Archives of Virology

Supplement 10

F. A. Murphy, C. M. Fauquet, D. H. L. Bishop,
S. A. Ghabrial, A. W. Jarvis, G. P. Martelli, M. A. Mayo,
M. D. Summers (eds.)

Virus Taxonomy

Classification and Nomenclature of Viruses

Sixth Report
of the International Committee
on Taxonomy of Viruses

Virology Division
International Union of Microbiological Societies

Springer-Verlag Wien New York

Dr. Frederick A. Murphy
School of Veterinary Science, University of California, Davis, CA, U.S.A.

Dr. Claude M. Fauquet
ORSTOM/The Scripps Research Institute, La Jolla, CA, U.S.A.

Dr. David H. L. Bishop
Natural Environment Research School, Institute of Virology, Oxford, U.K.

Dr. Said A. Ghabrial
Department of Plant Pathology, University of Kentucky, Lexington, KY, U.S.A.

Dr. Audrey W. Jarvis
New Zealand Dairy Research Institute, Palmerston North, New Zealand

Dr. Giovanni P. Martelli
Instituto di Patologia Vegetale, Bari, Italy

Dr. Mike A. Mayo
Scottish Crop Research Institute, Dundee, U.K.

Dr. Max D. Summers
Department of Entomology, Texas A&M University, College Station, TX, U.S.A.

Printed in Austria

Typesetting: Camera-ready by editors
Printing and Binding: Druckhaus Grasl, A-2540 Bad Vöslau

Printed on acid-free and chlorine-free bleached paper

With 185 Figures

Cover illustration: The three-dimensional surface-shaded map of rotavirus by cryoelectron microscopy and image processing (diameter ~1,000Å). The surface is characterized by 60 VP4 haemagglutinin spikes that bind to cell surface receptors and mediate infection (front cover). Three protein shells encapsidate the dsRNA genome, which has been removed for clarity: an outer capsid shell formed by 780 VP7 molecules; an inner capsid shell formed by 260, pillar-shaped, VP6 trimers; and a core shell formed primarily by VP2 as well as VP1 and VP3 (back cover) (with permission of Dr. M. Yeager, The Scripps Research Institute, La Jolla, California). From: The EMBO Journal, Volume 13, Number 7 (1 April 1994), front cover, by permission of Oxford University Press.

ISSN 0939-1983
ISBN 3-211-82594-0 Springer-Verlag Wien New York

Preface and Acknowledgments

Virus taxonomy is a polarizing subject when it comes up in hallway conversations. Some virologists tune out immediately, others tune in. In the end, after the skeptics walk off, down the hallway, the intensity of the conversation usually increases, because if there is one truism about virus taxonomy it is that it brings out among virologists "strongly held opinions" (a euphemism for "polite arguments"). The point is that virus taxonomy is based upon opinion rather than data, or better, it is based upon the opinionated usage of data. Since one opinion is usually as valid as the next, chaos can reign, but starting in 1966, chaos started to give way to order as the International Committee on Nomenclature of Viruses, later changed to the International Committee on Taxonomy of Viruses (ICTV), set out to provide a single universal system for the classification and nomenclature of all viruses. The system has been based upon true international consensus building, and true pragmatism — and it has been successful. The work of the Committee has been published in a series of reports, the *Reports of the International Committee on Taxonomy of Viruses, The Classification and Nomenclature of Viruses.* These Reports have become part of the history and infrastructure of modern virology:

ICTV Report	Editors	Reporting ICTV Proceedings at the International Congresses of Virology held in
The First Report, 1971	P. Wildy	Helsinki, 1968
The Second Report, 1976	F. Fenner	Budapest, 1971 and Madrid, 1975
The Third Report, 1979	R. E. F. Mathews	The Hague, 1978
The Fourth Report, 1982	R. E. F. Mathews	Strasbourg, 1981
The Fifth Report, 1991	R. I. B. Francki, C. M. Fauquet, D. L. Knudson, F. Brown	Sendai, 1984, Edmonton, 1987 and Berlin, 1990

This Report, the Sixth Report of the ICTV, adds to the accumulated taxonomic construction "in progress" since 1966. It records the proceedings of the Committee since 1990 and includes decisions reached at mid-term meetings in 1991 and 1992 and at the Ninth International Congress of Virology held in Glasgow in August of 1993.

The work of the Committee is far from complete — in fact, it would seem that as virus research continues to grow in breadth and depth and discoveries of the nature and diversity of the viruses become more and more amazing, the naïve goal of a "complete" taxonomy recedes into the distance. Virus taxonomy is a dynamic enterprise — to remain useful, it must continue to draw upon the wisdom and efforts of many virologists around the world, virologists representing all of the specialty disciplines that make up virology, overall. In this regard, this Report represents the work of about 400 virologists, the members of the Study Groups, Subcommittees and the Executive Committee of the ICTV for the term 1990-1993. The compilers of the Report wish to express their gratitude to all these virologists. We also wish to acknowledge, belatedly, the financial contribution of the Mayne Bequest Fund, the University of Queensland.

We also wish to express our gratitude to Ella Blanc, dedicated secretary to Dr. Claude M. Fauquet, and especially to Usha Padidam, responsible for the typing, formatting and layout of the VIth ICTV Report.

for the Committee

Frederick A. Murphy
President of the International Committee on Taxonomy of Viruses

VIRUS TAXONOMY

Sixth Report
of the
International Committee on Taxonomy of Viruses

Editors

F.A. Murphy
School of Veterinary Medicine
University of California, Davis
Davis, CA 95616
USA

C.M. Fauquet
ORSTOM/The Scripps Research Institute
Division of Plant Biology-MRC7
10666 North Torrey Pines Rd.
La Jolla, CA 92037 USA

D.H.L. Bishop
Natural Environment Research
School
Institute of Virology
Mansfield Road
Oxford OX1 3SR, UK

S.A. Ghabrial
University of Kentucky
Dept. of Plant Pathology
S-305 Ag. Sc. Bl. N
Lexington, KY 40546
USA

A.W. Jarvis
New Zealand Dairy Research
Institute
Private Bag
Palmerston North
New Zealand

G.P. Martelli
Instituto di Patologia Vegetale
Via Giovanni Amendola 165A
70126 Bari
Italy

M.A. Mayo
Scottish Crop Research Institute
Invergowrie
Dundee DD2 5DA
UK

M.D. Summers
Texas A & M University
Dept. of Entomology
College Station, TX 77843-2475
USA

Contributors

Ackermann, H.-W.
Adam, G.
Adrian, T.
Alexander, D.J.
Atabekov, J.G.
Baldwin, M.
Bamford, D.H.
Barbanti-Brodano, G.
Barnett, O.W.
Bartha, A.
Baxby, D.
Beaty, B.J.
Beckage, N.E.
Bergoin, M.
Berns, K.I.
Berthiaume, L.
Billeter, M.A.
Bishop, D.H.L.
Black, D.N.
Blissard, G.W.
Bloom, M.
Boccardo, G.
Bozarth, R.
Bradley D
Brain, D.A.
Briddon, R.W.
Brinton, M.A.
Brown, F
Bruenn, J.
Brunt, A.A.
Buchmeier, M.J.
Buck, K.W
Burrell, C.J.
Calisher, C.H.
Candresse, T.
Carter, M.J.
Cavanagh, D.
Chiba, S.
Clegg, J.C.S.
Coffin, J.M.
Collett, M.
Collinge, J.
Collins, P.L.
Dalrymple, J.M.
Ghabrial, S.
Gibbs, M.J.
Gingery, R.E.
Ginsberg, H.
Goldbach, R.
Goorha, R.
Graf, T.M.
Granados, R.R.
Gust, I.D.
Hamilton, R.I.
Hammond, J.
Hanzlik, T.
Heinz, F.X.
Hendry, D.
Herrmann, J.E.
Hierholzer, J.C.
Hill, J.H.
Hillman, B.I.
Hinuma, Y
Hoey, E.
Holmes, K.V
Horzinek, M.C.
Hoshino, Y.
Howard, C.
Howley, P.M.
Hull, R.
Hunter, E.
Incardona, N.L.
Jackson, A.O.
Jaenisch, R.
Jahrling, P.B.
Johnson, J.
Joklik, W.K.
Jordan, R.

Kaper, J.M.
Karabatsos, N.
Kashiwazaki, S.
Keddie, B.A.
Keese, P.
Lee, H.W.
Meshi, T.
Milne, R.G.
Minor, P.D.
Minson, A.C.
Monroe, S.S.
Morales, F.
Moss, B.
Moyer, R.W.
Muller, H.
Murant, A.F.
Muzyczka, N.
Nagai, Y.
Nakamura, K.
Namba, S.
Nasz, I.
Neurath, A.R.
Newbold, J.
Nichol, S.T.
Nicholson, B.L.
Noordaa, J.V.
Nuss, D.L.
Nusse, R.
Nuttall, P.A.
Ohki, S.T.
Olszewski, N.E.
Oroszlan, S.
Orth, G.
Örvell, C.
Palese, P.
Palmenberg, A.
Patterson, J.
Payment, P.
Peters, C.J.
Peters, D.
Petterson, R.F.
Pickup, D.J.
Plagemann, P.G.W.
Possee, R.
Pringle, C.R.
Prusiner, S.B.
Purcifull, D.
Randles, J.W.
Rathjen, J.P.
Reavy, B.
Rice, C.M.
Rima, B.
Robinson, A.J.
Robinson, D.J.
Robinson, W.
Rock, D.
Roizman, B.
Romaine, C.P.
Rott, R.
Rueckert, R.
Russell, W.C.
Russo, M.
Rybicki, E.P.
Saif, L.
Samal, K.S.K.
Sanchez, A.
Schaffer, F.
Schaller, H.
Schleper, C
Schlesinger, R.W.
Schmaljohn, C.S.
Scotti, P.D.
Shah, K.V.
Shikata, E.
Shope, R.E.
Shukla, D.D.
Siddell, S.G.
Siegl, G.
Smith, A.
Smith, J.S.
Southern, P.J.
Spaan, W.J.M.
Stanway, G.
Stoltz, D.B.
Strauss, J.H.
Stuart, K.
Studdert, M.J.
Summers, M.D.
Svoboda, J.
Swanepoel, R.
Taguchi, F.
Tal, J.
Talbot, P.J.
Tateishi, J.
Tattersall, P.
Taylor, J.M.
Teich, N.
ter Meulen, V.
Theilmann, D.A.
Thiel, H.J.
Tiollais, P.
Tischer, I.
Tomaru, K.
Toriyama, S.
Toyoshima, K.
Tripathy, D.N.
Turnbull-Ross, A.D.
Uyeda, I.
Van Alfen, N.K.
Van Duin, J.
Van Etten, J.L.
Van Regenmortel, M.H.V.
Varmus, H.
Vinuela, E.
Volkman, L.E.
Wadell, G.
Walker, P.J.
Wang, A.
Wang, C.
Webb, B.A.
Weissmann, C.
Wen, Y-M.
Wengler, G.
Wickner, R.
Will, H.
Wimmer, E.
Winton, J.R.
Wunner, W.H.
Yamashita, S.
Yin-Murphy, M.
Zillig, W.
zur Hausen, H.

Contents

Part I: Introduction to the Universal System of Virus Taxonomy

The History of Virus Taxonomy

The earliest experiments involving viruses were designed to separate them from microbes that could be seen in the light microscope and that usually could be cultivated on rather simple media. In the experiments that led to the first discoveries of viruses, by Beijerinck and Ivanovski (tobacco mosaic virus), Loeffler and Frosch (foot-and-mouth disease virus), and Reed and Carroll (yellow fever virus) at the turn of the century, one single physicochemical characteristic was measured, that being their small size as assessed by filterability (Waterson and Wilkinson, 1978). No other physicochemical measurements were made at that time, and most studies of viruses centered on their ability to cause infections and diseases. The earliest efforts to classify viruses, therefore, were based upon perceived common pathogenic properties, common organ tropisms, and common ecological and transmission characteristics. For example, viruses that share the pathogenic property of causing hepatitis (e.g., hepatitis A virus, hepatitis B virus, hepatitis C virus, yellow fever virus, and Rift Valley fever virus) would have been brought together as "the hepatitis viruses," and plant viruses causing mosaics (e.g., cauliflower mosaic virus, ryegrass mosaic virus, brome mosaic virus, alfalfa mosaic virus, and tobacco mosaic virus) would have been brought together as "the mosaic viruses."

Although the first studies of viruses were begun at the turn of the century, it was not until the 1930s that evidence of the structure and composition of virions started to emerge. This prompted Bawden (1941, 1950) to propose for the first time that viruses be grouped on the basis of shared virion properties. Among the first taxonomic groups constructed on this basis were the herpesvirus group (Andrewes, 1954), the myxovirus group (Andrewes, Bang and Burnet, 1955), the poxvirus group (Fenner and Burnet, 1957), and several groups of plant viruses with rod-shaped or filamentous virions (Brandes and Wetter, 1959). In the 1950s and 1960s, there was an explosion in the discovery of new viruses. Prompted by a rapidly growing mass of data, several individuals and committees independently advanced classification schemes. The result was confusion over competing, conflicting schemes, and for the first but not the last time it became clear that virus classification and nomenclature are topics that give rise to very strongly held opinions.

The International Committee on Taxonomy of Viruses (ICTV)

Against this background, in 1966 the International Committee on Nomenclature of Viruses (ICNV)[1] was established at the International Congress of Microbiology in Moscow. At that time, virologists already sensed a need for a single, universal taxonomic scheme. There was little dispute that the hundreds of viruses being isolated from humans, animals, plants, invertebrates, and bacteria, should be classified in a single system, and that this system should separate the viruses from all other biological entities. Nevertheless, there was much dispute over the taxonomic system to be used. Lwoff, Horne and Tournier (1962) argued for the adoption of an all-embracing scheme for the classification of viruses into subphyla, classes, orders, suborders, and families. Descending hierarchical divisions were to be based, arbitrarily and monothetically, upon nucleic acid type, capsid symmetry, presence or absence of an envelope, etc. Opposition to this scheme was based upon its arbitrariness in deciding the relative importance of virion characteristics to be used and upon the argument that not enough was known about the characteristics of most viruses to warrant an elaborate hierarchy. An alternative proposal was set forth in 1966 by Gibbs *et al.* (1966); in this system, divisions were based upon multiple criteria (polythetic criteria). The system was illustrated by the use of "cryptograms" (coded notations of eight virus characters). These early efforts succeeded well in stimulating interest in the development of the universal taxonomy system that evolved in the 1970s and has been built upon ever since (Wildy, 1971; Matthews, 1983).

[1] The International Committee on Nomenclature of Viruses (ICNV) became the International Committee on Taxonomy of Viruses (ICTV) in 1973. Today, the ICTV operates under the auspices of the Virology Division of the International Union of Microbiological Societies. The ICTV has six Subcommittees, 45 Study Groups, and over 400 participating virologists.

In the universal scheme developed by the ICTV, virion characteristics are considered and weighted as criteria for making divisions into families, in some cases subfamilies, and genera (until recently, the scheme did not use any hierarchical level higher than that of family, but now one order, the order *Mononegavirales*, has been approved). In each case, the relative hierarchy and weight assigned to each characteristic used in defining taxa is set arbitrarily and is still influenced by prejudgments of relationships that "we would like to believe (from an evolutionary standpoint), but are unable to prove" (Fenner, 1974). As the species taxon has been developed in the 1990s, it has become clearer that families and genera might best be defined monothetically (or by just a few characters), but species are better defined polythetically (Van Regenmortel, 1990).

At its meeting in Mexico City in 1970, the ICTV approved the first two families and 24 floating genera (Wildy, 1971; Matthews, 1983). At that time, 16 plant virus groups were also designated (Harrison, *et al.*, 1966). Since then, the ICTV has published five Reports entitled *The Classification and Nomenclature of Viruses* (Wildy, 1971; Fenner, 1976; Matthews, 1979; Matthews, 1982; Francki et al., 1991). Additionally, the Study groups of the ICTV published over the years detailed descriptions of the characteristics of the member viruses of many taxa (e.g., Melnick *et al.*, 1974; Pfau *et al.*, 1974; Dowdle *et al.*, 1975; Cooper *et al.*, 1978; Kingsbury *et al.*, 1978; Porterfield *et al.*, 1978; Brown *et al.*, 1979; Schaffer *et al.*, 1980; Bishop *et al.*, 1980; Roizman *et al.*, 1982, 1992; Kiley *et al.*, 1982; Wigand *et al.*, 1982; Gust *et al.*, 1983; Siddell *et al.*, 1983; Siegl *et al.*, 1985; Westaway *et al.*, 1985; Gust *et al.*, 1986; Brown, 1986). This, the Sixth Report of the ICTV, records a universal taxonomy scheme comprising one order, 71 families, 9 subfamilies, and 164 genera, including 24 floating genera, and more than 3,600 virus species. The system still contains hundreds of unassigned viruses, largely because of a lack of data.

The Universal System of Virus Taxonomy

Today, there is a sense that a significant fraction of all existing viruses of humans, domestic animals and economically important plants have already been isolated and entered into the taxonomic system. This sense is based upon the infrequency in recent years of discoveries of viruses that do not fit into present taxa. Of course, this sense does not extend to the viruses infecting the myriad of other species populating the Earth. This present sense of the diversity of the viruses, however imperfect, does point once again to the need for a universal, usable taxonomic system — a system to keep track of the large numbers of different viruses being isolated and studied throughout the world, a system to tie viral characteristics to virus names. The present universal system of virus taxonomy is useful and usable. It is set arbitrarily at hierarchical levels of order, family, subfamily, genus, and species. Lower hierarchical levels, such as subspecies, strain, variant, etc., are established by international specialty groups and by culture collections.

Virus Orders

Virus orders represent groupings of families of viruses that share common characteristics and are distinct from other orders and families. Virus orders are designated by names with the suffix *-virales*. To date, one order has been approved by the ICTV, the order *Mononegavirales*, comprising the families *Paramyxoviridae, Rhabdoviridae* and *Filoviridae*. It is ICTV's intention to move slowly in the approval of orders, limiting use to those instances where there is good evidence of phylogenetic relationship among the viruses of member families.

Virus Families and Subfamilies

Virus families represent groupings of genera of viruses that share common characteristics and are distinct from the member viruses of other families. Virus families are designated by names with the suffix *-viridae*. Despite concerns about the arbitrariness of early criteria for creating these taxa, most of the original families have stood the test of time and are still intact. This level in the taxonomic hierarchy now seems stable, and, indeed, is the benchmark of the entire universal taxonomy system. Most of the families of viruses have distinct virion morphology, genome structure, and/or strategies of replication, indicating phylogenetic independence or great phylogenetic separation. At the same time, the virus family is being recognized as a taxon uniting viruses with a common, even if distant phylogeny. In four families, namely the families *Poxviridae, Herpesviridae, Parvoviridae,* and *Paramyxoviridae,* subfamilies have been introduced to allow for a more complex hierarchy of taxa, in keeping with the apparent intrinsic complexity of the relationships among member viruses. Subfamilies are designated by terms with the suffix *-virinae.*

VIRUS GENERA

Virus genera represent groupings of species of viruses that share common characteristics and are distinct from the member viruses of other genera. Virus genera are designated by terms with the suffix *-virus.* This level in the hierarchy of taxa also seems stable and in many cases may be considered a benchmark for setting definitions of other taxa, especially species. The criteria used for creating genera differ from family to family. As more viruses are discovered and studied, there is pressure in many families to use smaller and smaller genetic, structural or other differences to create new genera. Since evidence of common phylogeny has entered the definition of many families, it is logical that even more such evidence will become the basis for defining genera. In fact, it might be said that in the future genera will not stand where evidence is obtained of distinct phylogenies among member species.

VIRUS SPECIES

The species taxon has always been regarded as the most important hierarchical level in classification, but with the viruses it has proved to be the most difficult to deal with. After years of controversy, in 1991, the ICTV accepted the definition of a virus species proposed by van Regenmortel (1990), as follows: *"A virus species is defined as a polythetic class of viruses that constitutes a replicating lineage and occupies a particular ecological niche."* Members of a polythetic class are defined by more than one property and no single property is essential or necessary. One major advantage in this definition is that it can accommodate the inherent variability of viruses and it does not depend on the existence of a single unique characteristic. Similarly, it can accommodate the different traditions of virologists working in different areas of virology, in some cases accommodating "the lumpers" and in others "the splitters."

The ICTV Study Groups are now determining the specific properties to be used to define species in the taxon for which they are responsible. It seems clear that the species term will eventually be defined somewhat similarly to the term *virus* — although in some cases the term virus matches best with subspecies, strain or even variant. Just as the term virus is defined differently in different virus families, so, species will be defined differently, in some cases with emphasis on genome properties, and in others on structural, physicochemical or serological properties. Some viruses have already been designated as species, for example: Sindbis virus, Newcastle disease virus, poliovirus 1, vaccinia virus, Fiji disease virus, tomato spotted wilt virus. These examples, however, do not reflect the difficulty that is being encountered in deciding whether a particular virus should be designated as a species or as a subspecies or strain or variant.

VIRUS NOMENCLATURE

USAGE OF FORMAL TAXONOMIC NOMENCLATURE

In formal taxonomic usage, the first letters of virus family, subfamily, and genus names are capitalized and the terms are printed in italics (underlined when typewritten). Species designations are not capitalized (unless they are derived from a place name or a host family or genus name), nor are they italicized. In formal usage, the name of the taxon should precede the term for the taxonomic unit; for example: ..."the family *Paramyxoviridae*" ..."the genus *Morbillivirus.*" Furthermore, it was decided years ago that virus nomenclature would not involve the use of latinized binomial terms. For example, terms such as *Flavivirus fabricis*, *Orthopoxvirus variolae* and *Herpesvirus varicellae*, which were used at one time, have been abandoned. The following represent examples of full formal taxonomic terminology:

1. Family *Poxviridae*, subfamily *Chordopoxvirinae*, genus *Orthopoxvirus*, vaccinia virus.
2. Family *Herpesviridae*, subfamily *Alphaherpesvirinae*, genus *Simplexvirus*, human herpes virus 2 (herpes simplex virus 2).
3. Family *Picornaviridae*, genus *Enterovirus*, poliovirus 1.
4. Order *Mononegavirales*, Family *Rhabdoviridae*, genus *Lyssavirus*, rabies virus.
5. Family *Bunyaviridae*, genus *Tospovirus*, tomato spotted wilt virus.
6. Family *Bromoviridae*, genus *Bromovirus*, brome mosaic virus.
7. Genus *Sobemovirus*, Southern bean mosaic virus.
8. Family *Totiviridae*, genus *Totivirus*, Saccharomyces cerevisiae virus L-A.
9. Family *Tectiviridae*, genus *Tectivirus*, enterobacteria phage PRD1.
10. Family *Plasmaviridae*, genus *Plasmavirus*, Acholeplasma phage L2.

Vernacular Usage of Virus Nomenclature

In informal vernacular usage, virus family, subfamily, genus and species names are written in lower case Roman script; they are not capitalized, nor are they printed in italics or underlined. In informal usage, the name of the taxon should not include the formal suffix, and the name of the taxon should follow the term for the taxonomic unit; for example, ..."the picornavirus family" ..."the enterovirus genus."

The use of vernacular terms for virus taxonomic units and virus names should not lead to unnecessary ambiguity or loss of precision in virus identification. The formal family, subfamily, and genus terms and standard ICTV vernacular species terms, rather than any synonyms or transliterations, should be used as the basis for choosing vernacular terms.

One particular source of ambiguity in vernacular nomenclature lies in the common use of the same root terms in formal family and genus names. Imprecision stems from not being able to easily identify in vernacular usage which hierarchical level is being cited. For example, the vernacular name "paramyxovirus" might refer to the family *Paramyxoviridae*, the genus *Paramyxovirus*, or one of the species in the genus *Paramyxovirus*, such as one of the human parainfluenza viruses. Some virologists have suggested that this problem be solved by renaming taxa so that the same root term is never used at multiple hierarchical levels; however, there is no consensus for this, in fact, as plant virus taxonomy switches away from groups and toward families and genera, this problem will be exacerbated. The solution in vernacular usage is to avoid "jumping" hierarchical levels and to add taxon identification wherever needed. For example, when citing the taxonomic placement of human parainfluenza virus 1, the term "paramyxovirus" should refer firstly to the genus, not the subfamily or family, and taxon identification should always be added: "human parainfluenza virus 1 is a member of the paramyxovirus genus," rather than " human parainfluenza virus 1 is a paramyxovirus." Most examples like this exemplify the advantage of switching, where necessary, into formal nomenclature usage: "human parainfluenza virus 1 is a species in the genus *Paramyxovirus*, family *Paramyxoviridae*." In this example, as is usually the case, adding the information that this virus is also a member of the subfamily *Paramyxovirinae* and the order *Mononegavirales* is unnecessary.

Structural, Genomic, Physicochemical and Replicative Properties of Viruses Used in Taxonomy

The way by which viruses are characterized, for taxonomic and other purposes, is changing rapidly. In the past, laboratory techniques have included characterizations of virion morphology (by electron microscopy), virion stability (by varying pH and temperature, adding lipid solvents and detergents, etc.), virion size (by filtration through fibrous and porous microfilters), and virion antigenicity (by many different serologic methods). These means worked because after large numbers of viruses had been studied and their characteristics placed into the universal taxonomic scheme, it was necessary in most cases to only measure a few characteristics to place a new virus, especially a new variant from a well studied source, in its proper taxonomic niche. For example, a new adenovirus, isolated from the human respiratory tract and identified by serologic means, was easy to place in its niche in the family *Adenoviridae*, genus *Mastadenovirus*. The exceptions occurred when a new virus was found that did not have a familiar set of properties. Such a virus became a candidate prototype for a new taxon, generally a new family or genus. In such cases, comprehensive characterization of all virion properties was called for.

One particularly important technological advance underpinning the development of modern virus taxonomy was the invention by Brenner and Horne (1959) of the negative staining technique for electron microscopic examination of virions. The impact of this technique was immediate: (a) virions could be characterized with respect to size, shape, surface structure, presence or absence of an envelope, and, often, symmetry; (b) the method could be applied simply and universally; and (c) virions could be characterized in unpurified material, including diagnostic specimens. Negative staining has facilitated the rapid accumulation of data about the physical properties of many viruses. Thin-section electron microscopy of virus-infected cell cultures and tissues of infected humans, animals (including experimental animals), and plants has provided complementary data on virion morphology, mode and site of virion morphogenesis (e.g., site of budding), etc. Thus, in many cases, viruses were placed in their appropriate family, and in some instances in their appropriate genus, after simple visualization and measurement by negative-stain and/or thin-section electron microscopy (Murphy, 1987).

The fundamental molecular bases for many of the empirical virion property measurements that were originally used to construct virus families and genera are now rather well understood. Many of the characteristics that have been used in deciding taxonomic constructions are listed in the following table:

Some Properties of Viruses Used in Taxonomy

Virion Properties

Morphology

virion size
virion shape
presence or absence and nature of peplomers
presence or absence of an envelope
capsid symmetry and structure

Physicochemical and Physical Properties

virion molecular mass (Mr)
virion buoyant density (in CsCl, sucrose, etc.)
virion sedimentation coefficient
pH stability
thermal stability
cation stability (Mg^{++}, Mn^{++})
solvent stability
detergent stability
irradiation stability
genome
type of nucleic acid (DNA or RNA)
size of genome in kb/kbp
strandedness — (single) stranded or (double) stranded
linear or circular
sense (positive-sense, negative-sense, ambisense)
number and size of segments
nucleotide sequence, or partial sequence

presence of repetitive sequence elements
presence of isomerization
G+C ratio
presence or absence and type of 5'-terminal cap
presence or absence of 5'-terminal covalently-linked protein
presence or absence of 3'-terminal poly (A) tract

Proteins

number, size and functional activities of structural proteins
number, size and functional activities of non-structural proteins
details of special functional activities of proteins: especially transcriptase, reverse transcriptase, hemagglutinin, neuraminidase, and fusion activities
amino acid sequence or partial sequence
glycosylation, phosphorylation, myristylation
epitope mapping

Lipids

content, character, etc.

Carbohydrates

content, character, etc.
genome organization and replication
genome organization
strategy of replication
number and position of open reading frames
transcriptional characteristics
translational characteristics
post-translational processing
site of accumulation of virion proteins
site of virion assembly
site and nature of virion maturation and release

Antigenic Properties

serologic relationships, especially as obtained in reference centers

Biologic Properties

natural host range
mode of transmission in nature
vector relationships
geographic distribution
pathogenicity, association with disease
tissue tropisms, pathology, histopathology

Through the use of monoclonal antibodies, synthesized peptides, and epitope mapping, there is new understanding of the molecular bases for those serological reactions that were originally used to construct families and genera. Today, genome sequencing, or partial sequencing, is often done very early in virus identification, and even in diagnostic activities. For comparison, genome sequences are available in readily accessible databases for the prototype viruses of nearly all taxa. Sequence data are even driving consideration for the construction of new families and new genera before other data are available. It is likely because of their absolute nature that genome sequence data will become the base for further refinement and expansion of the universal taxonomic scheme. Genome sequence data are also the key to the advance of "phylogenetic taxonomy" (see below). In addition, the derivatives of sequencing are advancing as taxonomic criteria: for example, genome organization, gene order, strategy of replication and other genetic considerations have been added to the taxonomic decision process (Murphy, 1985, 1987, 1988, 1994; Murphy and Kingsbury, 1990).

TAXONOMY AND UNAMBIGUOUS VIRUS IDENTIFICATION

Unambiguous virus identification is a major virtue of the universal system of taxonomy (Murphy, 1983), and of particular value when the editor of a journal requires precise naming of viruses cited in a publication. At a minimum, precise naming avoids problems caused by synonyms, transliterated vernacular names, and local laboratory jargon. Precise virus identification includes taxonomic status, such as name of family, genus, and species, as well as strain designation terms. The matter of deciding how type species and strains are chosen and designated remains the responsibility of international specialty groups, some of which operate under the auspices of the World Health Organization and other agencies. In the past, some confusion was caused by the ICTV's identification of type species as part of descriptions of taxa; however, the ICTV has never tried to identify type species with the kind of precision that must be used by specialty groups and culture collections. There has also been some confusion caused by conflicting claims of individuals having personal interests in the choice of prototypes. This has occurred where prototype strains have become valuable as substrates for vaccines, diagnostic reagents, etc.

One of the best models for the kind of description necessary to avoid ambiguity in virus strain identification is that of the American Type Culture Collection in its frequently updated *Catalogue of Animal Viruses and Antisera, Chlamydiae and Rickettsiae* (1990). For example, St. Louis encephalitis virus is listed as:

St. Louis encephalitis virus Class III   ATCC VR-80
Strain: Hubbard. Original source: Brain of patient, Missouri, 1937. Reference: McCordock, H. A., *et al.*, Proc. Soc. Exp. Biol. Med. 37:288, 1937. Preparation: 20% SMB in 50% NIRS infusion broth; supernatant of low speed centrifugation. Host of Choice: sM (i.c.); M (i.c.). Incubation: 3-4 days. Effect: Death. Host Range: M, Ha, CE, HaK, CE cells. Special Characteristics: Infected brain tissue will have a titer of about 107. Agglutinates goose and chicken RBC. Cross reacts with many or all members of group B arboviruses....

TAXONOMY AND THE ADEQUATE DESCRIPTION OF NEW VIRUSES

As thousands of viruses have been isolated from human animal and plant specimens, there have been many errors and duplications over the years; that is, viruses isolated in different laboratories have been given different names and the chance for coexistence in various virus lists (Murphy, 1983). Viruses have also been placed in the wrong lists, the most notable instance involving the emergence of the family *Reoviridae* (genus *Reovirus*) from the initial placement of its prototype viruses in the list of human enteroviruses. Recently, several named but serologically "ungrouped" viruses listed in the International Catalogue of Arboviruses (Karabatsos, 1985) were found to actually be unrecognized isolates of Lassa virus and Rift Valley fever virus, viruses that must be handled under maximum containment conditions. The reasons for these and similar problems have been: (1) inadequate characterization and description of viruses by those who isolate them, and (2) inadequate review of data by international specialty groups. Such problems seem to be declining in frequency, but continuous attention is warranted because there can be serious consequences. Assuring the adequacy of characterization and description of new viruses is a particular responsibility of reference laboratories, international reference centers, international specialty groups, and culture collections.

When an "unknown" is first studied in a laboratory, its initial characterization may involve only standardized protocols. That is, only a few characteristics may be determined before the application of specific identification procedures. Only when an "unknown" fails to yield to routine procedures is there call for more extensive study. One key to simplifying and rationalizing such study is to set useful techniques into a proper sequence based upon taxonomic characteristics. This sequence of procedures should include logical short-cuts, so as to avoid extra effort and expense. For example, negative-stain electron microscopy represents a logical short-cut for the initial placement of unknowns that emerge from characterization protocols. If an "unknown" is shown by electron microscopy to be a rhabdovirus, there is little value of checking whether its genome is DNA or RNA. Likewise, there is little value of doing serology against other than the known rhabdoviruses (except perhaps in testing for the presence of contaminant viruses). Comprehensive characterization is the key to discovery of novel viruses, but fully characterizing usual isolates rarely contributes to such discovery.

Taxonomy in Diagnostic Virology

A universal system for taxonomy and nomenclature of viruses is a practical necessity whenever large numbers of distinct isolates are being dealt with, as in a reference diagnostic laboratory. The clinician usually makes a preliminary diagnosis of a viral disease on the basis of four kinds of evidence: (1) clinical features, which allow recognition with varying certainty in typical cases of many viral diseases (e.g., varicella exanthem, measles exanthem); (2) epidemic behavior, which in a typical population may allow recognition (e.g., epidemic influenza, arbovirus diseases such as dengue and yellow fever, enterovirus exanthems); (3) circumstances of occurrence, which may indicate probable etiology (e.g., respiratory syncytial virus as the primary cause of croup and bronchiolitis in infants, hepatitis B or hepatitis C as the likely cause of hepatitis following blood transfusion); and (4) organ involvement, which may suggest a probable etiology (e.g., mumps virus as the cause of parotitis, viruses in general as the cause of 80-90 percent of acute respiratory infections). Shortcomings in the predictive value of these kinds of evidence suggest that the laboratory diagnostician as well as the clinician must appreciate the range of possible etiologic agents in particular disease syndromes. There is value in initially assembling an inclusive "long list" of possible etiologic agents, so that no candidate agent is overlooked. In most cases this is done informally, and the process is adjusted to the complexity of the case.

The universal system of virus taxonomy may be used as the source of the "long list" of candidate etiologic agents. The system serves to organize the "long list" logically, and because the system is comprehensive, it is unlikely that known viruses will be overlooked. The observations that serve to place an etiologic agent in its proper family and genus should also play a major role in shortening the list, and should in most cases provide etiologic information needed to select immunologic (serologic) identification techniques. For example, the "long list" of possible etiologies for a slowly progressive central nervous system disease would include many viruses that are difficult or impossible to cultivate. However, the identification of spherical, 45 nm, nonenveloped virions in the nuclei of cells in a brain biopsy from a patient with such a disease would go far toward shortening the list to the family *Papovaviridae*, genus *Polyomavirus*, thereby suggesting a diagnosis of JC or SV 40 virus-induced progressive multifocal leukoencephalopathy. In this case, many viruses known to invade the brain and cause slowly progressive neurologic disease would be eliminated from the differential diagnostic list; e.g., member viruses of the families *Herpesviridae, Adenoviridae, Togaviridae, Flaviviridae, Paramyxoviridae, Rhabdoviridae, Bunyaviridae, Arenaviridae, Retroviridae,* and *Picornaviridae.*

The Future of Virus Taxonomy

The Advance of Phylogenetic Taxonomy and the Use of Higher Taxa

Until recently, one of the rules of virus taxonomy stated that the system was not meant to imply any phylogenetic relationships. Since viruses leave no fossils (*except perhaps within arthropods and other creatures embedded in amber!*), it was presumed that there never would be enough evidence to prove whether or not different taxa had common evolutionary roots. In fact, since the prevailing concept was that each kind of virus was derived separately from its host, it was considered foolish to consider any idea of a single evolutionary "tree" for the viruses (Goldbach, 1986, Gibbs, 1987). The generally very different morphological and physicochemical characteristics of the member viruses of many different taxa supported this view.

Now, as genome sequencing of many viruses, and many organisms from archaebacteria to humans, is revealing many conserved functional or "fossil" domains of ancient lineage, for the first time the archeology of the viruses is being explored from the perspective of data, not just "armchair theory." We now know that many viruses have gained some functional genes from their hosts (and hosts have gained some genes from viruses) and have gained other genes from other viruses — that is, viral genomes seem to represent more-or-less ancient "grab-bags" of genes, fine-tuned by the Darwinian forces of selection into replicative machines with extraordinary functional economy. We now know that the genomes of viruses in different families, in most cases, are extremely different from each other, but we also know that in some cases seemingly unrelated viruses (and taxa) are similar — similar in gene order and arrangement, fine points of strategy of replication, and even in conserved sequence domains encoding similarly functioning proteins. Overall, the differences between most taxa are so great that it still seems foolish to think of building a monophylogenous "tree" uniting all the viruses. On the other hand, the unexpected similarities have prompted some consideration of a partial phylogenetic taxonomy (Goldbach, 1986, 1987; Gibbs, 1987; Goldbach and Wellink, 1988; Kingsbury, 1988; Morse, 1993).

As these evidences of phylogenetic relationships between families have been studied, there has been a wish to reflect these relationships in the universal taxonomic scheme. There has been no wish to combine families exhibiting distant phylogenetic relationships and thereby have them lose their practical identities. As one virologist stated: "the family is the fixed point, the benchmark, in virus taxonomy, so let's not do anything to change this". Instead, there has been increasing interest in capturing these relationships by uniting distantly related families in higher taxa, namely orders. The order *Mononegavirales*, comprising the families *Paramyxoviridae, Rhabdoviridae* and *Filoviridae,* was formed in recognition that the member viruses had common sequences in their nucleocapsid genes and similar gene arrangements and gene products (Pringle, 1991). At this point, ICTV is committed to reserving the hierarchical level of order solely for recognizing phylogenetic relationships.

There is a further occasion for considering the grouping of families together; this involves many of the positive sense, single-stranded RNA viruses. Similarities in genome organization, gene arrangements and sequence similarities in particular domains among viruses in several taxa, representing diverse vertebrate, invertebrate, plant and bacterial viruses, have been studied for the past 10 years. Kamer and Argos (1984) first aligned RNA-dependent RNA polymerase gene sequences of several plant, animal and bacterial viruses. In 1986, Goldbach greatly broadened this approach and used the data to explore the possible paths of evolution of the many positive sense RNA viruses. He proposed the formation of several "supergroups" to formalize the recognition of similarities. Gibbs (1987) and Strauss, Strauss and Levine (1988, 1991) have explored the possible mechanisms underpinning such evolutionary relationships. Today, work is centered on assessment of many characteristics of the RNA viruses: (1) genome organization and gene order; (2) presence of a 5'-terminal covalently-linked polypeptide, a 3'-terminal poly (A) tract, a 5'-terminal cap; (3) presence of subgenomic RNA; (4) polyprotein processing and enzymology; etc. At the heart of the matter is the assessment of conserved sequences in genes encoding the RNA-dependent RNA polymerases, helicases, and proteases. These characteristics are being used as the basis for several different ideas for constructing higher taxa (Koonin and Gorbalenya, 1989; Koonin, 1991; Mahy, 1991; Goldbach *et al.*, 1991; Strauss, Strauss and Levine, 1991; Koonin and Dolja, 1993; Ward, 1994).

In their most recent proposals, Goldbach and de Haan (1993), Koonin and Dolja (1993) and Ward (1994) have proposed three major clusters of positive sense, single-stranded RNA viruses, one for the "picorna-like viruses," one for the "toga-like viruses," and one for the "flavi-like viruses" (although Goldbach calls attention to the major differences among the viruses in the latter cluster). Goldbach and de Haan (1993) continue to use the terms "supergroups" and "superfamilies" for these clusters, but Koonin and Dolja (1993) and Ward (1994) have taken this a step further in using the taxonomic hierarchical levels of class and order to denote the same groupings (it must be reminded that these usages do not have

ICTV sanction). It was thought for a time that the approach could also been extended to the double-stranded RNA viruses, but recent evidence suggests a polyphyletic origin of double-stranded RNA viruses from different groups of positive sense RNA viruses (Bruenn, 1991; Koonin, 1992). This matter will be debated by the ICTV over the next few years, but already it is clear that there is no general wish by most virologists to abandon the present system which is based upon assessment and weighting of multiple virion characteristics. As one virologist has stated: "I am alarmed by the idea of erecting higher taxa upon a scheme that assumes that the polymerase *is* the virus." Clearly, there will be interest in melding phylogenetic considerations and traditional approaches into a unifying system.

Within the subject of phylogenetic taxonomy, one of the most interesting debates centers on which characteristics of an organism are most ancient, which are most recent, which are most stable and which are most changing. With increasing knowledge of which characteristics are conserved through evolutionary divergence, arbitrary cladistic taxonomy seemingly must be melded with phylogenetic taxonomy. Many virologists have over the years considered that virion structural elements were most ancient; after all, cumulative mutations in icosahedral capsids could only lead to lethal instability. Similarly, many virologists have considered that viral genome expression strategies were most ancient; again, even if a virus figured out how to reinvent its capsid, changes in integrated multigenic replication steps would certainly be lethal. At present, however, the finding of conserved sequence domains in polymerases, helicases and proteases, but not in structural or other genes of the positive sense, single-stranded RNA viruses, suggests that the whole subject must be revisited. Of course, horizontal gene transfer between viruses and dynamic gene acquisition from host cells adds to the sense of phylogenetic complexity. It is unfortunate that viruses have such small genomes, so that there will not be an opportunity to find confirmatory evidences of relationships by analyzing additional genes.

The ICTV Database (*ICTVdB*)

The number of viruses occupying geographic and/or host niches as pathogens or silent passengers of humans, animals, plants, invertebrates, protozoa, fungi, and bacteria is very large. Our lists are increasing regularly as we search in new niches and as the sensitivity and specificity of our techniques for detection get better and better. Today, the ICTV recognizes more than 3,600 virus species. Specialty groups keep track of far more viruses, virus strains, and subtypes, each having particular health or economic distinction and importance. It has been estimated that more that 30,000 viruses, virus strains, and subtypes are being tracked in various specialty laboratories, reference centers, and culture collections communicating with the WHO, FAO, and other international agencies. Further, the development of the viral quasispecies concept, with its prediction of rapid evolution of variants that may become fixed in nature as new species, portends future needs to track even more viral entities (Holland, *et al.*, 1982; Zimmern, 1988; Holland, de la Torre and Steinhauer 1992; Dolja and Carrington, 1992; Eigen, 1993). It has been estimated that to describe a virus comprehensively, approximately 500-1,000 characters must be determined (Atherton, Holmes and Jobbins, 1983; Boswell, *et al.*, 1986). This means that to comprehensively describe all the known viruses of the world, we must "fill in the blanks" for 3 to 21 million data points (of course, many of the data points are the same when entering related viruses). This situation is even more complex; as we add more and more genome sequence information, our data systems will become truly enormous.

A major goal of the ICTV is to design, build and make available to all virologists, worldwide, a universal virus database, the *ICTVdB*. This database will encompass data that are now used in developing and managing the universal system for virus taxonomy. The *ICTVdB* will first describe viruses down to the species level, in keeping with the level of responsibility of the ICTV, but it will then go further, interfacing with the databases of international specialty groups which are cataloguing data down to subspecies, strain, variant and isolate levels, that is, levels important in medicine, agriculture, and other scholarly fields. ICTV's goal is to design the database to feed directly into user friendly programs that will be directly accessible to users (Pankhurst and Aitchison, 1975). Particular products, tailored to particular users, will be compressed to fit into equipment and software that can be readily accessed around the world.

There are several virus databases in operation in the world which will be integrated into the *ICTVdB*, including: (1) the plant virus database operated at the Australian National University in Canberra, Australia (the VIDE Project; AJ Gibbs, personal communication, 1994); (2) the veterinary virus database operated by the CSIRO/Australian Animal Health Laboratory in Geelong, Australia (the VIREF Project; A Della Porta, personal communication, 1994); and (3) an arbovirus database operated for the American Committee on Arthropod-borne Viruses (ACAV) by the Centers for Disease Control in Ft. Collins, Colorado, USA (C.H. Calisher, personal communication, 1994). Additionally, a human virus/human disease database, in planning stage in the Division of Viral and Rickettsial Diseases, Centers for Disease Control in Atlanta, Georgia, USA, will be integrated into the *ICTVdB* (B.W.J. Mahy, personal communication, 1994). The most advanced of these databases is the VIDE (Virus Identification Data Exchange) project on plant viruses (Dallwitz, 1974, 1980; Boswell *et al.*, 1983, 1986; Dallwitz and Paine, 1986; Partridge Dallwitz and Watson, 1988). This project, centered in Canberra, Australia, interfaces with Horticulture Research International (Littlehampton, U.K.) and C.A.B. International (Wallingford, U.K.), and involves a worldwide network of more than 200 collaborating plant virologists (Buchen-Osmond *et al.*, 1988, 1993; Brunt, Crabtree and Gibbs, 1990, 1992). The VIDE database now contains 569 characters for more than 890 plant virus species in 55 genera (A.J. Gibbs, personal communication, 1994). Using these databases as a foundation, the ICTV has laid out a plan to develop the universal virus database, the *ICTVdB*, over the next 10 years.

References for Part I

American Type Culture Collection (1990) Catalogue of Animal Viruses and Antisera, *Chlamydiae* and *Rickettsiae*, Sixth Edition. American Type Culture Collection, Rockville, Maryland

Andrewes CH (1954) Nomenclature of Viruses. Nature 173: 260-261

Andrewes CH, Bang FB, Burnet FM (1955) A short description of the Myxovirus group (influenza and related viruses). Virology 1: 176-180

Atherton JG, Holmes IH, Jobbins EH (1983) ICTV code for the description of virus characters. Monogr Virol 14: 1-154

Bawden FC (1941) Plant Viruses and Virus Diseases, First Edn. Chronica Botanica Company, Waltham, Massachusetts

Bawden FC (1950) Plant Viruses and Virus Diseases, Third Edn. Chronica Botanica Company, Waltham, Massachusetts

Bishop DHL, Calisher CH, Casals J, Chumakov MP, Gaidamovich SY, Hanoun C, Lvov DK, Marshall ID, Oker Blom N, Pettersson R, Porterfield JS, Russell PK, Shope RE, Westaway EG (1980) *Bunyaviridae*. Intervirology 14: 125-143

Boswell KF, Gibbs AJ (eds) (1983) Viruses of Legumes. Australian National University, Canberra

Boswell KF, Dallwitz MJ, Gibbs AJ, Watson L (1986) The VIDE (Virus Identification Data Exchange) project: a data bank for plant viruses. Rev Plant Pathol 65: 221-231

Brandes J, Wetter C (1959) Classification of elongated plant viruses on the basis of particle morphology. Virology 8:99-109

Brenner S, Horne RW (1959) A negative staining method for high resolution electron microscopy of viruses. Biochim Biophys Acta 34: 103-110

Brown F, Bishop DHL, Crick J, Francki RIB, Holland JJ, Hull R, Johnson KM, Martelli GP, Murphy FA, Obijeski JF, Peters D, Pringle CR, Reichmann ME, Schneider LG, Shope RE, Simpson DIH, Summers DF, Wagner RR (1979) *Rhabdoviridae*. Intervirology 12: 1-17

Brown F (1986) The classification and nomenclature of viruses: summary of results of meetings of the ICTV in Sendai, 1984. Intervirology 25:141-143

Bruenn JA (1991) Relationship among the positive strand and double-strand RNA viruses as viewed through their RNA-dependent RNA polymerases. Nucl Acids Res 19: 217-225

Brunt AA, Crabtree K, Gibbs AJ (eds) (1990) Viruses of Tropical Plants. C.A.B. International, London, pp 1-707

Brunt AA, Crabtree K, Gibbs AJ, Watson L (eds) (1994) Viruses of Plants, 2 volumes. C.A.B. International

Buchen-Osmond C, Blaine LD, Gibbs AJ (1993) Towards a comprehensive virus data base. Proceedings of the Ninth International Congress of Virology, Glasgow W76-1:116

Buchen-Osmond C, Crabtree K, Gibbs AJ, McLean GD (eds) (1988) Viruses of Plants in Australia. Australian National University, Canberra

Cooper PD, Agol VI, Bachrach HL, Brown F, Ghendon Y, Gibbs AJ, Gillespie JH, Lonberg-Holm K, Mandel B, Melnick JL, Mohanty SB, Povey RC, Rueckert RR, Schaffer FL, Tyrrell DAJ (1978) *Picornaviridae*: Second Report. Intervirology 10: 165-180

Dallwitz MJ (1974) A flexible computer program for generating identification keys. Systematic Zoology 23: 50-57

Dallwitz MJ (1980) A general system for coding taxonomic descriptions. Taxon 29: 41-69

Dallwitz MJ, Paine TA (1986) User's guide to the DELTA system - a general system for processing taxonomic descriptions. Third Edn. CSIRO Aust Div Entomol Rep. 13: 1-106

Dolja VV, Carrington JC (1992) Evolution of positive-strand RNA viruses. Sem Virol 3: 315-326

Dowdle WR, Davenport FM, Fukumi H, Schild GC, Tumova B, Webster RG, Zakstelskaja GE (1975) *Orthomyxoviridae*. Intervirology 5: 245-251

Eigen M (1993) Viral quasispecies. Scientific American 269: 42-49

Fenner F (1974) The classification of viruses; why, when and how. Aust J Experimental Biol Med Sci 52: 223-231

Fenner F (1976) The Classification and Nomenclature of Viruses. Second Report of the International Committee on Taxonomy of Viruses. Intervirology 7: 1-115

Fenner F, Burnet FM (1957) A short description of the poxvirus group (vaccinia and related viruses). Virology 4: 305-310

Francki RIB, Fauquet CM, Knudson DL, Brown F (1991) Classification and Nomenclature of Viruses. Fifth Report of the International Committee on Taxonomy of Viruses. Springer-Verlag, Wien, New York

Gibbs AJ (1987) Molecular evolution of viruses: "trees, clocks and modules." J Cell Sci, Supplementum 7: 319-337

Gibbs AJ, Harrison BD (1966) Realistic approach to virus classification and nomenclature. Nature 218: 927-929

Gibbs AJ, Harrison BD, Watson DH, Wildy P (1966) What's in a virus name? Nature 209: 450-454

Goldbach R (1986) Molecular evolution of plant RNA viruses. Annu Rev Phytopathol 24: 289-310

Goldbach R (1987) Genome similarities between plant and animal RNA viruses. Microbiol Sci 4: 197-202

Goldbach R, de Haan PT (1993) RNA viral supergroups and the evolution of RNA viruses. In: Morse SS (ed) The Evolutionary Biology of Viruses. Raven Press, New York, pp 105-119

Goldbach R, Le Gall O, Wellink J (1991) Alpha-like viruses in plants. Sem Virol 2: 19-25

Goldbach R, Wellink J (1988) Evolution of plus-strand RNA viruses. Intervirology 29: 260-268

Gust ID, Burrell CJ, Coulepis AG, Robinson WS, Zuckerman AJ (1986) Taxonomic classification of human hepatitis B virus. Intervirology 25: 14-29

Gust ID, Coulepis AG, Feinstone SM, Locarnini SA, Moritsugu Y, Najera R, Siegl G (1983) Taxonomic classification of hepatitis A virus. Intervirology 20: 1-7

Harrison BD, Finch JT, Gibbs AJ, Hollings M, Shepherd RJ, Valenta V, Wetter C (1966) Sixteen groups of plant viruses. Virology 45: 356-363

Holland JJ, de la Torre JC, Steinhauer DA (1992) RNA virus populations as quasispecies. Curr Top Microbiol Immunol 176: 1-20

Holland JJ, Spindler K, Horodyski F, Grabau E, Nichol ST, van de Pol S (1982) Rapid evolution of RNA genomes. Science 215: 1577-1585

Kamer G, Argos P (1984) Primary structural comparison of RNA-dependent polymerases from plant, animal and bacterial viruses. Nucl Acids Res 12: 7269-7282

Karabatsos N (ed) (1985) International Catalogue of Arboviruses, Third Edn. American Society of Tropical Medicine and Hygiene, San Antonio, Texas

Kiley MP, Bowen ETA, Eddy GA, Isaacson M, Johnson KM, Murphy FA, Pattyn SR, Peters D, Prozesky OW, Regnery RL, Simpson DIH, Slenczka W, Sureau P, van der Groen G, Webb PA (1982) *Filoviridae*: A taxonomic home for Marburg and Ebola viruses? Intervirology 18: 24-32

Kingsbury DW (1988) Biological concepts in virus classification. Intervirology 29: 242-253

Kingsbury DW, Bratt MA, Choppin PW, Hanson RP, Hosaka Y, ter Meulen V, Norrby E, Plowright W, Rott R, Wunner WH (1978) *Paramyxoviridae*. Intervirology 10: 137-152

Koonin EV (1991) The phylogeny of RNA-dependent RNA polymerases of positive-strand RNA viruses. J Gen Virol 72: 2197-2206

Koonin EV (1992) Evolution of double-stranded RNA viruses: a case for polyphyletic origin from different groups of positive-stranded RNA viruses. Sem Virol 3: 327-339

Koonin EV, Dolja VV (1993) Evolution and taxonomy of positive-strand RNA viruses: implications of comparative analysis of amino acid sequences. CRC Crit Rev Biochem Mol Biol 28: 375-430

Koonin EV, Gorbalenya AE (1989) Evolution of RNA genomes: does the high mutation rate necessitate a high rate of evolution of viral proteins? J Mol Evol 28: 524-527

Lwoff A, Horne RW, Tournier P (1962) A system of viruses. Cold Spring Harb Series Quant Biol 27: 51-55

Mahy BW (1991) Related Viruses of the Plant and Animal Kingdoms. Sem Virol 2: 1-77

Matthews REF (ed) (1979) Classification and nomenclature of viruses. Third Report of the International Committee on Taxonomy of Viruses. Intervirology 12: 132-296

Matthews REF (ed) (1982) Classification and Nomenclature of Viruses. Fourth Report of the International Committee on Taxonomy of Viruses. Intervirology 17: 1-199

Matthews REF (ed) (1983) A Critical Appraisal of Viral Taxonomy. CRC Press, Boca Raton, FL

Melnick JL, Allison AC, Butel JS, Eckhart W, Eddy BE, Kit S, Levine AJ, Miles JAR, Pagano JS, Sachs L, Vonka V (1974) *Papovaviridae*. Intervirology 3: 106-120

Morse SS (ed) (1993) The Evolutionary Biology of Viruses. Raven Press, New York

Murphy FA (1995) Virus taxonomy. In: Fields BN, Knipe DM (eds) Fundamental Virology, Third Edn. Raven Press, New York

Murphy FA (1983) Current problems in vertebrate virus taxonomy. In: Matthews REF (ed) A Critical Appraisal of Viral Taxonomy. CRC Press, Boca Raton, FL, pp 37-62

Murphy FA (1985) Virus taxonomy. In: Fields BN, Knipe DM (eds) Fundamental Virology, First Edn. Raven Press, New York, pp 7-26

Murphy FA (1987) Taxonomy of animal viruses. In: Nermut MV, Steven AC (eds) Animal Virus Structure. Elsevier, Amsterdam, pp 99-106

Murphy FA (1988) Virus taxonomy and nomenclature. In: Lennette EH, Halonen P, Murphy FA (eds) Laboratory Diagnosis of Infectious Diseases, Principles and Practices, Vol II. Springer-Verlag, New York, pp 153-176

Murphy FA, Kingsbury DW (1991) Virus taxonomy. In: Fields BN, Knipe DM (eds) Fundamental Virology, Second Edn. Raven Press, New York, pp 9-36

Pankhurst RJ, Aitchison RR (1975) An on-line identification program. In: Pankhurst RJ (ed) Biological Identification with Computers. Academic Press, London, pp 181-185

Partridge TR, Dallwitz MJ, Watson L (1988) A primer for the DELTA system on MS-DOS and VMS. 2nd edition. CSIRO Aust Div Entomol Rep. 38, 1-17

Pfau CJ, Bergold GH, Casals J, Johnson KM, Murphy FA, Pedersen IR, Rawls WE, Rowe WP, Webb PA, Weissenbacher MC (1974) *Arenaviridae.* Intervirology 4: 207-218

Porterfield JS, Casals J, Chumakov MP, Gaidamovich SY, Hanoun C, Holmes IH, Horzinek MC, Mussgay M, Oker Blom N, Russell PK, Trent DW (1978) *Togaviridae.* Intervirology 9: 129-148

Pringle CR (1991) The order *Mononegavirales.* Arch Virol 117: 137-140

Roizman B, Carmichael LE, Deinhardt F, de The G, Nahmias AJ, Plowright W, Rapp F, Sheldrick P, Takahashi M, Wolf K (1982) *Herpesviridae.* Definition, provisional nomenclature and taxonomy. Intervirology 16: 201-217

Roizman B, Desrosiers RC, Fleckenstein B, Lopez C, Minson AC, Studdert MJ (1992) The family *Herpesviridae*: an update. Arch Virol 123: 425-449

Schaffer FL, Bachrach HL, Brown F, Gillespie JH, Burroughs JN, Madin SH, Madeley CR, Povey RC, Scott F, Smith AW, Studdert MJ (1980) *Caliciviridae.* Intervirology 14:1-6

Siddell SG, Anderson R, Cavanagh D, Fujiwara K, Klenk H-D, MacNaughton MR, Pensaert MB, Stohlman SA, Sturman L, van der Zeijst BAM (1983) *Coronaviridae.* Intervirology 20: 181-190

Siegl G, Bates RC, Berns KI, Carter BJ, Kelly DC, Kurstak E, Tattersall P (1985) Characteristics and Taxonomy of *Parvoviridae.* Intervirology 23: 61-73

Strauss JH, Strauss EG (1988) Evolution of RNA viruses. Annu Rev Microbiol 42: 657-683

Strauss JH, Strauss EG, Levine AJ (1991) Virus evolution. In: Fields BN, Knipe DM (eds) Fundamental Virology, second edn. Raven Press, New York, pp 167-190

van Regenmortel MHV (1990) Virus species, a much overlooked but essential concept in virus classification. Intervirology 31:241-254

Ward CW (1993) Progress towards a higher taxonomy of viruses. Res Virol 144: 419-453

Waterson AP, Wilkinson L (eds) (1978) An Introduction to the History of Virology. Cambridge University Press, London

Westaway EG, Brinton MA, Gaidamovich SY, Horzinek MC, Igarashi A, Kääriäinen L, Lvov DK, Porterfield JS, Russell PK, Trent DW (1985) *Flaviviridae.* Intervirology 24:183-192

Wigand R, Bartha A, Dreizin RS, Esche H, Ginsberg HS, Green M, Hierholzer JC, Kalter SS, McFerran JB, Pettersson U, Russell WC, Wadell G (1982) *Adenoviridae*: Second report. Intervirology 18: 169-176

Wildy P (1971) Classification and Nomenclature of Viruses. First Report of the International Committee on Taxonomy of Viruses. Monogr Virol 5: 1-65

Zimmern D (1988) Evolution of RNA viruses. In: Holland JJ, Domingo E, Ahlquist P (eds) RNA Genetics. CRC Press, Boca Raton, FL, pp 211-240

PART II: THE VIRUSES

This report describes the taxa and member viruses approved by the ICTV between 1970 and 1993. Descriptions of the most important characteristics of these taxa are provided, together with a list of members and selected references. These descriptions represent the work of the chairpersons and members of the Subcommittees and Study Groups of the ICTV. A glossary of abbreviations and terms is provided first; followed by a set of virus diagrams and listings of the taxa, alphabetically, then by host, and then by nucleic acid and genome characteristics. A key to the placement of the viruses in the taxa is provided. Descriptions of the taxa and a listing of unassigned viruses follow.

The names of orders, families and genera approved by ICTV are printed in italics. Names that have not yet been approved are printed in quotation marks in standard type. Vernacular species names, whether approved or not, are printed in standard type.

Throughout the Report, three categories of member viruses of the various taxa have been defined: (1) *Type species:* pertains to the type species used in defining the taxon. As noted above, the choice of the type species by ICTV is not made with the kind of precision that must be used by international specialty groups and culture collections or when choosing substrates for vaccines, diagnostic reagents, etc. In this regard, the designation of prototype viruses and strains must be seen as a primary responsibility of international specialty groups. (2) *Other species:* pertains to those viruses which on the basis of all present evidence definitely belong to the taxon. (3) *Tentative species:* pertains to those viruses for which there is presumptive but not conclusive evidence favoring membership of the taxon.

The ICTV has approved one order, 50 families, 9 subfamilies and 164 genera. Descriptions of virus satellites, viroids and the agents of spongiform encephalopathies (prions) of humans and several animal species are included. Finally a list of unassigned viruses is provided with a pertinent reference for each.

GLOSSARY OF ABBREVIATIONS AND VIROLOGICAL TERMS

Note: These terms were approved by the Coordination Subcommittee of ICTV for use in ICTV Report but have no official status.

ABBREVIATIONS

bp	basepair
CF	complement fixing
CPE	cytopathic effect
D	diffusion coefficient
DI	defective interfering
ds	double-stranded
HI	hemagglutination inhibition
kbp	kilo base pair
kDa	kilo Dalton
Mr	molar ratio
ORF	open reading frame
RF	replicative from
RI	replicative intermediate
RNP	ribonucleoprotein
ss	single-stranded

RNA REPLICASES, TRANSCRIPTASES AND POLYMERASES

In the synthesis of viral RNA, the term polymerase has been replaced in general by two somewhat more specific terms: RNA replicase and RNA transcriptase. The term transcriptase has become associated with the enzyme involved in messenger RNA synthesis, most recently with those polymerases which are virion-associated. However, it should be borne in mind that for some viruses it has yet to be established whether or not the replicase and transcriptase activities reflect distinct enzymes rather than alternative activities of a single enzyme. Confusion also arises in the case of the small positive-sense RNA viruses where the term replicase (e.g., Qβ replicase) has been used for the enzyme capable both of transcribing the genome into messenger RNA via an intermediate negative-sense strand and of synthesizing the genome strand from the same template. In the text, the term replicase will be restricted as far as possible to the enzyme synthesizing progeny viral strands of either polarity. The term transcriptase is restricted to those RNA polymerases that are virion-associated and synthesize mRNA. The generalized term RNA polymerase (i.e., RNA-dependent RNA polymerase) is applied where no distinction between replication and transcription enzymes can be drawn (e.g., Qβ, R 17, poliovirus and many plant viruses).

OTHER DEFINITIONS

Enveloped: possessing an outer (bounding) lipoprotein bilayer membrane

Positive-sense (= plus strand, message strand); for RNA, the strand that contains the coding triplets which can be translated by ribosomes. For DNA, the strand that contains the same base sequence as the mRNA. However, in some dsDNA viruses mRNAs are transcribed from both strands and the transcribed regions may overlap. For such viruses this definition is inappropriate.

Negative sense(= minus strand); for RNA or DNA, the negative strand is the strand with base sequence complementary to the positive-sense strand.

Pseudotypes: Enveloped virus particles in which the envelope is derived from one virus and the internal constituents from another.

Transcriptase: found as part of the reverse transcribed viruses.

Reverse virus-encoded RNA-dependent DNA polymerase

Surface projections (= spikes, peplomers, knobs); morphological: features, usually consisting of glycoproteins, that protrude from the lipoprotein envelope of many enveloped viruses.

Virion: Morphologically complete virus particle.

Viroplasm: (= virus factory, virus inclusion, X-body); a modified region within the infected cell in which virus replication occurs, or is thought to occur.

Virus Diagrams

The following pages provide line drawings for the virus families and genera according to their given major host; bacteria (and mycoplasma), algae, fungi and protozoa, plants, invertebrates, and vertebrates. In case of virus families comprising viruses infecting several hosts we have indicated the genera for which it is the primary host. For example the *Togaviridae, Flaviviridae, Rhabdoviridae, Bunyaviridae, Tospovirus* for the families of viruses infecting plants. When all the genera have viruses affecting several hosts we only indicated the family name. For example *Bunyaviridae* and *Picornaviridae* for the families of viruses infecting Invertebrates and Vertebrates. All the diagrams have been drawn similarly: there are frames to separate taxa containing double stranded (ds) and single stranded (ss) genomes and horizontal grey blocks to separate taxa containing DNA and RNA viruses. Taxa containing reverse transcribing (RT) viruses and the negative (-) and positive (+) ssRNA genomes are also indicated. When no virus has been identified in a category, the box has been left empty or not shown.

All the diagrams have been drawn approximately to the same scale to provide an indication of the relative sizes of the viruses; but this cannot be taken as definitive for the following reasons: (i) Different viruses within a family or genus may vary somewhat in size and shape. In general the size and shape have been taken from the type member of the taxon. (ii) Dimensions of some viruses have not been determined with precision. (iii) Some viruses, particularly the larger enveloped ones, are pleomorphic. Only the outlines of most of the smallest viruses are shown, with an indication of the icosahedral structure shown whenever appropriate. The large viruses are shown schematically in surface outline, or in section, as appropriate to display major morphological characteristics.

Most of the diagrams are reproduced from the Fourth ICTV Report (Matthews, 1982) and from the Fifth ICTV Report (Francki *et al*., 1991), updated according to the suggestions of the chairmen of ICTV Subcommittees and Study Groups. In some cases individual virologists provided drawings. We would like to thank all the persons having contributed to help to draw these virus diagrams.

Contributed by

Fauquet CM, Berthiaume L, Ackermann H-W, Calisher CH, Goldbach R, Payment P

FAMILIES OF VIRUSES INFECTING BACTERIA

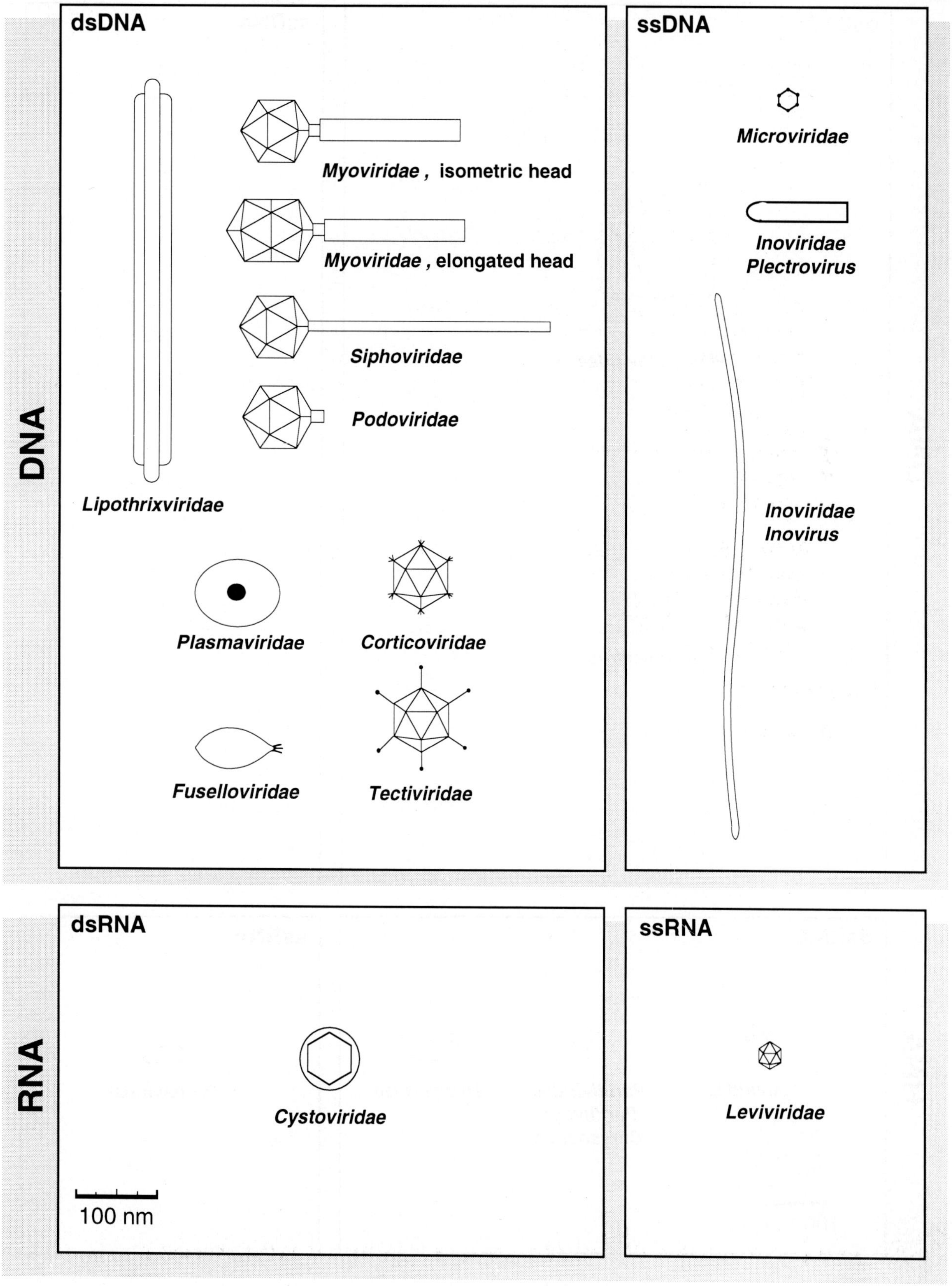
DNA
dsDNA
Myoviridae , isometric head
Myoviridae , elongated head
Siphoviridae
Podoviridae
Lipothrixviridae
Plasmaviridae
Corticoviridae
Fuselloviridae
Tectiviridae
ssDNA
Microviridae
Inoviridae
Plectrovirus
Inoviridae
Inovirus
RNA
dsRNA
Cystoviridae
100 nm
ssRNA
Leviviridae

FAMILIES OF VIRUSES INFECTING ALGAE, FUNGI AND PROTOZOA

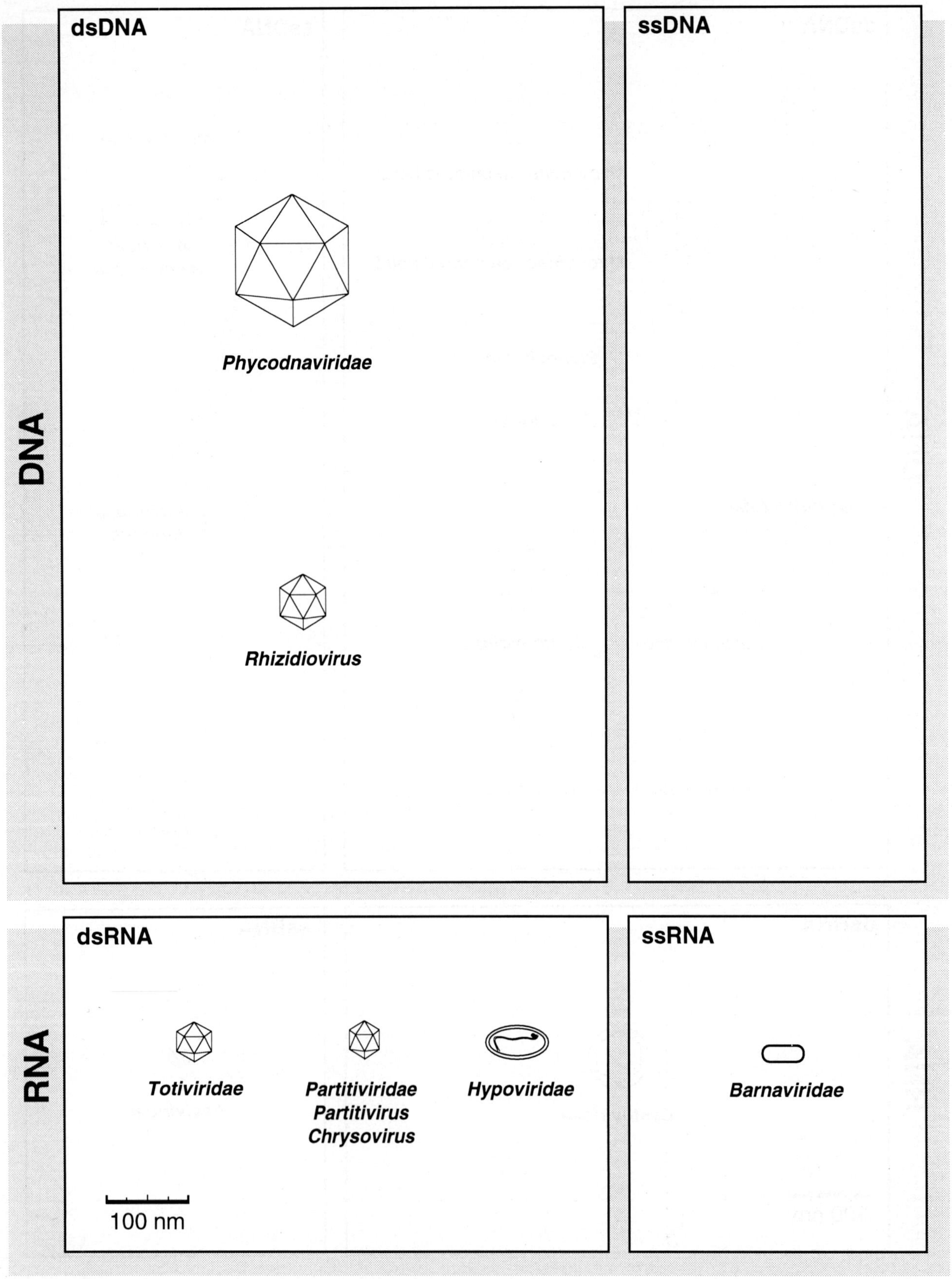

FAMILIES AND GENERA OF VIRUSES INFECTING PLANTS

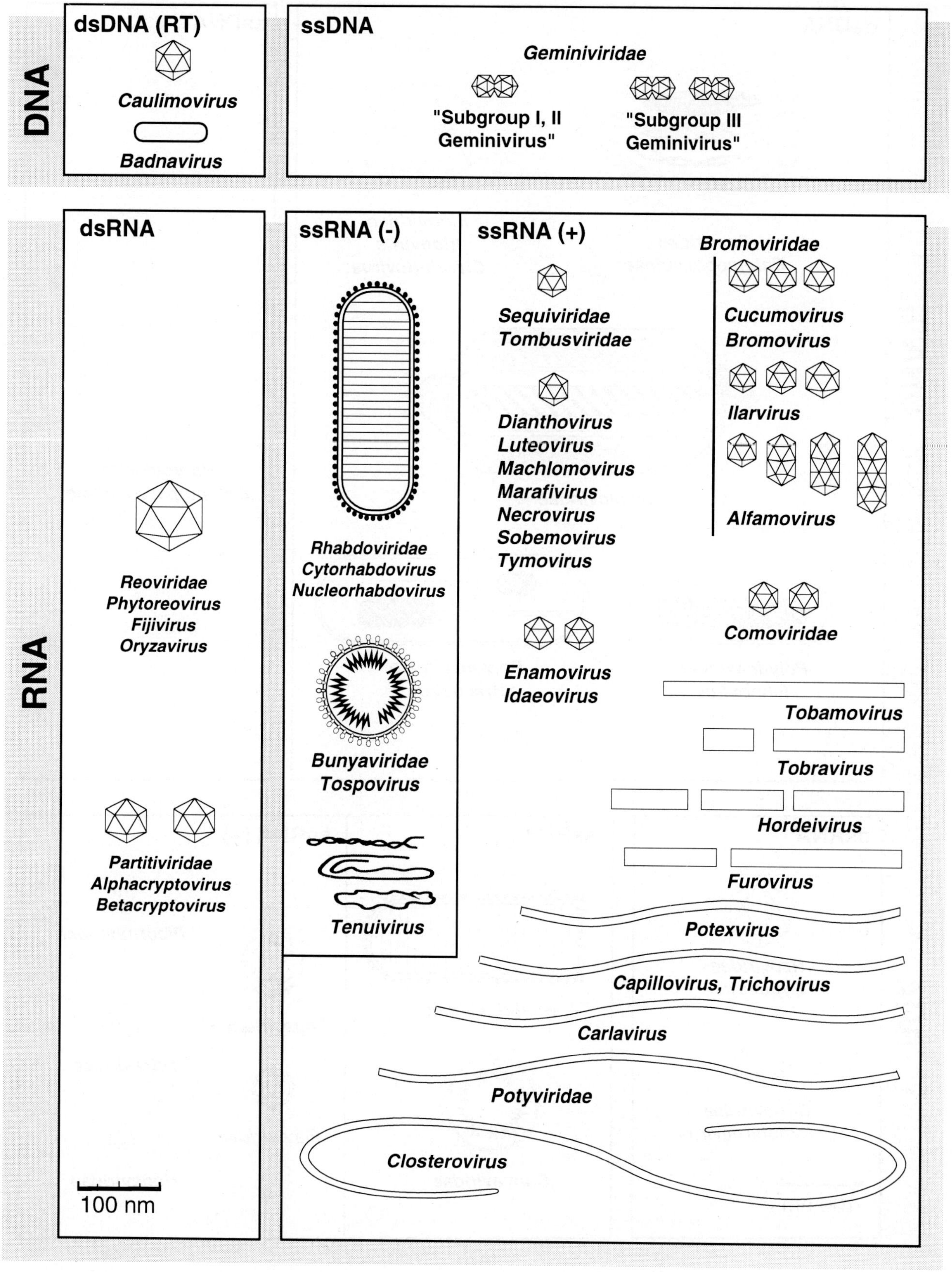
DNA
dsDNA (RT)
Caulimovirus
Badnavirus
ssDNA
Geminiviridae
"Subgroup I, II Geminivirus"
"Subgroup III Geminivirus"
RNA
dsRNA
Reoviridae
Phytoreovirus
Fijivirus
Oryzavirus
Partitiviridae
Alphacryptovirus
Betacryptovirus
100 nm
ssRNA (-)
Rhabdoviridae
Cytorhabdovirus
Nucleorhabdovirus
Bunyaviridae
Tospovirus
Tenuivirus
ssRNA (+)
Sequiviridae
Tombusviridae
Dianthovirus
Luteovirus
Machlomovirus
Marafivirus
Necrovirus
Sobemovirus
Tymovirus
Enamovirus
Idaeovirus
Bromoviridae
Cucumovirus
Bromovirus
Ilarvirus
Alfamovirus
Comoviridae
Tobamovirus
Tobravirus
Hordeivirus
Furovirus
Potexvirus
Capillovirus, Trichovirus
Carlavirus
Potyviridae
Closterovirus

FAMILIES OF VIRUSES INFECTING INVERTEBRATES

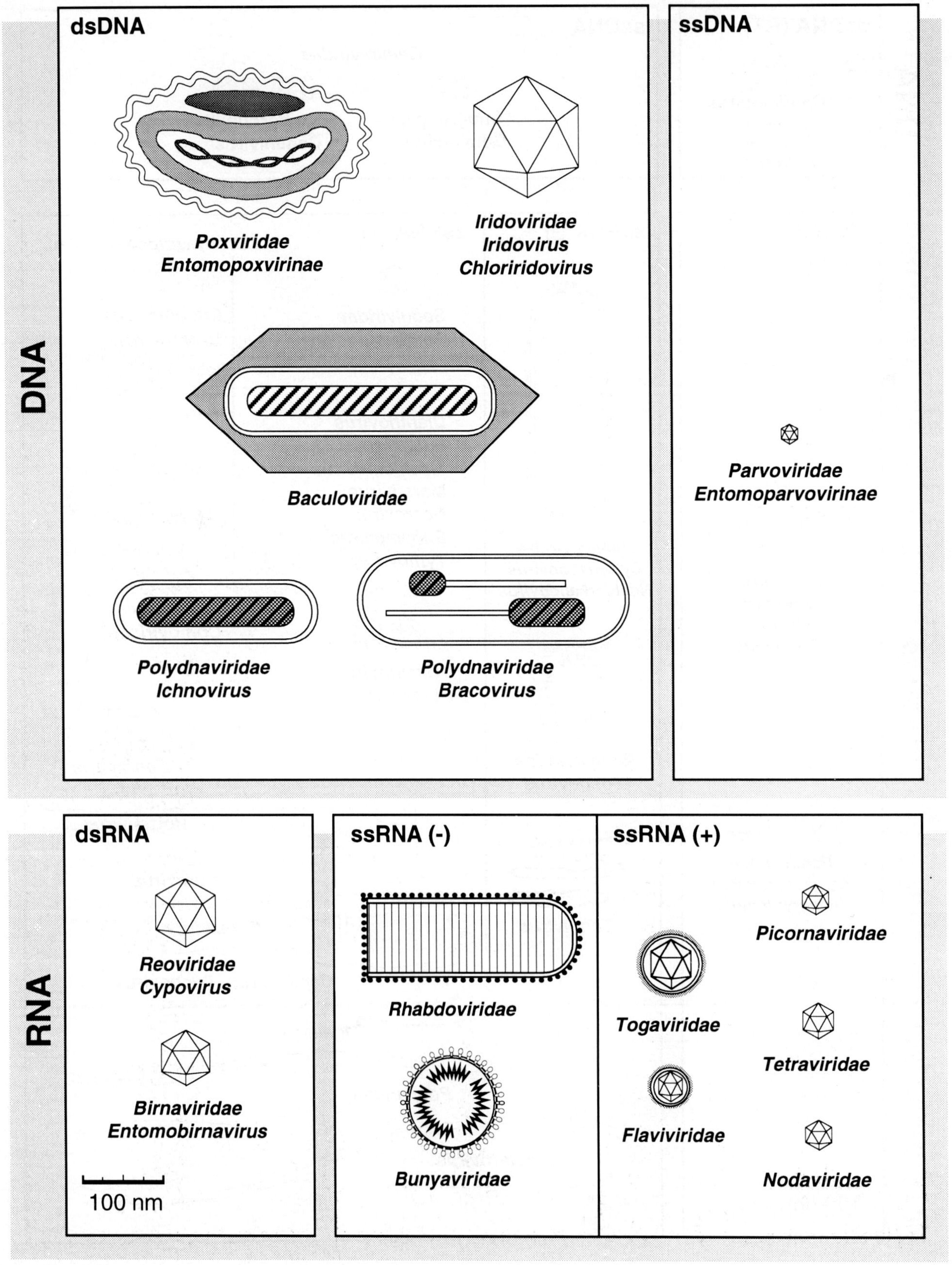

FAMILIES OF VIRUSES INFECTING VERTEBRATES

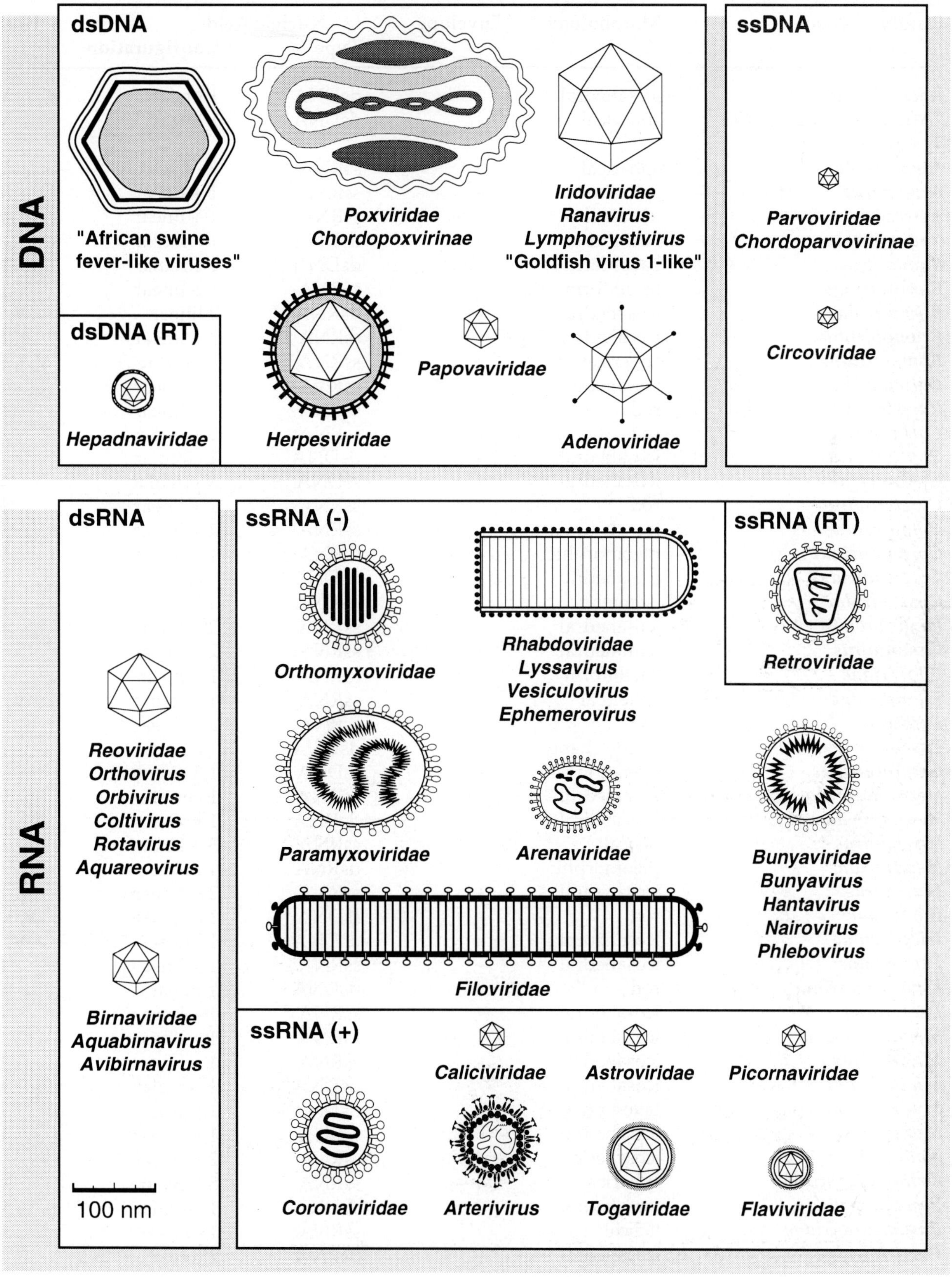
DNA
dsDNA
"African swine fever-like viruses"
Poxviridae
Chordopoxvirinae
Iridoviridae
Ranavirus
Lymphocystivirus
"Goldfish virus 1-like"
dsDNA (RT)
Hepadnaviridae
Herpesviridae
Papovaviridae
Adenoviridae
ssDNA
Parvoviridae
Chordoparvovirinae
Circoviridae
RNA
dsRNA
Reoviridae
Orthovirus
Orbivirus
Coltivirus
Rotavirus
Aquareovirus
Birnaviridae
Aquabirnavirus
Avibirnavirus
100 nm
ssRNA (-)
Orthomyxoviridae
Rhabdoviridae
Lyssavirus
Vesiculovirus
Ephemerovirus
Paramyxoviridae
Arenaviridae
Bunyaviridae
Bunyavirus
Hantavirus
Nairovirus
Phlebovirus
Filoviridae
ssRNA (RT)
Retroviridae
ssRNA (+)
Caliciviridae
Astroviridae
Picornaviridae
Coronaviridae
Arterivirus
Togaviridae
Flaviviridae

Listing of Virus Families and Floating Genera

Table I: Alphabetical Listing of Families and Floating Genera

Family or Genus	Morphology	Envelope	Nucleic Acid Type	Nucleic Acid Configuration	Host
Adenoviridae	icosahedral	-	dsDNA	1 linear	V
"African swine fever-like viruses"	spherical	+	dsDNA	1 linear	V
Arenaviridae	spherical	+	ssRNA	2 - linear	V
Arterivirus	spherical	+	ssRNA	1 + linear	V
Astroviridae	icosahedral	-	ssRNA	1 + linear	V
Baculoviridae	bacilliform	+	dsDNA	1 circular	I
Badnavirus	bacilliform	-	dsDNA	1 circular	P
Barnaviridae	bacilliform	-	ssRNA	1 + linear	F
Birnaviridae	icosahedral	-	dsRNA	2 linear	V, I
Bromoviridae	icosahedral	-	ssRNA	3 + linear	P
Bunyaviridae	spherical	+	ssRNA	3 - linear	V, I, P
Caliciviridae	icosahedral	-	ssRNA	1 + linear	V
Capillovirus	rod	-	ssRNA	1 + linear	P
Carlavirus	rod	-	ssRNA	1 + linear	P
Caulimovirus	icosahedral	-	dsDNA	1 circular	P
Circoviridae	icosahedral	-	ssDNA	X circular	V
Closterovirus	rod	-	ssRNA	1 + linear	P
Comoviridae	icosahedral	-	ssRNA	2 + linear	P
Coronaviridae	pleomorphic	+	ssRNA	1 + linear	V
Corticoviridae	icosahedral	-	dsDNA	1 circular	B
Cystoviridae	isometric	+	dsRNA	3 linear	B
Dianthovirus	icosahedral	-	ssRNA	2 + linear	P
Enamovirus	icosahedral	-	ssRNA	2 + linear	P
Filoviridae	bacilliform	+	ssRNA	1 - linear	V
Flaviviridae	spherical	+	ssRNA	1 + linear	V, I
Furovirus	rod	-	ssRNA	2 + linear	P
Fuselloviridae	lemon shape	+	dsDNA	1 circular	B
Geminiviridae	isometric	-	ssDNA	1,2 circular	P
Hepadnaviridae	icosahedral	-	ssDNA	1 circular	V
Herpesviridae	icosahedral	+	dsDNA	1 linear	V
Hordeivirus	helical	-	ssRNA	3 + linear	P
Hypoviridae	pleomorphic	+	dsRNA	1 linear	F
Idaeovirus	icosahedral	-	ssRNA	2 + linear	P
Inoviridae	rod	-	ssDNA	1 circular	B, M
Iridoviridae	icosahedral	+	dsDNA	1 linear	V, I
Leviviridae	icosahedral	-	ssRNA	1 + linear	B
Lipothrixviridae	rod	+	dsDNA	1 linear	B
Luteovirus	icosahedral	-	ssRNA	1 + linear	P
Machlomovirus	icosahedral	-	ssRNA	1 + linear	P
Marafivirus	icosahedral	-	ssRNA	1 + linear	P
Microviridae	icosahedral	-	dsDNA	1 circular	B
Myoviridae	tailed phage	-	dsDNA	1 linear	B
Necrovirus	icosahedral	-	ssRNA	1 + linear	P
Nodaviridae	icosahedral	-	ssRNA	2 + linear	I
Orthomyxoviridae	spherical	+	ssRNA	8 - linear	V
Papovaviridae	icosahedral	-	dsDNA	1 circular	V
Paramyxoviridae	helical	+	ssRNA	1 - linear	V
Partitiviridae	icosahedral	-	dsRNA	2 linear	F, P

Family or Genus	Morphology	Envelope	Nucleic Acid Type	Nucleic Acid Configuration	Host
Parvoviridae	icosahedral	-	ssDNA	1 - linear	V, I
Phycodnaviridae	icosahedral	-	dsDNA	1 + linear	A
Picornaviridae	icosahedral	-	ssRNA	1 + linear	V, I
Plasmaviridae	pleomorphic	+	dsDNA	1 circular	M
Podoviridae	tailed phage	-	dsDNA	1 linear	B
Polydnaviridae	rod, fusiform	+	dsDNA	X supercoiled	I
Potexvirus	rod	-	ssRNA	1 + linear	P
Potyviridae	rod	-	ssRNA	1 + linear	P
Poxviridae	ovoid	+	dsDNA	1 linear	V, I
Reoviridae	icosahedral	-	dsRNA	10 - 12 linear	V, I, P
Retroviridae	spherical	+	ssRNA	dimer 1 + linear	V
Rhabdoviridae	bacilliform	+	ssRNA	1 - linear	V, I, P
Rhizidiovirus	icosahedral	-	dsDNA	1 linear	F
Sequiviridae	icosahedral	-	ssRNA	1 + linear	P
Siphoviridae	tailed phage	-	dsDNA	1 linear	B
Sobemovirus	icosahedral	-	ssRNA	1 + linear	P
Tectiviridae	icosahedral	-	dsDNA	1 linear	B
Tenuivirus	amorphic	?	ssRNA	4-5 +/- linear	P
Tetraviridae	icosahedral	-	ssRNA	1, 2 + linear	I
Tobamovirus	rod	-	ssRNA	1 + linear	P
Tobravirus	rod	-	ssRNA	2 + linear	P
Togaviridae	spherical	+	ssRNA	1 + linear	V, I
Tombusviridae	icosahedral	-	ssRNA	1 + linear	P
Totiviridae	icosahedral	-	dsRNA	1 + linear	F, Pr
Trichovirus	helical	-	ssRNA	1 + linear	P
Tymovirus	icosahedral	-	ssRNA	1 + linear	P
Umbravirus	?	?	ssRNA	1 + linear	P

TABLE II: FAMILIES AND FLOATING GENERA LISTED BY HOST

Family or Genus	Morphology	Envelope	Nucleic Acid Type	Nucleic Acid Configuration	Host
Phycodnaviridae	icosahedral	-	dsDNA	1 + linear	A
Corticoviridae	icosahedral	-	dsDNA	1 circular	B
Cystoviridae	isometric	+	dsRNA	3 linear	B
Fuselloviridae	lemon shape	+	dsDNA	1 circular	B
Leviviridae	icosahedral	-	ssRNA	1 + linear	B
Lipothrixviridae	rod	+	dsDNA	1 linear	B
Microviridae	icosahedral	-	dsDNA	1 circular	B
Myoviridae	tailed phage	-	dsDNA	1 linear	B
Podoviridae	tailed phage	-	dsDNA	1 linear	B
Siphoviridae	tailed phage	-	dsDNA	1 linear	B
Tectiviridae	icosahedral	-	dsDNA	1 linear	B
Inoviridae	rod	-	ssDNA	1 circular	B, M
Barnaviridae	bacilliform	-	ssRNA	1 + linear	F
Hypoviridae	pleomorphic	+	dsRNA	1 linear	F
Rhizidiovirus	icosahedral	-	dsDNA	1 linear	F
Partitiviridae	icosahedral	-	dsRNA	2 linear	F, P
Totiviridae	icosahedral	-	dsRNA	1 + linear	F, Pr
Baculoviridae	bacilliform	+	dsDNA	1 circular	I
Nodaviridae	icosahedral	-	ssRNA	2 + linear	I
Polydnaviridae	rod, fusiform	+	dsDNA	X supercoiled	I
Tetraviridae	icosahedral	-	ssRNA	1, 2 + linear	I
Plasmaviridae	pleomorphic	+	dsDNA	1 circular	M
Badnavirus	bacilliform	-	dsDNA	1 circular	P
Bromoviridae	icosahedral	-	ssRNA	3 + linear	P
Capillovirus	rod	-	ssRNA	1 + linear	P
Carlavirus	rod	-	ssRNA	1 + linear	P
Caulimovirus	icosahedral	-	dsDNA	1 circular	P
Closterovirus	rod	-	ssRNA	1 + linear	P
Comoviridae	icosahedral	-	ssRNA	2 + linear	P
Dianthovirus	icosahedral	-	ssRNA	2 + linear	P
Enamovirus	icosahedral	-	ssRNA	2 + linear	P
Furovirus	rod	-	ssRNA	2 + linear	P
Geminiviridae	isometric	-	ssDNA	1,2 circular	P
Hordeivirus	helical	-	ssRNA	3 + linear	P
Idaeovirus	icosahedral	-	ssRNA	2 + linear	P
Luteovirus	icosahedral	-	ssRNA	1 + linear	P
Machlomovirus	icosahedral	-	ssRNA	1 + linear	P
Marafivirus	icosahedral	-	ssRNA	1 + linear	P
Necrovirus	icosahedral	-	ssRNA	1 + linear	P
Potexvirus	rod	-	ssRNA	1 + linear	P
Potyviridae	rod	-	ssRNA	1 + linear	P
Sequiviridae	icosahedral	-	ssRNA	1 + linear	P
Sobemovirus	icosahedral	-	ssRNA	1 + linear	P
Tenuivirus	amorphic	?	ssRNA	4-5 +/- linear	P
Tobamovirus	rod	-	ssRNA	1 + linear	P
Tobravirus	rod	-	ssRNA	2 + linear	P
Tombusviridae	icosahedral	-	ssRNA	1 + linear	P
Trichovirus	helical	-	ssRNA	1 + linear	P
Tymovirus	icosahedral	-	ssRNA	1 + linear	P
Umbravirus	?	?	ssRNA	1 + linear	P
Adenoviridae	icosahedral	-	dsDNA	1 linear	V

Family or Genus	Morphology	Envelope	Nucleic Acid Type	Nucleic Acid Configuration	Host
"African swine fever-like viruses"	spherical	+	dsDNA	1 linear	V
Arenaviridae	spherical	+	ssRNA	2 - linear	V
Arterivirus	spherical	+	ssRNA	1 + linear	V
Astroviridae	icosahedral	-	ssRNA	1 + linear	V
Caliciviridae	icosahedral	-	ssRNA	1 + linear	V
Circoviridae	icosahedral	-	ssDNA	X circular	V
Coronaviridae	pleomorphic	+	ssRNA	1 + linear	V
Filoviridae	bacilliform	+	ssRNA	1 - linear	V
Hepadnaviridae	icosahedral	-	ssDNA	1 circular	V
Herpesviridae	icosahedral	+	dsDNA	1 linear	V
Orthomyxoviridae	spherical	+	ssRNA	8 - linear	V
Papovaviridae	icosahedral	-	dsDNA	1 circular	V
Paramyxoviridae	helical	+	ssRNA	1 - linear	V
Retroviridae	spherical	+	ssRNA	dimer 1+linear	V
Birnaviridae	icosahedral	-	dsRNA	2 linear	V, I
Flaviviridae	spherical	+	ssRNA	1 + linear	V, I
Iridoviridae	icosahedral	+	dsDNA	1 linear	V, I
Parvoviridae	icosahedral	-	ssDNA	1 - linear	V, I
Picornaviridae	icosahedral	-	ssRNA	1 + linear	V, I
Poxviridae	ovoid	+	dsDNA	1 linear	V, I
Togaviridae	spherical	+	ssRNA	1 + linear	V, I
Bunyaviridae	spherical	+	ssRNA	3 - linear	V, I, P
Reoviridae	icosahedral	-	dsRNA	10 - 12 linear	V, I, P
Rhabdoviridae	bacilliform	+	ssRNA	1 - linear	V, I, P

TABLE III: FAMILIES AND FLOATING GENERA LISTED BY NUCLEIC ACID

Family or Genus	Morphology	Envelope	Nucleic Acid Type	Nucleic Acid Configuration	Host
Phycodnaviridae	icosahedral	-	dsDNA	1 + linear	A
Baculoviridae	bacilliform	+	dsDNA	1 circular	I
Badnavirus	bacilliform	-	dsDNA	1 circular	P
Caulimovirus	icosahedral	-	dsDNA	1 circular	P
Corticoviridae	icosahedral	-	dsDNA	1 circular	B
Fuselloviridae	lemon shape	+	dsDNA	1 circular	B
Microviridae	icosahedral	-	dsDNA	1 circular	B
Papovaviridae	icosahedral	-	dsDNA	1 circular	V
Plasmaviridae	pleomorphic	+	dsDNA	1 circular	M
Adenoviridae	icosahedral	-	dsDNA	1 linear	V
"African swine fever-like viruses	spherical	+	dsDNA	1 linear	V
Herpesviridae	icosahedral	+	dsDNA	1 linear	V
Iridoviridae	icosahedral	+	dsDNA	1 linear	V, I
Lipothrixviridae	rod	+	dsDNA	1 linear	B
Myoviridae	tailed phage	-	dsDNA	1 linear	B
Podoviridae	tailed phage	-	dsDNA	1 linear	B
Poxviridae	ovoid	+	dsDNA	1 linear	V, I
Rhizidiovirus	icosahedral	-	dsDNA	1 linear	F
Siphoviridae	tailed phage	-	dsDNA	1 linear	B
Tectiviridae	icosahedral	-	dsDNA	1 linear	B
Polydnaviridae	rod, fusiform	+	dsDNA	X supercoiled	I
Totiviridae	icosahedral	-	dsRNA	1 + linear	F, Pr
Hypoviridae	pleomorphic	+	dsRNA	1 linear	F
Birnaviridae	icosahedral	-	dsRNA	2 linear	V, I
Partitiviridae	icosahedral	-	dsRNA	2 linear	F, P
Cystoviridae	isometric	+	dsRNA	3 linear	B
Reoviridae	icosahedral	-	dsRNA	10 - 12 linear	V, I, P
Parvoviridae	icosahedral	-	ssDNA	1 - linear	V, I
Hepadnaviridae	icosahedral	-	ssDNA	1 circular	V
Inoviridae	rod	-	ssDNA	1 circular	B, M
Geminiviridae	isometric	-	ssDNA	1,2 circular	P
Circoviridae	icosahedral	-	ssDNA	X circular	V
Arterivirus	spherical	+	ssRNA	1 + linear	V
Astroviridae	icosahedral	-	ssRNA	1 + linear	V
Barnaviridae	bacilliform	-	ssRNA	1 + linear	F
Caliciviridae	icosahedral	-	ssRNA	1 + linear	V
Capillovirus	rod	-	ssRNA	1 + linear	P
Carlavirus	rod	-	ssRNA	1 + linear	P
Closterovirus	rod	-	ssRNA	1 + linear	P
Coronaviridae	pleomorphic	+	ssRNA	1 + linear	V
Flaviviridae	spherical	+	ssRNA	1 + linear	V, I
Leviviridae	icosahedral	-	ssRNA	1 + linear	B
Luteovirus	icosahedral	-	ssRNA	1 + linear	P
Machlomovirus	icosahedral	-	ssRNA	1 + linear	P
Marafivirus	icosahedral	-	ssRNA	1 + linear	P
Necrovirus	icosahedral	-	ssRNA	1 + linear	P
Picornaviridae	icosahedral	-	ssRNA	1 + linear	V, I
Potexvirus	rod	-	ssRNA	1 + linear	P
Potyviridae	rod	-	ssRNA	1 + linear	P
Sequiviridae	icosahedral	-	ssRNA	1 + linear	P
Sobemovirus	icosahedral	-	ssRNA	1 + linear	P

Family or Genus	Morphology	Envelope	Nucleic Acid Type	Nucleic Acid Configuration	Host
Tobamovirus	rod	-	ssRNA	1 + linear	P
Togaviridae	spherical	+	ssRNA	1 + linear	V, I
Tombusviridae	icosahedral	-	ssRNA	1 + linear	P
Trichovirus	helical	-	ssRNA	1 + linear	P
Tymovirus	icosahedral	-	ssRNA	1 + linear	P
Umbravirus	?	?	ssRNA	1 + linear	P
Filoviridae	bacilliform	+	ssRNA	1 - linear	V
Paramyxoviridae	helical	+	ssRNA	1 - linear	V
Rhabdoviridae	bacilliform	+	ssRNA	1 - linear	V, I, P
Tetraviridae	icosahedral	-	ssRNA	1, 2 + linear	I
Comoviridae	icosahedral	-	ssRNA	2 + linear	P
Dianthovirus	icosahedral	-	ssRNA	2 + linear	P
Enamovirus	icosahedral	-	ssRNA	2 + linear	P
Furovirus	rod	-	ssRNA	2 + linear	P
Idaeovirus	icosahedral	-	ssRNA	2 + linear	P
Nodaviridae	icosahedral	-	ssRNA	2 + linear	I
Tobravirus	rod	-	ssRNA	2 + linear	P
Arenaviridae	spherical	+	ssRNA	2 - linear	V
Bromoviridae	icosahedral	-	ssRNA	3 + linear	P
Hordeivirus	helical	-	ssRNA	3 + linear	P
Bunyaviridae	amorphic	?	ssRNA	4-5 +/- linear	V,I, P
Orthomyxoviridae	spherical	+	ssRNA	8 - linear	V
Retroviridae	spherical	+	ssRNA	dimer 1+linear	V

A: algae; B: bacteria; F: fungi; I: invertebrates; M: mycoplasma; P: plants; Pr: protozoa; V: vertebrates

KEY TO THE PLACEMENT OF VIRUSES IN TAXA

1.	Genome DNA	2
	Genome RNA	49
2.	Virion DNA is continuous; reverse transcriptase not used during replication	3
	Virion DNA contains discontinuities; reverse transcriptase used during replicaton	46
3.	DNA double-stranded	4
	DNA single-stranded	35

THE DS DNA VIRUSES

4.	Host a prokaryote	5
	Host a eukaryote	12
5.	Virion tailed	6
	Virion not tailed	8
6.	Tail contractile > 15 nm in diameter	*Myoviridae* / "T4-like phages"
	Tail not contractile < 12 nm in diameter	7
7.	Tail long (65 - 600 nm)	*Siphoviridae* / "λ-like phages"
	Tail short (10 - 20 nm)	*Podoviridae* / "T7-like phages"
8.	Virion not enveloped	9
	Virion enveloped	10
9.	DNA linear > 10 kbp; inner capsid can form a tail-like appendage	***Tectiviridae* / *Tectivirus***
	DNA circular < 10 kbp; no tail-like appendage is formed	***Corticoviridae* / *Corticovirus***
10.	Host a mycoplasma	***Plasmaviridae* / *Plasmavirus***
	Host an archaebacterium	11
11.	Virion rod-shaped	***Lipothrixviridae* / *Lipothrixvirus***
	Virion lemon-shaped	***Fuselloviridae* / *Fusellovirus***
12.	Virion contains one or more fusiform or cylindrical nucleocapsids and multiple DNA molecules	***(Polydnaviridae)*** 13
	Virion contains a single DNA molecule	14
13.	Nucleocapsid 85 x 330 nm with 2 envelopes	***Polydnaviridae* / *Ichnovirus***
	Nucleocapsid cylindrical, 40 nm diameter x 30-150 nm, with 1 envelope	***Polydnaviridae* / *Bracovirus***
14.	DNA ≥ 90 kbp	15
	DNA < 90 kbp	31
15.	DNA > 300 kbp; virion not enveloped; host an alga	***Phycodnaviridae* / *Phycodnavirus***
	DNA usually < 300 kbp; virion enveloped; host an animal	16
16.	Genome covalently closed circular DNA; nucleocapsid rod-shaped	***(Baculoviridae)*** 17
	Genome linear DNA; nucleocapsid not rod-shaped	18
17.	Inclusions typically contain numerous virions	***Baculoviridae* / *Nucleopolyhedrovirus***
	Inclusions typically contain a single virion	***Baculoviridae* / *Granulovirus***

18. Virion ovoid or brick-shaped — ***(Poxviridae)*** 19
Virion not ovoid or brick-shaped — 25

19. Host a vertebrate — ***(Poxviridae / Chordopoxvirinae)*** 20
Host an invertebrate — ***(Poxviridae / Entomopoxvirinae)*** 24

20. Virion ovoid — ***Poxviridae / Chordopoxvirinae / Parapoxvirus***
Virion brick-shaped — 21

21. Largest virion dimension > 320 nm; DNA > 250 kbp; host a bird — ***Poxviridae / Chordopoxvirinae / Avipoxvirus***
Largest virion dimension < 290 nm; DNA < 250 kbp — ***Poxviridae / Chordopoxvirinae /Orthopoxvirus***
Largest virion dimension > 290 nm; DNA < 250 kbp — 22

22. DNA 175 kbp; largest virion dimension 300 nm — ***Poxviridae / Chordopoxvirinae / Suipoxvirus***
DNA 188 kbp; largest virion dimension 320 nm — ***Poxviridae / Chordopoxvirinae / Molluscipoxvirus***
DNA < 170 kbp — 23

23. Virion 300 x 270 x 200 nm; DNA about 145 kbp — ***Poxviridae / Chordopoxvirinae / Capripoxvirus***
Virion 300 x 250 x 200 nm; DNA 160 kbp; GC content about 40% — ***Poxviridae / Chordopoxvirinae / Leporipoxvirus***
Virion 300 x 250 x 200 nm; DNA 146 kbp, GC content about 33% — ***Poxviridae / Chordopoxvirinae / Yatapoxvirus***

24. Virion ovoid, 450 x 250 nm; host from *Coleoptera* — ***Poxviridae / Entomopoxvirinae / Entomopoxvirus A***
Virion ovoid, 350 x 250 nm; DNA about 225 kbp; host from *Lepidoptera* or *Orthoptera* — ***Poxviridae / Entomopoxvirinae / Entomopoxvirus B***
Virion brick-shaped, 320 x 230 x 110 nm; DNA > 240 kbp; host from *Diptera* — ***Poxviridae / Entomopoxvirinae / Entomopoxvirus C***

25. Virion icosahedral with 70 - 100 nm diameter cores; virus multiplies in ticks and swine — "African swine fever-like viruses"
Virion icosahedral; genome circularly permutated and terminally redundant; multiplies only in poikilothermic animals — ***(Iridoviridae)*** 26
Virion quasi-spherical with 100 - 110 nm diameter cores; genome not circularly permutated; multiplies only in vertebrates — ***(Herpesviridae)*** 30

26. Host an invertebrate — 27
Host a vertebrate — 28

27. Virion 120 nm in diameter — ***Iridoviridae / Iridovirus***
Virion 180 nm in diameter — ***Iridoviridae / Chloriridovirus***

28. Host an amphibian — ***Iridoviridae / Ranavirus***
Host a fish — 29

29. Virion ≥ 200 nm in diameter — ***Iridoviridae / Lymphocystivirus***
Virion < 200 nm in diameter — ***Iridoviridae*** / "Goldfish virus 1-like viruses"

30. Reproductive cycle short, spread in culture rapid; infection often induces epithelial lesions; gene complement characteristic of human herpesvirus 1 — ***Herpesviridae / Alphaherpesvirinae / Simplexvirus / Varicellovirus***

Reproductive cycle long, spread in culture slow; gene complement characteristic of human herpesvirus 5 — ***Herpesviridae / Betaherpesvirinae / Cytomegalovirus / Muromegalovirus / Roseolovirus***

Infection often latent in lymphocytes and may cause lymphoproliferative disease; gene complement characteristic of human herpesvirus 4 — ***Herpesviridae / Gammaherpesvirinae / Lymphocryptovirus / Rhadinovirus***

31. DNA < 30 kbp — ***(Papovaviridae)*** 32
DNA > 30 kbp — 33

32. Virion 45 nm in diameter; DNA about 5 kbp with proteins encoded on both strands — ***Papovaviridae / Polyomavirus***
Virion about 55 nm in diameter; DNA about 8 kbp with proteins encoded on one strand — ***Papovaviridae / Papillomavirus***

33. Host a fungus — ***Rhizidiovirus***
Host a vertebrate — ***(Adenoviridae)*** 34

34. Host a mammal — ***Adenoviridae / Mastadenovirus***
Host a bird — ***Adenoviridae / Aviadenovirus***

THE ssDNA VIRUSES

35. Host a prokaryote — 36
Host a eukaryote — 39

36. Virion has helical symmetry — ***(Inoviridae)*** 37
Virion icosahedral — ***(Microviridae)*** 38

37. Virion filamentous, 700 - 2000 nm in length — ***Inoviridae / Inovirus***
Virion short, rod-shaped, 70 - 280 nm in length — ***Inoviridae / Plectrovirus***

38. Host an enterobacterium — ***Microviridae / Microvirus***
Host *Spiroplasma sp.* — ***Microviridae / Spiromicrovirus***
Host *Bdellovibrio bacteriovorus* — ***Microviridae / Bdellomicrovirus***
Host *Chlamydia psittaci* — ***Microviridae / Chlamydiamicrovirus***

39. Host a plant — ***(Geminiviridae)*** 40
Host not a plant — 41

40. Genome monopartite; host graminaceous; vector a leafhopper — ***Geminiviridae*** / "Subgroup I Geminivirus"
Genome monopartite; host dicotyledonous; vector a leafhopper — ***Geminiviridae*** / "Subgroup II Geminivirus"
Genome mono or bipartite; vector a whitefly — ***Geminiviridae*** / "Subgroup III Geminivirus"

41. DNA circular — ***Circoviridae / Circovirus***
DNA linear — ***(Parvoviridae)*** 42

42. Host a vertebrate — ***(Parvoviridae / Parvovirinae)*** 43
Host an invertebrate — ***(Parvoviridae / Densovirinae)*** 45

43. A helper virus (adenovirus or herpesvirus) needed for productive multiplication — ***Parvoviridae / Parvovirinae / Dependovirus***
Virus multiplies autonomously — 44

44. DNA contains 2 mRNA promoters — *Parvoviridae / Parvovirinae / Parvovirus*
DNA contains 1 mRNA promoter — *Parvoviridae / Parvovirinae / Erythrovirus*

45. DNA 6 kb, structural and non-structural proteins encoded on different strands — *Parvoviridae / Densovirinae / Densovirus*
DNA 5 kb; proteins all encoded on one strand; virion contains similar amounts of each sense DNA — *Parvoviridae / Densovirinae / Iteravirus*
DNA 4 kb; proteins all encoded on one strand; virion contains mainly negative sense DNA — *Parvoviridae / Densovirinae / Contravirus*

THE DNA AND RNA REVERSE TRANSCRIBING VIRUSES

46. DNA < 5 kbp; host a vertebrate — *(Hepadnaviridae)* 47
DNA > 7 kbp; host a plant — 48

47. Virion < 45 nm in diameter; nucleocapsid about 27 nm in diameter; host a mammal — *Hepadnaviridae / Orthohepadnavirus*
Virion > 45 nm in diameter; nucleocapsid about 35 nm in diameter; host a bird — *Hepadnaviridae / Avihepadnavirus*

48. Virion bacilliform — *Badnavirus*
Virion icosahedral — *Caulimovirus*

49. Genome encodes reverse transcriptase; DNA copies integrate in host genome — *(Retroviridae)* 50
Genome does not encode reverse transcriptase; virus genome does not integrate — 56

50. RNA > 8.5 kb — 51
RNA < 8.5 kb — 53

51. RNA < 10 kb; nucleocapsid bar-shaped or cone-shaped — *Retroviridae / Lentivirus*
RNA > = 10 kb; nucleocapsid not bar- or cone-shaped — 52

52. RNA 10 kb; nucleocapsid spherical and centrally located; gag, pro and pol encoded in different reading frames — *Retroviridae* / "Mammalian type B retroviruses"
RNA 11 kb; nucleocapsid eccentric; gag, pro and pol encoded in the same reading frame — *Retroviridae / Spumavirus*

53. RNA < 8 kb; LTR about 350 nt in length — 54
RNA 8.3 kb; LTR about 600 nt in length — 55

54. RNA 7.2 kb; gag and pol encoded in the same reading frame; host a bird — *Retroviridae* / "Avian type C retroviruses"
RNA 8 kb; gag and pro encoded in different reading frames; host a mammal — *Retroviridae* / "Type D retroviruses"

55. gag, pro and pol encoded in the same reading frame; R sequence in the LTR about 60 nt — *Retroviridae* / "Mammalian type C retroviruses"
pro encoded in a reading frame different from that encoding gag and pol; R sequence in the LTR >130 nt — *Retroviridae* / "HTLV-BLV retroviruses"

56. RNA double-stranded — 57
RNA single-stranded — 77

THE dsRNA VIRUSES

57. Host a prokaryote — *Cystoviridae / Cystovirus*
Host a eukaryote — 58

58.	Genome in > 9 segments	*(Reoviridae)* 59
	Genome in < 9 segments	67
59.	Host an animal	60
	Host a plant	65
60.	Genome in 10 segments	61
	Genome in > 10 segments	63
61.	Virion lacks an outer capsid and is < 70 nm in diameter	*Reoviridae / Cypovirus*
	Virion comprises cores and outer capsid and is > 70 nm in diameter	62
62.	Outer capsid distinct; virion sediments at > 600 S	*Reoviridae / Orthoreovirus*
	Outer capsid indistinct; virion sediments at < 600 S	*Reoviridae / Orbivirus*
63.	Genome in 12 segments	*Reoviridae / Coltivirus*
	Genome in 11 segments	64
64.	Virion appears wheel-like; 9 RNA segments are > 2 kbp; host a mammal or a bird	*Reoviridae / Rotavirus*
	Virion not wheel-like; 6 RNA segments are > 2 kbp; host a fish or a shellfish	*Reoviridae / Aquareovirus*
65.	Genome in 12 segments; virion lacks spikes	*Reoviridae / Phytoreovirus*
	Genome in 10 segments; virion bears spikes	66
66.	Virion 65 - 70 nm in diameter, with an outer capsid	*Reoviridae / Fijivirus*
	Virion 57 - 65 nm in diameter, lacks an outer capsid	*Reoviridae / Oryzavirus*
67.	Host an animal	*(Birnaviridae)* 68
	Host not an animal	70
68.	Host an invertebrate	*Birnaviridae / Entomobirnavirus*
	Host a vertebrate	69
69.	Host an aquatic animal, usually a fish	*Birnaviridae / Aquabirnavirus*
	Host a bird	*Birnaviridae / Avibirnavirus*
70.	No virions are formed in diseased tissue	*Hypoviridae / Hypovirus*
	RNA is encapsidated	71
71.	Genome monopartite	*(Totiviridae)* 72
	Genome multipartite	*(Partitiviridae)* 74
72.	Virion 40 - 43 nm in diameter; host a fungus	*Totiviridae / Totivirus*
	Virion < 40 nm in diameter; host a protozoa	73
73.	RNA > 6 kbp; host *Giardia* sp.	*Totiviridae / Giardiavirus*
	RNA < 6 kbp; host *Leishmania* sp.	*Totiviridae / Leishmaniavirus*
74.	Host a fungus	75
	Host a plant	76
75.	Virions 30 - 35 nm in diameter; genome bipartite	*Partitiviridae / Partitivirus*
	Virions 35 - 40 nm in diameter; genome tri- or quadripartite	*Partitiviridae / Chrysovirus*
76.	Virion 30 nm in diameter	*Partitiviridae / Alphacryptovirus*
	Virion 38 nm in diameter	*Partitiviridae / Betacryptovirus*

77.	RNA negative sense or ambisense	78
	RNA positive sense	96

THE NEGATIVE SENSE ssRNA VIRUSES

78.	RNA circular; productive multiplication is helper virus-dependent	***Deltavirus***
	RNA linear	7
79.	Genome monopartite	(order ***Mononegavirales***) 80
	Genome multipartite	87
80.	Virion filamentous and/or pleomorphic; RNA 18-19 kb	***Filoviridae / Filovirus***
	Virion pleomorphic, usually spherical; RNA 15-16 kb	***(Paramyxoviridae)*** 81
	Virion bullet-shaped or bacilliform, not pleomorphic; RNA 11 to 15 kb	***(Rhabdoviridae)*** 84
81.	RNA contains 10 transcriptional elements	***Paramyxoviridae / Pneumovirinae / Pneumovirus***
	RNA contains < 10 transcriptional elements	***(Paramyxoviridae / Paramyxovirinae)*** 82
82.	Virion lacks a neuraminidase	***Paramyxoviridae / Paramyxovirinae / Morbillivirus***
	Virion contains a neuraminidase	83
83.	RNA encodes a C protein	***Paramyxoviridae / Paramyxovirinae / Paramyxovirus***
	RNA does not encode a C protein	***Paramyxoviridae / Paramyxovirinae / Rubulavirus***
84.	Host an animal	85
	Host a plant	86
85.	RNA about 11 kb; virion assembles by budding from the plasma membrane	***Rhabdoviridae / Vesiculovirus***
	RNA about 12 kb; virion assembles by budding from intracytoplasmic membranes	***Rhabdoviridae / Lyssavirus***
	RNA > 13 kb	***Rhabdoviridae / Ephemerovirus***
86.	Virions accumulate in the cytoplasm	***Rhabdoviridae / Cytorhabdovirus***
	Virions accumulate in the perinuclear space	***Rhabdoviridae / Nucleorhabdovirus***
87.	Genome in > 5 segments	***(Orthomyxoviridae)*** 88
	Genome in < 5 segments	90
88.	Genome in 8 segments	***Orthomyxoviridae / Influenzavirus A, B***
	Genome in 7 segments	89
89.	Nucleoprotein Mr 64 x 10^3; infects only vertebrates	***Orthomyxoviridae / Influenzavirus C***
	Nucleoprotein Mr 54 x 10^3; infects ticks and vertebrates	***Orthomyxoviridae*** / "Thogoto-like viruses"
90.	Virion about 8 nm filaments, host a plant	***Tenuivirus***
	Virion not filamentous	91
91.	Genome bipartite; virion contains host ribosomes	***Arenaviridae / Arenavirus***
	Genome tripartite; virion does not contain host ribosomes	***(Bunyaviridae)*** 92
92.	All RNA segments negative sense	93
	S RNA ambisense	95
93.	S RNA < 1 kb; S RNA encodes NSS protein + N protein	***Bunyaviridae / Bunyavirus***
	S RNA > 1 kb; S RNA encodes only N protein	94

94. L RNA > 10 kb; G2 protein Mr < 50 x 10^3 — *Bunyaviridae / Nairovirus*
L RNA < 10 kb; G2 protein Mr > 50 x 10^3 — *Bunyaviridae / Hantavirus*

95. Host an animal — *Bunyaviridae / Phlebovirus*
Host a plant — *Bunyaviridae / Tospovirus*

THE POSITIVE SENSE ssRNA VIRUSES

96. Host a prokaryote — *(Leviviridae)* 97
Host a eukaryote — 98

97. RNA < 4 kb; genome encodes a protein for cell lysis — *Leviviridae / Levivirus*
RNA > 4 kb; genome does not encode a cell lysis protein — *Leviviridae / Allolevivirus*

98. No specific virions identified; RNA can be encapsidated in heterologous coat protein; host a plant — *Umbravirus*
Virus-specific capsids formed in infected cells — 99

99. Virion not enveloped — 100
Virion enveloped — 127

100. Coat protein(s) are expressed by proteolysis of a large (Mr > 100 x 10^3) polyprotein — 101
Coat protein(s) expressed by translation of a small genome segment or a sub-genomic RNA — 112

101. Host an animal; structural proteins formed from the sequence at or within about 300 residues of the N-terminus of the polyprotein — *(Picornaviridae)* 102
Host a plant; structural proteins preceded upstream in the polyprotein by > 400 residues of non-structural protein — 106

102. Polyprotein contains a 'leader' protein — 103
Polyprotein does not contain a 'leader' protein — 104

103. Virion buoyant density in CsCl < 1.35 g/cm^3; 'leader' protein is not a protease — *Picornaviridae / Cardiovirus*
Virion buoyant density in CsCl > 1.35 g/cm^3; 'leader' protein is a protease — *Picornaviridae / Aphthovirus*

104. Virion not stable at acid pH; virion buoyant density in CsCl > 1.35 g/cm^3 — *Picornaviridae / Rhinovirus*
Virion stable at acid pH; virion buoyant density in CsCl < 1.35 g/cm^3 — 105

105. Protein 1A (VP4) small (< 2 kDa) or absent — *Picornaviridae / Hepatovirus*
Protein 1A > 3 kDa — *Picornaviridae / Enterovirus*

106. Virion filamentous — *(Potyviridae)* 107
Virion isometric — 108

107. Genome monopartite; vector an aphid — *Potyviridae / Potyvirus*
Genome monopartite; vector a mite — *Potyviridae / Rymovirus*
Genome genome bipartite; vector a fungus — *Potyviridae / Bymovirus*

108. Genome monopartite — *(Sequiviridae)* 109
Genome bipartite — *(Comoviridae)* 110

109. Virus transmitted by aphids — *Sequiviridae / Sequivirus*
Virus phloem-limited, not mechanically transmissible — *Sequiviridae / Waikavirus*

110. Larger RNA species > 7 kb; virion usually contains 1 coat protein with Mr of about 57 x 10^3; virus usually transmitted by nematodes — *Comoviridae / Nepovirus*
Larger RNA species < 7 kb; virion contains 2 coat proteins — 111

111. Vector a beetle — *Comoviridae / Comovirus*
Vector an aphid — *Comoviridae / Fabavirus*

112. Host a vertebrate — 113
Host an invertebrate — 114
Host a plant or a fungus — 116

113. Virion 30 nm or more in diameter and with cup-shaped depressions; virion contains one structural protein — *Caliciviridae / Calicivirus*
Virion 30 nm or less in diameter, often appearing star-shaped; virion contains 2 or 3 structural proteins — *Astroviridae / Astrovirus*

114. Structural protein Mr < 40 x 10^3 — *Nodaviridae / Nodavirus*
Structural protein Mr > 60 x 10^3 — *(Tetraviridae)* 115

115. Genome monopartite — *Tetraviridae* / "Nudaurelia capensis β-like viruses"
Genome bipartite — *Tetraviridae* / "Nudaurelia capensis ω-like viruses"

116. Virus circulates in the bodies of the vectors — 117
No vector known or transmission non-circulative — 119

117. Vector a leafhopper — *Marafivirus*
Vector an aphid — 118

118. Virion contains 1 RNA; virus not transmissible mechanically — *Luteovirus*
Virion contains 2 RNA; virus readily transmissible mechanically — *Enamovirus*

119. Virion isometric or bacilliform — 120
Virion has helical symmetry — 138

120. Host a fungus; virion bacilliform — *Barnaviridae / Barnavirus*
Host not a fungus — 121

121. RNA about 6 kb; coat protein Mr about 20 x 10^3; vector a beetle — *Tymovirus*
RNA < 5.5 kb; coat protein Mr > 20 x 10^3 — 122

122. Genome bipartite; virion contains both genome segments — *Dianthovirus*
Genome monopartite — 123
Genome multipartite; genome contained in > one virion — 134

123. Coat protein Mr > 35 x 10^3 — *(Tombusviridae)* 124
Coat protein Mr < 35 x 10^3 — 125

124. RNA 4 kb; coat protein Mr < 40 x 10^3 — *Tombusviridae / Carmovirus*
RNA > 4 kb; coat protein Mr > 40 x 10^3 — *Tombusviridae / Tombusvirus*

125. RNA < 4 kb; vector a fungus — *Necrovirus*
RNA > 4 kb; vector an insect — 126

126. RNA has a VPg at the 5'-end; coat protein Mr 30 x 10^3 — *Sobemovirus*
RNA is 5'-capped; coat protein Mr 25 x 10^3 — *Machlomovirus*

127. Genome expressed as a polyprotein, no sub-genomic RNA are formed in infected cells — *(Flaviviridae)* 128
Sub-genomic RNA are formed in infected cells — 130

128. RNA > 12 kb; RNA encodes 3 envelope proteins and 1 nucleocapsid protein — ***Flaviviridae / Pestivirus***
RNA < 12 kb; RNA encodes 2 envelope proteins and 1 core protein — 129

129. RNA > 10 kb; host a vertebrate and often also an invertebrate — ***Flaviviridae / Flavivirus***
RNA < 10 kb; man is the only host — ***Flaviviridae*** / "Hepatitis C-like viruses"

130. Infected cells contain 1 species of sub-genomic RNA — ***(Togaviridae)*** 131
Infected cells contain > 1 species of sub-genomic RNA — 132

131. Virion 70 nm in diameter; infects vertebrates and insects — ***Togaviridae / Alphavirus***
Virion 60 nm in diameter; host a vertebrate — ***Togaviridae / Rubivirus***

132. RNA < 20 kb; virion spherical — ***Arterivirus***
RNA > 20 kb; virion pleomorphic — ***(Coronaviridae)*** 133

133. Virion spherical or pleomorphic with club-shaped surface projections — ***Coronaviridae / Coronavirus***
Virion biconcave disk-, kidney- or rod-shaped with a peplomer-bearing envelope — ***Coronaviridae / Torovirus***

134. Genome bipartite; largest RNA > 5 kb — ***Idaeovirus***
Genome tripartite; largest RNA < 4kb — ***(Bromoviridae)*** 135

135. Virions isometric, sedimenting as 1 component — 136
Virions not isometric, sedimenting as > 1 component — 137

136. Coat protein Mr about 20 x 10^3; virus not aphid-transmitted — ***Bromoviridae / Bromovirus***
Coat protein Mr > 24 x 10^3; virus aphid-transmitted — ***Bromoviridae / Cucumovirus***

137. Some virions bacilliform; virus aphid-transmitted — ***Bromoviridae / Alfamovirus***
Virions slightly pleomorphic; virus not aphid-transmitted — ***Bromoviridae / Ilarvirus***

138. Virion rod-shaped — 139
Virion filamentous — 142

139. Genome monopartite — ***Tobamovirus***
Genome multipartite — 140

140. Virion > 20 nm in diameter; vector a nematode — ***Tobravirus***
Virion < 20 nm in diameter — 141

141. Some virions > 250 nm in length; largest RNA > 5 kb; vector a fungus — ***Furovirus***
Virions < 200 nm long; largest RNA < 5 kb — ***Hordeivirus***

142. Virion > 700 nm in length — ***Closterovirus***
Virion < 700 nm in length — 143

143. Virion < 600 nm; coat protein Mr < 25 x 10^3 — ***Potexvirus***
Virion > 600 nm; coat protein Mr > 25 x 10^3 — 144

144. Virion with prominent banding; genome lacks a triple gene block — 145
Virion without obvious banding; genome contains a triple gene block — ***Carlavirus***

145. Replicase and coat protein encoded in the same open reading frame — ***Capillovirus***
Non-structural proteins and the coat protein encoded in different open reading frames — ***Trichovirus***

THE ORDER OF PRESENTATION OF THE VIRUSES

The order of presentation of virus families and genera does not reflect any hierarchical or phylogenetic classification, but only a convenient order of presentation. Since a taxonomic structure above the level of family or genus has not been developed, (with the exception of the order *Mononegavirales*) any sequence of listing must be arbitrary. The order of presentation of virus families and genera follows four criteria: (i) the nature of the viral genome, (ii) the strandedness of the viral genome, (iii) the fact that some viruses are reverse transcribed, and (iv) the polarity of the virus genome. As there are no known ssDNA, nor dsRNA reverse transcribed viruses, and there are negative sense viruses only for ssRNA viruses, these four criteria give rise to seven clusters comprising the 51 families and 24 genera of viruses. In addition, subviral agents, namely the satellites, viroids and agents of spongiform encephalopathies (prions) are included, in most cases without official taxonomic status. Finally, a list of unassigned viruses is provided.

Order Family	Subfamily	Genus	Type Species	Host	Page
The DNA Viruses					
The dsDNA Viruses					
Myoviridae		"T4-like phages"[1]	coliphage T4	Bacteria	51
Siphoviridae		"λ-like phages"	coliphage λ	Bacteria	55
Podoviridae		"T7-like phages"	coliphage T7	Bacteria	60
Tectiviridae		*Tectivirus*	enterobacteria phage PRD1	Bacteria	64
Corticoviridae		*Corticovirus*	Alteromonas phage PM2	Bacteria	67
Plasmaviridae		*Plasmavirus*	Acholeplasma phage L2	Mycoplasma	70
Lipothrixviridae		*Lipothrixvirus*	Thermoproteus virus 1	Bacteria	73
Fuselloviridae		*Fusellovirus*	Sulfolobus virus 1	Bacteria	76
Poxviridae					79
	Chordopoxvirinae				83
		Orthopoxvirus	vaccinia virus	Vertebrates	83
		Parapoxvirus	orf virus	Vertebrates	84
		Avipoxvirus	fowlpox virus	Vertebrates	85
		Capripoxvirus	sheeppox virus	Vertebrates	85
		Leporipoxvirus	myxoma virus	Vertebrates	86
		Suipoxvirus	swinepox virus	Vertebrates	86
		Molluscipoxvirus	Molluscum contagiosum virus	Vertebrates	87
		Yatapoxvirus	Yaba monkey tumor virus	Vertebrates	87
	Entomopoxvirinae				88
		Entomopoxvirus A	Melolontha melolontha entomopoxvirus	Invertebrates	88
		Entomopoxvirus B	Amsacta moorei entomopoxvirus	Invertebrates	89
		Entomopoxvirus C	Chironomus luridus entomopoxvirus	Invertebrates	89
		"African swine fever-like viruses"	African swine fever virus	Vertebrates[2]	92

[1] Quotes are used to denote taxa without ICTV international approved names.
[2] Vertebrate arthropod-borne viruses are listed according to their vertebrate hosts.

Order Family	Subfamily	Genus	Type Species	Host	Page
Iridoviridae					95
		Iridovirus	Chilo iridescent virus	Invertebrates	96
		Chloriridovirus	mosquito iridescent virus	Invertebrates	97
		Ranavirus	frog virus 3	Vertebrates	97
		Lymphocystivirus	flounder virus	Vertebrates	97
		"Goldfish virus 1-like viruses"	goldfish virus 1	Vertebrates	98
Phycodnaviridae		*Phycodnavirus*	Paramecium bursaria Chlorella virus 1	Algae	100
Baculoviridae					104
		Nucleopolyhedrovirus	Autographa californica nucleopolyhedrovirus	Invertebrates	107
		Granulovirus	Plodia interpunctella granulovirus	Invertebrates	111
Herpesviridae					114
	Alphaherpesvirinae				119
		Simplexvirus	human herpesvirus 1	Vertebrates	119
		Varicellovirus	human herpesvirus 3	Vertebrates	120
	Betaherpesvirinae				121
		Cytomegalovirus	human herpesvirus 5	Vertebrates	121
		Muromegalovirus	mouse cytomegalovirus 1	Vertebrates	122
		Roseolovirus	human herpesvirus 6	Vertebrates	122
	Gammaherpesvirinae				123
		Lymphocryptovirus	human herpesvirus 4	Vertebrates	123
		Rhadinovirus	ateline herpesvirus 2	Vertebrates	123
Adenoviridae					128
		Mastadenovirus	human adenovirus 2	Vertebrates	131
		Aviadenovirus	fowl adenovirus 1	Vertebrates	132
		Rhizidiovirus	Rhizidiomyces virus	Fungi	134
Papovaviridae					136
		Polyomavirus	murine polyomavirus	Vertebrates	140
		Papillomavirus	cottontail rabbit papillomavirus (Shope)	Vertebrates	141
Polydnaviridae					143
		Ichnovirus	Campoletis sonorensis virus	Invertebrates	144
		Bracovirus	Cotesia melanoscela virus	Invertebrates	145

Order Family	Subfamily	Genus	Type Species	Host	Page
The ssDNA Viruses					
Inoviridae					148
		Inovirus	coliphage fd	Bacteria	150
		Plectrovirus	Acholeplasma phage L51	Mycoplasma	151
Microviridae					153
		Microvirus	coliphage ϕX174	Bacteria	155
		Spiromicrovirus	Spiroplasma phage 4	Spiroplasma	156
		Bdellomicrovirus	Bdellovibrio phage MAC1	Bacteria	156
		Chlamydiamicrovirus	Chlamydia phage 1	Bacteria	157
Geminiviridae					158
		"Subgroup I Geminivirus"	maize streak virus	Plants	159
		"Subgroup II Geminivirus "	beet curly top virus	Plants	160
		"Subgroup III Geminivirus "	bean golden mosaic virus	Plants	161
Circoviridae		*Circovirus*	chicken anemia virus	Vertebrates	166
Parvoviridae					169
	Parvovirinae				173
		Parvovirus	mice minute virus	Vertebrates	174
		Erythrovirus	B19 virus	Vertebrates	174
		Dependovirus	adeno-associated virus 2	Vertebrates	175
	Densovirinae				176
		Densovirus	Junonia coenia densovirus	Invertebrates	176
		Iteravirus	Bombyx mori densovirus	Invertebrates	176
		Contravirus	Aedes aegypti densovirus	Invertebrates	177

Order Family	Subfamily	Genus	Type Species	Host	Page

The DNA and RNA Reverse Transcribing Viruses

Family	Subfamily	Genus	Type Species	Host	Page
Hepadnaviridae					179
		Orthohepadnavirus	hepatitis B virus	Vertebrates	183
		Avihepadnavirus	duck hepatitis B virus	Vertebrates	184
		Badnavirus	Commelina yellow mottle virus	Plants	185
		Caulimovirus	cauliflower mosaic virus	Plants	189
Retroviridae					193
		"Mammalian type B retroviruses"	mouse mammary tumor virus	Vertebrates	196
		"Mammalian type C retroviruses"	murine leukemia virus	Vertebrates	197
		"Avian type C retroviruses"	avian leukosis virus	Vertebrates	198
		"Type D retroviruses"	Mason-Pfizer monkey virus	Vertebrates	199
		"BLV-HTLV retroviruses"	bovine leukemia virus	Vertebrates	200
		Lentivirus	human immunodeficiency virus 1	Vertebrates	201
		Spumavirus	human spumavirus	Vertebrates	203

Order Family	Subfamily	Genus	Type Species	Host	Page

The RNA Viruses
The dsRNA Viruses

Cystoviridae		*Cystovirus*	Pseudomonas phage $\phi 6$	Bacteria	205
Reoviridae					208
		Orthoreovirus	reovirus 3	Vertebrates	210
		Orbivirus	bluetongue virus 1	Vertebrates	214
		Rotavirus	simian rotavirus SA11	Vertebrates	219
		Coltivirus	Colorado tick fever virus	Vertebrates	223
		Aquareovirus	golden shiner virus	Vertebrates	225
		Cypovirus	Bombyx mori cypovirus 1	Invertebrates	227
		Fijivirus	Fiji disease virus	Plants	232
		Phytoreovirus	wound tumor virus	Plants	234
		Oryzavirus	rice ragged stunt virus	Plants	237
Birnaviridae					240
		Aquabirnavirus	infectious pancreatic necrosis virus	Vertebrates	242
		Avibirnavirus	infectious bursal disease virus	Vertebrates	242
		Entomobirnavirus	Drosophila X virus	Invertebrates	243
Totiviridae					245
		Totivirus	Saccharomyces cerevisiae virus L-A	Fungi	245
		Giardiavirus	Giardia lamblia virus	Protozoa	248
		Leishmaniavirus	Leishmania RNA virus 1-1	Protozoa	249
Partitiviridae					253
		Partitivirus	Gaeumannomyces graminis virus 019/6-A	Fungi	254
		Chrysovirus	Penicillium chrysogenum virus	Fungi	255
		Alphacryptovirus	white clover cryptic virus 1	Plants	257
		Betacryptovirus	white clover cryptic virus 2	Plants	258
Hypoviridae		*Hypovirus*	Cryphonectria hypovirus 1-EP713	Fungi	261

Order / Family	Subfamily	Genus	Type Species	Host	Page

The Negative Stranded ssRNA Viruses

Order / Family	Subfamily	Genus	Type Species	Host	Page
Mononegavirales					265
Paramyxoviridae					268
	Paramyxovirinae				271
		Paramyxovirus	human parainfluenza virus 1	Vertebrates	271
		Morbillivirus	measles virus	Vertebrates	271
		Rubulavirus	mumps virus	Vertebrates	272
	Pneumovirinae				273
		Pneumovirus	human respiratory syncytial virus	Vertebrates	273
Rhabdoviridae					275
		Vesiculovirus	vesicular stomatitis Indiana virus	Vertebrates	274
		Lyssavirus	rabies virus	Vertebrates	281
		Ephemerovirus	bovine ephemeral fever virus	Vertebrates	282
		Cytorhabdovirus	lettuce necrotic yellows virus	Plants	283
		Nucleorhabdovirus	potato yellow dwarf virus	Plants	284
Filoviridae					289
		Filovirus	Marburg virus	Vertebrates	289
Orthomyxoviridae					293
		Influenzavirus A, B	influenza A virus	Vertebrates	296
		Influenzavirus C	influenza C virus	Vertebrates	297
		"Thogoto-like viruses"	Thogoto virus	Vertebrates	298
Bunyaviridae					300
		Bunyavirus	Bunyamwera virus	Vertebrates	304
		Hantavirus	Hantaan virus	Vertebrates	308
		Nairovirus	Nairobi sheep disease virus	Vertebrates	309
		Phlebovirus	sandfly fever Sicilian virus	Vertebrates	311
		Tospovirus	tomato spotted wilt virus	Plants	313
		Tenuivirus	rice stripe virus	Plants	316
Arenaviridae		*Arenavirus*	lymphocytic choriomeningitis virus	Vertebrates	319

The Positive Stranded ssRNA Viruses

Order / Family	Subfamily	Genus	Type Species	Host	Page
Leviviridae					324
		Levivirus	enterobacteria phage MS2	Bacteria	325
		Allolevivirus	enterobacteria phage Qβ	Bacteria	326
Picornaviridae					329
		Enterovirus	poliovirus 1	Vertebrates	332
		Rhinovirus	human rhinovirus 1A	Vertebrates	333
		Hepatovirus	hepatitis A virus	Vertebrates	333
		Cardiovirus	encephalomyocarditis virus	Vertebrates	334
		Aphtovirus	foot-and-mouth disease virus O	Vertebrates	334
Sequiviridae					337
		Sequivirus	parsnip yellow fleck virus	Plants	338
		Waïkavirus	rice tungro spherical virus	Plants	339
Comoviridae					341
		Comovirus	cowpea mosaic virus	Plants	343
		Fabavirus	broad bean wilt virus 1	Plants	344
		Nepovirus	tobacco ringspot virus	Plants	345
Potyviridae					348
		Potyvirus	potato virus Y	Plants	350
		Rymovirus	ryegrass mosaic virus	Plants	355
		Bymovirus	barley yellow mosaic virus	Plants	356
Caliciviridae		*Calicivirus*	vesicular exanthema of swine virus	Vertebrates	359
Astroviridae		*Astrovirus*	human astrovirus 1	Vertebrates	364
Nodaviridae		*Nodavirus*	Nodamura virus	Invertebrates	368
Tetraviridae					372
		"Nudaurelia capensis β-like viruses"			
			Nudaurelia capensis β virus	Invertebrates	374
		"Nudaurelia capensis ω-like viruses"			
			Nudaurelia capensis ω virus	Invertebrates	374
		Sobemovirus	Southern bean mosaic virus	Plants	376
		Luteovirus	barley yellow dwarf virus	Plants	379

Order Family	Subfamily	Genus	Type Species	Host	Page
		Enamovirus	pea enation mosaic virus	Plants	384
		Umbravirus	carrot mottle virus	Plants	388
Tombusviridae					392
		Tombusvirus	tomato bushy stunt virus	Plants	394
		Carmovirus	carnation mottle virus	Plants	395
		Necrovirus	tobacco necrosis virus	Plants	398
		Dianthovirus	carnation ringspot virus	Plants	401
		Machlomovirus	maize chlorotic mottle virus	Plants	404
Coronaviridae					407
		Coronavirus	avian infectious bronchitis virus	Vertebrates	409
		Torovirus	Berne virus	Vertebrates	410
		Arterivirus	equine arteritis virus	Vertebrates	412
Flaviviridae					415
		Flavivirus	yellow fever virus	Vertebrates	416
		Pestivirus	bovine diarrhea virus	Vertebrates	421
		"Hepatitis C-like viruses"	hepatitis C virus	Vertebrates	424
Togaviridae					428
		Alphavirus	Sindbis virus	Vertebrates	431
		Rubivirus	rubella virus	Vertebrates	432
		Tobamovirus	tobacco mosaic virus	Plants	434
		Tobravirus	tobacco rattle virus	Plants	438
		Hordeivirus	barley stripe mosaic virus	Plants	441
		Furovirus	soil-borne wheat mosaic virus	Plants	445
Bromoviridae					450
		Alfamovirus	alfalfa mosaic virus	Plants	453
		Ilarvirus	tobacco streak virus	Plants	453
		Bromovirus	brome mosaic virus	Plants	454
		Cucumovirus	cucumber mosaic virus	Plants	455

Order Family	Subfamily	Genus	Type Species	Host	Page
		Idaeovirus	rasberry bushy dwarf virus	Plants	458
		Closterovirus	beet yellows virus	Plants	461
		Capillovirus	apple stem grooving virus	Plants	465
		Trichovirus	apple chlorotic leaf spot virus	Plants	468
		Tymovirus	turnip yellow mosaic virus	Plants	471
		Carlavirus	carnation latent virus	Plants	475
		Potexvirus	potato virus X	Plants	479
Barnaviridae		*Barnavirus*	mushroom bacilliform virus	Fungi	483
		Marafivirus	maize rayado fino virus	Plants	485

The Subviral Agents: Satellites, Viroids, and Agents of Spongiform Encephalopathies (Prions)

Subviral Agents	Genus	Example	Host	Page
Satellites		tobacco necrosis virus satellite	Plants Vertebrates Invertebrates Fungi	487
	Deltavirus	hepatitis delta virus	Vertebrates	493
Viroids		potato spindle tuber viroid	Plants	495
Prions		scrapie agent	Vertebrates	498

Unassigned Viruses 504

TAILED PHAGES

Tailed phages are an extremely large and differentiated group of viruses. About 4,000 descriptions have been published. Three families are distinguished by tail structure; most data on replication have been derived from a few well-studied viruses.

TAXONOMIC STRUCTURE

Tailed Phages	
Family	*Myoviridae*
Family	*Siphoviridae*
Family	*Podoviridae*

VIRION PROPERTIES

MORPHOLOGY

Virions consist of a head (capsid), a tail, and fixation organelles. They have no envelope. Heads are isometric or elongated and are icosahedra or derivatives thereof (proposed triangulation numbers T=1, T=7, T=9, T=12, T=13, T=16). Capsomers are seldom visible and heads usually appear smooth and thin-walled (2-3 nm). Estimated capsomer numbers vary between 17 and 812. Isometric heads are 45-170 nm in diameter. Elongated heads derive from icosahedra by addition of rows of capsomers and are bipyramidal antiprisms up to 230 nm long. The DNA forms a tightly packed coil inside the phage head. Tails are long and contractile, long and noncontractile, or short. They are helical or consist of stacked disks of subunits, varying between 3 and 570 nm in length, and are usually equipped with base plates, spikes, or terminal fibers. Some phages have collars, head or collar appendages, transverse tail disks, or other attachments.

PHYSICOCHEMICAL AND PHYSICAL PROPERTIES

Virion Mr ranges from 29 to 470 x 10^6; S_{20w} is 226-1230. Both values may be higher, as the largest phages have not been studied in this respect. Buoyant density in CsCl is about 1.49 g/cm^3. Most tailed phages are stable at pH 5-9; a few resist pH 2 or pH 11. Heat sensitivity is variable and resembles that to the host. Many phages are inactivated by heating at 56-60° C for 30 min. Tailed phages are rather resistant to UV irradiation. Heat and UV inactivation generally follow first-order kinetics. Many tailed phages are ether- and chloroform-sensitive. Inactivation by nonionic detergents is variable and partly concentration-dependent.

NUCLEIC ACID

Virions contain one molecule of linear dsDNA. Genome sizes range from 19 to about 700 kbp, corresponding to Mr values of 11-490 x 10^6. Relative DNA content is about 45%. G+C content ranges between 27 and 72% and usually resembles that of host DNA. The DNA of many viruses has particular features such as circular permutation, terminal repeats, cohesive ends, proteins covalently linked to 5'-termini, fragments of host DNA attached to the ends of the phage genome, single-stranded interruptions, unusual bases which partially or completely replace normal nucleotides (e.g. 5-hydroxy-methylcytosine), or are glycosylated or associated with internal proteins or basic polyamines. The DNA of only six viruses has been fully sequenced (T7, P2, λ, L5, ϕ29, PZA). Nucleotide sequence data are available from GenBank.

PROTEINS

The number of structural proteins varies between 7 and 42. The Mr range is 4-200 x 10^3. Lysozyme is located at the tail tip; the spikes of some capsule-specific phages have endoglycosidase activity. A few exceptional phages contain transcriptases, dihydrofolate reductase, or thymidylate synthetase.

Lipids

Most virions contain no lipid. Up to 15% lipid has been found in a few phages of *mycobacteria;* its presence in others is doubtful.

Carbohydrates

Glycoproteins, glycolipids, hexosamine, and a polysaccharide have been found in individual phages.

Genome Organization and Replication

Detailed functional genetic maps are available for 10 phages only. They show evidence for considerable gene rearrangement during evolution and few common features. Genes with related functions tend to cluster together. The number of genes varies between 17 and >100. Genomes seem to consist of interchangeable gene blocks or "modules".

Virions adsorb tail first to specific receptors located on the cell wall, capsule, flagella, or pili of bacteria. In some phages, the cell wall is digested by phage lysozyme. Phage DNA enters the cytoplasm by as yet unknown mechanisms. Phages are virulent or temperate and present several strategies of replication:
1. In virulent phages, infection normally results in production of progeny phages and destruction of the host; however, persistent infections exist. The infecting DNA remains linear.
2. In temperate phages, the infecting DNA is replicated and the infecting DNA becomes latent within the host (prophage state) or, alternatively, is replicated and prophages are produced. Prophage DNA must be activated (derepressed) before replication. Hosts are lysogenized in several ways.

The infecting DNA:
a. Circularizes and integrates into the host genome at a specific site, at several sites, or at random.
b. Circularizes and persists in the cytoplasm as a plasmid.
c. Remains linear and integrates into host DNA at random.

Gene expression is largely time-ordered and sequential. "Early" genes are involved into DNA replication and integration. "Late" genes mainly specify structural proteins. In most species, transcription depends fully on host polymerases. DNA replication generally starts at fixed sites, is semiconservative and bidirectional or unidirectional. It usually results in the formation of multimeric DNA molecules or concatemers. Translational control is poorly understood and no generalizations are possible with the present state of knowledge. Particle assembly is complex and includes separate pathways for each phage part (head, tail, fibers). Head assembly starts with a prohead stage at the periphery of the nucleoplasm. Phage DNA is cut to size and enters preformed capsids. Some phages form intracellular arrays. Progeny phages are liberated by lysis of the host cell. Many phages produce aberrant structures (polyheads, polytails, giant, multitailed, or misshapen particles).

Antigenic Properties

Viruses are antigenically complex and efficient immunogens, inducing the formation of neutralizing and complement-fixing antigens. The existence of group antigens is likely.

Biological Properties

Host Range

Tailed phages have been found in over 100 genera of eubacteria and archaebacteria. They are usually host genus-specific. Enterobacterial phages are specific for the family *Enterobacteriaceae*. Some species have a world-wide distribution.

Family *Myoviridae*

Distinguishing Features

Tails are contractile, more or less rigid, long and relatively thick (80-455 x 16-20 nm). They are complex, consisting of a central core surrounded by a contractile sheath, which is separated from the head by a neck. During contraction, sheath subunits slide over each other and the sheath becomes shorter and thicker. This brings the tail core in contact with the bacterial plasma membrane and is an essential stage of infection. With respect to other tailed phages, myoviruses tend to have larger heads and higher particle weights and DNA contents, and seem to be more sensitive to freezing and thawing and to osmotic shock.

Taxonomic Structure of the Family

Family *Myoviridae*
Genus "T4 -like phages"

Genus "T4-like Phages"

Type Species coliphage T4 (T4)

Virion Properties

Morphology

Phage heads are elongated, pentagonal bipyramidal antiprisms, measure about 111 x 78 nm, and consist of 152 capsomers (T=13). Tails measure 113 x 16 nm and have a collar, a base plate, 6 short spikes and 6 long fibers.

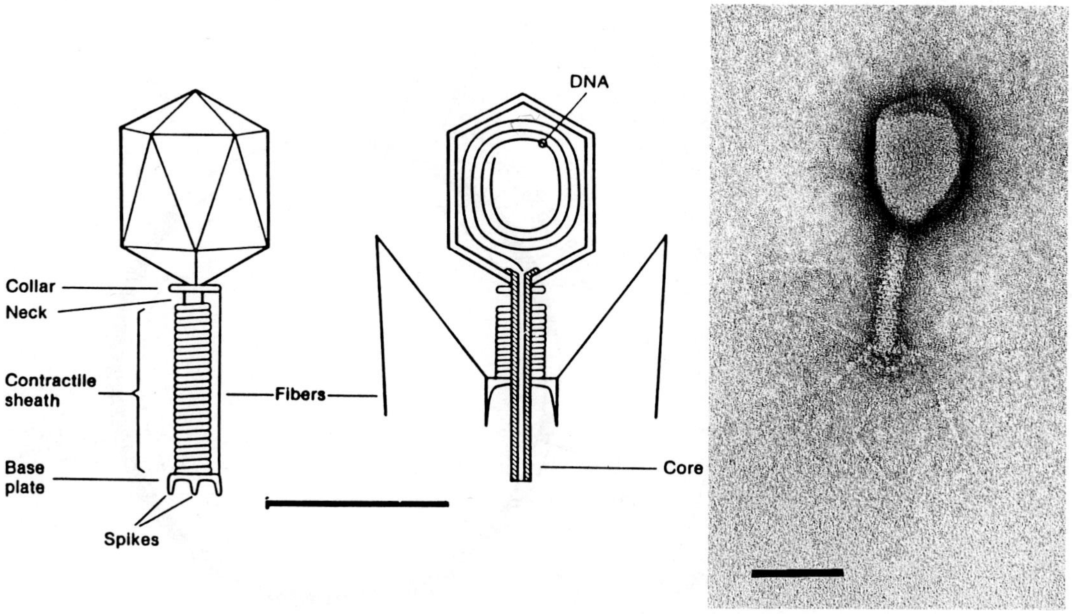

Figure 1: (left) Coliphage T4 in surface view (tail extended) and section (tail contracted). (From Ackermann H-W, DuBow MS (1987), with permission). (right) Negative contrast electron micrograph of coliphage T4, stained with uranyl acetate. Bars represent 100 nm.

Physicochemical and Physical Properties

Virion Mr is 210 x 10^6, S_{20w} is about 1030; buoyant density in CsCl is 1.51 g/cm^3. Infectivity is ether and chloroform resistant.

NUCLEIC ACID

Genomes have an Mr of about 175 x 10^6, corresponding to 48% of particle weight, contain 5-hydroxymethylcytosine (HMC) instead of thymine, have a G+C content of 35%, and are glycosylated, circularly permuted, and terminally redundant.

PROTEINS

Particles contain at least 42 polypeptides (Mr 8-155 x 10^3), including 1,600-2,000 copies of the major capsid protein (Mr 43 x 10^3); 3 proteins are located inside the head. Various enzymes are present, e.g. dehydrofolate reductase and lysozyme. ATP is present in the tail.

LIPIDS

None reported.

CARBOHYDRATES

Glucose is covalently linked to HMC in phage DNA. Gentobiose may be present.

GENOME ORGANIZATION AND REPLICATION

The genome is circular and includes 150-160 genes. Morphopoietic genes generally cluster together, but the whole genome appears disorganized, suggesting extensive translocation of genes during evolution. Phage adsorb to the cell wall and initiate a virulent infection. The host chromosome breaks down and viral DNA replicates as a concatemer, giving rise to forked replicative intermediates. Heads, tails, and tail fibers are assembled by 3 different pathways. Aberrant head structures (polyheads, isometric heads) are frequent.

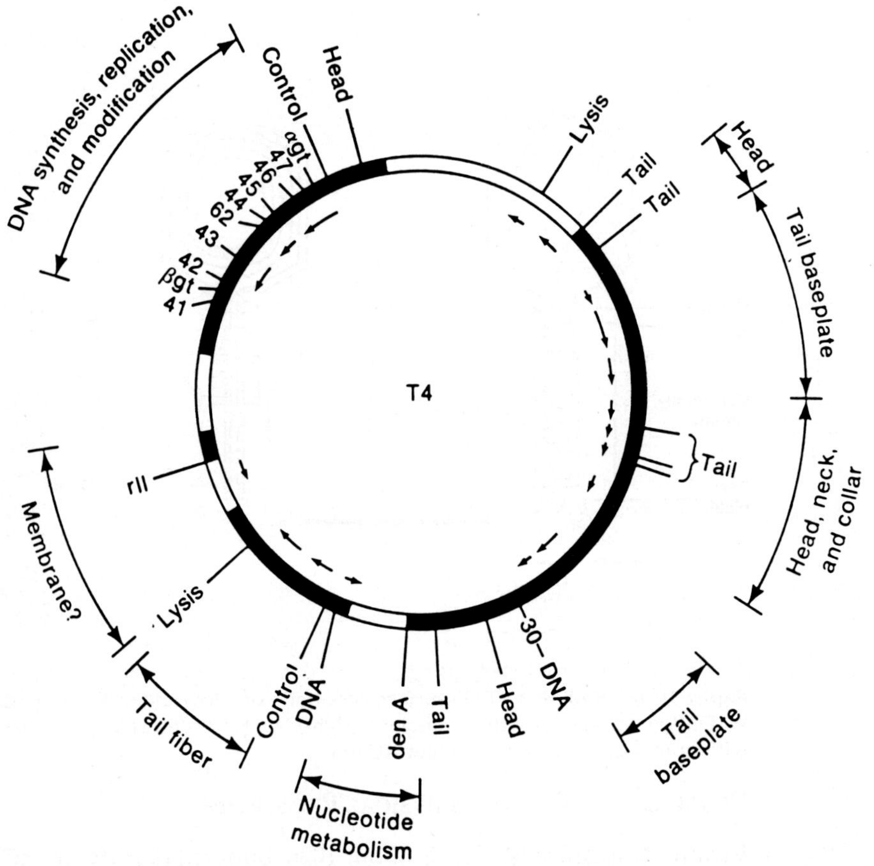

Figure 2: Simplified genetic map of coliphage T4 showing clustering of genes with related functions, location of essential genes (solid bars), and direction and origin of transcripts (arrows). (From Freifelder D (1983). Molecular Biology. Science Books International, Boston, and Van Nostrand Reynolds, New York, p 614, with permission).

Antigenic Properties

A group antigen and antigens defining 8 subgroups have been identified by complement fixation.

Biological Properties

Phages are specific for enterobacteria.

List of Species in the Genus

The viruses and their assigned abbreviations ( ) are:

Species in the Genus

The genus includes a large number of isolates of uncertain taxonomic status; these are either strains of the coliphage T4 species or represent independent species:

Aeromonas phage 44RR2.8t	(44RR2.8t)
Aeromonas phage 65	(65)
Aeromonas phage Aeh1	(Aeh1)
coliphage T2	(T2)
coliphage T4	(T4)
coliphage T6	(T6)
enterobacteria phage C16	(C16)
enterobacteria phage DdVI	(DdVI)
enterobacteria phage PST	(PST)
enterobacteria phage SMB	(SMB)
enterobacteria phage SMP2	(SMP2)
enterobacteria phage α1	(α1)
enterobacteria phage 3	(3)
enterobacteria phage 3T+	(3T+)
enterobacteria phage 9/0	(9/0)
enterobacteria phage 11F	(11F)
enterobacteria phage 50	(50)
enterobacteria phage 66F	(66F)
enterobacteria phage 5845	(5845)
enterobacteria phage 8893	(8893)
(and many other enterobacteria phages not well characterized).	
Vibrio phage nt-1	(nt-1)

Tentative Species in the Genus

None reported.

List of Unassigned Species in the Family

Actinomycetes phage SK1	(SK1)
Actinomycetes phage 108/016	(108/016)
Aeromonas phage Aeh2	(Aeh2)
Aeromonas phage 29	(29)
Aeromonas phage 37	(37)
Aeromonas phage 43	(43)
Aeromonas phage 51	(51)
Aeromonas phage 59.1	(59.1)
Agrobacterium phage PIIBNV6	(PIIBNV6)
Alcaligenes phage A6	(A6)
Bacillus phage G	(G)
Bacillus phage MP13	(MP13)
Bacillus phage PBS1	(PBS1)

Bacillus phage SP3	(SP3)
Bacillus phage SP8	(SP8)
Bacillus phage SP10	(SP10)
Bacillus phage SP15	(SP15)
Bacillus phage SP50	(SP50)
Bacillus phage SPy-2	(SPy-2)
Bacillus phage SST	(SST)
Clostridium phage HM3	(HM3)
Clostridium phage CEß	(CEß)
coryneforms phage A19	(A19)
cyanobacteria phage AS-1	(AS-1)
cyanobacteria phage N1	(N1)
cyanobacteria phage S-6(L)	(S-6(L))
enterobacteria phage Beccles	(Beccles)
enterobacteria phage FC3-9	(FC3-9)
enterobacteria phage K19	(K19)
enterobacteria phage Mu	(Mu)
enterobacteria phage 01	(01)
enterobacteria phage P1	(P1)
enterobacteria phage P2	(P2)
enterobacteria phage ViI	(ViI)
enterobacteria phage ϕ92	(ϕ92)
enterobacteria phage 121	(121)
enterobacteria phage 16-19	(16-19)
enterobacteria phage 9266	(9266)
Lactobacillus phage fri	(fri)
Lactobacillus phage hv	(hv)
Lactobacillus phage hw	(hw)
Lactobacillus phage 222a	(222a)
Listeria phage 4211	(4211)
mollicutes phage Br1	(Br1)
Mycobacterium phage I3	(I3)
Pasteurella phage AU	(AU)
Pseudomonas phage PB-1	(PB-1)
Pseudomonas phage PP8	(PP8)
Pseudomonas phage PS17	(PS17)
Pseudomonas phage ϕKZ	(ϕKZ)
Pseudomonas phage ϕW-14	(ϕW-14)
Pseudomonas phage ϕ1	(ϕ1)
Pseudomonas phage 12S	(12S)
Rhizobium phage CM_1	(CM_1)
Rhizobium phage CT4	(CT4)
Rhizobium phage m	(m)
Rhizobium phage WT1	(WT1)
Rhizobium phage ϕgal-1-R	(ϕgal-1-R)
Staphylococcus phage Twort	(Twort)
Xanthomonas phage XP5	(XP5)
Vibrio phage kappa	(kappa)
Vibrio phage 06N-22P	(06N-22P)
Vibrio phage VP1	(VP1)
Vibrio phage X29	(X29)
Vibrio phage II	(II)

FAMILY *SIPHOVIRIDAE*

DISTINGUISHING FEATURES

Virions have long, noncontractile, thin tails (65?-570 x 7-10 nm) which are often flexible. Tails are helical or built of stacked disks of subunits.

TAXONOMIC STRUCTURE OF THE FAMILY

Family	***Siphoviridae***	
Genus		"λ-like phages"

GENUS "λ-LIKE PHAGES"

Type Species coliphage λ (λ)

VIRION PROPERTIES

MORPHOLOGY

Phage heads are isometric, measure about 60 nm in diameter, and consist of 72 capsomers (T=7). Tails are flexible, measure 150 x 8 nm, and have short terminal and subterminal fibers.

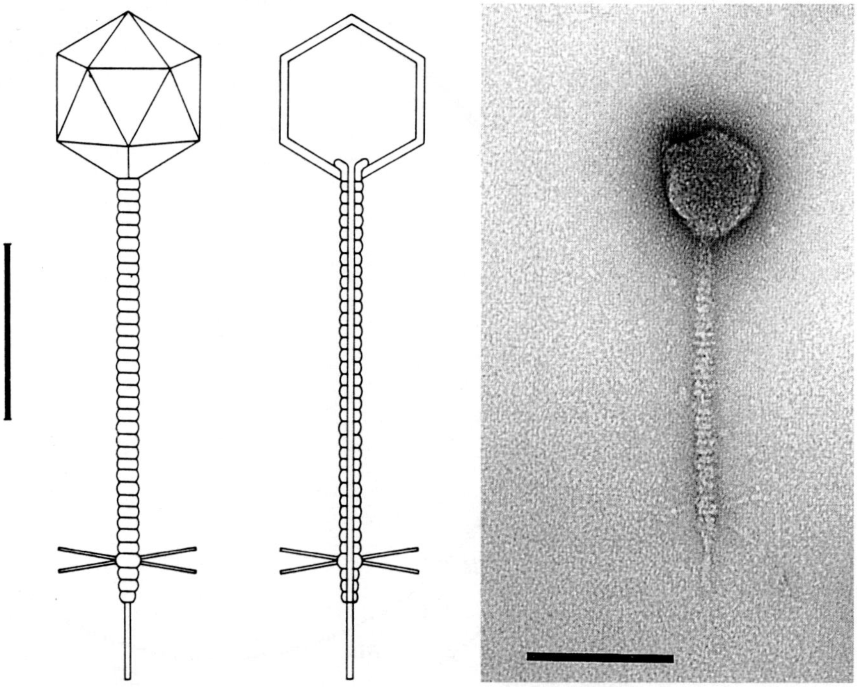

Figure 1: (left) Coliphage λ in surface view and section. (right) Negative contrast electron micrograph of coliphage λ stained with uranyl acetate. Bars represent 100 nm.

PHYSICOCHEMICAL AND PHYSICAL PROPERTIES

Virion Mr is about 60 x 10^6; S_{20w} is about 390; buoyant density in CsCl is 1.50 g/cm^3. Infectivity is ether-resistant.

NUCLEIC ACID

Genomes are about 48.5 kbp in size, corresponding to 54% of particle weight, have G+C contents of 52% and cohesive ends, and are nonpermuted.

PROTEINS

Virions contain 9 structural proteins (Mr 17-130 x 10^3), including 420 copies each of major capsid proteins D and E (Mr 38 and 53 x 10^3).

LIPIDS AND CARBOHYDRATES

None reported.

GENOME ORGANIZATION AND REPLICATION

The genome is linear and includes about 50 genes. Related functions cluster together. Phages adsorb to the cell wall and initiate a temperate infection. The infecting DNA circularizes and integrates into the host genome, generally at a preferred site, or is involved directly, without integration, in replication and transcription. Bidirectional DNA replication as a ϑ structure is followed by unidirectional replication via a rolling-circle mechanism. There is no breakdown of host DNA. Heads and tails assemble by 2 separate pathways. Proheads are frequent in lysates.

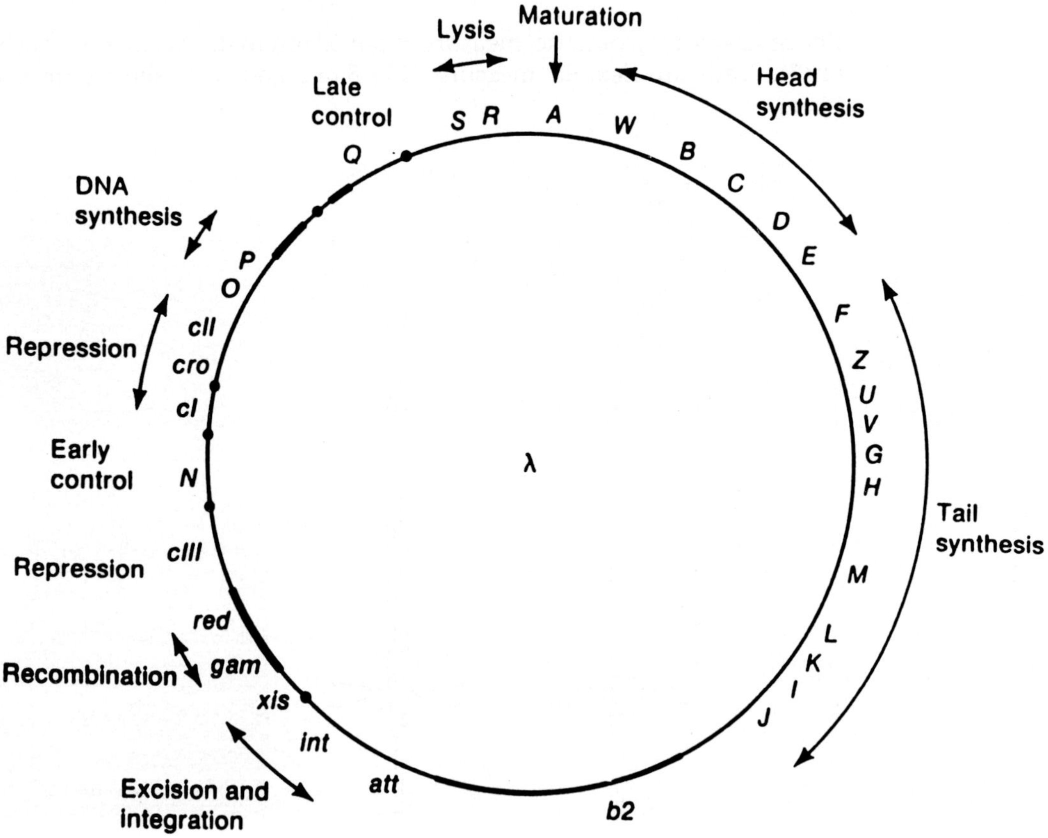

Figure 2: Simplified genetic map of coliphage λ. Solid lines indicate non-essential regions. (From Freifelder D (1983) Molecular Biology. Science Books International, Boston, and Van Nostrand Reynolds, New York, p 639, with permission).

BIOLOGICAL PROPERTIES

Phages are specific for enterobacteria.

LIST OF SPECIES IN THE GENUS

The viruses, their genomic sequence accession numbers [ ] and assigned abbreviations ( ) are:

SPECIES IN THE GENUS

coliphage λ	[V00636]	(λ)
The genus includes several isolates, called:		
lambdoid phage HK97		(HK97)
lambdoid phage HK022		(HK022)
lambdoid phage PA-2		(PA-2)
lambdoid phage ΦD328		(ΦD328)
lambdoid phage ø80		(ø80)

PA-2 and ø80 may represent independent species.

TENTATIVE SPECIES IN THE GENUS

None reported.

LIST OF UNASSIGNED SPECIES IN THE FAMILY

Actinomycetes phage A1-Dat	(A1-Dat)
Actinomycetes phage Bir	(Bir)
Actinomycetes phage M_1	(M_1)
Actinomycetes phage MSP8	(MSP8)
Actinomycetes phage P-a-1	(P-a-1)
Actinomycetes phage R_1	(R_1)
Actinomycetes phage R_2	(R_2)
Actinomycetes phage SV2	(SV2)
Actinomycetes phage VP5	(VP5)
Actinomycetes phage φC	(φC)
Actinomycetes phage φ31C	(φ31C)
Actinomycetes phage φUW21	(φUW21)
Actinomycetes phage φ115-A	(φ115-A)
Actinomycetes phage φ150A	(φ150A)
Actinomycetes phage 119	(119)
Agrobacterium phage PS8	(PS8)
Agrobacterium phage PT11	(PT11)
Agrobacterium phage ψ	(ψ)
Alcaligenes phage 8764	(8764)
Alcaligenes phage A5/A6	(A5/A6)
Bacillus phage α	(α)
Bacillus phage BLE	(BLE)
Bacillus phage IPy-1	(IPy-1)
Bacillus phage mor1	(mor1)
Bacillus phage MP15	(MP15)
Bacillus phage PBP1	(PBP1)
Bacillus phage SPP1	(SPP1)
Bacillus phage SPβ	(SPβ)
Bacillus phage type F	(type F)
Bacillus phage φ105	(φ105)
Bacillus phage 1A	(1A)
Bacillus phage II	(II)
Clostridium phage F1	(F1)
Clostridium phage HM7	(HM7)
coryneforms phage β	(β)
coryneforms phage φA8010	(φA8010)
coryneforms phage A	(A)
coryneforms phage Arp	(Arp)
coryneforms phage BL3	(BL3)
coryneforms phage CONX	(CONX)
coryneforms phage MT	(MT)

cyanobacteria phage S-2L	(S-2L)
cyanobacteria phage S-4L	(S-4L)
enterobacteria phage β4	(β4)
enterobacteria phage H-19J	(H-19J)
enterobacteria phage Jersey	(Jersey)
enterobacteria phage T5	(T5)
enterobacteria phage ViII	(ViII)
enterobacteria phage χ	(χ)
enterobacteria phage ZG/3A	(ZG/3A)
Lactobacillus phage 1b6	(1b6)
Lactobacillus phage 223	(223)
Lactobacillus phage ϕFSW	(ϕFSW)
Lactobacillus phage PL-1	(PL-1)
Lactobacillus phage y5	(y5)
Lactococcus phage 936	(936)
Lactococcus phage 949	(949)
Lactococcus phage 1358	(1358)
Lactococcus phage 1483	(1483)
Lactococcus phage BK5-T	(BK5-T)
Lactococcus phage c2	(c2)
Lactococcus phage PO87	(PO87)
Lactococcus phage P107	(P107)
Lactococcus phage P335	(P335)
Leuconostoc phage pro2	(pro2)
Listeria phage 2389	(2389)
Listeria phage 2671	(2671)
Listeria phage 2685	(2685)
Listeria phage H387	(H387)
Micrococcus phage N1	(N1)
Micrococcus phage N5	(N5)
Mycobacterium phage lacticola	(lacticola)
Mycobacterium phage Leo	(Leo)
Mycobacterium phage R1-Myb	(R1-Myb)
Pasteurella phage 32	(32)
Pasteurella phage C-2	(C-2)
Pseudomonas phage D3	(D3)
Pseudomonas phage Kf1	(Kf1)
Pseudomonas phage M6	(M6)
Pseudomonas phage PS4	(PS4)
Pseudomonas phage SD1	(SD1)
Rhizobium phage NM1	(NM1)
Rhizobium phage NT2	(NT2)
Rhizobium phage ϕ2037/1	(ϕ2037/1)
Rhizobium phage 5	(5)
Rhizobium phage 7-7-7	(7-7-7)
Rhizobium phage 16-2-12	(16-2-12)
Rhizobium phage 317	(317)
Staphylococcus phage 3A	(3A)
Staphylococcus phage 77	(77)
Staphylococcus phage 107	(107)
Staphylococcus phage 187	(187)
Staphylococcus phage 2848A	(2848A)
Staphylococcus phage B11-M15	(B11-M15)
Streptococcus phage 24	(24)
Streptococcus phage A25	(A25)
Streptococcus phage PE1	(PE1)

Streptococcus phage VD13	(VD13)
Streptococcus phage ω8	(ω8)
Vibrio phage α3a	(α3a)
Vibrio phage IV	(IV)
Vibrio phage OXN-52P	(OXN-52P)
Vibrio phage VP3	(VP3)
Vibrio phage VP5	(VP5)
Vibrio phage VP11	(VP11)

FAMILY PODOVIRIDAE

DISTINGUISHING FEATURES

Virions have short, noncontractile tails about 20 x 8 nm in dimension.

TAXONOMIC STRUCTURE OF THE FAMILY

Family	*Podoviridae*	
Genus		"T7-like Phages"

GENUS "T7-LIKE PHAGES"

Type Species coliphage T7 (T7)

VIRION PROPERTIES

MORPHOLOGY

Phage heads are isometric, measure about 60 nm in diameter, and consist of 72 capsomers (T=7). Tails measure 17 x 8 nm and have 6 short fibers.

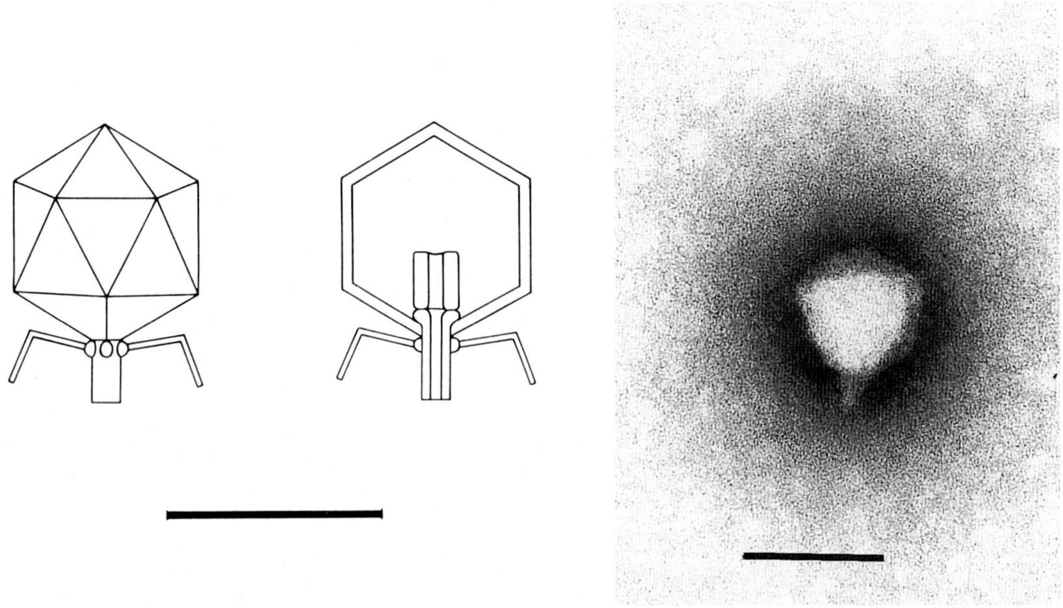

Figure 1: (left) Coliphage T7 in surface view and section. (Modified from Eiserling FA (1979) Bacteriophage structure. In: Fraenkel-Conrat H, Wagner RR (eds) Comprehensive Virology Vol 13. Plenum Press, New York, p. 553, with permission). (right) Negative contrast electron micrograph of coliphage T7; stained with phosphotungstate. Bars represent 100 nm.

PHYSICOCHEMICAL AND PHYSICAL PROPERTIES

Virion Mr is about 48×10^6; S_{20w} is about 510; buoyant density in CsCl is 1.50 g/cm^3. Infectivity is ether and chloroform resistant.

NUCLEIC ACID

Genomes are about 40 kbp in size, corresponding to 50% of particle weight, have G+C contents of 50%, and are nonpermuted and terminally redundant.

PROTEINS

Particles contain about 12 proteins (Mr 13-150 x 10^3), including about 40 copies of major capsid proteins (Mr 38 x 10^3); 3 proteins are located inside the head.

Lipids

None reported.

Carbohydrates

None reported.

Genome Organization and Replication

The genetic map is linear and codes for about 50 genes. Related functions cluster together. Phages adsorb to the cell wall and initiate a virulent infection with breakdown of the host chromosome. The viral DNA forms concatemers during replication. Tails assemble on preformed heads. Irregular polyheads are frequently observed.

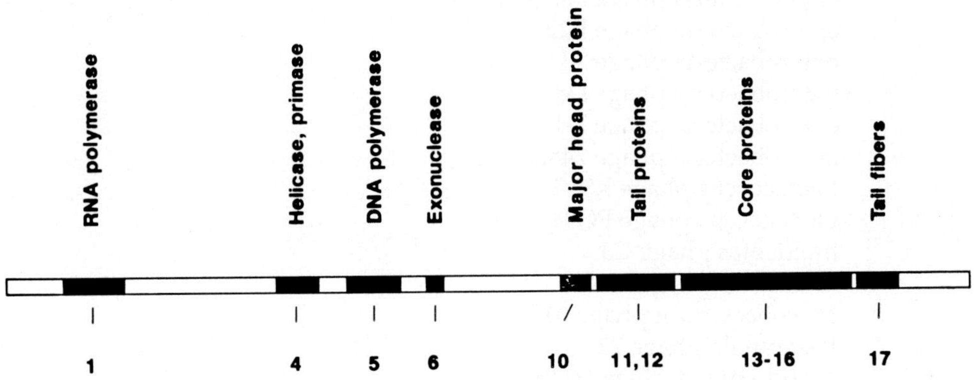

Figure 2: Simplified genetic map of coliphage T7. (Redrawn after Freifelder D (1983) Molecular Biology. Science Books International).

Biological Properties

Phages are specific for enterobacteria.

List of Species in the Genus

The viruses, their genomic sequence accession numbers [ ] and assigned abbreviations ( ) are:

Species in the Genus

The genus includes a number of isolates which may or may not represent independent species:

coliphage T7	[V01146]	(T7)
enterobacteria phage H		(H)
enterobacteria phage PTB		(PTB)
enterobacteria phage R		(R)
enterobacteria phage T3		(T3)
enterobacteria phage W31		(W31)
enterobacteria phage Y		(Y)
enterobacteria phage øI		(øI)
enterobacteria phage øII		(øII)
Pseudomonas phage gh-1		(gh-1)
Rhizobium phage 2		(2)
Vibrio phage III		(III)

Tentative Species in the Genus

None reported.

List of Unassigned Species in the Family

Bacillus phage GA-1	(GA-1)
Bacillus phage φ29	(φ29)
Brucella phage Tb	(Tb)
Clostridium phage HM2	(HM2)
coryneforms phage AN25S-1	(AN25S-1)
coryneforms phage 7/26	(7/26)
cyanobacteria phage AC-1	(AC-1)
cyanobacteria phage A-4(L)	(A-4(L))
cyanobacteria phage SM-1	(SM-1)
cyanobacteria phage LPP-1	(LPP-1)
enterobacteria phage Esc-7-11	(Esc-7-11)
enterobacteria phage N4	(N4)
enterobacteria phage P22	(P22)
enterobacteria phage sd	(sd)
enterobacteria phage Ω8	(Ω8)
enterobacteria phage 7-11	(7-11)
enterobacteria phage 7480b	(7480b)
Lactococcus phage KSY1	(KSY1)
Lactococcus phage PO34	(PO34)
mollicutes phage C3	(C3)
mollicutes phage L3	(L3)
Mycobacterium phage φ17	(φ17)
Pasteurella phage 22	(22)
Pseudomonas phage F116	(F116)
Rhizobium phage φ2042	(φ2042)
Staphylococcus phage 44AHJD	(44AHJD)
Streptococcus phage CP-1	(CP-1)
Streptococcus phage Cvir	(Cvir)
Streptococcus phage H39	(H39)
Streptococcus phage 2BV	(2BV)
Streptococcus phage 182	(182)
Xanthomonas phage RR66	(RR66)
Vibrio phage OXN-100P	(OXN-100P)
Vibrio phage 4996	(4996)
Vibrio phage I	(I)

Similarity with Other Taxa

See *Tectiviridae*.

Derivation of Names

myo: from Greek *mys, myos*, "muscle", relating to the contractile tail
podo: from Greek *pous*, "foot", for short tail
sipho: from Greek *siphon*, "tube"

References

Ackermann H-W (1992) Frequency of morphological phage descriptions. Arch Virol 124: 69-82
Ackermann H-W, DuBow MS (eds) (1987) Viruses of Prokaryotes. Vol I General Properties of Bacteriophages. CRC Press, Boca Raton FL
Ackermann H-W, DuBow MS (eds) (1987) Viruses of Prokaryotes. Vol II Natural Groups of Bacteriophages. CRC Press, Boca Raton FL
Casjens S, Hendrix RW (1988) Control mechanisms in dsDNA bacteriophage assembly. In: Calendar R (ed) The Bacteriophages Vol 1. Plenum Press, New York, pp 15-91
Hausmann R (1988) The T7 group. In: Calendar R (ed) The Bacteriophages Vol 1 Plenum Press, New York, pp 259-289
Hendrix RW, Roberts JW, Stahl FW, Weisberg RA (eds) (1983) Lambda II. Cold Spring Harbor, New York

Jarvis AW, Fitzgerald GF, Mata M, Mercenier A, Neve H, Powell IB, Ronda C, Saxelin M, Teuber M (1991) Species and type phages of lactococcal bacteriophages. Intervirology 32: 2-9
Klaus S, Krüger DH, Meyer J (1992) Bakterienviren. Gustav Fischer, Jena-Stuttgart, pp 133-247
Keppel F, Fayet O, Georgopoulos K (1988) Strategies of bacteriophage DNA replication. In: Calendar R (ed) The Bacteriophages Vol 2. Plenum Press, New York, pp 145-262
Mosig G, Eiserling F (1988) Phage T4 structure and metabolism. In: Calendar R (ed) The Bacteriophages Vol 2. Plenum Press, New York, pp 521-606

Contributed By

Ackermann H-W, Dubow MS

FAMILY *TECTIVIRIDAE*

TAXONOMIC STRUCTURE OF THE FAMILY

Family	***Tectiviridae***
Genus	*Tectivirus*

GENUS *TECTIVIRUS*

Type Species enterobacteria phage PRD1 (PRD1)

VIRION PROPERTIES

MORPHOLOGY

Virions exhibit icosahedral symmetry, have no envelope, and measure about 63 nm in diameter. Bacillus phage AP50 virions have 20 nm long spikes at their vertices. The capsid consists of two parts: a smooth, rigid, 3 nm thin protein outer shell and a flexible, 5 - 6 nm thick inner lipoprotein vesicle. The DNA is coiled within the vesicle. Virions are normally tail-less, but produce tail-like tubes of about 60 x 10 nm upon adsorption or after chloroform treatment.

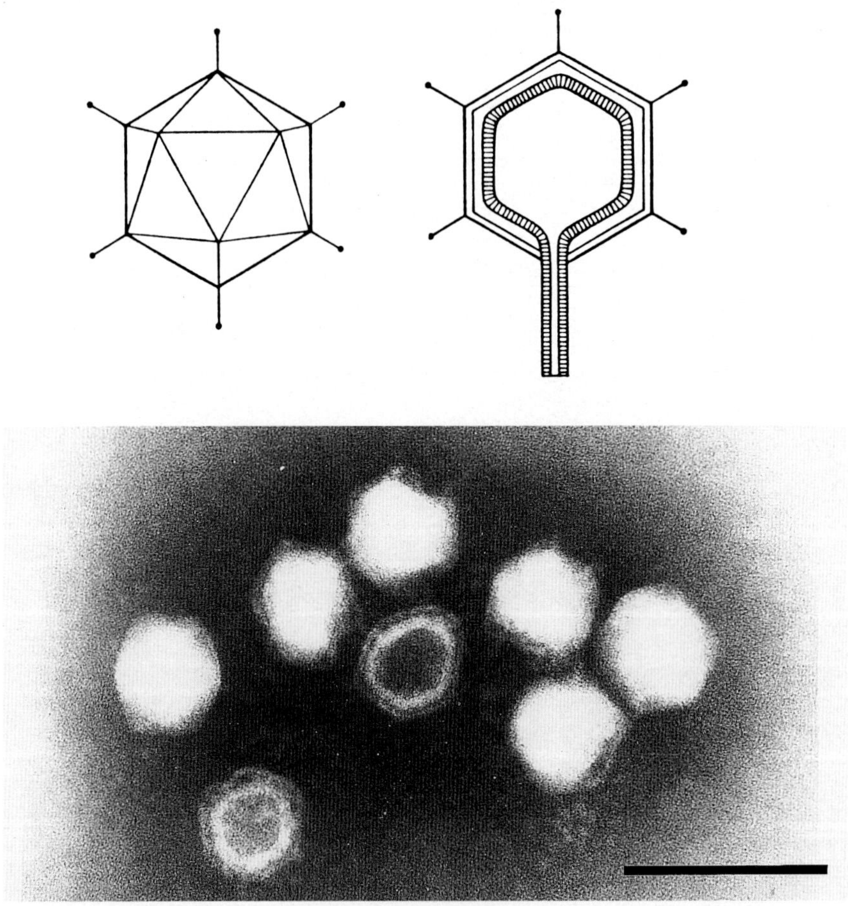

Figure 1: (upper) Diagram of virion in surface and in section; (lower) negative contrast electron micrograph of enterobacteria phage PRD1 particles. The bar represents 100 nm.

PHYSICOCHEMICAL AND PHYSICAL PROPERTIES

Virion Mr is about 70 x 10^6; S_{20w} is 357 - 416; buoyant density in CsCl is about 1.29 g/cm^3. Virions are usually stable at pH 5 - 8. Bacillus phage øNS11 has a pH optimum of 3.5. Infectivity is sensitive to ether, chloroform, and detergents.

Nucleic Acid

Virions contain a single molecule of linear dsDNA, 147 - 157 kbp in size (Mr 9.2 - 9.9 x 10^6). The DNA mass corresponds to 14 -15% of particle weight. The complete nucleotide sequence of enterobacteria phage PDR1 is available.

Proteins

Enterobacteria phage PRD1 is composed of at least 17 proteins (Mr 3-65 x 10^3). Two proteins constitute the outer shell and 15, mostly regulatory, are associated with the inner vesicle. The major capsid protein (Mr 43 x 10^3) is present in about 1,100 copies. Protein P1 is a DNA polymerase. Tectiviruses infecting members of the genus *Bacillus* contain at least 6 proteins (Mr 12.4 - 63 x 10^3). The major capsid protein has a Mr of 43-48 x 10^3 and is present in about 920 copies. Amino acid sequence data are available.

Lipids

Virions are composed of about 15% lipids by weight (5-6 species). Lipids constitute 60% of the inner vesicle. In PRD1 virus, lipids form a bilayer and seem to be in a liquid crystalline phase. Phospholipid contents (56% phosphatidylethanolamine and 37% phosphatidylglycerol) are higher than in the host, but vary according to the host strains. The fatty acid composition of the inner coat is identical to that of the host.

Carbohydrates

None reported.

Genome Organization and Replication

Enterobacteria phage PRD1 adsorbs to receptors coded by conjugative plasmids and tectiviruses of bacilli adsorb to the cell wall. Upon contact with the latter, the inner vesicle transforms itself into a tube and DNA is injected. Phages are virulent and liberated by lysis. The genome has inverted terminal repeats and a protein molecule covalently linked to each of its 5' ends. In enterobacteria phage PRD1, 33 ORFs have been identified. DNA is primed by the terminal protein. Transcription of early genes is bidirectional and directed toward the center of the genome. New phage DNA is packaged into preformed capsids.

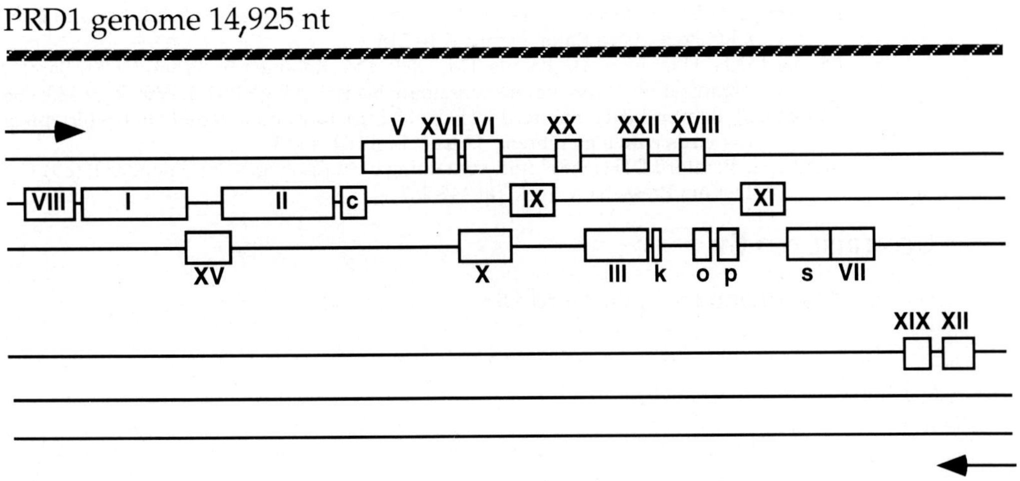

Figure 2: Diagram of enterobacteria phage PRD1 genome.

BIOLOGICAL PROPERTIES

Enterobacteria phage PRD1 multiplies in Gram-negative bacteria harboring P, N, or W incompatibility plasmids (enterobacteria, *Acinetobacter, Pseudomonas, Vibrio*). The phages AP50 and øNS11 are specific for the genus *Bacillus*.

LIST OF SPECIES IN THE GENUS

The viruses and their assigned abbreviations ( ) are:

SPECIES IN THE GENUS

Bacillus phage AP50	(AP50)
Bacillus phage øNS11	(øNS11)
enterobacteria phage PRD1	(PRD1)

TENTATIVE SPECIES IN THE GENUS

None reported.

LIST OF UNASSIGNED VIRUSES IN THE FAMILY

Vibrio phage 06N-58P (may be a corticovirus)

SIMILARITY WITH OTHER TAXA

Tectiviruses have morphological similarities to tailed phages (capsid size, tail) and corticoviruses (capsid size, thick inner component). They differ from tailed phages by their double capsid and the transitory nature of their "tail", and from corticoviruses by their ability to produce a "tail" or nucleic acid ejection device.

DERIVATION OF NAMES

tecti: from Latin *tectus*, 'covered'

REFERENCES

Ackermann H-W, DuBow MS, (eds) (1987) Viruses of Prokaryotes. Vol I General Properties of Bacteriophages. CRC Press, Boca Raton FL, pp 49 - 85

Ackermann H-W, DuBow MS (eds) (1987) Viruses of Prokaryotes. Vol II Natural Groups of Bacteriophages. CRC Press, Boca Raton FL, pp 171 - 218

Bamford JKH, Hanninen AL, Pakula TM, Ojala PM, Kalkinen N, Frilander M, Bamford DH (1991) Genome organization of membrane-containing bacteriophage PRD1. Virology 183: 658-676

Caldentey J, Bamford JKH, Bamford DH (1990) Structure and assembly of bacteriophage PRD1, an *Escherichia coli* virus with a membrane. J Struct Biol 104: 44-51

Mindich L, Bamford DH (1988) Lipid-containing bacteriophages. In: Calendar R (ed) The Bacteriophages Vol 2. Plenum Press, New York, pp 145-262

CONTRIBUTED BY

Ackermann H-W, Bamford DH

FAMILY *CORTICOVIRIDAE*

TAXONOMIC STRUCTURE OF THE FAMILY

Family *Corticoviridae*
Genus *Corticovirus*

GENUS *CORTICOVIRUS*

Type Species Alteromonas phage PM2 (PM2)

VIRION PROPERTIES

MORPHOLOGY

Virions exhibit icosahedral symmetry (T=12 or T=13) and are about 60 nm in diameter. They have no envelope. Capsids are complex and consist of an outer and an inner protein shell enclosing a lipid bilayer. Brush-like spikes protrude from each apex.

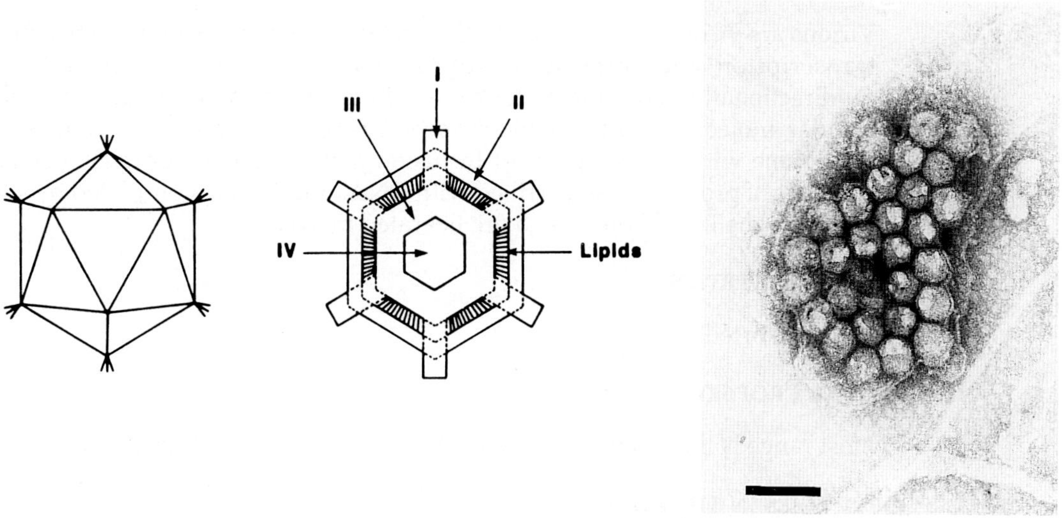

Figure 1: (left) Alteromonas phage PM2 in surface view and section (center), indicating locations of proteins I to IV and of lipid bilayer. (right) Negative contrast electron micrograph of Alteromonas phage PM2; phosphotungstate, the bar represents 100 nm. (From Schäfer R, Hinnen R, Franklin RM (1974) Eur J Biochem 50: 15-27, with permission).

PHYSICOCHEMICAL AND PHYSICAL PROPERTIES

Virion Mr is 49 x 10^6, S_{20w} is 230, buoyant density in CsCl is 1.28 g/cm^3. Virions are stable at pH 6-8, and are very sensitive to ether, chloroform, and detergents.

NUCLEIC ACID

Virions contain a single molecule of covalently closed, circular dsDNA about 9 kbp in size (Mr 5.8 x 10^6). The DNA contains a large number of superhelical turns. The DNA comprises 12.7% of virion weight and is coiled within the inner shell in association with protein IV. The G+C content is 43%. Parts of the Alteromonas phage PM2 genome have been sequenced.

PROTEINS

Four structural proteins are present. Protein II makes up 65% of the total protein. Proteins III and IV behave as lipoproteins. Transcriptase activity is associated with the virion.

Protein	Mr x 10^3	Location and function
I	43.6	Spikes, adsorption
II	27.7	Outer shell, major coat protein
III	13.0	Inner shell
IV	6.6	Transcriptase activity ?

LIPIDS

Particles are composed of 13% lipid by weight (5 species). Lipids form a bilayer between the outer and the inner shell. About 90% are phospholipids, mainly phosphatidylglycerol and phosphati-dylethanolmelamine. The lipid composition of phages differs from that of the host.

CARBOHYDRATES

None reported.

GENOME ORGANIZATION AND REPLICATION

Virions are virulent and adsorb to the bacterial cell wall. Genome structure and modes of transcription and translation are largely unknown. DNA replication proceeds unidirectionally and counterclockwise. Replicative intermediates include rings, nicked circular molecules, and double-branched rings. Phages are assembled at the plasma membrane without formation of inclusion bodies. The inner shell assembles first in the presence of protein IV, and is filled with DNA. Virions are completed by addition of lipids, the outer shell, and spikes, and are liberated by lysis.

ANTIGENIC PROPERTIES

None reported.

BIOLOGICAL PROPERTIES

Host range is limited to a marine bacterium of the genus *Alteromonas*.

LIST OF SPECIES IN THE GENUS

SPECIES IN THE GENUS

The viruses, and their assigned abbreviations ( ) are:

Alteromonas phage PM2 (PM2)

TENTATIVE SPECIES IN THE GENUS

None reported.

LIST OF UNASSIGNED VIRUSES IN THE FAMILY

Vibrio phage 06N-58P

SIMILARITY WITH OTHER TAXA

Corticoviruses have similarities to tectiviruses (capsid size, presence of lipids, sensitivity to ether, chloroform, and detergents). They differ from corticoviruses by the absence of an inner vesicle and a nucleic acid ejection device.

DERIVATION OF NAMES

cortico: from Latin *cortex, corticis*, "bark, crust"

References

Ackermann H-W, DuBow MS (1987) Viruses of Prokaryotes. Vol I General Properties of Bacteriophages. CRC Press, Boca Raton FL, pp 49-85

Ackermann H-W, DuBow MS (1987) Viruses of Prokaryotes. Vol II Natural Groups of Bacteriophages. CRC Press, Boca Raton FL, pp 171-218

Armour GA, Brewer GJ (1990) Membrane morphogenesis from cloned fragments of bacteriophage PM2 DNA that contain the sp6.6 gene. FASEB J 4: 1488-1493

Franklin RM, Marcoli R, Satake H, Schäfer R, Schneider D (1977) Recent studies on the structure of bacteriophage PM2. Med Microbiol Immunol 164: 87-95

Mindich L (1978) Bacteriophages that contain lipid. In: Fraenkel-Conrat H, Wagner RR (eds) Comprehensive Virology, Vol 12. Plenum Press, NY, pp 271-335

Contributed By

Ackermann H-W

FAMILY PLASMAVIRIDAE

TAXONOMIC STRUCTURE OF THE FAMILY

Family	***Plasmaviridae***
Genus	*Plasmavirus*

GENUS PLASMAVIRUS

Type Species Acholeplasma phage L2 (L2)

VIRION PROPERTIES

MORPHOLOGY

Virions are quasi-spherical, slightly pleomorphic, enveloped, and about 80 nm (range 50-125 nm) in diameter. Size varies due to virion heterogeneity; at least three distinct virion forms are produced during infection. Thin sections show virions with densely stained centers, seemingly containing condensed DNA, and particles with lucent centers. The absence of a regular capsid structure suggests the Acholeplasma phage L2 virion is an asymmetric nucleoprotein condensation bounded by a lipid-protein membrane.

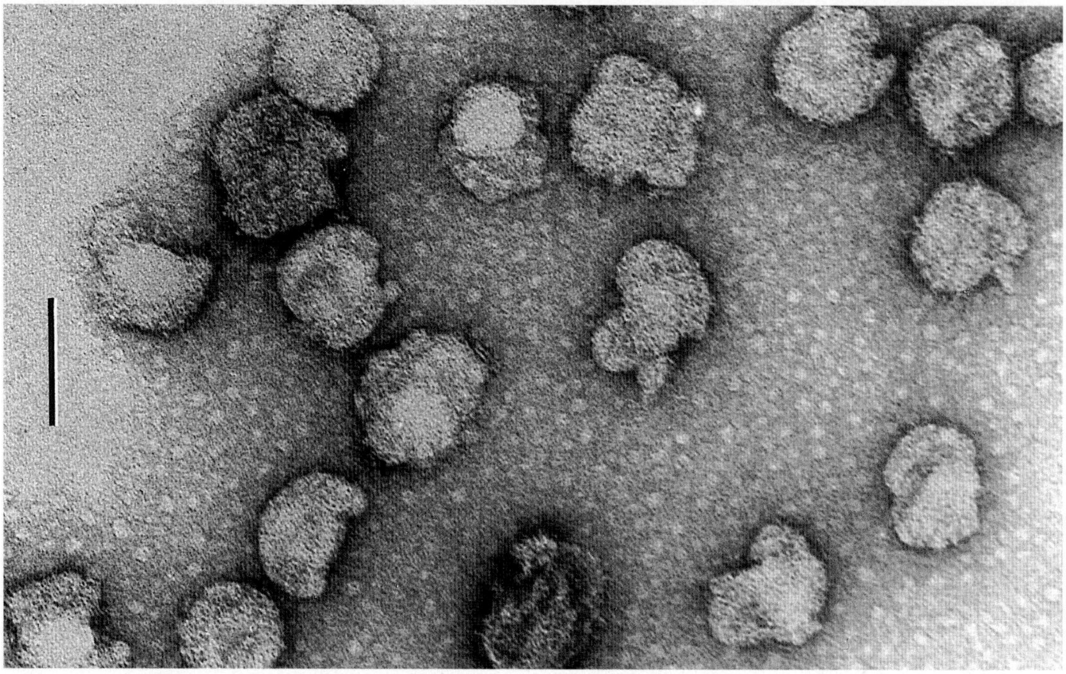

Figure 1: Negative contrast electron micrograph of Acholeplasma phage L2 virions. The pleomorphic virion appears as a core (perhaps a nucleoprotein condensation) within a baggy membrane. The bar represents 100 nm. (From Poddar SK, Cadden SP, Das J, Maniloff J (1985) with permission).

PHYSICOCHEMICAL AND PHYSICAL PROPERTIES

Virions are extremely heat sensitive, relatively cold stable, and inactivated by nonionic detergents (Brij-58, Triton X-100, and Nonidet P-40), ether, and chloroform. Viral infectivity is resistant to DNase I and phospholipase A, but sensitive to pronase and trypsin treatment. UV-irradiated virions can be reactivated in host cells by excision and SOS DNA repair systems. Virions are relatively resistant to photodynamic inactivation.

NUCLEIC ACID

Virions contain one molecule of infectious, circular, negative superhelical, dsDNA. The Acholeplasma phage L2 genome is 11,965 bp in size, with a G+C value of 32%. All ORFs are encoded in one strand. Several genes are translated from overlapping reading frames.

Proteins

Virions contain at least four major proteins, with Mr about 64, 61, 58, and 19 x 10^3. Several minor protein bands are also observed in virion preparations. DNA sequence analysis indicates 15 ORFs (encoding proteins of sizes from 7 to 81 kDa).

Lipids

Virions and host cell membranes have similar fatty acid compositions. Variation of host cell membrane fatty acid composition leads to virions with corresponding fatty acid composition variations. Data indicate viral membrane lipids are in a bilayer structure.

Carbohydrates

None reported.

Genome Organization and Replication

Acholeplasma phage L2 infection involves a noncytocidal productive infectious cycle followed by a lysogenic cycle in each infected cell. At least 11 overlapping mRNAs are transcribed from the DNA coding strand, from at least 8 promoters.

Noncytocidal infection involves progeny virus release by budding from the host cell membrane, with the host surviving as a lysogen. Lysogeny involves integration of the Acholeplasma phage L2 genome into a unique site in the host cell chromosome. Lysogens are resistant to superinfection by homologous virus but not by heterologous virus (apparently due to a repressor), and are inducible by UV-irradiation and mitomycin C.

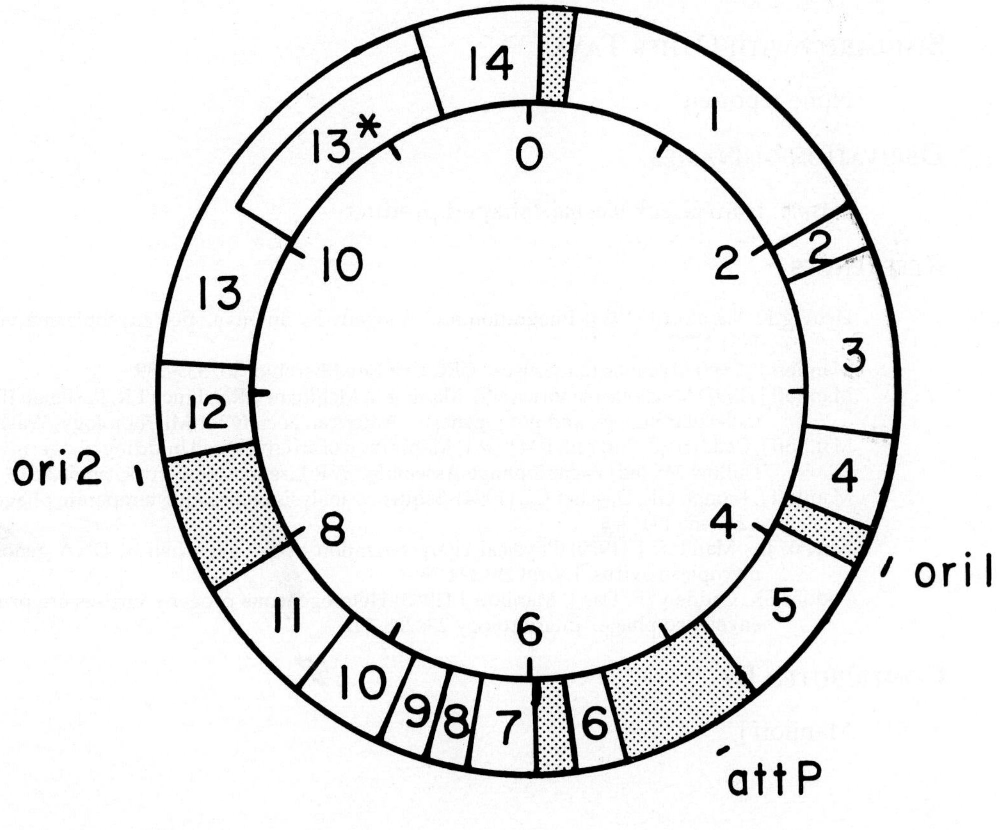

Figure 2: Map of genome organization, showing ORFs as determined from analysis of the 11,965 bp sequence. The base on the 3' -side of the single BstE II cleavage site is taken as the first base on the DNA sequence. The map also shows locations of the Acholeplasma phage L2 integration site (attP) and the two Acholeplasma phage L2 DNA replication origin sites (ori1 and ori2). (From Maniloff J, Kampo GJ, and Dascher CC. (1994)).

Biological Properties

Host Range

Acholeplasma phage L2 infects *Acholeplasma laidlawii* strains: other possible plasmaviruses have been reported to infect *A. modicum* and *A. oculi* strains.

List of Species in the Genus

The viruses, their genomic sequence accession numbers [ ], host { } and assigned abbreviations ( ) are:

Species in the Genus

Acholeplasma phage L2	{*A. laidlawii*} [L13696]	(L2)

Tentative Species in the Genus

Acholeplasma phage v1	{*A. laidlawii*}	(v1)
Acholeplasma phage v2	{*A. laidlawii*}	(v2)
Acholeplasma phage v4	{*A. laidlawii*}	(v4)
Acholeplasma phage v5	{*A. laidlawii*}	(v5)
Acholeplasma phage v7	{*A. laidlawii*}	(v7)
Acholeplasma phage M1	{*A. modicum*}	(M1)
Acholeplasma phage O1	{*A. oculi*}	(O1)

List of Unassigned Viruses in the Family

None reported.

Similarity with Other Taxa

None reported.

Derivation of Names

plasma: from Greek *plasma*, 'shaped product'

References

Dybvig K, Maniloff J (1983) Integration and lysogeny by an enveloped mycoplasma virus. J Gen Virol 64: 1781-1785

Maniloff J (1988) Mycoplasma viruses. CRC Crit Rev Microbiol 15: 339-389

Maniloff J (1992) Mycoplasma viruses In: Maniloff J, McElhaney RN, Finch LR, Baseman JB (eds) Mycoplasmas: molecular biology and pathogenesis. American Society for Microbiology, Washington DC, pp 41-59

Maniloff J, Cadden SP, Putzrath RM (1981) Maturation of an enveloped budding phage: mycoplasmavirus L2 In: DuBow MS (ed) Bacteriophage Assembly. A R Liss Inc, New York, pp 503-513

Maniloff J, Kampo GK, Dascher CC (1994) Sequence analysis of a unique temperate phage: mycloplasma virus L2. Gene 141: 1-8

Nowak JA, Maniloff J (1979) Physical characterization of the superhelical DNA genome of an enveloped mycoplasmavirus. J Virol 29: 374-380

Poddar SK, Cadden SP, Das J, Maniloff J (1985) Heterogeneous progeny viruses are produced by a budding enveloped phage. Intervirology 23: 208-221

Contributed By

Maniloff J

FAMILY LIPOTHRIXVIRIDAE

TAXONOMIC STRUCTURE OF THE FAMILY

Family	***Lipothrixviridae***
Genus	*Lipothrixvirus*

GENUS LIPOTHRIXVIRUS

Type Species Thermoproteus virus 1 (TTV-1)

VIRION PROPERTIES

MORPHOLOGY

Virions are rigid rods, 410 nm long and 38 nm in diameter, with protrusions arising asymmetrically from both ends. The envelope does not show structure in electron micrographs. It contains a helical core.

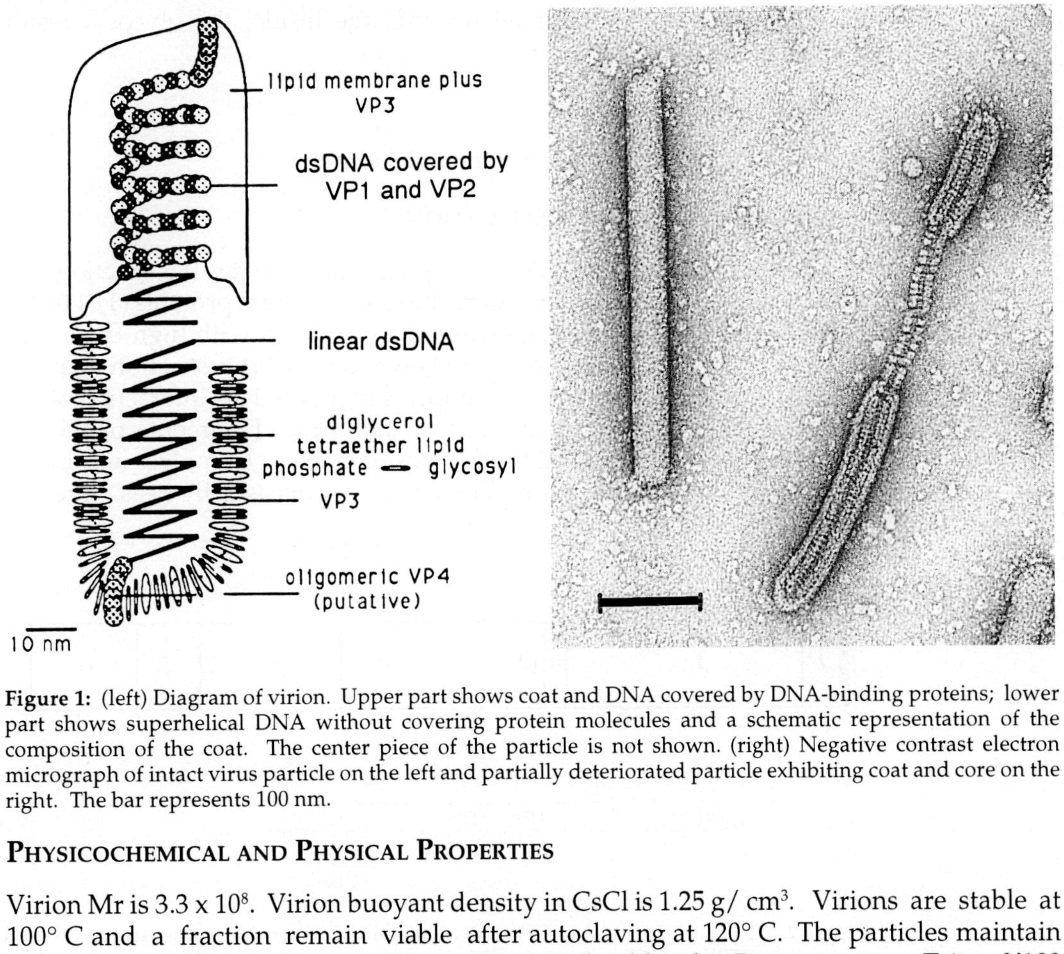

Figure 1: (left) Diagram of virion. Upper part shows coat and DNA covered by DNA-binding proteins; lower part shows superhelical DNA without covering protein molecules and a schematic representation of the composition of the coat. The center piece of the particle is not shown. (right) Negative contrast electron micrograph of intact virus particle on the left and partially deteriorated particle exhibiting coat and core on the right. The bar represents 100 nm.

PHYSICOCHEMICAL AND PHYSICAL PROPERTIES

Virion Mr is 3.3 x 10^8. Virion buoyant density in CsCl is 1.25 g/ cm^3. Virions are stable at 100° C and a fraction remain viable after autoclaving at 120° C. The particles maintain their structure in 6M urea and 7M guanidine hydrochloride. Detergents, e.g. Triton X100 and octylglycoside, dissociate virions into viral cores, containing the DNA plus DNA-binding proteins, and viral envelopes, containing isopranyl ether lipid and coat protein.

Virions contain about 3% (w/w) DNA, about 75% protein and about 22% isopranyl ether lipids.

NUCLEIC ACID

Virions contain one molecule of linear dsDNA; Mr 10 x 10^6 (15.9 kbp). About 85% (except the left most ClaI fragment) of the total sequence has been determined. The ends of the DNA molecule are masked in an unknown manner.

PROTEINS

Virions contain at least four proteins of the following sizes: TP1 12.9 kDa; TP2 16.3 kDa; TP3 18.1 kDa; and TP4 24.5 kDa. TP1 and TP2 are DNA-associated, TP3 is the envelope protein, and the location of TP4 in the virus particle is unknown. Only TP1 is a basic protein. TP3 is highly hydrophobic, TP4 hydrophilic. Additional minor proteins may be present. A fifth protein, TPX, carrying a C-terminal (thr, pro)$_n$ repeat, is present in infected cells in high amount, but absent in virus particle.

LIPIDS

The virion envelope contains the same lipids as the host's membrane, essentially diphytanyl tetraether-lipids. The envelope has a bilayer structure. The phosphate residues of the phospholipids are oriented towards the inside, the glycosyl residues towards the outside of the particles.

CARBOHYDRATES

Virions contain carbohydrate in their glycolipid.

GENOME ORGANIZATION AND REPLICATION

The genome contains several transcription units. So far, the function of only few genes is known, among them those encoding the four structural proteins (TP1 to TP4). There are two ORFs between which specific recombination occurs with high frequency, encoding (TP)$_n$ and (PT)$_n$. Their map positions are shown in the ClaI restriction map of the viral genome (Fig. 2). Adsorption and infection appears to proceed via interaction of the tips of pili of the host with the terminal protrusions of the virus. Fragments of the viral genome have sometimes been found integrated in host genomes. Complete non-integrated virus DNA exists in the cell in linear form. The virions are released by lysis. Infection may be latent.

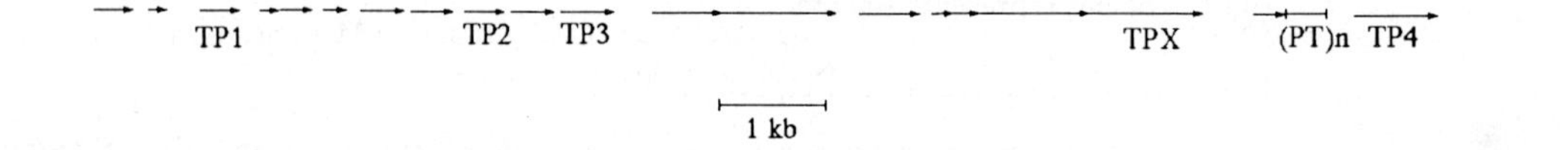

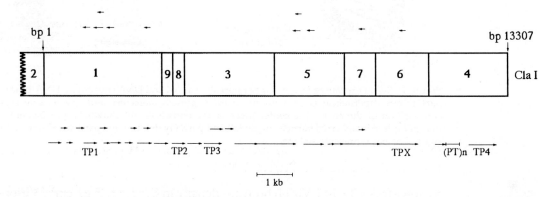

Figure 2: ClaI restriction map of TTV-1 DNA showing ORFs including the structural proteins of the virus and two regions, TPX and (PT)$_n$ containing (thr, pro)$_n$ repeats, between which recombination occurs frequently.

BIOLOGICAL PROPERTIES

HOST RANGE

The host range is limited to the archaeon *Thermoproteus tenax*. Other rod-shaped DNA-containing viruses of similar morphology but different dimensions have been found associated with *Thermoproteus* cultures or have been observed by electron microscopy in waters from Icelandic solfataras but virus has not been cultivated from these sources. TTV-2 and TTV-3 resemble TTV-1 in yielding a DNA containing core structure and a lipid-containing coat upon treatment with detergent. The coat of TTV-4 contains only one protein and no lipid.

Geographic Distribution

So far, these viruses have been found only in samples taken from solfataras at the Krafla in Iceland.

List of Species in the Genus

The viruses, their genomic sequence accession numbers [ ] and assigned abbreviations ( ) are:

Species in the Genus

Thermoproteus virus 1	[X14855]	(TTV-1)
Thermoproteus virus 2		(TTV-2)
Thermoproteus virus 3		(TTV-3)
Thermoproteus virus 4		(TTV-4)

Tentative Species in the Genus

None reported.

List of Unassigned Viruses in the Family

None reported.

Similarity with Other Taxa

None reported.

Derivation of Names

lipo: from Greek, *lipos*, 'fat'
thrix: from Greek, *thrix*, 'hair'

References

Neumann H, Schwass V, Eckerskorn C, Zillig W (1989) Identification and characterization of the genes encoding three structural proteins of the Thermoproteus tenax virus TTV1. Mol Gen Genet 217: 105-110

Neumann H, Zillig W (1989) Coat protein TP4 of the virus TTV1: primary structure of the gene and the protein. Nucl Acids Res 17: 9475

Neumann H, Zillig W (1990) The TTV1-encoded viral protein TPX: primary structure of the gene and the protein. Nucl Acids Res 18: 195

Neumann H, Zillig W (1990) Nucleotide sequence of the viral protein TPX of the TV1 variant VT3. Nucl Acids Res 18: 2171

Neumann H, Zillig W (1990) Structural variability in the genome of the Thermoproteus tenax virus TTV1. Mol Gen Genet 222: 435-437

Reiter W-D, Zillig W, Palm P (1988) Archaebacterial viruses. In Maramorosch K, Murphy FA, Shatkin AJ (eds) Adv Virus Res, Vol 34, pp 143-188

Rettenberger M (1990) Das Virus TTV1 des extrem thermophilen Schwefel-Archaebakteriums Thermoproteus tenax: Zusammensetzung und Struktur. Doctor's thesis. Ludwig-Maximilians-Universität München

Zillig W, Reiter W-D, Palm P, Gropp F, Neumann H, Rettenberger M (1988) Viruses of archaebacteria. In: Calendar R (ed) The Bacteriophages, Vol 1, Plenum Press, New York, pp 517-558

Contributed By

Zillig W

FAMILY *FUSELLOVIRIDAE*

TAXONOMIC STRUCTURE OF THE FAMILY

Family	***Fuselloviridae***
Genus	*Fusellovirus*

GENUS *FUSELLOVIRUS*

Type Species Sulfolobus virus 1 (SSV-1)

VIRION PROPERTIES

MORPHOLOGY

Virions are lemon-shaped, slightly flexible in appearance with short tail fibers attached to one pole. Virions are 60 x 100 nm in size; a small fraction of the SSV-1 population (up to 1 %) is larger with a particle length of about 300 nm. The virion envelope consists of host lipids and of two virus-encoded proteins; a third protein is DNA-associated (Fig. 1).

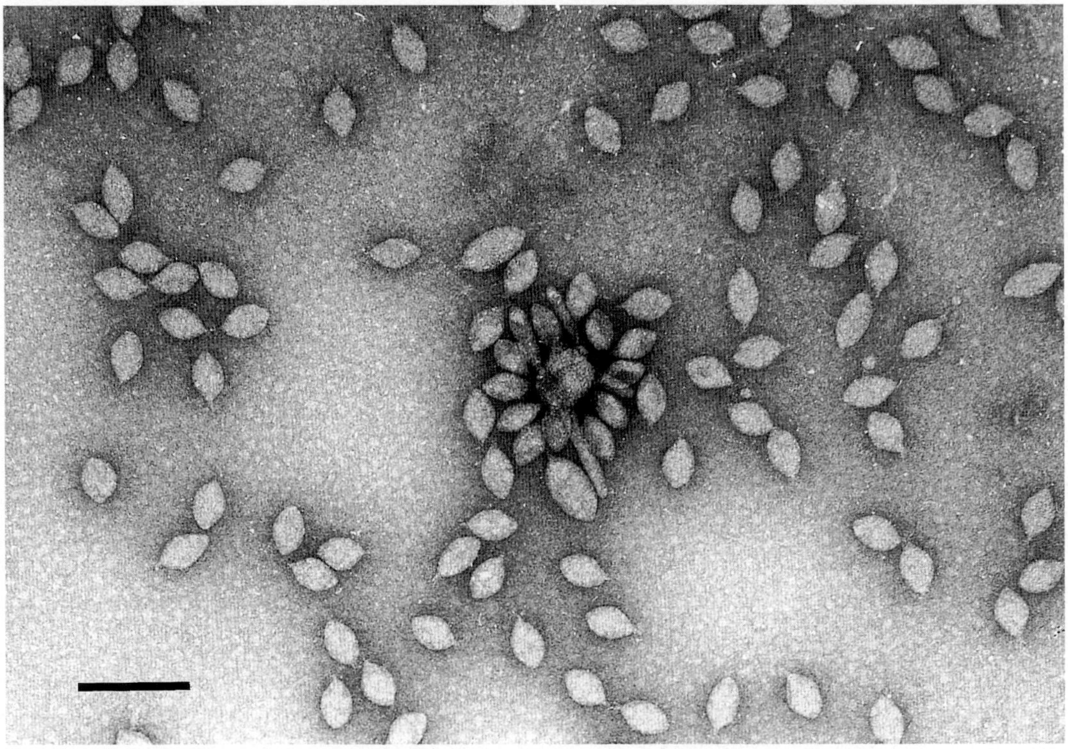

Figure 1: Negative contrast electron micrograph of SSV-1 virions, bar represents 200 nm.

PHYSICOCHEMICAL AND PHYSICAL PROPERTIES

Virion buoyant density in CsCl is 1.24 g/cm^3. The particles are stable at up to 85° C and are insensitive to urea and ether. Low pH (below 5) reduces viability due to degradation of the DNA; virions are sensitive to high pH (above 11) and trichloromethane.

NUCLEIC ACID

Virions contain circular, positively supercoiled dsDNA, of 15,465 bp in size. Virion DNA is associated with polyamines and a virus-coded basic protein. The nucleic acid sequence has been completely determined and the data are available from EMBL/GenBank.

PROTEINS

Two basic proteins (VP1 and VP3) are constituents of the virion envelope. They consist of 73 and 92 amino acid residues as deduced from the nucleic acid sequence. A very basic protein (VP2, 74 amino acids) is attached to the viral DNA. The genes encoding these three structural proteins are closely linked on the SSV-1 genome, in the order VP1, 3, 2.

The second largest ORF of SSV-1 (ORF d335, 335 amino acids) shows sequence homology to the integrase family of site-specific recombinases. This protein has been expressed in *E. coli* and recombines DNA fragments sequence-specifically *in vitro*.

LIPIDS

10 % of the SSV-1 coat consists of host lipids.

CARBOHYDRATES

None reported.

GENOME ORGANIZATION AND REPLICATION

The SSV-1 genome is present in the cells as cccDNA and also site-specifically integrated into a tRNA gene of the host chromosome. The integrated copy is flanked by a 44 bp direct repeat (attachment core) that occurs once in the circular SSV-1 DNA. Upon integration, ORF d335 is disrupted.

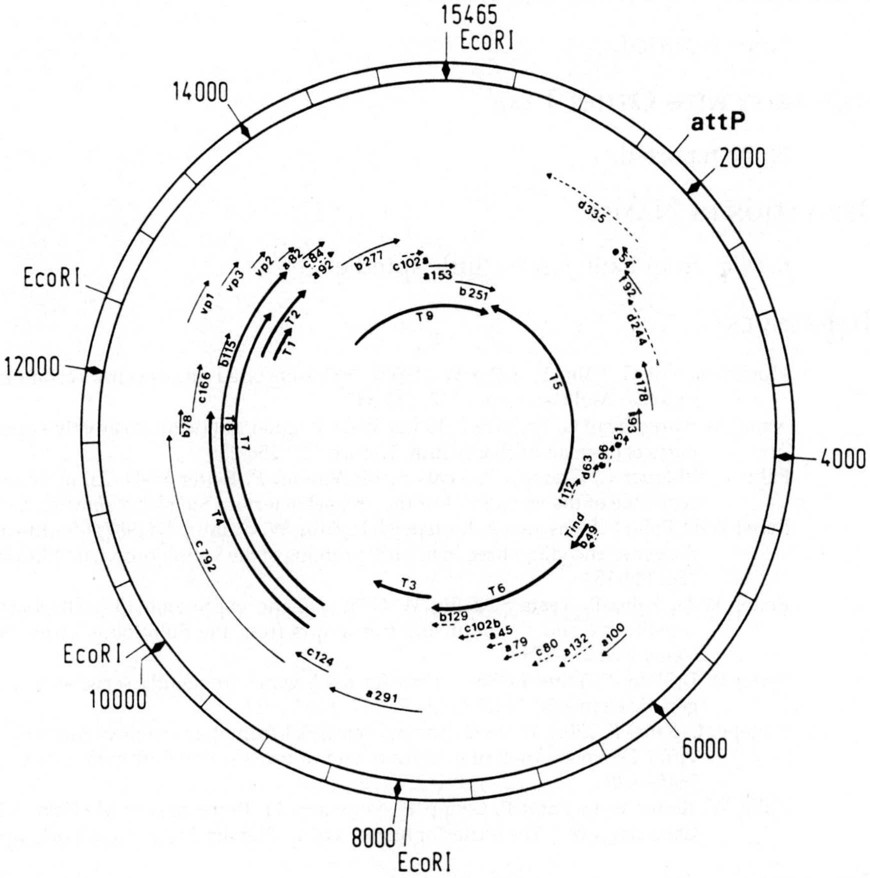

Figure 2: Genome organization of SSV-1. Numbers refer to base pairs relative to the start of the largest *Eco*RI fragment. AttP indicates the cleaving site for integration into the host genome. Bold arrows represent transcripts T1 to T9 and T_{ind}; thin arrows are protein genes VP1 to VP3 and ORFs, both without cysteine codons; dotted lines indicate ORFs that contain cysteine codons.

Eleven transcripts, initiated from 7 promoters, cover the SSV-1 genome. UV-irradiation is a stimulus for virus production and the particles are released without evident lysis of the host cells. A small transcript (T_{ind}) is strongly induced upon induction. Particles are probably assembled at the cell membrane, since no virus particles have been observed in host cells.

Biological Properties

Host range

Host range is limited to two extremely thermophilic archaeon, *Sulfolobus shibatae* and *Sulfolobus* isolates P1 and P2. Few phage particles are produced in cultures of lysogens. UV-irradiation strongly induces phage production without evident lysis of the host.

List of Species in the Genus

The viruses, their genomic sequence accession numbers [ ], and assigned abbreviations ( ) are:

Species in the Genus

Sulfolobus virus 1	[XO7234]	(SSV-1)

Tentative Species in the Genus

None reported.

Unassigned Viruses in the Family

None reported.

Similarity with Other Taxa

None reported.

Derivation of Names

fusello: from Latin *fusello*, 'little spindle'

References

Muskhelishvili G, Palm P, Zillig W (1993) SSV1-encoded site-specific recombination system in *Sulfolobus shibatae*. Mol Gen Genet 237: 334-342

Nadal M, Mirambeau G, Forterre P, Reiter W-D, Duguet M (1986) Positively supercoiled DNA in a virus-like particle of an archaebacterium. Nature 321: 256-258

Palm P, Schleper C, Grampp B, Yeats S, McWilliam P, Reiter W-D, Zillig W (1991) Complete nucleotide sequence of the virus SSV1 of the archaebacterium *Sulfolobus shibatae*. Virology 185: 242-250

Reiter W-D, Palm P, Henschen A, Lottspeich F, Zillig W, Grampp B (1987) Identification and characterization of the genes encoding three structural proteins of the Sulfolobus virus-like particle SSV1. Mol Gen Genet 206: 144-153

Reiter W-D, Palm P, Yeats S, Zillig W (1987) Gene expression in archaebacteria: Physical mapping of constitutive and UV-inducible transcripts from the Sulfolobus virus-like particle SSV1. Mol Gen Genet 209: 270-275

Reiter W-D, Palm P, Yeats S (1989) Transfer RNA genes frequently serve as integration sites for prokaryotic genetic elements. Nucl Acids Res 17: 1907-1914

Schleper C, Kubo K, Zillig W (1992) The particle SSV1 from the extremely thermophilic archaeon *Sulfolobus* is a virus: Demonstration of infectivity and of transfection with viral DNA. Proc Natl Acad Sci USA 89: 7645-7649

Zillig W, Reiter W-D, Palm P, Gropp F, Neumann H, Rettenberger M (1988) Viruses of archaebacteria In: Calendar R (ed) The bacteriophages, Vol 1. Plenum Press, New York, pp 517-558

Contributed By

Schleper C

FAMILY POXVIRIDAE

TAXONOMIC STRUCTURE OF THE FAMILY

Family	***Poxviridae***	
Subfamily	***Chordopoxvirinae***	
Genus		*Orthopoxvirus*
Genus		*Parapoxvirus*
Genus		*Avipoxvirus*
Genus		*Capripoxvirus*
Genus		*Leporipoxvirus*
Genus		*Suipoxvirus*
Genus		*Molluscipoxvirus*
Genus		*Yatapoxvirus*
Subfamily	***Entomopoxvirinae***	
Genus		*Entomopoxvirus A*
Genus		*Entomopoxvirus B*
Genus		*Entomopoxvirus C*

VIRION PROPERTIES

MORPHOLOGY

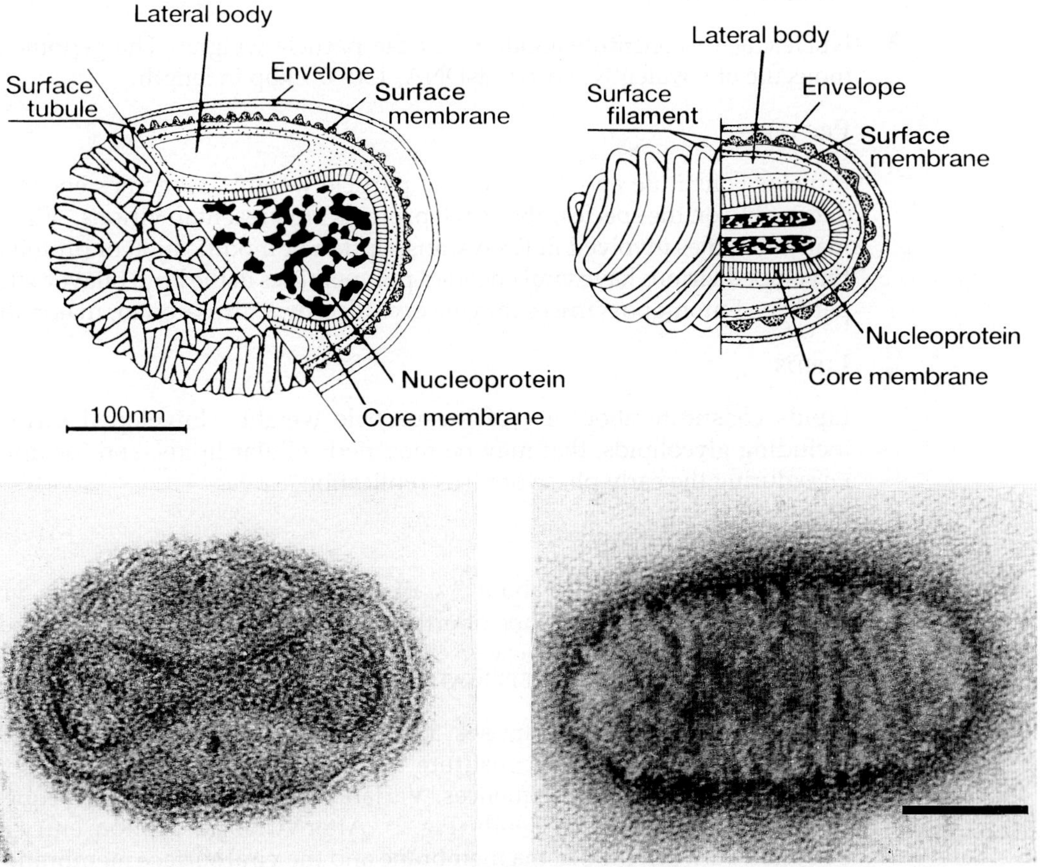

Figure 1: (upper left) Schematic of brick-shaped orthopoxvirus; the left side of the diagram shows the surface structure of a non-enveloped orthopoxvirus particle; on the right side is shown a cross-section of an enveloped form of the orthopoxvirus particle; (upper right) schematic of ovoid-shaped parapoxvirus; the left side of the diagram shows the surface of a non-enveloped parapoxvirus particle with a single long filament seemingly wound around the particle; on the right side is shown a section through the enveloped form of the virus; (lower left) thin section of non-enveloped vaccinia virus (lower right) negative contrast electron micrograph of, non-enveloped orf virus (the bar represents 100 nm). (Modified from Fenner and Nakano (1988) with permission).

Virions are somewhat pleomorphic, generally either brick-shaped (220-450 nm long x 140-260 nm wide x 140-260 nm thick) with a lipoprotein surface membrane displaying tubular or globular units (10-40 nm), or ovoid (250-300 nm long x 160-190 nm diameter) with a surface membrane possessing a regular spiral filament (10-20 nm in diameter).

Negative contrast images show the surface membrane encloses a biconcave or cylindrical core that contains the genome DNA and proteins organized in a nucleoprotein complex. One or two lateral bodies appear to be present in concavities between the core membrane and the surface membrane. Some virions are enclosed in an envelope derived from the cell and containing virus-specified proteins. Others (e.g., entomopoxviruses) may be occluded.

Physicochemical and Physical Properties

Particle weight is about 5×10^{-15} g. S_{20w} is about 5000. Buoyant density of virions is subject to osmotic influences: in dilute buffers it is about 1.16 g/cm^3, in sucrose about 1.25 g/cm^3, in CsCl and potassium tartrate about 1.30 g/cm^3. Virions tend to aggregate in high salt solution. Infectivity of some members is resistant to trypsin. Some members are insensitive to ether. Generally, virions are sensitive to common detergents, formaldehyde, oxidizing agents, and temperatures greater than 40° C. The virion surface membrane is removed by nonionic detergents and sulfhydryl reagents. Virions are relatively stable in dry conditions at room temperature; they can be lyophilized with little loss of infectivity.

Nucleic Acid

Nucleic acids constitute about 3% of the particle weight. The genome is a single, linear molecule of covalently-closed, dsDNA, 130-375 kbp in length.

Proteins

Proteins constitute about 90% of the particle weight. Genomes encode 150-300 proteins depending on the species; about 100 proteins are present in virions. Virus particles contain many enzymes involved in RNA transcription or modification of proteins or nucleic acids. Enveloped virions have viral encoded polypeptides in the lipid bilayer which surrounds the particle. Entomopoxviruses may be occluded by a virus-coded, major structural protein.

Lipids

Lipids constitute about 4% of the particle weight. Enveloped virions contain lipids, including glycolipids, that may be modified cellular lipids, and/or lipids synthesized *de novo* during the early phase of virus replication.

Carbohydrates

Carbohydrates constitute about 3% of the particle weight. Certain viral proteins, e.g., hemagglutinin in the envelope of orthopoxviruses, have N- and C-linked glycans.

Genome Organization and Replication

The poxvirus genome comprises a linear molecule of dsDNA with covalently closed termini; terminal hairpins constitute two isomeric "flip-flop" DNA forms consisting of inverted complementary sequences. Variably sized, tandem repeat sequence arrays may or may not be present near the ends (Fig. 2). After virion adsorption, entry into the host cell is by fusion between the plasma membrane and the viral surface membrane or, when present, the envelope, after which cores are released into the cytoplasm and uncoated further. Endocytosis, involving fusion between the plasma and vacuolar membranes, may also occur.

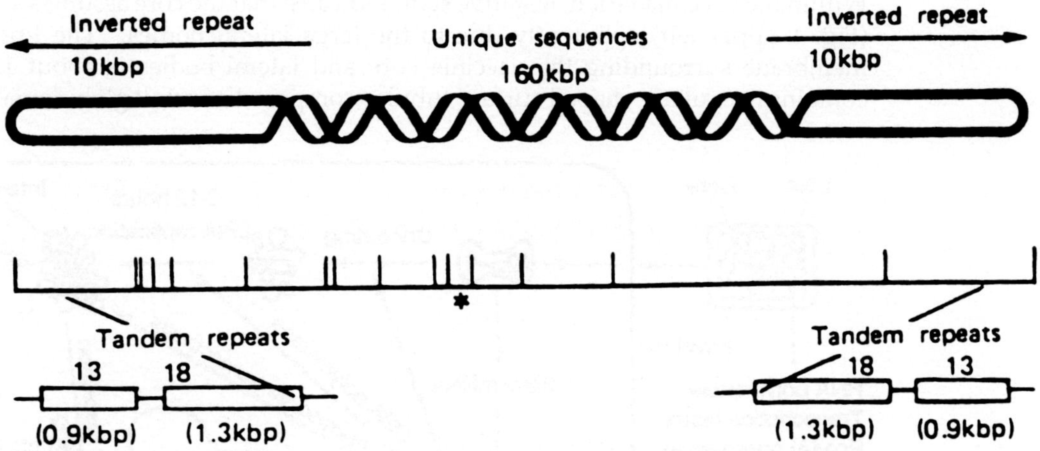

Figure 2: Schematic representation of the DNA of vaccinia virus (WR strain): (upper) Linear double-stranded molecule with terminal hairpins and inverted repeats (not to scale). The denatured DNA forms a single-stranded circular molecule. In (center) are shown the HindIII cleavage sites of the vaccinia virus (WR strain) genome, the asterisk indicates the fragment that contains the thymidine kinase gene. (lower) Each 10-kbp terminal portion includes two groups of tandem repeats of short sequences rich in AT. (From Fenner, Wittek, and Dumbell 1989 with permission).

Polyadenylated primary mRNA transcripts, representing about 15% of the genome, are synthesized from both DNA strands by enzymes within the core, including a multisubunit RNA polymerase; transcripts are extruded from the core for translation by host ribosomes. During synthesis of early proteins, host macromolecular synthesis is shut-off. Virus reproduction ensues in the host cell cytoplasm, producing basophilic (B-type) inclusions termed "viroplasms" (or "virus factories"). The genome contains closely spaced ORFs preceded by promoters that temporally regulate transcription of three classes of genes: early genes, expressed before and during genome uncoating (these encode many non-structural proteins, including enzymes involved in replicating the genome and modifying DNA, RNA, and proteins); intermediate genes, expressed during the period of DNA replication (these appear to regulate late gene transcription); and late genes that are expressed during the post-replicative phase (these mainly encode virion proteins). The mRNAs are capped, polyadenylated at the 3' termini, and not spliced. Many late mRNAs and some early mRNAs have 5'-polyadenylated sequences. Early protein synthesis is generally decreased during the switch-over to late gene expression, but some genes are expressed first from an early promoter and then from a late promoter. Certain proteins are modified post-translationally (e.g., by proteolytic cleavage, phosphorylation, glycosylation, ribosylation, sulfation, acylation, myristylation, by binding metal ions, by disulfide cross-linking, etc.). A summary of the infectious cycle is given in Fig. 3

The DNA genome appears to be replicated mainly by viral enzymes. Although incompletely understood, it may involve a self-priming, unidirectional, strand displacement mechanism in which concatemerized replicative intermediates are resolved into unit length DNAs that are subsequently covalently closed. Genetic recombination within genera has been shown, and may occur between daughter molecules during replication. Non-genetic genome reactivation generating infectious virus has been shown within and between genera of the *Chordopoxvirinae*.

Virus morphogenesis proceeds via coalescence of DNA within crescent-shaped, lipoprotein bilayers (nascent surface membranes) that are coated with spicules. Eventually, the lipoprotein encloses the genome to form an immature particle. Virus DNA and several proteins are organized as a nucleoprotein complex within the core. Particle maturation involves continued protein synthesis and the formation of intracellular naked virions (INVs) which contain an encompassing surface membrane, lateral bodies, and the nucleoprotein core complex. For vaccinia, the core has a 9 nm thick membrane with a regular subunit structure.

Within the vaccinia virion, negative stain indicates that the core assumes a biconcave shape (Fig. 1 upper left) apparently due to the large lateral bodies. The lipoprotein surface membrane surrounding the vaccinia core and lateral bodies is about 12 nm thick and contains irregularly shaped surface tubules composed of small globular subunits.

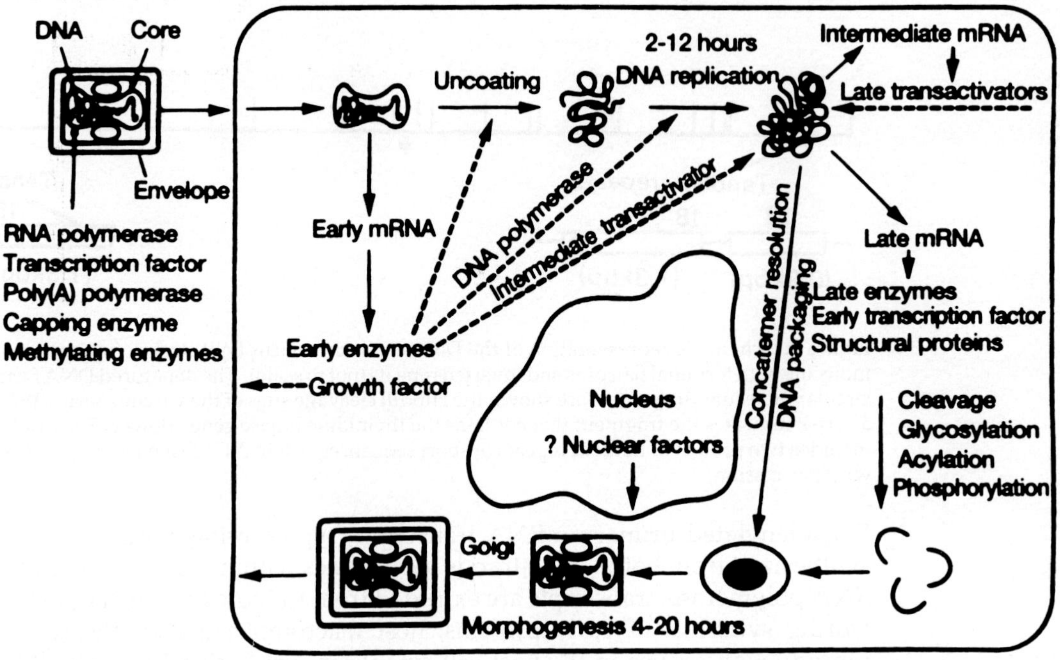

Figure 3: The infectious cycle of vaccinia virus (from Moss, Science 252:1662, 1991; Copyright AAAS, 1991, with permission).

Mature INVs are released by cellular disruption. A few may be enveloped on exocytosis following acquisition of modified Golgi membranes or following extrusion through microvilli. Enveloped virions thereby acquire host cell lipids and additional virus-specific proteins, including the virus hemagglutinin protein. The envelope is closely applied to the surface membrane. Although the internal structure of vaccinia is revealed in thin sections, the detailed internal structure of parapoxvirus particles is less evident. In negatively stained preparations of parapoxviruses, superimposition of dorsal and ventral views of the surface filament sometimes produces a distinctive criss-cross surface appearance. Both INVs and enveloped virions are infectious and contain different exterior antigens.

Antigenic Properties

Within each genus of the subfamily *Chordopoxvirinae* there is considerable serologic cross-protection and cross-reactivity. Neutralizing antibodies are genus-specific. Nucleoprotein antigen, obtained by treatment of virus suspensions with 0.04 M NaOH and 56° C treatment of virus suspensions, is highly cross-reactive among members. Orthopoxviruses have hemagglutinin antigens, although this is rare in other genera.

Biological Properties

Transmission of various member viruses of the subfamily *Chordopoxvirinae* occurs by (1) aerosol, (2) direct contact, (3) arthropods (via mechanical carriage), or (4) indirect contact via fomites; transmission of member viruses of the subfamily *Entomopoxvirinae* occurs between arthropods by mechanical means. Host range may be broad in laboratory animals and in tissue culture; however, in nature it is generally narrow . Many poxviruses of vertebrates produce dermal maculopapular, vesicular rashes after systemic or localized infections. Poxviruses infecting humans are zoonotic except for Molluscum contagiosum virus and the orthopoxvirus variola (smallpox, now eradicated). Members may or may not be occluded within proteinaceous inclusions (*Chordopoxvirinae*: acidophilic (A-type) inclusion bodies, or *Entomopoxvirinae*: occlusions or spheroids). Occlusions may protect such poxviruses in environments of low transmission opportunity.

Neutralizing antibodies and cell-mediated immunity play a major role in clearance of vertebrate poxvirus infections. Reinfection rates are generally low and usually less severe. *Molluscum contagiosum* infections may recur, especially by autoinoculation of other areas of the skin with virus derived from the original lesions (e.g., by scratching).

SUBFAMILY *CHORDOPOXVIRINAE*

TAXONOMIC STRUCTURE OF THE SUBFAMILY

Subfamily	***Chordopoxvirinae***
Genus	*Orthopoxvirus*
Genus	*Parapoxvirus*
Genus	*Avipoxvirus*
Genus	*Capripoxvirus*
Genus	*Leporipoxvirus*
Genus	*Suipoxvirus*
Genus	*Molluscipoxvirus*
Genus	*Yatapoxvirus*

DISTINGUISHING FEATURES

Includes brick-shaped or ovoid poxviruses of vertebrates with a low G+C content (30-40%), except for the parapoxviruses (64%). Extensive serologic cross-reaction and cross-protection is observed within genera, this is less obvious among the avipoxviruses. Some viruses produce pocks on the chorioallantoic membranes of embryonated chicken eggs.

GENUS *ORTHOPOXVIRUS*

Type Species vaccinia virus (VACV)

DISTINGUISHING FEATURES

Virions are brick-shaped, about 200 nm x 200 nm x 250 nm. Infectivity is ether-resistant. Extensive serologic cross-reactivity exists between the viruses. Virus-infected cells synthesize a hemagglutinin (HA) glycoprotein that contributes to the modification of cell membranes and enables hemadsorption and hemagglutination of certain avian erythrocytes and alteration of the envelope of extracellular enveloped viruses. Neutralization sites on enveloped viruses are distinct from those on INVs. The host range is broad in laboratory animals and in tissue culture; in nature it may be relatively narrow. DNA is 170-250 kbp, G+C content is about 36%. The DNAs cross-hybridize extensively between members of the genus and sometimes with DNA of members of other genera. By comparison to the American species, DNA restriction maps suggest independent evolution of the Eurasian-African species.

LIST OF SPECIES IN THE GENUS

The viruses, their alternative names ( ), host { }, genomic sequence accession numbers [ ], and assigned abbreviations ( ) are:

SPECIES IN THE GENUS

buffalopox virus (BPXV)
(vaccinia subspecies)
{buffalo, cattle, human}
camelpox virus (CMLV)
{camel}

cowpox virus	[M19531]	(CPXV)
{rodents, felines, bovines, human}		
ectromelia virus	[M83102]	(ECTV)
{mousepox, reservoir unknown}		
monkeypox virus	[K02025]	(MPXV)
{rodents, primates, human}		
rabbitpox virus	[M60387]	(RPXV)
(vaccinia subspecies)		
{colonized rabbit, no natural reservoir}		
raccoonpox virus	[M94169]	(RCNV)
{North America raccoon}		
taterapox virus		(GBLV)
{African gerbil}		
vaccinia virus	[M35027]	(VACV)
{no natural reservoir}		
variola virus	[K02031]	(VARV)
{human; eradicated from nature}		
volepox virus		(VPXV)
{California pinon mouse and voles}		

Tentative Species in the Genus

skunkpox virus	(SKPV)
{North American striped skunk}	
Uasin Gishu disease virus	(UGDV)
{Central African horses}	

Genus *Parapoxvirus*

Type Species orf virus (ORFV)

Distinguishing Features

Virions are ovoid, 220-300 nm x 140-170 nm in size, with a surface filament that may appear as a regular cross-hatched, spiral coil involving a continuous thread. Infectivity is ether-sensitive. DNA is 130-150 kbp in size, G+C is about 64%. Most species show extensive DNA cross-hybridization and serological cross-reactivity. Cross-hybridizations and DNA maps suggest extensive sequence divergence among members. Generally the member viruses come from ungulates and livestock. They exhibit a narrow cell culture host range.

List of Species in the Genus

The viruses, their alternative names ( ), host { }, genomic sequence accession numbers [ ] and assigned abbreviations ( ) are:

Species in the Genus

bovine papular stomatitis virus		(BPSV)
{bovines, human}		
orf virus	[M30023]	(ORFV)
(contagious pustular dermatitis virus)		
(contagious ecthyma virus)		
{sheep, goats, musk oxen, human, deer}		
parapoxvirus of red deer in New Zealand		(PVNZ)
pseudocowpox virus		(PCPV)
(Milker's nodule virus)		
(paravaccinia virus)		
{bovines, human}		

Tentative Species in the Genus

Auzduk disease virus
(camel contagious ecthyma virus)
chamois contagious ecthyma virus
sealpox virus

Genus *Avipoxvirus*

Type Species fowlpox virus (FWPV)

Distinguishing Features

Virions are brick-shaped, about 330 nm x 280 nm x 200 nm. Infectivity is usually ether-resistant. Genus includes viruses of birds that usually produce proliferative skin lesions (cutaneous form) and/or upper digestive tract lesions (diptheritic form). Cross-protection is variable. Viruses are primarily transmitted mechanically by arthropods or by direct contact. DNA is about 300 kbp. Viruses exhibit extensive serologic cross-reaction. Viruses produce A-type inclusion bodies with considerable amounts of lipid. Viruses grow productively in avian cell cultures, but abortively in mammals and mammalian cell lines that have been examined.

List of Species in the Genus

The viruses, their genomic sequence accession numbers[ ] and assigned abbreviations ( ) are:

Species in the Genus

canarypox virus		(CNPV)
fowlpox virus	[X17202, D00295]	(FWPV)
juncopox virus		(JNPV)
mynahpox virus		(MYPV)
pigeonpox virus	[M88588]	(PGPV)
psittacinepox virus		(PSPV)
quailpox virus		(QUPV)
sparrowpox virus		(SRPV)
starlingpox virus		(SLPV)
turkeypox virus		(TKPV

Tentative Species in the Genus

peacockpox virus	(PKPV)
penguinpox virus	(PEPV)

Genus *Capripoxvirus*

Type Species sheeppox virus (SPPV)

Distinguishing Features

Virions are brick-shaped, about 300 nm x 270 nm x 200 nm. Infectivity is sensitive to trypsin and ether. Genus includes viruses of sheep, goats and cattle. Viruses can be mechanically transmitted by arthropods and by direct contact, fomites. DNA is about 145 kbp. There is extensive DNA cross-hybridization between species. In addition, extensive serologic cross-reaction and cross-protection is observed among members.

List of Species in the Genus

The viruses, their genomic sequence accession numbers [ ] and assigned abbreviations ( ) are:

Species in the Genus

goatpox virus		(GTPV)
lumpy skin disease virus		(LSDV)
sheeppox virus	[M28823, M30039, D00423]	(SPPV)

Tentative Species in the Genus

None reported.

Genus *Leporipoxvirus*

Type Species myxoma virus (MYXV)

Distinguishing Features

Virions are brick-shaped, about 300 nm x 250 nm x 200 nm. Infectivity is ether-sensitive. Genus includes viruses of lagomorphs and squirrels with extended cell culture host range. Usually viruses are mechanically transmitted by arthropods; they are also transmitted by direct contact and fomites. Myxoma and fibroma viruses cause localized benign tumors in their natural hosts. Myxoma viruses cause severe generalized disease in European rabbits. DNA is about 160 kbp, G+C about 40%. Extensive DNA cross-hybridization is observed between member viruses. Serologic cross-reaction and cross-protection has been demonstrated between different species.

List of Species in the Genus

The viruses, their alternative names ( ), host { }, genomic sequence accession numbers [ ] and assigned abbreviations ( ) are:

Species in the Genus

hare fibroma virus		(FIBV)
{European hare}		
myxoma virus	[M93049]	(MYXV)
rabbit fibroma virus	[M14899]	(SFV)
(Shope fibroma virus)		
squirrel fibroma virus		(SQFV)

Tentative Species in the Genus

None reported.

Genus *Suipoxvirus*

Type Species swinepox virus (SWPV)

Distinguishing Features

Virions are brick-shaped, about 300 nm x 250 nm x 200 nm. DNA is about 175 kbp in size with inverted terminal repeats of about 5 kbp. Virus forms foci in pig kidney cell culture (one-step growth is about 3 days at 37° C) and plaques in swine testes cell cultures. Virus causes asymptomatic generalized skin disease in swine that appears to be localized to epithelial cells and draining lymph nodes. Virus neutralizing antibodies are not usually detected. Mechanical transmission by arthropods (probably lice) is suspected. Viruses have a worldwide distribution. Rabbits can be infected experimentally, however serial transmission in rabbits is unsuccessful.

LIST OF SPECIES IN THE GENUS

The viruses, their genomic sequence accession numbers [ ] and assigned abbreviation ( ) are:

SPECIES IN THE GENUS

swinepox virus	[M59931, M64000]	(SWPV)

TENTATIVE SPECIES IN THE GENUS

None reported.

GENUS *MOLLUSCIPOXVIRUS*

Type Species Molluscum contagiosum virus (MOCV)

DISTINGUISHING FEATURES

Virions are brick-shaped, about 320 nm x 250 nm x 200 nm. Their buoyant density in CsCl is about 1.288 g/cm^3. DNA is about 188 kbp in size, G + C content is about 60%. DNAs cross-hybridize extensively. Restriction maps suggest two major sequence divergences among the isolates examined. Molluscum contagiosum virus grows poorly or not at all in primary human and other cell cultures. It is transmitted mechanically by direct contact between children, or between young adults. It is often sexually transmitted. Sometimes the virus causes opportunistic infections of persons with eczyma or AIDS. Virus produces localized lesions containing enlarged cells with cytoplasmic inclusions. Infections can recur and lesions may be disfiguring when combined with bacterial infections.

LIST OF SPECIES IN THE GENUS

The viruses, their genomic sequence accession numbers [ ] and assigned abbreviations ( ) are:

SPECIES IN THE GENUS

Molluscum contagiosum virus	[M63487]	(MOCV)

TENTATIVE SPECIES IN THE GENUS

Unnamed viruses of horses, donkeys, chimpanzees

GENUS *YATAPOXVIRUS*

Type Species Yaba monkey tumor virus (YMTV)

DISTINGUISHING FEATURES

Virions are brick-shaped, about 300 nm x 250 nm x 200 nm. DNA is about 146 kbp in size, G+C is about 33%. Yaba monkey tumor virus in primates causes histiocytomas, tumor-like masses of mononuclear cells. Viruses have been isolated from captive monkeys, baboons, and experimentally infected rabbits. Laboratory infections of man have been reported. Although DNAs cross-hybridize extensively, DNA restriction maps suggest major sequence divergences between Tanapox and Yaba monkey tumor viruses. Tanapox virus produces localized lesions in primates that likely result from the mechanical transmission by insects generally during the rainy season in African rain forests. Lesions commonly contain virions with a double-layer envelope surrounding the viral surface membrane.

LIST OF SPECIES IN THE GENUS

The viruses, and their assigned abbreviations ( ) are:

SPECIES IN THE GENUS

tanapox virus	(TANV)
Yaba monkey tumor virus	(YMTV)

TENTATIVE SPECIES IN THE GENUS

None reported.

SUBFAMILY *ENTOMOPOXVIRINAE*

TAXONOMIC STRUCTURE OF THE SUBFAMILY

Subfamily	***Entomopoxvirinae***
Genus	*Entomopoxvirus A*
Genus	*Entomopoxvirus B*
Genus	*Entomopoxvirus C*

DISTINGUISHING FEATURES

The viruses infect insects. The viruses include different morphologic forms, e.g., brick-shaped, or ovoid. They are about 70-250 nm x 350 nm in size and chemically similar to other family members. Virions contain at least 4 enzymes equivalent to those found in vaccinia virus. Virions of several morphological types have globular surface units that give a mulberry-like appearance; some have one lateral body, others have two. The DNA G+C content is about 20%. No serologic relationships have been demonstrated between entomopoxviruses and chordopoxviruses. Entomopoxviruses replicate in the cytoplasm of insect cells (hemocytes and adipose tissue cells). Mature virions are usually occluded in spheroids comprised of a major crystalline occlusion body protein (termed "spheroidin"). The subdivision into genera is based on virion morphology, host range, and the genome sizes of a few isolates. The genetic basis for these different traits is unknown.

GENUS *ENTOMOPOXVIRUS A*

Type Species Melolontha melolontha entomopoxvirus (MMEV)

DISTINGUISHING FEATURES

The genus includes poxviruses of *Coleoptera*. Virions are ovoid, about 450 nm x 250 nm in size, with one lateral body and a unilateral concave core. Surface globular units are 22 nm in diameter. DNA is about 260-370 kbp in size.

LIST OF SPECIES IN THE GENUS

The viruses, and their assigned abbreviations ( ) are:

SPECIES IN THE GENUS

Anomala cuprea entomopoxvirus	(ACEV)
Aphodius tasmaniae entomopoxvirus	(ATEV)
Demodema boranensis entomopoxvirus	(DBEV)
Dermolepida albohirtum entomopoxvirus	(DAEV)
Figulus subleavis entomopoxvirus	(FSEV)
Geotrupes sylvaticus entomopoxvirus	(GSEV)
Melolontha melolontha entomopoxvirus	(MMEV)

TENTATIVE SPECIES IN THE GENUS

None reported.

GENUS *ENTOMOPOXVIRUS B*

Type Species Amsacta moorei entomopoxvirus (AMEV)

DISTINGUISHING FEATURES

The genus includes poxviruses of *Lepidoptera* and *Orthoptera*. Virions are ovoid, about 350 nm x 250 nm in size, with a sleeve-shaped lateral body and cylindrical core. Surface globular units are 40 nm in diameter. DNA is about 225 kbp in size with covalently closed termini and inverted terminal repetitions. The G+C content is about 18.5%. Viruses produce a 115 kDa occlusion body composed of spheroidin protein.

LIST OF SPECIES IN THE GENUS

The viruses, their origins 'L' = lepidopteran, 'O' = orthopteran and assigned abbreviations ( ) are:

SPECIES IN THE GENUS

Amsacta moorei entomopoxvirus 'L'	(AMEV)
Acrobasis zelleri entomopoxvirus 'L'	(AZEV)
Arphia conspersa entomopoxvirus 'O'	(ACOEV)
Choristoneura biennis entomopoxvirus 'L'	(CBEV)
Choristoneura conflicta entomopoxvirus 'L'	(CCEV)
Choristoneura diversuma entomopoxvirus 'L'	(CDEV)
Chorizagrotis auxiliars entomopoxvirus 'L'	(CXEV)
Locusta migratoria entomopoxvirus 'O'	(LMEV)
Melanoplus sanguinipes entomopoxvirus 'O'	(MSEV)
Oedaleus senegalensis entomopoxvirus 'O'	(OSEV)
Operophtera brumata entomopoxvirus 'L'	(OBEV)
Schistocerca gregaria entomopoxvirus 'O'	(SGEV)

TENTATIVE SPECIES IN THE GENUS

None reported.

GENUS *ENTOMOPOXVIRUS C*

Type Species Chironomus luridus entomopoxvirus (CLEV)

DISTINGUISHING FEATURES

The genus includes poxviruses of *Diptera*. Virions are brick-shaped, about 320 nm x 230 nm x 110 nm in size, with two lateral bodies and a biconcave core. DNA is about 250-380 kbp in size.

LIST OF SPECIES IN THE GENUS

The viruses isolated from *Diptera*, and their assigned abbreviations ( ) are:

SPECIES IN THE GENUS

Aedes aegypti entomopoxvirus	(AAEV)
Camptochironomus tentans entomopoxvirus	(CTEV)
Chironomus attenuatus entomopoxvirus	(CAEV)
Chironomus luridus entomopoxvirus	(CLEV)
Chironomus plumosus entomopoxvirus	(CPEV)

Goeldichironomus holoprasimus entomopoxvirus (GHEV)

Tentative Species in the Genus

None reported.

List of Unassigned Viruses in the Family

The viruses, their host { } and assigned abbreviations ( ) are:

California harbor sealpox virus	(SPV)
{may also infect dog, cat}	
cotia virus	(CPV)
{sentinel mice, Brazil}	
dolphinpox virus	(DOV)
{bottle-nose dolphin}	
embu virus	(ERV)
{mosquitoes, human blood}	
grey kangaroopox virus	(KXV)
marmosetpox virus	(MPV)
Molluscum-likepox virus	(MOV)
{horse, donkey, chimpanzee}	
Nile crocodilepox virus	(CRV)
Quokkapox virus	(QPV)
{marsupial, Australia}	
red kangaroopox virus	(KPV)
Salangapox virus	(SGV)
{*Aethomys medicatus*, Cent. Afr. Rep}	
spectacled caimanpox virus	(RPV)
volepox virus	(VPV)
{vole, Turkmenia}	
mule deerpox virus	(DPV)
{*Odocoileus hemionus*, Wyoming}	
yokapox virus	(YKV)
{*Aedes simpsoni* , Centr. Afr. Rep.}	

Similarity with Other Taxa

None reported.

Derivation of Names

avi: from Latin *avis*, "bird"
capri: from Latin *caper*, "goat"
entomo: from Greek *entomon*, "insect"
lepori: from Latin *lepus*, "hare"
molluscum: from Latin molluscum, "clam", "snail" related to appearance of lesion
orf: Scottish word based on Icelandic hrufa, "scab", "boil"
ortho: from Greek *orthos*, "straight"
para: from Greek *para*, "by side of"
pox: from old English poc, pocc, "pustule"
sui: from Latin *sus*, "swine"

References

Buller RM, Palumbo GJ (1991) Poxvirus pathogenesis. Microbiol Rev 55: 80-122

Dales S, Pogo BGT (1981) Biology of Poxviruses. In: Kingsbury DW, zur Hausen H (eds) Virology Monographs, No. 18. Springer-Verlag, New York

Esposito JJ, Nakano JH (1991) Poxvirus infections in humans. In: Balows A, Hausler WJ, *et al.* (eds) Manual of Clinical Microbiology, 5th edn. American Society for Microbiology, Washington DC, pp 858-867

Fenner F, Henderson DA, Arita I, Jezek Z, Ladnyi D (1988) Smallpox and its eradication. World Health Organization, Geneva

Fenner F, Nakano JH (1988) Poxviridae: The Poxviruses. In: Lennette EH, Halonen P, Murphy FA (eds) Laboratory Diagnosis of Infectious Diseases: Principles and Practice, Vol 2. Viral, Rickettsial and Chlamydial Diseases. Springer-Verlag, New York, pp 177-207

Fenner F, Wittek R, Dumbell KR (eds) (1989) The Orthopoxviruses. Academic Press, New York

Granados RR (1981) Entomopoxvirus infections in insects. In: Davidson I (ed) Pathogenesis of Invertebrate Microbial Diseases. Allenheld Osmu, Totowa New Jersey, pp 101-129

Moyer RW, Turner PC (eds) (1990) Poxviruses. In: Current Topics in Microbiology and Immunology, Vol 163. Springer-Verlag, New York

Moss B (1990) *Poxviridae* and their replication. In: Fields BN, Knipe DM (eds) Virology, 2nd edn. Vol 2. Raven Press, New York, pp 2079-2111

Robinson AJ, Lyttle DJ (1992) Parapoxviruses: their biology and potential as recombinant vaccine vectors. In: Binns MM, Smith GL (eds) Recombinant Poxviruses. CRC Press, Boca Raton FL, pp 285-327

Tripathy DN (1991) Pox. In: Calek BW, Barnes HJ et al (eds) Diseases of Poultry, 9th edn. Iowa State University Press, Ames Iowa, pp 583-596

Tripathy DN, Hanson LE, Crandell RA (1981) Poxviruses of veterinary importance: diagnosis of infections. In: Kurstak E, Kurstak C (eds) Comparative Diagnosis of Viral Diseases Vol 3. Academic Press, New York, pp 267-346

Contributed By

Esposito JJ, Baxby D, Black DN, Dales S, Darai G, Dumbell KR, Granados RR, Joklik WK, McFadden G, Moss B, Moyer RW, Pickup DJ, Robinson AJ, Tripathy DN

Genus "African swine fever-like viruses"

Type species African swine fever virus (ASFV)

Virion Properties

Morphology

Virions consist of a nucleoprotein core structure, 70-100 nm in diameter, surrounded by an icosahedral capsid, 172 to 191 nm in diameter, and a lipid-containing envelope. The capsid exhibits icosahedral symmetry (T=189-217) corresponding to 1,892-2,172 capsomers (each capsomer is 13 nm in diameter and appears as a hexagonal prism with a central hole; intercapsomeric distance is 7.4-8.1 nm). Extracellular virions (enveloped) have an overall diameter of 175 to 215 nm (Fig. 1).

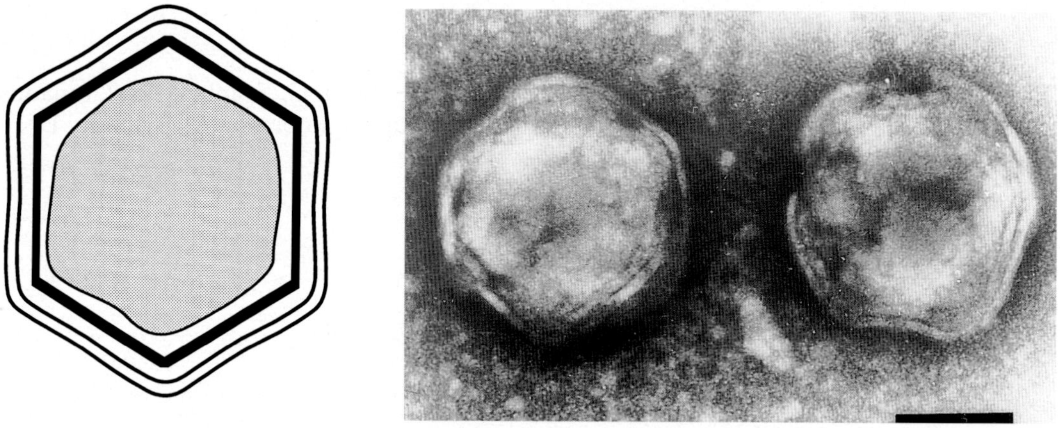

Figure 1: (left) Schematic representation of ASFV virion. (right) Negative contrast electron micrograph of ASFV. The bar represents 100 nm.

Physicochemical and Physical Properties

Virion buoyant density is 1.095 g/cm^3 in Percoll, 1.19-1.24 g/cm^3 in CsCl; S_{20w} is about 3,500. Virions are sensitive to ether, chloroform and deoxycholate and are inactivated at 60° C within 30 min., but survive for years at 20° C or 4° C. Infectivity is stable over a wide pH range. Some infectious virus may survive treatment at pH 4 or pH 13. Infectivity is destroyed by some disinfectants (1% formaldehyde in 6 days, 2% NaOH in 1 day); para-phenylphenolic disinfectants are very effective. Virus is sensitive to irradiation.

Nucleic Acid

The genome consists of a single molecule of linear, covalently close-ended, dsDNA, 170-190 kbp in size (varying among isolates). The end sequences are inverted, complementary, tandem repeats. Genes are encoded on both DNA strands and are generally closely spaced. At several intergenic locations there are short tandem repeat arrays. The genome may encode 150-200 proteins.

Proteins

Virions contain more than 54 proteins, including several virion-associated enzymes required for transcription and post-transcriptional modification of mRNA, including RNA polymerase, poly (A) polymerase and mRNA capping enzymes. Synthesis of more than 100 virus-induced proteins has been detected in infected cells.

Lipids

Enveloped virions contain lipids, including glycolipids.

Carbohydrates

No carbohydrates have been demonstrated in virions other than in the form of glycolipids.

Genome Organization and Replication

The virus enters cells by receptor mediated endocytosis. Virus cores contain enzymes required for early mRNA synthesis and processing which begins in the cytoplasm immediately following virus entry. Transcripts are 3'-polyadenylated and 5'-capped. The virus genome encodes many enzymes involved in mRNA transcription and DNA replication. DNA replication reaches a peak about 8 hours post-infection; head-to-head concatameric forms of DNA, which may be replicative intermediates, are found in cells at this time. DNA replication may proceed by a self-priming mechanism. Late genes are expressed after the onset of DNA replication; synthesis of some early genes continues throughout infection. Some virus proteins are post-translationally modified (proteolytic cleavage, phosphorylation, glycosylation, myristylation, etc.). The cell nucleus is required for productive infection. Virus morphogenesis takes place in a perinuclear area rich in fibrillar and membranous organelles; this area is often surrounded by an enlarged Golgi apparatus and many ribosomes. Virus is released by cell destruction or by budding through plasma membrane.

Antigenic Properties

Infected swine mount a protective immune response against non-fatal virus strains and produce antibodies. Antibodies can cause a reduction in virus infectivity but do not neutralize virus. Antigenic variation mainly involves the structural proteins p150, p14 and p12, as shown by monoclonal antibody analyses. Standard immunological tests fail to differentiate between virus isolates. However, isolates can be divided and grouped into distinct genotypes by restriction enzyme analyses. Hemadsorption of swine erythrocytes is obtained using swine bone marrow cells or leukocytes. Antibody can inhibit hemadsorption and inhibition can be used to differentiate isolates.

Biological Properties

The only animal species naturally infected are domestic and wild swine (*Sus scrofa domesticus* and *S. s. ferus*), warthogs, bushpigs and giant forest hogs. Soft ticks of the genus *Ornithodoros* are also infected by the virus (*O. moubata* that infests warthog burrows and domestic pig pens in parts of Africa south of the Sahara; *O. erraticus* in pig pens in parts of Portugal and south-west Spain). Virus can be transmitted in ticks trans-stadially, trans-ovarially and sexually. Warthogs and domestic swine may be infected by the bites of infected ticks. Warthogs show no signs of disease. Domestic and European wild pigs may exhibit disease. Neither vertical nor direct horizontal transmission between warthogs is believed to occur. However, transmission between domestic swine can occur by direct contact, or from infected pork, or fomites, or mechanically by biting flies. Both warthogs and *O. moubata* act as reservoirs for the virus. Disease is endemic in domestic swine in many African countries and in Europe in Portugal, Sardinia and south-west Spain. Sporadic outbreaks have occurred in and been eradicated from Belgium, Brazil, Cuba, the Dominican Republic, Haiti, Holland, Italy and Malta. Virus isolates differ in virulence and may produce a variety of signs ranging from inapparent, to acute, to chronic. Virulent isolates may cause 100% mortality in 7-10 days. Less virulent isolates may produce a mild disease from which a proportion of infected swine recover and become carriers. Viruses replicate in cells of the mononuclear phagocytic system and reticulo-endothelial cells in lymphoid tissues and organs of domestic swine. The main histological lesions in acute disease are seen in the antigen processing cells of the lymphoreticular system. Widespread lymphoid necrosis and damage to endothelial cells in arterioles and capillaries account for the lesions seen in acute disease.

List of Species in the Genus

The viruses, their genomic sequence accession numbers [ ] and assigned abbreviations ( ) are:

Species in the Genus

African swine fever virus	[X71982]	(ASFV)

Tentative Species in the Genus

None reported.

List of Unassigned Viruses in the Family

None reported.

Similarity with Other Taxa

Earlier, African swine fever virus was listed as a member virus of the family *Iridoviridae*, but as more information was obtained, it was removed from this family. Now, the virus has been placed as the only member of an, as yet unnamed, separate genus. Additionally, the virus exhibits some similarities in genome structure and strategy of replication to the poxviruses, but it has a quite different virion structure and many other properties that distinguish it from the member viruses of the family *Poxviridae*.

Derivation of Names

No defined taxonomic name.

References

Costa JV (1990) African swine fever. In: Darai G (ed) Molecular Biology of Iridoviruses. Kluwer Academic Publishers, Boston, pp 247-270

Dixon LK, Wilkinson PJ, Sumpton KJ, Ekue F (1990) Diversity of the African swine fever virus genome. In: Darai G (ed) Molecular Biology of Iridoviruses. Kluwer Academic Publishers, Boston, pp 271-296

Hess WR (1971) African swine fever virus. Virol Monogr 9: 1-33

Plowright W (1984) African swine fever. In: Brown F, Wilson G (eds) Principles of Bacteriology, Virology and Immunity. Vol 4. Virology, Edward Arnold, London, pp 538-554

Vinuela E (1985) African swine fever. Curr Top Microbiol Immunol 116: 151-170

Wilkinson PJ (1989) African swine fever virus family. In: Pensaert MB (ed) Virus Infections of Porcines. Elsevier Science Publications BV, New York, pp 15-36

Wilkinson PJ (1989) Iridoviridae and African swine fever virus. In: Porterfield JS (ed) Andrewes Viruses of Vertebrates, 5th edn. Balliere Tindall, London, pp 333-345

Wilkinson PJ (1990) African swine fever. In: Collier LH, Timbury MC (eds) Topley and Wilson's Principles of Bacteriology, Virology and Immunology, 8th edn, Vol 4. Edward Arnold, London, pp 623-629

Contributed By

Dixon LK, Rock D, Vinuela E

FAMILY *IRIDOVIRIDAE*

TAXONOMIC STRUCTURE OF THE FAMILY

Family	***Iridoviridae***
Genus	*Iridovirus*
Genus	*Chloriridovirus*
Genus	*Ranavirus*
Genus	*Lymphocystivirus*
Genus	"goldfish virus 1-like viruses"

VIRION PROPERTIES

MORPHOLOGY

Virions have icosahedral symmetry and are 130-170 nm in diameter. Tipula iridescent virus has 812 surface subunits. Animal iridoviruses have envelopes derived by budding through the plasma membrane. All iridoviruses contain an internal lipid membrane-like structure. Some iridoviruses have numerous fibers trailing from the icosahedron.

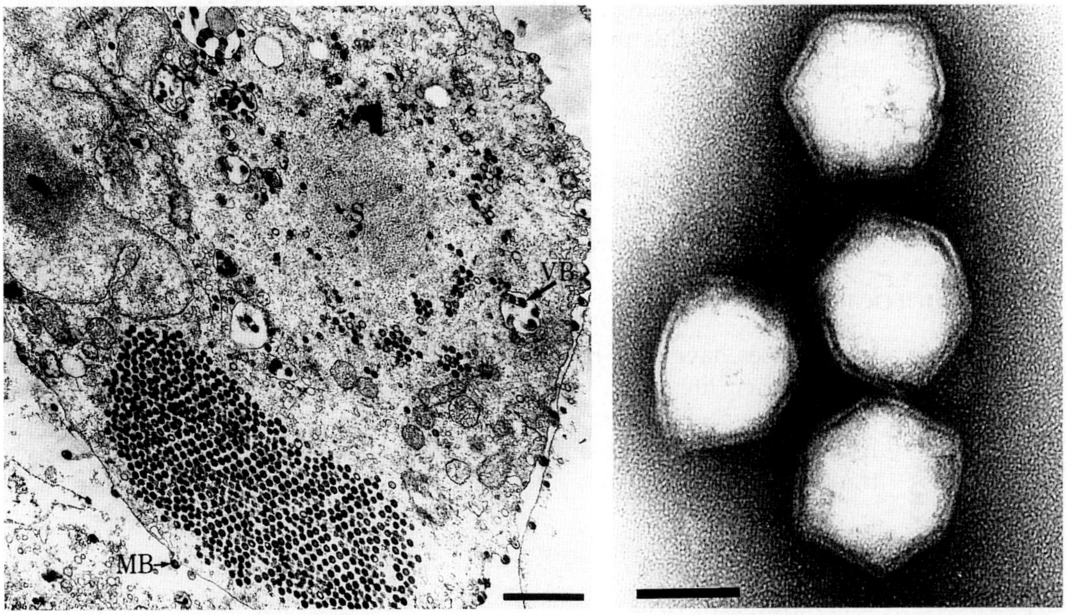

Figure 1: (left) Electron micrograph of a thin section of FV-3-infected cell; a crystalline array of virus particles is shown, the bar represents 1 µm. (right) Negative contrast electron micrograph of FV-3 virions, the bar represents 100 nm.

NUCLEIC ACID

Virions contain a single molecule of linear dsDNA, 170 to 200 kbp in size. Mosquito iridescent virus has been reported to have a genome of 440 kbp, the largest genome of any virus. The viruses contain circularly permuted and terminally redundant DNA. Genomic DNA of vertebrate iridoviruses is highly methylated (frog virus 3 DNA is methylated at all cytosines in the dinucleotide CpG by a virus-encoded DNA methyl-transferase). Insect iridovirus DNA presumably is not methylated since it is readily cleaved by cytosine sensitive restriction endonucleases.

PROTEINS

Virions contain approximately 9 to 36 polypeptides. Purified virions contain an assortment of enzyme activities such as protein kinase, nucleotide phosphohydrolase, ribonuclease which cleaves both single- and double-stranded RNA, deoxyribonuclease activities having pH optima of 5 and 7.5, and protein phosphatase.

LIPIDS

Purified virions contain approximately 3 to 14% lipids. The composition of lipids has led to the suggestion that viral membranes are not derived from host membranes.

CARBOHYDRATES

No carbohydrates are present in purified virions.

GENOME ORGANIZATION AND REPLICATION

The replication strategy of iridoviruses, as exemplified by frog virus 3 is strikingly different from other DNA viruses. The genome of an infecting virion reaches the nucleus where it is transcribed. Cellular RNA polymerase II, modified by a virion structural protein(s) is utilized for viral transcription at an early stage. The parental genome serves as the template for the first stage of DNA replication. DNA synthesized during this stage is often less than genome size. Viral DNA in the nucleus may be utilized as template for further transcription or be transported to the cytoplasm where it participates in the second stage of DNA replication. During the second stage, the newly synthesized DNA is in the form of a large concatamer. The concatamer is processed to produce mature viral DNA. Presumably, concatamer processing is intimately associated with DNA packaging into the virion. The consequence of this process is the generation of a circularly permuted and terminally redundant genome.

ANTIGENIC PROPERTIES

Antigenic relationships among the iridoviruses have not, as yet, been systematically investigated. There appears to be no serologic or nucleic acid sequence relationship between vertebrate and invertebrate iridoviruses.

BIOLOGICAL PROPERTIES

Iridoviruses have only been isolated from poikilothermic animals that have an aquatic stage in their life cycle. Most insect iridoviruses impart a blue or turquoise coloration to infected larvae. However, vertebrate iridoviruses do not cause any coloration. Mosquitoes iridoviruses can be transmitted transovarially, in contrast to the other viruses which are transmitted horizontally.

GENUS *IRIDOVIRUS*

Type Species Chilo iridescent virus (CIV)

DISTINGUISHING FEATURES

Virions are about 120 nm in diameter. The complex icosahedral shell contains lipid, but infectivity is not sensitive to ether. Infected larvae and concentrated purified virus produce a blue to purple iridescence. Chilo iridescent virus has circularly permuted and terminally redundant DNA.

LIST OF SPECIES IN THE GENUS

The viruses, and their assigned abbreviations ( ) are:

SPECIES IN THE GENUS

Chilo iridescent virus	(CIV)
insect iridescent virus 1	(IIV-1)
insect iridescent virus 2	(IIV-2)
insect iridescent virus 6	(IIV-6)
insect iridescent virus 9	(IIV-9)
insect iridescent virus 10	(IIV-10)
insect iridescent viruses 16 to 32	(IIV-16 to 32)

TENTATIVE SPECIES IN THE GENUS

None reported.

GENUS *CHLORIRIDOVIRUS*

Type Species mosquito iridescent virus (MIV)

DISTINGUISHING FEATURES

Virion diameter is about 180 nm. Infected larvae and virus pellets of most members iridesce with a yellow-green color.

LIST OF SPECIES IN THE GENUS

The viruses, and their assigned abbreviations ( ) are:

SPECIES IN THE GENUS

insect iridescent viruses 3 to 5	(IIV-3 to 5)
insect iridescent virus 7	(IIV-7)
insect iridescent virus 8	(IIV-8)
insect iridescent viruses 11 to 15	(IIV-11 to 15)
mosquito iridescent virus	(MIV)
(iridescent virus type 3, regular strain)	

TENTATIVE SPECIES IN THE GENUS

Chironomus plumosus iridescent virus

GENUS *RANAVIRUS*

Type Species frog virus 3 (FV-3)

DISTINGUISHING FEATURES

FV3 grows in piscine, avian, and mammalian cells and at 12° C to 32° C. Structural proteins cause rapid inhibition of host macromolecular synthesis. DNA contains a high proportion of 5-methyl cytosine and is circularly permuted and terminally redundant. DNA synthesis occurs in 2 stages: (1) synthesis of unit-length molecules in the nucleus and (2) synthesis of concatemers in the cytoplasm. mRNA lacks poly (A).

LIST OF SPECIES IN THE GENUS

The viruses and their assigned abbreviations ( ) are:

SPECIES IN THE GENUS

frog virus 1	(FV-1)
frog virus 2	(FV-2)
frog virus 3	(FV-3)
frog viruses 5 to 24	(FV-5 to 24)
frog virus L2	(FV-L2)
frog virus L4	(FV-L4)
frog virus L5	(FV-L5)
newt viruses T6 to T20	(NV-T6 to T20)
tadpole edema virus LT 1-4 (from Rana catesbriana)	(TEVLT-1 to 4)
Xenopus virus T21	(XV-T21)

TENTATIVE SPECIES IN THE GENUS

None reported.

Genus *Lymphocystivirus*

Type Species flounder virus (LCDV-1)

Distinguishing Features

Viruses grow in centrarchid fish, where they form giant cells in connective tissue at 25° C. Genomic DNA is circularly permuted, terminally redundant, and is highly methylated at cytosine residues.

List of Species in the Genus

The viruses and their assigned abbreviations ( ) are:

Species in the Genus

flounder virus (LCDV-1)
lymphocystis disease virus (dab isolate) (LCDV-2)

Tentative Species in the Genus

Octopus vulgaris disease virus

Genus "Goldfish virus 1-like viruses"

Type Species goldfish virus 1 (GFV-1)

Distinguishing Features

Viruses have a more restricted host range *in vitro* than amphibian viruses. Infection produces cytoplasmic vacuolization and cell rounding in the goldfish cell line, CAR, at 25° C. DNA is highly methylated at cytosine residues, not only at CpG sequences but most likely, also at CpT.

List of Species in the Genus

The viruses and their assigned abbreviations ( ) are:

Species in the Genus

goldfish virus 1 (GFV-1)
goldfish virus 2 (GFV-2)

Tentative Species in the Genus

None reported.

List of Unassigned Viruses in the Family

None reported.

Similarity with Other Taxa

None reported.

Derivation of Names

irido: from Greek *iris, iridos*, goddess whose sign was the rainbow, hence iridescent: 'shining like a rainbow,' from appearance of infected larval insects and centrifuged pellets of virions
chloro: from Greek *chloros*, 'green'
rana: from Latin *rana*, 'frog'
cyssti: from Greek *kystis*, 'bladder or sac'
lympho: from Latin *lympha*, 'water'

REFERENCES

Darai G (ed) (1990) Molecular Biology of Iridoviruses. Developments in Molecular Virology. Kluwer Academic Publishers, Boston Dordrecht London

Delius H, Darai G, Flugel RM (1984) DNA analysis of insect iridescent virus 6: evidence for circular permutation and terminal redundancy. J Virol 49: 609-614

Essani K, Granoff A (1989) Amphibian and piscine iridoviruses proposal for nomenclature and taxonomy based on molecular and biological properties. Intervirology 30: 187-193

Goorha R, Murti KG (1994) The genome of an animal DNA virus (frog virus 3) is circularly permuted and terminally redundant. Proc Natl Acad Sci USA (in press)

Willis DB (1985) *Iridoviridae*. Curr Topics Microbiol Immunol, Vol 116. Springer, Berlin Heidelberg New York Tokyo

CONTRIBUTED BY

Goorha R

FAMILY PHYCODNAVIRIDAE

TAXONOMIC STRUCTURE OF THE FAMILY

Family	***Phycodnaviridae***
Genus	*Phycodnavirus*

GENUS PHYCODNAVIRUS

Type Species Paramecium bursaria Chlorella virus 1 (PBCV-1)

VIRION PROPERTIES

MORPHOLOGY

Virions are polyhedral with a multilaminate shell surrounding an electron dense core. Virions do not have an external membrane and are 130-190 nm in diameter. Some electron micrographs indicate the virions have flexible hair like appendages with swollen structures at the end; these appendages extend from at least some of the vertices. One virion vertex may contain a 20-25 nm spike structure.

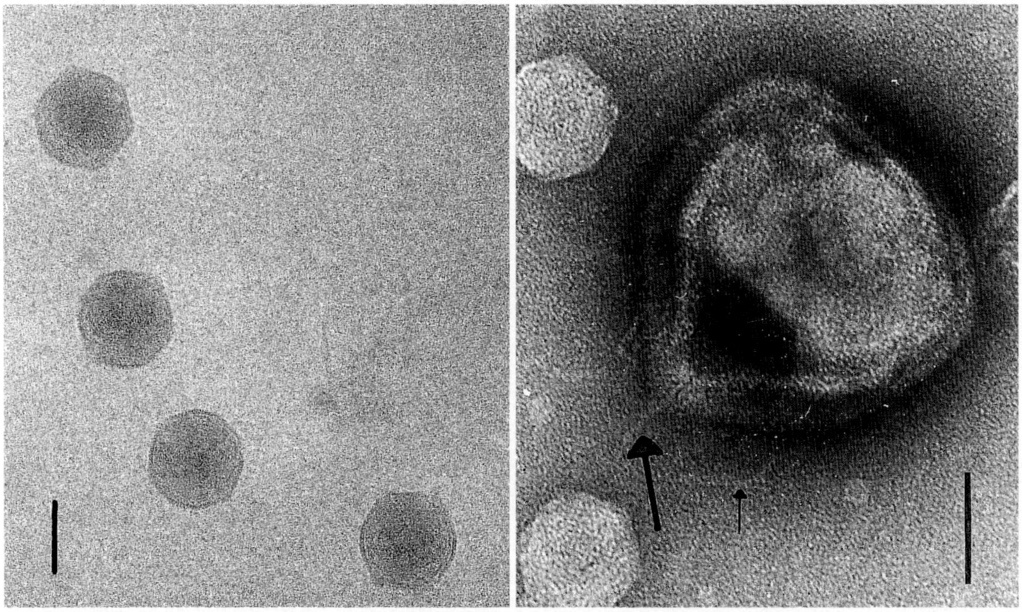

Figure 1: (left) Frozen hydrated PBCV-1 virions; (right) negative contrast electron micrograph of stained PBCV-1. Note that (i) long fibers are associated with the particles (small arrow), (ii) a distinctive 20- to 25- nm spike structure (large arrow) extends from one vertex of the particle. The bar represents 100 nm.

PHYSICOCHEMICAL AND PHYSICAL PROPERTIES

Virion Mr is about 1 x 10^9; S_{20w} is more than 2,000; some virions are disrupted in CsCl. Virions are insensitive to non-ionic detergents but are inactivated by organic solvents. Infectivity is lost after exposure to 5 mM dithiothreitol or dithioerythritol but not mercaptoethanol.

NUCLEIC ACID

Virions contain linear, nonpermuted dsDNA more than 300 kbp in size. The DNA has cross-linked hairpin ends. G + C content is 40-52%. The DNA termini, contain identical inverted 1-2.2 kbp repeats. The remainder of the genome appears to represent unique DNA sequences.

The DNA contains methylated bases, both 5-methyl-cytosine (5mC) and N^6-methyladenine (6mA). Proportions of methylated bases vary with the virus and range from no 6mA and 0.1% 5mC to 37% 6mA and 47% 5mC.

PROTEINS

Purified virions contain more than 50 proteins ranging in size from 10 to more than 200 kDa; at least three of the proteins are glycoproteins, including the major capsid protein, Vp54, which comprises 40% of the total virion protein. Four proteins, including Vp54, are located on the virus surface.

LIPIDS

Five to 10% of the virion is composed of lipid. The lipid component is located inside the glycoprotein shell and is required for virus infectivity.

CARBOHYDRATES

At least three of virus proteins are glycosylated including the major capsid protein Vp54. The glycan portion of Vp54 is on the external surface of the virion. Unlike any other known viruses, PBCV-1 appears to code for the enzymes involved in its glycosylation.

GENOME ORGANIZATION AND REPLICATION

The intracellular site of virion DNA replication and transcription is unknown. DNA packaging occurs in localized regions in the cytoplasm; however, recent evidence indicates that the nucleus may play an important role in virus replication.

A DNA restriction map of the prototype virus, PBCV-1, is available. Genes are rapidly being mapped on the PBCV-1 genome including DNA polymerase, DNA topoisomerase, a ser/thr protein kinase, both subunits of ribonucleotide reductase, the major capsid protein Vp54, a glycoprotein Vp260, a DNA methyltransferase M.CviAII, a DNA site-specific endonuclease CviAII, a translation elongation factor-3, and a DNA methyltransferase pseudogene. The viruses code for DNA methyltransferases and DNA site-specific (restriction) endonucleases of unknown biological function.

ANTIGENIC PROPERTIES

Antigenic variants of PBCV-1 virus can be isolated which are completely resistant to polyclonal antibody prepared against prototype PBCV-1. These variants occur at a frequency of about 1×10^{-6}-1×10^{-7}. Using polyclonal antibodies prepared against the mutants, four distinct PBCV-1 antigenic variants have been identified. The antibodies react primarily with the glycan portion of the major capsid protein.

Additional variants of these viruses can easily be isolated from natural sources.

BIOLOGICAL PROPERTIES

HOST RANGE

Nature: The viruses, which are ubiquitous in fresh water throughout the world, are extremely host specific and only attach rapidly and irreversibly to cell walls of certain unicellular, eukaryotic, exsymbiotic chlorella-like green algae. Virus attachment is followed by dissolution of the host wall at the point of attachment and entry of the viral DNA and associated proteins into the cell, leaving an empty capsid on the host surface. Beginning about 2-4 hr. after infection, progeny virions are assembled in the cytoplasm of the host. Infectious virions can be detected inside the cell about 30 to 40 min. prior to virus release; virus release occurs by cell wall lysis.

Laboratory: The hosts, *Chlorella* strains NC64A and Pbi, can easily be grown in the laboratory and the viruses can be plaque assayed. Thus large quantities of these viruses can easily be produced in the laboratory.

TRANSMISSION

The viruses are transmitted horizontally.

TAXONOMIC STRUCTURE OF THE GENUS

Three groups of viruses are delineated based on host specificity.

Group 1. Paramecium bursaria Chlorella NC64A viruses (NC64A viruses)
Group 2. Paramecium bursaria Chlorella Pbi viruses (Pbi viruses)
Group 3. Hydra viridis Chlorella viruses (HVC viruses)
Chlorella strains NC64A, ATCC 30562, and N1A (originally symbionts of the protozoan *P. bursaria*), collected in the United States, are the only known host for NC64A viruses. *Chlorella* strain Pbi (originally a symbiont of a European strain of *P. bursaria*) collected in Germany, is the only known host for Pbi viruses. Pbi viruses do not infect *Chlorella* strains NC64A, ATCC 30562, and N1A. *Chlorella* strain Florida (originally a symbiont of *Hydra viridis*) is the only known host for HVCV. NC64A viruses are placed in 16 subgroups based on plaque size, serological reactivity, resistance of the genome to restriction endonucleases, and nature and content of methylated bases.

LIST OF SPECIES IN THE GENUS

The viruses and their assigned abbreviations ( ) are:

SPECIES IN THE GENUS

1-Paramecium bursaria Chlorella NC64A virus group:

Paramecium bursaria Chlorella virus 1	(PBCV-1)
Paramecium bursaria Chlorella virus AL1A	(PBCV-AL1A)
Paramecium bursaria Chlorella virus AL2A	(PBCV-AL2A)
Paramecium bursaria Chlorella virus AL2C	(PBCV-AL2C)
Paramecium bursaria Chlorella virus BJ2C	(PBCV-BJ2C)
Paramecium bursaria Chlorella virus CA1A	(PBCV-CA1A)
Paramecium bursaria Chlorella virus CA1D	(PBCV-CA1D)
Paramecium bursaria Chlorella virus CA2A	(PBCV-CA2A)
Paramecium bursaria Chlorella virus CA4A	(PBCV-CA4A)
Paramecium bursaria Chlorella virus CA4B	(PBCV-CA4B)
Paramecium bursaria Chlorella virus IL2A	(PBCV-IL2A)
Paramecium bursaria Chlorella virus IL2B	(PBCV-IL2B)
Paramecium bursaria Chlorella virus IL3A	(PBCV-IL3A)
Paramecium bursaria Chlorella virus IL3D	(PBCV-IL3D)
Paramecium bursaria Chlorella virus IL5-2s1	(PBCV-IL5-2s1)
Paramecium bursaria Chlorella virus MA1D	(PBCV-MA1D)
Paramecium bursaria Chlorella virus MA1E	(PBCV-MA1E)
Paramecium bursaria Chlorella virus NC1A	(PBCV-NC1A)
Paramecium bursaria Chlorella virus NC1B	(PBCV-NC1B)
Paramecium bursaria Chlorella virus NC1C	(PBCV-NC1C)
Paramecium bursaria Chlorella virus NC1D	(PBCV-NC1D)
Paramecium bursaria Chlorella virus NE-8D	(PBCV-NE8D)
Paramecium bursaria Chlorella virus NE8A	(PBCV-NE8A)
Paramecium bursaria Chlorella virus NY2A	(PBCV-NY2A)
Paramecium bursaria Chlorella virus NY2B	(PBCV-NY2B)
Paramecium bursaria Chlorella virus NY2C	(PBCV-NY2C)
Paramecium bursaria Chlorella virus NY2F	(PBCV-NY2F)
Paramecium bursaria Chlorella virus NYb1	(PBCV-NYb1)

Paramecium bursaria Chlorella virus NYs	(PBCV-NYs)
Paramecium bursaria Chlorella virus SC1A	(PBCV-SC1A)
Paramecium bursaria Chlorella virus SC1B	(PBCV-SC1B)
Paramecium bursaria Chlorella virus SH6A	(PBCV-SH6A)
Paramecium bursaria Chlorella virus XY6E	(PBCV-XY6E)
Paramecium bursaria Chlorella virus XZ3A	(PBCV-XZ3A)
Paramecium bursaria Chlorella virus XZ4A	(PBCV-XZ4A)
Paramecium bursaria Chlorella virus XZ5C	(PBCV-XZ5C)
Paramecium bursaria Chlorella virus XZ4C	(PBCV-XZ4C)
2-Paramecium bursaria Chlorella Pbi virus group:	
Paramecium bursaria Chlorella virus A1	(PBCV-A1)
Paramecium bursaria Chlorella virus B1	(PBCV-B1)
Paramecium bursaria Chlorella virus G1	(PBCV-G1)
Paramecium bursaria Chlorella virus M1	(PBCV-M1)
Paramecium bursaria Chlorella virus R1	(PBCV-R1)
3-Hydra viridis Chlorella virus group:	
Hydra viridis Chlorella virus 1	(HVCV-1)
Hydra viridis Chlorella virus 2	(HVCV-2)
Hydra viridis Chlorella virus 3	(HVCV-3)

TENTATIVE SPECIES IN THE GENUS

None reported.

LIST OF UNASSIGNED VIRUSES IN THE FAMILY

None reported.

SIMILARITY WITH OTHER TAXA

Many large polyhedral virus-like particles have been observed in electron micrographs of eukaryotic algae. However, for the most part these particles have not been characterized. Particles isolated from three of these algae are reported to contain large dsDNA genomes of unknown structure.

DERIVATION OF NAMES

phyco: from Greek *phycos*, meaning algae
dna: sigla for *d*eoxyribo*n*ucleic *a*cid

REFERENCES

Grabherr R, Strasser P, van Etten JL (1992) The DNA polymerase gene from Chlorella viruses PBCV-1 and NY-2A contains an intron with nuclear splicing sequences. Virology 188: 721-731

Nelson M, Zhang Y, van Etten JL (1993) DNA methyltransferases and DNA site-specific endonucleases encoded by Chlorella viruses. In: Jost J, Saluz HP (eds) DNA Methylation: Molecular biology and Biological Significance. Birkhauser Publishers Ltd Basel, Switzerland pp 186-211

Reisser W, Burbank DE, Meints SM, Meints RH, Becker B, van Etten JL (1988) A comparison of viruses infecting two different Chlorella-like green alga. Virology 167: 143-149

Rohozinski J, Girton LE, van Etten JL (1989) Chlorella viruses contain linear nonpermuted double-stranded DNA genomes with covalently closed hairpin ends. Virology 168: 363-369

Strasser P, Zhang Y, Rohozinski J, van Etten JL (1991) The termini of the Chlorella virus PBCV-1 genome are identical 2.2 kbp inverted repeats. Virology 180: 763-769

van Etten JL, Lane LC, Meints RH (1991) Viruses and virus-like particles of eukaryotic algae. Microbiol Rev 55: 586-620

Wang I, Li Y, Que Q, Bhattacharya M, Lane LC, Chaney WG, van Etten JL (1993) Evidence for virus-encoded glycosylation specificity. Proc Natl Acad Sci USA 90: 3840-3844

CONTRIBUTED BY

van Etten JL

FAMILY BACULOVIRIDAE

TAXONOMIC STRUCTURE OF THE FAMILY

Family	*Baculoviridae*
Genus	*Nucleopolyhedrovirus*
Genus	*Granulovirus*

VIRION PROPERTIES

MORPHOLOGY

One or two virion phenotypes may be involved in baculovirus infections. The virion phenotype that initiates infections in the gut epithelium is occluded in a crystalline protein

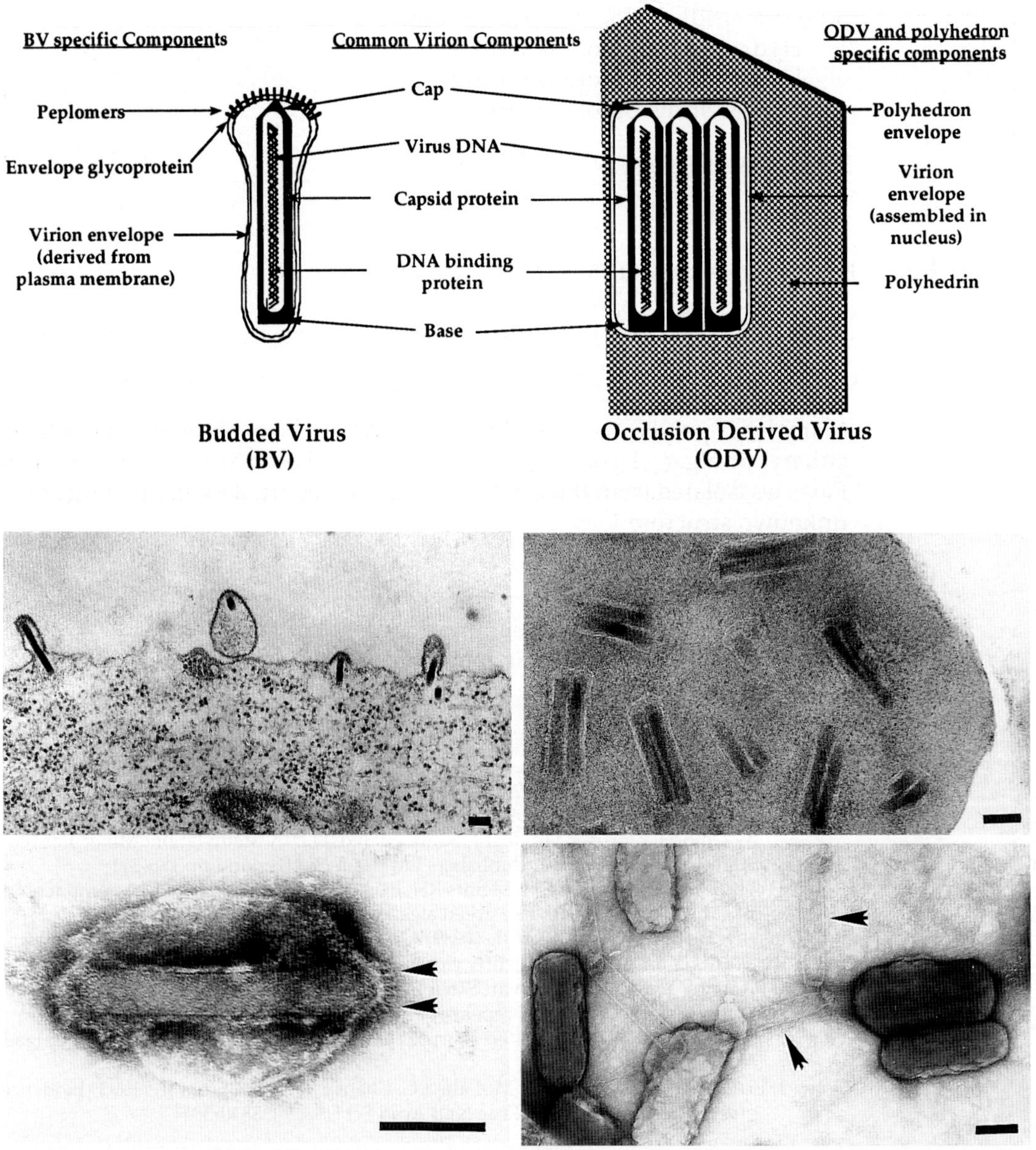

Figure 1: (upper) Diagram of the location of baculovirus structural components. The two baculovirus phenotypes are shown with shared and phenotype-specific components indicated. (center left) Transmission electron micrograph of AcMNPV budding from an infected TN-368 cell. (lower left) Negative contrast electron micrograph of AcMNPV BV with arrows indicating peplomers. (center right) Transmission electron micrograph of AcMNPV occlusion containing bundles of enveloped virions. (lower right) Negative contrast electron micrograph of AcMNPV ODV and empty capsids (arrows). Bars represent 100 nm.

matrix which may be polyhedral in shape. This occlusion may range in size from 0.15 to 15 µm and contain many virions (genus *Nucleopolyhedrovirus*), or may be ovicylindrical (about 0.3 x 0.5 µm) and contain only one, or rarely two or more virions (genus *Granulovirus*). Virions within occlusions consist of one or more rod-shaped nucleocapsids with distinct structural polarity enclosed within an envelope thought to be generated by *de novo* synthesis and assembled in the nucleus (genus *Nucleopolyhedrovirus*) or in the nuclear-cytoplasmic milieu after rupture of the nuclear membrane (genus *Granulovirus*). The nucleocapsids average 30-35 nm in diameter and 250-300 nm in length. The envelope of the occlusion derived virus (ODV) has no peplomers. If infection is not restricted to the gut epithelium, a second phenotype may infect other tissues. This second phenotype is characterized by virions that bud primarily as single nucleocapsids from the plasma membrane of infected cells. Envelopes of the budded virus (BV) are characteristically loose-fitting and contain terminal peplomers 14-15 nm in length with a single glycoprotein as the major component.

Physicochemical and Physical Properties

ODV buoyant density in CsCl is 1.18-1.25 g/cm^3, and that of the nucleocapsid is 1.47 g/cm^3. BV buoyant density in sucrose is 1.17-1.18 g/cm^3. Virions of both phenotypes are sensitive to organic solvents and detergents. BV is marginally sensitive to heat and pH 8-12, is inactivated by pH 3.0, and is stable in Mg^{++} (10^{-1} M to 10^{-5} M).

Nucleic Acid

Nucleocapsids contain a single molecule of circular supercoiled dsDNA, 90-160 kb in size.

Proteins

Virions contain approximately 12 to 30 different polypeptides. The major protein of the occlusion is a single polypeptide, viral encoded, Mr 25-33 x 10^3. This protein is called polyhedrin for polyhedroviruses and granulin for granuloviruses. Virions of both phenotypes contain a major capsid protein and a basic DNA binding protein, but only BV contains a major envelope protein (the peplomer protein) with fusogenic properties.

Lipids

Lipids are present in the envelopes of ODV and BV.

Carbohydrates

Carbohydrates are present as glycoproteins and glycolipids.

Genome Organization and Replication

Circular genomic DNA is infectious suggesting that no virion-associated proteins are essential for infection. Transcription of baculovirus genes is temporally regulated, and two main classes of genes are recognized, early and late. Some late genes are described as very late. The gene classes are not clustered on the baculovirus genome, and both strands of the genome are involved in coding functions. Early genes are transcribed by host RNA polymerase II, while late and very late genes are transcribed by an alpha-amanitin resistant RNA polymerase activity. Transcriptional activity throughout replication frequently results in nested transcripts, both with variable 5' and co-terminal 3' ends, and with co-terminal 5' and variable 3' ends. RNA splicing occurs, but is rare. BV production begins during the late phase, and occlusion production during the very late phase. Replication initiates in the midgut (insects) or digestive gland epithelium (shrimp) of its arthropod hosts following ingestion of viral occlusions. The occlusions are solubilized in the gut lumen releasing the enveloped virions which are thought to enter the target epithelium via fusion with the cell surface membrane. In lepidopteran insects, fusion occurs at in an alkaline environment, up to pH 12. Replication takes place in the nucleus. In granulovirus-infected cells, the nuclear membrane appears to lose its integrity during the replication process. With some baculoviruses, replication is restricted to the gut epithelium and progeny virions become enveloped and occluded within these cells, and may be shed into the gut lumen with sloughed epithelium, or released upon death of the host. Other baculoviruses produce a second phenotype which buds from the basolateral membrane of infected gut cells. This budded virus is thought to transmit the infection to internal organs and tissues. In secondarily infected tissues, BV is produced first and occluded virus second, with infected fat body being the primary location of occluded virus production. Occluded virus matures within nuclei of infected cells for nucleopolyhedroviruses (nuclear-cytoplasmic milieu for granuloviruses) and is released upon death, and usually liquification, of the host.

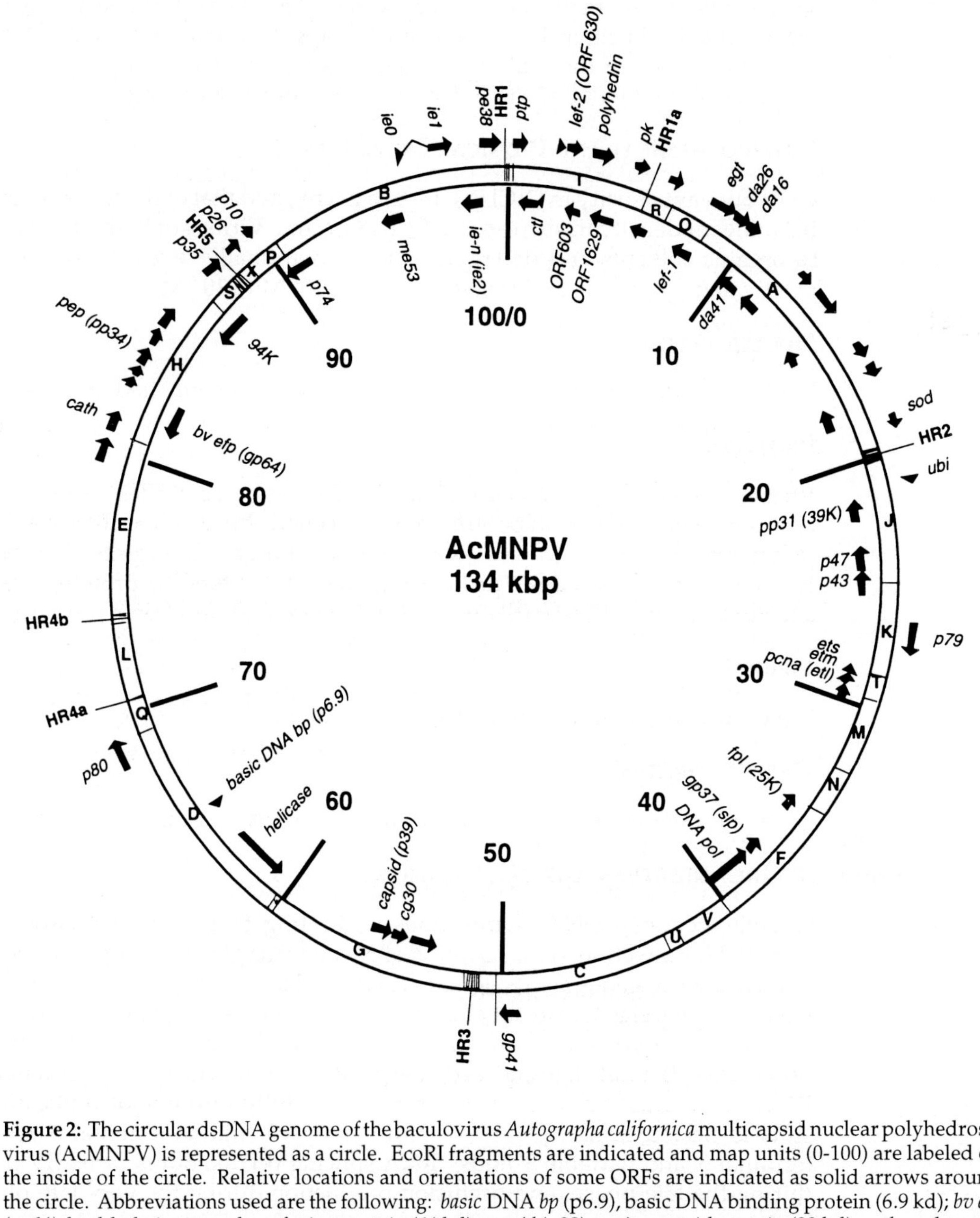

Figure 2: The circular dsDNA genome of the baculovirus *Autographa californica* multicapsid nuclear polyhedrosis virus (AcMNPV) is represented as a circle. EcoRI fragments are indicated and map units (0-100) are labeled on the inside of the circle. Relative locations and orientations of some ORFs are indicated as solid arrows around the circle. Abbreviations used are the following: *basic* DNA *bp* (p6.9), basic DNA binding protein (6.9 kd); *bv efp (gp64)*, budded virus envelope fusion protein (64 kd); *capsid* (p39), major capsid protein (39 kd); *cath*, cathepsin; *cg30*, HindIII-C/EcoRI-G 30 kd protein; *da16*, HindIII-D/EcoRI-A 16 kd protein; *da26*, HindIII-D/EcoRI-A 26 kd protein; *da41*, HindIII-D/EcoRI-A 41 kd protein; DNA pol, DNA polymerase; *egt*, ecdysterioid UDP-glucosylttransferase; *ets*, HindIII-E-EcoRI-T-small; *fpl* (25 kd), few polyhedra locus protein (25 kd); *gp37, glycoprotein 37 kd (slp*, spheroidin-like protein); HR, homologous repeat; *p35*, suppressor of apoptosis; *ie0*, immediate early gene 0; *ie1*, immediate early gene 1; *ie-n (ie2)*, immediate early gene 2; *lef-1*, late expression factor 1; *lef-2*, late expression factor 2; *me53*, major early 53 kd; *orf1629*, 1629 nt ORF; *orf603*, 603 nt ORF; *pcna (etl)*, proliferating cell nuclear antigen (HindIII-E-EcoRI-T-large); *pe38*, PstI-EciR1 38 kd; *pep (pp34)*, polyhedral envelope protein (phosphoprotein 34 kd); *pk*, protein kinase; *polyhedrin*, major occlusion protein; *pp31* (*39k*),

phosphoprotein 31 kd (originally named 39 kd protein); ptp, protein tyrosine/serine phosphatase; *sod*, superoxide dismutase; *ubi*, ubiquitin.

ANTIGENIC PROPERTIES

Antigenic determinants that cross-react exist on virion proteins and on the major subunit of polyhedrin and granulin polypeptides. Neutralizing antibodies react with the major surface glycoprotein of BV.

BIOLOGICAL PROPERTIES

Baculoviruses have been isolated only from arthropods; primarily from insects of the order *Lepidoptera*, but also *Hymenoptera*, *Diptera*, *Coleoptera*, *Neuroptera*, *Thysanura* and *Trichoptera* as well as from the crustacean order *Decapoda* (shrimp). Horizontal transmission occurs by contamination of food, egg surface, etc.; vertical transmission via the egg has been reported; experimental transmission can be accomplished by injection of intact hosts or by infection or transfection of cell cultures. Typically the infectious process in insects takes a week, and as an end result, the diseased insect liquifies, releasing occluded virus into the environment.

GENUS *NUCLEOPOLYHEDROVIRUS*

Type Species Autographa californica nucleopolyhedrovirus (AcMNPV)

DISTINGUISHING FEATURES

Two virion phenotypes may be characteristic of a virus species, but one is occluded within a polyhedral proteinic matrix composed primarily of a single protein. Each occlusion measures 0.15 to 15 µm in size, matures within nuclei of infected cells and characteristically contains many enveloped virions. The occluded phenotypes of species are packaged as one (S) or multiple (M) nucleocapsids within a single viral envelope. Factors that regulate nucleocapsid packaging are unknown and for some species packaging arrangements may be variable. S/M designations in common usage have been retained for species where variability has not been reported and for distinct viruses that would otherwise have identical designations under the current nomenclature. Nucleocapsids are rod-shaped (30-60 nm x 250-300 nm) and contain a single molecule of circular supercoiled dsDNA 90-160 kb in size. Nucleocapsids are thought to be transported through the nuclear pore into the nucleus to initiate replication. Species may infect any of seven orders of insects and an order of *Crustacea*.

LIST OF SPECIES IN THE GENUS

The viruses and their assigned abbreviations ( ) are:

SPECIES IN THE GENUS

Anticarisia gemmatalis MNPV	(AgMNPV)
Autographa californica MNPV	(AcMNPV)
Bombyx mori NPV	(BmNPV)
Choristoneura fumiferana MNPV	(CfMNPV)
Galleria mellonella MNPV	(GmMNPV)
Helicoverpa zea SNPV	(HzSNPV)
Lymantria dispar MNPV	(LdMNPV)
Mamestra brassicae MNPV	(MbMNPV)
Orgyia pseudosugata MNPV	(OpMNPV)
Orgyia pseudosugata SNPV	(OpSNPV)
Rachiplusia ou MNPV	(RoMNPV)
Spodoptera exigua MNPV	(SeMNPV)
Spodoptera frugiperda MNPV	(SfMNPV)
Trichoplusia ni MNPV	(TnMNPV)
Trichoplusia ni Single SNPV	(TnSNPV)

TENTATIVE SPECIES IN THE GENUS

Abraxas grossulariata NPV (AbgrNPV)
Achaea janata NPV (AcjaNPV)
Acidalia carticcaria NPV (AccaNPV)
Acleris variana NPV (AcvaNPV)
Actebia fennica NPV (AcfeNPV)
Adisura atkinsoni NPV (AdatNPV)
Aedes aegypti NPV (AeaeNPV)
Aedes atropalpus NPV (AeatNPV)
Aedes nigromaculis NPV (AeniNPV)
Aedes sollicitans NPV (AesoNPV)
Aedes tormentor NPV (AetoNPV)
Aedia leucomelas NPV (AeleNPV)
Agraulis vanillae NPV (AgvaNPV)
Agrotis ipsilon NPV (AgipNPV)
Alabama argillacea NPV (AlarNPV)
Alphaea phasma NPV (AlphNPV)
Amathes c-nigrum NPV (Amc-nNPV)
Amphidasis cognataria NPV (AmcoNPV)
Amsacta lactinea NPV (AmlaNPV)
Amyelois transitella NPV (AmtrNPV)
Anagasta kuehniella NPV (AnkuNPV)
Anaitis plagiata NPV (AnplNPV)
Anomis flava NPV (AnflNPV)
Anomogyna elimata NPV (AnelNPV)
Anthela varia NPV (AnvaNPV)
Antheraea paphia NPV (AnpaNPV)
Antheraea polyphemus NPV (AnpoNPV)
Anthonomus glandis PV (AnglNPV)
Apamea anceps NPV (ApanNPV)
Apocheima pilosaria NPV (AppiNPV)
Aproaerema modicella NPV (ApmoNPV)
Archips cerasivoranus NPV (ArceNPV)
Artica villica NPV (ArviNPV)
Arge pectoralis NPV (ArpeNPV)
Argyrogramma basigera NPV (ArbaNPV)
Autographa biloha NPV (AubiNPV)
Autographa gamma NPV (AugaNPV)
Autographa precationis NPV (AuprNPV)
Bellura gortynoides NPV (BegoNPV)
Biston betularia NPV (BibeNPV)
Biston hispidaria NPV (BihsNPV)
Biston robustum NPV (BiroNPV)
Boarmia bistortata NPV (BobiNPV)
Bucculatrix thurbeliella NPV (ButhNPV)
Buzura suppressaria NPV (BusuNPV)
Cadra cautella NPV (CacaNPV)
Calliphora vomitoria NPV (CavoNPV)
Canephora asiatica NPV (CaasNPV)
Carposina niponensis NPV (CaniNPV)
Catocala conjuncta NPV (CacoNPV)
Catocala nymphagoga NPV (CanmNPV)
Cephalcia abietis NPV (CeabNPV)
Ceramica pisi NPV (CepsNPV)
Cerura hermelina NPV (CeheNPV)
Chirono mustentans NPV (ChteNPV)
Choristoneura diversana NPV (ChdiNPV)
Choristoneura occidentalis NPV (ChooNPV)
Choristoneura rosaceana NPV (ChroNPV)
Chrysodeixis eriosoma NPV (CherNPV)
Cingilia caternaria NPV (CicaNPV)
Coleophora laricella NPV (ColaNPV)
Colias eurytheme NPV (CoeuNPV)
Colias philodice NPV (CophNPV)
Corcyrace phalonica NPV (CophNPV)
Cossus cossus NPV (CocoNPV)
Cryptothelea junodi NPV (CrjuNPV)
Culcuta panterinaria NPV (CupaNPV)
Culex salinarius NPV (CusaNPV)
Acantholyda erythrocephala NPV (AcerNPV)
Achroia grisella NPV (AcgrNPV)
Acleris gloverana NPV (AcglNPV)
Acronicta aceris NPV (AcaoNPV)
Actias selene NPV (AcseNPV)
Adoxophyes orana NPV (AdorNPV)
Aedes annandalei NPV (AeanNPV)
Aedes epactius NPV (AeepNPV)
Aedes scutellaris NPV (AescNPV)
Aedes taeniorhynchus NPV (AetaNPV)
Aedes triseriatus NPV (AetrNPV)
Aglais urticae NPV (AgurNPV)
Agrotis exclamationis NPV (AgexNPV)
Agrotis segetum NPV (AgseNPV)
Aletia oxygala NPV (AloxNPV)
Alsophila pometaria NPV (AlpoNPV)
Amphelophaga rubiginosa NPV (AmruNPV)
Amsacta albistriga NPV (AmalNPV)
Amsacta moorei NPV (AmmoNPV)
Anadevidia peponis NPV (AnpeNPV)
Anagrapha falcifera NPV (AnfaNPV)
Anisota senatoria NPV (AnseNPV)
Anomis sabulifera NPV (AnsaNPV)
Anopheles crucians NPV (AncrNPV)
Anthelia hyperborea NPV (AnhyNPV)
Antheraea pernyi NPV (AnpeNPV)
Antheraea yamamai NPV (AnyaNPV)
Anthrenus museorum NPV (AnmuNPV)
Apocheima cinerarius NPV (ApciNPV)
Aporia crataegi NPV (ApcrNPV)
Araschnia levana NPV (ArleNPV)
Arctia caja NPV (ArcaNPV)
Ardices glatignyi NPV (ArglNPV)
Argynnis paphia NPV (ArpaNPV)
Astero campaceltis NPV (AscaNPV)
Autographa bimaculata NPV (AubmNPV)
Autographa nigrisigna NPV (AuniNPV)
Batocera lineolata NPV (BaliNPV)
Bhima undulosa NPV (BhunNPV)
Biston hirtaria NPV (BihiNPV)
Biston marginata NPV (BimaNPV)
Biston strataria NPV (BistNPV)
Boarmia obliqua NPV (BoobNPV)
Bupalus piniarius NPV (BupiNPV)
Buzura thibtaria NPV (ButiNPV)
Cadra figulilella NPV (CafiNPV)
Calophasia lunula NPV (CaluNPV)
Caripeta divisata NPV (CadiNPV)
Catabena esula NPV (CaesNPV)
Catocala nymphaea NPV (CanyNPV)
Catopsilia pomona NPV (CapoNPV)
Ceramica picta NPV (CepiNPV)
Cerapteryx graminis NPV (CegrNPV)
Chilo suppressalis NPV (ChsuNPV)
Choristoneura conflictana NPV (ChcoNPV)
Choristoneura murinana NPV (ChmuNPV)
Choristoneura pinus NPV (ChpiNPV)
Chrysodeixis chalcites NPV (ChchNPV)
Chrysopa perla NPV (ChpeNPV)
Cnidocampa flavescens NPV (CnflNPV)
Colias electo NPV (CoelNPV)
Colias lesbia NPV (ColeNPV)
Coloradia pandora NPV (CopaNPV)
Cosmotriche podatoria NPV (CopoNPV)
Cryptoblabes lariciana NPV (CrlaNPV)
Cryptothelea variegata NPV (CrvaNPV)
Culex pipiens NPV (CupiNPV)
Cyclophragma undans NPV (CyunNPV)

Cyclophragma yamadai NPV	(CyyaNPV)	Cydia pomonella NPV	(CypoNPV)
Dasychira abietis NPV	(DaabNPV)	Dasychira argentata NPV	(DaarNPV)
Dasychira axutha NPV	(DaaxNPV)	Dasychira basiflava NPV	(DabaNPV)
Dasychira confusa NPV	(DacoNPV)	Dasychira glaucinoptera NPV	(DaglNPV)
Dasychira locuples NPV	(DaloNPV)	Dasychira mendosa NPV	(DameNPV)
Dasychira plagiata NPV	(DaplNPV)	Dasychira pseudabietis NPV	(DapsNPV)
Dasychira pudibunda NPV	(DapuNPV)	Deilephila elpenor NPV	(DeelNPV)
Deileptenia ribeata NPV	(DeriNPV)	Dendrolimus latipennis NPV	(DelaNPV)
Dendrolimus pini NPV	(DepiNPV)	Dendrolimus punctatus NPV	(DepuNPV)
Dendrolimus spectabilis NPV	(DespNPV)	Dermeste lardarius NPV	(DelaNPV)
Diachrysia orichalcea NPV	(DiorNPV)	Diacrisia obliqua NPV	(DiobNPV)
Diacrisia purpurata NPV	(DipuNPV)	Diacrisia virginica NPV	(DiviNPV)
Diaphora mendica NPV	(DimeNPV)	Diatraea grandiosella NPV	(DigrNPV)
Diatraea saccharalis NPV	(DisaNPV)	Dichocrocis punctiferalis NPV	(DipuNPV)
Dictyoploca japonica NPV	(DijaNPV)	Dicycla oo NPV	(DiooNPV)
Dilta hibernica NPV	(DihiNPV)	Dioryctria pseudotsugella NPV	(DipsNPV)
Diparopsis watersi NPV	(DiwaNPV)	Diprion hercyniae NPV	(DiheNPV)
Diprion leuwanensis NPV	(DileNPV)	Diprion nipponica NPV	(DiniNPV)
Diprion pallida NPV	(DipaNPV)	Diprion pindrowi NPV	(DipdNPV)
Diprion pini NPV	(DipiNPV)	Diprion polytoma NPV	(DipoNPV)
Diprion similis NPV	(DisiNPV)	Dirphia gragatus NPV	(DigrNPV)
Doratifera casta NPV	(DocaNPV)	Dryobota furva NPV	(DrfuNPV)
Dryobota protea NPV	(DrprNPV)	Dryobotodes monochroma NPV	(DrmoNPV)
Earias insulana NPV	(EainNPV)	Ecpantheria icasia NPV	(EcicNPV)
Ectropis crepuscularia NPV	(EccrNPV)	Ectropis obliqua NPV	(EcobNPV)
Ennomos quercaria NPV	(EnquNPV)	Ennomos quercinaria NPV	(EnquNPV)
Ennomos subsignarius NPV	(EnsuNPV)	Enypia venata NPV	(EnveNPV)
Epargyreus clarus NPV	(EpclNPV)	Ephestia elutella NPV	(EpelNPV)
Epiphyas postvittana NPV	(EppoNPV)	Erannis ankeraria NPV	(EranNPV)
Erannis defoliaria NPV	(ErdeNPV)	Erannis tiliaria NPV	(ErtiNPV)
Erannis vancouverensis NPV	(ErvaNPV)	Eratmapodites quinquevittatus NPV	(ErquNPV)
Erinnyis ello NPV	(ErelNPV)	Eriogyna pyretorum NPV	(ErpyNPV)
Estigmene acrea NPV	(EsacNPV)	Eupithecia annulata NPV	(EuanNPV)
Eupithecia longipalpata NPV	(EuloNPV)	Euproctis bipunctapex NPV	(EubiNPV)
Euproctis chrysorrhoea NPV	(EuchNPV)	Euproctis flava NPV	(EuflNPV)
Euproctis flavinata NPV	(EufvNPV)	Euproctis karghalica NPV	(EukaNPV)
Euproctis pseudoconspersa NPV	(EupsNPV)	Euproctis similis NPV	(EusiNPV)
Euproctis subflava NPV	(EusuNPV)	Euthyatira pudens NPV	(EupuNPV)
Euxoa auxiliaris NPV	(EuauNPV)	Euxoa messoria NPV	(EumeNPV)
Euxoa ochrogaster NPV	(EuocNPV)	Euxoa scandens NPV	(EuscNPV)
Feralia jacosa NPV	(FejaNPV)	Gastropacha quercifolia NPV	(GaquNPV)
Hadena sordida NPV	(HasoNPV)	Halisidota argentata NPV	(HaarNPV)
Halisidota caryae NPV	(HacaNPV)	Helicoverpa armisgera NPV	(HearNPV)
Helicoverpa assulta NPV	(HeasNPV)	Helicoverpa obtectus NPV	(HeobNPV)
Helicoverpa paradoxa NPV	(HepaNPV)	Helicoverpa peltigera NPV	(HepeNPV)
Helicoverpa phloxiphaga NPV	(HephNPV)	Helicoverpa punctigera NPV	(HepuNPV)
Helicoverpa rubrescens NPV	(HeruNPV)	Helicoverpa subflexa NPV	(HesuNPV)
Helicoverpa virescens NPV	(HeviNPV)	Hemerobius stigma NPV	(HestNPV)
Hemichroa crocea NPV	(HecrNPV)	Hemileuca eglanterina NPV	(HeegNPV)
Hemileuca maia NPV	(HemaNPV)	Hemileuca oliviae NPV	(HeolNPV)
Hemileuca tricolor NPV	(HetrNPV)	Hesperumia sulphuraria NPV	(HesuNPV)
Hippotion eson NPV	(HiesNPV)	Homona magnanima NPV	(HomaNPV)
Hoplodrina ambigua NPV	(HoamNPV)	Hyalophora cecropia NPV	(HyceNPV)
Hydriomena irata NPV	(HyirNPV)	Hydriomena nubilofasciata NPV	(HynuNPV)
Hyles euphorbiae NPV	(HyeuNPV)	Hyles gallii NPV	(HygaNPV)
Hyles lineata NPV	(HyliNPV)	Hylesia nigricans NPV	(HyniNPV)
Hyloicus pinastri NPV	(HypiNPV)	Hyperetis amicaria NPV	(HyamNPV)
Hyphantria cunea NPV	(HycuNPV)	Hyphorma minax NPV	(HymiNPV)
Hypocrita jacobeae NPV	(HyjaNPV)	Inachis io NPV	(InioNPV)
Ilragoides fasciata NPV	(IlfaNPV)	Ivela auripes NPV	(IvauNPV)
Ivela ochropoda NPV	(IvocNPV)	Jankowskia athleta NPV	(JaatNPV)
Junonia coenia NPV	(JucoNPV)	Lacanobia oleracea NPV	(LaolNPV)
Lambdina fiscellaria NPV	(LafiNPV)	Laothoe populi NPV	(LapoNPV)
Lasiocampa quercus NPV	(LaquNPV)	Lasiocampa trifolii NPV	(LatrNPV)
Lebeda nobilis NPV	(LenNPV)	Lechriolepis basirufa NPV	(LebaNPV)
Leucoma candida NPV	(LecaNPV)	Leucoma salicis NPV	(LesaNPV)
Lophopteryx camelina NPV	(LocaNPV)	Loxostege sticticalis NPV	(LostNPV)
Luehdorfia japonica NPV	(LujaNPV)	Lymantria dispar NPV	(LydiNPV)
Lymantria dissoluta NPV	(LydsNPV)	Lymantria fumida NPV	(LyfuNPV)
Lymantria incerta NPV	(LyinNPV)	Lymantria mathura NPV	(LymaNPV)
Lymantria monacha NPV	(LymoNPV)	Lymantria ninayi NPV	(LyniNPV)

Lymantria obfuscata NPV	(LyobNPV)	Lymantria violaswinhol NPV	(LyviNPV)
Lymantria xylina NPV	(LyxyNPV)	Macrothylacia rubi NPV	(MaruNPV)
Mahasena miniscula NPV	(MamiNPV)	Malacosoma alpicola NPV	(MaalNPV)
Malacosoma americanum NPV	(MaamNPV)	Malacosoma californicum NPV	(MacaNPV)
Malacsoma constrictum NPV	(MacoNPV)	Malacosoma disstria NPV	(MadiNPV)
Malacsoma fragile NPV	(MafrNPV)	Malacsoma lutescens NPV	(MaluNPV)
Malacsoma neustria NPV	(Mane NPV)	Malacsoma pluvia1e NPV	(MaplNPV)
Mamestra configurata NPV	(MacoNPV)	Mamestra suasa NPV	(MasuNPV)
Manduca sexta NPV	(MaseNPV)	Melanolophia imitata NPV	(MeimNPV)
Melitaea didyma NPV	(MediNPV)	Merophyas divulsana NPV	(MediNPV)
Mesonura rufonota NPV	(MeruNPV)	Moma champa NPV	(MochNPV)
Myrteta tinagmaria NPV	(MytiNPV)	Nacoleia octosema NPV	(NaocNPV)
Nadata gibbosa NPV	(NagiNPV)	Nematus olfaciens NPV	(NeolNPV)
Neodiprion abietis NPV	(NeabNPV)	Neodiprion excitans NPV	(NeexNPV)
Neodiprion leconti NPV	(NeleNPV)	Neodiprion nanultus NPV	(NenaNPV)
Neodiprion pratti NPV	(NeprNPV)	Neodiprion sertifer NPV	(NeseNPV)
Neodiprion swainei NPV	(NeswNPV)	Neodiprion taedae NPV	(NetaNPV)
Neodiprion tsugae NPV	(NetsNPV)	Neodiprion virginiana NPV	(NeviNPV)
Neophasia menapia NPV	(NemeNPV)	Neopheosia excurvata NPV	(NeexNPV)
Nephelodes emmedonia NPV	(NeemNPV)	Nepytia freemani NPV	(NefrNPV)
Nepytia phantasmaria NPV	(NephNPV)	Noctua pronuba NPV	(NoprNPV)
Nyctobia limitaria NPV	(NyliNPV)	Nymphalis antiopa NPV	(NyanNPV)
Nymphalis polychloros NPV	(NypoNPV)	Nymphula depunctalis NPV	(NydeNPV)
Ocinara varians NPV	(OcvaNPV)	Operophtera bruceata NPV	(OpbrNPV)
Operophtera brumata NPV	(OpbuNPV)	Opisina arenosella NPV	(OparNPV)
Opisthograptis luteolata NPV	(OpluNPV)	Oporinia autumnata NPV	(OpauNPV)
Opsiphanes cassina NPV	(OpcaNPV)	Oraesia emarginata NPV	(OremNPV)
Orgyia anartoides NPV	(OranNPV)	Orgyia antiqua NPV	(OratNPV)
Orgyia australis NPV	(OrauNPV)	Orgyia badia NPV	(OrbaNPV)
Orgyia gonostigma NPV	(OrgoNPV)	Orgyia leucostigma NPV	(OrleNPV)
Orgyia postica NPV	(OrpoNPV)	Orgyia turbata NPV	(OrtuNPV)
Orgyia vetusta NPV	(OrveNPV)	Orthosia hibisci NPV	(OrhiNPV)
Orthosia incerta NPV	(OrinNPV)	Ostrinia nubilalis NPV	(OsnuNPV)
Pachypasa capensis NPV	(PacaNPV)	Pachypasa otus NPV	(PaotNPV)
Paleacrita vernata NPV	(PaveNPV)	Panaxia dominula NPV	(PadoNPV)
Pandemis heparana NPV	(PaheNPV)	Pandemis lamprosana NPV	(PalaNPV)
Panolis flammea NPV	(PaflNPV)	Pantana phyllostachysae NPV	(PaphNPV)
Panthea portlandia NPV	(PapoNPV)	Parasa consocia NPV	(PacoNPV)
Parasa lepida NPV	(PaleNPV)	Parasa sinica NPV	(PasiNPV)
Parnara guttata NPV	(PaguNPV)	Parnara mathias NPV	(PamaNPV)
Papilio daunis NPV	(PadaNPV)	Papilio demoleus NPV	(PadeNPV)
Papilio podalirius NPV	(PapoNPV)	Papilio polyxenes NPV	(PaplNPV)
Papilio xuthus NPV	(PaxuNPV)	Pectinophora gossypiella NPV	(PegoNPV)
Peribatoides simpliciaria NPV	(PesiNPV)	Pericallia ricini NPV	(PeriNPV)
Peridroma saucia NPV	(PesaNPV)	Pero behrensarius NPV	(PebeNPV)
Pero mizon NPV	(PemiNPV)	Phalera assimilis NPV	(PhasNPV)
Phalera bucephala NPV	(PhbuNPV)	Phalera flavescens NPV	(PhflNPV)
Phauda flammans NPV	(PhfaNPV)	Phigalia titea NPV	(PhtiNPV)
Phlogophora meticulosa NPV	(PhmeNPV)	Phryganidia californica NPV	(PhcaNPV)
Phthonosema tendinosaria NPV	(PhteNPV)	Phthorimaea operculella NPV	(PhopNPV)
Pieris rapae NPV	(PiraNPV)	Pikonema dimmockii NPV	(PidiNPV)
Plathypena scabra NPV	(PlscNPV)	Platynota idaesalis NPV	(PlidNPV)
Plusia argentifera NPV	(PlarNPV)	Plusia balluca NPV	(PlbaNPV)
Plusia signata NPV	(PlsiNPV)	Plutella xylostella NPV	(PlxyNPV)
Polygonia c-album NPV	(Poc-aNPV)	Polygonia satyrus NPV	(PosaNPV)
Porthesia scintillans NPV	(PoscNPV)	Pristophora erichsonii NPV	(PrerNPV)
Pristophora geniculata NPV	(PrgeNPV)	Prodenia litosia NPV	(PrliNPV)
Prodenia praefica NPV	(PrprNPV)	Prodenia terricola NPV	(PrteNPV)
Protoboarmia porcelaria NPV	(PrpoNPV)	Pseudaletia convecta NPV	(PscoNPV)
Pseudaletia separata NPV	(PsseNPV)	Pseudoplusia includens NPV	(PsinNPV)
Psorophora confinnis NPV	(PscnNPV)	Psorophora ferox NPV	(PsfeNPV)
Psorophora varipes NPV	(PsvaNPV)	Pterolocera amplicornis NPV	(PtaeNPV)
Ptycholomoides aeriferana NPV	(PtaeNPV)	Ptychopoda seriata NPV	(PtseNPV)
Pygaera anastomosis NPV	(PyanNPV)	Pygaera fulgurita NPV	(PyfuNPV)
Pyrausta diniasalis NPV	(PydiNPV)	Rachiplusia nu NPV	(RanuNPV)
Rhyacionia duplana NPV	(RhduNPV)	Rhynchosciara angelae NPV	(RhanNPV)
Rhynchosciara hollaenderi NPV	(RhhoNPV)	Rhynchosciara milleri NPV	(RhmiNPV)
Rondiotia menciana NPV	(RomeNPV)	Samia cynthia NPV	(SacyNPV)
Samia pryeri NPV	(SaprNPV)	Samia ricini NPV	(SariNPV)
Saturnia pyri NPV	(SapyNPV)	Sceliodes cordalis NPV	(SccoNPV)
Scirpophaga incertulas NPV	(ScinNPV)	Scoliopteryx libatrix NPV	(ScliNPV)

Scopelodes contracta NPV	(SccoNPV)	Scopelodes venosa NPV	(ScveNPV)
Scopula subpunctaria NPV	(ScsuNPV)	Scotogramma trifolii NPV	(SctrNPV)
Selenephera lunigera NPV	(SeluNPV)	Selidosema suavis NPV	(SesuNPV)
Semidonta biloba NPV	(SebiNPV)	Sesamia calamistis NPV	(SecaNPV)
Sesamia inferens NPV	(SeinNPV)	Smerinthus ocellata NPV	(SmocNPV)
Sparganothis pettitana NPV	(SppeNPV)	Sphinx ligustri NPV	(SpligNPV)
Spilarctia subcarnea NPV	(SpsuNPV)	Spilonota ocellana NPV	(SpocNPV)
Spilosoma lubricipeda NPV	(SpluNPV)	Spodoptera exempta NPV	(SpexNPV)
Spodoptera exigua NPV	(SpeiNPV)	Spodoptera frugiperda NPV	(SpfrNPV)
Spodoptera latifascia NPV	(SplaNPV)	Spodoptera littoralis NPV	(SpliNPV)
Spodoptera litura NPV	(SpltNPV)	Spodoptera mauritia NPV	(SpmaNPV)
Spodoptera ornithogalli NPV	(SporNPV)	Synaxis jubararia NPV	(SyjuNPV)
Synaxis pallulata NPV	(SypaNPV)	Syngrapha selecta NPV	(SyseNPV)
Tetralopha scortealis NPV	(TescNPV)	Tetropium cinnamopterum NPV	(TeciNPV)
Thaumetopoea pityocampa NPV	(ThpiNPV)	Thaumetopoea processionea NPV	(ThprNPV)
Theophila mandarina NPV	(ThmaNPV)	Theretra japonica NPV	(ThjaNPV)
Thosea baibarana NPV	(ThbaNPV)	Thymelicus lineola NPV	(ThliNPV)
Thylidolpteryx ephemeraeformis NPV	(ThepNPV)	Ticera castanea NPV	(TicaNPV)
Tinea pellionella NPV	(TipeNPV)	Tineola hisselliella NPV	(TihiNPV)
Tipula paludosa NPV	(TipaNPV)	Tiracola plagiata NPV	(TiplNPV)
Tortrix loeflingiana NPV	(ToloNPV)	Tortrix viridana NPV	(ToviNPV)
Toxorhynchites brevipalpis NPV	(TobrNPV)	Trabala vishnou NPV	(TrviNPV)
Trichiocampus irregularis NPV	(TrirNPV)	Trichiocampus viminalis NPV	(TrvmNPV)
Ugymyia sericariae NPV	(UgseNPV)	Uranotaenia sapphirina NPV	(UrsaNPV)
Urbanus proteus NPV	(UrprNPV)	Vanessa atalanta NPV	(VaatNPV)
Vanessa cardui NPV	(VacaNPV)	Vanessa prorsa NPV	(VaprNPV)
Wiseana cervinata NPV	(WiceNPV)	Wiseana signata NPV	(WisiNPV)
Wiseana umbraculata NPV	(WiumNPV)	Wyeomyia smithii NPV	(WysmNPV)
Xylena curvimacula NPV	(XycuNPV)	Yponomeuta cognatella NPV	(YpcoNPV)
Yponomeuta evonymella NPV	(YpevNPV)	Yponomeuta malinellus NPV	(YpmaNPV)
Yponomeuta padella NPV	(YppaNPV)	Zeiraphera diniana NPV	(ZediNPV)
Zeiraphera pseudotsugana NPV	(ZepsNPV)		

NPV, nucleopolyhedrovirus; M, multiple; S, single.

GENUS *GRANULOVIRUS*

Type Species   Plodia interpunctella granulovirus   (PiGV)

DISTINGUISHING FEATURES

Two virion phenotypes may be characteristic of a virus species, but one is occluded within an ovicylindrical proteinic matrix composed primarily of a single protein. Each occlusion measures 0.13 x 0.5 µm in size and characteristically contains one enveloped nucleocapsid. One nucleocapsid generally is contained within a single envelope. Occluded virions may mature among nuclear-cytoplasmic cellular contents after rupture of the nuclear membrane of infected cells. Nucleocapsids are rod-shaped (30-60 nm x 250-300 nm) and contain a single molecule of circular supercoiled dsDNA 90-180 kb in size. Viral DNA is thought to be extruded into the nucleus through the nuclear pore to initiate infection; the capsid remains in the cytoplasm. Species of this genus have only been isolated from lepidopteran insects.

LIST OF SPECIES IN THE GENUS

The viruses and their assigned abbreviations ( ) are:

SPECIES IN THE GENUS

Trichoplusia ni granulovirus	(TnGV)
Pieris brassicae granulovirus	(PbGV)
Artogeia rapae granulovirus	(ArGV)
Cydia pomonella granulovirus	(CpGV)

TENTATIVE SPECIES IN THE GENUS

Amelia pallorana GV	(AmpaGV)	Amsacta lactinea GV	(AmlaGV)
Andraca bipunctata GV	(AnbiGV)	Apamea anceps GV	(ApanGV)
Apamea sordens GV	(ApsoGV)	Archippus breviplicanus GV	(ArbrGV)

Archippus packardianus GV	(ArpaGV)	Archips argyrospila GV	(ArarGV)
Archips longicellana GV	(ArloGV)	Argyrotaenia velutinana GV	(ArveGV)
Artona funeralis GV	(ArfuGV)	Athetis albina GV	(AtalGV)
Autographa californica GV	(AucaGV)	Cadra cautella GV	(CacaGV)
Cadra figulilella GV	(CafiGV)	Carposina niponensis GV	(CaniGV)
Cephalcia fascipennis GV	(CefaGV)	Chilo infuscatellus GV	(ChinGV)
Chilo sacchariphagus GV	(ChsaGV)	Chilo suppressalis GV	(ChsuGV)
Choristoneura conflictana GV	(ChcoGV)	Choristoneura fumiferana GV	(ChfuGV)
Choristoneura murinana GV	(ChmuGV)	Choristoneura occidentalis GV	(ChooGV)
Choristoneura retiniana GV	(ChreGV)	Choristoneura viridis GV	(ChviGV)
Clepsis persicana GV	(ClpeGV)	Cnaphalocrocis medinalis GV	(CnmeGV)
Cnidocampa flavescens GV	(CnflGV)	Coleotechnites milleri GV	(ComiGV)
Cryptophlebia leucotreta GV	(CrleGV)	Cydia nigricana GV	(CyniGV)
Darna trima GV	(DatrGV)	Dendrolimus sibiricus GV	(DesiGV)
Dendrolimus spectabilis GV	(DespGV)	Diacrisia obliqua GV	(DiobGV)
Diacrisia virginica GV	(DiviGV)	Diatraea saccharalis GV	(DisaGV)
Dionychopus amasis GV	(DiamGV)	Dioryctria abietella GV	(DiabGV)
Dryobota furva GV	(DrfuGV)	Ecpantheria icasia GV	(EcicGV)
Ectropis obliqua GV	(EcobGV)	Epinotia aporema GV	(EpapGV)
Estigmene acrea GV	(EsacGV)	Euplexia lucipara GV	(EuluGV)
Eupsilia satellitia GV	(EusaGV)	Euxoa auxiliaris GV	(EuauGV)
Euxoa messoria GV	(EumeGV)	Euxoa ochrogaster GV	(EuocGV)
Exartema appendiceum GV	(ExapGV)	Feltia subterranea GV	(FesuGV)
Glena bisulca GV	(GlbiGV)	Grapholitha molesta GV	(GrmoGV)
Griselda radicana GV	(GrraGV)	Hadena basilinea GV	(HabaGV)
Hadena sordida GV	(HasoGV)	Harrisina brillians GV	(HabrGV)
Helicoverpa armisgera GV	(HearGV)	Helicoverpa punctigera GV	(HepuGV)
Helicoverpa zea GV	(HezeGV)	Hemileuca eglanterina GV	(HeegGV)
Hemileuca oliviae GV	(HeolGV)	Homona coffearia GV	(HocoGV)
Homona magnanima GV	(HomaGV)	Hydria prunivora GV	(HyprGV)
Hyphantria cunea GV	(HycuGV)	Junonia coenia GV	(JucoGV)
Lacanobia oleracea GV	(LaolGV)	Lambdina fiscellaria GV	(LafiGV)
Lathronympha phaseoli GV	(LaphGV)	Lobesia botrana GV	(LoboGV)
Loxostege sticticalis GV	(LostGV)	Macroglossum bombylans GV	(MaboGV)
Malacsoma pluviale GV	(MaplGV)	Mamestra brassicae GV	(MabrGV)
Mamestra configurata GV	(MacoGV)	Manduca quinquemaculata GV	(MaquGV)
Manduca sexta GV	(MaseGV)	Megalopyge opercularis GV	(MeopGV)
Melanchra persicariae GV	(MepeGV)	Nacoleia diemenalis GV	(NadiGV)
Natada nararia GV	(NanaGV)	Nematocampa filamentaria GV	(NefiGV)
Nephelodes emmedonia GV	(NeemGV)	Nymphalis antiopa GV	(NyanGV)
Papaipema purpurifascia GV	(PapuGV)	Parasa bicolor GV	(PabiGV)
Parasa consocia GV	(PacoGV)	Parasa lepida GV	(PaleGV)
Parasa sinica GV	(PasiGV)	Pericallia ricini GV	(PeriGV)
Peridroma saucia GV	(PesaGV)	Persectania ewingii GV	(PeewGV)
Phragmatobia fuliginosa GV	(PhfuGV)	Phthorimaea operculella GV	(PhopGV)
Pieris melete GV	(PimeGV)	Pieris napi GV	(PinaGV)
Pieris rapae GV	(PiraGV)	Pieris virginiensis GV	(PiviGV)
Plathypena scabra GV	(PlscGV)	Plusia circumflexa GV	(PlciGV)
Plutella xylostella GV	(PlxyGV)	Pontia daplidice GV	(PodaGV)
Prodenia androgea GV	(PranGV)	Pseudaletia convecta GV	(PscoGV)
Pseudaletia separata GV	(PsseGV)	Pseudaletia unipuncta GV	(PsunGV)
Psilogramma menephron GV	(PsmeGV)	Pygaera anachoreta GV	(PyaaGV)
Pygaera anastomosis GV	(PyanGV)	Rheumaptera hastata GV	(RhhaGV)
Rhyacionia buoliana GV	(RhbuGV)	Rhyacionia duplana GV	(RhduGV)
Rhyacionia frustrana GV	(RhfrGV)	Sabulodes caberata GV	(SacaGV)
Sciaphila duplex GV	(ScduGV)	Scotogramma trifolii GV	(SctrGV)
Selepa celtis GV	(SeceGV)	Semiothisa sexmaculata GV	(SeseGV)
Sesamia cretica GV	(SecrGV)	Sesamia nonagrioides GV	(SenoGV)
Spodoptera exigua GV	(SpexiGV)	Spodoptera frugiperda GV	(SpfrGV)
Spodoptera littoralis GV	(SpliGV)	Spodoptera litura GV	(SpltGV)
Thaumetopoea pityocampa GV	(ThpiGV)	Thosea sinensis GV	(ThsiGV)
Wiseana cervinata GV	(WiceGV)	Wiseana umbraculata GV	(WiumGV)
Zeiraphera diniana GV	(ZediGV)		

GV, granulovirus.

LIST OF UNASSIGNED VIRUSES IN THE FAMILY

None reported.

Similarity with Other Taxa

None reported.

Derivation of Names

baculo: from *baculum*, 'stick', from morphology of virion
polyhedro: from polyhedron, shape of occlusions
granulo: from granule

References

Adams JR, Bonami JR (eds) (1991) Atlas of Invertebrate Viruses. CRC Press, Boca Raton FL
Blissard GW, Rohrmann GF (1990) Baculovirus diversity and molecular biology. Annu Rev Entomol 35: 127-155
Consigli RA, Russell DL, Wilson ME (1986) The biochemistry and molecular biology of the granulosis virus that infects *Plodia interpunctella*. Cur Topics Microbiol Immunol 131: 69-101
Doerfler W, Bohm P (1986) The Molecular Biology of Baculoviruses. Cur Topics Microbiol Immunol 131: 1-168
Fraser MJ (1986) Ultrastructural Observations of Virion Maturation in Autographa californica Nuclear Polyhedrosis Virus Infected *Spodoptera frugiperda* Cell Cultures. J Ultrastruct Mol Struct Res 95: 189-195
Granados RR, Federici BA (eds) (1986) The Biology of Baculoviruses. CRC Press, Boca Raton FL
Hull R, Brown F, Payne CC (eds) (1989) Dictionary and Directory of Animal, Plant and Bacterial Viruses. Macmillan, London
Rohrmann GF (1992) Baculovirus Structural Proteins. J Gen Virol 73: 749-761
Summers MD (1977) Baculoviruses (*Baculoviridae*). The Atlas of Insects and Plant Viruses Including Mycoplasma Viruses and Viroids. Academic Press Inc, New York pp 3-27
Volkman LE, Keddie BA (1992) Nuclear Polyhedrosis Virus Pathogenesis. Semin Virol 1: 249-256

Contributed By

Volkman LE, Blissard GW, Friesen P, Keddie BA, Possee R, Theilmann DA

FAMILY *HERPESVIRIDAE*

TAXONOMIC STRUCTURE OF THE FAMILY

Family	***Herpesviridae***
Subfamily	***Alphaherpesvirinae***
Genus	*Simplexvirus*
Genus	*Varicellovirus*
Subfamily	***Betaherpesvirinae***
Genus	*Cytomegalovirus*
Genus	*Muromegalovirus*
Genus	*Roseolovirus*
Subfamily	***Gammaherpesvirinae***
Genus	*Lymphocryptovirus*
Genus	*Rhadinovirus*

VIRION PROPERTIES

MORPHOLOGY

Virions range from 102 to 200 nm in diameter. They are quasi-spherical and enveloped with surface projections. Between the envelope and the capsid is the viral tegument. It consists of several proteins arranged in an amorphous, sometimes asymmetric, layer. The capsid is 100-110 nm in diameter, icosahedral in structure and contains 162 capsomers of which 150 are hexameric and 12 are pentameric. The viral DNA genome is located in the center. Although the size of the DNA varies in different species, the capsids of herpesviruses are of comparable size.

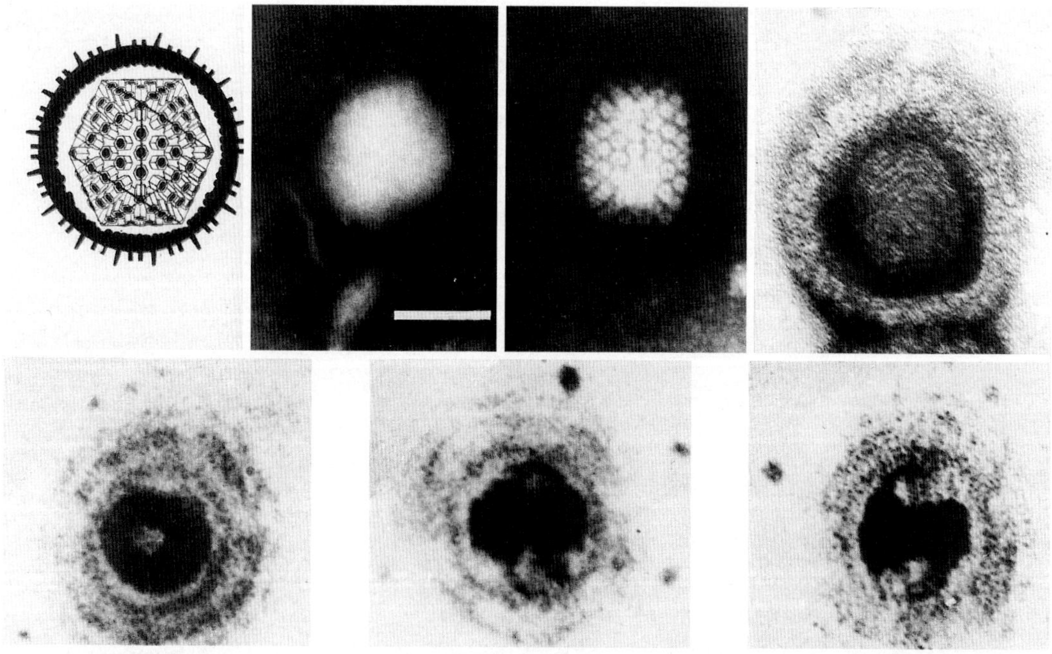

Figure 1: (upper left) Schematic representation of a herpesvirus virion [the outer envelope has projecting spikes; the capsid exhibits icosahedral symmetry; the irregular inner perimeter of the envelope represents the occasional asymmetrical arrangement of the tegument]; (upper center left) an intact, negative contrast electron micrograph of HHV-1 virion; the bar represents 100 nm; (upper center right) negative contrast electron micrograph of HHV-1 capsid, exhibiting icosahedral symmetry; (upper right) HHV-1 core permeated with uranyl acetate [the presence of thread-like structures, 4-5 nm wide are evident]; (bottom) electron micrographs of thin sections of HHV-1 virions showing the core cut at different angles [the preparation was stained with uranyl acetate and counterstained with lead citrate). The core preferentially takes up the stain and appears as a toroid with an outer diameter of 70 nm and lumen of 18 nm diameter. The micrographs show the toroid seen looking: (lower left) through the lumen, (lower center) in cross-section, (lower right) from the side (courtesy of Roizman B, 1990). Cryoelectron microscopy has provided further definition of the virion structure.

PHYSICOCHEMICAL AND PHYSICAL PROPERTIES

The dry weight of HHV-1, virions, full capsids, empty capsids, and cores are about 13.3 x 10^{-16} g, 7.5 x 10^{-16} g, 5.2 x 10^{-16} g, and 2.1 x 10^{-16} g, respectively. Virions contain 19.4 x 10^{-16} g of protein. The average mass ratio of a virion, or full or empty capsid, or core, to DNA is 8, 1, 4.6, and 1.25 to 1, respectively. The buoyant density of virions in CsCl is about 1.20-1.29 g/cm^3. Virions are unstable in detergents or other lipid solvents and less stable at low than at neutral pH values.

NUCLEIC ACID

The genome is composed of linear, double stranded DNA, ranging from 124 to 235 kbp in size, depending on the virus species. Individual genomes may be larger than the normal size of that species (usually by <10 kbp) due to a number of terminal and, or internal, reiterated sequences. The G+C base composition of herpesvirus DNAs range from 32 to 75 %.

Herpesvirus genomes contain a single nucleotide extension at the 3' ends of the genome. Terminally associated proteins have not been detected. Some herpesvirus genomes contain internal repeats of one or both terminal sequences which cause the sequences flanked by the repeats to invert relative to the remainder of the genome and therefore result in the formation of 2 or 4 isomeric forms. The different isomeric forms appear to have no biological consequence.

PROTEINS

The surface of virions contain both glycosylated and non-glycosylated proteins which vary in number depending on the virus species. HSV-1 contains 11 glycosylated and at least two non-glycosylated proteins in the virion envelope. A common feature of the envelope proteins is the presence of an Fc receptor specified by the virus. The precise number of structural proteins is not known. In the case of HHV-1, about half the proteins encoded by the virus are thought to be components of the virion.

LIPIDS

Lipids are located in the viral envelope. The exact composition is not known. They probably reflect the lipid composition of nuclear or other cellular membranes.

CARBOHYDRATES

Glycans associated with the viral envelope proteins are generally of the complex type. High mannose glycans are found on glycoproteins of infectious virions that are retained in cells.

GENOME ORGANIZATION AND REPLICATION

The number of ORFs contained in herpesvirus genomes range from about 70 to more than 200. Among the proteins specified by all herpesviruses are a DNA polymerase, DNA binding proteins and a protease. HSV possesses a helicase-primase. Additional proteins with enzymatic activities known to exist in at least some herpesviruses are thymidine kinase, thymidylate synthase, dUTPase, uracil glycosylase, ribonucleotide reductase, dihydrofolate reductase, alkaline DNase, and as many as three protein kinases. The list of viral proteins includes one or more factors which activate transcription; however, no RNA polymerases have been identified as viral-coded products.

The herpesvirus genomes have been assigned into one of six groups depending on the arrangement of the terminal and internal reiterated sequences (Fig. 2). However, a particular genome structure is not restricted to a single subfamily. In the genomes of viruses comprising group A, e.g., IgHV-1, EHV-2, HHV-6, a large sequence from one terminus is directly repeated at the other terminus. In the group B genomes, e.g., SaHV-2, the terminal sequence is directly repeated numerous times at both termini. Also, the number of

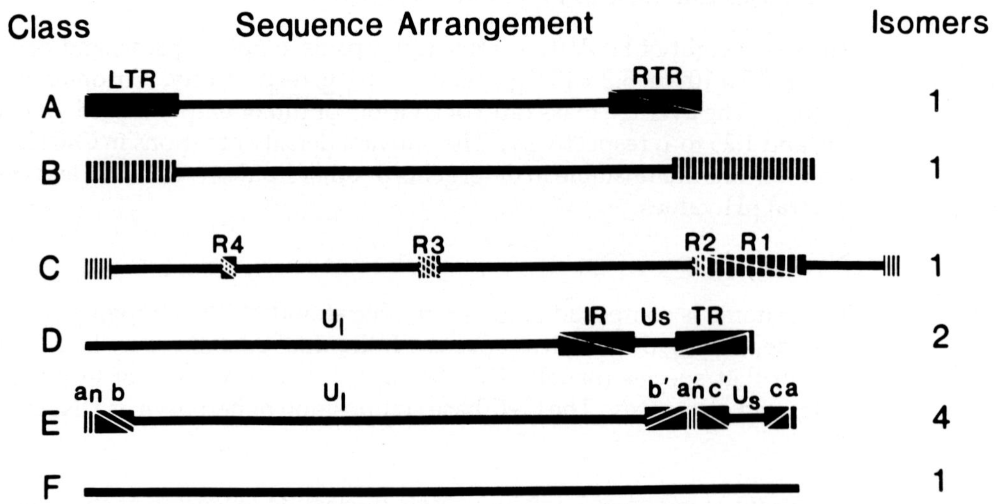

Figure 2: Schematic diagram of the sequence arrangements in the classes of genomes of the viruses comprising the family *Herpesviridae* (A-F, see text). In the diagrams the narrow boxes represent unique, or quasi-unique regions; the reiterated domains are shown as rectangles and are designated as Left and Right Terminal Repeats (LTR and RTR) for Group A, repeats R1 to R4 for internal repeats of Group C, and internal and terminal repeats (IR and TR) of Group D; the termini of Group E, e.g., HHV-1 consist of two elements: one contains n copies of sequence "a" next to a larger sequence designated "b", the other terminus has one directly repeated "a" sequence next to a sequence designated as "c", the terminal ab and ca sequences are inserted in an inverted orientation (denoted by primes) separating the unique sequences into long (Ul) and short (Us) domains; terminal reiterations in the genomes of group F have not been described; in group B, the terminal sequences are reiterated numerous times at both termini and the number of reiterations at each terminus may vary; the components of the genomes in classes D and E invert; in class D, the short component inverts relative to the long; although rarely the long component may also invert, most of the DNA forms two populations differing in the orientation of the short component; in the class E genomes, both the short and long components can invert and viral DNA consists of 4 equimolar isomers (from Roizman B, 1990). The number of isomers for each class is shown at the right.

reiterations at both termini may vary. In the group C genomes, e.g., HHV-4, the number of direct terminal reiterations is smaller, but there may be other, unrelated, sequences greater than 100 bp that are directly repeated and which subdivide the unique (or quasi-unique) sequences of genome into several well delineated stretches. In group D genomes, e.g., HHV-3, PRV, sequences at the termini are repeated in an inverted orientation internally. In these genomes, the domain consisting of the stretch of unique sequences flanked by inverted repeats (i.e., the short, or S component) can invert relative to the remaining sequences (i.e., the long, or L component) such that DNA extracted from virions (or infected cells) consists predominantly of two equimolar populations, differing solely in the relative orientation of the S component relative to the (fixed) orientation of the L component. In group E viral genomes, e.g., HHV-1, HHV-2 , HHV-5, sequences from both termini are repeated in an inverted orientation and juxtaposed internally dividing the genomes into two components (L and S), each of which consists of unique sequences flanked by inverted repeats. In this instance, both components may invert relative to each other and DNA extracted from virions (or infected cells) consists of four equimolar populations differing in the relative orientation of the two components. For the genomes comprising the F group, e.g., MCMV-1, the sequences at the termini have short repeats.

Herpesvirus genomes also differ in gene organization. Whereas in the genomes of HHV-4 and HHV-5 many mRNAs result from splicing of sequences that code for two or more exons, only 6 of about 70 different genes of HHV-1 and HHV-2 yield spliced mRNAs. All herpesviruses attach to one or more type of cellular receptor and enter by a pH-independent fusion of the envelope with the plasma membrane (Fig. 2, stage 1), releasing tegument proteins that for HHV-1 cause shut-off of host protein synthesis (Figure, VHS, stage 2). The HHV-1 α-TIF protein (VP16) is transported to the nucleus. The virus capsid is transported to the nuclear pore. The viral DNA enters the nucleus and is circularized without *de novo* protein synthesis (stage 3). At this point infections may become latent or productive. The decision depends on the type of cell infected by the virus, the combination of cell and viral gene expression (e.g., HHV-4), or cellular gene expression alone (HHV-1). In lytic infections, transcription of early genes by nuclear enzymes is induced (by α-TIF) (stage 4), and mRNAs (α-mRNAs) are transported to the cytoplasm and translated (stage 5). The expressed immediate early (or α) proteins are then transported to the nucleus and are involved in the synthesis of additional mRNAs (β mRNA, stage 6). At this stage of a lytic infection the chromatin (Fig. 2, c) is degraded and displaced toward the nuclear membrane, and the nucleoli (Fig 2, n) become disaggregated (stage 7). The β-proteins are involved in the replication of the viral DNA by the rolling circle mechanism (stage 8) yielding head-to-tail concatemers. β-proteins are also involved in the transcription of the late (γ) mRNAs that are translated (stage 9) mostly into the structural proteins that are required for virion morphogenesis and formation of empty capsids (stage 10) into which unit lengths of viral DNA are packaged (stage 11). The addition of further structural proteins occurs (stage 12). Particle envelopment takes place at nuclear membranes where, on the outer surface, virion surface proteins are located and together with inner tegument proteins particles are assembled (stage 13). The enveloped virions accumulate in the endoplasmic reticulum, the final processing of glycoproteins occurs in the Golgi and virions eventually reach the extracellular space by exocytosis (stage 14).

Among other proteins, common to all herpesviruses are an encoded DNA polymerase, a ssDNA binding protein, proteins which specify a helicase, a primase, and a DNA origin binding protein. The incorporation of one or more specific glycoproteins into the plasma membrane causes the cell to become refractory to superinfection by the same virus. Partially enveloped capsids in the cytoplasm have been variously interpreted as an irreversible de-envelopment as a result of fusion of the envelope with the transport vesicle membrane and as a naturally occurring process of serial envelopment and de-envelopment which culminates in the final envelopment of the capsid at the nuclear membrane. Depending on the virus, infected cells frequently round up and may fuse to form syncytia.

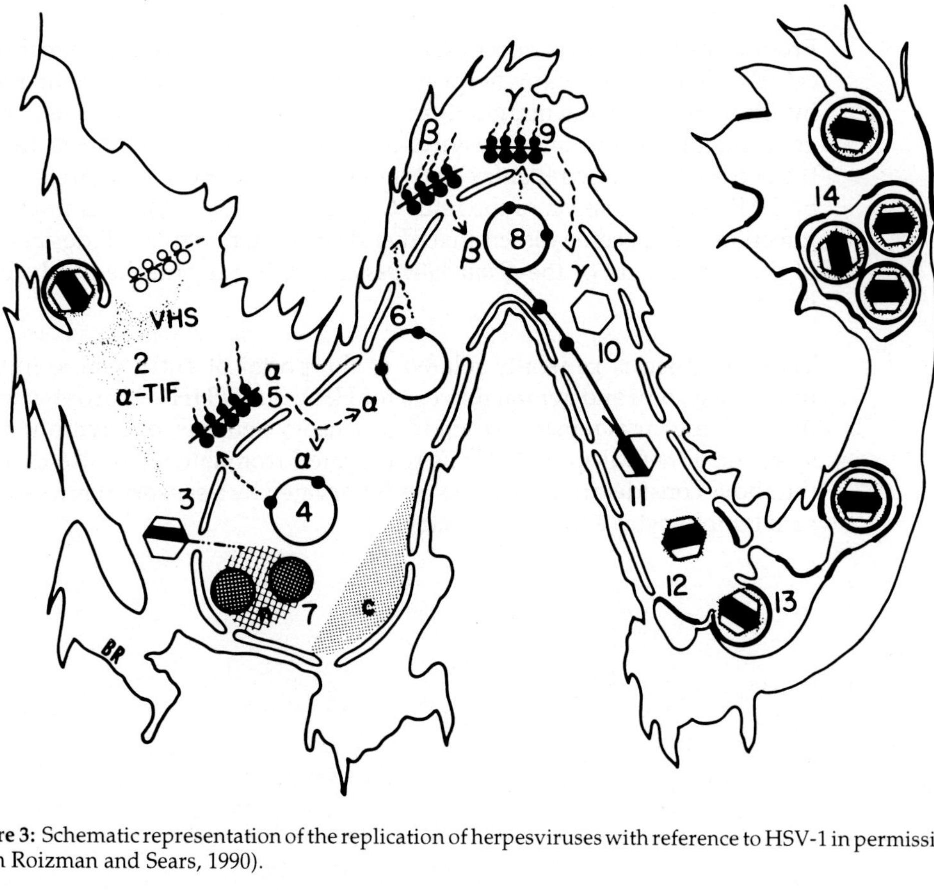

Figure 3: Schematic representation of the replication of herpesviruses with reference to HSV-1 in permissive cells (from Roizman and Sears, 1990).

Antigenic Properties

The antibody response that is protective against infection is usually directed against the virion glycoproteins. The number of virion glycoproteins capable of inducing protective immunity in the form of complement independent neutralizing antibody ranges up to 3 (HHV-1). T cell specific epitopes have been reported. They vary depending on the virus and the host species.

Biological Properties

As a general rule the natural host ranges of herpesviruses are restricted. Transmission from one host species to another can occur, e.g., the simian herpes B virus (CeHV-1) may be transmitted to humans. In experimental animal systems, transmission between host species varies considerably. It is greater for member viruses of the subfamily *Alphaherpesvirinae* (e.g., HHV-1) than for member families of the subfamilies *Betaherpesvirinae* (e.g., HHV-5, or HHV-6), or *Gammaherpesvirinae* (e.g., HHV-4). Natural transmission is usually by infected cells from an infected individual (e.g., HHV-1, HHV-2), or by free virus, (in saliva, urogenital excretions, etc.) (HHV-4, HHV-5, HHV-7), or by aerosol (HHV-3). The geographic distribution of herpesvirus in nature coincides with that of its natural host.

Herpesviruses are highly adapted to their hosts and except for very young or immunologically debilitated hosts, infection is seldom lethal. Herpesviruses normally remain latent in a specific cell type of the host and form a reservoir of virus available either frequently or constantly in excretions, or intermittently in recurrent lesions. For many members of the subfamily *Alphaherpesvirinae,* the site of latency is particular sensory ganglia. The sites of latency for member viruses of the subfamily *Betaherpesvirinae* are not known but macrophages and salivary glands have been implicated. B lymphocytes of the oropharynx maintain members of the *Lymphocryptovirus* genus in a latent state.

At the cellular level, host range varies from very wide (e.g., most *Alphaherpesvirinae*) to very narrow (e.g., lymphocryptoviruses such as HHV-4).

Productive herpesvirus infection results in cell death and this contributes to the pathological manifestation of many herpesvirus infections. A characteristic feature of herpesvirus infection of cells is the margination of the host chromatin. Serious, life-threatening pathogenic manifestations of herpesviruses in immunocompetent hosts are rare and usually are the consequence of viral entry and replication in a specific organ (e.g., encephalitis caused by HHV-1), or invasion of the fetus (e.g., EHV-1, HHV-5). In immunocompromised hosts infection may become disseminated and result in massive cell destruction, and, in the case of some members of the *Gammaherpesvirinae,* in uncontrolled polyclonal proliferation of lymphocytes.

Tissue tropism is generally related to the portal of entry where initial virus replication occurs (e.g., oral and genital mucosa for HHV-1 and HHV-2, oropharynx for HHV-4). Cells in which the virus remains latent (e.g., sensory neurons, or B lymphocytes) are infected via systemic or neural spread. Virus reactivated from latency is also distributed according to the above considerations (i.e., tissues innervated by a sensory neuron harboring latent virus, or B lymphocytes and the oropharynx).

Subfamily *Alphaherpesvirinae*

Taxonomic Structure of the Subfamily

Subfamily	***Alphaherpesvirinae***
Genus	*Simplexvirus*
Genus	*Varicellovirus*

Distinguishing Features

Viruses may exhibit a variable host range, a relatively rapid reproductive cycle, rapid spread in culture, efficient destruction of infected cells and capacity to establish latent infections in sensory ganglia. Common genetic attributes that characterize these viruses are not yet defined. As in other subfamilies, and as a general principle, related viruses are classified as distinct species if (a) their genomes differ in a readily assayed and distinctive manner across the entire genome and not merely at a specific site and (b) if the virus can be shown to have distinct epidemiologic and biologic characteristics. The numbers assigned to the viruses are not of taxonomic significance. They were assigned on the basis of the chronology of virus isolation. They do not refer to a common antigenic type (virus serotype).

Genus *Simplexvirus*

Type Species human herpesvirus 1 (HHV-1)

Distinguishing Features

Viruses assigned to this genus have a common genome structure and exhibit serologic relatedness.

List of Species in the Genus

The viruses, their alternative names ( ), genomic sequence accession numbers [ ] and assigned abbreviations ( ) are:

Species in the Genus

bovine herpesvirus 2		(BoHV-2)
(bovine mamillitis virus)		
(Allerton virus)		
(pseudolumpy skin disease virus)		
human herpesvirus 1	[X14112]	(HHV-1)
(herpes simplex virus 1)		
human herpesvirus 2		(HHV-2)
(herpes simplex virus 2)		
herpes virus B		(HBV)
(cercopithecine herpesvirus 1)		
(herpes simiae virus)		

Tentative Species in the Genus

None reported.

GENUS *VARICELLOVIRUS*

Type Species human herpesvirus 3 (HHV-3)

DISTINGUISHING FEATURES

The type virus has a distinctive genome structure and causes a distinctive disease, acutely varicella, and recrudescently zoster.

LIST OF SPECIES IN THE GENUS

The viruses, their alternative names ( ), genomic sequence accession numbers [ ] and assigned abbreviations ( ) are:

SPECIES IN THE GENUS

human herpesvirus 3 [X04370] (HHV-3)
(varicella-zoster virus 1)

TENTATIVE SPECIES IN THE GENUS

bovine herpesvirus 1 (BoHV-1)
(infectious bovine rhinotracheitis virus)
equid herpesvirus 1 [M86664] (EHV-1)
(equine herpesvirus 1)
(equine abortion herpesvirus)
equid herpesvirus 4 (EHV-4)
(equine herpesvirus 4)
(equine rhinopneumonitis virus)
pseudorabies virus (PRV)
(suid herpesvirus 1)
(Aujeszky's disease virus)

LIST OF UNASSIGNED SPECIES IN THE SUBFAMILY

anatid herpesvirus 1 (AnHV-1)
(duck plague herpesvirus)
ateline herpesvirus 1 (AtHV-1)
(spider monkey herpesvirus)
bovine herpesvirus 5 (BoHV-5)
(bovine encephalitis herpesvirus)
canid herpesvirus 1 (CaHV-1)
(canine herpesvirus)
caprine herpesvirus 1 (CpHV-1)
(goat herpesvirus)
cercopithecine herpesvirus 2 (CeHV-2)
(SA8 virus)
cercopithecine herpesvirus 6 (CeHV-6)
(Liverpool vervet monkey virus)
cercopithecine herpesvirus 7 (CeHV-7)
(patas monkey herpesvirus pH delta)
cercopithecine herpesvirus 9 (CeHV-9)
(Medical Lake macaque herpesvirus)
(simian varicella herpesvirus)
cervid herpesvirus 1 (CvHV-1)
(red deer herpesvirus)
cervid herpesvirus 2 (CvHV-2)
(reindeer herpesvirus)
(Rangifer tarandus herpesvirus)

equid herpesvirus 3 (EHV-3)
  (equine herpesvirus 3)
  (coital exanthema virus)
equid herpesvirus 6 (EHV-6)
  (asinine herpesvirus 1)
equid herpesvirus 8 (EHV-8)
  (asinine herpesvirus 3)
felid herpesvirus 1 (FeHV-1)
  (feline viral rhinotracheitis virus)
  (feline herpesvirus 1)
gallid herpesvirus 1 (GaHV-1)
  (infectious laryngotracheitis virus)
macropodid herpesvirus 1 (MaHV-1)
  (parma wallaby herpesvirus)
macropodid herpesvirus 2 (MaHV-2)
  (docropsis wallaby herpesvirus)
saimiriine herpesvirus 1 (SaHV-1)
  (marmoset herpesvirus)
  (herpesvirus M)
  (herpesvirus platyrrhinae type)
  (herpesvirus T)
  (herpesvirus tamarinus)

Subfamily *Betaherpesvirinae*

Taxonomic Structure of the Subfamily

Subfamily	***Betaherpesvirinae***
Genus	*Cytomegalovirus*
Genus	*Muromegalovirus*
Genus	*Roseolovirus*

Characteristics of the members of this subfamily are a restricted host range, a long reproductive cycle and slow spread of infection from cell to cell in culture. Infected cells frequently become enlarged (cytomegalia) and carrier cultures are readily established. Viruses can be maintained in latent form in lymphoreticular cells and possibly in secretory glands, kidneys and other tissues.

Genus *Cytomegalovirus*

Type Species human herpesvirus 5 (HHV-5)

Distinguishing Features

There is a single virus assigned to this genus with a genome structure that is different to those of other genera.

List of Species in the Genus

The viruses, their alternative names ( ), genomic sequence accession numbers [ ] and assigned abbreviations ( ) are:

Species in the Genus

human herpesvirus 5 [X17403] (HHV-5)
  (human cytomegalovirus)

Tentative Species in the Genus

None reported.

Genus *Muromegalovirus*

Type Species mouse cytomegalovirus 1 (MCMV-1)

Distinguishing Features

There is a single virus assigned to this genus with a genome structure that is different to those of other genera.

List of Species in the Genus

The viruses, their alternative names ( ) and assigned abbreviations ( ) are:

Species in the Genus

mouse cytomegalovirus 1 (MCMV-1)
(murid herpesvirus)

Tentative Species in the Genus

None reported.

Genus *Roseolovirus*

Type Species human herpesvirus 6 (HHV-6)

Distinguishing Features

The viruses assigned to this genus have a distinctive genome structure. They have been isolated from lymphocytes.

List of Species in the Genus

The viruses, and their assigned abbreviation ( ) are:

human herpesvirus 6 (HHV-6)

List of Unassigned Species in the Subfamily

The viruses, their alternative names ( ) and assigned abbreviations ( ) are:

aotine herpesvirus 1 (AoHV-1)
(herpesvirus aotus 1)
aotine herpesvirus 3 (AoHV-3)
(herpesvirus aotus 3)
callitrichine herpesvirus 2 (CaHV-2)
(marmoset cytomegalovirus)
caviid herpesvirus 2 (CaHV-2)
(guinea pig cytomegalovirus)
cebine herpesvirus 1 (CbHV-1)
(capuchin herpesvirus AL-5)
cebine herpesvirus 2 (CbHV-2)
(capuchin herpesvirus AP-18)
cercopithecine herpesvirus 3 (CeHV-3)
(SA6 virus)
cercopithecine herpesvirus 4 (CeHV-4)
(SA 15 virus)

cercopithecine herpesvirus 5 (CeHV-5)
(African green monkey cytomegalovirus)
cercopithecine herpesvirus 8 (CeHV-8)
(rhesus monkey cytomegalovirus)
cricetid herpesvirus (CrHV-1)
(hamster herpesvirus)
equid herpesvirus 2 (EHV-2)
(equine cytomegalovirus)
equid herpesvirus 5 (EHV-5)
(equine herpesvirus 5)
equid herpesvirus 7 (EHV-7)
(asinine herpesvirus 2)
murid herpesvirus 2 (MuHV-2)
(rat cytomegalovirus)
sciurid herpesvirus (ScHV-1)
(European ground squirrel cytomegalovirus)
(American ground squirrel herpesvirus)
suid herpesvirus 2 (SuHV-2)
(swine cytomegalovirus)
(inclusion body rhinitis virus)

SUBFAMILY *GAMMAHERPESVIRINAE*

TAXONOMIC STRUCTURE OF THE SUBFAMILY

Subfamily	***Gammaherpesvirinae***
Genus	*Lymphocryptovirus*
Genus	*Rhadinovirus*

The experimental host range of the members of this subfamily is frequently, but not exclusively, limited to the family or order to which the natural host belongs. *In vitro* all members replicate in lyphoblastoid cells and some also cause lytic infections in certain types of epithelioid and fibroblastic cells. Viruses in this group tend to be specific for either T or B lymphocytes, but exceptions occur. In the lymphocyte, infection often occurs without the production of infectious progeny. Latent virus is frequently demonstrated in lymphoid tissue.

GENUS *LYMPHOCRYPTOVIRUS*

Type Species human herpesvirus 4 (HHV-4)

DISTINGUISHING FEATURES

The viruses have a distinctive genome structure and produce latent infections in B lymphocytes.

LIST OF SPECIES IN THE GENUS

The viruses, their alternative names ( ), genomic sequence accession numbers [ ] and assigned abbreviations ( ) are:

SPECIES IN THE GENUS

cercopithecine herpesvirus 12 (CeHV-12)
(papio Epstein-Barr herpesvirus)
(herpesvirus papio)
(baboon herpesvirus)
cercopithecine herpesvirus 14 (CeHV-14)
(African green monkey HHV-4-like virus)

cercopithecine herpesvirus 15		(CeHV-15)
(rhesus HHV-4-like virus)		
human herpesvirus 4	[V01555]	(HHV-4)
(Epstein-Barr virus)		
pongine herpesvirus 1		(PoHV-1)
(chimpanzee herpesvirus)		
(pan herpesvirus)		
pongine herpesvirus 2		(PoHV-2)
(orangutan herpesvirus)		
pongine herpesvirus 3		(PoHV-3)
(gorilla herpesvirus)		

TENTATIVE SPECIES IN THE GENUS

None reported.

GENUS *RHADINOVIRUS*

Type Species ateline herpesvirus 2 (AtHV-2)

DISTINGUISHING FEATURES

There is a single virus assigned to this genus. It has a distinctive genome structure.

LIST OF SPECIES IN THE GENUS

The viruses, their alternative names ( ) and assigned abbreviations ( ), are:

SPECIES IN THE GENUS

ateline herpesvirus 2	(AtHV-2)
(herpes ateles 2)	

TENTATIVE SPECIES IN THE GENUS

None reported.

LIST OF UNASSIGNED SPECIES IN THE SUBFAMILY

The viruses, their alternative names ( ), and assigned abbreviations ( ) are:

alcelaphine herpesvirus 1	(AlHV-1)
(malignant catarrhal fever virus of European cattle)	
(wildbeest herpesvirus)	
alcelaphine herpesvirus 2	(AlHV-2)
(hartebeest herpesvirus)	
bovine herpesvirus 4	(BoHV-4)
(Movar herpesvirus)	
caviid herpesvirus 1	(CvHV-1)
(guinea pig herpesvirus 1)	
(hsiung Kaplow herpesvirus)	
herpesvirus saimiri 2	(SaHV-2)
(saimiriine herpesvirus 2)	
(squirrel monkey herpesvirus)	(SMHV-2)
leporid herpesvirus 1	(LeHV-1)
(cottontail herpesvirus)	
(herpesvirus sylvilagus)	
marmodid herpesvirus 1	(MaHV-1)
(woodchuck herpesvirus marmota 1)	
meleagrid herpesvirus 1	(MeHV-1)

(turkey herpesvirus 1)
murid herpesvirus 4 (MuHV-4)
(mouse herpesvirus strain 68)
ovine herpesvirus 2 (OvHV-2)
(sheep associated malignant catarrhal fever of cattle virus)

LIST OF UNASSIGNED VIRUSES IN THE FAMILY

The viruses, their alternative names ( ), genomic sequence accession numbers [ ] and assigned abbreviations ( ) are:

acciptrid herpesvirus 1		(AcHV-1)
(bald eagle herpesvirus)		
allitrich herpesvirus 1		(AlHV-1)
aotine herpesvirus 2		(AoHV-2)
ateline herpesvirus 3		(AtHV-3)
(herpesvirus ateles strain 73)		
boid herpesvirus 1		(BaHV-1)
callitrichine herpesvirus 1		(CAHV-1)
(herpesvirus sanguinus)		
caviid herpesvirus 3		(CvHV-3)
(guinea pig herpesvirus 3)		
cercopithecine herpesvirus 10		(CeHV-10)
(rhesus leukocyte associated herpesvirus strain 1)		
cercopithecine herpesvirus 13		(CeHV-13)
(herpesvirus cyclopsis)		
channel catfish herpesvirus	[M75136]	(CCHV)
(ictalurid herpesvirus)		
chelonid herpesvirus 1		(ChHV-1)
(gray patch disease agent of green sea turtle)		
chelonid herpesvirus 2		(ChHV-2)
(Pacific pond turtle herpesvirus)		
chelonid herpesvirus 3		(ChHV-3)
(painted turtle herpesvirus)		
(map turtle herpesvirus)		
chelonid herpesvirus 4		(ChHV-4)
(Geochelone chilensis herpesvirus)		
(Geochelone carbonaria herpesvirus)		
(Argentine turtle herpesvirus)		
ciconiid herpesvirus 1		(CiHV-1)
(black stork herpesvirus)		
columbid herpesvirus 1		(CoHV-1)
(pigeon herpesvirus)		
cyprinid herpesvirus 1		(CyHV-1)
(carp pox herpesvirus)		
elephantid herpesvirus		(EiHV-1)
(elephant loxondontal herpesvirus)		
elapid herpesvirus		(EpHV-1)
(Indian cobra herpesvirus)		
(banded krait herpesvirus)		
(siamese cobra herpesvirus)		
erinaceid herpesvirus 1		(ErHV-1)
(European hedgehog herpesvirus)		
esocid herpesvirus 1		(EsHV-1)
(Northern pike herpesvirus)		
falconid herpesvirus 1		(FaHV-1)
(falcon inclusion body disease)		

gallid herpesvirus 2 (GaHV-2)
(Marek's disease herpesvirus 1)
gallid herpesvirus 3 (GaHV-3)
(Marek's disease herpesvirus 2)
gruid herpesvirus (GrHV-1)
(crane herpesvirus)
human herpesvirus 7 (HHV-7)
iguanid herpesvirus 1 (IgHV-1)
(green iguana herpesvirus)
lorisine herpesvirus 1 (LoHV-1)
(kinkajou herpesvirus)
(herpesvirus pottos)
lacertid herpesvirus (LaHV-1)
(green lizard herpesvirus)
leporid herpesvirus 2 (LeHV-2)
(herpesvirus cuniculi)
(virus III)
murid herpesvirus 3 (MuHV-3)
(mouse thymic herpesvirus)
murid herpesvirus 5 (MuHV-5)
(field mouse herpesvirus)
(Microtus pennsylvanicus herpesvirus)
murid herpesvirus 6 (MuHV-6)
(sand rat nuclear inclusion agents)
murid herpesvirus 7 (MuHV-7)
(murine herpesvirus)
ovine herpesvirus 1 (OvHV-1)
(sheep pulmonary adenomatosis associated herpesvirus)
percid herpesvirus 1 (PeHV-1)
(walleye epidermal hyperplasia)
perdicid herpesvirus 1 (PdHV-1)
(bobwhite quail herpesvirus)
phalacrocoracid herpesvirus 1 (PhHV-1)
(cormorant herpesvirus)
(Lake Victoria cormorant herpesvirus)
phocid herpesvirus 1 (PoHV-1)
(harbor seal herpesvirus)
pleuronectid herpesvirus (PiHV-1)
(herpesvirus scophthalmus)
(turbot herpesvirus)
psittacid herpesvirus 1 (PsHV-1)
(parrot herpesvirus)
(Pacheco's disease virus)
ranid herpesvirus 1 (RaHV-1)
(Lucke frog herpesvirus)
ranid herpesvirus 2 (RaHV-2)
(frog herpesvirus 4)
salmonid herpesvirus 1 (SaHV-1)
(herpesvirus salmonis)
salmonid herpesvirus 2 (SaHV-2)
(Onchorhynchus masou herpesvirus)
sciurid herpesvirus 2 (ScHV-2)
sphenicid herpesvirus 1 (SpHV-1)
(black footed penguin herpesvirus)
strigid herpesvirus 1 (StHV-1)
(owl hepatosplenitis herpesvirus)
tupaiid herpesvirus 1 (TuHV-1)
(tree shrew herpesvirus)

Similarity with Other Taxa

None reported.

Derivation of Names

herpes: from Greek *herpes*, "creeping"
alpha: Greek letter α, "a"
beta: Greek letter β, "b"
gamma: Greek letter γ, "g"
simplex: from Latin *simplex*, "simple"
varicello: derived from Latin *varius*, "spotted", and its diminuitive variola, "smallpox"
cytomegalo: from Greek *kytos*, "cell" and *megas*, "large"
muromegalo: from Latin *mus*, "mouse" and Greek megas, "great"
roseolo: from Latin *rose* "rose, rosy"
lymphocrypto: from Latin *lympha*, "water" and Greek kryptos, "concealed"
rhadino: from Greek adjective *rhadinos*, "slender, taper"

References

Albrecht JC, Nicholas J, Biller D, Cameron KR, Biesinger B, Newman C, Wittmann S, Craxton MA, Coleman H, Fleckenstein B, Honess RW (1992) Primary structure of the herpesvirus saimiri genome. J Virol 66: 5047-5058

Baer R, Bankier AT, Biggin MD, Deininger PL, Farrell PJ, Gibson TJ, Hatfull G, Hudson GS, Satchwell SC, Seguin C, Tuffnell PS, Barrell BG (1984) DNA sequence and expression of the B95-8 Epstein-Barr virus genome. Nature 310: 207-211

Booy FP, Newcomb WW, Trus BL, Brown JC, Baker TS, Steven AC (1991) Liquid-crystalline phage-like packing of encapsidated DNA in hepres simplex virus. Cell 64: 1007-1015

Bornkamm GW, Delius H, Fleckenstein B, Werner F-J, Mulder C (1976) Structure of herpesvirus saimiri genomes: arrangment of heavy and light sequences in the M genome. J Virol 19: 154-161

Chee MS, Bankier AT, Beck S, Bohni R, Brown CM, Cerny R, Horsnell T, Hutchinson III CA, Kouzarides T, Martignetti JA, Preddie E, Satchwell SC, Tomlinson P, Weston KM, Barrell BG (1990) Analysis of the protein-coding content of the sequence of human cytomegalovirus strain AD169. Curr Top Micro Immunol 154: 125-169

Davison AJ (1992) Channel catfish virus: a new type of herpesvirus. Virology 186: 9-14

Davison AJ, Scott JE (1986) The complete DNA sequence of Varicella-zoster virus. J Gen Virol 67: 1759-1816

Fleckenstein B, Bornkamm GW, Mulder C, Werner F-J, Daniel MD, Falk LA, Delius H (1978) Herpesvirus ateles DNA and its homology with herpesvirus saimiri nucleic acid. J Virol 25: 361-373

McGeoch DJ, Dalrymple MA, Davison AJ, Dolan A, Frame MC, McNab D, Perry LJ, Scott JE, Taylor P (1988) The complete DNA sequence of the long unique region in the genome of herpes simplex virus type 1. J Gen Virol 69: 1531-1574

McGeoch DJ, Dolan A, Donald S, Brauer DHK (1986) Complete DNA sequence of the short repeat region in the genome of herpes simplex virus type 1. Nucl Acids Res 14: 1727-1745

McGeoch DJ, Dolan A, Donald S, Rixon FJ (1985) Sequence determination and genetic content of the short unique region in the genome of herpes simplex virus type 1. J Mol Biol 181: 1-13

Mosmann TR, Hudson JB (1973) Some properties of the genome of murine cytomegalovirus (MCV). Virology 54: 135-149

Roizman B (1990) The family *Herpesviridae*: a brief introduction. In: Fields BN, Knipe DM (eds) Virology 2nd edn. Raven Press, New York, pp 1787-1794

Roizman B. (1993) The Family *Herpesviridae*. In: Roizman B, Lopez C, Whitley R.J (eds) The human herpesvirus. Raven Press New York, N.Y pp 1-10

Roizman B, Sears AE (1993) The replication of Herpes simplex viruses. In: Roizman B, Lopez C, and Whitley RJ (eds) The human herpesviruses. Raven Press, New York pp 11-68

Roizman B, Carmichael LE, Deinhardt F, de The G, Plowright W, Rapp F, Sheldrick P, Takahashi M, Wolf K (1981) *Herpesviridae*: definition, provisional nomenclature, and taxonomy. Intervirology 16: 201-21

Telford EAR, Watson MS, McBride K, Davison AJ (1992) The DNA sequence of equine herpesvirus-1. Virology 189: 304-316

Contributed By

Roizman B, Desrosiers RC, Fleckenstein B, Lopez C, Minson AC, Studdert MJ

FAMILY ADENOVIRIDAE

TAXONOMIC STRUCTURE OF THE FAMILY

Family	***Adenoviridae***
Genus	*Mastadenovirus*
Genus	*Aviadenovirus*

VIRION PROPERTIES

MORPHOLOGY

Virions are non-enveloped, 80-110 nm in diameter and exhibit icosahedral symmetry. Virions have 240 non-vertex capsomers (hexons), 8-10 nm in diameter, and 12 vertex capsomers (pentons) with fibers that protrude 9-30 nm from the virion surface (Fig. 1).

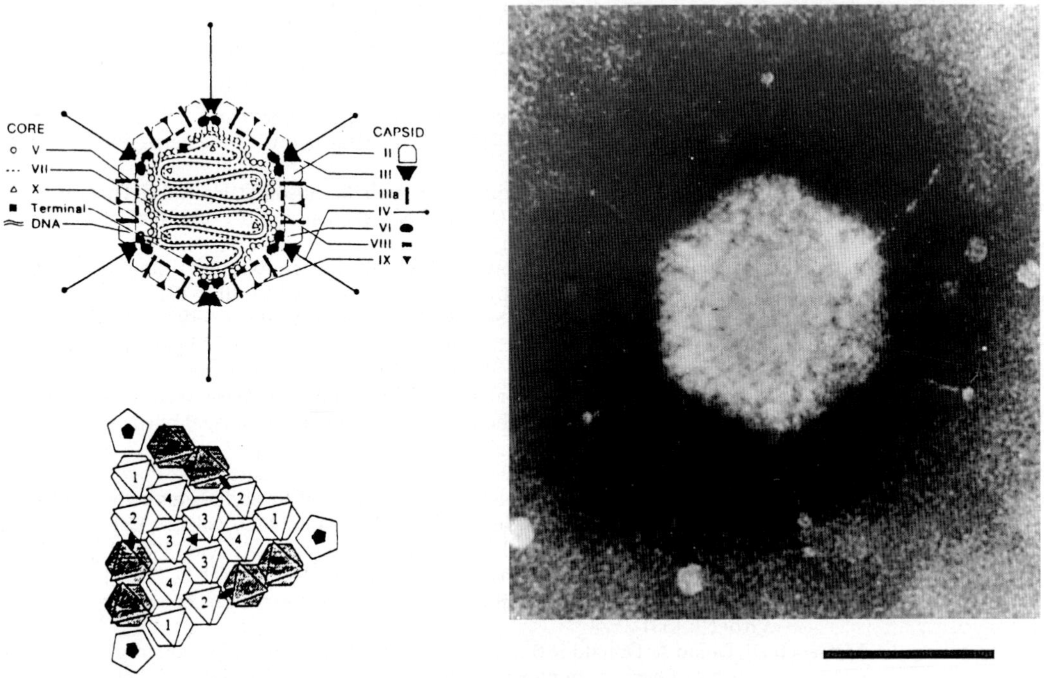

Figure 1: (upper left) Stylized section of the adenovirus particle. The 240 hexons are formed by the interaction of three identical polypeptides (designated II) and consist of two distinct parts - a triangular top with three "towers", and a pseudohexagonal base with a central cavity. The hexon bases are tightly packed together and form a protein shell that protects the inner components. The positions of hexons (II), penton bases (III), fibers (IV) and protein IX are well established. Twelve copies of polypeptides IX are found between 9 hexons in the center of each facet. The positions of proteins IIIa, VI and VIII are tentatively assigned. Two monomers of IIIa penetrate the hexon capsid at the edge of each facet. Multiple copies of VI form a ring underneath the peripentonal hexons. The 12 penton bases are each formed by the interaction of five polypeptides (III) and are tightly associated with one or two fibres each consisting of three polypeptides (IV) that interact to form a shaft of characteristic length with a distal knob. The 12 pentons (III and IV) are less tightly associated with the neighboring (peripentonal) hexons. Polypeptide VIII has been assigned to the inner surface of the hexon capsid. Other polypeptides (monomers of IIIa, trimers of IX, and multimers of VI) are in contact with hexons forming a continuous protein shell. Polypeptides VI and VIII appear to link the capsid to the virus core. The core consists of the DNA genome complexed with four polypeptides (V, VII, X, terminal). As the structure of the nucleoprotein core has not been established, the polypeptides associated with the DNA are shown in hypothetical locations. Two other structural proteins (IVa2 and protease) are not depicted because their location is unknown; (lower left) schematic diagram of the 12 hexons in one of the 20 facets (top view), each represented as a triangular "tower" superimposed on a pseudohexagonal base. There are four variants of hexon in the capsid, each with different environments (1-4). The two edge hexons from the three adjacent facets are shaded. Three vertex pentons are indicated in the diagram (upper left and lower left provided by Stewart PL and Burnett RM); (right) negative contrast electron micrograph of human adenovirus particle. The bar represents 100 nm.

PHYSICOCHEMICAL AND PHYSICAL PROPERTIES

Virion Mr is 150-180 x 10^6; buoyant density in CsCl is 1.32-1.35 g/cm^3. Viruses are stable on storage in the frozen state. They are stable to mild acid and insensitive to lipid solvents. Virus infectivity is inactivated after heating at 56° C for more than 10 min.

NUCLEIC ACID

Virions contain a single linear molecule of dsDNA of Mr about 20-25 x 10^6 for mastadenoviruses, or Mr about 30 x 10^6 for aviadenoviruses. A virus-coded terminal protein is covalently linked to the 5'-end of each DNA strand. The genome of human adenovirus 2 (HAdV-2) comprises 35,937 bp and contains an inverted terminal repetition (ITR) of 103 bp. ITR's of 50-200 bp have been found in all viruses so far analyzed. The DNA G+C content varies from 48-61% for mastadenoviruses and 54-55% for aviadenoviruses.

PROTEINS

About 40 different polypeptidesa are derived from the genome-mostly via complex splicing mechanisms (Fig. 2). Almost a third of these provide structural proteins as in Fig. 1. In general terms, the early gene products facilitate extensive modulation of the host cell's transcriptional machinery (E1 and E4), assemble the virus DNA replication complex (E2) and provide means for subverting host defence mechanisms (E3). Intermediate and late gene products (L1 - L5) are concerned with the assembly and maturation of the virion.

LIPIDS

None reported.

CARBOHYDRATES

Fiber proteins and some of the non-structural proteins are glycosylated.

GENOME ORGANIZATION AND REPLICATION

Virus entry is by attachment via the fiber, followed by endocytosis, uncoating and delivery of the virus core to the nucleus which is the site of mRNA transcription, virus DNA replication and assembly. Virus infection mediates the early shut-down of host DNA synthesis and, later, host RNA and protein synthesis. Transcription by the host RNA polymerase II involves both DNA strands and initiates from four early (E1-E4), two

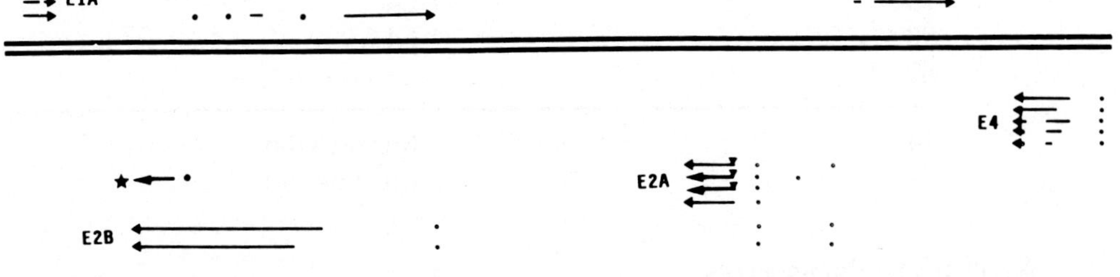

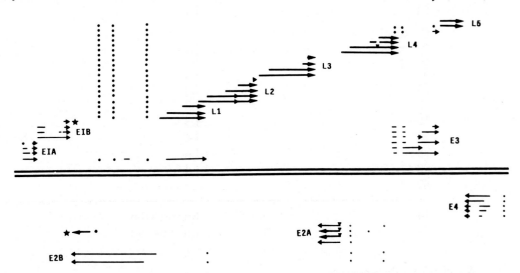

Figure 2: Schematic of the transcription pattern of human Ad2 virus. The parallel lines indicate the linear duplex genome of 36 kbp. The dots, broken lines and split arrows indicate the spliced structures of the mRNAs. EIA, E3, etc., refer to early transcription units. Most (but not all) late genes are in the major late transcription unit which initiates at map position 16 of the indicated top strand, and which includes the L1, L2, L3, L4 and L5 families of mRNAs. Other (intermediate) genes include those starred (adapted from Wold and Gooding, 1991).

intermediate, and one major late (L) promoter in a pattern as shown in Fig. 2. All primary transcripts are capped and polyadenylated. There are complex splicing patterns to produce families of mRNAs. There are also one or two VA RNA genes which are transcribed by cellular RNA polymerase III and these encode RNA products which facilitate translation of late mRNAs.

There are many non-structural proteins in addition to the structural proteins (Table). A number of polypeptides are modified by phosphorylation, some by glycosylation. Proteolysis of some structural polypeptides by the virus-coded protease is an essential prerequisite for virion maturation (Table). DNA replication is by strand-displacement using a protein priming mechanism (terminal protein) together with a virus-coded DNA polymerase and DNA binding protein in concert with cellular factors. Virions are assembled in the nucleus sometimes in paracrystalline arrays along with similar arrays of virus structural proteins. Release is achieved following disintegration of the host cell.

Table: Deduced proteins encoded by human adenovirus serotype 2 (HAdV-2). Mr, rounded to nearest k, are presented as unmodified and uncleaved gene products. NS = non-structural; S = structural; p-protein = phosphoprotein; DBP = DNA binding protein; DNA pol = DNA polymerase; Term = terminal protein; * = Mr are significantly different from those obtained by SDS-PAGE; † = cleaved by viral protease; other ORFs are not yet identified.

Mr (x 10^3)	Transcription class	Description
13, 27, 32	EIA	NS
16, 21, 55	EIB	NS
59	E2A	NS; 72kDa* DBP
120	E2B	NS; 140kDa* DNA pol.
75	E2B	S; Term†, 80kDa* pTP
4, 7, 8, 10, 12 13, 15, 15, 19	E3	NS
7, 13, 13, 14, 15, 17	E4	NS
47	L1	NS; maturation 52/55kDa*
64	L1	S (IIIa); p-protein
10	L2	S (X)†; and μ
22	L2	S (pVII); major core†
42	L2	S (V); minor core
63	L2	S (III); penton*
23	L3	S; protease
27	L3	S (pVI)†;
109	L3	S (II); hexon
25	L4	NS; 33kDa* p-protein
25	L4	S (pVIII)†;
90	L4	NS; 100kDa*
62	L5	S (IV); fiber
14	Intermediate	S (IX);
51	Intermediate	S (IVa2);

Antigenic Properties

Adenovirus serotypes are defined on the basis of neutralization assays. A serotype is defined as one which either exhibits no cross-reaction with others, or shows an homologous : heterologous titer ratio greater than 16 (in both directions). For homologous : heterologous titer ratios of 8 or 16, a serotype assignment is made if either the viral hemagglutinins are unrelated (as shown by lack of cross-reaction in hemagglutination-inhibition tests), or if substantial biophysical or biochemical differences exist. Antigens at the surface of the virion are mainly type-specific. Hexons are involved in neutralization, fibers in neutralization and hemagglutination-inhibition. Soluble antigens associated with virus infections include surplus capsid proteins which have not been assembled. As defined with monoclonal antibodies, hexons and other soluble antigens carry numerous epitopes, some that are genus-specific, others that are type-specific and others that group viruses within the genus. Free hexon protein mainly reacts as a genus-specific antigen (*Mastadenovirus* or *Aviadenovirus*). The hexon genus-specific antigen is located on the basal surface of the hexon, whereas hexon serotype-specific antigens are located mainly on the 'tower' region of the hexon.

BIOLOGICAL PROPERTIES

The natural host range of adenoviruses is mostly confined to one species, or to closely related species. This also applies for cell cultures. Some human adenoviruses cause productive infection in rodent cells but with low efficiency. Several viruses cause tumors in newborn hosts of heterologous species. Subclinical infections are frequent in various virus-host systems. Direct or indirect transmission occurs from throat, feces, eye, or urine, depending on the virus serotype. Human adenovirus infections are mostly asymptomatic but can be associated with diseases of the respiratory, ocular and gastrointestinal systems. Human adenovirus types 1, 2, 3, 5, 6 and 7 cause respiratory infections in children. Enteric infection, as indicated by fecal shedding, is predominant in all serotypes. Human serotypes 40 and 41 can be isolated in high yield from feces of young children with acute gastroenteritis and are second only to rotaviruses as a major cause of infantile viral diarrhea. Human adenovirus type 11 is associated with hemorrhagic cystitis. Canine adenoviruses are responsible for hepatitis as well as respiratory disease. Canine adenoviruses have caused epizotics in foxes, bears, wolves, cyotes and skunks. Avian adenoviruses have been associated with diverse disease patterns eg. hemorrhagic enteritis, 'marble spleen' disease, pulmonary congestion and edema. Adenoviruses infecting susceptible cells cause similar gross pathology e.g., early rounding of cells and aggregation of chromatin followed by the later appearance of characteristic basophilic nuclear inclusions.

GENUS *MASTADENOVIRUS*

Type Species human adenovirus 2 (HAdV-2)

DISTINGUISHING FEATURES

The adenoviruses that infect mammals are serologically distinct from those that infect birds.

TAXONOMIC STRUCTURE OF THE GENUS

There are 10 groups of adenoviruses that infect mammals. The serotypes assigned to the groups are given numbers.

LIST OF SPECIES IN THE GENUS

The viruses, their genomic sequence accession numbers [ ] and assigned abbreviations ( ) are:

SPECIES IN THE GENUS

bovine adenoviruses 1 to 9	[K01264]	(BAdV-1 to 9)
canine adenovirus 1	[J04368]	(CAdV-1)
canine adenovirus 2		(CAdV-2)
caprine adenovirus 1		(GAdV-1)
equine adenovirus 1	[M14895]	(EAdV-1)

human adenoviruses 1 to 47	[J01903, J01915, J01917, J01993, M14785, M14918, M15952, M1954, M62712, M73260, M86665, X03000]	(HAdV-1 to 47)
murine adenovirus 1	[M22245]	(MAdV-1)
murine adenovirus 2		(MAdV-2)
ovine adenoviruses 1 to 6		(OAdV-1 to 6)
porcine adenoviruses 1 to 6		(PAdV-1 to 6)
simian adenoviruses 1 to 27	[X01027]	(SAdV-1 to 27)
tree shrew adenovirus 1	[M10054]	(TSAdV-1)

Tentative Species in the Genus

None reported.

Genus *Aviadenovirus*

Type Species fowl adenovirus 1 (FAdV-1)

Distinguishing Features

The adenoviruses that infect birds are serologically distinct from those that infect mammals.

Taxonomic Structure of the Genus

There are 5 groups of adenoviruses that infect birds. The serotypes assigned to the groups are given numbers.

List of Species in the Genus

The viruses, their genomic sequence accession numbers [ ] and assigned abbreviations are:

Species in the Genus

duck adenovirus 1		(DAdV-1)
duck adenovirus 2		(DAdV-2)
fowl adenoviruses 1 to 12	[M12738, X17217]	(FAdV-1 to 12)
goose adenoviruses 1 to 3		(GoAdV-1 to 3)
pheasant adenovirus 1		(PhAdV-1)
turkey adenoviruses 1 to 3		(TAdV-1 to 3)

Tentative Species in the Genus

None reported.

List of Unassigned Viruses in the Family

None reported.

Similarity with Other Taxa

None reported.

Derivation of Names

adeno: from Greek *aden, adenos,* "gland"; in recognition of the fact that adenoviruses were first isolated from human adenoid tissue
avi: from Latin *avis*, "bird"
mast: from Greek *mastos*, "breast"

REFERENCES

Chroboczek J, Bieber F, Jacrot B (1992) The sequence of the genome of adenovirus type 5 and its comparison with the genome of adenovirus type 2. Virology 186: 280-285

Furcinitti PS, van Oostrum J, Burnett RM (1989) Adenovirus polypeptide IX revealed as capsid cement by difference images from electron microscopy and crystallography. EMBO J 8: 3563-570

Ginsberg HS (ed) (1984) The Adenoviruses. Plenum Press, New York

Hasson TB, Soloway PD, Ornelles DA, Doerfler W, Shenk T (1989) Adenovirus L1 52- and 55-kilodalton proteins are required for assembly of virions. J Virol 63: 3612-3621

Hierholzer JC, Wigand R, Anderson LJ, Adrian T, Gold JWM (1988) Adenoviruses from patients with AIDS; a plethora of serotypes and a description of five new serotypes of subgenus D (types 43-47). J Infect Dis 158: 804-813

Horwitz MS (1990) Adenoviruses and their replication. In: Fields BN, Knipe DM (eds) Virology 2nd edn. Raven Press, New York, pp 1679-1722

Mautner V (1989) *Adenoviridae.* In: Porterfield JS (ed) Andrewes Viruses of Vertebrates, 5th edn. Balliere Tindall, London, pp 249-282

Roberts RJ, Akusjarvi G, Alestrom P, Gelinas RE, Gingeras TR, Sciaky D, Pettersson U (1986) A consensus sequence for the adenovirus 2 genome. In: Doerfler W (ed) Adenovirus DNA. The viral genome and its expression. Martinus Nijhoff Publishing, Boston, pp 1-51

Russell WC, Bartha A, de Jong JC, Fujinaga K, Ginsberg HS, Hierholzer JC, Li QG, Mautner V, Nasz I, Wadell G (1991) *Adenoviridae.* In: Francki RIB, Fauquet CM, Knudson DL, Brown F (eds) Classification and Nomenclature of Viruses, Fifth Report ICTV Arch Virol Suppl 2. Springer-Verlag, Wein NewYork, pp 140-144

Stewart PL, Burnett RM, Cyrklaff M, Fuller SD (1991) Image reconstruction reveals the complex molecular organization of adenovirus. Cell 67: 145-154

Wold WS, Gooding LR (1991) Region E3 of adenovirus: a casette of genes involved in host immuno surveilllance and virus-cell interactions. Virology 184: 1-8

CONTRIBUTED BY

Russell WC, Adrian T, Bartha A, Fujinaga K, Ginsberg HS, Hierholzer JC, de Jong JC, Li QG, Mautner V, Nasz I, Wadell G

Genus *Rhizidiovirus*

Type Species Rhizidiomyces virus (RZV)

Virion Properties

Morphology

Virions are isometric, 60 nm in diameter.

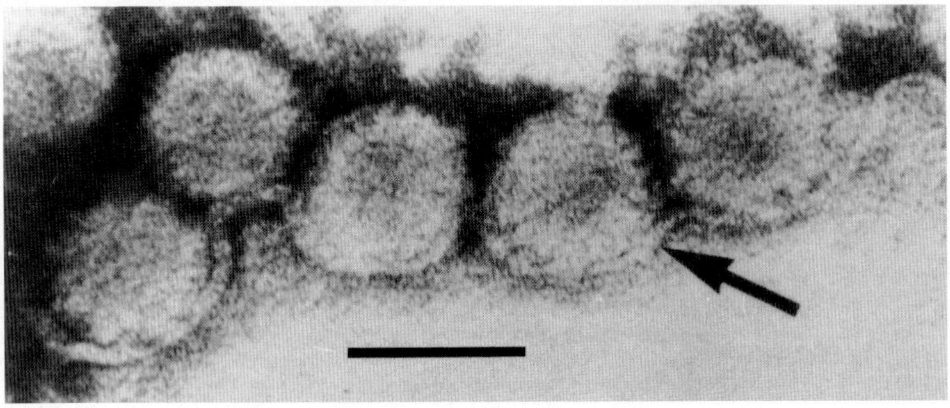

Figure 1: Negative contrast electron micrograph of RZV particles which have been physically separated from the fungus are observed attached on a membrane-like structure (arrow) (from Dawe and Kuhn, 1983 Virology 130: 10-20). The bar represents 50 nm.

Physicochemical and Physical Properties

The buoyant density of virions in CsCl is 1.31 g/cm^3; S_{20w} is 625. Virions contain 10% nucleic acid.

Nucleic Acid

Virions contain a single molecule of dsDNA with an Mr of 16.8 x 10^6 and a G+C ratio of 42%.

Proteins

Virions contain at least 14 polypeptides with Mr in the range of 26-84.5 x 10^3.

Lipids

None reported.

Carbohydrates

None reported.

Genome Organization and Replication

Particles appear first in the nucleus.

Biological Properties

The virus appears to be transmitted in a latent form in the zoospores of the fungus. Activation of the virus, which occurs under stress conditions such as heat, poor nutrition, or aging, results in cell lysis.

List of Species in the Genus

The viruses, their host { } and assigned abbreviation ( ) are:

Species in the Genus

Rhizidiomyces virus {from *Rhizidiomyces* sp isolate F} (RZV)

Tentative Species in the Genus

None reported.

Similarity with Other Taxa

None reported.

Derivation of Names

Rhizidio: from name of the host *Rhizidiomyces* sp

References

Dawe VH, Kuhn CW (1983) Virus-like particles in the aquatic fungus, *Rhizidiomyces*. Virology 130: 10-20

Dawe VH, Kuhn CW (1983) Isolation and characterization of a double-stranded DNA mycovirus infecting the aquatic fungus, *Rhizidiomyces*. Virology 130: 21-28

Contributed By

Ghabrial SA, Buck KW

FAMILY PAPOVAVIRIDAE

TAXONOMIC STRUCTURE OF THE FAMILY

Family	***Papovaviridae***
Genus	*Polyomavirus*
Genus	*Papillomavirus*

VIRION PROPERTIES

MORPHOLOGY

Virions are non-enveloped, 40 nm (*Polyomavirus*) and 55 nm (*Papillomavirus*) in diameter. The icosahedral capsid is composed of 72 capsomers in skewed (T= 7) arrangement. Filamentous and tubular forms are observed as a result of aberrant maturation.

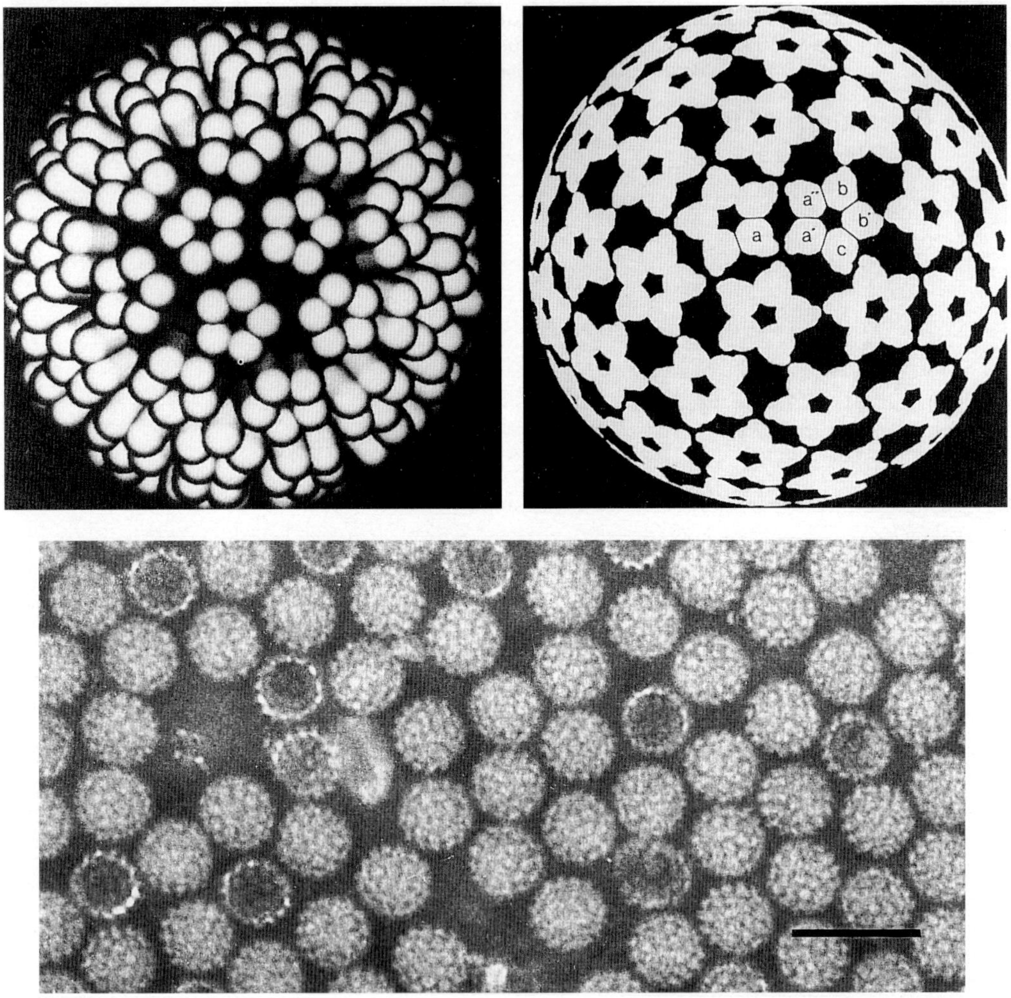

Figure 1: Computer graphics representation of: (upper left) the surface of the mouse polyomavirus capsid (the icosahedral structure includes 360 VP1 subunits arranged in 12 pentavalent and 60 hexavalent capsomers); (upper right) capsomer bonding relations (there are six VP1 molecules in each icosahedral asymmetric unit, which include one subunit of a pentavalent pentamer. The six symmetrically different subunits are designated a, a', a'', b, b' and c, corresponding to three different bonding states) (from Eckhart, 1991; adapted from Salunke et al., 1986; with permission). (lower) Negative contrast electron micrograph of HPV-1 virions. The bar represents 100 nm.

PHYSICOCHEMICAL AND PHYSICAL PROPERTIES

Virion Mr is 25×10^6 (*Polyomavirus*) and 47×10^6 (*Papillomavirus*). Buoyant density of virions in sucrose and CsCl gradients is 1.20 and 1.34-1.35 g/cm^3, respectively. Virion S_{20w} is 240

(*Polyomavirus*) and 300 (*Papillomavirus*). Virions are resistant to ether, acid and heat treatment (50° C, 1 hr.). Virions are unstable at 50° C for 1 hr in the presence of 1 M $MgCl_2$.

NUCLEIC ACID

Virions contain a single molecule of circular dsDNA. The genomic size is fairly uniform within each genus; for members of the genus *Polyomavirus* it is about 5 kbp (e.g., SV-40 [strain 776] has 5,243 bp, JCV[Mad1] has 5,130 bp, BKV[Dun] has 5,153 bp, murine polyomavirus [A2] has 5,297 bp, BPyV has 4,697 bp); for members of the genus *Papillomavirus* it is about 8 kbp (e.g., BPV-1 has 7,946 bp, DPV has 8,374 bp, CRPV has 7,868 bp, HPV-1a has 7,815 bp, HPV-16 has 7,905 bp). The Mr of the genome is 3-5 x 10^6 and the DNA constitutes about 10-13% of the virion by weight. The G+C content is 40-50%. A 5' terminal cap or 5' terminal covalently-linked polypeptide is absent from the genome. In the mature virion the viral DNA is associated with host cell histone proteins H2a, H2b, H3 and H4 in a chromatin-like complex.

PROTEINS

The virus genomes encode at least 5-10 proteins with Mr ranging from 3-88 x 10^3 (Table 1). Three structural proteins, VP1, VP2 and VP3 make up the polyomavirus capsid; of these, VP1 is the major component. A fourth protein, agnoprotein, or LP1, may be produced and may facilitate the assembly of the polyomavirus capsid. It is not a structural component of the mature virion.

Table 1: Deduced polyomavirus proteins (kDa), (N: none), ELP: Early Leader Protein predicted from the DNA sequence in the case of JCV and BKV.

Virus:	PyV	SV-40	JCV	BKV	KV	LPV	BPyV
Structural proteins:							
VP1	42.4	39.9	39.6	40.1	41.7	40.2	40.5
VP2	34.8	38.5	37.4	38.3	37.4	39.3	39.1
VP3	22.9	27.0	25.7	26.7	25.2	27.3	26.9
Non-structural proteins:							
T	88.0	81.6	79.3	80.5	72.3	79.9	66.9
mT	48.6	N	N	N	N	N	N
t	22.8	20.4	20.2	20.5	18.8	22.2	14.0
ELP	N	2.7	4.3	4.3	N	N	N
LP1	N	7.3	8.1	7.4	N	N	13.1

The capsids of the papillomaviruses are composed of structural proteins encoded by the L1 and L2 ORFs, (Table 2).

Table 2: Deduced papillomavirus proteins (kDa).

Virus:	CRPV	BPV-1	HP-1
Structural proteins:			
L1	57.9	55.5	59.6
L2	52.8	50.1	50.7
Non-structural proteins:			
E1	67.9	68.0	73.0
E2	44.0	48.0	41.8
E4	25.8	12.0	10.4
E5	11.3	7.0	9.4
E6	29.7	15.1	19.2
E7	10.5	14.0	11.0

Genetic evidence has not been presented that associates specific viral proteins with the E3 and E8 ORFs.

LIPIDS

None present.

CARBOHYDRATES

None present.

GENOME ORGANIZATION AND REPLICATION

Virions that attach to cellular receptors are engulfed by the cell and are transported to the nucleus. During a productive infection, transcription of the viral genome is divided into an early and late stage. Transcription of the early and late coding regions is controlled by separate promoters, and occurs on opposite DNA strands in the case of the polyomaviruses and on the same strand for the papillomaviruses (Fig. 2).

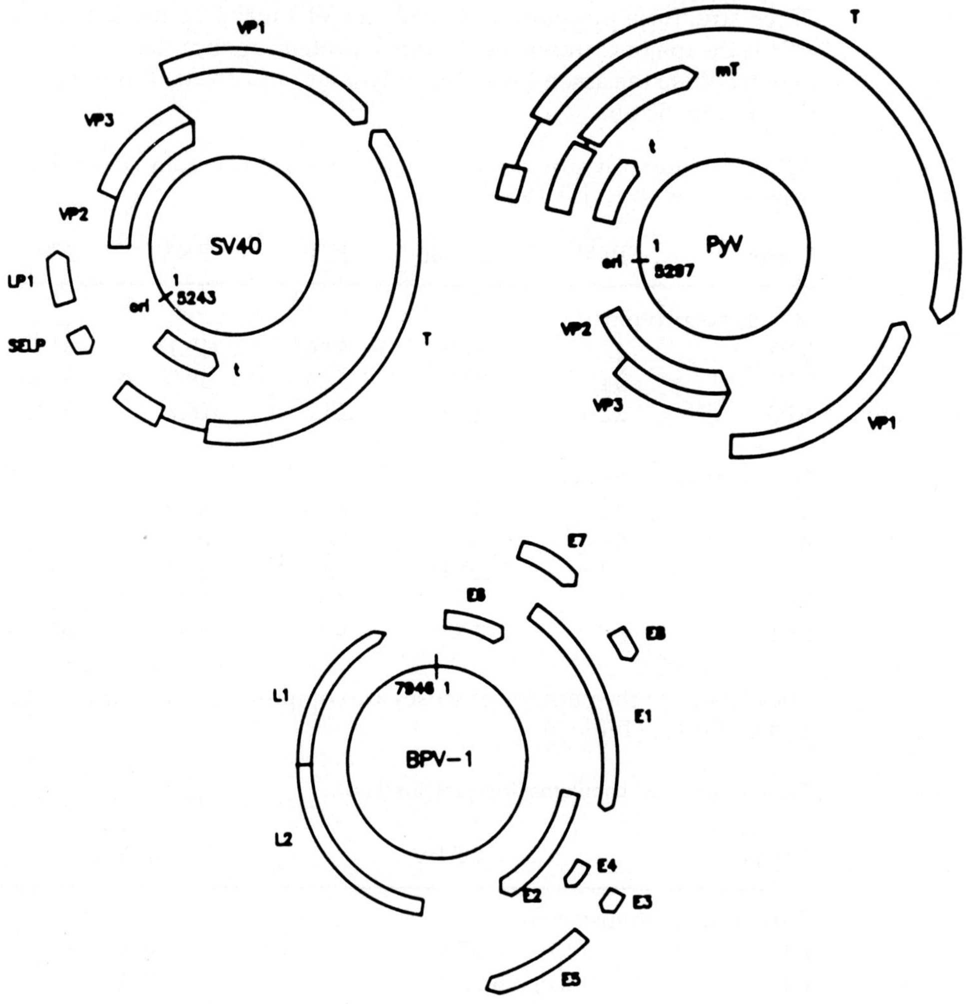

Figure 2: Diagram of (upper left) the SV-40, (upper right) PyV and (lower) BPV-1 genomes and encoded proteins. Inner circles represent the viral dsDNAs (sizes in bp, origin of replication: ori), the outer arrows indicate the encoded viral proteins, or ORFs, as well as the direction of transcription. Introns are denoted by a single line.

Precursor mRNAs undergo post-transcriptional processing that includes capping and polyadenylation of the 5' and 3' termini, respectively, as well as splicing. Efficient use of coding information involves differential splicing of the messages and use of overlapping ORFs. Early mRNAs encode regulatory proteins that may exhibit trans-activating properties. These include proteins that are required for viral DNA replication. Their expression leads to de-repression of some host cell enzymes and stimulation of cell DNA synthesis. Prior to the start of the late events, viral DNA replication is initiated in the nucleus. Translation of most of the late transcripts produces structural proteins that are involved in capsid assembly. Post-translational modifications of some early and late viral proteins include phosphorylation, N-acetylation, fatty acid acylation, ADP-ribosylation, methylamination, adenylation, glycosylation and sulphation. Several of the viral proteins contain sequences, termed nuclear localization signals, which facilitate transport of the proteins to the host cell nucleus where virion maturation occurs. Virions are released by lysis of infected cells.

Members of the genus *Polyomavirus* express 2-3 non-structural proteins which include large T, middle (m)T and small t for mouse and hamster polyomaviruses, and large T and small t for the other species (e.g., SV-40, JCV, and BKV, Table 1). An exception is BPyV for which no mRNA encoding a protein of a size comparable to the small t proteins of other viruses has been identified. An ORF for a third protein, ELP (Early Leader Protein) has been identified in the SV-40 genome; ORFs with the potential to encode a similar protein are present within the JCV and BKV genomes (Table 1). The function(s) of this polypeptide is unknown whereas the T proteins, first named for their involvement in Tumorigenicity and Transformation, play key roles in the regulation of transcription and DNA replication. The best characterized of these, the SV-40 large T protein, exhibits multiple functions that can be mapped to discrete domains.

The genomes of most members of the genus *Papillomavirus* that have been sequenced contain 9-10 ORFs called E1-8 and L1-2 (Fig. 2). Some members lack the E3 and E8 ORFs and have an L3 ORF. Proteins encoded by the E ORFs may represent non-structural polypeptides involved in transcription, DNA replication and transformation, whereas those encoded by the L ORFs appear to represent structural proteins.

Replication of the viral genome is initiated by the specific binding of one or more viral proteins (the polyomavirus T protein; the papillomavirus E1 and E2 proteins) at a unique origin of replication and their interaction with host DNA polymerase(s). Due to the limited amount of genetic information encoded by the viral genomes, the papovaviruses rely heavily upon host cell machinery to replicate their DNA. Replication proceeds bi-directionally via a "Cairns" structure and terminates about 180° from the origin of replication. Late in the replication cycle, rolling circle-type molecules have been identified. The viral proteins involved in intitiation may also promote elongation through helicase and ATPase activities.

Antigenic Properties

Antisera prepared against disrupted virions detect antigens shared with other species in the genus. Members of the genus *Polyomavirus* can be distinguished antigenically by neutralization, hemagglutination inhibition and immuno-electron microscopy tests. Polyclonal and monoclonal antibodies can be used to demonstrate cross-reactivity between the T proteins of the primate polyomaviruses.

Biological Properties

Each virus has a specific host range in nature and in cell culture. The host range is often highly restricted, although cells which fail to support viral replication may be transformed via the action of the early viral gene products. Replication of papillomaviruses *in vivo* is dependent upon the terminal differentiation of keratinocytes.

Virus spread occurs by reactivation of persistent infections in the mother during pregnancy, low-level shedding of virus in urine, and rarely by tissue transplantation (humans). Transmission may also involve contact and air-borne infection; some human papillomaviruses are transmitted sexually. Vectors do not appear to play a role in transmission. The papovaviruses are distributed worldwide, and persistent infections are frequently established, usually early in life. The papillomaviruses cause benign tumors (warts, papillomas) in their natural host and in related species. Papillomas are induced in the skin and in mucous membranes, often at specific sites on the body. Warts may progress to malignant tumors, and certain types of human papillomaviruses (HPV) have been associated with specific tumors (e.g., HPV-16 and HPV-18 are associated with cervical carcinoma). The viral DNA is often present in an integrated form in cervical cancer cell lines which is in contrast to other papillomavirus-infected cells in which the DNA is maintained in an episomal state. The polyomaviruses often demonstrate highly tissue-specific expression. Involvement of the kidney is frequently observed and viruria may be noted, especially in immunodeficient hosts. Infection of humans has been associated with some pathologic changes in the urinary tract. One of the human polyomaviruses, JCV, may infect and destroy oligodendrocytes of the central nervous system, thereby leading to a fatal demyelinating disease termed progressive multifocal leukoencephalopathy (PML). SV-40 causes a PML-like disease in rhesus monkeys. Most polyomaviruses have oncogenic potential in rodents. JCV induces tumors in primates. Under some conditions mouse polyomavirus produces a wide variety of tumors in its natural host. Transformation and oncogenicity result from expression of virus-specific early proteins and their interaction with products of cellular tumor suppressor genes. In transformed and tumor cells the polyomavirus genomes are usually integrated into the host cell DNA.

Genus *Polyomavirus*

Type Species murine polyomavirus (strain A2) (PyV)

Distinguishing Features

In contrast to the papillomaviruses, viral proteins are coded on both strands of the DNA genome (Fig. 2).

List of Species in the Genus

The viruses, their alternative names ( ), genomic sequence accession numbers [ ] and assigned abbreviations ( ) are:

Species in the Genus

African green monkey polyomavirus		(LPV)
(B-lymphotropic papovavirus strain K38)	[K02562]	
baboon polyomavirus 2		(PPV-2)
BK virus (strain Dun)	[J02038]	(BKV)
bovine polyomavirus		(BPyV)
(stump-tailed macaque virus)		
(fetal rhesus kidney virus)	[D00755]	
budgerigar fledgling disease virus		(BFDV)
hamster polyomavirus	[X02449]	(HaPV)
JC virus (strain Mad1)	[J02226]	(JCV)
murine polyomavirus	[M55904]	(KV)
(mice pneumotropic virus)		
(Kilham strain, or K virus)		
murine polyomavirus (strain A2)	[J02288]	(PyV)
rabbit kidney vacuolating virus		(RKV)
simian agent virus 12		(SAV-12)
simian virus 40 (strain 776)	[J02400]	(SV-40)

Tentative Species in the Genus

None reported.

GENUS *PAPILLOMAVIRUS*

Type Species cottontail rabbit papillomavirus (Shope) (CRPV)

DISTINGUISHING FEATURES

The proteins are coded on only one of the two strands of DNA. The genomes are larger than those of polyomaviruses.

LIST OF SPECIES IN THE GENUS

Members of this genus are known from humans (more than 63 types, HPV-1, etc.), chimpanzee, colobus and rhesus monkeys, cow (6 types), deer, dog, horse, sheep, elephant, elk, opossum, multimammate and European harvest mouse, turtle, chaffinch and parrot.

The viruses, their alternative names ( ), genomic sequence accession numbers [ ] and assigned abbreviations ( ) are:

SPECIES IN THE GENUS

bovine papillomavirus 1	[X02346]	(BPV-1)
bovine papillomavirus 2	[M20219]	(BPV-2)
bovine papillomavirus 4	[X05817]	(BPV-4)
canine oral papillomavirus		(COPV)
chaffinch papillomavirus		(ChPV)
cottontail rabbit papillomavirus (Shope)	[K02708]	(CRPV)
deer papillomavirus (deer fibroma virus)	[M11910]	(DPV)
elephant papillomavirus		(EPV)
equine papillomavirus		(EqPV)
European elk papillomavirus	[M15953]	(EEPV)
human papillomavirus 1a	[V01116]	(HPV-1a)
human papillomavirus 5		(HPV-5)
human papillomavirus 6b		(HPV-6b)
human papillomavirus 8		(HPV-8)
human papillomavirus 11	[M14119]	(HPV-11)
human papillomavirus 16	[K02718]	(HPV-16)
human papillomavirus 18	[X05015]	(HPV-18)
human papillomavirus 31	[J04353]	(HPV-31)
human papillomavirus 33	[M12732]	(HPV-33)
multimammate mouse papillomavirus		(MnPV)
rabbit oral papillomavirus		(ROPV)
reindeer papillomavirus		(RePV)
rhesus monkey papillomavirus		(RMPV)
sheep papillomavirus		(SPV)

TENTATIVE SPECIES IN THE GENUS

None reported.

LIST OF UNASSIGNED VIRUSES IN THE FAMILY

None reported.

SIMILARITY WITH OTHER TAXA

None reported.

Derivation of Names

papova: sigla from *pa*pilloma, *po*lyoma, and *va*cuolating agent (early name for SV-40)
papilloma: from Latin *papilla*, "nipple, pustule", also Greek suffix *-oma*, used to form nouns denoting "tumors"
polyoma: from Greek *poly*, "many", and *-oma*, denoting "tumors"

References

Boroweic JA, Dean FB, Bullock PA, Hurwitz J (1990) Binding and unwinding - how T antigen engages the SV40 origin of DNA replication. Cell 60: 181-184
Buchman AR, Burnett L, Berg P (1981) The SV40 nucleotide sequence. In: Tooze J (ed) DNA tumor viruses, 2nd edn. Cold Spring Harbor Laboratory, New York pp 799-841
Chan S-Y, Bernard H-U, Ong C-K, Chan S-P, Hofmann B, Delius H (1992) Phylogenetic analysis of 48 papillomavirus types and 28 subtypes and variants: a showcase for the molecular evolution of DNA viruses. J Virol 66: 5714-5725
Chen EY, Howley PM, Levinson AD, Seeburg PH (1982) The primary structure and genetic organization of the bovine papillomavirus type 1 genome. Nature 299: 529-534
Eckhart W (1991) *Polyomavirinae* and their replication. In Fields & Knipe (eds) Fundamental virology, 2nd edn. Raven Press, New York, pp 727-741
Fanning E (1992) Simian virus 40 large T antigen: the puzzle, the pieces, and the emerging picture. J Virol 66: 1289-1293
Frisque RJ, Bream GL, Cannella MT (1984) Human polyomavirus JC virus genome. J Virol 51: 458-469
Giri I, Danos O, Yaniv M (1985) Genomic structure of the cottontail rabbit (Shope) papillomavirus. Proc Natl Acad Sci USA 82: 1580-1584
Griffin BE, Soeda E, Barrell BG, Staden R (1981) Sequences and analysis of polyoma virus DNA. In Tooze J (ed) DNA tumor viruses, 2nd edn. Cold Spring Harbor Laboratory, New York, pp 843-910
Lambert PF (1991) Papillomavirus DNA replication. J Virol 65:3417-3420
Meyers C, Frattini MG, Hudson JB, Laimins LA (1992) Biosynthesis of human papillomavirus from a continuous cell line upon epithelial differentiation. Science 257: 971-973
Salunke DM, Caspar DLD, Garcea RL (1986) Self-assembly of purified polyomavirus capsid protein VP_1. Cell 46: 895-904
Salzman NP (ed) (1986) The *Papovaviridae*, the polyomaviruses, vol 1. Plenum Press, New York
Salzman NP, Howley PM (eds) (1987) The *Papovaviridae*, the papillomaviruses, vol 2. Plenum Press, New York
Villarreal LP (ed) (1989) Common mechanisms of transformation by small DNA tumor viruses. American Society for Microbiology, Washington, DC
zur Hausen H (1991) Human papillomaviruses in the pathogenesis of anogenital cancer. Virology 184: 9-13

Contributed By

Frisque RJ, Barbanti-Brodano G, Crawford LV, Gardner SD, Howley PM, Orth G, Shah KV, va der Noordaa J, zur Hausen H

FAMILY *POLYDNAVIRIDAE*

TAXONOMIC STRUCTURE OF THE FAMILY

Family	***Polydnaviridae***
Genus	*Ichnovirus*
Genus	*Bracovirus*

VIRION PROPERTIES

MORPHOLOGY

Ichnovirus virions consist of nucleocapsids of uniform size (approximately 85 nm x 330 nm), having the form of a prolate ellipsoid, surrounded by 2 unit-membrane envelopes. The inner envelope appears to be assembled *de novo* within the nucleus of infected calyx cells, while the outer envelope is acquired by budding through the plasma membrane into the oviduct lumen. *Bracovirus* virions consist of enveloped cylindrical electron-dense nucleocapsids of uniform diameter but of variable length (40 nm diameter by 30-150 nm length) and may contain one or more nucleocapsids within a single envelope; the latter appears to be assembled *de novo* within the nucleus. *Bracovirus* nucleocapsids in some cases possess long unipolar tail-like appendages.

PHYSICOCHEMICAL AND PHYSICAL PROPERTIES

None reported.

NUCLEIC ACID

Genomes consist of multiple supercoiled dsDNAs of variable size ranging from approximately 2.0 to more than 28 kbp. No aggregate size for any polydnavirus genome has as yet been determined. Estimates of genome size and complexity are complicated by the presence of related DNA sequences shared among two or more DNA genome segments.

PROTEINS

Virions are structurally complex and contain at least 20-30 polypeptides, with Mr ranging from 10-200 x 10^3.

LIPIDS

Lipids are present, but uncharacterized.

CARBOHYDRATES

Carbohydrates are present, but uncharacterized.

GENOME ORGANIZATION AND REPLICATION

Unique among the dsDNA viruses, polydnaviruses have segmented genomes (see above). Chromosomally integrated sequences homologous to viral DNAs are located within the parasitoid genome; this proviral DNA form is responsible for the transmission of viral genomes within parasitoid populations.

The polydnavirus genome appears to be unusual in other respects as well: some viral genes contain introns; several viral gene families exist, members of which are distributed on one or more genome segments; transcriptional activity is host-specific, in the sense that some genes are expressed in the wasp ovary while others are expressed only in the parasitized host animal; families of viral genome segments exist in some cases; polydnavirus genomes, at least potentially, are genetically redundant (e.g., they would appear to be diploid).

Polydnavirus replication is nuclear, begins during wasp pupation, and is very likely induced by a change in ecdysone titre. Virus morphogenesis occurs in the calyx epithelium of the ovaries of all female wasps belonging to all affected species. Ichnovirus particles bud directly from the calyx epithelial cells into the lumen of the oviduct. The mode of release of bracovirus particles is presently unclear, but probably involves lysis of affected calyx epithelial cells. Extrachromosomal, circular DNAs are present both in male wasps and in non-ovarian female tissues (but viral morphogenesis has not been demonstrated). Viral replication does not occur in parasitized host insects.

ANTIGENIC PROPERTIES

Cross-reacting antigenic determinants are shared by a number of different *Ichnovirus* isolates; in some cases, viral nucleocapsids share at least one major conserved epitope. It has recently been shown that CsPDV and *C. sonorensis* venom protein display common epitopes. Antigenic relationships among the bracoviruses have not as yet been investigated.

BIOLOGICAL PROPERTIES

Polydnaviruses have been isolated only from endoparasitic hymenopteran insects (wasps) belonging to the families *Ichneumonidae* and *Braconidae*. In nature, polydnavirus genomes are apparently transmitted as proviruses. Polydnavirus particles are injected into host animals during oviposition; virus-specific expression leads to significant changes in host physiology, some of which are assumed to be responsible for successful parasitism.

GENUS *ICHNOVIRUS*

Type Species Campoletis sonorensis virus

DISTINGUISHING FEATURES

Ichnoviruses have been found only in the wasp family *Ichneumonidae*. *Ichnovirus* nucleocapsids are fusiform in shape, and are enveloped by two unit membranes. Typically, virus particles each contain a single nucleocapsid (viruses from the wasp genera *Glypta* and *Dusona* are the only known exceptions).

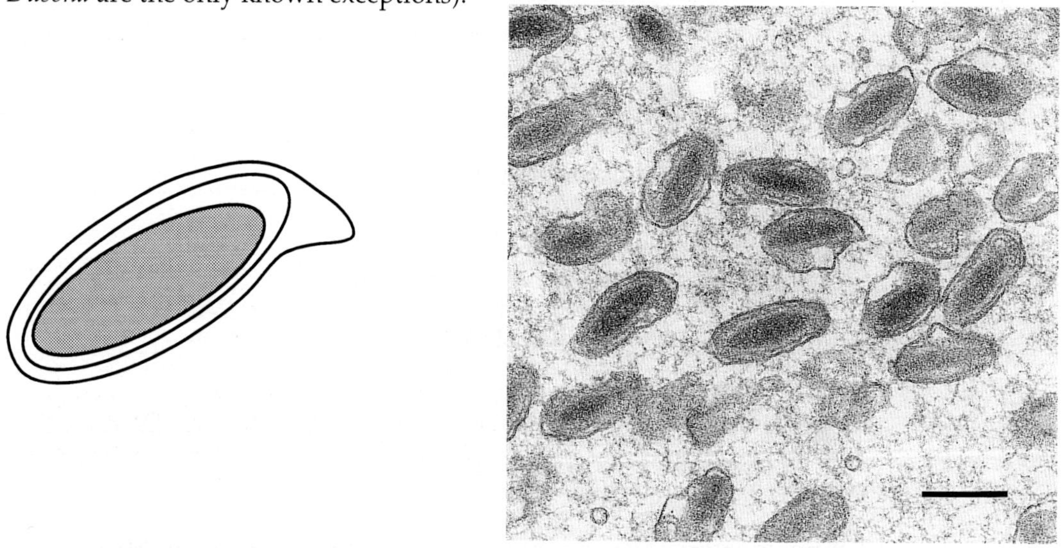

Figure 1: Sectional diagram (left) and electron micrograph (right) of *Ichnovirus* from *Hyposoter exiguae*. The bar represents 200 nm.

LIST OF SPECIES IN THE GENUS

SPECIES IN THE GENUS

Campoletis aprilis virus
Campoletis flavicincta virus

Campoletis sonorensis virus
Campoletis sp. virus
Casinaria arjuna virus
Casinaria forcipata virus
Casinaria infesta virus
Casinaria sp. virus
Diadegma acronyctae virus
Diadegma interruptum virus
Diadegma terebrans virus
Dusona sp. virus
Eriborus terebrans virus
Enytus montanus virus
Glypta fumiferanae virus
Glypta sp. virus
Hyposoter annulipes virus
Hyposoter exiguae virus
Hyposoter fugitivus virus
Hyposoter lymantriae virus
Hyposoter pilosulus virus
Hyposoter rivalis virus
Lissonota sp. virus
Olesicampe benefactor virus
Olesicampe geniculatae virus
Synetaeris tenuifemur virus
Tranosema sp. virus

TENTATIVE SPECIES IN THE GENUS

None reported.

GENUS *BRACOVIRUS*

Type Species Cotesia melanoscela virus

DISTINGUISHING FEATURES

Bracoviruses are found only in certain species of braconid wasps. *Bracovirus* nucleocapsids are cylindrical, of variable length, and are surrounded by only a single unit membrane envelope.

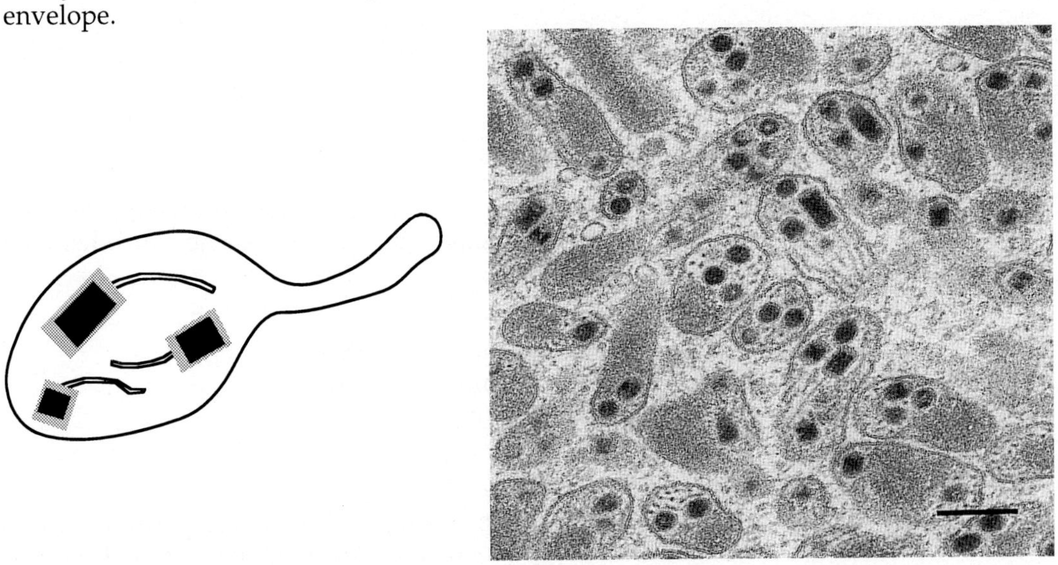

Figure 2: Sectional diagram (left) and electron micrograph (right) of Protapanteles paleacritae virus. The bar represents 200 nm.

List of Species in the Genus

Species in the Genus

Apanteles crassicornis virus
Apanteles fumiferanae virus
Ascogaster argentifrons virus
Ascogaster quadridentata virus
Cardiochiles nigriceps virus
Chelonus altitudinis virus
Chelonus blackburni virus
Chelonus nr. curvimaculatus virus
Chelonus insularis virus
Chelonus texanus virus
Cotesia congregata virus
Cotesia flavipes virus
Cotesia glomerata virus
Cotesia hyphantriae virus
Cotesia kariyai virus
Cotesia marginiventris virus
Cotesia melanoscela virus
Cotesia rubecula virus
Cotesia schaeferi virus
Diolcogaster facetosa virus
Glyptapanteles flavicoxis virus
Glyptapanteles indiensis virus
Glyptapanteles liparidis virus
Hypomicrogaster canadensis virus
Hypomicrogaster ectdytolophae virus
Microplitis croceipes virus
Microplitis demolitor virus
Phanerotoma flavitestacea virus
Pholetesor ornigis virus
Protapanteles paleacritae virus

Tentative Species in the Genus

None reported.

List of Unassigned Viruses in the Family

None reported.

Similarity with Other Taxa

Occasionally, very long *Bracovirus* nucleocapsids are observed; at least superficially, these resemble baculovirus nucleocapsids. Ichnoviruses resemble no other known type of virus.

Derivation of Names

polydna: from *poly* (meaning several), and *DNA*
ichno: from *Ichn*eumonidae, a family of wasps
braco: from *Braco*nidae, a family of wasps

References

Blissard GW, Fleming JGW, Vinson SB, Summers MD (1986) Campoletis sonorensis virus: expression in *Heliothis virescens* and identification of expressed sequences. J Insect Physiol 32: 352-359

de Buron I, Beckage NE (1992) Characterization of a polydnavirus (PDV) and virus-like filamentous particle (VLFP) in the braconid wasp *Cotesia congregata* (*Hymenoptera: Braconidae*). J Invert Path 59: 315-327

Fleming JGW, Summers MD (1986) *Campoletis sonorensis* endoparasitic wasps contain forms of C. sonorensis virus DNA suggestive of integrated and extrachromosomal polydnavirus DNAs. J Virol 57: 552-562
Fleming JGW, Summers MD (1991) Polydnavirus DNA is integrated in the DNA of its parasitoid wasp host. Proc Natl Acad Sci 88: 9770-9774
Fleming JGW (1992) Polydnaviruses: mutualists and pathogens. Ann Rev Entomol 37: 401-25
Krell PJ, Stoltz DB (1979) Unusual baculovirus of the parasitoid wasp *Apanteles melanoscelus:* isolation and preliminary characterization. J Virol 29: 1118-1130
Krell PJ, Stoltz DB (1980) Virus-like particles in the ovary of an ichneumonid wasp: purification and preliminary characterization. Virology 101: 408-418
Krell PJ, Summers MD, Vinson SB (1982) Virus with a multipartite superhelical DNA genome from the ichneumonid parasitoid *Campoletis sonorensis*. J Virol 43: 859-870
Stoltz DB, Vinson SB (1979) Viruses and parasitism in insects. Adv Virus Res 24: 125-171
Stoltz DB (1990) Evidence for chromosomal transmission of polydnavirus DNA. J Gen Virol 71: 1051-1056
Theilmann DA, Summers MD (1986) Molecular analysis of Campoletis sonorensis virus DNA in the lepidopteran host *Heliothis virescens*. J Gen Virol 67: 1961-1969
Theilmann DA, Summers MD (1987) Physical analysis of tandemly repeated elements. J Virol 61: 2589-2598
Theilmann DA, Summers MD (1988) Identification and comparison of Campoletis sonorensis virus transcripts in the insect hosts *Campoletis sonorensis* and *Heliothis virescens*. Virology 167: 329-341
Webb BA, Summers MD (1990) Venom and viral expression products of the endoparasitic wasp *Campoletis sonorensis* share epitopes and related sequences. Proc Natl Acad Sci USA 87: 4961-4965
Webb BA, Summers MD (1992) Stimulation of polydnavirus replication by 20-hydroxyecdysone. Experientia 48: 1018-1022
Xu D, Stoltz DB (1991) Evidence for a chromosomal location of polydnavirus DNA in the ichneumonid parasitoid, *Hyposoter fugitivus*. J Virol 65: 6693-6704

CONTRIBUTED BY

Stoltz DB, Beckage NE, Blissard GW, Fleming JGW, Krell PJ, Theilmann DA, Summers MD, Webb BA

FAMILY *INOVIRIDAE*

TAXONOMIC STRUCTURE OF THE FAMILY

Family	***Inoviridae***
Genus	*Inovirus*
Genus	*Plectrovirus*

VIRION PROPERTIES

MORPHOLOGY

Virions are nonenveloped, helical, and filamentous or rod-shaped. Particles of abnormal length are frequently observed. *Inovirus* virions are usually flexible rods, 760 to 1,950 nm long and 6 to 8 nm in diameter. *Plectrovirus* virions are filamentous with one rounded end: Acholeplasma phage L51 virions are 71 to 90 nm long and 14 to 16 nm in diameter, and Spiroplasma phage SpV1 virions are 230 to 280 nm long and 10 to 15 nm in diameter.

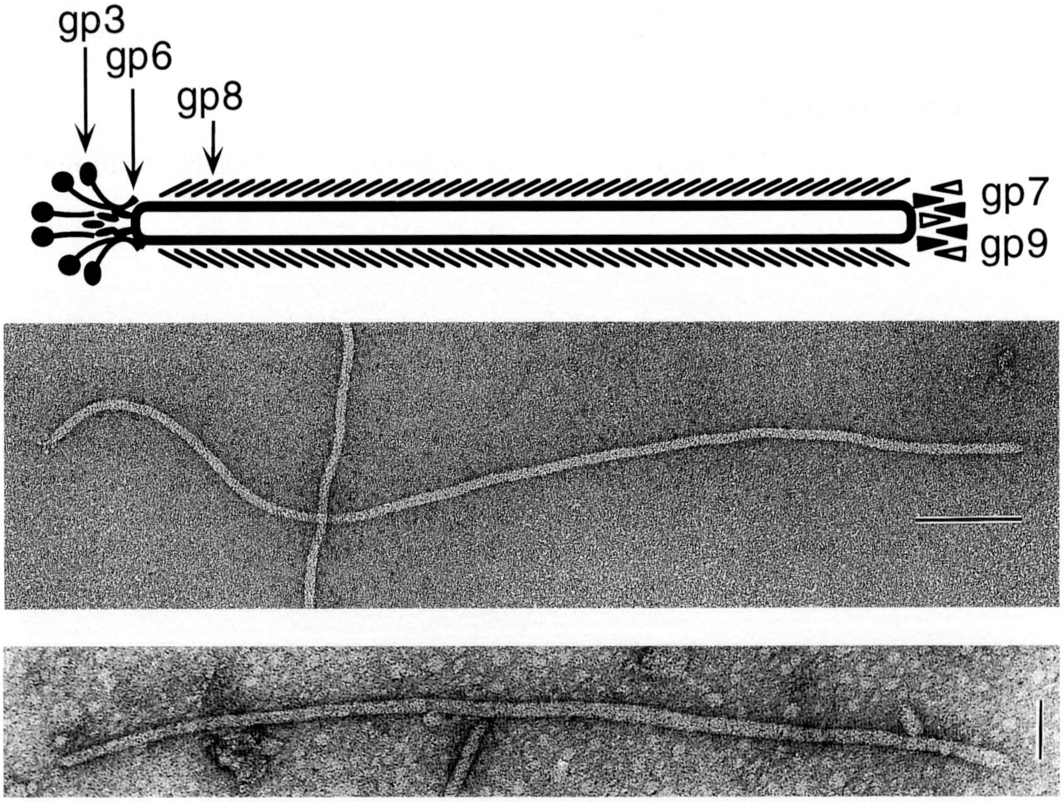

Figure 1: *Inoviridae* virions: (upper) diagram of the coat proteins and ssDNA of an *Inovirus* F pilus-specific coliphage. (From Kornberg A, Baker TA (1991) DNA replication, 2nd ed. WH Freeman and Co., New York, p. 562). (center) *Inovirus* fd virion, showing adsorption proteins at one end. The virus contains a molecule of circular ssDNA of 6408 bp, ensheathed in a protein coat. The bar represents 100 nm. (From Gray CW, Brown RS, Marvin DA (1981) Adsorption complex of filamentous fd virus. J Mol Biol 146: 621-627, courtesy of Gray CW). (lower) Negative contrast electron micrograph of Acholeplasma phage L51 virion preparation, showing rod-shaped virion and long abnormal length particle. The bar represents 50 nm. (From Maniloff J, Das J, Putzrath).

PHYSICOCHEMICAL AND PHYSICAL PROPERTIES

Virion buoyant density in CsCl is 1.3-1.4 g/cm^3, depending on the genus. Virions are sensitive to chloroform and detergents, and resistant to heat. The Mr of *Inovirus* virions is 12-23 x 10^6 and the S_{20w} is 41-45.

Nucleic Acid

Virions contain one molecule of infectious, circular, positive sense ssDNA, 4.4 to 8.5 kb in size. *Inovirus* genome sizes range from 5833 bases for Pseudomonas phage Pf3, to 6407 to 6883 bases for the coliphages, to 7308 bases for Xanthomonas phage Cf1. *Plectrovirus* genome sizes are 4.3 to 4.5 kb for Acholeplasma phage L51 and 8273 bases for Spiroplasma phage SpV1. Several genes are translated from overlapping reading frames. Intergenic regions contain the complementary- and viral-strand replication origins and the DNA packaging signal. The complete DNA sequences of *Inovirus* fd, M13, f1, Ike, Pf3 and Cf1, and *Plectrovirus* SpV1 are available from either GenBank or EMBL database.

Proteins

The *Inovirus* F pilus-specific coliphage virion contains about 2700 copies of gp8 (Mr 5.2 x 10^3), 5 copies each of gp 3 (Mr 43 x 10^3) and gp6 (Mr 12 x 10^3) forming the adsorption end, and 5 copies each of gp 7 (Mr 3.5 x 10^3) and gp9 (Mr 3.3 x 10^3) forming the other end. Five nonstructural proteins have been identified: gp 1 (Mr 35 x 10^3) and gp4 (Mr 50 x 10^3) are involved in morphogenesis, gp2 (Mr 46 x 10^3) and gpX (Mr 12 x 10^3) are involved in DNA replication, and gp5 (Mr 9.8 x 10^3) is a ssDNA binding protein. *Plectrovirus* L51 virions contain at least four proteins, with Mr of 70, 53, 30, and 19 x 10^3.

Lipids

None reported.

Carbohydrates

None reported.

Genome Organization and Replication

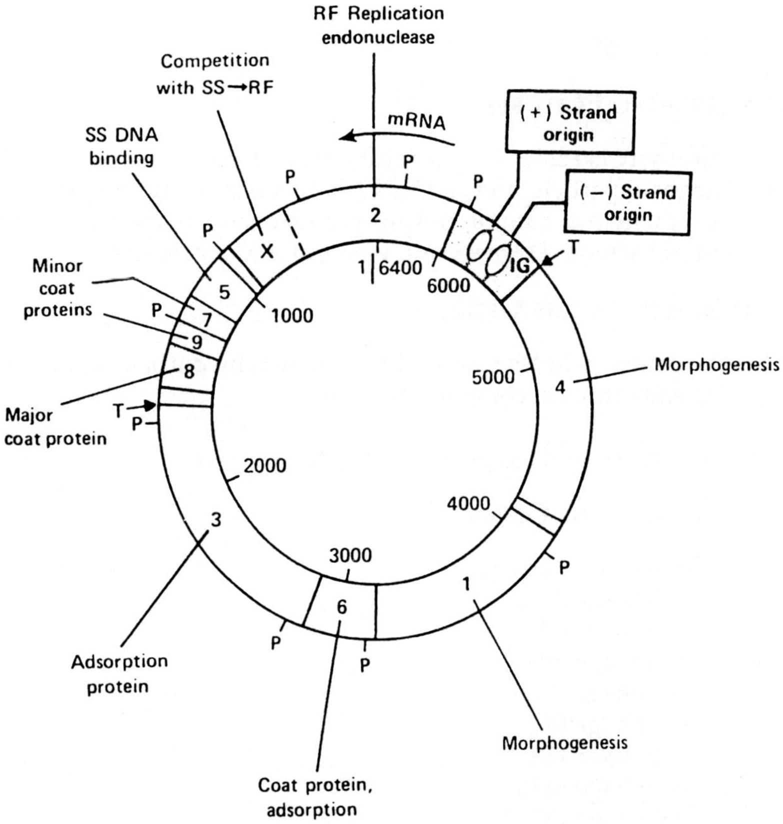

Figure 2: Genetic map of *Inovirus* F pilus-specific coliphages with functions of gene products. DNA replication origins in intergenic region (IG) are shown. P = promoter, T = transcription terminator. (From Kornberg A, Baker TA (1991) DNA replication, 2nd edn. WH Freeman and Co., New York, p. 561).

Infection involves conversion of parental ssDNA into a dsDNA replicative form (RF), semiconservative RF replication, synthesis of progeny ssDNA, and release by extrusion through host membranes without cell lysis. Infected cells continue to grow slowly, producing and releasing progeny virus. Replication of *Inovirus* F pilus-specific coliphages begins with transfer of parental ssDNA into the cell and its conversion to ds RF by host cell proteins. Messenger RNA is transcribed from several promoters on the complementary strand by host cell RNA polymerase. One of the proteins (gp2) made from this mRNA is an endonuclease-topoisomerase and makes a specific cleavage in the parental RF, leading to replication to form progeny ds RF. Progeny RF molecules act as templates for mRNA synthesis and further RF replication (via ssDNA intermediates formed by rolling circle replication). When sufficient amounts of gp5 and gpX are made, complementary strand synthesis is blocked and complexes of gp5-progeny viral ssDNA molecules accumulate. GpX may down-regulate the activity of gp2. Assembly is at adhesion zones between the inner and outer membranes. Gp1 is involved in adhesion zone formation, and gp4 also participates (in an unknown way) in assembly. Assembly involves extrusion of progeny viral ssDNA, with gp5 being replaced by gp8. Since intracellular ds RF has been detected for other *Inovirus* and *Plectrovirus* species, they presumably follow a replication pathway similar to that of the *Inovirus* F pilus-specific coliphages.

BIOLOGICAL PROPERTIES

The host range of the *Inovirus* coliphages is determined by the type of host cell pilus; i.e., phages fd, f1, and M13 require the F pilus for adsorption, and phage Ike requires the N pilus. Similar specificity determinants are presumed for other *Inovirus* species, which also infect Gram-negative eubacteria (i.e., *Pseudomonas, Vibrio,* and *Xanthomonas*); but the nature of the specificity determinants for *Plectrovirus* species, which infect wall-less *Acholeplasma* and *Spiroplasma*, is not known.

GENUS *INOVIRUS*

Type Species coliphage fd (fd)

DISTINGUISHING FEATURES

Infectivity is sensitive to sonication; ether sensitivity is variable. Nucleic acid is 6-21% by weight of particle, and G+C is 40-60%. Virions have no carbohydrate. Host range is certain genera in the gamma-purple phylogenetic branch of Gram-negative eubacteria; i.e., Enterobacteria, *Pseudomonas, Vibrio,* and *Xanthomonas*.

LIST OF SPECIES IN THE GENUS

The genus includes species differentiated by particle length, host range, antigenic properties and chemical composition.

The viruses, and assigned abbreviations ( ) are:

SPECIES IN THE GENUS

1-Coliphage fd group:

coliphage AE2	(AE2)
coliphage δA	(δA)
coliphage Ec9	(Ec9)
coliphage f1	(f1)
coliphage fd	(fd)
coliphage HR	(HR)
coliphage M13	(M13)
coliphage ZG/2	(ZG/2)
coliphage ZJ/2	(ZJ/2)

2-Other enterobacteria phages:	
enterobacteria phage C-2	(C-2)
enterobacteria phage If1	(If1)
enterobacteria phage If2	(If12)
enterobacteria phage Ike	(Ike)
enterobacteria phage I_2-2	(I_2-2)
enterobacteria phage PR64FS	(PR64FS)
enterobacteria phage SF	(SF)
enterobacteria phage tf-1	(tf-1)
enterobacteria phage X	(X)
3-Pseudomonas phages:	
Pseudomonas phage Pf1	(Pf1)
Pseudomonas phage Pf2	(Pf2)
Pseudomonas phage Pf3	(Pf3)
4-Vibrio phages:	
Vibrio phage v6	(v6)
Vibrio phage Vf12	(Vf12)
Vibrio phage Vf33	(Vf33)
5-Xanthomonas phages:	
Xanthomonas phage Cf	(Cf)
Xanthomonas phage Cf1t	(Cf1t)
Xanthomonas phage Xf	(Xf)
Xanthomonas phage Xf2	(Xf2)

TENTATIVE SPECIES IN THE GENUS

None reported.

GENUS *PLECTROVIRUS*

Type Species Acholeplasma phage L51 (L51)

DISTINGUISHING FEATURES

Virions are resistant to nonionic detergents (Nonidet P-40 and Triton X-100) and slightly sensitive to ether. Genome of Spiroplasma phage SpV1 is 23% G+C. No data on carbohydrates have been reported. Adsorption is to cell membrane of wall-less mycoplasma host cells. Host range of the Acholeplasma phage L51 is some *Acholeplasma laidlawii* strains, and of the Spiroplasma phage SpV1 is some *Spiroplasma citri* strains.

LIST OF SPECIES IN THE GENUS

The viruses, their genomic sequence accession numbers [ ] and assigned abbreviations ( ) are:

SPECIES IN THE GENUS

Acholeplasma phage L51		(L51)
Acholeplasma phage MV-L1		(MV-L1)
Acholeplasma phage MVG51		(MVG51)
Acholeplasma phage 0c1r		(0c1r)
Acholeplasma phage 10tur		(10tur)
Spiroplasma phage 1	[X51344]	(SpV1)
Spiroplasma phage aa		(SpVaa)

TENTATIVE SPECIES IN THE GENUS

Spiroplasma phage C1/TS2 (C1/TS2)

List of Unassigned Viruses in the Family

None reported.

Similarity with Other Taxa

None reported.

Derivation of Names

ino: from Greek *nos*, 'muscle'
plectro: from Greek *plektron*, 'small stick'

References

Ackermann H-W, DuBow MS (1987) Viruses of prokaryotes, Vol 2. CRC press, Boca Raton FL, pp 171-218

Baas PD (1985) DNA replication of single-stranded *Escherichia coli* DNA phages. Biochim Biophys Acta 825: 111-139

Day LA, Marzec CJ, Reisberg SA, Casadevall A (1988) DNA packing in filamentous bacteriophages. Annu Rev Biophysics Biophys Chem 17: 509-539

Kuo TT, Lin YH, Huang CM, Chang SF, Dai H, Feng TY (1987) The lysogenic cycle of the filamentous phage Cflt from *Xanthomonas campestris* pv *citri*. Virology 156: 305-312

Kuo TT, Tan MS, Su MT, Yang MK (1991) Complete nucleotide sequence of filamentous phage Cflt from *Xanthomonas campestris* pv. *citri*. Nucl Acids Res 19: 2498

Maniloff J (1992) Mycoplasma viruses. In: Maniloff J, McElhaney RN, Finch LR, Baseman JB (eds) American Society for Microbiology, Mycoplasmas: molecular biology and pathogenesis. Washington DC, pp 41-59

Model P, Russel M (1988) Filamentous bacteriophage. In: Calendar R (ed) The bacteriophages, Vol 2. Plenum Press, New York, pp 375-456

Rasched H, Oberer E (1986) Ff coliphages:structural and functional relationships. Microbiol Rev 50: 401-427

Renaudin J, Aullo P, Vignault JC, Bove JM (1990) Complete nucleotide sequence of the genome of Spiroplasma citri virus SpV1-R8A2 B. Nucl Acids Res 18: 1293

Renaudin J, Bodin-Ramiro C, Vignault JC, Bove JM (1990) Spiroplasmavirus 1: presence of viral sequences in the spiroplasma genome. Zbl Bakteriol Suppl 20: 125-130

Contributed By

Maniloff J

FAMILY MICROVIRIDAE

TAXONOMIC STRUCTURE OF THE FAMILY

Family	***Microviridae***
Genus	*Microvirus*
Genus	*Spiromicrovirus*
Genus	*Bdellomicrovirus*
Genus	*Chlamydiamicrovirus*

VIRION PROPERTIES

MORPHOLOGY

Virions exhibit icosahedral symmetry (T = 1) with projections at each of the 12 vertices. There is no envelope, and the diameter of unstained hydrated particles is 22 nm between the depressions at the 2-fold axes and 33 nm between the outermost edges of the projections at the 5-fold axes. Thus, reported diameters from electron micrographs of negative stained preparations vary from 26-32 nm, depending on the orientation chosen for measurement.

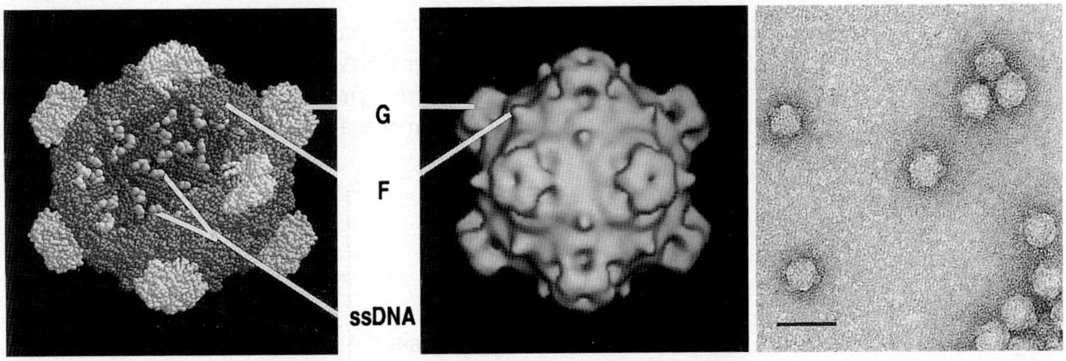

Figure 1: Coliphage ϕX174 virions: (left) molecular model; (center) image reconstruction of frozen-hydrated virion; (right) negative contrast electron micrograph. The bar represents 50 nm.

PHYSICOCHEMICAL AND PHYSICAL PROPERTIES

Virion buoyant density in CsCl is 1.36-1.41 g/cm^3, depending on the genus. Infectivity is chloroform and detergent resistant and stable in the pH range of 6-9, but highly sensitive to radiation. Virion Mr (genus *Microvirus)* is 6-7 x 10^6, and the S_{20w} is 83-121.

NUCLEIC ACID

Virions contain one molecule of circular, positive sense ssDNA; genome sizes are as follows:

Genus	Phage	Number of bases
Microvirus	ϕX174	5,386
	St-1	6,050
Spiromicrovirus	SpV4	4,421
Bdellomicrovirus	MAC-1	about 4,600
Chlamydiamicrovirus	Chp1	4,877

Several genes are translated from overlapping reading frames. The complete sequences of the genomes of ϕX174, S13, and G4 viruses (genus *Microvirus*), SV4 virus (genus *Spiromicrovirus)*, and Chp1 virus (genus *Chlamydiamicrovirus*), are available from either GenBank or EMBL database.

PROTEINS

Virions (genus *Microvirus*) contain 60 copies of three proteins (gp J, F, and G) with Mr of 4, 48, and 19 x 10^3, respectively, and 12 copies of one protein (gp H) with an Mr of 34 x 10^3. The atomic structure of ϕX174 virus has been determined; its F capsid protein contains an eight-stranded antiparallel beta barrel similar to that found in picornaviruses and many icosahedral plant viruses. The C-terminal end of each J protein is bound to the inner surface of each F protein near the 3-fold axis, forming a binding pocket for segments of the ssDNA. Seven nonstructural proteins have been identified: gp B and D are components of the procapsid, gp A and C are involved in synthesis of RF and progeny DNA, gp A* suppresses host DNA synthesis, gp E functions in cell lysis, and gp K increases the progeny yield.

LIPIDS

None reported.

CARBOHYDRATES

None reported.

GENOME ORGANIZATION AND REPLICATION

Virus replication begins with transfer of the parental ssDNA into the cell and its conversion to ds RF by host cell proteins. Messenger RNA is transcribed from this template by host cell RNA polymerase. One of the proteins (protein A) made from this mRNA becomes covalently bound to the parental RF and leads to progeny RF replication. Progeny RF

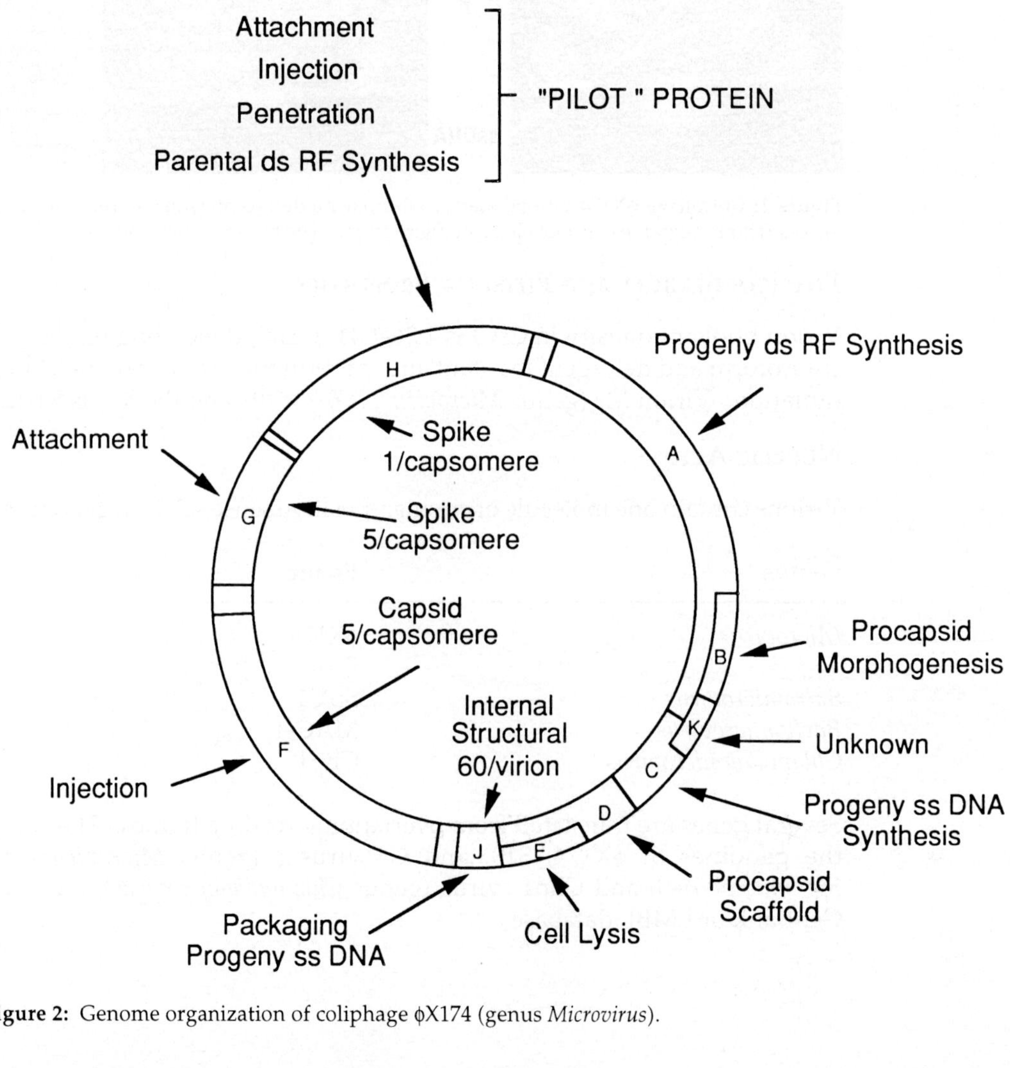

Figure 2: Genome organization of coliphage ϕX174 (genus *Microvirus*).

molecules act as templates for mRNA synthesis and further RF replication (via ssDNA intermediates formed by rolling circle replication), until sufficient levels of viral structural proteins and two additional procapsid proteins are made and assembled into procapsids. Procapsids then bind to some RF molecules and C protein switches DNA synthesis from ds RF replication to synthesis of progeny ssDNA. As nascent viral ssDNA is synthesized, it interacts with procapsid-associated J proteins and is packaged into proheads. Maturation of filled procapsids involves loss of procapsid-associated B and D proteins, and occurs as the cell is lysed by E protein. Since intracellular ds RF has been detected for other genera, they presumably follow a replication pathway similar to that of the genus *Microvirus*, but there are only limited data on the replication details of other genera of the family *Microviridae*.

ANTIGENIC PROPERTIES

Native virions (genus *Microvirus*) generate both non-neutralizing and neutralizing monoclonal antibodies. Polyclonal antisera produce first-order inactivation kinetics. Members of the genus *Microvirus* can be assigned to at least three main groups based on serologic cross-reactivity patterns.

BIOLOGICAL PROPERTIES

The host range of member viruses (genus *Microvirus*) is determined by the carbohydrate structure of the host cell outer membrane lipopolysaccharide receptor. Thus, various species of the *Enterobacteriaceae* constitute the host range for individual viruses. Similar specificity determinants are presumed for the MAC-1 and Chp1 phage (genera *Bdellomicrovirus* and *Chlamydiamicrovirus*), which also infect Gram-negative eubacteria (genus *Bdellovibrio* and *Chlamydia*, respectively); but the nature of the specificity determinants for the genus *Spiromicrovirus*, which infects wall-less *Spiroplasma*, is not known.

GENUS *MICROVIRUS*

Type Species coliphage ϕX174 (ϕX174)

DISTINGUISHING FEATURES

Members have different temperature ranges for plaque formation; e.g., 10-39° C for G4, 22-43° C for S-13, and 33-43° C for St-1 virus. In addition, the host cell enzyme requirement for viral DNA replication is not identical for all members of this genus, and three major groups arise based on this criterion. Host range: *Enterobacteriaceae* species and strains.

LIST OF SPECIES IN THE GENUS

The viruses, their genomic sequence accession numbers [ ] and assigned abbreviations ( ) are:

SPECIES IN THE GENUS

coliphage 1ϕ1	(1ϕ1)
coliphage 1ϕ3	(1ϕ3)
coliphage 1ϕ7	(1ϕ7)
coliphage 1ϕ9	(1ϕ9)
coliphage 2D/13	(2D/13)
coliphage α10	(α10)
coliphage α3	(α3)
coliphage BE/1	(BE/1)
coliphage δ1	(δ1)
coliphage dϕ3	(dϕ3)
coliphage dϕ4	(dϕ4)
coliphage dϕ5	(dϕ5)
coliphage ϕA	(ϕA)
coliphage ϕB	(ϕB)
coliphage ϕC	(ϕC)
coliphage ϕK	(ϕK)

coliphage ϕR		(ϕR)
coliphage ϕX174	[J02482]	(ϕX174)
coliphage G13		(G13)
coliphage G14		(G14)
coliphage G4		(G4)
coliphage G6		(G6)
coliphage η8		(η8)
coliphage M20		(M20)
coliphage o6		(o6)
coliphage S13		(S13)
coliphage St-1		(St-1)
coliphage U3		(U3)
coliphage WA/1		(WA/1)
coliphage WF/1		(WF/1)
coliphage WW/1		(WW/1)
coliphage ζ3		(ζ3)

Tentative Species in the Genus

None reported.

Genus *Spiromicrovirus*

Type Species Spiroplasma phage 4 (SpV-4)

Distinguishing Features

Virus and host cells use TGA as tryptophan codon instead of "universal" stop codon. Host range: *Spiroplasma melliferum* strains.

List of Species in the Genus

The viruses, their genomic sequence accession numbers [ ] and assigned abbreviations ( ) are:

Species in the Genus

Spiroplasma phage 4	[M17988]	(SpV-4)

Tentative Species in the Genus

None reported.

Genus *Bdellomicrovirus*

Type Species Bdellovibrio phage MAC 1 (MAC-1)

Distinguishing Features

Host range: *Bdellovibrio bacteriovorus* strains

List of Species in the Genus

The viruses and their assigned abbreviations ( ) are:

Species in the Genus

Bdellovibrio phage MAC 1	(MAC-1)
Bdellovibrio phage MAC 1'	(MAC-1')
Bdellovibrio phage MAC 2	(MAC-2)
Bdellovibrio phage MAC 4	(MAC-4)
Bdellovibrio phage MAC 4'	(MAC-4')
Bdellovibrio phage MAC 5	(MAC-5)
Bdellovibrio phage MAC 7	(MAC-7)

Tentative Species in the Genus

None reported.

Genus *Chlamydiamicrovirus*

Type Species Chlamydia phage 1 (Chp-1)

Distinguishing Features

Host range: *Chlamydia psittaci* strains

List of Species in the Genus

The viruses and their assigned abbreviations ( ) are:

Species in the Genus

Chlamydia phage 1 [D00624] (Chp-1)

Tentative Species in the Genus

None reported.

List of Unassigned Viruses in the Family

None reported.

Similarity with Other Taxa

The ssDNA genome is similar to that of the members of the family *Inoviridae*, in organization and existence of overlapping genes and in many aspects of DNA replication.

Derivation of Names

micro: from Greek *mikros*, 'small'

References

Ackermann H-W, DuBow MS (1987) Viruses of Prokaryotes Vol II. CRC Press, Boca Raton FL, pp 171-218

Althauser M, Samsonoff WA, Anderson C, Conti SF (1972) Isolation and preliminary characterization of bacteriophages for *Bdellovibrio bacteriovorus*. J Virol 10: 516-523

Hayashi MN, Aoyama A, Richardson Jr DL, Hayashi MN (1988) Biology of bacteriophage fX174. In: Calendar R (ed) The Bacteriophages Vol 2. Plenum Press, New York, pp 1-72

Maniloff J (1992) Mycoplasma viruses. In: Maniloff J, McElhaney RN, Finch LR, Baseman JB (eds) Mycoplasmas: molecular biology and pathogenesis. American Society for Microbiology, Washington DC, pp 41-59

McKenna R, Xia D, Willingmann P, Ilag LL, Krishnaswamy S, Rossmann MG, Olson NH, Baker TS, Incardona NL (1992) Atomic structure of single-stranded DNA bacteriophage fX174 and its functional implications. Nature 355: 137-143

Renaudin J, Pascarel MC, Bove JM (1987) Spiroplasma virus 4: nucleotide sequence of the viral DNA, regulatory signals, and proposed genome organization. J Bacteriol 169: 4950-4961

Renaudin J, Pascarel MC, Garnier M, Carle-Junca P, Bove JM (1984) SpV4, a new spiroplasma virus with circular single-stranded DNA. Ann Virol 135E: 343-361

Roberts RC, Keefer MA, Ranu RS (1987) Characterization of Bdellovibrio bacteriovorus bacteriophage MAC-1. J Gen Microbiol 133: 3065-3070

Storey CC, Lusher M, Richmond SJ (1989) Analysis of the complete nucleotide sequence of Chp1, a phage which infects avian *Chlamydia psittaci*. J Gen Virol 70: 3381-3390

Storey CC, Lusher M, Richmond SJ, Bacon J (1989) Further characterization of a bacteriophage recovered from an avian strain of *Chlamydia psittaci*. J Gen Virol 70: 1321-1327

Contributed By

Incardona NL, Maniloff J

FAMILY GEMINIVIRIDAE

TAXONOMIC STRUCTURE OF THE FAMILY

Family	***Geminiviridae***
Genus	"Subgroup I Geminivirus"
Genus	"Subgroup II Geminivirus"
Genus	"Subgroup III Geminivirus"

VIRION PROPERTIES

MORPHOLOGY

Virions are geminate (about 18 x 30 nm), consisting of two incomplete icosahedra (T=1) with a total of 22 capsomers.

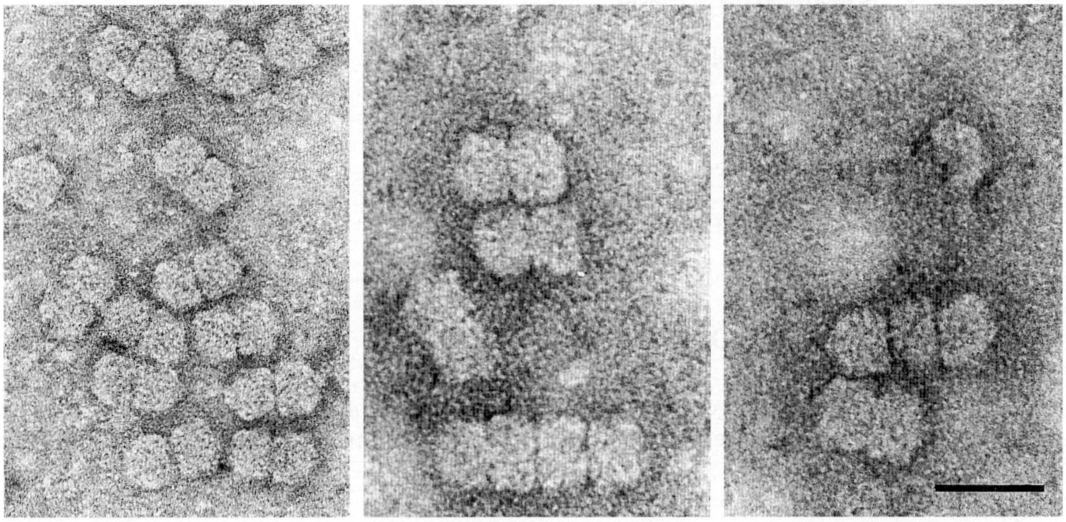

Figure 1: Typical geminiviruses consist of two quasi-isometric subunits (left), however sometimes three (right) or four (center) subunits are joined. The bar represents 50 nm.

PHYSICOCHEMICAL AND PHYSICAL PROPERTIES

S_{20w} is approximately 70.

NUCLEIC ACID

Virions contain a single molecule of circular ssDNA, 2.5-3.0 kb in size.

PROTEINS

Virions contain a single structural protein (coat protein; Mr 28-34 x 10^3). A virion consists of 22 copolymers with each capsomer estimated to contain 5 coat protein molecules.

LIPIDS

None reported.

CARBOHYDRATES

None reported.

GENOME ORGANIZATION AND REPLICATION

Both the viral (encapsidated) and complementary strands of the viral genome encode genes. Coding regions diverge from an intergenic region. Replication occurs through a double-stranded replicative intermediate via a rolling circle mechanism. ssDNA synthesis is initiated at a conserved TAATATTAC sequence within the intergenic region. Transcription of the viral genome is bidirectional with transcripts initiating within the intergenic region.

GENUS "SUBGROUP I GEMINIVIRUS"

Type Species maize streak virus (MSV)

DISTINGUISHING FEATURES

GENOME ORGANIZATION AND REPLICATION

The genomes of subgroup I geminiviruses consist of a single component, 2.6-2.8 kb in size. The presence of a small complementary senseprimer-like molecule, bound to the genome within the small intergenic region, has been shown for five species (CSMV, DSV, MSV, TYDV, WDV). The nucleotide sequences of the genomes of eight species (CSMV, DSV, MSV, MiSV, PanSV, SSV, TYDV, WDV) have been determined. The genomes of subgroup I geminiviruses encode four genes, two each on the viral and complementary strands. For most species, ORF C2 lacks a methionine start codon. For several species (DSV, MSV, TYDV, WDV) translation of this ORF has been shown to occur by splicing of the C1 and C2 transcripts.

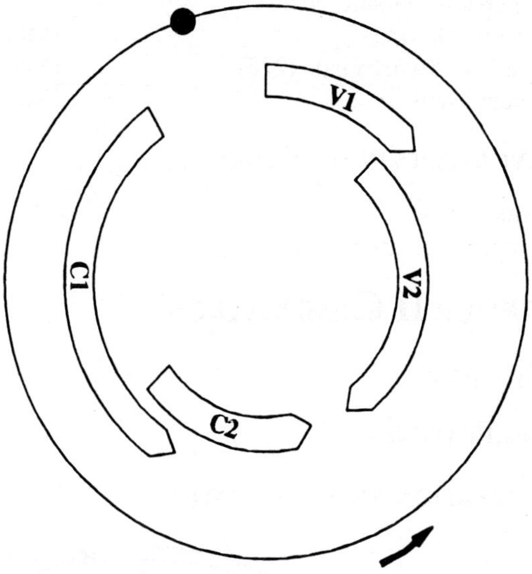

Figure 2: Typical genomic organization of subgroup I geminiviruses. Genes are denoted as either being encoded on the viral (V) or complementary (C) strand. Gene V2 encodes the coat protein. The positions of the conserved TAATATTAC sequence (●) and the encapsidated, complementary sense primer-like molecule (—>) are shown.

ANTIGENIC PROPERTIES

Serological analyses show close interrelationships between viruses originating from the same continent, although DSV (originating from Vanuatu) is closely related serologically to the African subgroup I geminiviruses. Viruses originating from different continents are either unrelated or distantly related.

BIOLOGICAL PROPERTIES

HOST RANGE

Subgroup I geminiviruses have narrow host ranges. With the exception of TYDV (which infects dicotyledonous plants) subgroup I geminivirus host ranges are limited to members of the *graminae*.

TRANSMISSION

Transmitted in nature by leafhoppers (Homoptera: *Cicadellidae*), in most cases by a single species. Mechanism of transmission is persistent (circulative, non-propagative). Subgroup I geminiviruses are not transmissible by mechanical inoculation. Experimentally some members have been transmitted by *Agrobacterium*-mediated transfer using recombinant DNA methods (CSMV, DSV, MSV, MiSV, PanSV, TYDV, WDV).

LIST OF SPECIES IN THE GENUS

The viruses, their genomic sequence accession numbers [ ], CMI/AAB description # ( ) and assigned abbreviations ( ) are:

SPECIES IN THE GENUS

Bromus striate mosaic virus		(BrSMV)
Chloris striate mosaic virus (221)	[M20021]	(CSMV)
Digitaria streak virus	[M23022]	(DSV)
Digitaria striate mosaic virus		(DiSMV)
maize streak virus (133)	[X01089, X01633]	(MSV)
Miscanthus streak virus (348)	[D00800, D01030]	(MiSV)
Panicum streak virus	[X60168]	(PanSV)
Paspalum striate mosaic virus		(PSMV)
sugarcane streak virus	[M82918]	(SSV)
tobacco yellow dwarf virus (278)	[M81103]	(TYDV)
wheat dwarf virus	[X02869]	(WDV)

TENTATIVE SPECIES IN THE GENUS

bajra streak virus	(BaSV)
chickpea chlorotic dwarf virus	(CpCDV)

GENUS "SUBGROUP II GEMINIVIRUS"

Type Species beet curly top virus (BCTV)

DISTINGUISHING FEATURES

GENOME ORGANIZATION AND REPLICATION

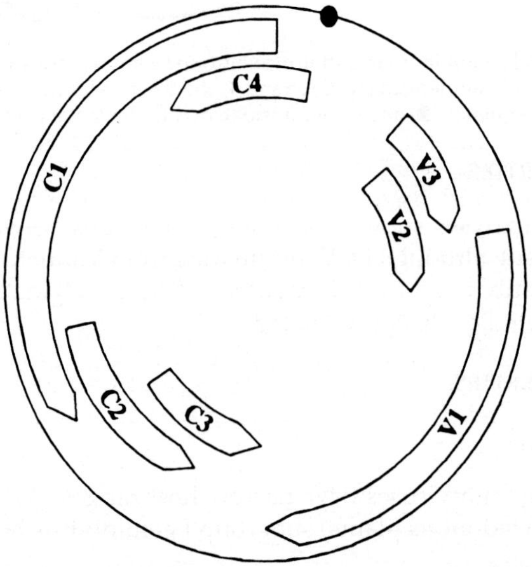

Figure 3: Genomic organization of BCTV. Genes are denoted as either being encoded on the viral (V) or complementary (C) strand. The coat protein is encoded by gene V1. The position of the conserved nanonucleotide sequence (TAATATTAC) is shown (●).

The genomes of subgroup II geminiviruses consist of a single component (2.7-3.0 kb). The nucleotide sequence of the genome of BCTV has been determined. The genome of BCTV encodes six genes, two on the viral strand and four on the complementary strand.

ANTIGENIC PROPERTIES

Serological tests show BCTV, TPCTV and TLRV to be relatively closely related. Distant relationships between subgroup II geminiviruses and subgroup III geminiviruses have been shown in serological tests.

BIOLOGICAL PROPERTIES

HOST RANGE

Type species BCTV has a very wide host range, over 300 species in 44 plant families.

TRANSMISSION

Transmitted in nature by leafhoppers (Homoptera: *Cicadellidae*), with the exception of TPCTV, which is transmitted by a treehopper (Homoptera: *Membracidae*). Mechanism of transmission is persistent (circulative, non-propagative). BCTV may be transmitted with difficulty by mechanical inoculation. Experimentally some species have been transmitted by *Agrobacterium*-mediated transfer using recombinant DNA methods (BCTV, TPCTV).

LIST OF SPECIES IN THE GENUS

The viruses, their genomic sequence accession numbers [ ], CMI/AAB description # ( ) and assigned abbreviations ( ) are:

SPECIES IN THE GENUS

beet curly top virus (210)	[U02311, X04144]	(BCTV)

TENTATIVE SPECIES IN THE GENUS

tomato leafroll virus	(TLRV)
tomato pseudo-curly top virus	(TPCTV)

GENUS "SUBGROUP III GEMINIVIRUS"

Type Species bean golden mosaic virus (BGMV)

DISTINGUISHING FEATURES

GENOME ORGANIZATION AND REPLICATION

The genomes of the majority subgroup III geminiviruses consist of two components (each 2.5-2.8 kb) although subgroup III geminiviruses with only a single DNA component have recently been identified (TYLCV, TLCV). The larger of the two components (DNA A) encodes four genes (one on the viral and three on the complementary strand whereas the smaller component (DNA B) encodes two genes (one on each strand). The DNA A component encodes the coat protein and all functions required for replication. The products of DNA B are involved in spread within plants. The two genomic components have an approximately 200 bp block (encompassing the conserved TAATATTAC sequence) within the intergenic region with near sequence identity which is termed the "common region". The genomes of 18 subgroup III geminiviruses have been sequenced (AbMV, ACMV, BDMV, BGMV, ICMV, MYMV, PHV, PYMV, SLCV, TGMV, TMoV, ToLCV-Au, ToLCV-In, TYLCV-Ls, TYLCV-Sr, TYLCV-Th, TYLCV-Yem, TLCrV).

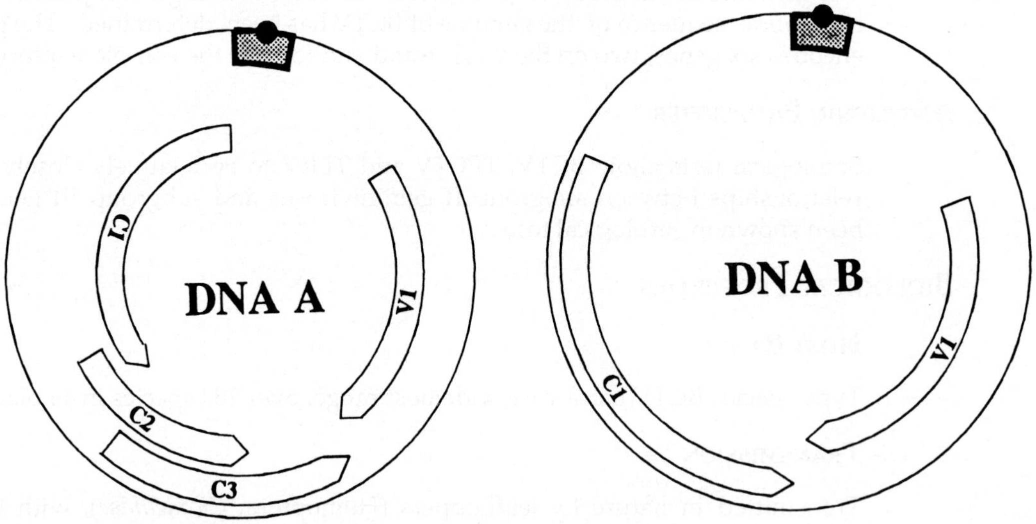

Figure 4: Typical genomic organization of bipartite subgroup III geminiviruses. Genes are denoted as either being encoded on the viral (V) or complementary (C) strand. The coat protein is encoded by gene V1. The position of the conserved TAATATTAC sequence (●) and the "common region" between the two genomic components (shaded boxes) are shown.

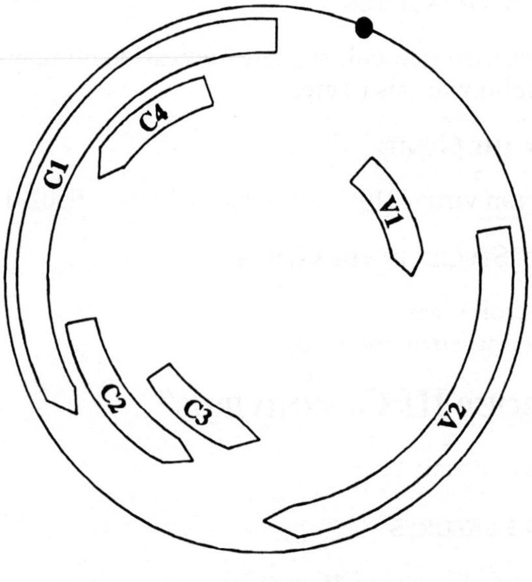

Figure 5: Genomic organization of monopartite subgroup III geminiviruses. Genes are denoted as either being encoded on the viral (V) or complementary (C) strand. The coat protein is encoded by gene V2. The position of the conserved TAATATTAC sequence is shown (●).

Antigenic Properties

Serological tests show all subgroup III geminiviruses to be relatively closely related. The use of monoclonal antisera has show that subgroup III geminiviruses may be grouped geographically based on shared epitopes.

Biological Properties

Host Range

Individual subgroup III geminiviruses generally have narrow host ranges amongst dicotyledonous plants.

TRANSMISSION

Transmitted in nature exclusively by the whitefly *Bemisia tabaci* (Genn.) (Homoptera: *Aleyrodidae*). Some species are transmissible by mechanical inoculation. Experimentally some species have been transmitted by either *Agrobacterium*-mediated transfer or biolistics using recombinant DNA methods (AbMV, ACMV, BDMV, BGMV, PHV, PYMV, SLCV, TGMV, TMoV, TLCV, TYLCV).

LIST OF SPECIES IN THE GENUS

The viruses, their genomic sequence accession numbers [ ], CMI/AAB description # ( ) and assigned abbreviations ( ) are:

SPECIES IN THE GENUS

Abutilon mosaic virus	[X15983, X15984]	(AbMV)
Acalypha yellow mosaic virus		(AYMV)
African cassava mosaic virus	[X17095, X17096, J02058, J02057]	(ACMV)
Ageratum yellow vein virus		(AYVV)
Asystasia golden mosaic virus		(AGMV)
bean calico mosaic virus	[L27264, L27266]	(BCaMV)
bean dwarf mosaic virus	[M88179, M88180]	(BDMV)
bean golden mosaic virus (192)	[M10070, M91604, M10080, L01635, L01636, M91605, D00200, D00201, M88686]	(BGMV)
Bhendi yellow vein mosaic virus		(BYVMV)
Chino del tomate virus		(CdTV)
cotton leaf crumple virus		(CLCrV)
cotton leaf curl virus		(CLCuV)
Croton yellow vein mosaic virus		(CYVMV)
Dolichos yellow mosaic virus		(DoYMV)
Eclipta yellow vein virus		(EYVV)
Euphorbia mosaic virus		(EuMV)
honeysuckle yellow vein mosaic virus		(HYVMV)
horsegram yellow mosaic virus		(HgYMV)
Indian cassava mosaic virus	[Z24758, Z24759]	(ICMV)
Jatropha mosaic virus		(JMV)
limabean golden mosaic virus		(LGMV)
Malvaceous chlorosis virus		(MCV)
melon leaf curl virus		(MLCV)
Macrotyloma mosaic virus		(MaMV)
mungbean yellow mosaic virus (323)	[D14703, D14704]	(MYMV)
okra leaf curl virus		(OLCV)
pepper huasteco virus	[X70418, X70419]	(PHV)
pepper mild tigré virus		(PepMTV)
potato yellow mosaic virus	[D00940, D00941]	(PYMV)
Pseuderanthemum yellow vein virus		(PYVV)
Rhynchosia mosaic virus		(RhMV)
Serrano golden mosaic virus		(SGMV)
sida golden mosaic virus		(SiGMV)
squash leaf curl virus	[M38182, M38183, M63155] M63156, M63157, M63158]	(SLCV)
Texas pepper virus		(TPV)
tobacco leaf curl virus (232)		(TLCV)
tomato golden mosaic virus (303)	[K02029, K02030]	(TGMV)
tomato leaf curl virus - Au	[S53251]	(ToLCV-Au)
tomato leaf curl virus - In	[L12739, L11746]	(ToLCV-In)

tomato mottle virus	[L14460, L14461]	(TMoV)
tomato yellow dwarf virus		(ToYDV)
tomato yellow leaf curl virus - Is	[X15656]	(TYLCV-Is)
tomato yellow leaf curl virus - Sr	[X61153, Z25751, L27708]	(TYLCV-Sr)
tomato yellow leaf curl virus - Th	[M59838, M59839]	(TYLCV-Th)
tomato yellow leaf curl virus - Ye	[X79429]	(TYLCV-Yem)
tomato leaf crumple virus	[L27267 to L27269]	(TLCrV)
tomato yellow mosaic virus		(ToYMV)
watermelon chlorotic stunt virus	[X79430]	(WmCSV)
watermelon curly mottle virus		(WmCMV)

TENTATIVE SPECIES IN THE GENUS

cowpea golden mosaic virus	(CpGMV)
eggplant yellow mosaic virus	(EYMV)
Eupatorium yellow vein virus	(EpYVV)
lupin leaf curl virus	(LLCV)
papaya leaf curl virus	(PaLCV)
sida yellow vein virus	(SiYVV)
Solanum apical leaf curl virus	(SALCV)
soybean crinkle leaf virus	(SCLV)
Wissadula mosaic virus	(WiMV)

LIST OF UNASSIGNED VIRUSES IN THE FAMILY

None reported.

SIMILARITY WITH OTHER TAXA

None reported.

DERIVATION OF NAMES

Gemini: from latin, "twins" describing the characteristic twinned particle morphology

REFERENCES

Bennett CW (1935) Studies on properties of the curly top virus. J Agri Res 50: 211-241

Boulton MI, King DI, Donson J, Davies JW (1991) Point substitutions in a promoter-like region and the V1 affect the host range and symptoms of maize streak virus. Virology 183: 114-121

Briddon RW, Lunness P, Chamberlin LCL, Pinner MS, Brundish H, Markham PG (1992) The nucleotide sequence of an infectious insect-transmissible clone of the geminivirus Panicum streak virus. J Gen Virol 73: 1041-1047

Dry IB, Rigden JE, Krake LR, Mullineaux PM, Rezaian MA (1993) Nucleotide sequence and genome organization of tomato leaf curl geminivirus. J Gen Virol 74: 147-151

Faria JC, Gilbertson RI, Hanson SFG, Morales FJ, Ahlquist P, Loniello AO, Maxwell DP (1994) Bean golden mosaic geminivirus type II isolates from the Dominican Republic and Guatamala: Nucleotide sequences, infectious pseudorecombinants, and phylogenetic relationships. Mol Plant Pathol 84: 321-329

Howarth AJ, Caton J, Bossert M, Goodman RM (1985) Nucleotide sequence of bean golden mosaic virus and a model for gene regulation in geminiviruses. Proc Natl Acad Sci USA 82: 3572-3576

Kheyr-Pour A, Bendahmane M, Matzeit V, Accotto GP, Crespi S, Gronenborn B (1991) Tomato yellow leaf curl virus from Sardinia is a whitefly transmitted monopartite geminivirus. Nucl Acids Res 19: 6763-6769

Lazarowitz SG (1992) Geminiviruses: genome structure and gene function. Crit Rev Plant Sci 11: 327-349

Morris BAM, Richards KA, Haley A, Zhan X, Thomas JE (1992) The nucleotide sequence of the infectious cloned DNA component of tobacco yellow dwarf virus reveals features of geminiviruses infecting monocotyledonous plants. Virology 187: 633-642

Navot N, Pichersky E, Zeidan M, Zamir D, Czosneck H (1991) Tomato yellow leaf curl virus: A whitefly-transmitted geminivirus with a single genomic component. Virology 185: 151-161

Pinner MS, Markham PG, Markham RH, Dekker EL (1988) Characterization of maize streak virus: description of strains; symptoms. Plant Pathol 37: 74-87

Pinner MS, Markham PG (1990) Serotyping and strain identification of maize streak virus isolates. J Gen Virol 71: 1635-1640

Pinner MS, Markham PG, Rybicki E, Greber RS (1992) Serological relationships of geminivirus isolates from *Gramineae* in Australia. Plant Pathol 41: 618-625

Saunders K, Lucy A, Stanley J (1991) DNA forms of the geminivirus African cassava mosaic virus consistent with a rolling circle mechanism of replication. Nucl Acids Res 19: 2325-2330

Stanley J, Markham PG, Callis RJ, Pinner MS (1986) The nucleotide sequence of an infectious clone of the geminivirus beet curly top virus. EMBO J 15: 1761-1767

Stanley J (1991) The molecular determinants of geminivirus pathogenesis. Semin Virol 2: 139-149

Stanley J, Latham JR, Pinner MS, Bedford ID, Markham PG (1992) Mutational analysis of the monopartite geminivirus beet curly top virus. Virology 191: 396-405

Thomas JE, Massalski PR, Harrison BD (1986) Production of monoclonal antibo-dies to African cassava mosaic virus and differences in their reactivities with other whitefly transmitted geminiviruses. J Gen Virol 67: 2739-2748

CONTRIBUTED BY

Briddon RW, Markham PG

FAMILY *CIRCOVIRIDAE*

TAXONOMIC STRUCTURE OF THE FAMILY

Family *Circoviridae*
Genus *Circovirus*

GENUS *CIRCOVIRUS*

Type Species chicken anemia virus (CAV)

VIRION PROPERTIES

MORPHOLOGY

Virions are 17-22 nm in diameter, icosahedral in structure, and do not possess an envelope. Chicken anemia virus (CAV), has a defined surface structure, whereas porcine circovirus (PCV) and beak and feather disease virus (BFDV) exhibit no surface structure.

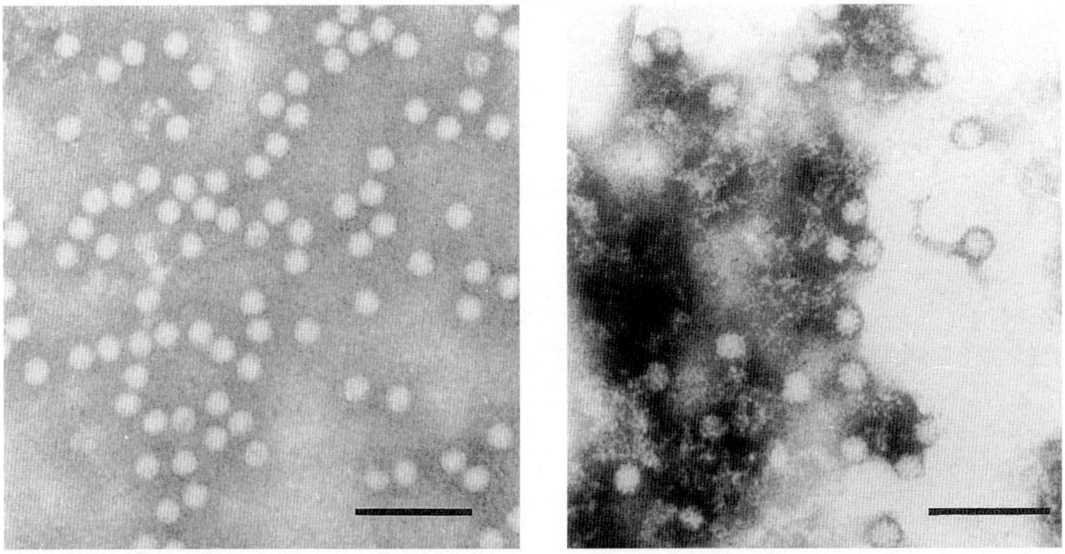

Figure 1: (left) Negative contrast electron micrograph of BFDV virions; (right) negative contrast electron micrograph of CAV virions. Bars represent 100 nm.

PHYSICOCHEMICAL AND PHYSICAL PROPERTIES

The buoyant density of virions in CsCl is 1.33 - 1.37 g/cm^3. Virion Mr, sedimentation coefficient, pH stability, heat sensitivity and other characteristics have not been reported.

NUCLEIC ACID

Virions contain circular ssDNA, 1.7-2.3 kb in size. Possible plant virus members have ssDNA 0.85-1 kb in size.

PROTEINS

CAV and PCV are composed of one protein, Mr 50 x 10^3, and 36 x 10^3, respectively. BFDV is composed of three proteins, Mr 26.3, 23.7 and 15.9 x 10^3. Possible plant virus members have one protein, Mr 19-20 x 10^3 in size.

LIPIDS

None reported.

CARBOHYDRATES

None reported.

GENOME ORGANIZATION AND REPLICATION

CAV has 3 ORFs but only 1 protein has been associated with the virion. BFDV has 3 proteins. PCV depends on cellular enzymes that are expressed during the S growth phase of host cells. Details of replication and morphogenetic strategies are not known.

ANTIGENIC PROPERTIES

No common antigens have been reported between CAV, PCV and BFDV. BFDV exhibits hemagglutination.

BIOLOGICAL PROPERTIES

Viruses appear to be specific for species of origin. Modes of transmission and possible vectors are not known. The viruses have a worldwide distribution. CAV causes transient anemia and immunosuppression in baby chicks. BFDV causes chronic and ultimately fatal disease in large psittacine birds. No disease has been associated with PCV infection. Cells of the hematopoietic system are infected by CAV and BFDV.

LIST OF SPECIES IN THE GENUS

The viruses, their genomic sequence accession numbers [ ] and assigned abbreviations ( ) are:

SPECIES IN THE GENUS

beak and feather disease virus		(BFDV)
chicken anemia virus	[M55918]	(CAV)
porcine circovirus		(PCV)

TENTATIVE SPECIES IN THE GENUS

None reported.

LIST OF UNASSIGNED VIRUSES IN THE FAMILY

Unassigned viruses, and their abbreviations ( ) that are considered possible members of the family are:

banana bunchy top virus	(BBTV)
coconut foliar decay virus	(CFDV)
subterranean clover stunt virus	(SCSV)

SIMILARITY WITH OTHER TAXA

None reported.

DERIVATION OF NAMES

circo: sigla to indicate that the viral DNA has a *cir*cular *co*nformation

REFERENCES

Chu PNG, Helms K (1988) Novel virus-like particles containing circular single-stranded DNAs associated with subterranean clover stunt disease. Virology 167: 38-49

Harding RM, Burns TM, Dale JL (1991) Virus-like particles associated with banana bunchy top disease contain single-stranded DNA. J Gen Virol 72: 225-230

Noteborn N, de Boer GF, van Roozelaar D, Karreman C, Karenburg O, Vos J, Jeurissen S, Hoeben R, Zantema Z, Koch G, van Ormondt H, van der Eb A (1991) Characterization of cloned chicken anemia virus DNA that contains all elements for the infectious replication cycle. J Virol 65: 3131-3139

Randles JW, Harold D, Julia J (1987) Small circular single-stranded DNA associated with foliar decay disease of coconut palm in Vanuatu. J Gen Virol 68: 272-280
Ritchie B, Niagro F, Lukert P, Steffens W, Latimer K (1989) Characterization of a new virus from cockatoos with psittacine beak and feather disease. Virology 171: 83-88
Tischer I, Gelderblom H, Vetterman W, Koch M (1982) A very small porcine virus with circular single-stranded DNA. Nature 295: 64-66
Todd D, Creelan J, Mackie D, Rixon F, McNulty MS (1990) Purification and biochemical characterization of chicken anemia agent. J Gen Virol 71: 819-823
Todd D, Niagro F, Ritchie B, Curran W, Alan G, Lukert P, Latimer K, McNulty MS (1991) Comparison of three animal viruses with circular single-stranded DNA genomes. Arch Virol 117: 129-135
Yuasa N, Taniguchi T, Yoshida I (1971) Isolation and some characteristics of an agent inducing anemia in chicks. Avian Dis 23: 366-385

CONTRIBUTED BY

Lukert P, de Boer GF, Dale JL, Keese P, McNulty MS, Randles JW, Tischer I

FAMILY *PARVOVIRIDAE*

TAXONOMIC STRUCTURE OF THE FAMILY

Family	*Parvoviridae*		
Subfamily		*Parvovirinae*	
Genus			*Parvovirus*
Genus			*Erythrovirus*
Genus			*Dependovirus*
Subfamily		*Densovirinae*	
Genus			*Densovirus*
Genus			*Iteravirus*
Genus			*Contravirus*

VIRION PROPERTIES

MORPHOLOGY

Virions are unenveloped, 18-26 nm in diameter, and exhibit icosahedral symmetry. The particles are composed of 60 copies of the capsid protein. The principal protein appears to be either VP2 or VP3 although 12 of the copies may be VP1.

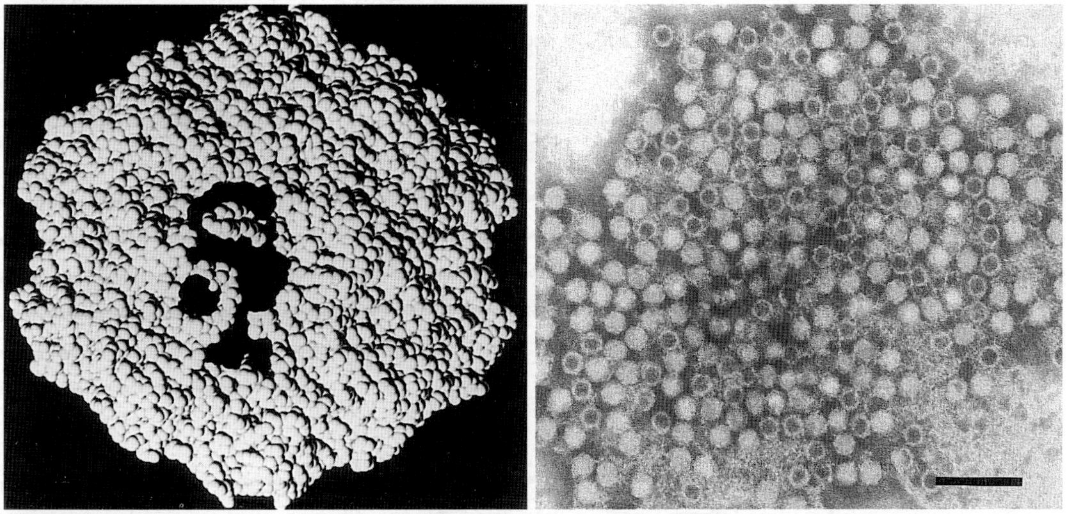

Figure 1: (left) Canine parvovirus capsid structure using a space-filling model, where each amino acid is represented by a 4Å sphere. One VP2 molecule is shown using darker spheres to illustrate the contribution of each VP2 protein to the structure and the intertwined arms of the VP2 molecules. (right) Negative contrast electron micrograph of canine parvovirus, the bar represents 100 nm, (courtesy of Parrish CR).

PHYSICOCHEMICAL AND PHYSICAL PROPERTIES

Virion Mr is about 5.5-6.2 x 10^6. Virion buoyant density is 1.39-1.42 g/cm^3 in CsCl. The S_{20w} is 110-122. Infectious particles are composed of about 80% protein and about 20% DNA. Infectious particles with buoyant densities about 1.45 g/cm^3 may represent conformational or other variants, or precursors to the mature particles. Defective particles with deletions in the genome occur and exhibit lower densities. Mature virions are stable in the presence of lipid solvents, or on exposure to pH 3-9 or, for most species, incubation at 56° C for at least 60 min. Viruses can be inactivated by treatment with formalin, β-propriolactone, hydroxylamine, or oxidizing agents.

NUCLEIC ACID

The genome is a linear, molecule of ssDNA, 4-6 kb in size with a (Mr 1.5-2.0 x 10^6). The G+C content is 41-53%. Some members preferentially encapsidate ssDNA of negative polarity (i.e., complementary to the viral mRNA species; e.g., MMV), others may encapsidate ssDNA species of either polarity in equivalent (e.g., AAV), or different proportions (BPV).

The percentage of particles encapsidating the positive strand can vary from 1 to 50% and may be influenced by the host cell in which the virus is produced (e.g., LUIII virus). After extraction, and depending on the amounts present, the complementary strands may hybridize *in vitro* to form dsDNA.

PROTEINS

Viruses generally have 2-4 virion proteins species (VP1-4). Depending on the species, the Mr of VP1 species is 80-96 x 10^3, the VP2 species is 64-85 x 10^3, the VP3 species is 60-75 x 10^3 and the VP4 species 49-52 x 10^3. The viral proteins represent alternative forms of the same gene product. Enzymes are lacking. The principal protein species is VP2 or VP3. Spermidine, spermine, and putrescine have been identified in some virus particles.

LIPIDS

Virions lack lipids.

CARBOHYDRATES

None of the viral proteins is glycosylated.

GENOME ORGANIZATION AND REPLICATION

Parvoviruses possess 2 major genes, the REP (or NS) ORF that encodes functions required for transcription and DNA replication, and the CAP (or S) ORF that encodes the coat

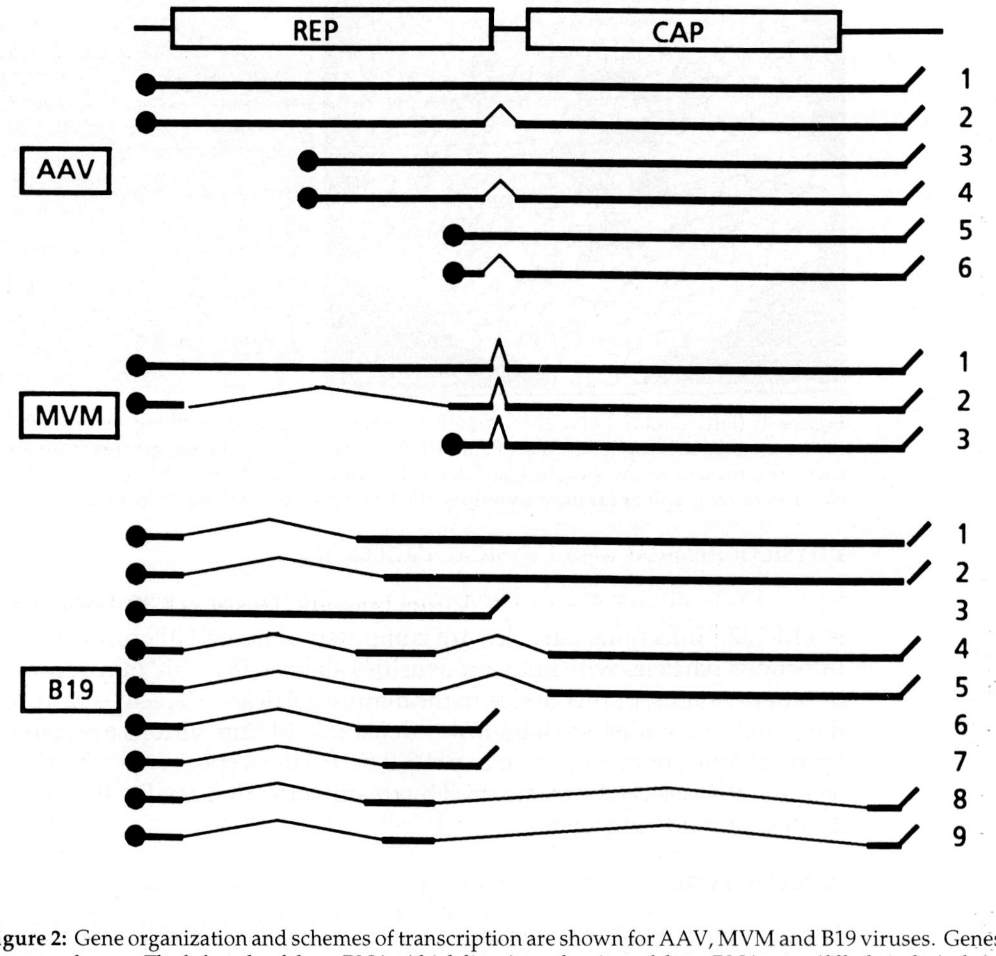

Figure 2: Gene organization and schemes of transcription are shown for AAV, MVM and B19 viruses. Genes are shown as boxes. The left ends of the mRNAs (thick lines) are the sites of the mRNA caps (filled circles), the right ends are the polyadenylation sites (oblique lines); introns are indicated by thin lines (adapted from Berns, 1990).

proteins. Both genes are present on the same DNA strand in the cases of the vertebrate parvoviruses (Fig. 2) and some densoviruses (e.g., *Densovirinae* genera *Iteravirus* and *Contravirus*, Fig. 3 lower). In the case of *Densovirus*, the REP function and the coat proteins are encoded on complementary strands (Fig. 3 upper). Other minor ORFs have been detected in some viruses. For some of these a protein product has been identified (e.g., the ORF for the amino terminus of VP1). The MMV REP ORF produces 2 major non-structural proteins, NS1, NS2.

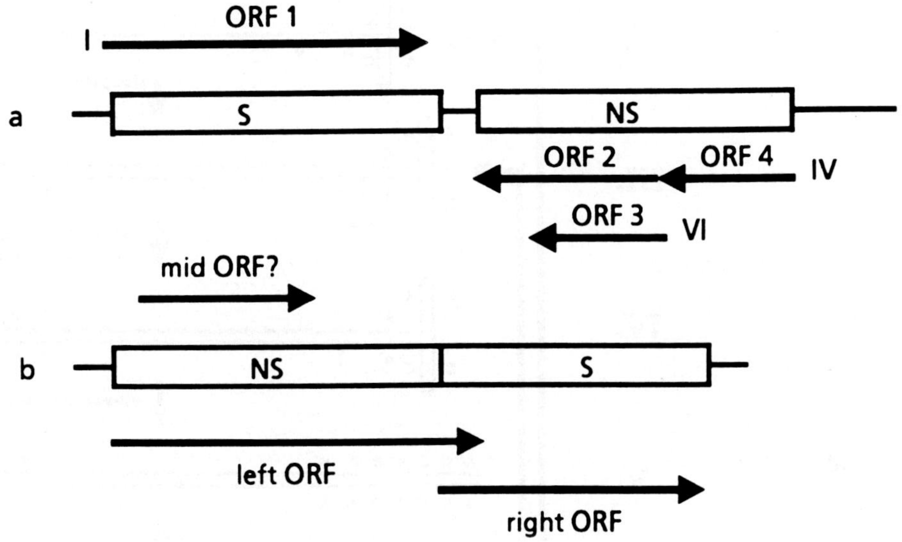

Figure 3: The genetic organization of (upper) the invertebrate Junonia coenia densovirus (*Densovirus*), and (lower) Aedes albopictus densovirus (*Contravirus*). S = structural proteins; NS = non-structural proteins; I, IV, VI are reading frames. The arrowed lines indicate the possible transcription products that have been deduced from DNA sequence analyses.

Mutations within the REP (NS) ORF of MMV block virus replication and gene expression. For some viruses alternative splicing allows different forms of the REP gene products to be produced. The coat (CAP) ORF of MMV produces up to 3 proteins. MMV VP3 is generated in the intact capsid by proteolytic cleavage of VP2. VP1 and VP2 are identical except for their amino termini. Synthesis of VP1 derives from a spliced mRNA that brings an upstream small ORF with basic amino acids motifs to the 5' of the VP2-coding sequence. Parvoviruses use an alternative splice donor, while dependoviruses use an alternative splice acceptor for this purpose. VP1, by virtue of its particular position in the capsid structure may facilitate DNA binding. Mutants in REP or CAP can be complemented in trans. The palindromic sequences (at both termini) are required in *cis* for DNA replication to occur.

The processes of adsorption and uncoating are poorly understood. Viral replication takes place in the cell nucleus and appears to require the cell to go through its S phase, indicating a close association between the host and viral replication processes, and probably involving host DNA polymerase(s) (e.g., α, δ, or others). Rendering the viral genome into a dsDNA is thought to be required before mRNA transcription occurs. DNA synthesis derives from a self-priming mechanism and the existence of palindromic sequences (Fig. 4). The replicative intermediate is a linear duplex molecule covalently linked at one end by a hairpin primer. The covalent link is broken by the REP protein(s) and the hairpin is transferred to the progeny strand. The resulting 3' terminal gap in the parental strand is repaired using the transferred sequence as a template. In the case of MMV, NS1 (REP equivalent) is covalently bound to the 5' end of the progeny strand. Other replicative intermediates include concatameric structures. Mature ssDNA genome equivalents are removed from the replicative complex in a manner that seems to be dependent on the availability of some species of NS protein and empty capsid assembly.

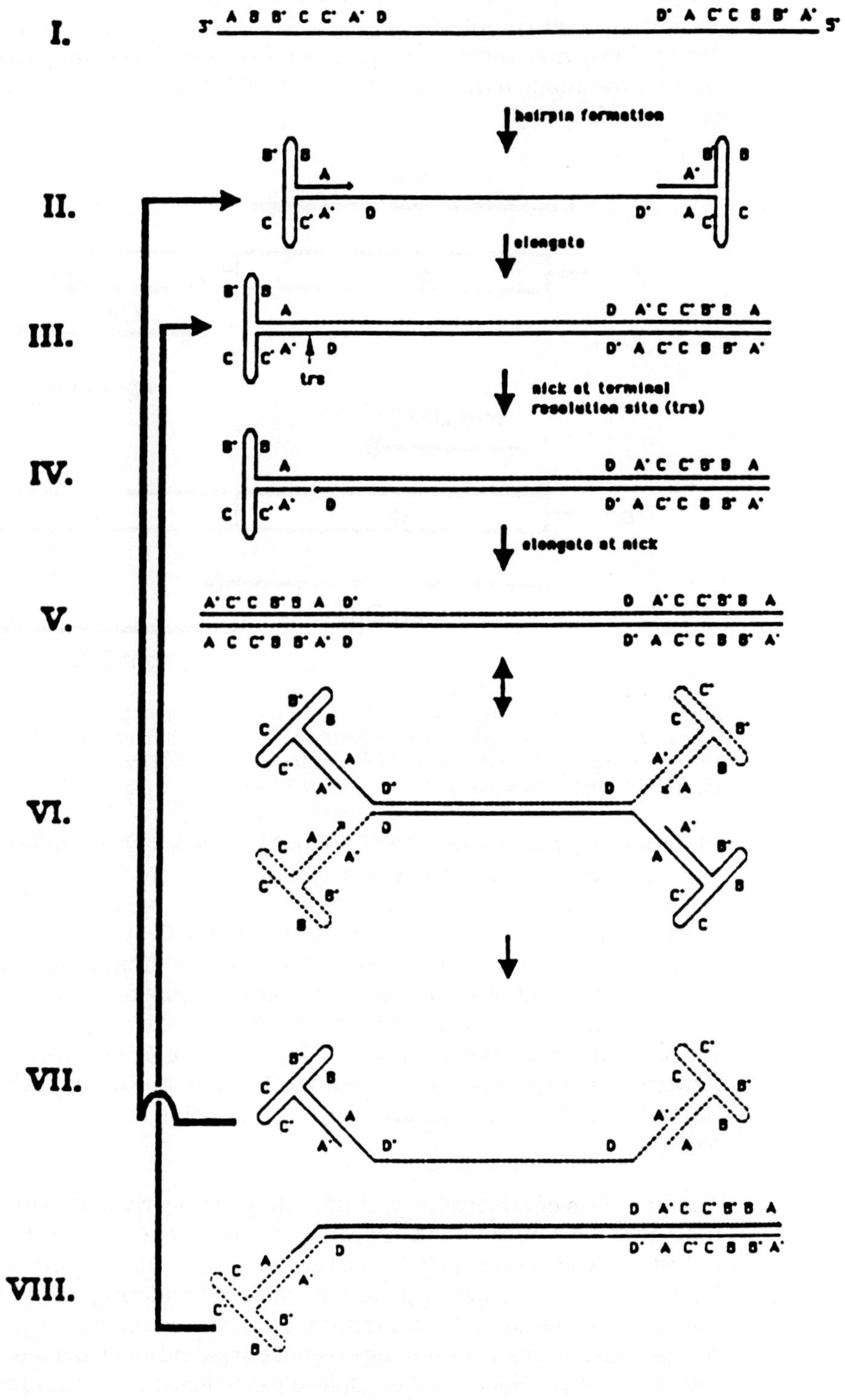

Figure 4: DNA replication model for AAV. The terminal repeats of AAV are self-complementary and capable of forming hairpins shown in structure II. This allows for self-primed DNA synthesis from the 3' hydroxyl group. The site and strand specific nick shown in IV is made by Rep 68 or Rep 78. The two large Rep proteins also possess helicase activity, as required for the isomerization to convert structure V to VI. Structures VII and VIII are equivalent to structures II and III. Structure VII can either be encapsidated during strand displacement (resulting in net virion production) or it can enter the template amplification pathway as shown.

Rep68 or Rep78, the two large Rep proteins also possess helicase activity, as required for the isomerization to convert structure V to VI. Structures VII and VIII are equivalent to structures II and III. Structure VII can either be encapsidated during strand displacement (resulting in net virion production) or it can enter the template amplification pathway as shown. Depending on the virus there may be 1 (B19 virus, *Iteravirus* and *Contravirus*), 2 (MMV, *Densovirus*), or 3 (AAV) promoters for mRNA transcription (Fig. 2). Some of the mRNAs are spliced allowing alternate forms of the protein products to be produced. The mRNA species are capped and polyadenylated either at a common 3' site near the end of the genome (MMV, AAV), or at an alternative polyadenylation site in the centre of the genome as well as at a site near the end of the genome (B19, ADV).

Depending on the species, viruses may benefit from co-infection with other viruses, such as adenoviruses, or herpesviruses, or from the effects of chemical or other treatments of the host. Viral proteins accumulate in the nucleus in the form of empty capsid structures. Progeny infectious virions accumulate in the cell nucleus.

ANTIGENIC PROPERTIES

Some, but not all, species in a genus may be antigenically related by epitopes in the NS proteins.

BIOLOGICAL PROPERTIES

Autonomous parvoviruses require host cell passage through S-phase. Certain parvoviruses replicate efficiently in the presence of helper viruses (e.g., adenoviruses, herpesviruses). These helper functions involve the adenovirus or herpes early gene products and *trans*-activation of parvovirus replication. The helper functions appear to relate to effects of the helper virus upon the host cell rather than direct involvement of helper virus gene products in parvovirus replication.

Association of parvoviruses with tumor cell lines appears to relate to increased DNA replication and/or the state of differentiation in such cells rather than previous involvement as an etiologic agent of oncogenesis. Co-infection involving certain parvoviruses and selected oncogenic adenoviruses (or other viruses) may reduce the oncogenic effect of those viruses, possibly by promoting cell death.

In certain circumstances parvovirus DNA may integrate into the host genome from which it may be activated by subsequent helper virus infection. The site of integration may be specific in certain hosts (e.g., the q arm of human chromosome 19 for AAV-2).

SUBFAMILY *PARVOVIRINAE*

TAXONOMIC STRUCTURE OF THE GENUS

Subfamily	***Parvovirinae***
Genus	*Parvovirus*
Genus	*Erythrovirus*
Genus	*Dependovirus*

DISTINGUISHING FEATURES

Viruses assigned to the subfamily *Parvovirinae* infect vertebrates and vertebrate cell cultures, frequently in association with other viruses.

GENUS *PARVOVIRUS*

Type Species mice minute virus (MMV)

DISTINGUISHING FEATURES

For some members of the genus, mature virions contain negative-strand DNA of 5 kb. In other members, positive-strand DNA occurs in variable proportions (1-50%). The linear molecule of ssDNA has hairpin structures at both the 5'- and 3'-ends. The 3'-terminal hairpin is 115-116 nt in length, the 5' structure is 200-242 nt long. There are two mRNA promoters (map units 4 and 39) and a single polyadenylation site at the 3' end. Characteristic cytopathic effects are induced by the viruses during replication in cell culture. Many species exhibit hemagglutination with red blood cells of one or more species. Under experimental conditions the host range may be extended to a large number of vertebrate species (e.g., rodent viruses and LUIII replicate in Syrian hamsters). Transplacental transmission has been detected for a number of species. Goose parvovirus is transmitted vertically through the ovary.

LIST OF SPECIES IN THE GENUS

The viruses, their alternative names ( ), genomic sequence accession numbers [ ] and assigned abbreviations ( ) are:

SPECIES IN THE GENUS

Aleutian mink disease virus (Aleutian disease virus)	[M20036]	(AMDV)
bovine parvovirus	[M14363]	(BPV)
canine minute virus		(CMV)
canine parvovirus	[M19296]	(CPV)
chicken parvovirus		(ChPV)
feline panleukopenia virus	[M75728]	(FPV)
feline parvovirus		
goose parvovirus		(GPV)
HB virus		(HBPV)
H-1 virus	[X01457]	(H-1PV)
Kilham rat virus (rat virus, R)		(KRV)
lapine parvovirus		(LPV)
LUIII virus	[M81888]	(LUIIIV)
mink enteritis virus		(MEV)
mice minute virus	[J02275]	(MMV)
porcine parvovirus	[D00623]	(PPV)
raccoon parvovirus	[M24005]	(RPV)
RT parvovirus		(RTPV)
tumor virus X		(TVX)

TENTATIVE SPECIES IN THE GENUS

rheumatoid arthritis virus (RAV-1)

GENUS *ERYTHROVIRUS*

Type Species B19 virus (B19V)

DISTINGUISHING FEATURES

Populations of mature virions contain equivalent numbers of positive and negative sense ssDNA, 5 kb in size. The DNA molecules contain inverted terminal repeats of 383 nucleotides, the first 365 of which form a palindromic sequence. Upon extraction, the complementary DNA strands usually form dsDNA. There is a single mRNA promoter (map unit 6) and two polyadenylation signals, one near the middle of the genome, the other near the 3' end. Efficient replication occurs in primary erythrocyte precursors. There have also been reports of productive infection of primary umbilical cord erythrocytes and of a continuous line of megakaryoblastoid cells.

LIST OF SPECIES IN THE GENUS

The viruses, their genomic sequence accession numbers [ ] and assigned abbreviations ( ) are:

SPECIES IN THE GENUS

B19 virus	[M13178, M24682]	(B19V)

TENTATIVE SPECIES IN THE GENUS

None reported.

GENUS *DEPENDOVIRUS*

Type Species adeno-associated virus 2 (AAV-2)

DISTINGUISHING FEATURES

Populations of mature virions contain equivalent numbers of positive or negative strand ssDNA 4.7 kb in size. The DNA molecules contain inverted terminal repeats of 145 nucleotides, the first 125 of which form a palindromic sequence. Upon extraction, the complementary DNA strands usually form dsDNA. The are three mRNA promoters (map units 5, 19, 40). Efficient virus replication is dependent upon helper adenoviruses or herpes viruses. Under certain conditions (presence of mutagens, synchronization of cell replication with hydroxyurea), replication can also be detected in the absence of helper viruses. All AAV isolates share a common antigen as demonstrated by fluorescent antibody staining. Transplacental transmission has been observed for AAV-1 and vertical transmission has been reported for avian AAV.

LIST OF SPECIES IN THE GENUS

The viruses, their genomic sequence accession numbers [ ] and assigned abbreviations ( ) are:

SPECIES IN THE GENUS

adeno-associated virus 1		(AAV-1)
adeno-associated virus 2	[J01901]	(AAV-2)
adeno-associated virus 3		(AAV-3)
adeno-associated virus 4		(AAV-4)
adeno-associated virus 5		(AAV-5)
avian adeno-associated virus		(AAAV)
bovine adeno-associated virus		(BAAV)
canine adeno-associated virus		(CAAV)
equine adeno-associated virus		(EAAV)
ovine adeno-associated virus		(OAAV)

TENTATIVE SPECIES IN THE GENUS

None reported.

SUBFAMILY *DENSOVIRINAE*

TAXONOMIC STRUCTURE OF THE SUBFAMILY

Subfamily	***Densovirinae***
Genus	*Densovirus*
Genus	*Iteravirus*
Genus	*Contravirus*

DISTINGUISHING FEATURES

Viruses assigned to the subfamily *Densovirinae* infect arthropods. The ssDNA genome of virions is either of positive or negative sense. Upon extraction, the complementary DNA strands usually form dsDNA. There are four structural proteins. Viruses multiply efficiently in most of the tissues of larvae, nymphs, and adult host species without the involvement of helper viruses. Cellular changes consist of hypertrophy of the nucleus with accumulation of virions therein to form dense, voluminous intranuclear masses. The known host range includes members of the Dictyoptera, Diptera, Lepidoptera, Odonata and Orthoptera. There is evidence that densovirus-like viruses also infect and multiply in crabs and shrimps.

GENUS *DENSOVIRUS*

Type Species Junonia coenia densovirus (JcDNV)

DISTINGUISHING FEATURES

The ssDNA genome is about 6 kb in size. Populations of virions encapsidate equal amounts of positive and negative strands. On one strand there are 3 ORFs which encode NS proteins using a single mRNA promoter (7 map units from the end). The four structural proteins are encoded on the complementary strand, using an mRNA promoter that is 9 map units from the end of that strand. JaDNV has an inverted terminal repeat of 517 bases, the first 96 of which can fold to form a T-shaped structure of the type found in the ITR of AAV DNA.

LIST OF SPECIES IN THE GENUS

The viruses, and their assigned abbreviations ( ) are:

SPECIES IN THE GENUS

Galleria mellonella densovirus (GmDNV)
Junonia coenia densovirus (JcDNV)

TENTATIVE SPECIES IN THE GENUS

None reported.

GENUS *ITERAVIRUS*

Type Species Bombyx mori densovirus (BmDNV)

DISTINGUISHING FEATURES

The ssDNA genome is about 5 kb in size. Populations of virions encapsidate equal amounts of plus and minus strands. ORFs for both the structural and NS proteins are located on the same strand. There is apparently one mRNA promoter upstream of each ORF. There is a small ORF on the complementary strand of unknown function. The DNA has an inverted terminal repeat of 225 bases, the first 175 are palindromic but do not form a T-shaped structure when folded.

List of Species in the Genus

The viruses, their genomic sequence accession numbers [ ] and assigned abbreviations ( ) are:

Species in the Genus

Bombyx mori densovirus	[M15123, M60583, M60584]	(BmDNV)

Tentative Species in the Genus

None reported.

Genus *Contravirus*

Type Species Aedes aegypti densovirus (AaDNV)

Distinguishing Features

The genome is about 4 kb in size. Populations of virions encapsidate positive and negative strands, a majority of which are of negative polarity (85%). ORFs for the structural and NS proteins are on the same strand. There are mRNA promoters at map units 7 and 60. There is a small ORF of unknown function on the complementary strand. A palindromic sequence of 146 bases is found at the 3' end of the genome and a different palindromic sequence of 164 bases at the 5' end. Both terminal sequences can fold to form a T-shaped structure.

List of Species in the Genus

The viruses, and their assigned abbreviations ( ) are:

Species in the Genus

Aedes aegypti densovirus	(AaDNV)
Aedes albopictus densovirus	(AlDNV)

Tentative Species in the Genus

Acheta domestica densovirus	(AdDNV)
Aedes pseudoscutellaris densovirus	(ApDNV)
Agraulis vanillae densovirus	(AvDNV)
Casphalia extranea densovirus	(CeDNV)
Diatraea saccharalis densovirus	(DsDNV)
Euxoa auxiliaris densovirus	(EaDNV)
Leucorrhinia dubia densovirus	(LdDNV)
Lymantria dubia densovirus	(LdDNV)
Periplanata fuliginosa densovirus	(PfDNV)
Pieris rapae densovirus	(PrDNV)
Pseudaletia includens densovirus	(PiDNV)
Sibine fusca densovirus	(SfDNV)
Simulium vittatum densovirus	(SvDNV)

List of Tentative Species in the Subfamily

hepatopancreatic parvo-like virus of shrimps	(HPPLV)
parvo-like virus of crabs	(PCV84)

List of Unassigned Viruses in the Family

None reported.

Similarity with Other Taxa

None reported.

Derivation of Names

adeno: from Greek *aden*, "gland"
contra: from Latin *contra*, "opposite"
dependo: from Latin *dependeo*, "to hang down"
entomo: from Greek *entomon*, "insect"
erythro: from Greek *erythros*, "red"
denso: from Latin *densus*, "thick, compact"
parvo: from Latin *parvus*, "small"

References

Afanasiev BN, Galyov EE, Buchatsky LP, Kozlov YV (1991) Nucleotide sequence and genomic organization of Aedes densonucleosis virus. Virology 185: 323-336
Bando H, Choi H, Ito Y, Kawase S (1990) Terminal structure of a densovirus implies a hairpin transfer replication which is similar to the model for AAV. Virology 179: 57-63
Bando H, Kusuda J, Kawase S (1987) Molecular cloning and characterization of Bombyx densovirus genomic DNA. Arch Virol 98: 139-146
Berns KI (ed) (1984) The parvoviruses. Plenum Press, New York London
Berns KI (1990) Parvovirus replication. Microbiol Rev 54: 316-329
Dumas B, Jourdan M, Pascaud A-M, Bergoin M (1992) Complete nucleotide sequence of the cloned infectious genome of Junonia coenia densovirus reveals an organization unique among parvoviruses. Virology 191: 202-222
Lightner DV, Redman RM (1985) A parvo-like virus disease of penaeid shrimp. J Invert Pathol 45: 47-53
Mari J, Bonami JR (1988) PC84, a parvo-like virus from crab *Carcinus mediterraneus*: pathological aspects, ultrastructure of the agent, and first biochemical characterization. J Invert Pathol 51: 145-156
Pattison JR (ed) (1988) Parvoviruses and human disease. CRC Press, Boca Raton FL
Siegl G, Bates RC, Berns KI, Carter BJ, Kelly DC, Kurstak E, Tattersall P (1985) Characteristics and taxonomy of *Parvoviridae*. Intervirology 23: 61-73
Tijssen P (ed) (1990) Handbook of parvoviruses, Vols I and II. CRC Press, Boca Raton FL

Contributed By

Berns KI, Bergoin M, Bloom M, Lederman M, Muzyczka N, Siegl G, Tal J, Tattersall P

FAMILY HEPADNAVIRIDAE

TAXONOMIC STRUCTURE OF THE FAMILY

Family	***Hepadnaviridae***	
Genus		*Orthohepadnavirus*
Genus		*Avihepadnavirus*

VIRION PROPERTIES

MORPHOLOGY

Hepadnaviruses are spherical, occasionally pleomorphic, 40-48 nm in diameter but with no evident surface projections. The outer 7 nm thick, detergent-sensitive envelope contains the surface antigens and surrounds an icosahedral, 27-35 nm diameter nucleocapsid core with 180 capsomers arranged in a T= 3 symmetry. The core is composed of one major protein species, the core antigen, and encloses the viral genome (DNA) and associated minor protein(s). For some viruses, variable length, 22 nm diameter, filamentous forms and spherical 16-25 nm structures occur that lack cores (HBsAg).

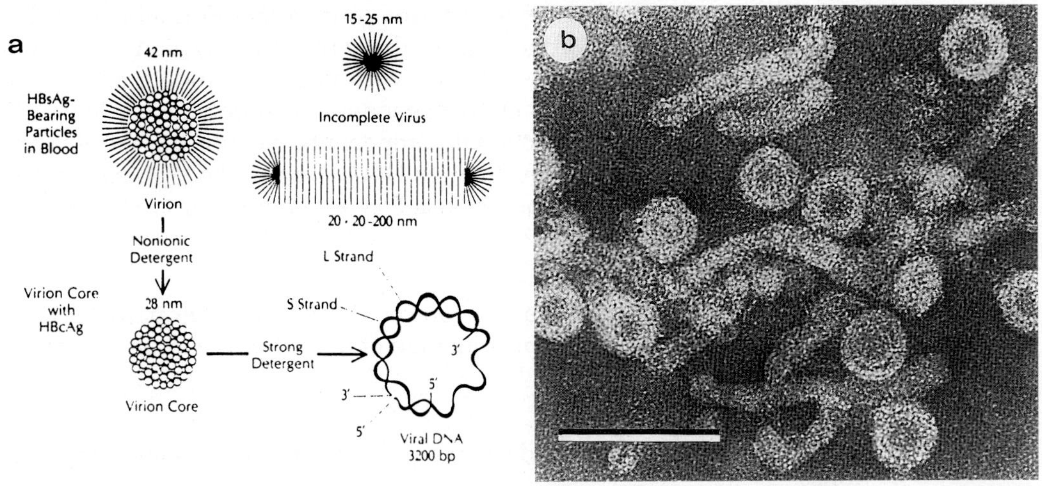

Figure 1: (a) Diagram of virion and virus-associated particles in section (from Hollinger, 1990); (b) negative contrast electron micrograph of virions and filamentous forms of HBsAg. The bar represents 100 nm.

PHYSICOCHEMICAL AND PHYSICAL PROPERTIES

The virion S_{20w} is about 280. The buoyant density in CsCl is about 1.25 g/cm^3. The buoyant density of particles lacking cores is about 1.18 g/cm^3. Virus-derived cores (lacking envelopes) have densities of about 1.36 g/cm^3. Viruses are unstable at acid pH. Generally, the virus infectivity is retained for 6 months at 30-32° C and neutral pH.

NUCLEIC ACID

The genome consists of a single molecule of non-covalently closed, circular DNA that is partially double-stranded and partially single-stranded. Virion Mr is about 1.6-1.8 x 10^6; S_{20w} is about 15 and G+C content is about 48%. One strand (negative sense, i.e., complementary to the viral mRNAs) is full-length (3.0-3.3 kb), the other varies in size. In orthohepadnaviruses, the full-length negative sense DNA has a nick at a unique site corresponding to a position 242 nucleotides downstream from the 5' end of the positive sense strand (Fig. 2). The ssDNA may represent up to 60% of the circle. For avihepadnaviruses the nick in the negative sense DNA is about 50 nt from the end and genomes may be fully double-stranded. The uniquely-located 5'-ends of the two strands overlap by about 240 nt so that the circular configuration is maintained by base-pairing of cohesive ends. The 5' end of the negative sense DNA has a covalently attached terminal protein. The 5' end of the positive sense DNA has a covalently attached 19-nt, 5' capped oligoribonucleotide primer. The DNA sequence of HBV has an enhancer region (ENH), two 11-base direct repeat sequences, DR1, DR2, (not always conserved among viruses), a U5-like sequence, a polyadenylation signal (TATAAA) and a putative glucocorticoid-responsive element (GRE, Fig. 2). The 5' end of the negative strand is located within DR1, the 5' end of the positive strand is at the 3' boundary of DR2.

Proteins

In orthohepadnaviruses, the envelope (surface antigen) proteins of virions consists of three groups of antigenically complex proteins: S-proteins (p24, GP27), M-proteins (P33, GP36) and L-proteins (P39, GP42). All the envelope proteins have common carboxy termini and differ in amino termini (due to different sites of translation initiation) and in the presence and form of glycosylation. For HBV, the major S proteins appear to have the same amino acid composition, however, GP27 has a single glycosylation site (complex glycan type) that is shared by the M-proteins GP33 and GP36 which are composed of P24 with an additional 55 amino acids and glycosylation site (high mannose glycan type). The L proteins contain a further about 120 amino acids and their N-termini are myristylated.

The 20-25 nm particles (HBsAg) contain predominantly S-proteins (p24, GP27) and occasionally M-proteins. Filamentous forms contain these proteins and the L-proteins. The virion core is composed principally of the core antigen (HBcAg), Mr about 22 kDa. It is a phosphoprotein. Enzymes associated with virions include a protein kinase and reverse transcriptase with RNA- and DNA-dependent DNA polymerase and RNase H activities (P gene products). Other functional components include the terminal protein covalently attached to the 5'-end of the full-length DNA strand. The terminal protein has been shown to be a component of the about 90 kDa P gene product.

Lipids

Lipids are components of the envelope of virions and other particles and are derived from host cell membranes. The lipids include phospholipids, sterols and fatty acids.

Carbohydrates

Demonstrated in particles and virions as N-linked glycans of the complex and high mannose types.

Genome Organization and Replication

The HBV genome DNA has four partially overlapping genes (S, C, P, X), all orientated in the same direction (Fig. 2). For DHBV there are three genes (S, C, P). There appear to be no intervening sequences. The S gene ORF codes for the surface antigens. In the S gene, the p24 protein (for HBV, the HBsAg) is preceded by pre-S2, which, in turn, is preceded by pre-S1. Each has an in-frame ATG codon for the initiation of protein synthesis. For different mammalian hepadnaviruses the pre-S1 and S sequences may vary in size, otherwise the genes are similar for the different viruses. The C gene ORF specifies the major core protein (for HBV, the HBcAg). It is preceded by a short pre-C region that varies in size between different viruses. The C gene of avihepadnaviruses is larger than its mammalian counterpart. The P-gene covers 80% of the genome and overlaps the other three ORFs (Fig. 2). It codes for the reverse transcriptase, with DNA polymerase and RNAse H activities, and the genome-linked terminal protein. The X gene specifies a protein with a probable transactivation function. It varies in size among the HBV serotypes, being largest for HBV adr.

Virus enters hepatocytes by an unknown mechanism. The virus polymerase uses the 3' end of the positive sense DNA strand as a primer and repairs ss regions to make full-length dsDNA molecules. The DNA is converted into a covalently closed circular DNA species by removal of the terminal protein of the negative strand, the oligoribonucleotide of the positive strand and the terminally redundant region of the negative strand. Closed DNA is then achieved by ligation. dsDNA is located in the nucleus of infected cells. Transcription of viral mRNAs by host RNA polymerase II yields predominantly 3.4, 2.4 and 2.1 kb mRNA species (Fig 2). Transcription is enhanced by the X protein. The 3.4 kb polyadenylated mRNA is greater in size than the genome length due to terminally redundant sequences (Fig. 2). Following transcription, translation of the viral gene products ensues. For HBV, the p39 protein (GP42) is translated from a 2.4 kb polyadenylated mRNA, the p33 protein (GP39) from a 2.1 kb polyadenylated mRNA and the p24 protein (GP27) from a 2.1 kb polyadenylated mRNA and possibly others that are about 2 kb in size. The C antigens are translated from the 3.4 kb species. The mRNAs are unspliced and are made from distinct promoters. The C promoter may have tissue specificity. Two regions of the HBV genome have transcription enhancer activities, another is similar to glucocorticoid responsive elements. P protein may be translated from the 3.4 kb mRNA by an unknown mechanism. The X protein may be translated from a minor mRNA that has yet to be identified, or from the other mRNAs by an unknown mechanism. The 3.4, 2.4 and 2.1 kb mRNAs terminate at the same polyadenylation signal. The greater-than-genome length 3.4 kb mRNA is initiated near the start of the pre-C ORF and terminates about 100 nucleotides downstream of the pre-C initiation site after making a complete copy of the genome. The polyadenylation signal for all the mRNAs is located within the C coding region.

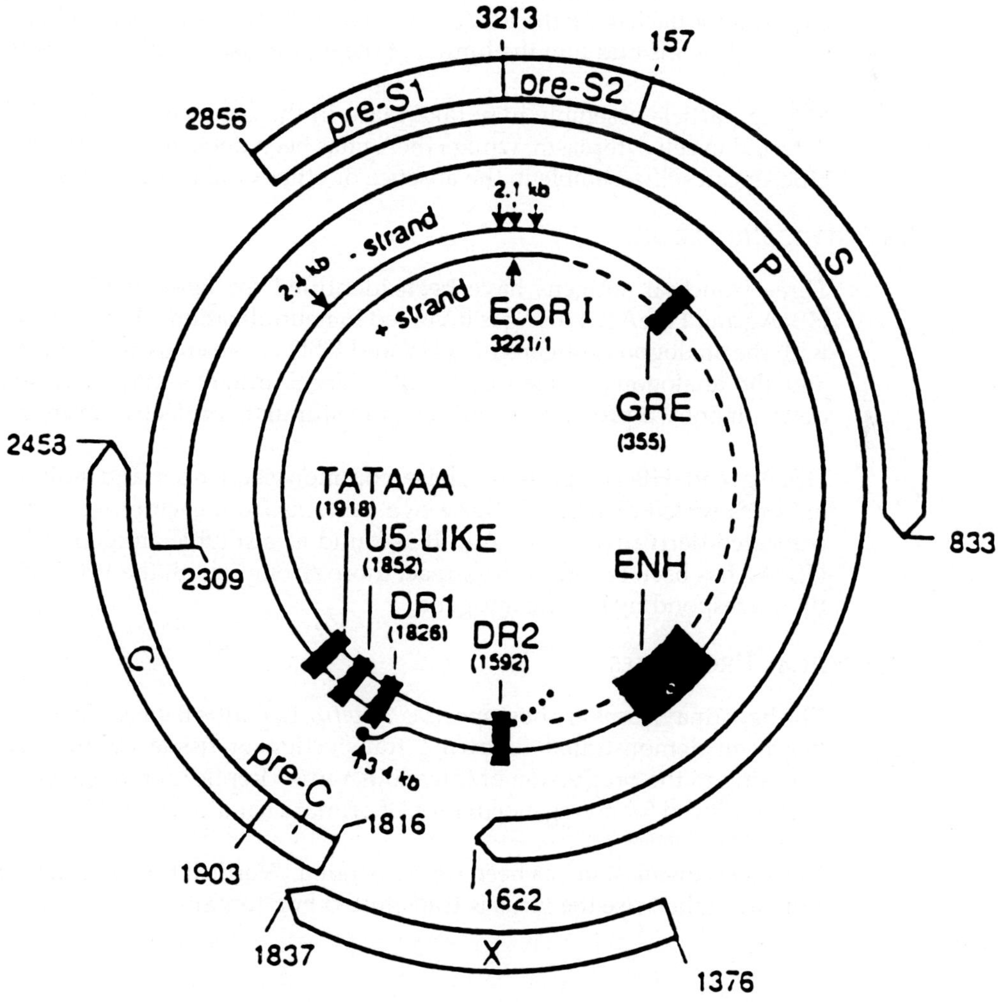

Figure 2: Diagram of genome organization of HBV (adw2) indicating the DNA arrangement, the positions of 4 ORFs (C, P, S, X), mRNA initiation sites and other sequence elements (courtesy of Robinson WS and Plenum Press).

HBcAg may regulate its mRNA synthesis. The S mRNAs are found in cells expressing the HBsAg only. Both S and C mRNAs are found in cells supporting virus replication. The X protein has been postulated to have tissue-specific transactivation properties for viral and cellular genes, however, its role in natural infections is unknown. At least for GSHV and WHV the X gene appears to be essential for virus replication. While integration of viral DNA into the host genome is not required for replication, integrated and deletion derivatives of viral species occur in hepatocellular carinoma (HCC) cells in culture and in hepatocytes of HCC patients. Singly integrated forms cannot serve as templates for the synthesis of the 3.4 kb mRNA (which requires circularized or concatenated copies of integrated DNA), which may account for the observation that predominantly subgenomic mRNAs and HBsAg are synthesized in HCC cell lines with integrated HBV DNA. However in such cells defective HBV sequences often occur. Current evidence indicates that following the generation of a covalently closed circular DNA and synthesis of the 3.4 kb mRNA, this RNA associates with viral core particles where it serves as a template for synthesis of minus strand DNA by reverse transcription using a protein primer. Reverse transcription is initiated in the vicinity of DR1 (near the mRNA poly [A] tail) and proceeds to the 5' end of the mRNA. The minus DNA strand serves as template for plus strand DNA synthesis and is primed by transposition of the 5'-end of the plus strand RNA that remains after RNase H digestion (i.e., transposition from the 5' proximal DR1 position to the 3' proximal DR2). The plus strand DNA strand is incomplete in most core particles at the time of virion assembly and release from infected cells. The carboxy-terminal domain of C protein probably is required for packaging the RNA. Cytoplasmic core particles attached to the p39- and p33-related proteins bud into the lumen of the endoplasmic reticulum as HBV particles.

HBsAg particle assembly may take place in the absence of cores. HBsAg has only been detected in cell cytoplasm, while HBcAg has been detected in both cytoplasm and nucleus. HBcAg can self-assemble in the absence of other viral components.

Antigenic Properties

Three principal antigens have been identified for hepadnaviruses. These are HBsAg, HBcAg and HBeAg. HBsAg is involved in neutralization. It cross-reacts to a limited extent with the analogous antigens of WHV and GSHV. No cross-reaction exists between HBsAg and the analogous antigen of DHBV. Pre-S antigens may bear specific neutralization determinants. S proteins are sufficient to stimulate protective immunity.

HBeAg and HBcAg proteins share common sequences and epitopes but also contain epitopes which distinguish these two proteins from each other. The HBeAg is a 16 kDa truncated derivative of HBcAg. It is found as a soluble antigen in the serum of patients. HBcAg has been found to cross-react more strongly with the WHV core antigen than with the corresponding surface antigens.

Biological Properties

The hepadnaviruses are host specific. *In vitro,* hepatitis B virus, GSHV and WHV replication has been demonstrated following transfection of tissue culture cells by cloned DNA, resulting in the production of infectious virus. Replication of several hepadnaviruses has been achieved following inoculation of primary hepatocytes with serum that contains virus.

Vertical transmission has been demonstrated. Vertical transmission of HBV may occur in humans, otherwise the virus is transmitted horizontally.

GENUS *ORTHOHEPADNAVIRUS*

Type Species hepatitis B virus (HBV)

DISTINGUISHING FEATURES

Viruses infect mammals. The only known natural host of HBV is humans, although chimpanzees and gibbons may be infected experimentally. Experimental transmission of HBV has also been reported in African monkeys, rhesus and woolly monkeys. Virions are spherical particles, 40-42 nm in diameter with an internal nucleocapsid that is 27 nm in diameter. Virus DNA is mostly partially double-stranded. Virus genomes contain the X gene.

BIOLOGICAL PROPERTIES

HBV may cause acute and chronic hepatitis, cirrhosis, hepatocellular carcinoma, immune complex disease, polyarteritis, glomerulonephritis, infantile papular acrodematitis and aplastic anemia. Horizontal transmission of HBV can be by perinatal, percutaneous, sexual and other routes of close contact, e.g., intravenous drug abuse, and by use of infected blood and blood products. HBV can survive on surfaces which may come into contact with mucous membranes or open skin breaks (e.g., toothbrushes, baby bottles, toys, eating utensils, razors or hospital equipment such as respirators, endoscopes or laboratory equipment). Although some populations of mosquitoes and bedbugs have been shown to contain HBsAg, there has been no direct demonstration of HBV transmission to humans by insect vectors. Hepatitis occurs in woodchucks and squirrels infected with their respective viruses.

At least 5 antigenic specificities have been identified for HBV. A group determinant (a) is shared by all S antigen preparations. Two pairs of subtype determinants (d, y and w, r) have been demonstrated which are generally mutually exclusive (and thus usually behave as alleles). Antigenic heterogeneity of the w determinants and additional determinants, such as q and x or g, have also been described. To date, eight S antigen subtypes have been identified, namely ayw, ayw2, ayw3, ayw4, ayr, adw4 and adr. Unusual combinations of S subtype determinants such as awr, adwr, adyw, adyr and adywr have been reported. The S subtypes have an uneven geographical distribution. The subtype specificity of S antigen can be affected by mutations. HBV variants with amino acid mutations in the group determinant (a) have been identified.

LIST OF SPECIES IN THE GENUS

The viruses, their genomic sequence accession numbers [ ] and assigned abbreviations ( ) are:

SPECIES IN THE GENUS

ground squirrel hepatitis B virus	[K02715]	(GSHV)
hepatitis B virus	[M12906, J02202-3, J02205 X01587, X02763, X65257]	(HBV)
woodchuck hepatitis B virus	[J02442, J04514, M11082 M18752, M60764, M90520]	(WHV)

TENTATIVE SPECIES IN THE GENUS

None reported.

Genus *Avihepadnavirus*

Type Species duck hepatitis B virus (DHBV)

Distinguishing Features

Virions are spherical, 46-48 nm in diameter, with a nucleocapsid that is 35 nm in diameter and exhibit projections. The viral DNA is often nearly or completely full length. Viruses lack the X gene. Virus particles have only the largest (Mr 36 x 10^3) and smallest (Mr 17 x 10^3) S antigens. Transmission is predominantly vertical.

List of Species in the Genus

The viruses, their genomic sequence accession numbers [ ] and assigned abbreviations ( ) are:

Species in the Genus

duck hepatitis B virus	[K01834, X58567-9, M21953 M32990-1, M60677]	(DHBV)
heron hepatitis B virus	[M22056]	(HHBV)

Tentative Species in the Genus

None reported.

List of Unassigned Viruses in the Family

None reported.

Similarity with Other Taxa

The involvement of reverse transcription in the replication of hepadnaviruses is similar to that of retroviruses and cauliflower mosaic virus.

Derivation of Names

avi: from Latin *avis*, "bird"
dna: sigla for *d*eoxyribo*n*ucleic *a*cid)
hepa: from Greek *hepar*, "liver"
ortho: from Greek *orthos*, "straight"

References

Galibert F, Mandart E, Fitoussi F, Tiollais P, Charnay P (1979) Nucleotide sequence of the hepatitis B virus genome (subtype ayw) cloned in *E. coli*. Nature 281: 646-650

Ganem D, Varmus HE (1987) The molecular biology of the hepatitis B viruses. Ann Rev Biochem 56: 651-693

Gust ID, Burrell CJ, Coulepis AG, Robinson WS, Zuckerman AJ (1986) Taxonomic classification of human hepatitis B virus. Intervirology 25: 14-29

Hollinger FB (1990) Hepatitus B virus In: Fields BN, Knipe DM (eds) Virology 2nd edn. Raven Press, New York, pp 2171-2236

Howard C (1986) The biology of hepadnaviruses. J Gen Virol 67: 1215-1235

Howard C, Melnick JL (1991) Classification and taxonomy of hepatitis viruses. In: Hollinger FB, Lemon S, Margolis HS (eds) Viral hepatitis and liver disease. Williams and Williams, Baltimore, pp 890-892

Marion PL, Robinson WS (1983) Hepadnaviruses: Hepatitis B and related viruses. Curr Top Microb Immun 105: 99-121

Schodel F, Sprengel R, Weimer T, Fernholz D, Schneider R, Will H (1989) Animal hepatitis B viruses. Adv Viral Oncol 8: 73-108

Summers J, Mason WS (1982) Replication of the genome of a hepatitis B-like virus by reverse transcription of an RNA intermediate. Cell 29: 403-415

Tiollais P, Pourcel C, Dejean A (1985) The hepatitis B virus. Nature 317: 489-495

Contributed By

Howard C, Burrell CJ, Gerin JL, Gerlich WH, Gust ID, Koike K, Marion PL, Mason WS, Neurath AR, Newbold J, Robinson W, Schaller H, Tiollais P, Wen Y-M, Will H

GENUS *BADNAVIRUS*

Type Species Commelina yellow mottle virus (ComYMV)

VIRION PROPERTIES

MORPHOLOGY

Virions are bacilliform, with parallel sides and rounded ends. There is no envelope. Virions are uniformly 30 nm in width. Modal particle length is 130 nm, but particles ranging in length from 60-900 nm are commonly observed. No projections or other capsid surface features have been observed by electron microscopy. Virions have an electron-transparent central core, but there is no information on the nature of nucleic acid-capsomer interaction. The tubular portion of the virion has a structure based on an icosahedron cut across its 3-fold axis, with a structural repeat of 10 nm and 9 rings of hexamer subunits.

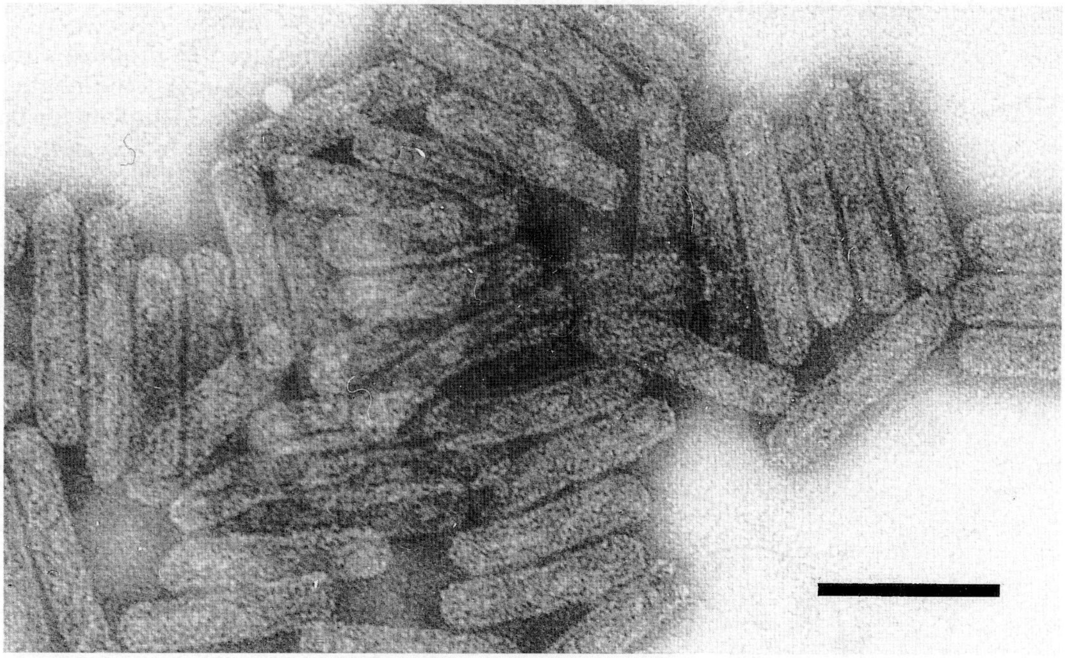

Figure 1: Negative contrast electron microscopy of Commelina yellow mottle virus virions, stained with sodium phosphotungstate, 2%, pH 7.0. The bar represents 100 nm.

PHYSICOCHEMICAL AND PHYSICAL PROPERTIES

Virions have a buoyant density of 1.31 g/cm^3 in CsCl and a S_{20w} of approximately 200. There are no data on Mr. Virions are stable at pH 6-9, and in 4M NaCl, 100 mM EDTA and Cs_2SO_4, but not CsCl. Virions are stable at room temperature for several weeks; infectivity is lost on exposure to 53-55° C for 10 minutes. Virions are unaffected by chloroform, ether, carbon tetrachloride and non-ionic detergents, but are sensitive to n-butanol.

NUCLEIC ACID

Virions contain a single molecule of circular dsDNA, 7.5-8.0 kbp in size, depending on the species, and a buoyant density in CsCl-ethidium bromide of 1.57 g/cm^3. Each strand of the genome is interrupted by one site-specific discontinuity.

PROTEINS

The CoYMV and ScBV genomes both contain three ORFs (I-III) that are capable of encoding proteins of Mr 23 or 22, 15 or 13 and 216 or 215 x 10^3, respectively. RTBV genome contains four ORFs (I-IV) that are capable of encoding proteins of Mr 24, 12, 194 and 46 x 10^3, respectively. The largest ORF product of each virus is believed to encode a polyprotein that is proteolytically processed to produce a protein of unknown function (U), the virus coat and RNA binding protein(s) (RB), an aspartic protease (PR), reverse transcriptase (RT) and ribonuclease H (RH).

Virions contain a major structural protein of 35-37 kd. A second polypeptide varying in size from 33 to 39 kd, and a series of other less prevalent species are detected by both SDS-PAGE and immunoblotting.

LIPIDS

None reported.

CARBOHYDRATES

Virions contain no carbohydrate detectable by periodic acid - Schiff's staining.

GENOME ORGANIZATION AND REPLICATION

Following entry into the cell, the genome is transcribed to produce a transcript that is, depending on the virus, 120 to 268 nucleotides greater than genome length. This transcript presumably serves both as a polycistronic mRNA and as a template for replication of the negative strand. Negative strand synthesis is primed by the host cytosolic $tRNA^{Met}$ and performed by virally encoded reverse transcriptase. Positive strand synthesis is carried out by the viral reverse transcriptase and ribonuclease H. The site specific discontinuities that are present in both the negative (D-) and positive (D+) strands occur because the strands are not ligated to form a closed circle following the completion of synthesis. There is no information on the cellular sites of synthesis of viral proteins and nucleic acid. Virions occur and accumulate only in the cytoplasm.

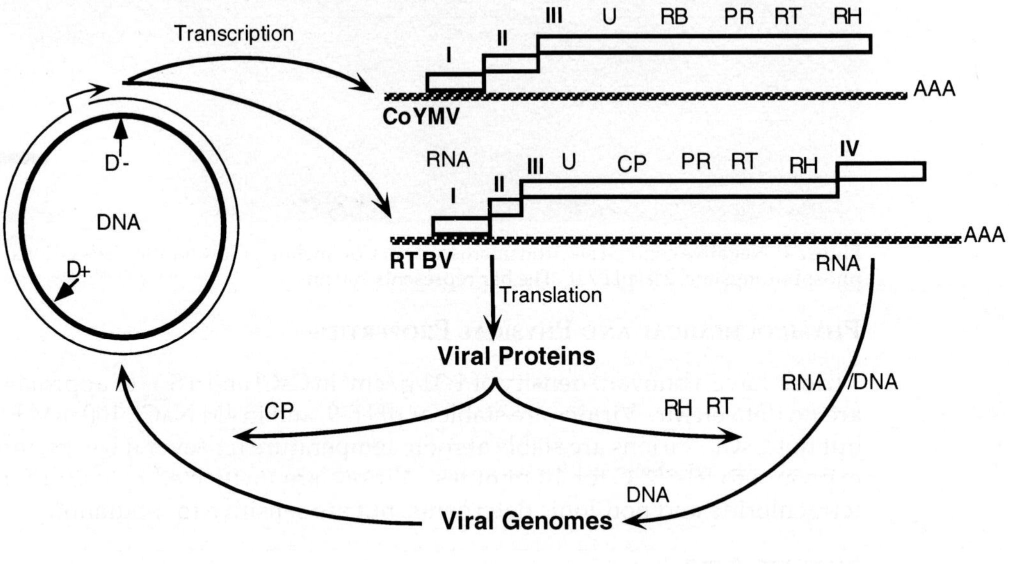

Figure 2: Genome organization and strategy of replication of ComYMV and RTBV.

ANTIGENIC PROPERTIES

Virions are only moderately antigenic. Polyclonal rabbit sera with immunodiffusion titres of 1/128-1/512 can be obtained. Pronounced antigenic variability occurs within several badnavirus species (e.g. BSV). A limited degree of inter-specific, though not group-specific, cross-reactivity can be demonstrated by enzyme immunosorbent assay or immunoelectron microscopy.

BIOLOGICAL PROPERTIES

NATURAL HOST RANGE

The viruses generally have a very restricted natural host range, often limited to a few species within a given plant genus. Experimental transmission outside the natural host range is generally unsuccessful.

MODE OF TRANSMISSION IN NATURE

The majority of badnaviruses occur in clonally-propagated plant hosts and are therefore spread by vegetative propagation of infected plant materials. The majority are transmitted in nature by mealybugs (*Homoptera, Pseudococcidae*), and several are also seed- and/or pollen-transmitted. Rice tungro badnavirus is transmitted by cicadellid leafhopper vectors.

VECTOR RELATIONSHIPS

The viruses are transmitted in a semi-persistent manner by mealybug or leafhopper (RTBV) vectors. Vectors can transmit virus after a 5 minutes acquisition feeding, but transmission efficiency increases with longer acquisition feeds. Vectors retain ability to transmit virus for up to 72 hours. Virus does not multiply in vectors and there is no transovarial transmission. All life stages of vectors can acquire and transmit virus.

GEOGRAPHIC DISTRIBUTION

The viruses occur worldwide, primarily in the tropics and subtropics. The majority of badnavirus-infected, clonally-propagated host plants have their centers of origin/diversity in Southeast Asia and Australasia.

PATHOGENICITY

Pathogenicity is variable, ranging from latency to plant mortality. The most frequent symptom type is interveinal chlorotic mottling. Symptoms are most severe in the primary stage following inoculation. Symptoms then become less pronounced, and may disappear for extended periods before reappearing.

HISTOPATHOLOGY

Virions occur only in the cytoplasm. They occur singly or in large groups, randomly distributed or arranged in palisade-like arrays. They do not occur within inclusion bodies or membrane-bound structures. Most badnaviruses are not tissue-limited, and occur in all tissue types. Rice tungro badnavirus is exceptional in being phloem-limited. Apart from changes in the internal organization of mitochondria there are no data on other histopathological effects.

LIST OF SPECIES IN THE GENUS

The viruses, their host { }, genomic sequence accession numbers [ ] and assigned abbreviations ( ) are:

SPECIES IN THE GENUS

banana streak virus		(BSV)
cacao swollen shoot virus	[L14546]	(CSSV)
Canna yellow mottle virus		(CaYMV)
Commelina yellow mottle virus	[X7924]	(ComYMV)
Dioscorea bacilliform virus		(DBV)
kalanchoe top-spotting virus		(KTSV)
piper yellow mottle virus		(PYMoV)
rice tungro bacilliform virus	[X57924, M65026]	(RTBV)
Schefflera ringspot virus		(SRV)

sugarcane bacilliform virus	[M89923]	(SCBV)

TENTATIVE SPECIES IN THE GENUS

Aucuba bacilliform virus	{*Aucuba japonica*}	(AuBV)
mimosa bacilliform virus	{*Albizzia julibrissin*}	(MBV)
taro bacilliform virus	{*Colocasia esculenta*}	(TaBV)
Yucca bacilliform virus	{*Yucca elephantipes*}	(YBV)

SIMILARITY WITH OTHER TAXA

Badnaviruses are similar to caulimoviruses in genome type (dsDNA). They differ from caulimoviruses in genome size, particle morphology, vector taxa, and histopathology.

DERIVATION OF NAMES

ba: from *ba*cilliform, relating to virion morphology
dan: from *d*eoxyribo*n*ucleic *a*cid (DNA), referring to genome type

REFERENCES

Bao Y, Hull R (1992) Characterization of the discontinuities in rice tungro bacilliform virus DNA. J Gen Virol 73: 1297-1301

Brunt AA (1970) Cacao swollen shoot virus. Descriptions of Plant Viruses. CMI/AAB Kew, Surrey, England, No. 10, 4pp.

Bouhida M, Lockhart BEL, Olszewski NE (1993) An analysis of the complete sequence of a sugarcane bacilliform virus genome infectious to banana and rice. J Gen Virol 74: 15-22

Hagen LS, Jacquemond M, Lepingle A, Lot H, Tepfer M (1993) Nucleotide sequence and genomic organization of cacao swollen shoot virus. J Gen Virol (in press)

Harrison BD, Roberts IM (1973) Association of virus-like particles with internal brown spot of yam (*Dioscorea alata*). Trop Agric, Trinidad 50: 335-340

Hay M, Jones MC, Blakebrough ML, Dasgupta I, Davies JW, Hull R (1991) An analysis of the sequence of an infectious clone of rice tungro bacilliform virus; a plant pararetovirus. Nucl Acids Res 19: 2615-2621

Hearon SS, Locke JC (1984) Graft, pollen, and seed transmission of an agent associated with top spotting in *Kalanchoe blossfeldiana*. Plant Dis 68: 347-350

James M, Kenten RH, Woods RD (1973) Virus-like particles associated with two diseases of *Colocasia esculenta* (L.) Schott in the Solomon Islands. J Gen Virol 21: 145-153

Lockhart BEL (1986) Purification and serology of a bacilliform virus associated with a streak disease of banana. Phytopathology 76: 995-999

Lockhart BEL (1988) Occurrence of canna yellow mottle virus in North America. Acta Hort 234: 72-78

Lockhart BEL (1990) Evidence for a double-stranded circular genome in a second group of plant viruses. Phytopathology 80: 127-131

Lockhart BEL, Autrey JC (1988) Occurence in sugarcane of a bacilliform virus related serologically to banana streak virus. Plant Dis 72: 230-233

Lockhart BEL, Ferji Z (1988) Purification and mechanical transmission of kalanchoe top-spotting associated virus. Acta Hort 234: 72-78

Lot H, Kjiekpor E, Jacquemond M (1991) Characterization of the genome of cacao swollen shoot virus. J Gen Virol 72: 1735-1739

Medberry SL, Lockhart BEL, Olszewski NE (1990) Properties of Commelina yellow mottle virus's complete DNA sequence, genomic discontinuities and transcript suggest that it is a pararetovirus. Nucl Acids Res 18: 5505-5513

Qu R, Bhattacharya M, Laco GS, de Kochko A, Subba Rao BL, Kaniewska MB, Elmer JS, Rochester DE, Smith CE, Beachy RN (1991) Characterization of the genome of rice tungro bacilliform virus: Comparison with Commelina yellow mottle virus and caulimoviruses. Virology 185: 354-364

CONTRIBUTED BY

Lockhart BEL, Olszewski NE, Hull R

Genus *Caulimovirus*

Type Species cauliflower mosaic virus (CaMV)

Virion Properties

Morphology

Isometric particles are about 50 nm in diameter with a T = 7 (420 subunits) multilayered structure. Virions have no envelope.

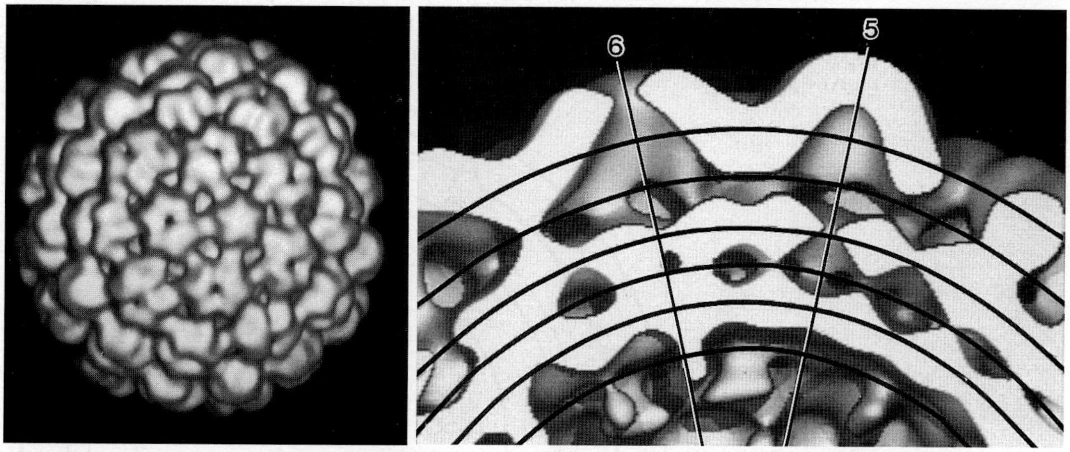

Figure 1: (left) Reconstruction of cauliflower mosaic virus surface structure showing T = 7 symmetry. (right) Cutaway surface reconstruction showing multilayer structure. (From Cheng *et al.*, 1992).

Physicochemical and Physical Properties

Virion Mr is about 20 x 10^6; S_{20w} is about 208. D is about 0.75 x 10^{-7} cm^2/s; apparent partial specific volume is about 0.704; buoyant density in CsCl is about 1.37 g/cm^3; particles are very stable.

Nucleic Acid

Virions contain one molecule of dsDNA in the form of an open circle with single-strand discontinuities at specific sites, the transcribed (α) strand with one and the non-transcribed (β) strand with two discontinuities; some other members have three discontinuities in the β strand. DNAs of five CaMV isolates have been sequenced.

Proteins

Capsid protein is translated from ORF IV, and is assembled into capsids as a 57 x 10^3 phosphorylated polypeptide. Rapid degradation occurs *in vivo* (and perhaps also during purification) to give several polypeptide forms, Mr predominantly about 42 x 10^3 and 37 x 10^3. The product of ORF I is involved in the cell-to-cell spread of the virus and that of ORF II is the aphid transmission helper factor. The function of ORF III protein is unknown. ORF V protein is the reverse transcriptase and ORF VI protein forms the matrix of the major virus inclusion body and transactivates the translation of the 35S RNA.

Lipids

None reported.

Carbohydrates

The coat protein is glycosylated.

Genome Organization and Replication

Six or possibly 8 ORFs (putative genes) are present on the α strand. The β strand is noncoding.

Transcription occurs in the nucleus from a DNA template with properties of a minichromosome. Two major transcripts (19S and 35S) are found. The 19S transcript is from ORF VI, and translates to a protein (Mr 62 x 10^3) found in cytoplasmic viral inclusion bodies in which most mature virus particles accumulate. These electron-dense inclusion bodies are characteristic of the group. The 35S transcript has not been translated *in vitro* but is thought to be the mRNA of several of the ORFs. The 35S transcript is 180 nt longer than the full length viral DNA (i.e., it contains a 180 nt terminal repeat), and is also thought to be a template for replication of the viral genome by reverse transcription. ORF V may code for the replication enzyme.

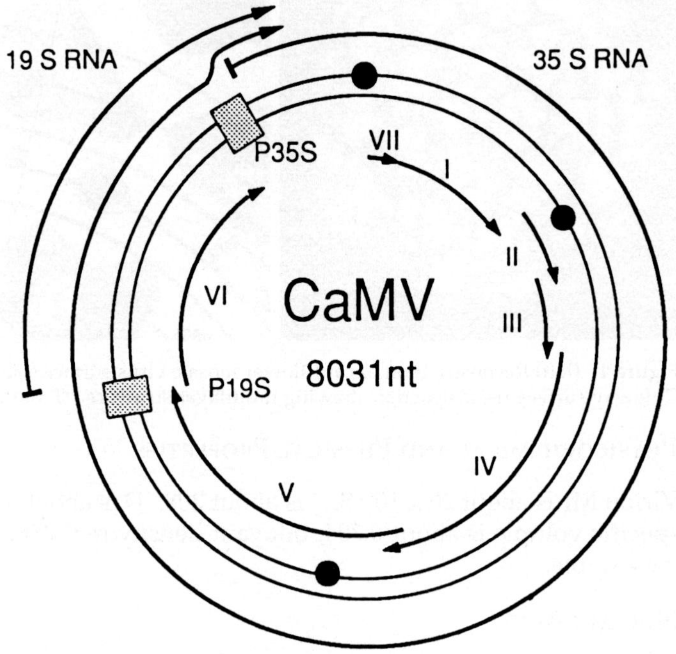

Figure 2: Genome map of cauliflower mosaic virus showing as outer arc and circle the two RNA transcripts (arrow head 3'-end). The double circle represents the genomic DNA with the discontinuities shown as spots and the promoter sites as boxes. The inner arcs are the ORFs.

Antigenic Properties

The viruses serve as efficient immunogens. There are serological relationships among some members.

Biological Properties

Host Range

The natural host range of most members is narrow. Disease symptoms are usually mosaics and mottles. Infection is systemic with most cell types being infected.

Transmission

The viruses are transmissible experimentally by mechanical inoculation; in nature they are transmitted by aphids in a semipersistent manner. Transmission of CaMV requires a virus-coded protein (the product of ORF II) which forms separate inclusion bodies.

GEOGRAPHICAL DISTRIBUTION

The geographic distribution varies between members, some of which have a very restricted distribution while others are distributed worldwide. Most occur in temperate regions.

CYTOPATHIC EFFECTS

Cells infected with caulimoviruses contain inclusion bodies which can be seen with the light microscope or, in thin sections with the electron microscope. Two sorts of inclusion bodies have been recognized: electron-dense vacuolar ones which have a matrix composed mainly of gene VI product and electron-lucent ones which are composed mainly of gene II product. Virus particles are found mainly in the electron-dense inclusions and, in some members, in the nucleus.

LIST OF SPECIES IN THE GENUS

The viruses, their genomic sequence accession numbers [ ] CMI/AAB description # ( ) and assigned abbreviations ( ) are:

SPECIES IN THE GENUS

blueberry red ringspot virus (327)		(BRRV)
carnation etched ring virus (182)	[EM_VI:CERVDNA]	(CERV)
cauliflower mosaic virus (24; 243)	[EM_VI:CAMVG2, EM_VI:MCACOMGEN, EM_VI:CAMVG1, EM_VI:MCACDH]	(CaMV)
dahlia mosaic virus (51)	[EM_VI:MCA1841]	(DMV)
figwort mosaic virus	[EM_VI:CAFMCXX]	(FMV)
horseradish latent virus		(HRLV)
Mirabilis mosaic virus		(MiMV)
peanut chlorotic streak virus		(PCSV)
soybean chlorotic mottle virus (331)	[EM_VI:CASCMVX]	(SbCMV)
strawberry vein banding virus (219)		(SVBV)
thistle mottle virus		(ThMoV)

TENTATIVE SPECIES IN THE GENUS

Aquilegia necrotic mosaic virus	(ANMV)
cassava vein mosaic virus	(CsVMV)
Cestrum virus	(CV)
petunia vein clearing virus	(PVCV)
Plantago virus 4	(PlV-4)
Sonchus mottle virus	(SMoV)

SIMILARITY WITH OTHER TAXA

Caulimoviruses are one of the two genera of reverse-transcribing viruses which infect plants, the other being the badnaviruses. The two genera differ from one another in virion morphology, details of genome organization, host range and vector.

DERIVATION OF NAMES

caulimo: sigla from *cauli*flower *mo*saic

REFERENCES

Cheng RH, Olson NH, Baker TS (1992) Cauliflower mosaic virus: a 420 subunit (T = 7), multilayer structure. Virology 186: 655-668

Covey SN, Hull R (1985) Advances in cauliflower mosaic virus research. Oxford Surveys of Plant Mol Cell Biol 2: 339-346

Covey SN (1985) Organization and expression of the cauliflower mosaic virus genome, In : Davies JW (ed), Molecular Plant Virology, Replication and Gene Expression, Vol II. CRC Press, Boca Raton FL, pp 121-160

Donson J, Hull R (1983) Physical mapping and molecular cloning of caulimovirus DNA. J Gen Virol 64: 2281-2288

Francki RIB, Milne RG, Hatta T (1985) Caulimovirus group, In: Atlas of plant viruses, Vol I. CRC Press, Boca Raton FL, pp 17-32

Frank A, Guilley H, Jonard G, Richards KE, Hirth L (1980) Nucleotide sequence of cauliflower mosaic virus DNA. Cell 21: 285-294

Gardner RC, Howarth AJ, Hahn P, Brown-Luedi M, Shepherd RJ, Messing J (1981) The complete nucleotide sequence of an infectious clone of cauliflower mosaic virus by M13mp7 shotgun sequencing. Nucl Acids Res 9: 2871-2888

Hull R, Covey SN (1983) Does cauliflower mosaic virus replicate by reverse transcription? Trends in Biochem Sci 8: 119-121

Hull R, Donson J (1982) Physical mapping of the DNAs of carnation etched ring and figwort mosaic virus. J Gen Virol 60: 125-134

Kruse J, Timmins P, Witz J (1987) The spherically averaged structure of a DNA isometric plant virus: cauliflower mosaic virus. Virology 159: 166-168

Maule AJ (1985) Replication of caulimoviruses in plants and protoplasts, In: Davies JW (ed) Molecular Plant Virology, Replication and Gene Expression, Vol II. CRC Press, Boca Raton FL, pp 161-190

Pfeiffer P, Hohn T (1983) Involvement of reverse transcription in the replication of cauliflower mosaic virus: A detailed model and test of some aspects. Cell 33: 781-789

Richins RD, Shepherd RJ (1983) Physical maps of the genomes of dahlia mosaic virus and mirabilis mosaic virus - two members of the caulimovirus group. Virology 124: 208-214

CONTRIBUTED BY

Hull R

FAMILY RETROVIRIDAE

TAXONOMIC STRUCTURE OF THE FAMILY

Family	***Retroviridae***
Genus	"Mammalian type B retroviruses"
Genus	"Mammalian type C retroviruses"
Genus	"Avian type C retroviruses"
Genus	"Type D retroviruses"
Genus	"BLV-HTLV retroviruses"
Genus	*Lentivirus*
Genus	*Spumavirus*

VIRION PROPERTIES

MORPHOLOGY

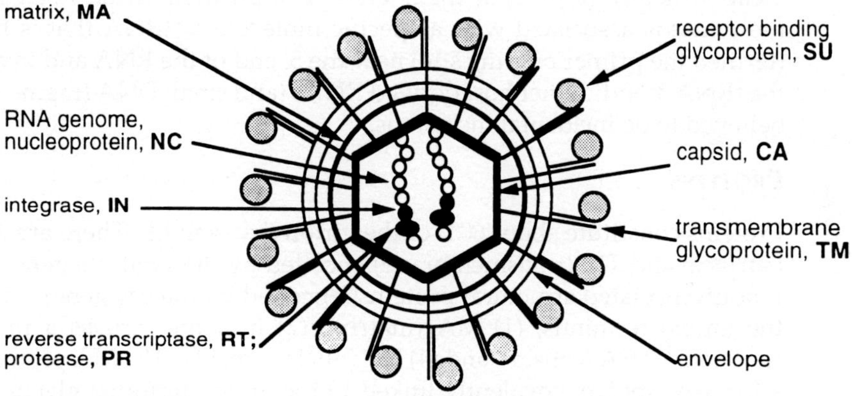

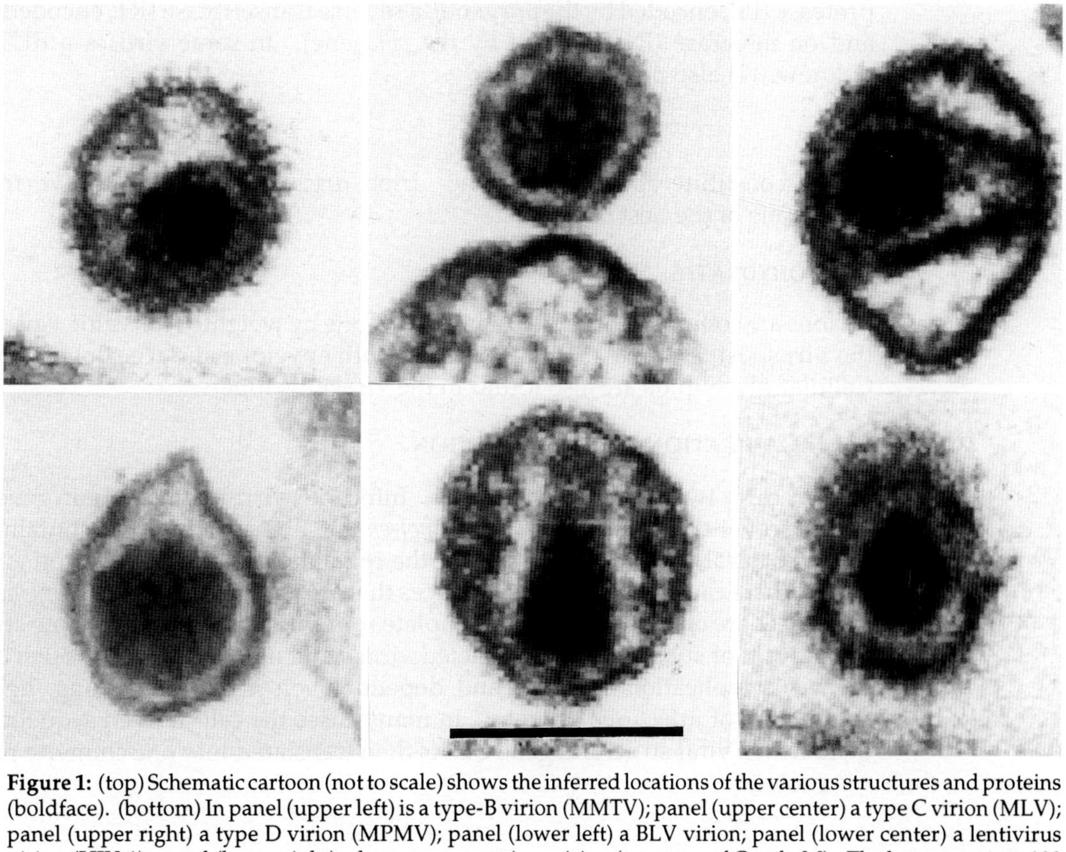

Figure 1: (top) Schematic cartoon (not to scale) shows the inferred locations of the various structures and proteins (boldface). (bottom) In panel (upper left) is a type-B virion (MMTV); panel (upper center) a type C virion (MLV); panel (upper right) a type D virion (MPMV); panel (lower left) a BLV virion; panel (lower center) a lentivirus virion (HIV-1); panel (lower right) a human spumavirus virion (courtesy of Gonda M). The bar represents 100 nm.

Virions are spherical, enveloped and 80-100 nm in diameter. Glycoprotein surface projections are about 8 nm in length. The internal core is spherical or icosahedral and encapsidates the viral nucleocapsid. The nucleocapsid (nucleoid) is eccentric in type B virions, concentric in type C, HTLV-BLV, and spumavirus virions, and is in the shape of a rod or truncated cone in lentivirus virions.

PHYSICOCHEMICAL AND PHYSICAL PROPERTIES

Virion buoyant density is 1.16-1.18 g/cm^3 in sucrose. Virions are sensitive to heat, detergents and formaldehyde. The surface glycoproteins may be partially removed by proteolytic enzymes. Virions are relatively resistant to UV light.

NUCLEIC ACID

The viral genome consists of a dimer of linear, positive sense, ssRNA, each monomer 7 to 11 kb in size. The RNA constitutes about 2% of the virion dry weight. The monomers are held together by hydrogen bonds. Each monomer of RNA is polyadenylated at the 3' end and has a cap structure (type 1) at the 5' end. The purified virion RNA is not infectious. Each monomer is associated with a specific molecule of tRNA that is base-paired to a region (termed the primer binding site) near the 5' end of the RNA and involves about 18 bases of the tRNA 3' end. Other host derived RNAs (and small DNA fragments) found in virions are believed to be incidental inclusions.

PROTEINS

Proteins constitute about 60% of the virion dry weight. There are 2 envelope proteins: SU (surface) and TM (transmembrane) encoded by the viral *env* gene. There are 3-6 internal, non-glycosylated structural proteins (encoded by the *gag* gene). These are, in order from the amino terminus, (1) MA (matrix), (2) in some viruses a protein of undetermined function, (3) CA (capsid), and (4) NC (nucleocapsid). The MA protein is often acylated with a myristyl moiety covalently linked to the amino terminal glycine. Other proteins are a protease (PR, encoded by the *pro* gene), a reverse transcriptase (RT, encoded by the *pol* gene) and on integrase (IN, encoded by the *pol* gene). In some viruses a dUTPase (DU, role unknown) is also present.

LIPIDS

Lipids constitute about 35% of the virion dry weight. They derive from the plasma membrane of the host cell.

CARBOHYDRATES

Virions are composed of about 3% carbohydrate by weight. This value varies, depending on the virus. At least one (SU), but usually both envelope surface proteins are glycosylated. Cellular glycolipids are also found in the viral envelope.

GENOME ORGANIZATION AND REPLICATION

Virions carry two copies of the genome. Infectious viruses have 4 main genes coding for the virion proteins in the order: 5'-*gag-pro-pol-env*-3'. Some retroviruses contain genes encoding non-structural proteins important for the regulation of gene expression and virus replication. Others carry cell-derived sequences that are important in pathogenesis. These cellular sequences are either inserted in a complete retrovirus genome (e.g., some strains of RSV), or in the form of substitutions for deleted viral sequences (e.g., MSV). Such deletions render the virus replication-defective and dependent on non-transforming, helper viruses for production of infectious progeny. In many cases the cell-derived sequences form a fused gene with a viral structural gene that is then translated into one chimeric protein (e.g., *gag-onc* protein).

Entry into the host cell is mediated by interaction between a virion glycoprotein and specific receptors at the host cell surface, resulting in fusion of the viral envelope with the plasma membrane, either directly or following endocytosis. Receptors are cell surface proteins. Four have been identified: CD4 protein (a receptor for HIV), which is an immunoglobulin-like molecule with a single transmembrane region; two others (receptors for ecotropic MLV and GALV), which are involved in the transport of small molecules and have a complex structure with multiple transmembrane domains; and an ALV receptor, which is a small molecule with a single transmembrane domain, distantly related to a cell receptor for low-density lipoprotein.

The process of intracellular uncoating of viral particles is not understood. Subsequent early events are carried out in the context of a nucleoprotein complex derived from the capsid.

Replication starts with reverse transcription (by RT) of virion RNA into cDNA using the 3' end of the tRNA as primer for synthesis of a negative-sense cDNA transcript. The initial short product (to the 5' end of the genome) transfers and primes further cDNA synthesis from the 3' end of the genome by virtue of duplicated end sequences at the ends of the viral RNA species. cDNA synthesis involves the concomitant digestion of the viral RNA (RNAse H activity of the RT protein). The products of this hydrolysis serve to prime virus-sense cDNA synthesis on the negative sense DNA transcripts. In its final form, the linear dsDNA transcripts derived from the viral genome contain long terminal repeats (LTRs) composed of sequences from the 3' (U3) and 5' (U5) ends of the viral RNA flanking a sequence (R) found near both ends of the RNA. The process of reverse transcription is characterized by a high frequency of recombination due to the transfer of the RT from one template RNA to the other.

Retroviral DNA becomes integrated into the chromosomal DNA of the host to form a provirus by a mechanism involving the viral IN protein. The ends of the virus DNA are joined to cell DNA, involving the removal of two bases from the ends of the linear viral DNA and generating a short duplication of cell sequences at the integration site. Virus DNA can integrate at many sites in the cellular genome. However, once integrated, a sequence is apparently incapable of further transposition within the same cell. The map of the integrated provirus is co-extensive with that of unintegrated linear viral DNA. Integration appears to be a prerequisite for virus replication.

The integrated provirus is transcribed by cellular RNA polymerase II into virion RNA and mRNA species in response to transcriptional signals in the viral LTRs. In some genera, transcription is also regulated by viral encoded transactivators. There are several classes of mRNA depending on the virus and the genetic map of the retrovirus. An mRNA comprising the whole genome serves for the translation of the *gag*, *pro*, and *pol* genes (positioned in the 5' half of the RNA). This results in the formation of polyprotein precursors which are cleaved to yield the structural proteins, protease, RT and IN, respectively. A smaller mRNA consisting of the 5' end of the genome spliced to sequences from the 3' end of the genome and including the *env* gene and the U3 and R regions, is translated into the precursor of the envelope proteins. In viruses that contain additional genes, other forms of spliced mRNA are also made, however, all mRNAs share a common sequence at their 5' ends. Most primary translational products in retrovirus infections are polyproteins which require proteolytic cleavage before becoming functional. The *gag*, *pro* and *pol* products are produced from a nested set of primary translation products. For *pro* and *pol*, translation involves bypassing translational termination signals by ribosomal frameshifting or by read-through at the *gag-pro* and/or the *pro-pol* boundaries.

Capsids assemble either at the plasma membrane (type C and most other viruses), or as intracytoplasmic (type A) particles and are released from the cell by a process of budding. Polyprotein processing of the internal proteins occurs concomitant with or just subsequent to the maturation of virions.

ANTIGENIC PROPERTIES

Virion proteins contain type-specific and group-specific determinants. Some type-specific determinants of the envelope glycoproteins are involved in antibody-mediated virus neutralization. Group-specific determinants are shared by members of a serogroup and may be shared between members of different serogroups within a particular genus. There is no evidence for cross-reactivities between members of different genera. Epitopes that elicit T-cell responses are found on many of the structural proteins. Antigenic properties are infrequently used in classification of Retroviridae.

BIOLOGICAL PROPERTIES

Retroviruses are widely distributed as exogenous infectious agents of vertebrates. Endogenous proviruses that have resulted at some time from infection of germ line cells are inherited as Mendelian genes. They occur widely among vertebrates.

Retroviruses are associated with a variety of diseases. These include: malignancies including certain leukemias, lymphomas, sarcomas and other tumors of mesodermal origin; mammary carcinomas and carcinomas of liver and kidney; immunodeficiencies (such as AIDS); autoimmune diseases; lower motor neuron diseases; and several acute diseases involving tissue damage. Some retroviruses are non-pathogenic. Transmission of retroviruses is horizontal via a number of routes, including blood, saliva, sexual contact, etc., and vertical via direct infection of the developing embryo, or via milk or perinatal routes. Endogenous retroviruses are transmitted by inheritance of proviruses.

GENUS "MAMMALIAN TYPE B RETROVIRUSES"

Type Species mouse mammary tumor virus (MMTV)

DISTINGUISHING FEATURES

Virions exhibit a B-type morphology with prominent surface spikes and an eccentric condensed core. Capsid assembly occurs within the cytoplasm as A-type particles prior to transport to, and budding from the plasma membrane. Protein sizes are: MA: about 10 kDa; p8: 8 kDa; p21: 21 kDa; CA: about 27 kDa; NC: about 14 kDa; DU; PR: about 13 kDa; RT; IN; SU: about 52 kDa; TM: about 36 kDa. The genome is about 10 kb in size (one monomer); its organization is illustrated in Fig. 2.

There is an additional gene (*sag*) whose product functions as a superantigen and is located at the 3' end of the genome. The tRNA primer is tRNA^{Lys-3}. The LTR is about 1300 nt long of which the U3 region is 1200, the R sequence 15 and the U5 region some 120 nt in length.

The recognized viruses in this genus are exogenous, vertically-transmitted (milk) and endogenous viruses of mice. Viruses are associated with mammary carcinoma and T-lymphomas. No oncogene-containing members are known.

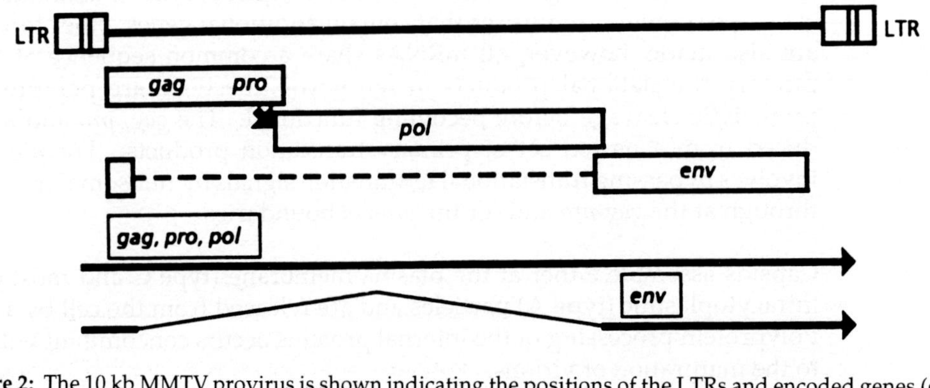

Figure 2: The 10 kb MMTV provirus is shown indicating the positions of the LTRs and encoded genes (*gag, pro, pol, env, sag*), their relative reading frames (ribosomal frame-shift sites: arrow heads; individual mRNAs: arrows).

TAXONOMIC STRUCTURE OF THE GENUS

Only one virus is recognized in this genus, although related endogenous proviruses have been identified in other mammalian species (rodents, primates).

LIST OF SPECIES IN THE GENUS

The viruses, their genomic sequence accession numbers [ ] and assigned abbreviations ( ) are:

SPECIES IN THE GENUS

mouse mammary tumor virus	[M1552]	(MMTV)

TENTATIVE SPECIES IN THE GENUS

None reported.

GENUS "MAMMALIAN TYPE C RETROVIRUSES"

Type Species murine leukemia virus (MuLV)

DISTINGUISHING FEATURES

Virions exhibit a C-type morphology with barely visible surface spikes. They have a centrally located, condensed core. Virus assembly occurs at the inner surface of the membrane at the same time as budding. Protein sizes are: MA: about 15 kDa; p12: 12 kDa; CA: about 30 kDa; NC: about 10 kDa; PR: about 14 kDa; RT: about 80 kDa; IN: about 46 kDa; SU: about 70 kDa; TM: about 15 kDa. The genome is about 8.3 kb in size (one monomer); its organization is illustrated in Fig. 3. There are no known additional genes. The tRNA primer is $tRNA^{Pro}$, ($tRNA^{Glu}$ is found in a few endogenous mouse viruses). The LTR is about 600 nt long of which the U3 region is 500, the R sequence 60 and the U5 region some 75 nt in size.

The viruses are widely distributed; exogenous (vertical and horizontal transmission) and endogenous viruses are found in many mammals. The reticuloendotheliosis viruses comprise a few isolates from birds with no known corresponding endogenous relatives. Related endogenous sequences are found in mammals. The viruses are associated with a variety of diseases including malignancies, immunosuppression, neurological disorders, and others. Many oncogene-containing members of the mammalian and reticuloendotheliosis virus groups have been isolated.

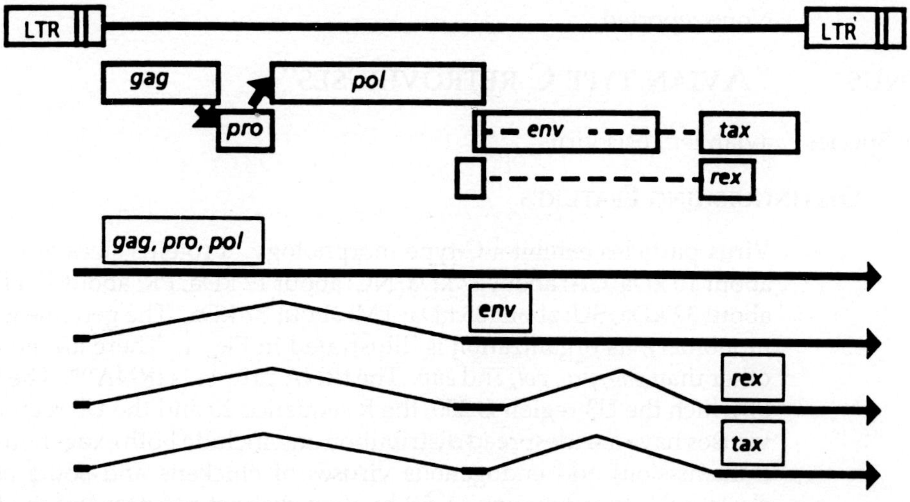

Figure 3: The 8.3 kb MLV provirus is shown indicating the positions of the LTRs and encoded genes (*gag, pro, pol, env*), their relative reading frames (ribosomal readthrough site, arrow head; individual mRNAs: arrows).

Taxonomic Structure of the Genus

Three serogroups are recognized, the mammalian type C oncoviruses, the reticuloendotheliosis viruses and the reptilian type C viruses.

List of Species in the Genus

The groups, viruses, their genomic sequence accession numbers [ ] and assigned abbreviations ( ) are:

Species in the Genus

1-Mammalian type C virus group:		
Abelson murine leukemia virus	[J02009]	(AbMLV)
AKR (endogenous) murine leukemia virus	[J01998]	(AKRMLV)
feline leukemia virus	[M18247]	(FeLV)
Finkel-Biskis-Jinkins murine sarcoma virus	[K02712]	(FBJVMSV)
Friend murine leukemia virus	[M93134, Z11128]	(FrMLV)
Gardner-Arnstein feline sarcoma virus		(GAFeSV)
gibbon ape leukemia virus	[M26927]	(GALV)
guinea pig type C oncovirus		(GPCOV)
Hardy-Zuckerman feline sarcoma virus		(HZFeSV)
Harvey murine sarcoma virus		(HaMSV)
Kirsten murine sarcoma virus		(KiMSV)
Moloney murine sarcoma virus	[J02266]	(MoMSV)
murine leukemia virus (Moloney virus)	[J02255]	(MoMLV)
porcine type C oncovirus		(PCOV)
Snyder-Theilen feline sarcoma virus		(STFeSV)
woolly monkey sarcoma virus (simian sarcoma virus)	[J02394]	(WMSV)
2-Reptilian type C oncovirus virus group:		
viper retrovirus		(VRV)
3-Reticuloendotheliosis virus group:		
chick syncytial virus		(CSV)
reticuloendotheliosis virus (strain T, A)		(REV)
Trager duck spleen necrosis virus		(TDSNV)

Tentative Species in the Genus

None reported.

Genus "Avian type C retroviruses"

Type Species avian leukosis virus (ALV)

Distinguishing Features

Virus particles exhibit a C-type morphology. Proteins sizes are: MA: about 19 kDa; p10: about 10 kDa; CA: about 27 kDa; NC: about 12 kDa; PR: about 15 kDa; RT: about 68 kDa; IN: about 32 kDa; SU: about 85 kDa; TM: about 37 kDa. The genome is about 7.2 kb in size (one monomer); its organization is illustrated in Fig. 4. There are no known additional genes other than *gag*, *pro*, *pol*, and *env*. The tRNA primer is $tRNA^{Trp}$. The LTR is about 350 nt long, of which the U3 region is 250, the R sequence 20 and the U5 region some 80 nt in size. The viruses have a widespread distribution and include both exogenous (vertical and horizontal transmission) and endogenous viruses of chickens and some other birds. Isolates are classified into subgroups (A-G) by their distinct receptor usage. Distantly related endogenous sequences are found in birds and mammals. Virus infections are associated with malignancies and some other diseases such as wasting, and osteopetrosis. Many oncogene-containing members of the genus have been isolated.

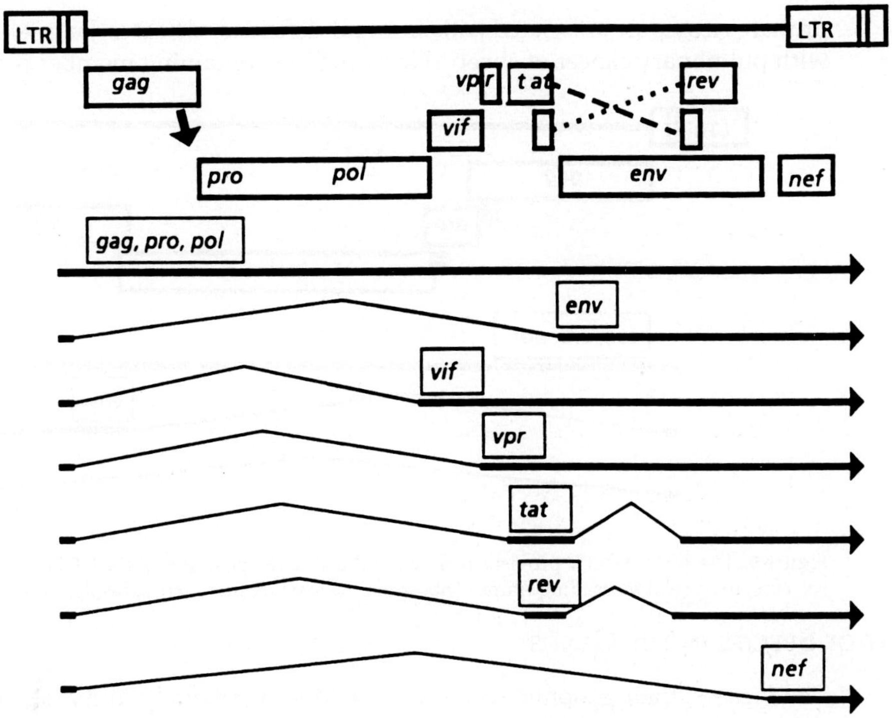

Figure 4: The 7.2 kbp ALV provirus is shown, indicating the positions of the LTRs and encoded genes (*gag*, *pro*, *pol*, *env*), their relative reading frames (ribosomal frameshift site: arrowhead; individual mRNAs: thin arrows).

List of Species in the Genus

The viruses, their genomic sequence accession numbers [ ] and assigned abbreviations ( ) are:

Species in the Genus

avian carcinoma, Mill Hill virus 2	[K02082]	(MHV-2)
avian leukosis virus - RSA	[M37980]	(ALV)
avian myeloblastosis virus	[J02013]	(AMV)
avian myelocytomatosis virus 29	[J02019]	(MCV-29)
Fujinami sarcoma virus	[J02194]	(FSV)
Rous sarcoma virus (Prague strain)	[J02342]	(RSV)
UR2 sarcoma virus	[M10455]	(UR2SV)
Y73 sarcoma virus	[J02027]	(Y73SV)

Tentative Species in the Genus

None reported.

Genus "Type D retroviruses"

Type Species Mason-Pfizer monkey virus (MPMV)

Distinguishing Features

Viruses exhibit a D-type morphology. They lack prominent surface spikes. Proteins sizes are: MA: about 10 kDa; p18: 18 kDa; p12: 12 kDa; CA: about 27 kDa; NC: about 14 kDa; p4: about 4 kDa; DU; PR: about 11 kDa; RT; IN; SU: about 70 kDa; TM: about 22 kDa. The genome is about 8.0 kb in size (one monomer); its organization is illustrated in Fig. 5. There are no known additional genes to *gag*, *pro*, *pol*, and *env*. The tRNA primer is $tRNA^{Lys\,1,2}$. The LTR is about 350 nucleotides long of which the U3 region is 240, the R sequence 15 and the U5 region some 95 nt in size. Viruses assigned to the genus include exogenous, horizontally transmitted and endogenous viruses of new and old world primates and sheep. Exogenous primate viruses are associated with immuno-deficiency diseases, Jaagsiekte virus is associated with pulmonary cancer of sheep. No oncogene-containing member is known.

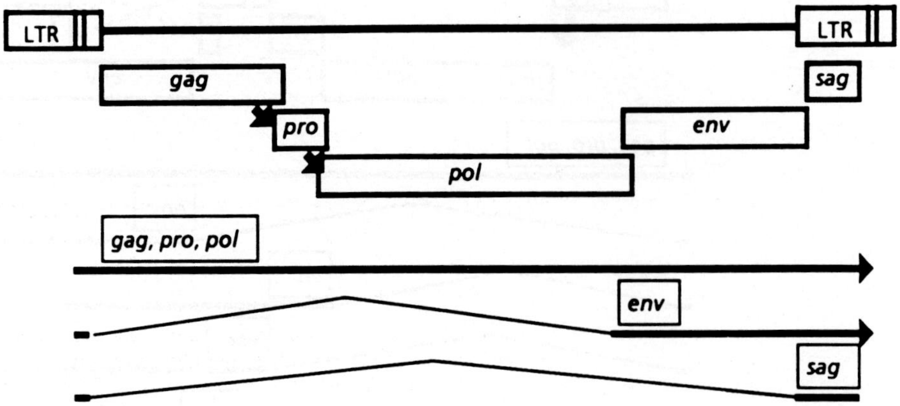

Figure 5: The 8.0 kb MMPV provirus is shown indicating the positions of the LTRs and encoded genes (*gag, pro, pol, env*), their relative reading frames (ribosomal frameshift sites, arrow heads; individual mRNAs: arrows).

LIST OF SPECIES IN THE GENUS

The viruses, their genomic sequence accession numbers [ ] and assigned abbreviations ( ) are:

SPECIES IN THE GENUS

Langur virus (Po-1-LU)		(LNGV)
Mason-Pfizer monkey virus	[M12349]	(MPMV)
ovine pulmonary adenocarcinoma virus (Jaagsiekte virus)	[M80216]	(OPAV)
simian type D virus 1	[M11841]	(SRV-1)
squirrel monkey retrovirus	[M23385]	(SMRV)

TENTATIVE SPECIES IN THE GENUS

None reported.

GENUS "BLV-HTLV RETROVIRUSES"

Type Species bovine leukemia virus (BLV)

DISTINGUISHING FEATURES

Virions are similar to C-type retroviruses in terms of morphology and assembly. Proteins sizes are: MA: about 19 kDa; CA about 24 kDa; NC about 12-15 kDa; PR about 14 kDa; RT; IN; SU about 60 kDa; TM about 21 kDa. The genome is about 8.3 kb in size (one monomer); its organization is illustrated in Fig. 6. There are non-structural genes, designated *tax*, and *rex* which are involved in regulation of synthesis and processing of virus RNA, in addition to *gag, pro, pol* and *env*. The tRNA primer is $tRNA^{Pro}$. The LTR is about 550-750 nt long, of which the U3 region is 200-300, the R sequence 135-235 and the U5 region 100-200 nt in size.

The exogenous viruses (horizontal transmission) in this genus are found in only a few groups of mammals. No related endogenous viruses are known. Virus infections are associated with B or T cell leukemias or lymphomas as well as neurological disease (tropical spastic paraparesis, or HTLV-associated myopathy) and exhibit a long latency with an incidence of less than 100%. No oncogene-containing members of this genus have been identified.

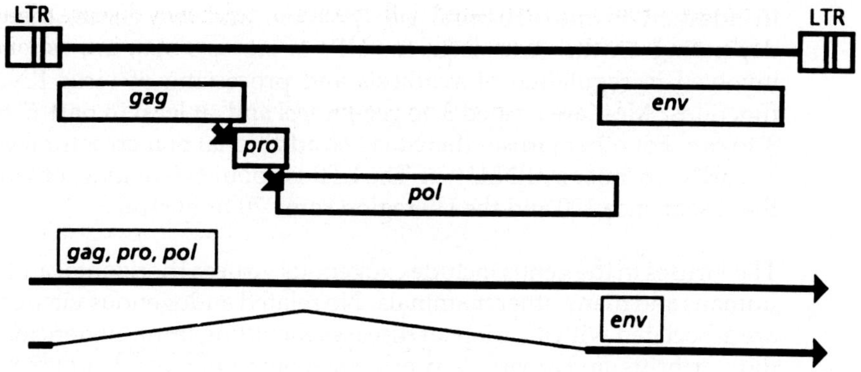

Figure 6: The 8.3 kbp HTLV-1 provirus genome is shown indicating the positions of the LTRs and encoded structural genes (*gag, pro, pol, env*) and certain other non-structural genes (*tax, rex*), their reading frames (ribosomal frameshift sites: arrow heads; individual mRNAs: arrows). The genes in other members of the genus may occupy different reading frames.

List of Species in the Genus

The viruses, their genomic sequence accession numbers [ ] and assigned abbreviations ( ) are:

Species in the Genus

bovine leukemia virus	[K02120]	(BLV)
human T-lymphotropic virus 1	[D00294]	(HTLV-1)
human T-lymphotropic virus 2	[M10060]	(HTLV-2)
simian T-lymphotropic virus		(STLV)

Tentative Species in the Genus

None reported.

Genus *Lentivirus*

Type Species human immunodeficiency virus 1 (HIV-1)

Distinguishing Features

Virions have a distinctive morphology with a bar, or cone-shaped core (nucleoid). Viruses assemble at the cell membrane. Proteins sizes are: MA: about 17 kDa; CA: about 24 kDa; NC: about 7-11 kDa; PR: about 14 kDa; RT: about 66 kDa; DU (in all except the primate lentiviruses); IN: about 32 kDa; SU: about 120 kDa; TM: about 41 kDa. The genome is about 9.2 kb in size (one monomer); its organization is illustrated in Fig. 7.

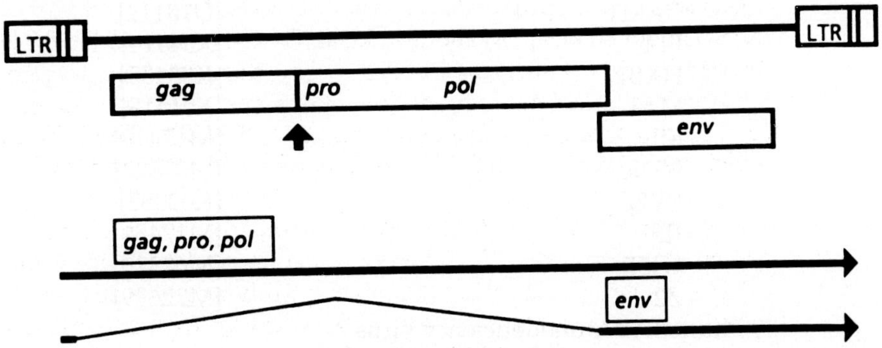

Figure 7: The 9.2 kbp HIV-1 provirus is shown indicating the positions of the LTRs and encoded structural genes (*gag, pro, pol, env*) and certain non-structural genes (*vif, vpr, tat, rev, nef*), their reading frames (ribosomal frameshift site, arrow head; individual mRNAs: arrows). The genes in other members of the genus may occupy different reading frames.

In addition to the structural *gag, pro, pol,* and *env* genes, there are additional genes, depending on the virus (e.g., for HIV-1: *vif, vpr, vpu, tat, rev, nef*) whose products are involved in regulation of synthesis and processing of virus RNA and other replicative functions. Most are located 3' to *gag-pro-pol* and, at least in part, 5' to *env*, one (*nef* in HIV) is 3' to *env*. For other viruses there may be additional non-structural genes (e.g., *vpx* in HIV-2). The tRNA primer is $tRNA^{Lys\,1,2}$. The LTR is about 600 nt long, of which the U3 region is 450, the R sequence 100 and the U5 region some 70 nt in size.

The viruses in the genus include exogenous viruses (horizontal and vertical transmission) of humans and many other mammals. No related endogenous viruses are known. The viruses are associated with a variety of diseases including immunodeficiencies, neurological disorders, arthritis, and others. No oncogene-containing member of this genus has been isolated.

Taxonomic Structure of the Genus

Five serogroups of lentiviruses are recognized, reflecting the hosts with which they are associated (primates, sheep and goats, horses, cats, and cattle). The primate lentiviruses are distinguished by the use of CD4 protein as receptor and the absence of DU. Some groups have cross-reactive gag antigens (e.g., the ovine, caprine and feline lentiviruses). Antibodies to *gag* antigens in lions and other large felids indicate the existence of other viruses related to FIV and the ovine/caprine lentiviruses.

List of Species in the Genus

The viruses, their genomic sequence accession numbers [ ] and assigned abbreviations ( ) are:

Species in the Genus

1-Bovine lentivirus group:		
bovine immunodeficiency virus	[M32690]	(BIV)
2-Equine lentivirus group:		
equine infectious anemia virus	[M16575]	(EIAV)
3-Feline lentivirus group:		
feline immunodeficiency virus (Petuluma)	[M25381]	(FIV)
4-Ovine/caprine lentivirus group:		
caprine arthritis encephalitis virus	[M33677]	(CAEV)
visna/maedi virus (strain 1514)	[M60609, M60610]	
5-Primate lentivirus group:		
human immunodeficiency virus 1		(HIV-1)
reference strains:		
ARV-2/SF-2	[K02007]	
BRU (LAI)	[K02013]	
CAM1	[D10112]	
ELI	[X04414]	
HXB2	[K03455]	
MAL	[X04415]	
MN	[M17449]	
NDK	[M27323]	
PV22	[K02083]	
RF	[M17451]	
U455	[M62320]	
Z2	[M22639]	
human immunodeficiency virus 2		(HIV-2)
reference strains:		
BEN	[M30502]	
C194	[J04542]	
GH-1	[M30895]	

ROD	[M15390]	
SBLISY	[J04498]	
ST	[M31113]	
simian immunodeficiency virus		(SIV)
reference strains:		
African green monkey (agm) 155	[M29975]	
African green monkey 3	[M30931]	
African green monkey TYO	[X07805]	
African green monkey AA	[M66437]	
chimpanzee (cpz)	[X52154]	
Grived (gr-1)	[M29973]	
mandrill (mnd)	[M27470]	
pig-tailed macaque (mne)	[M32741]	
Rhesus (Maccaca mulatta) (mac)	[M195499]	
sooty mangabey H4 (sm)	[X14307]	
stump-tailed macaque (stm)	[M83293]	

TENTATIVE SPECIES IN THE GENUS

None reported.

GENUS *SPUMAVIRUS*

Type Species human spumavirus (HSV)

DISTINGUISHING FEATURES

Virions exhibit a distinctive morphology with prominent surface spikes and a central condensed core. Capsid assembly occurs in the cytoplasm prior to budding. Proteins sizes and ranges are not well defined. The genome is about 11 kb in size (one monomer); its organization is illustrated in Fig. 8. There are several genes (designated *bel* 1, 2, 3, 4) some of which have a transactivation function, in additional to *gag*, *pro*, *pol*, and *env*. The tRNA primer is $tRNA^{Lys\ 1,2}$. The LTR is about 1150 nt long, of which the U3 region is 800, the R sequence 200 and the U5 region some 150 nt in size. Viruses have a widespread distribution. Exogenous viruses are found in many mammals. No related endogenous viruses are known. Many isolates cause characteristic "foamy" cytopathology in cell culture. No diseases have been associated with spumavirus infections. No oncogene-containing member of the genus has been found.

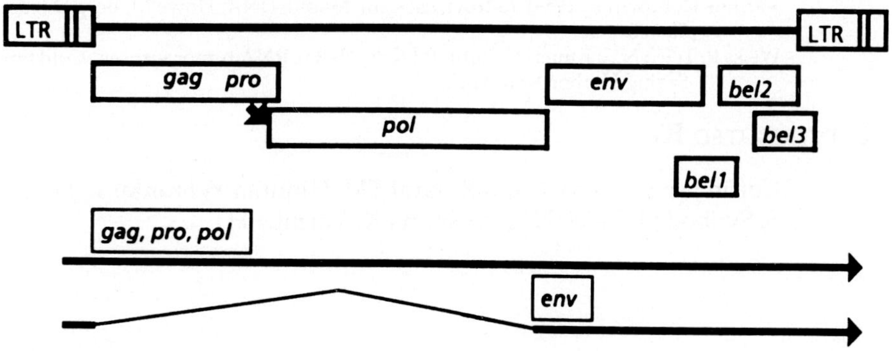

Figure 8: The 11 kb SFV provirus is shown indicating the positions of the LTRs and encoded structural genes (*gag, pro, pol, env*), certain other non-structural genes (*bel*), their relative reading frames (ribosomal frameshift site, arrow head; and known individual mRNAs: arrows, others not yet characterized).

LIST OF SPECIES IN THE GENUS

The viruses, their genomic sequence accession numbers [ ] and assigned abbreviations ( ) are:

Species in the Genus

bovine syncytial virus		(BSV)
feline syncytial virus		(FSV)
human spumavirus (human foamy virus)		(HSRV)
simian foamy virus	[X54482]	(SFV)

Tentative Species in the Genus

None reported.

List of Unassigned Viruses in the Family

None reported.

Similarity with Other Taxa

None reported.

Derivation of Names

retro: from Latin *retro*, "backwards", refers to the activity of reverse transcriptase and the transfer of genetic information from RNA to DNA
onco: from Greek *onkos*, "tumor"
spuma: from Latin *spuma*, "foam"
lenti: from Latin *lentus*, "slow"

References

Coffin JM (1990) *Retroviridae* and their replication. In: Fields BN, Knipe DM (eds) Virology, 2nd edn. Raven Press, New York, pp 1437-1500
Coffin JM (1992) Structure and classification of retroviruses. In: Levy J (ed) The *Retroviridae*, Vol 1. Plenum Press, New York, pp 19-50
Doolittle RF, Feng D-F, Johnson MS, McClure MA (1989) Origins and evolutionary relationships of retroviruses. Quart Rev Biol 64: 1-30
Gallo R, Wong-Staal F, Montagnier L, Haseltine WA, Yoshida M (1988) HIV/HTLV gene nomenclature. Nature 333: 504
Leis J, Baltimore D, Bishop JM, Coffin JM, Fleissner E, Goff SP, Roszlan S, Robinson H, Skalka AM, Temin HM, Vogt V (1988) Standardized and simplified nomenclature for proteins common to all retroviruses. J Virol 62: 1808-1809
Myers G, Rabson AB, Josephs SF, Smith TF, Berzofsky JA, Wong-Staal F (eds) (1992) Human retroviruses and AIDS 1992. Los Alamos National Laboratory, Los Alamos New Mexico
Varmus H, Brown P (1989) Retroviruses. In: Mobile DNA, Howe M, Berg D (eds) ASM Press, Washington, pp 53-108
Weiss R, Teich N, Varmus H, Coffin JM (eds) (1985) RNA tumor viruses. Cold Spring Harbor Laboratory, Cold Spring Harbor New York

Contributed By

Coffin JM, Essex M, Gallo R, Graf TM, Hinuma Y, Hunter E, Jaenisch R, Nusse R, Oroszlan S, Svoboda J, Teich N, Toyoshima K, Varmus H

FAMILY *CYSTOVIRIDAE*

TAXONOMIC STRUCTURE OF THE FAMILY

Family *Cystoviridae*
Genus *Cystovirus*

GENUS *CYSTOVIRUS*

Type Species Pseudomonas phage φ6 (φ6)

VIRION PROPERTIES

MORPHOLOGY

Virions are 86 nm in diameter, spherical, with an envelope covered by 8 nm long spikes. The envelope surrounds an icosahedral nucleocapsid which is about 58 nm in diameter. The removal of the nucleocapsid surface protein reveals a dodecahedral polymerase complex which is about 43 nm in diameter.

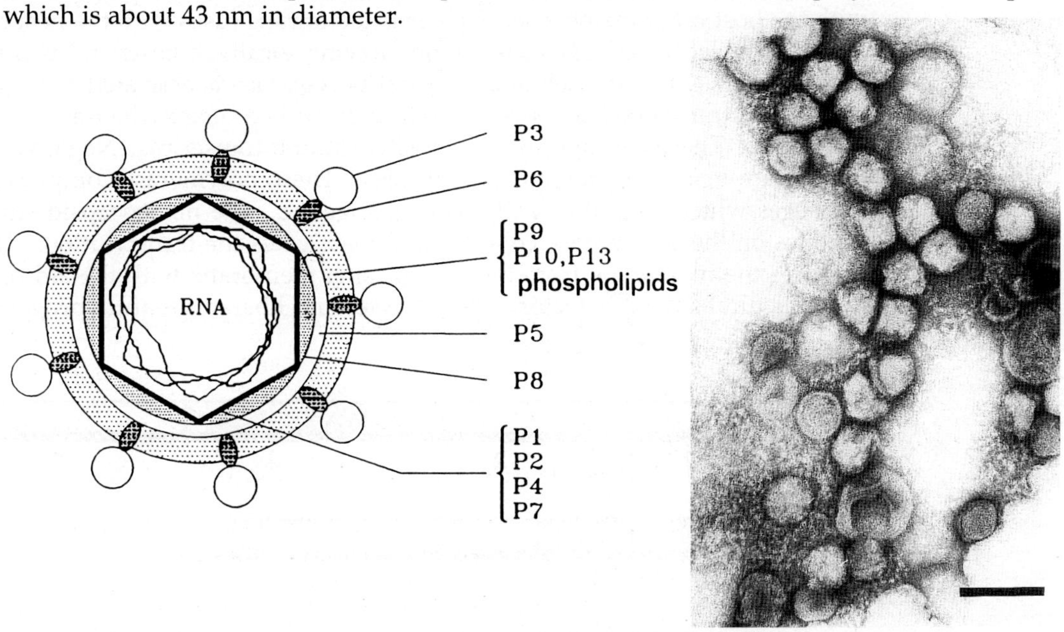

Figure 1: (left) Schematic of a cystovirus, Pseudomonas phage φ6 and indication of its proteins. (right) Negative contrast electron micrograph of Pseudomonas phage φ6. The bar represents 50 nm.

PHYSICOCHEMICAL AND PHYSICAL PROPERTIES

Virion Mr is about 99 x 10^6; and that of the nucleocapsid is about 40 x 10^6. Virion S_{20w} is about 405. The buoyant density of the virion is 1.27 g/cm^3 in CsCl and 1.24 g/cm^3 in sucrose. Pseudomonas phage φ6 is stable at pH 6 - 9 and very sensitive to ether, chloroform and detergents.

NUCLEIC ACID

Virions contains three linear dsRNA segments L (6374 bp), M (4057 bp), and S (2948 bp). The segments have a base composition of 55.2, 56.7, and 55.5 % G+C, respectively. Virions contain about 10% RNA. Nucleic acid sequence data are available from GenBank and EMBL.

PROTEINS

The genome codes for twelve proteins. The early proteins P1, P2, P4, and P7 are coded from the L segment and form the viral polymerase complex. The association of protein P8, the NC surface protein, and the viral lytic enzyme, P5, with the polymerase complex forms the NC.

These proteins are coded from the genome segment S. Proteins P9, P10, and P13 reside in the envelope. The absorption and fusion complex is formed by proteins P3 and P6. P3 is the spike protein recognizing the receptor, whereas P6 is a membrane protein with membrane fusion activity. P3 is associated with the virion though protein P6. There is so far only one identified nonstructural protein, P12, which is needed in the membrane assembly inside the host cell. Virions are composed of about 70% protein.

LIPIDS

Virions contain about 20% phospholipid. This is located in the envelope. There is enough lipid to cover about one-half of the envelope surface area (the rest being protein).

CARBOHYDRATES

None reported.

GENOME ORGANIZATION AND REPLICATION

Virions absorb to *Pseudomonas syringae* pili which retract bringing the virion into contact with the host outer membrane. The virus membrane fuses with the host outer membrane and the nucleocapsid associated lytic enzyme locally digests the peptidoglycan. The nucleocapsid enters the cell and the viral polymerase is activated to produce early transcripts. The translated L transcripts produce the early proteins which assemble to polymerase complexes. These package all three positive strand transcripts. Negative strand synthesis takes place inside the polymerase complex. These polymerase complexes transcribe late messages which code the synthesis of late genes. The nucleocapsid surface protein assembles on the polymerase complex and inactivates the transcription. The nucleocapsid acquires the membrane from the host plasma membrane with the aid of a virus specific nonstructural assembly factor. The cell lyses and liberates mature progeny particles.

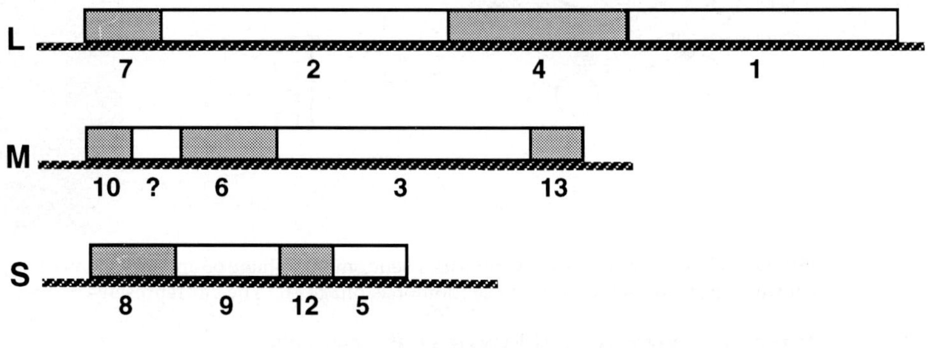

Figure 2: Genome organization of Pseudomonas phage $\phi 6$, the legend of the numbers in the figure are the following:

Segment	Gene	Protein function
L	1	Structural framework (dodecahedron)
	2	RNA polymerase active site
	4	Nucleoside triphosphate phosphohydrolase
	7	?
M	3	Spikes, host attachment
	6	Membrane, anchor for p3
	10	Membrane, lysis
	13	Membrane
S	8	Major capsid protein
	12	Envelopment of capsid, nonstructural
	5	Membrane assembly
	5	Endopeptidase, lysis and entry

Biological Properties

Pseudomonas phage ø6 infects many phytopathogenic *Pseudomonas* species. In addition, some *Pseudomonas pseudoalcaligenes* strains are sensitive to this virus.

List of Species in the Genus

The viruses, their genomic sequence accession numbers [ ] and assigned abbreviations are:

Species in the Genus

Pseudomonas phage ϕ6	[M17461, M17462, M12921]	(ϕ6)

Tentative Species in the Genus

None reported.

List of Unassigned Viruses in the Family

None reported.

Derivation of Names

cysto: from Greek *kystis*, 'bladder, sack'

References

Ackermann H-W, DuBow MS (1987) Viruses of Prokaryotes, Vol II. CRC Press, Boca Raton FL, pp 171-218

Mindich L (1988) Bacteriophage f6: A unique virus having a lipid-containing membrane and a genome composed of three dsRNA segments. Adv Virus Res 35: 137-176

Olkkonen VM, Gottlieb P, Strassman J, Qiao X, Bamford DH, Mindich L (1990) *In vitro* assembly of infectious nucleocapsid of bacteriophage f6: Formational a recombinant double-stranded RNA virus. Proc Natl Acad Sci USA 87: 9173-9177

Contributed By

Bamford DH

FAMILY REOVIRIDAE

TAXONOMIC STRUCTURE OF THE FAMILY

Family	***Reoviridae***
Genus	*Orthoreovirus*
Genus	*Orbivirus*
Genus	*Rotavirus*
Genus	*Coltivirus*
Genus	*Aquareovirus*
Genus	*Cypovirus*
Genus	*Fijivirus*
Genus	*Phytoreovirus*
Genus	*Oryzavirus*

VIRION PROPERTIES

MORPHOLOGY

Virions are icosahedral in structure, but many appear spherical in shape. They are 60-80 nm in diameter and consist of an inner core surrounded by several protein layers (Fig. 1). The precise morphology varies, depending on the genus. Some cypoviruses are occluded by a crystalline matrix of protein that forms a large polyhedron entrapping many virions.

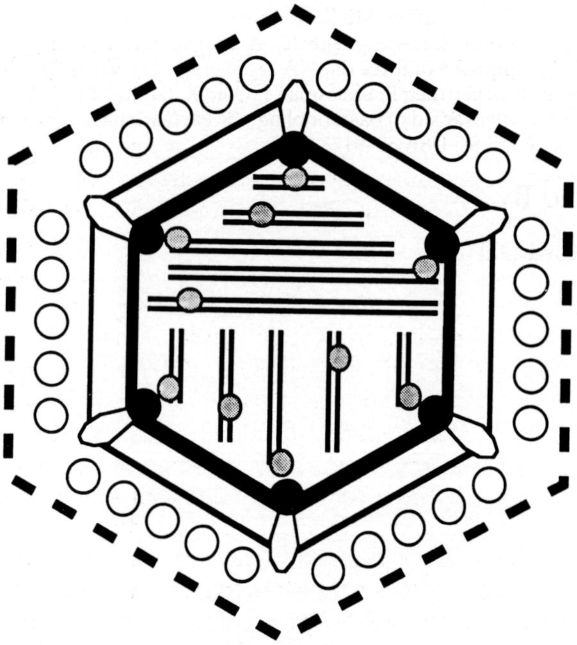

Figure 1: Schematic of an orbivirus particle showing 4 shells of protein forming the capsid, 10 internal dsRNA segments and associated minor proteins. The positions of the internal components are hypothetical. Members of other genera have different arrangements.

PHYSICOCHEMICAL AND PHYSICAL PROPERTIES

The virion Mr is about 120 x 10^6. The buoyant density in CsCl is 1.36-1.39 g/cm^3. Virus infectivity is moderately resistant to heat, organic solvents (e.g., ether) and to non-ionic detergents. The pH stability of virions varies among genera.

NUCLEIC ACID

Virions contain 10, 11 or 12 segments of linear dsRNA depending on the genus. The individual Mr of these RNAs range from 0.2 to 3.0 x 10^6. The total Mr of the genome is 12 -

20×10^6. The RNA constitutes about 15-20% of the virion dry weight. The positive strands of each duplex have 5' terminal caps (type 1 structure), the negative strands have phosphorylated 5' termini. RNAs lack 3' poly (A) tracts. The viral dsRNA species are present in equimolar proportions.

PROTEINS

At least 3 internal proteins constitute the virion RNA polymerase and associated enzymes involved in mRNA synthesis (including initiation of RNA synthesis, elongation, nucleotide phospho-hydrolase, capping, methylation and possible helicase activities). Some of the minor proteins may be integral components of the virion structure together with at least 3 major capsid proteins. The proteins range in size from Mr 15 -155 x 10^3. The proteins constitute about 80-85% of the dry weight of virions.

LIPIDS

Mature virions lack a lipid envelope. Depending on the genus, a myristyl residue may be covalently attached to one of the virion proteins. For rotaviruses and orbiviruses, an intermediate in virus morphogenesis has a lipid envelope that is subsequently removed.

CARBOHYDRATES

In some genera one of the outer virion proteins may be glycosylated with high mannose glycans, or O-linked N-acetylglucosamine.

GENOME ORGANIZATION AND REPLICATION

The viral RNA species are mostly monocistronic. Protein is encoded on one strand of each duplex (mRNA species). Some of the viral dsRNA species code for non-structural (NS) proteins. The mode of entry of viruses into cells varies between genera but often involves the loss of some components of the outer capsid. Virus-derived particles reside in the cell cytoplasm. Repetitive asymmetric transcription of full-length mRNA species from each dsRNA segment occurs within these particles throughout the infection course. The mRNA products are extruded from the icosahedral apices of the particles. Structures, termed viroplasms or virus inclusion bodies, occur in localized areas of the cytoplasm. They have a granular and moderately electron dense appearance by electron microscopy. The process of dsRNA synthesis is unknown. Evidence has been obtained for orthoreoviruses that sets of capped mRNAs and certain NS proteins are incorporated into "assortment complexes" that are considered to be the precursors of progeny virus particles. It is believed that such complexes, together with structural proteins, are encapsidated into sub viral particles and that the mRNAs are transcribed into minus strands with which they remain associated (dsRNA). In addition to the parental virus-derived particle, progeny sub viral particles synthesize mRNA species. Depending on the genus, some NS proteins are involved in the translocation of virus particles within cells and in virus egress. Some cypoviruses also form polyhedra, large protein matrices that occlude virus particles. The steps involved in virion morphogenesis and virus egress from cells vary according to the genus. Genome segment reassortment occurs readily in cells co-infected with closely related viruses.

ANTIGENIC PROPERTIES

Viruses generally possess type- and group-specific antigens. No antigenic relationship has been found between viruses in different genera. Some viruses hemagglutinate red blood cells.

BIOLOGICAL PROPERTIES

The biological properties of the viruses vary according to the genus. Some viruses replicate only in certain vertebrate species and are transmitted between hosts by respiratory or oral-fecal routes. Other vertebrate viruses replicate both in arthropod vectors (e.g., gnats, mosquitoes, or ticks, etc., - orbiviruses, coltiviruses) and vertebrate hosts. Plant viruses replicate both in plants and arthropod vectors. Viruses that are pathogens of insects (cypoviruses) are transmitted by contact.

Genus *Orthoreovirus*

Type Species reovirus 3 (REOV-3)

Distinguishing Features

Orthoreoviruses only infect vertebrates and are spread by the respiratory or oral-fecal routes. Virions have a well defined capsid structure and contain 10 dsRNA species.

Virion Properties

Morphology

Orthoreoviruses possess a double capsid shell. The diameter of intact REOV-3 particles is 81 nm (for avian reoviruses the size may be slightly different). The diameter of REOV-3 cores (i.e., virus particles from which the outer capsid has been removed) is 60 nm. The diameter of the central compartment where the dsRNA genome is located is 49 nm. Core particles have projections located at each of the 12 capsid vertices (Fig. 2).

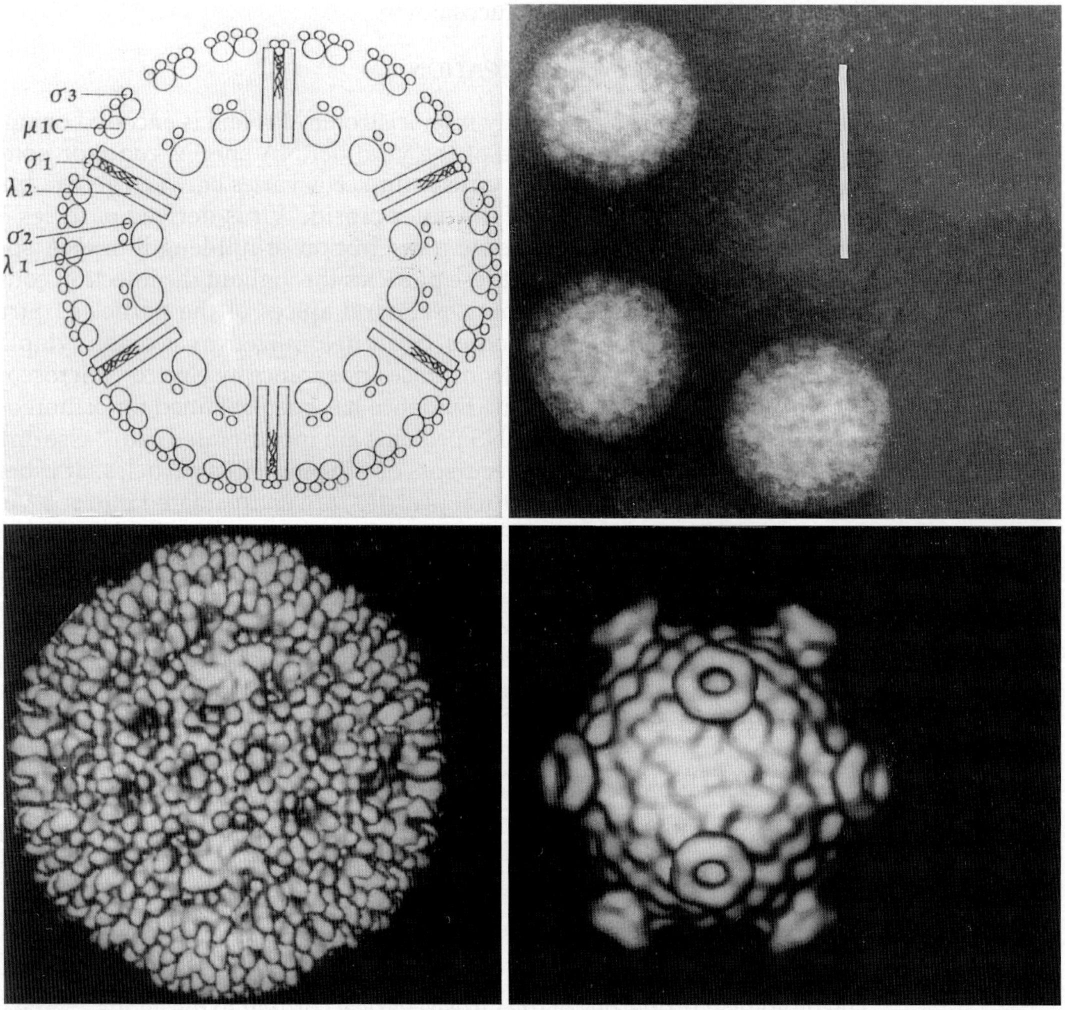

Figure 2: (upper left) Schematic arrangement of the capsid proteins of orthoreovirus (the dsRNA genome and minor proteins λ3 and μ2 are not shown); (upper right) electron micrograph of REOV-3 virions stained with uranyl formate (bar represents 100 nm); (lower left) computer image of REOV-3 virion constructed from cryoelectron micrographs; (lower right) computer image (cryoelectron microscopy) of a REOV-3 core particle showing projections at the icosahedral vertices.

These extend almost to the surface of the virion. For REOV-3 the projections are composed of trimers of the 49 kDa σ1 protein overlaying pentamers of the 144 kDa λ2 protein (a total of 36 molecules of σ1 and 60 molecules of λ2 per virion). The σ1 protein is in the form of an extended fiber topped with a knob. It has hemagglutinin activity and reacts with neutralizing antibodies. The other major structural proteins of the core are the 142 kDa λ1 (120 copies) and the 47 kDa σ2 proteins (240 copies). These form the principal components of the core shell. They are arranged in a T=13 (l) lattice. It is estimated that enclosed within the core are 12 copies of both the 137 kDa l3 and 83 kDa μ2 proteins, in addition to the 10 dsRNA species. How the l3 and μ2 proteins, or the dsRNA species, are arranged within the core is not known. The inner and outer capsids exhibit fivefold, threefold and twofold axes of rotational symmetry. The surface arrangement of the capsomers on the outer surface includes pentagonal and hexagonal arrays, 11-20 nm in diameter with central 4-6 nm cavities. Like the core, the capsomers which form these rings are arranged in a T=13 (l) lattice. They are composed of dimers of the 76 kDa μ1 protein (72 kDa μ1C and 4 kDa factor viii) associated with, and overlayed by two molecules of protein σ3. There are estimated to be some 720 molecules (each) of μ1C and σ3. The avian reovirus σC protein has a size of 35 kDa.

Physicochemical and Physical Properties

The Mr of orthoreovirus (e.g., REOV-3) is about 130 x 10^6, its buoyant density in CsCl is 1.36 g/cm^3. The virion S_{20w} is about 630. Virions are relatively stable to temperature changes, or treatment with cations, lipid solvents, detergents, or radiation.

Nucleic Acid

Orthoreoviruses have 10 dsRNA segments with Mr that range from 0.6 to 2.7 x 10^6. Based on their resolution by gradient centrifugation or by gel electrophoresis they are categorized into 3 size classes commonly referred to as large (L1-L3, about 3.9-3.8 kbp, S_{20w} of 14), medium (M1-M3, about 2.3-2.2 kbp, S_{20w} of 12) and small (S1-S4, about 1.4-1.2 kbp, S_{20w} of 10.5), although the individual sizes and relative electrophoretic mobilities may vary between viruses (e.g., the avian reovirus S1 dsRNA is significantly slower on gels than other S dsRNA species; the avian S1 species has a size of 1.6 kbp). The total Mr of REOV-3 is about 16 x 10^6 (23.7 kbp). Virions also contain numerous oligonucleotides. Defective virus particles may lack particular dsRNA species, or contain abnormal dsRNA sequences.

Proteins

The mammalian orthoreovirus structural proteins (e.g., REOV-3) are designated in terms of their relative sizes and size classes (λ1-3, μ1-2, σ1-3). The nomenclature used for avian reoviruses is similar (λA-C, μA-B, σA-C). The mammalian reovirus μ1 protein in the outer capsid is myristylated at its amino terminus. The μ1 protein is cleaved to μ1C and factor viii when it is complexed with σ3. The λ3 protein is the RNA polymerase, the λ2 protein is a guanylyl transferase involved in mRNA capping, the function of μ2 is not known. The λ1 and other proteins may also be involved in RNA transcription in addition to their structural roles.

Genome Organization and Replication

For other orthoreoviruses the coding assignments of the comparable RNAs are similar - when the differences in relative migrations of the dsRNA segments and different sizes of the encoded proteins are taken into account.

The overall course of infection involves adsorption, penetration, particle uncoating, asymmetric mRNA transcription and translation, assembly of progeny sub viral particles, further rounds of mRNA transcription and translation followed by virion assembly. Virions accumulate in the cell cytoplasm and are released when infected cells lyse.

The attachment of virions to cells involves components of the outer capsid. The σ1 protein mediates cell attachment and determines the cell and tissue tropism of the virus strain. The M2 gene product (μ1) of different strains of orthoreovirus determines the *in vitro* susceptibility of particles to proteolytic digestion and subsequent transcriptase activation. Cell penetration involves endocytosis and is subject to the effects of lysosomotropic agents.

The efficiency of translation of the various orthoreovirus mRNA species varies over a 100-fold range. The proportions of the mRNA species found in infected cells also vary. σNS and μNS proteins are produced in high abundance during an infection and, together with σ3, associate with mRNA to form virus mRNA-containing complexes. Complexes containing equimolar proportions of the dsRNA species are also formed and include μNS, σNS, σ3 and λ2. The σ3 protein has the ability to bind dsRNA near its carboxy terminus. The protein is a metalloprotein with a zinc-binding domain near the amino terminus. Although the μNS protein is a phosphoprotein, the 70 kDa μNSC is not phosphorylated. The roles of the latter and that of the basic σ1S protein that is made in low abundance during an infection are not known. During the later stages of infection, host macromolecular synthesis is inhibited. σ1 protein is somehow involved in the inhibition of host cell DNA replication. M2 gene products modulate the neurovirulence of different orthoreovirus strains.

Table 1: List of the dsRNA segments of REOV-3 with their respective size (bp) and their encoded proteins for which the name, calculated size (kDa) and function and/or location are indicated.

dsRNA #	Size (bp)	Proteins	Size (kDa)	Function (location)
L1	3854	λ3	142	RNA polymerase (core)
L2	3916	λ2	144	Guanylyl transferase [capping enzyme] (core spike)
L3	3896	λ1	137	(core)
M1	2304	μ2	83	(core)
M2	2203	μ1	76	u1C precursor
		μ1C	72	(outer capsid)
M3	2235	μNS	80	ssRNA-binding, phosphoprotein
		μNSC	75	Unknown
S1	1416	σ1	49	Cell attachment protein, HA, type-specific antigen (outer capsid)
		σ1S	16	Unknown
S2	1331	σ2	47	(core)
S3	1189	σNS	41	ssRNA-binding
S4	1196	σ3	41	dsRNA-binding (outer capsid)

ANTIGENIC PROPERTIES

The type-specific antigen of orthoreoviruses is protein σ1 (σC of avian species). It has hemagglutinin activity and reacts with neutralizing antibodies. σ1 and other proteins elicit cytotoxic T-cell activities. σ1 also reacts with neutralizing antibodies and has hemagglutinin activity. The avian orthoreovirus σC protein, however, lacks hemagglutinin activity. Proteins λ2 and σ3 are group-specific antigens. Depending on the species, orthoreovirus proteins exhibit considerable sequence homology between different virus serotypes. The most conserved are the structural and minor proteins of the core.

BIOLOGICAL PROPERTIES

The host range of orthoreoviruses includes a variety of vertebrate species (birds, cattle, humans, monkeys, sheep, swine, and bats). Transmission is horizontal. No arthropod vectors are involved.

Orthoreovirus distribution is ubiquitous and worldwide. Disease associated with human orthoreoviruses may include upper respiratory tract infections, enteritis in infants and children (albeit rare), and possibly biliary atresia in neonates. Orthoreovirus disease in mice includes diarrhea, runting, the so-called oily hair effect, jaundice, and neurologic symptoms. In horses, orthoreoviruses cause upper and lower respiratory illness (laryngitis, rhinitis, conjunctivitis, and cough). In cattle, sheep and swine, orthoreoviruses cause respiratory and diarrheal illnesses. In dogs, they cause conjunctivitis, rhinitis, pneumonia and diarrhea. In monkeys, orthoreoviruses cause hepatitis, extrahepatic biliary atresia, meningitis, and necrosis of ependymal and choroid plexus epithelial cells. Certain mammalian orthoreoviruses infect the M cells of Peyer's patches and cells of the central nervous system.

Avian orthoreoviruses do not infect mammalian species. They induce syncytia in cell culture. Several pathotypes of avian orthoreoviruses are recognized. The outcome of infection of birds may range from inapparent to lethal. The severity of orthoreovirus disease has been correlated with the age of the host bird. Disease presentations in chickens include: arthritis, feathering abnormalities, gastro-enteritis, hepatitis, malabsorption, mortality, myocarditis, paling, pneumonia, stunted growth, tenosynovitis, and weight loss. In turkeys, avian orthoreoviruses cause an infectious enteritis. Tissues associated with avian orthoreovirus infections include the bursa of Fabricius, the intestine, heart, kidney, liver, pancreas, Peyer's patches, spleen, tendons, thymus, and tonsils. Birds may have obvious joint and tendon disorders. In embryonated eggs avian orthoreoviruses infect the chorioallantoic membrane and yolk sac.

Taxonomic Structure of the Genus

There are at least 2 recognized antigenic groups of orthoreoviruses. One group infects mammals, the other infects birds.

List of Species in the Genus

The viruses, their genomic sequence accession numbers [ ] and assigned abbreviations ( ) are:

Species in the Genus

1-Mammalian orthoreoviruses:		
reovirus 1 (strain Lang)		(REOV-1)
reovirus 2 (strain D5/Jones)		(REOV-2)
reovirus 3 (strain Dearing)	[L1-M24734, L2-J03488, L3-M13139, M1-M27261, M2-M19408,M3-M27262, S1-M10262, S2-M25780, S3-X01627, S4-K02739]	(REOV-3)
2-Avian orthoreoviruses:		
avian reovirus 1 (Uchida, TS-17)		(AVREOV-1)
avian reovirus 2 (TS-17)		(AVREOV-2)
avian reovirus 3 (TS-142)		(AVREOV-3)
avian reovirus 4 (CS-108)		(AVREOV-4)
avian reovirus 5 (OS-161)		(AVREOV-5)
avian reovirus 6 (R24)		(AVREOV-6)
avian reovirus 7 (R25)		(AVREOV-7)
avian reovirus 8 (Fahey-Crawley)		(AVREOV-8)
avian reovirus 9 (59)		(AVREOV-9)
Nelson Bay virus		(NBV)
Somerville virus 4	[S1-L07069]	
WVU virus 71 to 212		
WVU virus 2937		

Tentative Species in the Genus

None reported.

GENUS ORBIVIRUS

Type Species bluetongue virus 1 (BTV-1)

DISTINGUISHING FEATURES

Virions have an indistinct outer capsid and a genome composed of 10 segments of dsRNA. They are transmitted between vertebrate hosts by a variety of hematophagous arthropods.

VIRION PROPERTIES

MORPHOLOGY

Virions are about 80 nm in diameter, core particles are about 60 nm. No lipid envelope is present on virions, although unpurified virus is often associated with cellular membranes. Surface projections are only observed on virions where the particle structure is maintained (e.g., using cryoelectron microscopy). Otherwise, by conventional electron microscopy, the surface of virions is indistinct (Fig. 3).

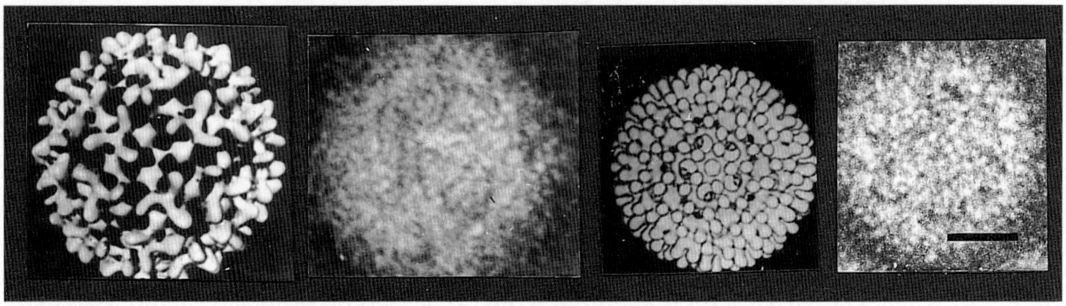

Figure 3: (left) Image of the surface arrangement of BTV as deduced by cryoelectron microscopy (courtesy of Hewat E); (center left) electron micrograph of BTV (courtesy of Booth T); (center right) image of the BTV core particle (courtesy of Prasad BVV); (right) electron micrograph of BTV-derived core (courtesy of Mertens PPC). The bar represents 20 nm.

PHYSICOCHEMICAL AND PHYSICAL PROPERTIES

The virion Mr is about 79 x 10^6, the core Mr is about 52 x 10^6. The buoyant density in CsCl are 1.36 g/cm^3 (virions) and 1.40 g/cm^3 (cores). The S_{20w} is 550 (virions) and 470 (cores). Virus infectivity is stable at pH 8-9 and virions exhibit a marked decrease in infectivity outside the pH range 6.5 -10.2. In part, this may be related to the loss of outer coat proteins. The sensitivity of the outer capsid proteins and their removal by cation treatment varies markedly with both pH and virus strain. Virus cores are about 100 times less infectious for mammalian cell cultures than intact viruses. At low pH values (less than 5.0), virions and cores are both disrupted. Unlike orthoreoviruses, at pH 3.0 virus infectivity is abolished. In blood samples, or serum, or albumin, viruses held *in vitro* may remain infectious for decades at less than 15° C. They are rapidly inactivated on heating to 60° C. In general, viruses are considered to be relatively resistant to treatment with solvents, or detergents. Freezing reduces virus infectivity by about 90%. When held at -70° C virus infectivity remains stable. Core particles are very stable when kept at 4° C.

NUCLEIC ACID

Virus particles and cores contain 16% and 25% RNA, respectively. The genome is composed of 10 dsRNA segments that, for bluetongue viruses, range in size from 3,954 to 822 bp (total size is 19.2 kbp). For bluetongue viruses, the RNAs are classified as 3 large (L1-3, about 3.9-2.8 kbp), 3 medium (M4-6, about 2.0-1.8 kbp) and 4 small segments (S6-10, 1.1-0.8 kbp). For other members of the genus, different sizes and size classes exist. For a particular virus species the dsRNA sizes of different isolates, or different serotypes are generally comparable. For BTV-10, the 5' non-coding sequences range from 8 to 34 bp, for the 3' ends they are 31 to 116 bp in length. For other serotypes and other viruses the lengths differ; in general, however, the 5' non-coding regions are shorter than the 3' non-coding sequences. The non-coding regions of BTVs and EDHVs include terminal sequences of 6 bp that are identical for all 10 dsRNA segments (so far reported) and conserved between virus isolates. For the mRNA sense strands these sequences are 5' GUUAAA....ACUUAC 3'. Other orbiviruses have end sequences comparable to those of BTVs, but which are not always identical and which may not be conserved in all 10 segments.

PROTEINS

There are 7 viral proteins (VP1-7). Proteins constitute 84% and 75% of the dry weight of virions and cores, respectively. For BTVs, the outer capsid consists of 180 copies of the 111 kDa VP2 protein arranged as triskellion structures, and 120 copies of an interdispersed and underlying VP5 protein (59 kDa). The surface of the icosahedral core consists of capsomers that are arranged in ring-like patterns. The surface is composed of 260 trimers of the 39 kDa VP7 protein. The core exhibits a T = 13 (l) surface arrangement. Both VP2 and VP5 are attached to VP7. Underlying VP7 are 12 pentamers of the 103 kDa VP3 protein that form the subcore. This encloses the 10 segment dsRNA genome and the three minor proteins, viz: the 150 kDa VP1 which is an RNA polymerase, the 76 kDa VP4 which is a guanylyl transferase and the 36 kDa VP6 (VP6A), whose function is not known but which binds ssRNA and dsRNA and may be an helicase. The arrangement of the minor proteins is not known, some or all may have structural roles. Other members of the genus may have different protein sizes.

CARBOHYDRATES

VP5 protein may be glycosylated.

GENOME ORGANIZATION AND REPLICATION

Table 2: List of the dsRNA segments of BTV-10 with the corresponding proteins with name, calculated size, and function and /or location.

dsRNA #	Size (bp)	Protein	Size (kDa)	Function (location)
L1	3 954	VP1	150	polymerase (core)
L2	-	VP2	111	Type-specific (outer capsid)
L3	2.8	VP3	103	(core)
M4	2.0	VP4	76	guanylyl transferase (core)
M5	-	NS1	59	(outer capsid)
M6	1.8	VP5	64	Unknown (tubules, inclusion)
S7	1.1	VP7	39	Group antigen (core surface)
S8	-	NS2	41	Binds mRNA (inclusion)
S9	-	VP6(VP6A)	36	Helicase? (core)
S10	0.8	NS3	25	Virus release from cell

For BTV-10, the coding assignments based on the dsRNA migration in 1% agarose are: L1-VP1, L2-VP2, L3-VP3, M4-VP4, M5-the 64 kDa NS1 protein, M6-VP5, S7-VP7, S8-the 34 kDa NS2 protein, S9-VP6, and S10-the 25 kDa NS3 glycoprotein. Cognate genes of other strains are similar. The S9 and S10 mRNA are translated from either of 2 in-frame AUG codons. The significance of the 2 forms of the S9 and S10 gene products (NS3, NS3A; VP6, VP6A) is not known. The NS3 proteins are glycosylated and associate with intracellular and plasma membranes. At the latter site they aid virus egress from the cell. In this process the NS3 proteins are also released. The NS2 protein is a phosphoprotein that binds ssRNA but not dsRNA. NS2 in conjunction with other virus proteins is believed to be involved in the recruitment of viral mRNA for encapsidation. NS2 and virus core proteins are major components of cytoplasmic inclusion bodies that are observed in orbivirus infections. NS1 forms tubules of unknown function. In some cases other virus proteins form morphologically defined structures in infected cells (e.g., the VP7 protein of AHSVs), but of unknown functional significance. Virus adsorption involves components of the outer capsid. The outer capsid layer is lost during the early stages of replication. The mRNA transcription frequency of individual genes varies with more copies produced from the smaller segments. Details of the process of virus replication are lacking. The inclusion bodies are considered to be the sites of morphogenesis of transcriptionally active virus cores containing dsRNA. The outer capsid proteins are added at the periphery of these inclusion bodies. Virus particles are transported within the cell by specific interaction with the cellular cytoskeleton and can be released prior to cell lysis through interaction with membrane-associated NS3 proteins. In mammalian cells, replication of orbiviruses leads to shut-off of host protein synthesis and contributes to cell lysis and the further release of virus particles. In insect cells there is no evidence for shut-off of host protein synthesis, or for extensive cell lysis. NS3 is particularly abundant in insect cells. Continuous release from infected cells and reinfection appears to be a feature of orbivirus replication.

Antigenic Properties

The main serogroup-specific antigen of orbiviruses such as BTVs is the VP7 protein, although other viral antigens are conserved between virus serotypes (in particular core antigens and certain NS proteins). Some of these antigens are cross-reactive with viruses in certain other serogroups. The BTV VP2 and VP5 proteins exhibit the greatest antigenic and sequence variation. BTV VP2 protein has hemagglutinin activity. Although 14 orbivirus serogroups are recognized, some exhibit close antigenic relationships (e.g., African horse sickness, bluetongue, epizootic hemorrhagic disease, equine encephalosis, Eubenangee serogroups). Virus serotype is determined by serum neutralization tests. The specificity of these reactions is determined by the 2 outer capsid proteins. In BTVs, VP2 is the main neutralization antigen while VP5 is also involved, possibly by imposing conformational constraints on VP2. In other viruses (Kemerovo complex viruses) these roles may be reversed.

Biological Properties

Depending on the virus, the vertebrate hosts that orbiviruses infect include ruminants (domesticated and wild), equids, rodents, bats, marsupials, birds, sloths, and primates, including humans. Orbiviruses replicate in, and are primarily transmitted by, arthropod vectors (gnats, mosquitoes, phlebotomies, or ticks, depending on the virus). Trans-stadial transmission in ticks has been demonstrated for some viruses. Infection of vertebrates *in utero* may also occur. Orbiviruses, particularly those transmitted by short-lived vectors (gnats, mosquitoes, phlebotomines), are only enzootic in areas where adults of the competent vector species persist and are present all, or most of the year. For example, BTV and EHDV serogroup viruses are distributed worldwide between about 50° North and about 30° South in the Americas and between 40° North and 35° South in the rest of the world. Virus distribution also depends on the initial introduction into areas containing susceptible vertebrate hosts and competent vector species. For this reason not all serotypes of each serogroup (e.g., BTV serogroup) are present at locations where some serotypes are endemic.

Orbivirus infection of arthropods has no evident effect. In vertebrates, infection can be inapparent to fatal, depending on the virus and the host. Some BTV strains cause death in sheep, others cause a variety of pathologies, including hemorrhagic conditions, lameness, oedema, a transitory cyanotic appearance of the tongue, nasal and mouth lesions, etc.; still others cause no overt pathology. BTV infection of cattle may show no signs of disease but involve long-lived viremias. AHSVs, EHDVs (deer) and EEVs can cause severe pathology in their respective vertebrate hosts.

Taxonomic Structure of Genus

Fourteen groups of orbiviruses are recognized in addition to a number of unclassified viruses. The groups include a number of serotypes and antigenic complexes. From the reported data, reassortment can occur between at least some member viruses of a group or antigenic complex, but not between members representing different groups. Sequence analyses indicate that some genes are more conserved across the genus than others.

LIST OF SPECIES IN THE GENUS

The Kemerovo group consists of at least 3 gene pools with reassortment potential (KEMV-GIV-BRDV; CNUV; MONOV), however these do not correspond to the recognized antigenic complexes listed below:

The viruses, their host { }, antigenic complexes (+), serotypes, genomic sequence accession numbers [ ] and assigned abbreviations ( ), are:

SPECIES IN THE GENUS

1-African horse sickness group: {Culicoides}
African horse sickness viruses 1 to 10 (AHSV-1 to 10)
[L2:M94680, L3:M94681, M5:D11390, M6:M94682, S7:D12533, S8:M69090, S10:D12479]
2-bluetongue viruses 1 to 24 (BTV-1 to 24)
[L1:X12819, L2:M11787, L3:M22096, M4:Y00421, M5:D12532, M6:Y00422, S7:X06463, S8:D00500, S9:D00509, S10:M28981]
3-Changuinola virus group: {phlebotomines}
Almeirim virus (ALMV)
Altamira virus (ALTV)
Caninde virus (CANV)
Changuinola virus (CGLV)
Gurupi virus (GURV)
Irituia virus (IRIV)
Jamanxi virus (JAMV)
Jari virus (JARIV)
Monte Dourado virus (MDOV)
Ourem virus (OURV)
Purus virus {culicine mosquitoes} (PURV)
Saraca virus (SRAV)
4-Corriparta virus group: {culicine mosquitoes}
Acado virus (ACDV)
Corriparta virus (CORV)
Jacareacanga virus (JACV)
5-Epizootic hemarrhogic disease virus group {Culicoides}
epizootic hemorrhagic disease viruses 1 to 10 (EHDV-1 to 10)
[L2:D10767, L3:M76616, M5:X55782, M6:X59000, S7:D10766, S8:M69091]
Ibaraki virus (IBAV)
6-Equine encephalosis virus group: {Culicoides}
equine encephalosis viruses 1 to 7 (EEV-1 to 7)
7-Eubenangee virus group:
{Culicoides, anopheline and culicine mosquitoes}
Eubenangee virus (EUBV)
Ngoupe virus (NGOV)
Pata virus (PATAV)
Tilligerry virus (TILV)
8-Lebombo virus group: {culicine mosquitoes}
Lebombo virus (LEBV)
9-Orungo virus group: {culicine mosquitoes}
Orungo virus 1 to 4 (ORUV-1 to 4)

10-Palyam virus group: {Culicoides, culicine mosquitoes}
- Abadina virus (ABAV)
- Bunyip creek virus (BCV)
- CSIRO village virus (CVGV)
- D'Aguilar virus (DAGV)
- Kasba virus (KASV)
- Kindia virus (KINV)
- Marrakai virus (MARV)
- Nyabira virus (NYAV)
- Palyam virus (PALV)
- Petevo virus (PETV)
- Vellore virus (VELV)

11-Umatilla virus group: {culicine mosquitoes}
- Llano Seco virus (LLSV)
- Minnal virus (MINV)
- Umatilla virus (UMAV)

12-Wallal virus group: {Culicoides}
- Mudjinbarry virus (MUDV)
- Wallal virus (WALV)

13-Warrego virus group:
{Culicoides, anopheline and culicine mosquitoes}
- Mitchell river virus (MRV)
- Warrego virus (WARV)

14-Kemerovo virus group:
{ticks}

14a+Kemerovo complex: {Ixodes; rodents, man}
- Kemerovo virus (KEMV)
- Kharagysh virus (KHAV)
- Lipovnik virus (LIPV)
- Tribec virus (TRBV)

14b+Chenuda complex:
{*Argas, Ornithodoros;* land-, seabirds}
- Baku virus (BAKUV)
- Chenuda virus (CNUV)
- Essaouira virus (ESSV)
- Huacho virus (HUAV)
- Kala Iris virus (KIRV)
- Mono Lake virus (MLV)
- Sixgun city virus (SCV)

14c+Great Island complex:
{*Argas, Ixodes, Ornithodoros;* seabirds}
- Arbroath virus (ABRV)
- Bauline virus (BAUV)
- Broadhaven virus [L2:M87875, M5:M36394, S7: M87876, S10:M83197] (BRDV)
- Cape Wrath virus (CWV)
- Ellidaey virus (ELLV)
- Foula virus (FOUV)
- Great Island virus (GIV)
- Great Saltee Island virus (GSIV)
- Grimsey virus (GSYV)
- Inner Farne virus (INFV)
- Kenai virus (KENV)
- Lundy virus (LUNV)
- Mill Door virus (MDRV)
- Mykines virus (MYKV)
- North Clett virus (NCLV)

North End virus	(NEDV)
Nugget virus	(NUGV)
Okhotskiy virus	(OKHV)
Poovoot virus	(POOV)
Rost Islands virus	(RSTV)
Saint Abb's Head virus	(SAHV)
Shiant Islands virus	(SHIV)
Thormódseyjarklettur virus	(THRV)
Tindholmur virus	(TDMV)
Vaeroy virus	(VAEV)
Wexford virus	(WEXV)
Yaquina Head virus	(YHV)
14d+Wad Medani complex:	
{*Boophilus, Rhipicephalus, Hyalomma, Argas*; domestic animals}	
Seletar virus	(SELV)
Wad Medani virus	(WMV)

TENTATIVE SPECIES IN THE GENUS

Andasibe virus	(ANDV)
Arkonam virus	(ARKV)
Chobar Gorge virus	(CGV)
Fromede virus	(FOMV)
Gomoka virus	(GOMV)
Ieri virus	(IERIV)
Ife virus	(IFEV)
Itupiranga virus	(ITUV)
Japanaut virus	(JAPV)
Kammavanpettai virus	(KMPV)
Lake Clarendon virus	(LCV)
Matucare virus	(MATV)
Ndelle virus	(NDEV)
Paroo river virus	(PRV)
Picola virus	(PIAV)
Tembe virus	(TMEV)
Wongorr virus	(WGRV)

GENUS *ROTAVIRUS*

Type Species simian rotavirus SA11 (SA11)

DISTINGUISHING FEATURES

Viruses only infect vertebrates and are transmitted by the fecal-oral route. They have a typical structure that appears wheel-like by negative contrast electron microscopy. Rotaviruses possess 11 dsRNA segments and undergo a process of morphogenesis that involves the temporary acquisition of a lipid envelope and the deposition of viral-coded glycoprotein.

VIRION PROPERTIES

MORPHOLOGY

Virions consist of a core (about 50 nm in diameter), inner capsid (about 60 nm) and outer capsid (about 70 nm). Cryoelectron microscopy and image processing reveals that both inner and outer capsids have T = 13 (l) icosahedral symmetry, with 132 channels superimposed and extending inwards from the surface to the core, and 60 short spikes extending 4.5 - 6 nm from the surface of the virus particle (Fig. 4).

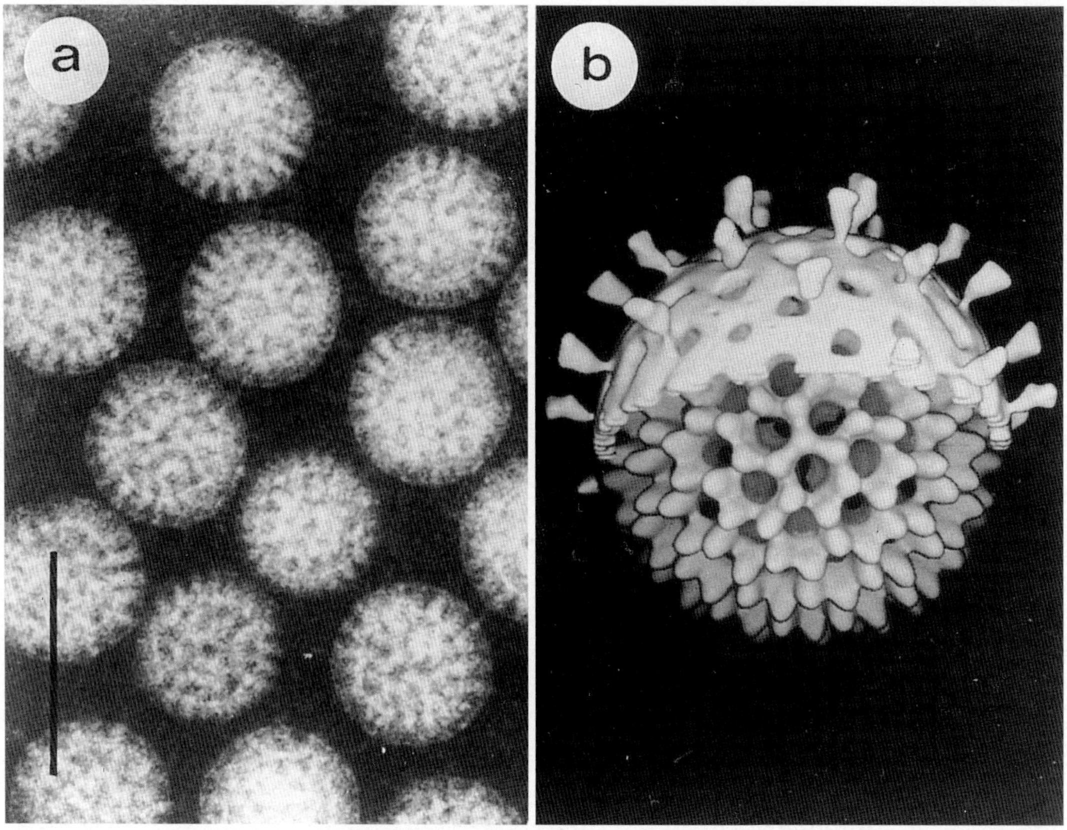

Figure 4: (a) Rotavirus particles visualized by negative staining. Particle forms include complete, infectious, triple-shelled particles with spikes, and incomplete, double-shelled particles that lack the outer shell (bar represents 100 nm); (b) representation (from cryoelectron micrographs) of the three dimensional structure of a complete rotavirus particle in which a portion of the outer shell has been removed to show the second shell. The outer shell is composed of the glycoprotein VP7 from which dimers of VP4 extend. The second shell consists of trimers of VP6. The innermost (third) shell is composed of VP2 and is visible through holes in the second shell (courtesy of Prasad BVV).

Physicochemical and Physical Properties

Infectivity is stable to pH 3.0 and relatively stable to heat.

Nucleic Acid

The rotavirus genome consists of 11 segments of dsRNA (size range: 0.6 - 3.3 kbp). Although the dsRNA sizes may be broadly categorized into 4 large, 5 medium, and 2 small, the RNA sizes vary significantly between the rotavirus groups and consequently the dsRNA species are numbered 1-11. RNA sizes and size classes are frequently species specific and some can be used to distinguish rotaviruses of different groups and from the 11-segment genomes of aquareoviruses. Aberrant dsRNA forms and sizes also may be present in a virus population, presumably representing rearrangements (usually duplications) within a segment.

Proteins

The structural proteins of rotaviruses include both primary gene products and those that are derived by post-translational modification (proteolytic cleavage, glycosylation).

Genome Organization and Replication

The coding assignments (primary translation product Mr) of the group A rotavirus dsRNAs (e.g., SA11) are: 1:VP1, 2:VP2, 3:VP3, 4:VP4 (VP5, VP8), 5: NS53 (59 kDa), 6: VP6, 7: NS34 (35 kDa), 8: NS35 (38 kDa), 9: VP7, 10: NS28 (a 20 kDa precursor protein to a high mannose glycoprotein of about 28 kDa), 11: NS26 (a 22 kDa, O-linked, phosphorylated protein of about 28 kDa). The individual sizes and relative electrophoretic mobilities of the RNAs and proteins may vary between viruses; however, cognate genes can be identified by sequence comparisons.

Table 3: List of the dsRNA segments of SA 11 with their respective size (bp) and the corresponding proteins for which name, calculated size (kDa) and function and/or location are indicated.

dsRNA #	Size (kbp)	Protein	Size (kDa)	Function (location)
1	3302	VP1	125	Polymerase (core)
2	2690	VP2	102	(core)
3	2591	VP3	98	guanylyl transferase (core)
4	2362	VP4	87	cleaved by trypsin to:
		VP5*	60	Cell attachment & entry,
		VP8*	28	HA, type-specific (outer capsid spike)
5	1611	N SP1	59	Unknown
6	1356	VP6	45	Group antigen (inner capsid)
7	1104	NSP3	35	Unknown
8	1059	NS35	37	Unknown
9	1062	VP7	37	Type Specific (outer capsid)
10	751	NSP4	20	Particle entry to RER and assembly
11	667	NS26	22	Unknown

Virus binding involves epitopes present on VP4 and requires sialic acid residues on cell surface components. Viruses may penetrate the plasma membrane directly. This penetration depends on the cleavage of VP4 that produces VP5 (alternatively designated VP5*) and VP8 (designated VP8*). Penetration following phagocytosis may occur although, since lysosomotropic agents exert little inhibitory effect, this mechanism of entry appears unlikely. The processes of synthesis of mRNA species and their translation (etc.), have not been studied in detail. Like other reoviruses, mRNAs are capped and not polyadenylated. They are produced by the endogenous RNA-directed RNA polymerase present in particles containing VP1, VP2, VP3 and VP6. Rotaviruses such as SA11 synthesize 5 NS proteins (NS53, NS35, NS34, NS28, NS26) whose functions probably include roles in mRNA recruitment into progeny particles, dsRNA synthesis and virus morphogenesis. Genetic studies indicate that VP2 and VP6 proteins also have roles in dsRNA synthesis. NS53 has a zinc finger domain. Two of the NS proteins are glycosylated (NS28, NS26), one of these (NS26) is phosphorylated. Translation of VP7 occurs on membrane associated ribosomes and appears to initiate at either of two in-frame AUGs that are separated by about 30 codons. Whether the 2 forms of the protein have different roles in the infection process is not known. The VP7 glycoprotein has signal sequences proximal to each AUG codon. These sequences are cleaved co-translationally, so that for SA11 the amino terminus of the protein is at residue 51 (glutamine). Depending on the virus, VP7 possesses one or more N-linked, high mannose glycans that are partially trimmed during virus maturation.

The process of morphogenesis of rotaviruses involves the translocation of progenitor particles (that accumulate in viroplasms), and their budding into the cisternae of the rough endoplasmic reticulum (RER). They thereby acquire a temporary envelope. Viruses are not subsequently translocated to the Golgi apparatus. NS28 mediates the translocation of partially assembled particles across the RER membrane. NS28 (NSP4) has a signal sequence that is not removed and a carboxy terminal half that extends into the cytoplasm. It also has a role in the eventual removal of the envelope that is acquired by the progenitor particles.

ANTIGENIC PROPERTIES

The rotavirus VP4 protein has type-specific antigens and elicits neutralizing antibodies. Most, but not all, rotavirus strains hemagglutinate red blood cells. VP4 is the hemagglutinin. The VP7 outer capsid protein also has type specific antigens that play a role in virus neutralization. Although all rotavirus proteins contain group-specific determinants, VP6, the major capsid protein, is most often considered the group-specific antigen. It is the antigen most easily detected in diagnostic tests. Six serogroups of rotaviruses are recognized (designated A-F). Within the rotavirus A group some 14 serotypes have been defined based on their VP7 antigens (designated G1-14) and 8 serotypes based on VP4 (designated P1-8). Distinct serotypes within the other rotavirus groups probably exist.

Biological Properties

Most rotaviruses are difficult to cultivate *in vitro*. They require epithelial cells of intestinal or kidney origin and media containing trypsin. Rotaviruses infect a variety of vertebrates. They cause diarrhea due to infection and lysis of intestinal enterocytes and consequent loss of the ability of the intestine to absorb water. Rotaviruses that affect humans include the Group A, B and C viruses. The A and C viruses are primarily associated with pediatric disease, often with initial infection occurring in the first few years of life. Probably infections by Group A and C viruses occur throughout life. The Group B viruses have caused epidemics of infection in adults as well as the young. All six groups of rotaviruses infect a variety of other vertebrates, including cats, cattle, horses, pigs, primates, rabbits, rodents, turkeys, etc. Several rotavirus genes contribute to virus virulence in model animal systems.

Taxonomic Structure of the Genus

There are 6 antigenic groups of rotaviruses (A-F).

List of Species in the Genus

Note, the cognate genes do not necessarily correspond to the RNA segments with the same number (e.g., Cowden rotavirus segments 5-8 correspond to SA11 segments 6, 7, 5, and 9, respectively).

The groups, viruses, their genomic sequence accession numbers [ ] and assigned abbreviations ( ) are:

Species in the Genus

group A rotaviruses (simian rotavirus SA11) (ROTAV-A)
[1-X16830, 2-X16831, 3-X16062, 4-X14204, 5-X14914, 6-X00421, 7-X00355, 8-J02353, 9-K02028, 10-KO1138, 11-X07831]
group B rotaviruses (ROTAV-B)
[5-M55982, 6-M84456, 9-M33872, D00911, 11-M34380, D00912]
group C rotaviruses (porcine Cowden strain) (ROTAV-C)
[1-M74216, 2-M74217, 3-M74218, 4-M74219, 5-M29287, 6-M69115, 7-X60546, 8-M61100, 10-M81488]
group D rotaviruses (chicken 132 strain) (ROTAV-D)
group E rotaviruses (porcine DC-9 strain) (ROTAV-E)
group F rotaviruses (avian) (ROTAV-F)

Tentative Species in the Genus

None reported.

GENUS *COLTIVIRUS*

Type Species Colorado tick fever virus (CTFV)

DISTINGUISHING FEATURES

MORPHOLOGY

Coltivirus particles are about 80 nm in diameter with a double layered capsid. Electron microscopic studies, using negative staining have shown that particles have a relatively smooth surface capsomer structure and icosahedral symmetry. Particles are frequently observed associated with membranes, but do not acquire a membrane envelope.

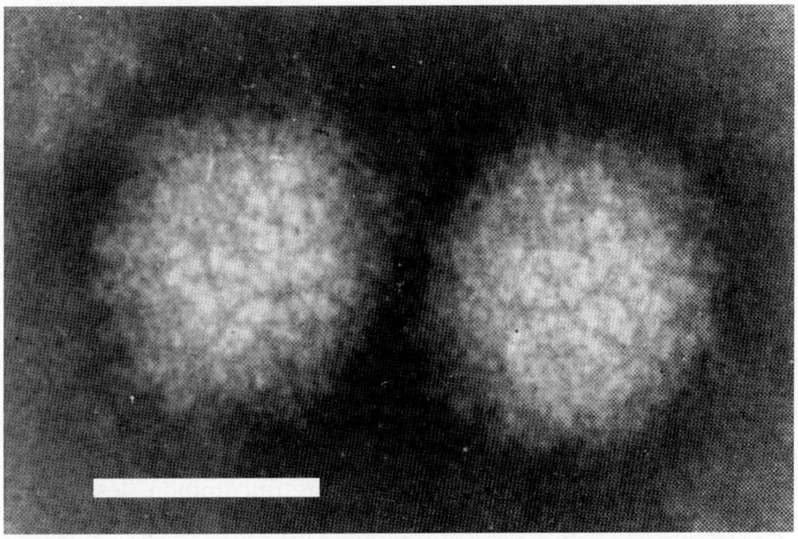

Figure 5: Negative contrast electron micrograph of Colorado tick fever virions (courtesy of Murphy FA). The bar represents 50 nm.

PHYSICOCHEMICAL AND PHYSICAL PROPERTIES

Virus infectivity is lost at pH 3.0 and is abolished by treatment with sodium deoxycholate. Viruses are stable between pH 7 and 8.

NUCLEIC ACID

The genome consists of 12 dsRNA segments with estimated Mr sizes ranging from 2.53×10^6 to 0.24×10^6 (total: Mr about 18×10^6).

PROTEINS

Viral proteins have not been characterized.

LIPIDS

None reported.

CARBOHYDRATES

None reported.

GENOME ORGANIZATION AND REPLICATION

In infected cells granular matrices are produced which contain virus-like particles. These structures are similar to the viral inclusion bodies produced during orbivirus infections. In addition, bundles of filaments, characterized by cross-striations, are found in the cytoplasm and, in some cases, in the nucleus of infected cells. There is no evidence for virus release prior to cell death and disruption.

Antigenic Properties

CTLV from North America and Eyach virus from Europe show little cross-reaction in neutralization tests. An isolate, S6-14-03, obtained from a hare (*Lepus californicus*) in Northern California, is related to Eyach virus, and is considered to be a third coltivirus.

Biological Properties

Coltiviruses have been isolated from several mammalian species (including humans) and from ticks which serve as vectors. The tick species include *Dermacentor andersoni, D occidentales, D. albipictus, D. parumapertus, Haemaphysalis leporispalustris, Otobius lagophilus, Ixodes sculptus, I. spinipalpis, I. ricinus and I. ventalloi.* Mosquito species may also act as vectors.

Although CTFV is not transmitted trans-ovarially in ticks it is transmitted trans-stadially. Ticks become infected on ingestion of a blood meal from an infected host. Adult and nymphal ticks become persistently infected and provide an overwintering mechanism for the virus. Some rodent species have prolonged viraemias (more than 5 months) which may also facilitate virus persistence. Humans become infected with CTLV when bitten by the wood tick *D. andersoni*, however humans probably do not act as a source of infection for other ticks. Transmission from person to person has been recorded as the result of blood transfusion. The prolonged viraemia observed in humans and rodents is thought to be due to the intra-erythrocytic location of virions, protecting them from immune clearance.

Colorado tick fever is characterized in humans by an abrupt onset of fever, chills, headache, retro-orbital pains, photophobia, myalgia and generalized malaise. Abdominal pain occurs in about 20% of patients. Rashes are uncommon (less than 10%). A diphasic, or even triphasic, febrile pattern has been observed, usually lasting for 5-10 days. Severe forms of the disease, involving infection of the central nervous system, or haemorrhagic fever, or both, have been infrequently observed (nearly always in children under 12 years of age). Three such cases were fatal. Congenital infection with CTFV may occur, although the risk of abortion and congenital defects remains uncertain. Antibodies to Eyach virus have been found in patients with meningoencephalitis and polyneuritis but a causal relationship to the virus has not been established.

Colorado tick fever virus causes leukopaenia in adult hamsters and in about two-thirds of infected humans. Suckling mice, which usually die at 6-8 days post-infection, suffer myocardial necrosis, necrobiotic cerebellar changes, widespread focal necrosis and perivascular inflammation in the cerebral cortex, degeneration of skeletal myofibers, hepatic necrosis, acute involution of the thymus, focal necrosis in the retina and in brown fat. The pathologic changes in mice due to CTFV infection (in skeletal muscle, heart and brain), are consistent with the clinical features of human infection which may include meningitis, meningo-encephalitis, encephalitis, gastro-intestinal bleeding, pneumonia and myocarditis.

Colorado tick fever occurs in forest habitats at 4,000 - 10,000 ft. elevation in the Rocky Mountain region of North America. Antibodies to the virus have been detected in hares in Ontario and a virus isolate has been reported from Long Island, New York. Eyach virus appears to be widely distributed in Europe.

List of Species in the Genus

Isolate S6-14-03 from a hare collected in California in 1976 shows some one-way cross-reaction in serum neutralization tests with Eyach virus, but is clearly distinguishable and has been reported as a distinct serotype. Serological variants of Eyach virus (AR 577 and AR 578) have also been reported. Recently, several Indonesian (JKT6423, JKT6969, JKT7041,

JKT7075) and Chinese (HN59, HN131, HN191, HN295) virus isolates have been made which may include serologically distinct coltiviruses.

The viruses, and their assigned abbreviations ( ) are:

Species in the Genus

Colorado tick fever virus	(CTFV)
Eyach virus (also AR 577, AR 578)	(EYAV)
S6-14-03 virus	(S6-14-03V)

Tentative Species in the Genus

None reported.

Genus *Aquareovirus*

Type Species golden shiner virus (GSV)

Distinguishing Features

Viruses physically resemble orthoreoviruses but possess 11 dsRNA segments. They infect certain aquatic organisms, including fish and clams. In fish cell culture lines they produce syncytia.

Virion Properties

Morphology

Viruses have a diameter of about 75 nm (core about 50 nm) (Fig. 6).

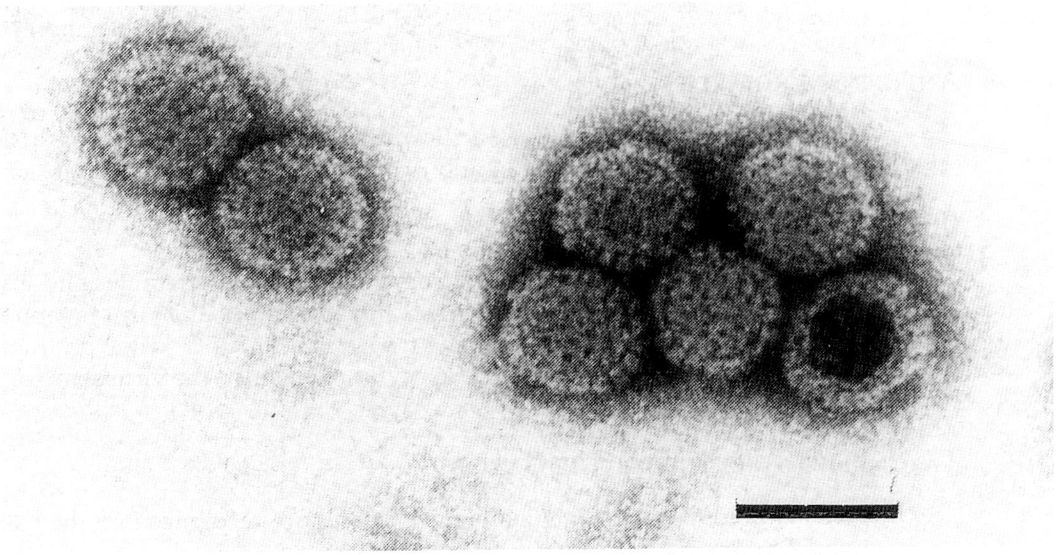

Figure 6: Negative contrast electron micrograph of GSV virions . The bar represents 100 nm.

Physicochemical and Physical Properties

Virion density in CsCl is 1.36 g/cm^3. Virus infectivity is not affected by treatment with ether or proteolytic enzymes.

NUCLEIC ACID

Viruses possess 11 segments of dsRNA (Mr 0.3 - 2.5 x 10^6; total about 15 x 10^6), 3 large, 3 medium and 5 small segments. Cross-hybridization studies indicate that many aquareoviruses are closely related.

PROTEINS

Virions contain 7 structural proteins (VP1: 130 kDa, inner core; VP2: 127 kDa, inner core; VP3: 126 kDa, inner core; VP4: 73 kDa, inner capsid; VP5: 71 kDa, inner capsid; VP6: 46 kDa, minor outer capsid; VP7: 35 kDa, major outer capsid); and five non-structural proteins (NS1:97 kDa; NS2:39 kDa; NS3: 29 kDa; NS4:28 kDa; NS5: 15kDa).

GENOME ORGANIZATION AND REPLICATION

Table 4: List of the dsRNA segments of SBR (*Aquareovirus*), with their estimated size (kbp), and corresponding proteins with name, size (estimated), and function and/or location.

dsRNA #	Size (kbp)	Protein	Size (kDa)	Function (location)
L1	3.8	VP 1	130	core
L2	3.6	VP 2	127	core
L3	3.3	VP 3	126	core
M4	2.5	VP 4	97	non-structural
M5	2.4	VP 5	71	inner capsid
M6	2.2	VP 4	73	inner capsid
S7	1.5	NS4	28	non-structural
S8	1.4	VP 6	46	minor outer capsid
S9	1.2	NS2	39	non-structural
S10	0.9	VP 7	34	major outer capsid
S11	0.8	NS3	29	non-structural
S11	0.8	NS5	15	non-structural

ANTIGENIC PROPERTIES

Viruses have type and group-specific antigenic determinants. Cross-reactivity has been demonstrated only between 2 (A and B) of the 5 recognized serogroups of aquareoviruses.

BIOLOGICAL PROPERTIES

Aquareoviruses have been isolated from poikilotherm vertebrates and invertebrates (fish, molluscs, etc.) obtained from both fresh and sea water. The viruses replicate efficiently in fish cell lines at temperatures ranging from 15° C to 30° C. They produce a characteristic cytopathic effect consisting of large syncytia. Generally, the viruses are of low pathogenicity in their host species.

TAXONOMIC STRUCTURE OF THE GENUS

Five genogroups and some unassigned viruses are recognized on the basis of RNA-RNA hybridization.

LIST OF SPECIES IN THE GENUS

The groups, viruses, and their assigned abbreviations ( ) are:

SPECIES IN THE GENUS

1-group A:

American oyster reovirus	(13p2)
angel fish reovirus	(AFRV)
Atlantic salmon reovirus USA	(HBRV)
Atlantic salmon reovirus Canada	(ASV)

Atlantic salmon reovirus Australia	(TSV)
Chinook salmon reovirus	(DRCV)
chum salmon reovirus	(CSV)
Masou salmon reovirus	(MSV)
smelt reovirus	(SRV)
striped bass reovirus	(SBRV)
2-group B:	
Chinook salmon reovirus	(GRC, LBS, YRC, ICR)
Coho salmon reovirus	(SCSV)
3-group C:	
golden shiner reovirus	(GSV)
4-group D:	
channel catfish reovirus	(CRV)
5-group E:	
turbot reovirus	(TRV)

TENTATIVE SPECIES IN THE GENUS

chub reovirus Germany	(CHRV)
grass carp reovirus	(GCRV)
hard clam reovirus	(HCRV)
landlocked salmon reovirus	(LSRV)
tench reovirus	(TNRV)

GENUS *CYPOVIRUS*

Type Species Bombyx mori cypovirus 1 (BmCPV-1)

DISTINGUISHING FEATURES

Cypovirus virions lack a double-shelled structure and that may be occluded by a virus-coded polyhedrin protein to form polyhedra in the cytoplasm of infected cells. Also cypoviruses only infect and are pathogenic for particular arthropod species.

VIRION PROPERTIES

MORPHOLOGY

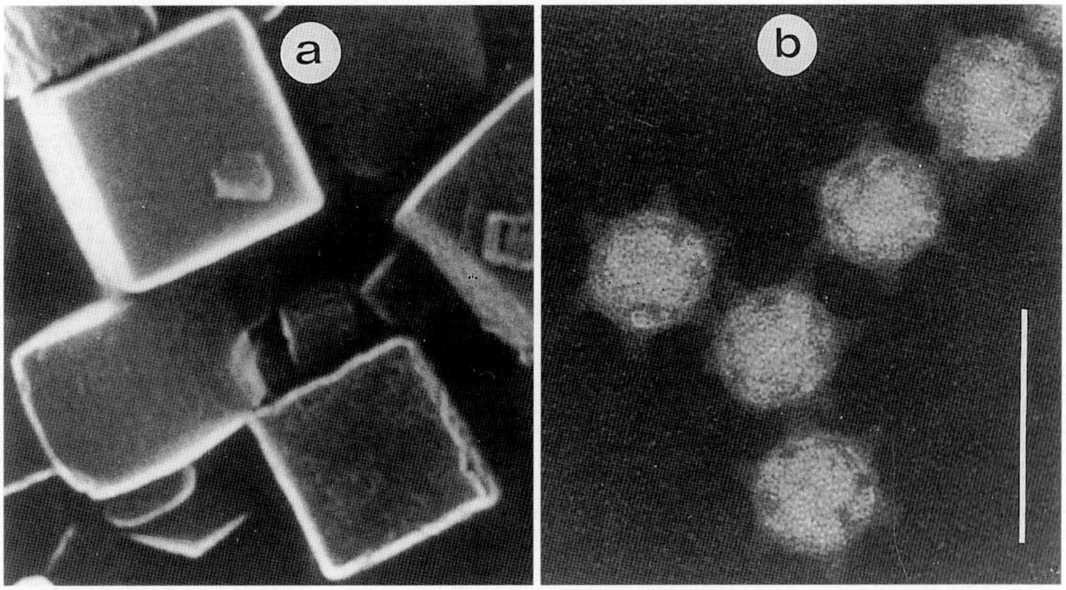

Figure 7: (a) Scanning electron micrograph of BmCPV-1 polyhedra (x 6,480); (b) negative contrast electron micrograph of BmCPV-1 virions stained with lithium tungstate, (courtesy of Bishop DHL). The bar represents 100 nm.

Virions have a single shelled capsid (55-69 nm in diameter, Fig. 7 left) with icosahedral symmetry and hollow surface spikes at the vertices (about 20 nm in length and 15-23 nm wide) and a central compartment about 35 nm in diameter. *Cypovirus* virions are structurally equivalent to the core particles of other members of the family *Reoviridae.*

Virus particles are also occluded by a crystalline matrix of polyhedrin protein forming a polyhedral inclusion body (Fig. 7 left). These structures have a symmetry (e.g., cubic, icosahedral, or irregular) which is dependent on both the virus strain and the host. The polyhedrin protein appears to be arranged as a face-centered cubic lattice with center to center spacing varying between 4.1 and 7.4 nm.

Physicochemical and Physical Properties

The virion Mr is about 54 x 10^6. The buoyant density in CsCl is 1.44 g/cm^3 (virions), about 1.30 g/cm^3 for empty particles, and about 1.28 g/cm^3 for polyhedra. The S_{20w} is about 420 (virions) and about 260 for empty particles. Polyhedra vary considerably in Mr and size and do not have a characteristic S value. Polyhedra may occlude many virus particles or only single particles.

Cypoviruses retain infectivity for several weeks at -15° C, 5° C, or 25° C, and retain full enzymatic activity after repeated freeze-thawing. Cations have relatively little effect on the virus structure. Heat treatment of virions at 60° C for 1 hr leads to degradation and release of genomic RNA. Under some conditions released RNA-protein complexes exhibit polymerase and capping activities. Viruses are resistant to treatment with trypsin, chrymotrypsin, ribonuclease A, deoxyribonuclease, and phospholipase C. Cypovirus particles are resistant to detergents such as sodium deoxycholate (0.5-1%) but are disrupted by 0.5-1% SDS. One or two fluorocarbon treatments have little effect on virus infectivity, however treatment with ethanol leads to release of RNA from virions. Viruses and polyhedra are readily inactivated by UV-irradiation. Polyhedra remain infectious for years on less than 20° C. Virions can be released from polyhedra by treatment with carbonate buffer at pH greater than 10.5. As in permissive insects' midguts, this pH treatment completely dissolves the polyhedral protein matrix.

Nucleic Acid

Cypoviruses contain 10 dsRNA genome segments with Mr that range from 0.3 to 2.6 x 10^6 and with a total genome Mr of 13.6 to 15.6 x 10^6. The pattern of size distribution of the genome segments varies widely between different cypoviruses (e.g., for some cypoviruses the smallest dsRNA Mr is about 0.8 x 10^6). At present, these size differences are the basis for cypovirus classification (12 different electrophoretypes by 1% agarose or 3% SDS-PAGE). Polyhedra contain significant amounts of adenylate-rich oligonucleotides. The termini of the coding strands are common for different genome segments of type 1 cypoviruses (5' AGUAAA...GUUAGCC 3'), but differ from those reported for type 5 cypoviruses (5' AGUUU...GAGUUGC 3'), suggesting that different cypovirus groups vary in this respect.

Proteins

Cypoviruses generally contain five distinct proteins, 2-3 with Mr of more than 100 kDa. For BmCPV-1 the structural proteins are 146 kDa, 138 kDa, 125 kDa, 70 kDa and 31 kDa. Polyhedra also contain a 25-37 kDa polyhedrin protein (27 kDa for BmCPV-1) that constitutes about 95% of the polyhedra protein dry weight.

Carbohydrates

The polyhedrin protein is glycosylated.

GENOME ORGANIZATION AND REPLICATION

For BmCPV-1 the coding assignments are indicated in table 5. The origin of a 31 kDa structural protein is not known, it may represent a processed product. The cognate genes of other cypoviruses are not known.

Table 5: List of dsRNA segments of BmCPV-1 (*Cypovirus*), with their respective size (kbp) and the corresponding proteins for which name, size (kDa) and function are indicated

dsRNA #	Size (kbp)	Protein	Size (kDa)	Function (location)
1	2.2-2.6		146	polymerase methyltransferase
2	2.3-2.6		138	structural protein
3	2.2-2.5		138	structural protein
4	2.0-2.2		125	structural protein
5	1.1-2.1	NS	107(80+23)	
6	1.0-1.3		70	structural protein
7	0.7-1.3	NS	58-61	
8	0.6-1.0	NS	55	
9	0.4-0.8	NS	39	
10	0.3-0.8		27	polyhedrin proteins

Unlike reoviruses, cypovirus uptake by insect cells does not require modification of the virions for activation of the core-associated enzymes. Virus replication and assembly occur in the host cell cytoplasm, although there is some evidence for virus RNA synthesis within the nucleus. Replication is accompanied by the formation of viroplasm (or virogenic stroma) within the cytoplasm. Viroplasm contain large amounts of virus proteins and virus particles. How genome segments are selected for packaging and assembly into progeny particles is not known. The importance of the terminal regions in this process is indicated by the packaging and transcription of a mutant segment 10 of a type 1 CPV that contained only 121 base pairs from the 5' end and 200 base pairs from the 3' end. Particles are occluded within polyhedra apparently at the periphery of the virogenic stroma, from about 15 hr post-infection. Polyhedrin protein is produced late in infection and in large excess compared to the other viral proteins. How polyhedrin protein synthesis is regulated is not known.

Many virus particles remain non-occluded. Following cell lysis virions spread infection between cells in culture, or within an individual host. Polyhedra serve to spread viruses between hosts.

ANTIGENIC PROPERTIES

Serological cross-comparisons of viral structural and polyhedrin proteins support the electrophoretype classification of cypoviruses with little or no cross-reaction evident for viruses representing different electrophoretypes, except for members of types 1 and 12. Depending on the virus, members assigned to an electrophoretype exhibit antigenic cross-reactions.

BIOLOGICAL PROPERTIES

Cypoviruses have only been isolated from arthropods. Attempts to infect vertebrates, or vertebrate cell lines, have failed. Also, cypovirus replication is inhibited at 35° C. Even susceptible insect larvae treated with the virus fail to develop infections at 35° C.

Cypoviruses are normally transmitted by ingestion of polyhedra on contaminated food materials. The polyhedra dissolve within the high pH environment of the insect gut releasing the virus particles which then infect the cells lining the gut wall. Virus infection is generally restricted in larvae to the columnar epithelial cells of the midgut, although goblet cells may also become infected. Cypovirus replication in the fat body has been reported. In larva, virus infection spreads throughout the midgut region. In some species the entire gut is occasionally infected. The production of very large numbers of polyhedra give the gut a characteristically creamy-white appearance. In the infected cell the endoplasmic reticulum is progressively degraded, mitochondria enlarge and the cytoplasm becomes highly vacuolated. In most cases the nucleus shows few pathological changes. An exception is a cypovirus strain which produces inclusion bodies within the nucleus. In the later stages of infection cellular hypertrophy is common and microvillae are reduced or completely absent. Very large numbers of polyhedra are released by cell lysis into the gut lumen and excreted. The gut pH is lowered during infection and this prevents dissolution of progeny polyhedra in the gut fluid.

The majority of cypovirus infections produce chronic disease often without extensive larval mortality. Consequently, many individuals reach the adult stage even though heavily diseased. Cypovirus infections do, however, produce symptoms of starvation due to changes in the gut cell structure and reduced adsorptive capacity. Infected larvae stop feeding as early as two days post-infection. Larval body size and weight are often reduced and diarrhea is common. The host larval stage can be significantly increased (about by 1.5 times the normal generation time).

The size of infected pupae is frequently reduced and the majority of diseased adults are malformed. They may not emerge correctly, and may be flightless. Infected females may exhibit a reduced egg laying capacity. Virus can be transmitted on the surface of eggs, producing high levels of infection in the subsequent generation. However, no transovarial transmission has been observed provided the egg surface is disinfected. The infectious dose increases dramatically with later larval instars. Different virus strains vary significantly in virulence. Larvae can recover from cypovirus infection, possibly because the gut epithelium has considerable regenerative capacity and because infected cells are shed at each larval moult.

Taxonomic Structure of the Genus

It is the custom in the literature to refer to cypoviruses by the name of the insect host species (e.g., Bombyx mori cypovirus 1). Although some host insect species appear to have an exclusive relationship to a particular virus type (e.g., BmCPV-1), other insect species support a wide range of different cypoviruses (e.g., *Spodoptera exempta* supports cypovirus types 3, 5, 8, 11 and 12). Also, many virus strains replicate in more than one insect species. Although prevalent, the use of host species names is inadequate for the purposes of taxonomy.

Cypoviruses are currently classified within 12 distinctive dsRNA electrophoretypes. Cross-hybridization analyses of dsRNA and serological comparisons of cypovirus proteins so far confirm the validity of this classification. However, only a few cypoviruses have been analyzed in this way.

The current classification system takes account of both the dsRNA electrophoretype and the host species from which viruses were originally isolated. The relationships at the molecular level of different cypoviruses within an electrophoretype, or to other cypoviruses, is not known. Only electrophoretypes 1 and 12 show any significant similarity in their overall genome profiles and levels of RNA cross-hybridization and serological cross-reaction.

List of Species in the Genus

Below is provided a list of some of the lepidopteran cypoviruses for which the RNA electrophoretypes have been deduced. In addition to many other lepidopteran cypoviruses that have been described (but are otherwise uncharacterized), there are dipteran and hymenopteran cypoviruses. One isolate from a freshwater daphnid has been reported. In total, more than 230 cypoviruses have been described, however the number of species is unknown. The recognized cypovirus electrophoretype groups (RNA sizes x 10^6) and certain recognized hosts (including the original and other members of the species from which the virus was isolated) genomic sequence accession numbers [ ] and assigned abbreviations ( ) are:

Species in the Genus

1-Cypovirus type 1 : (2.55, 2.42, 2.32, 2.03, 1.82, 1.12,0.84, 0.62, 0.56, 0.35)
- Bombyx mori cypovirus 1 (BmCPV-1)
- Dendrolimus spectabilis cypovirus 1 (DsCPV-1)
- Lymantria dispar cypovirus 1 (LdCPV-1)

2-Cypovirus type 2 : (2.29, 2.29, 2.16, 2.06, 1.25, 1.09, 1.01, 0.88, 0.78, 0.55)
- Aglais urticae cypovirus 2 (AuCPV-2)
- Agraulis vanillae cypovirus 2 (AvaCPV-2)
- Arctia caja cypovirus 2 (AcCPV-2)
- Arctia villica cypovirus 2 (AviCPV-2)
- Boloria dia cypovirus 2 (BdCPV-2)
- Dasychira pudibunda cypovirus 2 (DpCPV-2)
- Eriogaster lanestris cypovirus 2 (ElCPV-2)
- Hyloicus pinastri cypovirus 2 (HpCPV-2)
- Inachis io cypovirus 2 (IiCPV-2)
- Lacanobia oleracea cypovirus 2 (LoCPV-2)
- Malacosoma neustria cypovirus 2 (MnCPV-2)
- Mamestra brassicae cypovirus 2 (MbCPV-2)
- Operophtera brumata cypovirus 2 (ObCPV-2)
- Papilio machaon cypovirus 2 (PmCPV-2)
- Phalera bucephala cypovirus 2 (PbCPV-2)
- Pieris rapae cypovirus 2 (PrCPV-2)

3-Cypovirus type 3: (2.42, 2.32, 2.32, 2.08, 2.03, 1.29, 1.21, 0.61, 0.47, 0.34)
- Anaitis plagiata cypovirus 3 (ApCPV-3)
- Arctia caja cypovirus 3 (AcCPV-3)
- Danaus plexippus cypovirus 3 (DpCPV-3)
- Gonometa rufibrunnea cypovirus 3 (GrCPV-3)
- Malacosoma neustria cypovirus 3 (MnCPV-3)
- Operophtera brumata cypovirus 3 (ObCPV-3)
- Phlogophera meticulosa cypovirus 3 (PmCPV-3)
- Pieris rapae cypovirus 3 (PrCPV-3)
- Spodoptera exempta cypovirus 3 (SexmCPV-3)

4-Cypovirus type 4: (2.35, 2.35, 2.35, 2.20, 1.37, 1.22, 1.10, 0.97, 0.81, 0.81)
- Actias selene cypovirus 4 (AsCPV-4)
- Antheraea mylitta cypovirus 4 (AmCPV-4)
- Antheraea pernyi cypovirus 4 (ApCPV-4)

5-Cypovirus type 5: (2.35, 2.35, 2.35, 2.08, 1.82, 1.22, 1.16, 0.68, 0.50, 0.34)
- Euxoa scandens cypovirus 5 [dsRNA10 J04338] (EsCPV-5)
- Heliothis armigera cypovirus 5 (HaCPV-5)
- Orgyia pseudosugata cypovirus 5 (OpCPV-5)
- Spodoptera exempta cypovirus 5 (SexmCPV-5)
- Trichoplusia ni cypovirus 5 (TnCPV-5)

6-Cypovirus type 6: (2.35, 2.29, 2.23, 2.10, 1.54, 1.33, 1.26, 0.92, 0.79, 0.51)
- Aglais urticae cypovirus 6 (AuCPV-6)
- Agrochola helvolva cypovirus 6 (AhCPV-6)
- Agrochola lychnidis cypovirus 6 (AlCPV-6)
- Anaitis plagiata cypovirus 6 (ApCPV-6)
- Antitype xanthomista cypovirus 6 (AxCPV-6)
- Biston betularia cypovirus 6 (BbCPV-6)
- Eriogaster lanestris cypovirus 6 (ElCPV-6)
- Lasiocampa quercus cypovirus 6 (lqCPV-6)

7-Cypovirus type 7: (2.44, 2.34, 2.27, 2.15, 1.43, 1.28, 1.14, 0.61, 0.48, 0.30)
- Mamestra brassicae cypovirus 7 (MbCPV-7)
- Noctua pronuba cypovirus 7 (NpCPV-7)

8-Cypovirus type 8: (2.56, 2.56, 2.48, 2.21, 2.08, 1.07, 0.73, 0.67, 0.50, 0.37)
- Abraxas grossulariata cypovirus 8 (AgCPV-8)
- Heliothis armigera cypovirus 8 (HaCPV-8)
- Malacosoma disstria cypovirus 8 (MdCPV-8)
- Nudaurelia cytherea cypovirus 8 (NcCPV-8)
- Phlogophora meticulosa cypovirus 8 (PmCPV-8)
- Spodoptera exempta cypovirus 8 (SexmCPV-8)

9-Cypovirus type 9: (2.44, 2.36, 2.30, 2.04, 1.32, 0.97, 0.97, 0.44, 0.39, 0.39)
- Agrotis segetum cypovirus 9 (AsCPV-9)

10-Cypovirus type 10: (2.43, 2.43, 2.27, 2.27, 1.41, 1.29, 1.29, 0.95, 0.68, 0.56)
- Aporophyla lutulenta cypovirus 10 (AlCPV-10)

11-Cypovirus type 11: (2.59, 2.48, 2.48, 2.16, 1.12, 1.12, 0.76, 0.72, 0.55, 0.40)
- Heliothis armigera cypovirus 11 (HaCPV-11)
- Heliothis zea cypovirus 11 (HzCPV-11)
- Lymantria dispar cypovirus 11 (LdCPV-11)
- Mamestra brassicae cypovirus 11 (MbCPV-11)
- Pectinophora gossypiella cypovirus 11 (PgCPV-11)
- Pseudaletia unipuncta cypovirus 11 (PuCPV-11)
- Spodoptera exempta cypovirus 11 (SexmCPV-11)
- Spodoptera exigua cypovirus 11 (SexgCPV-11)

12-Cypovirus type 12 (2.50, 2.32, 2.32, 2.07, 1.86, 1.13, 0.81, 0.72, 0.64, 0.36)
- Autographa gamma cypovirus 12 (AgCPV-12)
- Mamestra brassicae cypovirus 12 (MbCPV-12)
- Pieris rapae cypovirus 12 (PrCPV-12)
- Spodoptera exempta cypovirus 12 (SexmCPV-12)

TENTATIVE SPECIES IN THE GENUS

None reported.

GENUS *FIJIVIRUS*

Type Species Fiji disease virus (FDV)

DISTINGUISHING FEATURES

Fijiviruses have a fragile structure and contain 10 dsRNA segments. They replicate in and are transmitted by delphacid planthoppers infecting phloem cells of susceptible plants.

VIRION PROPERTIES

MORPHOLOGY

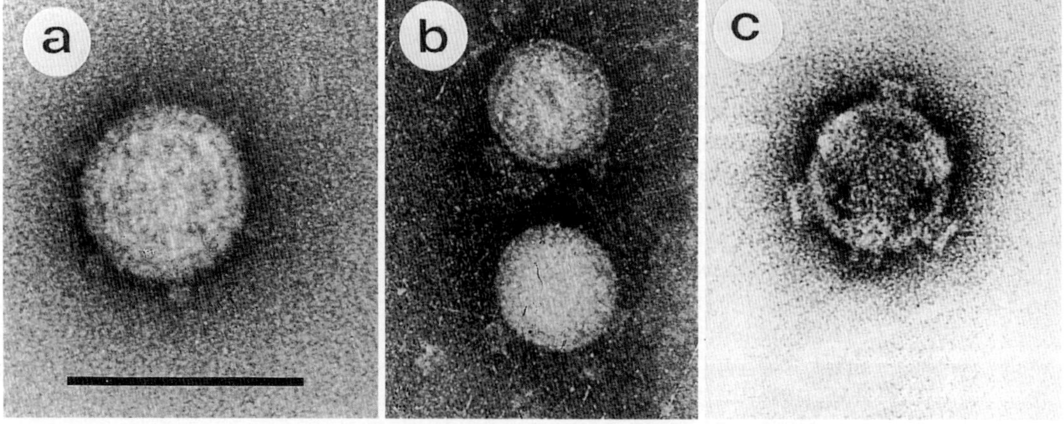

Figure 8: (a) Negative contrast electron micrograph of maize rough dwarf virus virions stained with uranyl acetate showing A-type spikes; (b) smooth subcores derived from MRDV on staining with neutral

phosphotungstate; (c) B-type spikes on virus-derived cores stained with uranyl acetate; (courtesy of Milne RG). The bar represents 100 nm.

Virions are double-shelled, spherical, 65-70 nm in diameter with "A"-type spikes of about 11 nm length and breadth at the 12 vertices on the icosahedral (Fig. 8 left). Unless prefixed, viruses readily break down *in vitro* to give cores, about 55 nm in diameter, with 12 "B"-type spikes, about 8 nm long and 12 nm in diameter (Fig. 8 right). Some treatments produce smooth subcores (Fig. 8 center) containing 2 proteins of 126 and 139 kDa.

PHYSICOCHEMICAL AND PHYSICAL PROPERTIES

The physicochemical properties of the virions have not been established.

NUCLEIC ACID

Fijiviruses have 10 dsRNA segments (S1-10) with Mr in the range 1.0-2.9 x 10^6 (total Mr 18-20 x 10^6). The coding strand of each segment of MRDV or RBSDV contains terminal nucleotides with the sequence: 5' AAGUUUUU....(U)GUC 3'. These are genus-specific terminal sequences, and, adjacent to them are segment-specific inverted repeats, similar to those of phytoreoviruses and oryzaviruses, although the sequences involved differ in these other genera. The sizes and groupings of the 10 dsRNA species are characteristic and distinctive for the three serogroups of Fijiviruses that are recognized.

PROTEINS

Fijiviruses have at least six structural proteins with Mr of 64-139 x 10^3.

GENOME ORGANIZATION AND REPLICATION

Most of the viral RNA segments are monocistronic. Segments S6 and S8 of MRDV (C. Marzachì, G Boccardo, unpublished) and S7 of RBSDV (I Uyeda, E. Shikata, unpublished) each possess 2 ORFs. These ORFs are in the same reading frame. On *in vitro* translation of their coding strands only the first ORFs of MRDV S6 and S8 are translated, forming NS proteins. When these segments and S7 of RBSDV are inserted in *Escherichia coli* cells, both ORFs are expressed but whether the second ORFs are expressed *in vivo* in insect or plant cells is not known. Although the coding assignments are not fully established, S10 RNA codes for a protein, probably structural, that is highly homologous between MRDV and RBSDV. Virus replication occurs in the cytoplasm of phloem-related cells in association with viroplasms composed partly of fine filaments. During infection, tubules, about 90 nm in diameter, accumulate. Sometimes these are incompletely closed and in the form of scrolls.

ANTIGENIC PROPERTIES

Three groups of Fijiviruses have been recognized based on the antigens associated with core particles.

BIOLOGICAL PROPERTIES

All Fijiviruses induce hypertrophy of the phloem (both expansion and multiplication of cells) leading to vein swellings and sometimes galls (enations, tumors) derived from phloem cells, especially on the backs of leaves. MRDV in maize induces longitudinal splitting of the roots. Other effects include the suppression of flowering, plant stunting, increased production of side shoots, and induction of a dark green coloring.

In insect hosts, no particular tissue tropism or severe disease is recognized. Viruses are transmitted propagatively by delphacid planthoppers (*Hemiptera, Delphacidae*, e.g., *Laodelphax, Javesella, Delphacodes, Sogatella, Perkinsiella* and *Unkanodes*). Virus is acquired from plants after some hours of feeding. The latent period is about 2 weeks and leads to a lifelong capacity for virus transmission to plants. No transovarial transmission or seed transmission of virus has been identified. Mechanical transmission from plant to plant can only be demonstrated with difficulty. Virus is spread by offsets in vegetatively propagated crops (e.g., pangolagrass and sugarcane). Viruses over-winter in diapausing planthoppers, in certain weed species and in autumn-sown cereals.

Generally, Fijiviruses are widespread in nature although apparently absent from North America and not reported from Africa, or confirmed from India. FDV has been reported from Australasia and the Pacific islands. RBSDV occurs in Japan and China, PaSV in northern countries of South America, and ODSV from northern Europe. MRDV is found in Scandinavia and in areas bordering the northern and eastern Mediterranean. There is a distinct variant found in Argentina (Conci L, Marzachì C, unpublished).

TAXONOMIC STRUCTURE OF THE GENUS

There are 3 antigenic groups of Fijiviruses.

LIST OF SPECIES IN THE GENUS

It is not clear whether MRDV and RBSDV should be classified as separate species since they are serologically closely related. Also, their host ranges, symptoms, vectors and, in part, their geographical distributions overlap.

The groups, viruses, their genomic sequences accession numbers [ ], CMI/AAB description # ( ) and assigned abbreviations ( ) are:

SPECIES IN THE GENUS

1-Fijivirus group 1:		
Fiji disease virus (119)		(FDV)
2-Fijivirus group 2:		
maize rough dwarf virus (72)	[S6:X55701]	(MRDV)
Pangola stunt virus (175)		(PaSV)
rice black streaked dwarf virus (135)		(RBSDV)
3-Fijivirus group 3:		
oat sterile dwarf virus (217)		(OSDV)

TENTATIVE SPECIES IN THE GENUS

None reported.

GENUS *PHYTOREOVIRUS*

Type Species wound tumor virus (WTV)

DISTINGUISHING FEATURES

Phytoreoviruses have distinctive angular particles and possess 12 dsRNA species. They are transmitted by cicadellid leafhoppers to susceptible plant species, replicating in both hosts.

VIRION PROPERTIES

MORPHOLOGY

Virions are 65-70 nm in diameter, more angular than spherical in uranyl acetate (Fig. 9), surviving intact in neutral phophotungstate negative stain. WTV possesses three protein shells, an outer amorphous layer, a layer of distinct capsomers, and a smooth core that is about 50 nm in diameter but lacks spikes.

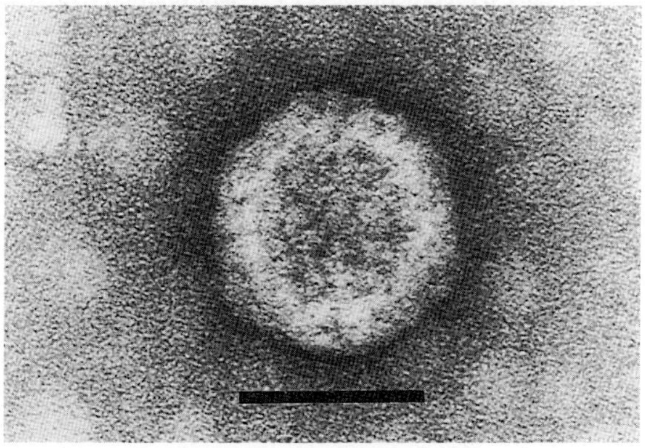

Figure 9: Negative contrast electron micrograph of rice gall dwarf virus virions stained with uranyl acetate. The bar represents 50 nm.

Physicochemical and Physical Properties

The Mr of phytoreoviruses is about 75 x 10^6. The virion S_{20w} is about 510. The optimal stability of particles is at pH 6.6. Viruses are resistant to Freon, CCl_4, and CsCl density gradient centrifugation.

Nucleic Acid

Phytoreoviruses have 12 segments of dsRNA (S1-12) with characteristic sizes for each virus. The dsRNA Mr is in the range 0.3 to 3.0 x 10^6 and G+C content is 38-44%. The RNA constitutes about 22% of the virion dry weight. The coding species of each genome segment of all viruses in the genus contains the conserved sequence: 5' GG(U/C)AUU...(U/C)GAU 3'. Adjacent to this genus-specific sequence, segments possess inverted repeats, 6-14 bases long. These sequences differ for each RNA segment. The mRNA 5' non-coding region is 14-63 nucleotides long, the 3' non-coding region is 56-495 nucleotides in length (I Uyeda, E. Shikata, unpublished). RDV particles encapsidate the genomic RNA in supercoiled form.

Proteins

Phytoreoviruses have seven structural proteins with Mr in the range 45 to 160 x 10^3. For WTV these are organized in three shells consisting of an amorphous outer shell of 2 species, an inner shell of 2 species and a core of three species. Protein constitutes about 78% of the particle dry weight. Removal of the outer shell is not required for activation of the virus transcriptase and associated enzymes.

Genome Organization and Replication

The coding strand of each dsRNA encodes a single ORF except for the S12 segment of WTV which has a second, small ORF downstream. No evidence has yet been obtained for the expression of this second ORF. Five structural and five NS WTV proteins have been assigned to their respective genome segments. For RDV, S1 encodes the putative transcriptase. The genus-specific and segment-specific sequence motifs appear necessary for successful replication, translation and encapsidation. Laboratory strains having internal deletions in some segments, but intact termini, replicate and compete favorably with wild-type virus, although the proteins expressed are aberrant, and the ability of the viruses to be transmitted by vectors ma be lost. Virus replication occurs in the cytoplasm of infected cells in association with viroplasms. WTV and RGDV are confined to phloem tissues of the plant host, whereas RDV can also multiply elsewhere. In the insect vector, there are no particular tissue tropisms. RDV induces abnormalities in fat body cells and mycetocytes.

Table 6: List of dsRNA segments of WTV (*Phytoreovirus*) with their respective size (bp) and their corresponding proteins for which the name, size (kDa) (calculated), and function and/or location are indicated.

dsRNA #	Size (bp)	Protein	Size (kDa)	Function (location)
S1		P1	(estim) 155	(core)
S2		P2	(estim) 130	(outer coat)
S3		P3	(estim) 108	(core)
S4	2565	Pns4	81	Unknown
S5	2613	P5	91	(outer coat)
S6	1700	Pns7	59	Unknown
S7	1726	P6	58	(core)
S8	1472	P8	48	(capsid)
S9	1182	pns10	39	Unknown
S10	1172	Pns11	39	Unknown
S11	1128	P9	36	(capsid)
S12	851	Pns12	19	Unknown

ANTIGENIC PROPERTIES

The three recognized phytoreoviruses are antigenically unrelated.

BIOLOGICAL PROPERTIES

Plant hosts are either dicotyledons, or the family *Gramineae*. WTV was originally identified in northeastern USA in the leafhopper *Agalliopsis novella*. The virus was recently found in New Jersey USA in a single periwinkle (*Catharanthus*) plant set out as bait for mycoplasmas in a blueberry (*Vaccinium*) field. The experimental plant host range of WTV is wide and encompasses many dicotyledons. The name of this virus derives from the fact that infected plants develop phloem-derived galls (tumors) at wound sites, notably at the emergence of side roots. RDV and RGDV have narrow and overlapping host ranges among *Gramineae*. RDV and RGDV cause severe disease in rice crops in south-east Asia, China, Japan and Korea. RDV is also found in Nepal. RDV induces white flecks and streaks on leaves, with stunting and excessive production of side shoots. RDV is the only plant reovirus that is not limited to the phloem. Also, RDV does not provoke enlargement and division of infected cells. RGDV induces stunting, shoot proliferation, dark green color and enations. In insect vectors phytoreoviruses induce no marked disease. They are transmitted propagatively by cicadellid leafhoppers (*Hemiptera, Cicadellidae,* e.g., *Agallia, Agalliopsis* and *Nephotettix*). Virus is acquired from plants shortly after feeding. The latent period in leafhoppers is about 2 weeks. Thereafter, infected insects have a lifelong ability to transmit virus to plants. Phytoreoviruses are also transmitted transovarially in their insect vectors. Experimentally, not mechanically transmissible from plant to plant. No seed transmission occurs.

TAXONOMIC STRUCTURE OF THE GENUS

Epitopes representing the inner surface of the outer capsid of RDV and RGDV are shared, however the outer surface epitopes of the 3 viruses are distinct.

LIST OF SPECIES IN THE GENUS

The viruses, their genomic sequence accession numbers [ ], CMI/AAB description # ( ) and assigned abbreviations ( ) are:

SPECIES IN THE GENUS

rice dwarf virus (102)	[S1:D90198, D10222, S3:X17203, D00607, S4:X51432, S5:D90033, X16017, S6 M31298, S11:D10249, D90199, S12:D90200]	(RDV)
rice gall dwarf virus (296)		(RGDV)

wound tumor virus (34)	[S4:M24117, S5:J03020, S6:M24116, S7:X14218, S8:J04344, S9:M24115, S10:M24114, S11:X14219]	(WTV)

TENTATIVE SPECIES IN THE GENUS

None reported.

GENUS *ORYZAVIRUS*

Type Species rice ragged stunt virus (RRSV)

DISTINGUISHING FEATURES

Oryzaviruses appear to lack an outer capsid and possess a genome consisting of 10 dsRNA species. They are transmitted by viruliferous planthoppers to plants in the family *Gramineae*, replicating in both hosts.

VIRION PROPERTIES

MORPHOLOGY

The particle diameter is in the range of 57-65 nm (Fig. 10). Particles possess 12 "B"-type spikes, 8-10 nm in height, 23-26 nm wide at the base and 14-17 nm at the top, that overlie the core. The cores are about 50 nm in diameter. The particle morphology is distinct from that of phytoreoviruses or Fijiviruses.

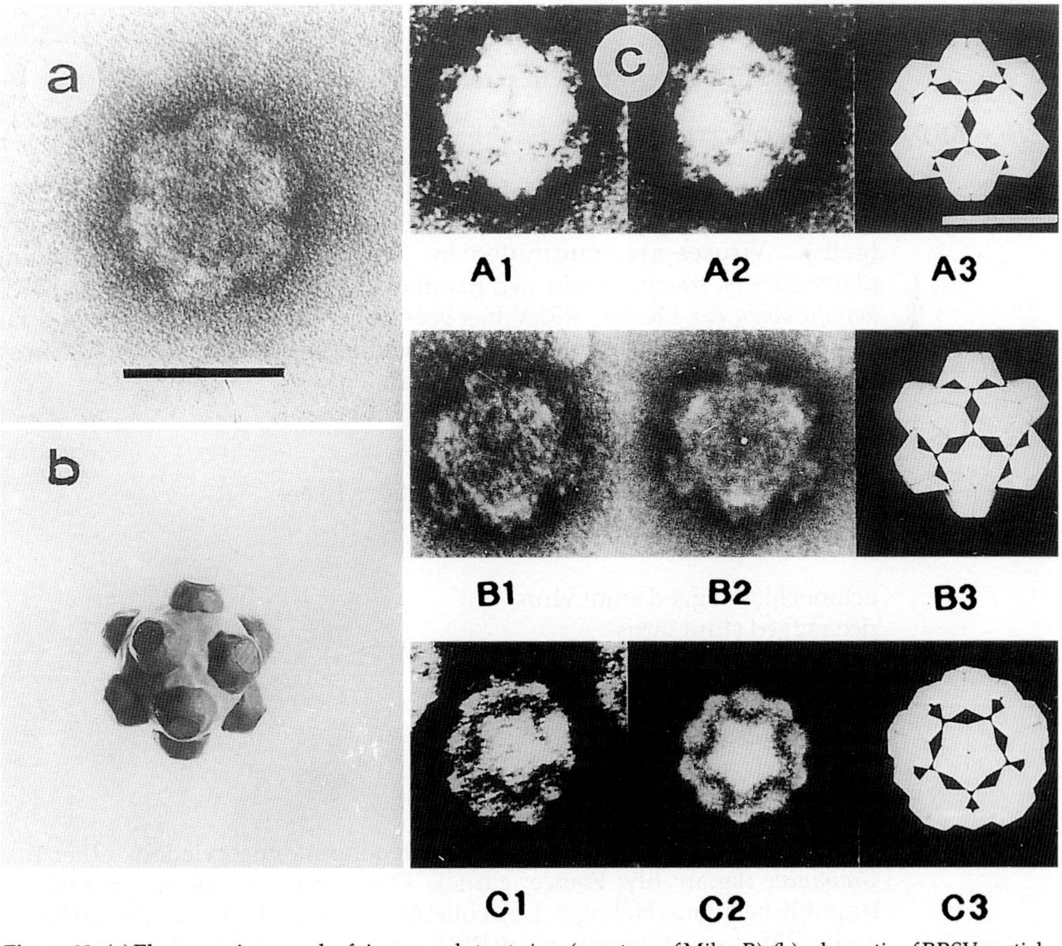

Figure 10: (a) Electron micrograph of rice ragged stunt virus (courtesy of Milne R); (b) schematic of RRSV particle; (c) micrographs of the virus arranged at 2-, 3- and 5-fold symmetries (A1, B1 and C1, respectively), images of the same rotated by increments of 180° (A2), or 120° (B2), or 72° (C2) and proposed models of the 2-, 3- and 5-fold symmetries (A3, B3, and C3 respectively); (courtesy of Shikata E). The bar represents 50 nm.

Nucleic Acid

The virus genome consists of 10 dsRNA segments (S1-10) with Mr values ranging from 0.8 to 2.5 x 10^6 and a total Mr of about 18 x 10^6 (about 27 kbp). For RRSV the S1 and S2 dsRNAs have similar sizes (about 3,900 bp), as do the S3 and S4 species (about 3,800 bp), the S5 is about 2,750 bp, the S6 about 2,300 bp, the S7 about 1,950 bp, the S8 about 1,900 bp, the S9 about 1,200 bp and the S10 about 1,160 bp. The end sequences of the mRNA strands (5' GAUAAA...GUGC 3') differ from those of phytoreoviruses or Fijiviruses.

Proteins

Up to eight structural proteins with sizes of about 125, 97, 66, 64, 48, 43, 36, and 32 kDa have been identified in RRSV particles.

Genome Organization and Replication

Limited information is available concerning the genome organization and replication strategy of oryzaviruses. The S1 RNA appears to encode two proteins, one (68 kDa) in the first half of the S1 segment, the second (about 70 kDa) in the second half, but partially overlapping the first. If correct, how both are translated (separate mRNAs?) is not known, frameshift sites typical of other viruses (retroviruses, coronaviruses) have not been identified. From sequencing data S5 encodes a 91 kDa protein. The S7 and S8 RNAs each encodes a about 67 kDa protein (Uyeda I, Shikata E, Waterhouse P, unpublished). How these and others relate to the structural and NS proteins of RRSV remains to be elucidated. The S10 RNA encodes the 36 kDa spike protein. The viruses induce viroplasms in the cytoplasm of infected cells.

Antigenic Properties

RRSV and ERSV cross-react in serological tests.

Biological Properties

Viruses infect plants in the family *Gramineae*, causing disease of rice (RRSV) and *Echinocloa* (ERSV). As in other plant reovirus infections induces phloem cell are induced to proliferate (galls). Viruses are transmitted by, and replicate in phloem-feeding, viruliferous planthoppers, specifically brown planthoppers (*Nilaparvata lugens* for RRSV and *Sogatella longifurcifera* for ERSV). RRSV has been reported in southeastern and far-eastern Asian countries where it affects rice yields (generally 10-20%, but up to 100% in severely affected areas). ERSV has been reported in Taiwan.

List of Species in the Genus

The viruses, and their assigned abbreviations ( ) are:

Species in the Genus

Echinochloa ragged stunt virus	(ERSV)
rice ragged stunt virus	(RRSV)

Tentative Species in the Genus

None reported.

Unassigned Species in the Family

Plant reoviruses have been observed infecting monocotyledons other than in the family *Gramineae* (Japan: lily; France: garlic). One report describes a reo-like virus in the lily. Unpublished data (H. Lot, B. Delecolle, G. Boccardo, R. Milne) identified a reo-like virus in garlic with distinctive genomic dsRNA sizes, antigenically distinct from other plant-infecting reoviruses but morphologically similar to Fijiviruses. No vectors have been identified. Reo-like viruses infecting *Liliaceae*

Derivation of Names

aqua: from Latin *aqua*, "water"
colti: sigla from *Co*lorado *ti*ck fever
cypo: sigla from *cy*toplasmic *po*lyhedrosis
Fiji: from name of country where virus was first isolated
orbi: from Latin *orbis*, "ring" or "circle" in recognition of the ring-like structures observed in micrographs of the surface of BTV cores
oryza: from Latin *oryza*, "rice"
phyto: from Greek *phyton*, "plant"
reo: sigla from *r*espiratory *e*nteric *o*rphan, due to the early recognition that the viruses caused respiratory and enteric infections, and (incorrect) belief that they were not associated with disease, hence they were considered "orphan" viruses
rota: from Latin *rota*, "wheel"

References

Boccardo G, Milne RG (1984) Plant reovirus group. CMI/AAB Descriptions of Plant Viruses. No 294, 4pp

Chen CC (1989) Comparison of proteins and nucleic acids of Echinochloa ragged stunt and rice ragged stunt viruses. Intervirology 30: 278-284

Estes MK (1991) Rotaviruses and their replication. In: Fields BN, Knipe DM (eds) Fundamental Virology, 2nd edn Raven Press, New York, pp 619-642

Francki RIB, Milne RG, Hatta T (1985) Atlas of Plant Viruses Vol 1. CRC Press, Boca Raton FL

Hetrick F, Samal KSK, Lupiana B, Dopazo C, Subramanian K, Mohanty SB (1992) Members of the family Reoviridae found in aquatic animals. In: Kimura T (ed) Salmonid Diseases (Proceedings of the Oji International Symposium on Salmonid Diseases). Hokkaido University Press, Sapporo Japan, pp 33-40

Hull R, Brown F, Payne CC (eds) (1989) Virology, directory and dictionary of animal, bacterial and plant viruses. Macmillan Press Ltd, London

Karabatsos N, Poland JD, Emmons RW, Mathews JH, Calisher CH, Wolf KL (1987) Antigenic variants of Colorado tick fever virus. J Gen Virol 68: 1463-1469

Lee SY, Uyeda I, Shikata E (1987) Characterization of RNA polymerase associated with rice ragged stunt virus. Intervirology 27: 189-195

Marzachi C, Boccardo G, Nuss DL (1991) Cloning of the maize rough dwarf virus genome: molecular confirmation of the plant reovirus classification scheme and identification of two large non-overlapping coding domains within a single genome segment. Virology 180: 518-526.

Mertens PPC, Crook NE, Rubinstein R, Pedley S, Payne CC (1989) Cytoplasmic polyhedrosis virus classification by electrophoretype; validation by serological analysis and agarose gel electrophoresis. J Gen Virol 70: 173-185

Nuss DL, Dall DJ (1990) Structural and functional properties of plant reovirus genomes. Adv Virus Res 38: 249-306

Rosenberger JK, Sterner FJ, Botts S, Lee KP, Margolin A (1989) In vitro and in vivo characterization of avian reoviruses. I Pathogenicity and antigenic relatedness of several avian reovirus isolates. Avian Dis 33: 535-544

Roy P, Gorman BM (eds) (1990) Bluetongue viruses. Curr Top Micro Immunol 162: 1-200

Saif LJ (1990) Nongroup A rotaviruses. In: Saif LJ, Theil KW (eds) Viral diarrhoeas of man and animals. CRC Press, Boca Raton FL, pp 73-95

Schnitzer TJ (1985) Protein coding assignments of the S genes of the avian reovirus S1133. Virology 141: 167-170

Winton JR, Lannan CN, Fryer JL, Hedrick RP, Meyers TR, Plumb JA, Yamamoto T (1987) Morphological and biochemical properties of four members of a novel group of reoviruses isolated from aquatic animals. J Gen Virol 68: 353-364

Yan J, Kudo H, Uyeda I, Lee SY, Shikata E (1992) Conserved terminal sequences of rice ragged stunt virus genomic RNA. J Gen Virol 73: 785-789

Contributed By

Holmes IH, Boccardo G, Estes MK, Furuichi MK, Hoshino Y, Joklik WK, McCrae M, Mertens PPC, Milne RG, Samal KSK, Shikata E, Winton JR, Uyeda I, Nuss DL

Family *Birnaviridae*

Taxonomic Structure of the Family

Family	***Birnaviridae***
Genus	*Aquabirnavirus*
Genus	*Avibirnavirus*
Genus	*Entomobirnavirus*

Virion Properties

Morphology

Virions are about 60 nm in diameter, single-shelled, non-enveloped icosahedrons. About 132 morphological subunits make up the viral capsid.

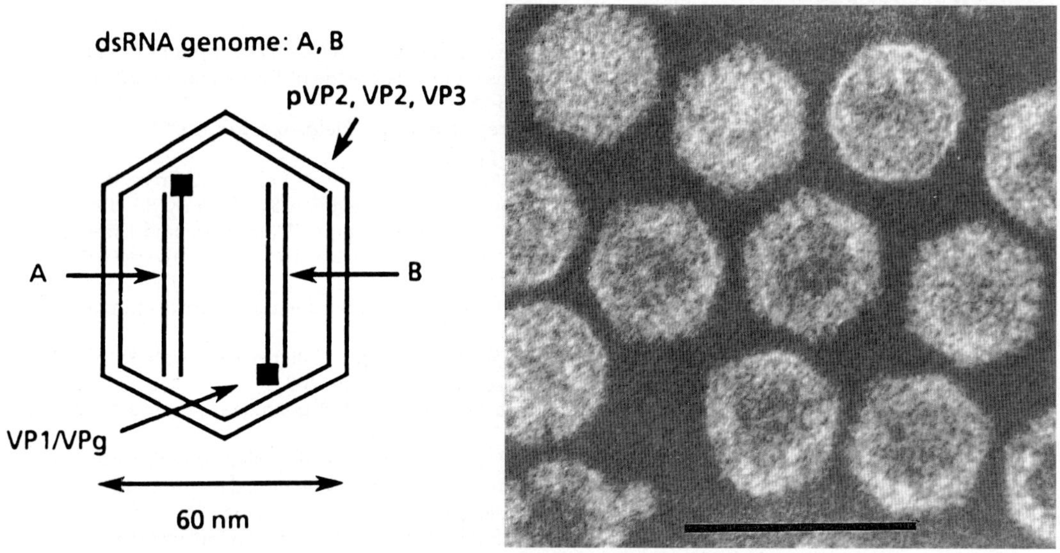

Figure 1: (left) Diagram of infectious pancreatic necrosis virus (IPNV); (right) negative contrast electron micrograph of virions. The bar represents 100 nm.

Physicochemical and Physical Properties

Virion Mr is about 55 x 10^6, S_{20w} is 435; buoyant density in CsCl is 1.33 g/cm^3. Viruses are stable at pH 3-9, resistant to heat (60° C, 1 hr), ether and 1% SDS at 20° C, pH 7.5 for 30 min.

Nucleic Acid

Virions contain two segments (A, B) of dsRNA which constitute about 9-10% of the particle by weight. The sizes of segments for infectious pancreatic necrosis virus (IPNV, strain Jasper) are: 3,092 bp (A) and 2,784 bp (B). For infectious bursal disease virus (IBDV) they are 3,129 and 2,795 bp, respectively. Both genome segments contain a 94 kDa 5' genome-linked protein (VPg). There are no poly (A) tracts at the 3' ends of the RNA segments.

Proteins

Virions contain five polypeptides: VP1 (94 kDa) which is the RNA-dependent RNA polymerase as well as the genome-linked protein; pre-VP2 (62 kDa) and VP2 (54 kDa), the major capsid polypeptides and type specific antigens; VP3 (30 kDa), an internal capsid protein and group specific antigen; and NS or VP4 (29 kDa), the virus coded protease. An additional 17 kDa, positively charged, minor polypeptide may also be present in virions. Guanylyl transferase and methyl transferase activities have been shown to be associated with the VP1 of IBDV.

LIPIDS

None present.

CARBOHYDRATES

The VP2 of IPNV may be glycosylated.

GENOME ORGANIZATION AND REPLICATION

Genome segment A contains two ORFs, encoding a 17 kDa protein (ORF 1) and a large 106 kDa polyprotein (ORF 2) in an overlapping reading frame. Genome segment B contains one large 94 kDa product (Fig. 2, ORF 3).

A single cycle of replication takes about 18-22 hr. After entry into the host cell, the virion RNA-dependent RNA polymerase becomes activated and produces two genome length (24S) mRNA molecules from each of the 14S dsRNA genome segments. It has not been determined whether these mRNAs are capped or have a VPg attached to their 5' ends; they lack 3' poly (A) tracts. Replicative intermediates have been identified in infected cells. Virus RNA is transcribed by a semi-conservative strand displacement mechanism *in vitro*; however, reinitiation of RNA synthesis *in vitro* has not been observed. There is no information on minus strand RNA synthesis. The two mRNAs can be detected in infected cells by 3-4 hr post-infection and are synthesized in the same relative proportions throughout the replicative cycle (i.e., about twice as many A as B mRNA species). Virus-specific polypeptides can be detected at 4-5 hr post-infection and are present in the same relative proportions to each other until the end of the replication cycle. There are no specifically early or late proteins. The segment A mRNA is translated to a 106 kDa polyprotein which contains (5' to 3') the pre-VP2, NS (VP4) and VP3 polypeptides (Fig. 2). The NS (VP4) protease co-translationally cleaves the polyprotein to generate the three polypeptides. Pre-VP2 is later processed by a slow maturation cleavage to produce VP2. This cleavage is incomplete since both pre-VP2 and VP2 are found in purified virus, although VP2 predominates. The polyprotein has been detected in *in vitro* translation systems, and the active site of the protease has been mapped to the carboxy end of NS (VP4). The exact cleavage sites on the polyprotein are not known. The product of the 17 kDa ORF has not been detected in infected cells.

The mRNA from segment B is translated to a 94 kDa polypeptide which is the viral RNA-dependent RNA polymerase (VP1, Fig. 2). It is found in virions both in a free, and genome-linked form. Virus particles assemble and accumulate in the cytoplasm. Subviral particles have not been found. The mechanism of virus release is unknown. In tissue culture about half of the progeny virions remains cell-associated.

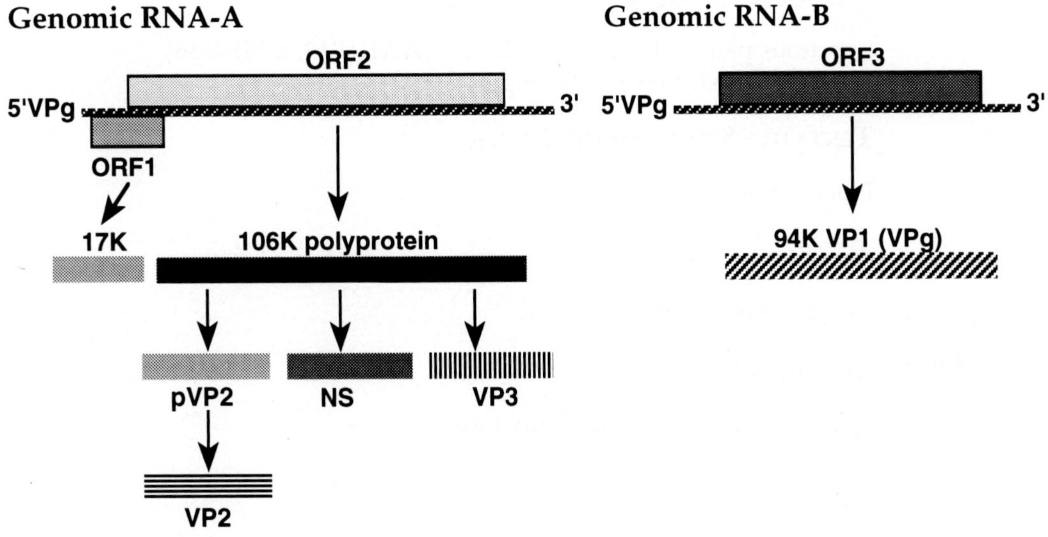

Figure 2: Schematic of infectious pancreatic necrosis virus genome organization.

Antigenic Properties

The major capsid protein VP2 is the type-specific antigen and contains the virus neutralizing epitopes. Anti-VP3 antibodies do not neutralize virus infectivity. There is no serological cross-reaction betweeen the fish, avian and insect birnaviruses.

Biological Properties

The natural hosts of IPNV are salmonids, although the virus has also been isolated from other fresh-water and marine fishes, as well as from bivalve molluscs. The virus is transmitted both vertically and horizontally. There are no known vectors. The geographic distribution is world-wide. IPNV can cause epizootics resulting in high mortality in hatchery-reared salmonid fries and fingerlings. The virus causes necrotic lesions in the pancreas and is also found, without lesions, in other organs such as kidney, gonad, intestine, brain etc. Infected adult fish become life-long carriers without exhibiting overt signs of infection.

The natural hosts of IBDV are chickens, ducks, turkeys and other domestic fowl. The mode of transmission is horizontal. There are no known vectors. IBDV has a world-wide distribution. The virus destroys the bursa of Fabricius of young chicks (less than 3 weeks old) causing B lymphocyte deficiency. Mortality occurs between 3 to 6 weeks of age and is associated with inflammation of the bursa Fabricius, formation of immune complexes, depletion of complement and clotting abnormalities.

Drosophila melanogaster is the natural host of Drosphila X virus (DXV). The mode of transmission is horizontal and there are no known vectors. The geographic distribution is unknown. Infected fruitflies become sensitive to CO_2. The target organs and histopathology are not known. DXV has also been isolated from populations of *Culicoides spp*.

Genus *Aquabirnavirus*

Type Species infectious pancreatic necrosis virus (IPNV)

Distinguishing Features

Species of the genus only infect fish, molluscs and crustaceans.

List of Species in the Genus

The viruses their genomic sequence accession numbers [ ] and assigned abbreviations ( ) are:

Species in the Genus

infectious pancreatic necrosis virus [A:M18049, B:M58756] (IPNV)
(reference strain VR 299, Jasper)

Tentative Species in the Genus

None reported.

Genus *Avibirnavirus*

Type Species infectious bursal disease virus (IBDV)

Distinguishing Features

Species of the genus infect only birds.

LIST OF SPECIES IN THE GENUS

The viruses, their genomic sequence accession numbers [ ] and assigned abbreviations ( ) are:

SPECIES IN THE GENUS

infectious bursal disease virus		(IBDV)
(reference strain 002-73)	[AM64738, BM19336]	
(reference strain STC)	[D00499]	

TENTATIVE SPECIES IN THE GENUS

None reported.

GENUS *ENTOMOBIRNAVIRUS*

Type Species Drosophila X virus (DXV)

DISTINGUISHING FEATURES

Species of genus infect only insects.

LIST OF SPECIES IN THE GENUS

The viruses, and their assigned abbreviations ( ) are:

SPECIES IN THE GENUS

Drosophila X virus (DXV)

TENTATIVE SPECIES IN THE GENUS

None reported.

LIST OF UNASSIGNED VIRUSES IN THE FAMILY

rotifer birnavirus (*Brachiorus plicatilis*) (RBV)

DERIVATION OF NAMES

aqua: from Latin *aqua*, "water"
avi: from Latin *avis*, "bird"
bi: from Latin prefix *bi*, "two", signifies the bisegmented nature of the viral genome as well as the presence of dsRNA
entomo: from Greek *entomon*, "insect"
rna: sigla from *r*ibo *n*ucleic *a*cid, indicating the nature of the viral genome

REFERENCES

Bayliss CD, Spies U, Shaw K, Peters RW, Papageorgiou A, Muller H, Boursnell MEG (1990) A comparison of the sequences of segment A of four infectious bursal disease virus strains and identification of a variable region in VP2. J Gen Virol 71: 1303-1312
Calvert JG, Nagy E, Soler M, Dobos P (1991) Characterization of the VPg-dsRNA linkage of infectious pancreatic necrosis virus. J Gen Virol 72: 2563-2567
Comps M, Mari J, Poisson F, Bonami JR (1991) Biophysical and biochemical properties of the RBV, and unusual birnavirus pathogenic for rotifers. J Gen Virol 72: 1229-1236
Dobos P, Hill BJ, Hallett R, Kells DTC, Becht H, Teninges D (1979) Biophysical and biochemical characterization of five animal viruses with bisegmented double-stranded RNA genomes. J Virol 32: 593-605
Dobos P, Roberts TE (1983) The molecular biology of infectious pacreatic necrosis virus; a review. Can J Microbiol 29: 377-384 (1983)
Duncan R, Dobos P (1986) The nucleotide sequence of infectious pancreatic necrosis virus dsRNA segment A reveals one large ORF encoding a precursor polyprotein. Nucl Acids Res 14: 5934

Duncan R, Mason CL, Nagy E, Leong JA, Dobos P (1991) Sequence analysis of infectious pancreatic necrosis virus genome segment B and its encoded VP1 protein: a putative RNA dependent RNA polymerase lacking the gly-asp-asp motif. Virology 181: 541-552

Duncan R, Nagy E, Krell PJ, Dobos P (1987) Synthesis of the infectious pancreatic necrosis virus polyprotein, detection of a virus encoded protease, and fine structure mapping of genome segment A coding regions. J Virol 61: 3655-3664

Morgan MM, Macreadie IG, Harley VR, Hudson PJ, Azad A (1988) Sequence of the small double stranded RNA genomic segment of infectious bursal disease virus and its deduced 90-kDa product. Virology 163: 240-242

Nagy E, Dobos P (1974) Coding assignments of Drosophila X virus genome segments: *in vitro* translation of native and denatured virion dsRNA. Virology 137: 58-66

Contributed By

Dobos P, Berthiaume L, Leong JA, Kibenge FSB, Muller H, Nicholson BL

FAMILY TOTIVIRIDAE

TAXONOMIC STRUCTURE OF THE FAMILY

Family	***Totiviridae***
Genus	*Totivirus*
Genus	*Giardiavirus*
Genus	*Leishmaniavirus*

VIRION PROPERTIES

MORPHOLOGY

Virions exhibit isometric symmetry, and are 30-40 nm in diameter, with no envelope or surface projections.

PHYSICOCHEMICAL AND PHYSICAL PROPERTIES

Virion buoyant density in CsCl is 1.33-1.43 g/cm^3. Additional components with different sedimentation coefficients are found in preparations of some viruses in the genus *Totivirus*. These consist of particles containing satellite or defective dsRNA.

NUCLEIC ACID

Virions contain a single molecule of linear uncapped dsRNA, 4.6-7.0 kbp in size.

PROTEINS

Virions contain a single major capsid polypeptide, with an Mr of 70-100 x 10^3. Virion-associated RNA polymerase activity is present.

LIPIDS

None reported.

CARBOHYDRATES

None reported.

GENOME ORGANIZATION AND REPLICATION

The virion-associated RNA-dependent RNA polymerase catalyzes *in vitro* end-to-end transcription of dsRNA to produce mRNA for capsid protein, by a conservative mechanism. The polymerase is expressed as a gag-pol-like fusion protein involving two ORFs.

BIOLOGICAL PROPERTIES

The viruses are associated with latent infections of their fungal or protozoal hosts.

GENUS TOTIVIRUS

Type Species Saccharomyces cerevisiae virus L-A (ScV-L-A)

VIRION PROPERTIES

MORPHOLOGY

Virions are isometric, 40-43 nm in diameter, with no envelope. Symmetry of particles has not been determined.

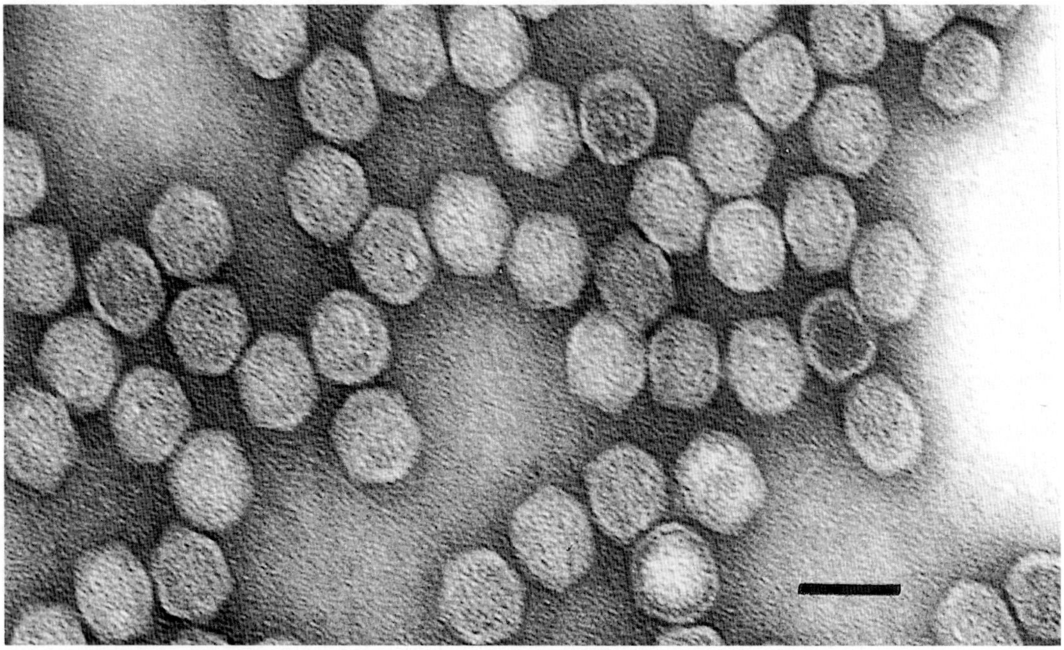

Figure 1: Negative contrast electron micrograph of Helminthosporium victoriae virus 190S (HvV-190S) virions, a representative species in the genus *Totivirus*. The bar represents 50 nm.

PHYSICOCHEMICAL AND PHYSICAL PROPERTIES

Virion Mr is estimated as 12.3 x 10^6. Buoyant density in CsCl is 1.40-1.43 g/cm^3 and S_{20w} is 160-190. Additional components with different sedimentation coefficients and buoyant densities are present in virus isolates with satellite or defective RNAs. Particles lacking nucleic acid have an S_{20w} of 98-113.

NUCLEIC ACID

Virions contain a single linear molecule of uncapped dsRNA (4.6-6.7 kbp in size). Some virus isolates contain additional satellite dsRNAs which encode "killer" proteins; these satellites are encapsidated separately in capsids encoded by the helper virus genome. Some virus isolates may contain (additionally or alternatively) defective dsRNAs which arise from the satellite dsRNAs; these additional dsRNAs are also encapsidated separately in capsids encoded by the helper virus genome. The complete nt sequence (4,579 bp) of ScV-L-A (L1) is available. The positive strand (4,580 nt; contains unpaired A at the 3' terminus) has two large ORFs that overlap by 130 nt. The first ORF encodes the viral major capsid polypeptide with a predicted size of 76 x 10^3. The two reading frames together encode, via translational frameshift, the putative RNA-dependent RNA polymerase as a fusion protein (analogous to gag-pol fusion proteins of the retroviruses) with a predicted size of 170 x 10^3. Sites essential for encapsidation and replication have been defined.

PROTEINS

Virions contain a single major capsid polypeptide species with an Mr of 73-88 x 10^3. Protein kinase activity is associated with HvV190S virions; capsids contain phosphorylated forms of the coat protein. RNA polymerase (replicase-transcriptase) is present. In ScV-L-A virions, RNA polymerase occurs as 1-2 molecules of the 170 kDa fusion protein. The *pol* domain of the *gag-pol* fusion protein has a single stranded RNA binding activity.

LIPIDS

Virions contain no lipids.

CARBOHYDRATES

None reported.

GENOME ORGANIZATION AND REPLICATION

ScV-L-A virus has a single 4.6 kbp dsRNA segment with two ORFs. The 5' ORF is *gag* and encodes the major capsid protein, while the 3' ORF, *pol*, encodes the RNA-dependent RNA polymerase, and has ssRNA binding activity. *Pol* is expressed only as a *gag-pol* fusion protein formed by a (-)1 frameshift in the 130 bp overlap region between the two ORFs. The (-)1 ribosomal frameshift is produced by a 72 b region that has a 7 base slippery site and an essential pseudoknot structure. The efficiency of frameshifting is critical for viral replication.

The virion-associated RNA polymerase catalyzes *in vitro* end-to-end transcription of dsRNA by a conservative mechanism to produce mRNA for capsid polypeptides. In the case of ScV-L-A, all of the positive strand transcripts are extruded from the particles. The positive strand of satellite RNA M_1, or deletion mutants of L-A or M_1, on the other hand, often remain within the particle where they are replicated to give two or more dsRNA molecules per particle (headful replication). The positive ssRNA of ScV-L-A is the species encapsidated to form progeny virus particles. The encapsidation signal on ScV-L-A or M_1 positive sense ssRNA is a 24 b stem-loop sequence located 400 b from the 3' end in each case. The *gag* protein must be acetylated for assembly and packaging to proceed. These particles have a replicase activity that synthesizes the negative strand on the positive strand template to produce dsRNA, thus completing the replication cycle. Replication requires an internal site and specific 3' end sequence and secondary/tertiary structure. Virions accumulate in the cytoplasm.

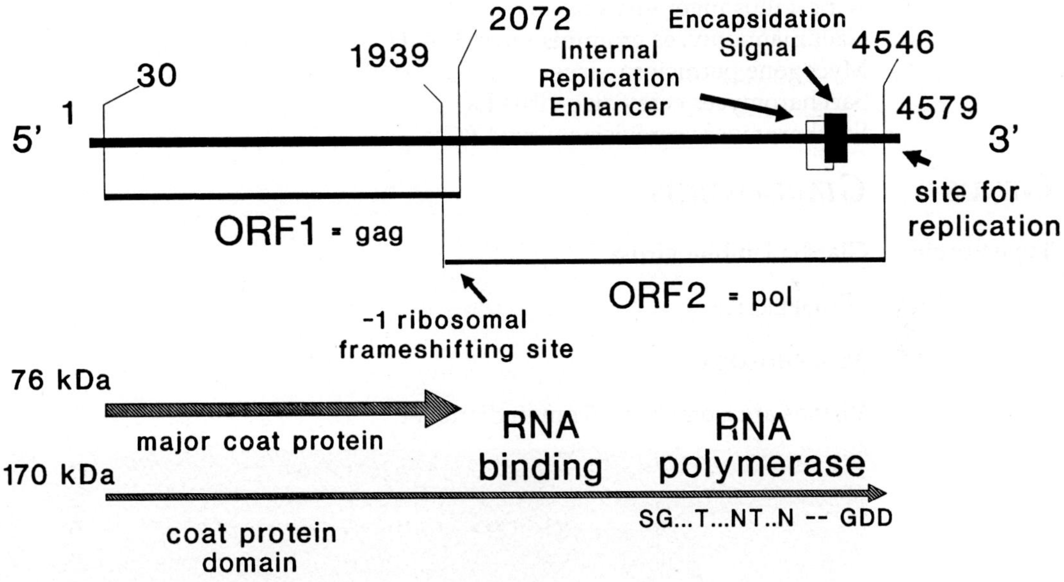

Figure 2: Genome organization of ScV-L-A.

ANTIGENIC PROPERTIES

Virions serve as efficient immunogens.

BIOLOGICAL PROPERTIES

TRANSMISSION

Virions remain intracellular and are transmitted during cell division, sporogenesis and cell fusion. In some ascomycetes, e.g. *Gaeumannomyces graminis*, virus is usually eliminated during ascospore formation.

Host Range

Saccharomyces cerevisiae L-A virus depends for its multiplication on the host genes, *MAK3, MAK10, MAK31* and *MAK32*. The *MAK3* gene encodes an N-acetyltransferase that acetylates the N-terminus of the major coat protein. Over 30 chromosomal genes are necessary for the replication of M_1 dsRNA. *S. cerevisiae* has an antiviral system, the *SKI* genes, whose only essential role is to repress the replication of ScV-L-A, M and, ScV-L-BC dsRNAs. If the *SKI* genes are defective, ScV-L-A becomes pathogenic; but only the M dsRNA causes a cytopathogenic effect. Cells become cold sensitive for growth.

List of Species in the Genus

The viruses, their alternative names ( ), genomic sequence accession numbers [ ] and assigned abbreviations ( ) are:

Species in the Genus

Helminthosporium victoriae virus 190S		(HvV-190S)
Saccharomyces cerevisiae virus L-A	[J04692, X13426]	(ScV-L-A)
Ustilago maydis virus 1		(UmV-P1)
Ustilago maydis virus 4		(UmV-P4)
Ustilago maydis virus 6		(UmV-P6)

Tentative Species in the Genus

Aspergillus foetidus virus S	(AfV-S)
Aspergillus niger virus S	(AnV-S)
Gaeumannomyces graminis virus 87-1-H	(GgV-87-1-H)
Mycogone perniciosa virus	(MpV)
Saccharomyces cerevisiae virus La	(ScV-La)
Saccharomyces cerevisiae virus LBC	(ScV-LBC)

Genus *Giardiavirus*

Type Species Giardia lamblia virus (GLV)

Virion Properties

Morphology

Virions are isometric, 36 nm in diameter.

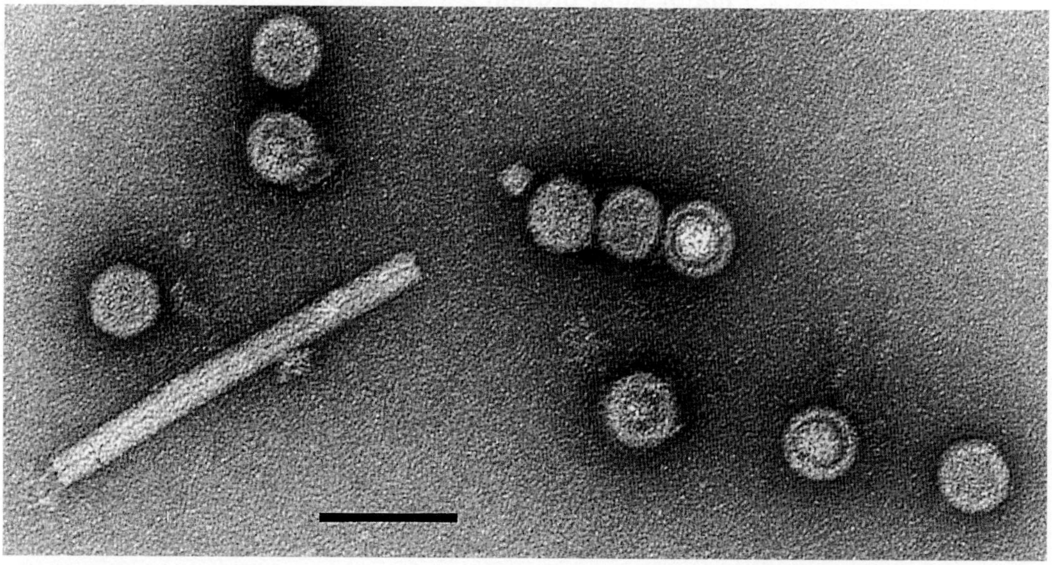

Figure 3: Negative contrast electron micrograph of Giardia lamblia virions. TMV is included as an internal size marker. The bar represents 100 nm.

Physicochemical and Physical Properties

Virion buoyant density in CsCl is 1.368 g/cm^3.

Nucleic Acid

Virions contain a single molecule of dsRNA, 7.0 kbp in size.

Proteins

Virions contain a single major capsid species, Mr of 100 x 10^3.

Lipids

None reported.

Carbohydrates

None reported.

Genome Organization and Replication

The virus is found in the nuclei of infected *G. lamblia*. Virus replicates without inhibiting the growth of *G. lamblia* trophozoites. Virus is also extruded into the culture medium and the extruded virus can infect many virus-free isolates of the protozoan host. There are isolates of the protozoan parasite, however, that are resistant to infection by GLV. A single-stranded copy of the viral dsRNA genome is present in infected cells. The concentration of the ssRNA observed during the time course of GLV infection is consistent with a role as a viral replicative intermediate or mRNA. The ssRNA does not appear to be polyadenylated.

Biological Properties

The virus infects many isolates of *G. lamblia*, a flagellated protozoan human parasite. The virus does not seem to be associated with the virulence of the parasite. It is not observed in the cyst form of the parasite and it is not known whether it can be carried through the transformation between cyst and trophozoite. The virus is infectious as purified particles and can infect uninfected *G. lamblia*.

List of Species in the Genus

The viruses, their genomic sequence accession numbers [ ] and assigned abbreviations ( ) are:

Species in the Genus

Giardia lamblia virus	[L13218]	(GLV)

Tentative Species in the Genus

Trichomonas vaginalis virus	(TVV)

Genus *Leishmaniavirus*

Type Species Leishmania RNA virus 1 - 1 (LRV1-1)

Virion Properties

Morphology

Virions are isometric, 33 nm in diameter, with no envelope or surface projections.

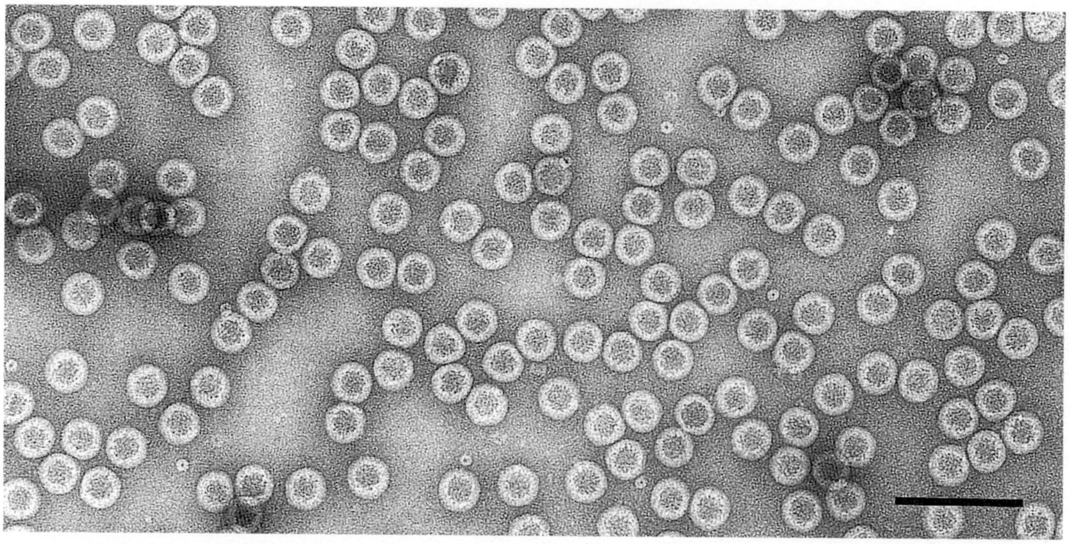

Figure 4: Negative contrast electron micrograph of Leishmania RNA virus 1 - 1 (LRV1-1) virions. The bar represents 100 nm.

Physicochemical and Physical Properties

Virion buoyant density in CsCl is 1.33 g/cm^3.

Nucleic Acid

Virions contain a single molecule of linear uncapped dsRNA, 5.3 kbp in size. The complete 5,284 nt sequence is available.

Proteins

Virions contain a single major capsid polypeptide of Mr 82 x 10^3.

Lipids

None reported.

Carbohydrates

None reported.

Genome Organization and Replication

The positive strand contains three ORFs. The predicted amino acid sequence of ORF 3 has motifs characteristic of viral RNA-dependent RNA polymerase. ORF 2 encodes the major capsid protein and overlaps ORF 3 by 71 nt, suggesting a +1 translational frameshift to produce a *gag-pol*-like fusion protein of predicted size of 176 x 10^3. Sequencing data support the idea that the abundant ssRNA found in infected cells is the message sense RNA.

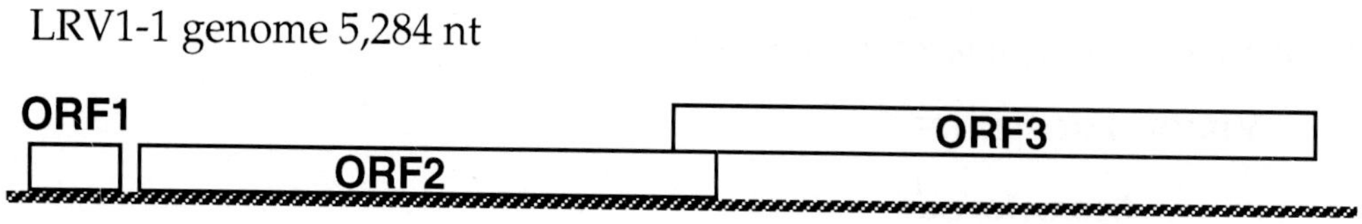

Figure 5: Genome organization of LRV1-1

BIOLOGICAL PROPERTIES

LRV1-1 is found in infected *Leishmania brasiliensis* strain CUMC1. Viruses infecting several other strains of *L. brasiliensis* and *L. guyanensis* are possibly strains of LRV1-1. A single strain of *L. major* is known to be infected with LRV1-1-like virus. The latter is designated LRV2-1 in order to distinguish it from the viruses infecting new world strains of *Leishmania*.

LIST OF SPECIES IN THE GENUS

The viruses, their host { }, genomic sequence accession numbers [ ] and assigned abbreviations ( ) are:

SPECIES IN THE GENUS

Leishmania RNA virus 1 - 1 {CUMC1}	[M92355]	(LRV1-1)
Leishmania RNA virus 1 - 2 {CUMC3} (formerly LR2)		(LRV1-2)
Leishmania RNA virus 1 - 3 {M2904}		(LRV1-3)
Leishmania RNA virus 1 - 4 {M4147} (formerly LBV)	[U01899]	(LRV1-4)
Leishmania RNA virus 1 - 5 {M1142}		(LRV1-5)
Leishmania RNA virus 1 - 6 {M1176}		(LRV1-6)
Leishmania RNA virus 1 - 7 {BOS12}		(LRV1-7)
Leishmania RNA virus 1 - 8 {BOS16}		(LRV1-8)
Leishmania RNA virus 1 - 9 {M6200}		(LRV1-9)
Leishmania RNA virus 1 - 10 {LC76}		(LRV1-10)
Leishmania RNA virus 1 - 11 {LH77}		(LRV1-11)
Leshmania RNA virus 1 - 12 {LC56}		(LRV1-12)
Leishmania RNA virus 2 - 1		(LRV2-1)

TENTATIVE SPECIES IN THE GENUS

None reported.

LIST OF UNASSIGNED VIRUSES IN THE FAMILY

None reported.

SIMILARITY WITH OTHER TAXA

None reported.

DERIVATION OF NAMES

totus: from *totus*, Latin for 'whole' or 'undivided'

REFERENCES

Bruenn JA (1980) Virus-like particles of yeast. Ann Rev Microbiol 34:49-68

Buck KW (1986) Fungal virology-an overview. In: Buck KW (ed) Fungal virology. CRC Press, Boca Raton FL, pp 1-84

Dinman JD, Icho T, Wickner RB (1991) A -1 ribosomal frameshift in a double stranded RNA virus of yeast forms a gag-pol fusion protein. Proc Natl Acad Sci USA 88: 174-178

Furfine ES, Wang CC (1990) Transfection of the Giardia lamblia double-stranded RNA virus into *Giardia lamblia* by electroporation of a single-stranded RNA copy of the viral genome. Mol Cell Biol 10: 3659-3663

Ghabrial SA (1994) New developments in Fungal Virology. Adv Virus Res

Ghabrial SA, Havens WM (1989) Conservative transcription of Helminthosporium victoriae 190S virus double-stranded RNA *in vitro*. J Gen Virol 70: 1025-1035

Ghabrial SA, Havens WM (1992) The Helminthosporium victoriae 190S mycovirus has two forms distinguishable by capsid composition and phosphorylation state. Virology 88: 657-665

Koltin Y (1988) The killer systems of Ustilago maydis. In: Koltin Y, Leibowitz M (eds) Viruses of fungi and simple enkaryotes. Marcel Dekker, New York pp 209-243

Patterson JL (1990) Viruses of protozoan parasites. Exper parasitol 70: 111-113

Shelbourn SL, Day PR, Buck KW (1988) Relationships and functions of virus double-stranded RNA in a P4 killer strain of Ustilago maydis. J Gen Virol 69: 975-982

Stuart KD, Weeks R, Guilbride L, Myler PJ (1992) Molecular organization of Leishmania RNA virus. Proc Natl Acad Sci USA 89: 8596-8600

Tarr PI, Aline RF, Smiley BL, Sholler J, Keithly J, Stuart KD (1988) LR1: a candidate RNA virus of Leishmania. Proc Natl Acad Sci USA 85: 9572-9575

Tzeng T-H, Tu C-L, Bruenn JA (1992) Ribosomal frameshifting requires a pseudoknot in the Saccharomyces cerevisiae double-stranded RNA virus. J Virol 66: 999-1006

Wang AL, Wang CC (1991) Viruses of the protozoa. Ann Rev Microbial 45: 251-263

White TC, Wang CC (1990) RNA dependent RNA polymerase activity associated with the double-stranded RNA virus of *Giardia lamblia*. Nucl Acids Res 18: 553-559

Widmer G, Patterson JL (1991) Genome structure and RNA polymerase activity in Leishmania virus. J Virol 65: 4211-4215

Wickner RB (1989) Yeast virology. FASEB J 3: 2257-2265

Wickner RB (1992) Double-stranded and single-stranded RNA viruses of *Saccharomyces Cerevisiae*. Ann Rev Microbiol 46: 347-375

CONTRIBUTED BY

Ghabrial SA, Bruenn JA, Buck KW, Wickner RB, Patterson JL, Stuart KD, Wang AL, Wang CC

FAMILY *PARTITIVIRIDAE*

TAXONOMIC STRUCTURE OF THE FAMILY

Family	***Partitiviridae***
Genus	*Partitivirus*
Genus	*Chrysovirus*
Genus	*Alphacryptovirus*
Genus	*Betacryptovirus*

VIRION PROPERTIES

MORPHOLOGY

Virions are isometric, nonenveloped, 30-40 nm in diameter. Symmetry of particles has not been determined.

PHYSICOCHEMICAL AND PHYSICAL PROPERTIES

Virion buoyant density in CsCl is in the range of 1.34-1.39 g/cm^3. Virions are stable in butanol and chloroform.

NUCLEIC ACID

Virions contain two unrelated linear dsRNA segments (1.4 - 3.0 kbp in size). The two segments of the individual viruses are usually of similar size. No nucleic acid sequencing data are available for any member of the family.

PROTEINS

Single major capsid polypeptide. Virion-associated RNA polymerase activity is present.

LIPIDS

None reported.

CARBOHYDRATES

None reported.

GENOME ORGANIZATION AND REPLICATION

The genome is comprised of two linear dsRNA segments, the smaller codes for the capsid polypeptide and the larger codes for an unrelated protein, probably the virion-associated RNA polymerase. Each dsRNA is probably monocistronic. *In vitro* transcription/replication occurs by a semi-conservative mechanism. Virions accumulate in the cytoplasm.

ANTIGENIC PROPERTIES

Virions are efficient immunogens. A single precipitin line is formed in gel diffusion tests. Members that are serologically related may be strains of a single virus. No serological relationships between the fungal viruses and the plant viruses in the family *Partitiviridae* have been detected.

BIOLOGICAL PROPERTIES

The viruses are associated with latent infections of their fungal and plant hosts. There are no known natural vectors. The fungal viruses are transmitted intracellularly during cell division, sporogenesis and cell division. In some ascomycetes, e.g. *Gaeumannomyces graminis*, virus is usually eliminated during ascospore formation. Experimental transmission of purified fungal partitiviruses has been reported by fusing virions with fungal protoplasts. The plant cryptoviruses are transmitted by ovule and by pollen to the seed embryo. There is no graft transmission and apparently no cell-to-cell transport, except at cell division; seed transmission is the only known mode for the transmission of cryptoviruses.

GENUS *PARTITIVIRUS*

Type Species Gaeumannomyces graminis virus 019/6-A (GgV-019/6-A)

VIRION PROPERTIES

MORPHOLOGY

Virions are 30-35 nm in diameter. Negatively stained particles are often penetrated by stain giving the appearance of empty particles even though physical data indicate that they contain dsRNA.

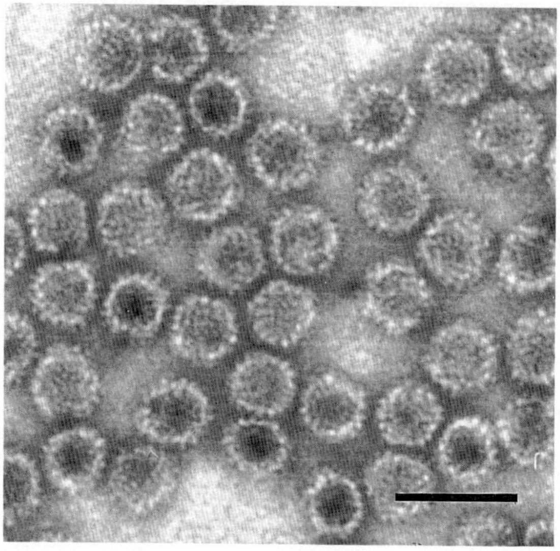

Figure 1: Negative contrast electron micrograph of virions of Penicillium stoloniferum virus S (PsV-S), a representative species of the genus *Partitivirus*. The bar represents 50 nm.

PHYSICOCHEMICAL AND PHYSICAL PROPERTIES

Mr of virions is estimated to range from 6 to 9 x 10^6. S_{20w} values range from 101-145. Particles lacking nucleic acid have an S_{20w} of 66-100. Virion buoyant density in CsCl is 1.29-1.30 and 1.34-1.36 g/cm^3 for particles without and with nucleic acid respectively. Additional density and sedimenting components are found in preparations of some viruses and are believed to comprise replicative intermediates. These consist of particles containing ssRNA and particles with both ssRNA and dsRNA. Virus purification is usually carried out at neutral pH.

NUCLEIC ACID

Virions contain two unrelated linear dsRNA segments, 1.4-2.2 kbp in size, which are separately encapsidated. The dsRNA segments of the individual viruses are of similar size. Additional segments of dsRNA (satellite or defective) may be present.

PROTEINS

Virions contain a single major capsid polypeptide, Mr 42-73 x 10^3. Virion-associated RNA polymerase activity is present.

GENOME ORGANIZATION AND REPLICATION

The virion-associated RNA polymerase catalyzes *in vitro* end-to-end transcription of each dsRNA to produce mRNA, by a semi-conservative mechanism. Virions accumulate in the cytoplasm.

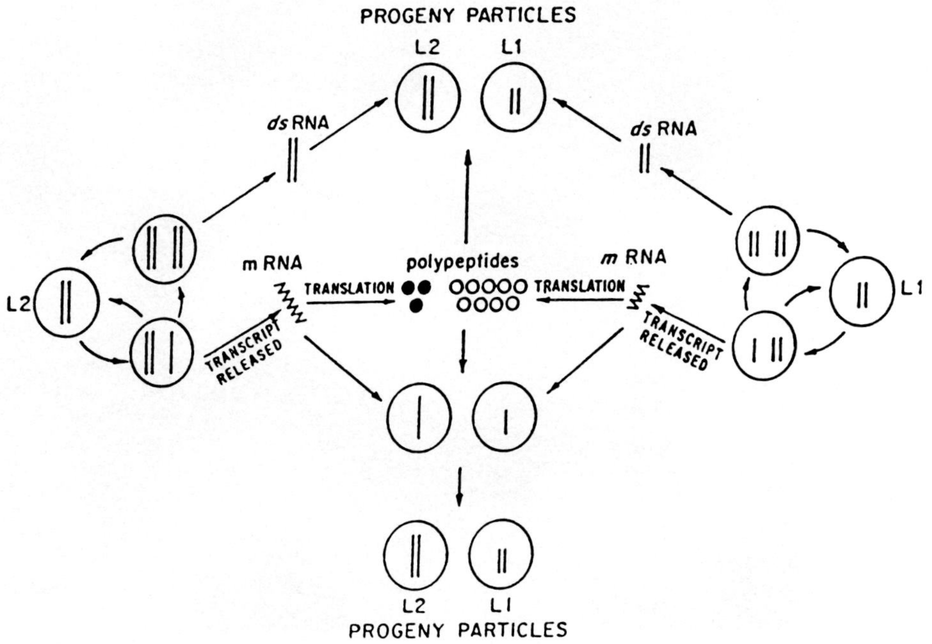

Figure 2: Model for replication of Penicillium stoloniferum S virus (PsV-S). The open circles represent capsid protein subunits and the closed circles represent RNA polymerase subunits. Solid lines represent RNA strands whereas wavy lines represent mRNA.

List of Species in the Genus

The viruses, their alternative names ( ) and assigned abbreviations ( ) are:

Species in the Genus

Agaricus bisporus virus 4 (mushroom virus 4)	(AbV-4)
Aspergillus ochraceous virus	(AoV)
Gaeumannomyces graminis virus 019/6-A	(GgV-019/6-A)
Gaeumannomyces graminis virus T1-A	(GgV-T1-A)
Penicillium stoloniferum virus S	(PsV-S)
Rhizoctonia solani virus	(RsV)

Tentative Species in the Genus

Diplocarpon rosae virus	(DrV)
Penicillium stoloniferum virus F	(PsV-F)
Phialophora radicicola virus 2-2-A	(PrV-2-2-A)

Genus *Chrysovirus*

Type Species Penicillium chrysogenum virus (PcV)

Virion Properties

Morphology

Virions are isometric and 35-40 nm in diameter.

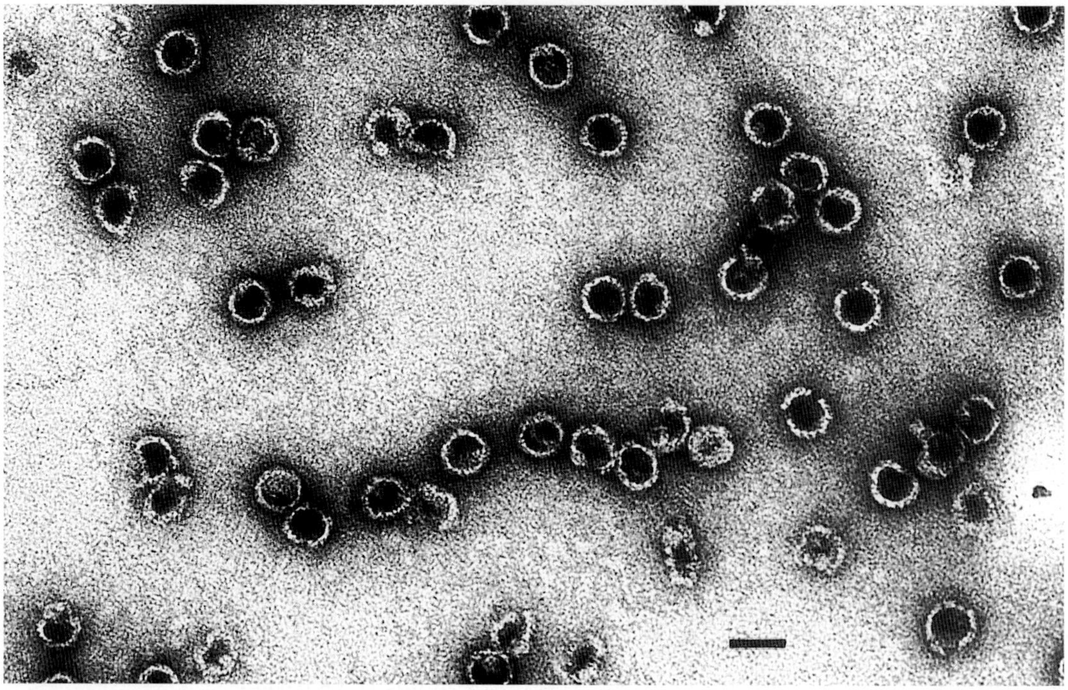

Figure 3: Negative contrast electron micrograph of Penicillium chrysogenum virus (PcV) virions, the type species of the genus *Chrysovirus*. The bar represents 50 nm.

Physicochemical and Physical Properties

Virion density in CsCl is 1.35 g/cm^3 and S_{20w} is 145-150.

Nucleic Acid

The virions typically contain three unrelated and separately encapsidated dsRNA segments of about 3 kbp each. Some virus isolates contain four dsRNA segments. The number of dsRNA species required for replication is not known. Because the genomes of members in the family *Partitiviridae* are bipartite in nature, the additional dsRNA segments that may be present in preparations of viruses in the genus *Chrysovirus* are tentatively considered satellite or defective dsRNAs.

Proteins

The capsids are made up of single polypeptide species (Mr 125 x 10^3). RNA polymerase activity is present.

List of Species in the Genus

The viruses, and their assigned abbreviations ( ) are:

Species in the Genus

Penicillium brevicompactum virus	(PbV)
Penicillium chrysogenum virus	(PcV)
Penicillium cyaneo-fulvum virus	(Pc-fV)

Tentative Species in the Genus

Helminthosporium victoriae virus 145S	(HvV-145S)

Genus *Alphacryptovirus*

Type Species white clover cryptic virus 1 (WCCV-1)

Virion Properties

Morphology

Virions are isometric, 30 nm in diameter. Particles lack fine structural detail, appearing rounded, usually penetrated by the stain to give a ring-like appearance.

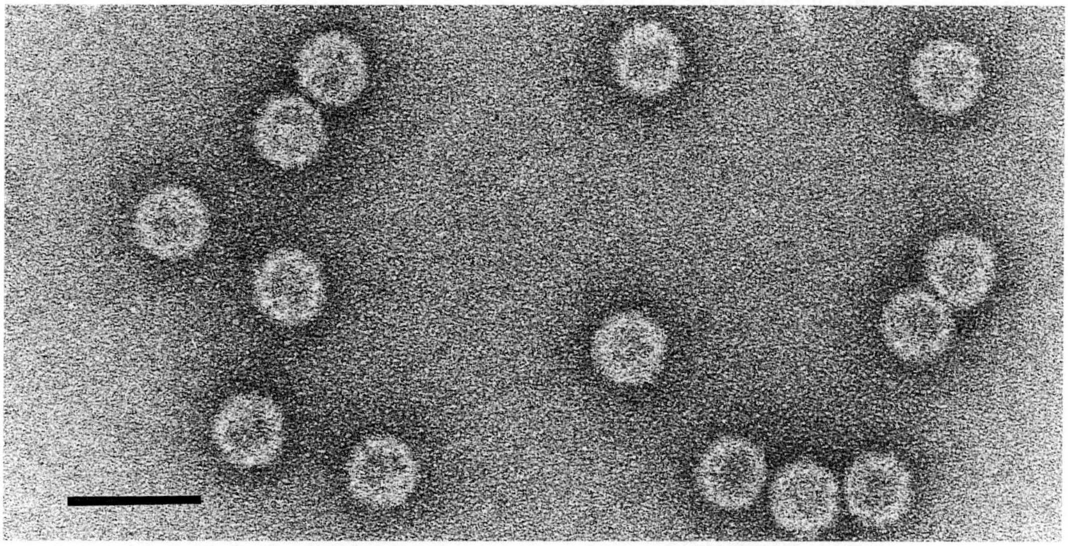

Figure 4: Negative contrast electron micrograph of white clover cryptic virus 1 (WCCV-1) virions, the type species of the genus *Alphacryptovirus*. The bar represents 50 nm.

Physicochemical and Physical Properties

Density in CsCl is 1.392 g/cm^3 .

Nucleic Acid

The virions typically contain two dsRNA segments, 1.7 and 2.0 kbp in size. The larger dsRNA segment codes for the virion-associated RNA polymerase and the smaller codes for the capsid polypeptide. It is not known whether the dsRNA segments are packaged together or separately.

Proteins

The capsids are made up of single polypeptide species (Mr 55 x 10^3). RNA polymerase activity is present.

Lipids

None reported.

Carbohydrates

None reported.

Antigenic Properties

Some viruses in the genus are serologically related; none are related to viruses in the genus *Betacryptovirus*. There are no serological relationships with mycoviruses in the genera *Partitivirus* and *Chrysovirus*.

LIST OF SPECIES IN THE GENUS

The viruses and their assigned abbreviations ( ) are:

SPECIES IN THE GENUS

alfalfa cryptic virus 1	(ACV-1)
beet cryptic virus 1	(BCV-1)
beet cryptic virus 2	(BCV-2)
beet cryptic virus 3	(BCV-3)
carnation cryptic virus 1	(CCV-1)
carrot temperate virus 1	(CTeV-1)
carrot temperate virus 3	(CTeV-3)
carrot temperate virus 4	(CTeV-4)
hop trefoil cryptic virus 1	(HTCV-1)
hop trefoil cryptic virus 3	(HTCV-3)
radish yellow edge virus	(RYEV)
ryegrass cryptic virus	(RGCV)
spinach temperate virus	(SpTV)
Vicia cryptic virus	(VCV)
white clover cryptic virus 1	(WCCV-1)
white clover cryptic virus 3	(WCCV-3)

TENTATIVE SPECIES IN THE GENUS

carnation cryptic virus 2	(CCV-2)
cucumber cryptic virus	(CuCV)
fescue cryptic virus	(FCV)
garland chrysanthemum temperate virus	(GCTV)
Mibuna temperate virus	(MTV)
poinsettia cryptic virus	(PnCV)
red pepper cryptic virus 1	(RPCV-1)
red pepper cryptic virus 2	(RPCV-2)
rhubarb temperate virus	(RTV)
Santosai temperate virus	(STV)

GENUS *BETACRYPTOVIRUS*

Type Species white clover cryptic virus 2 (WCCV-2)

VIRION PROPERTIES

MORPHOLOGY

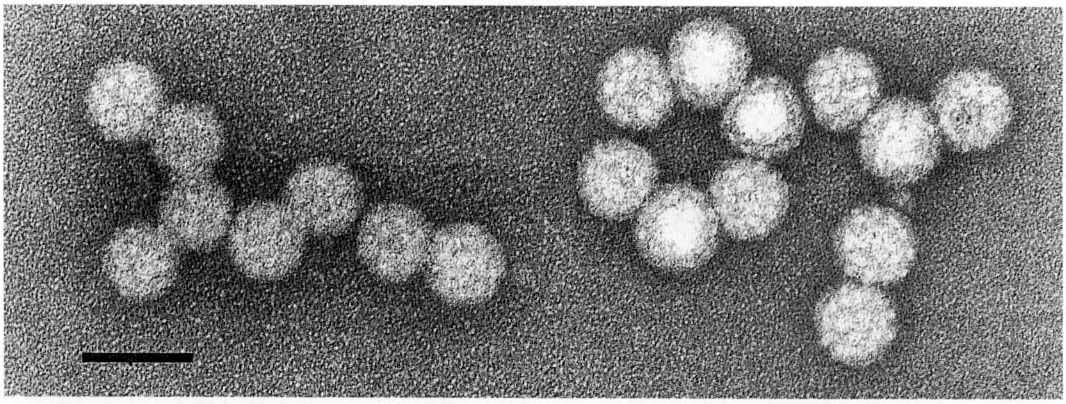

Figure 5: Negative contrast electron micrograph of white clover cryptic virus 2 virions (WCCV-2), the type species of the genus *Betacryptovirus*. The bar represents 50 nm.

Virions are isometric, 38 nm in diameter. Particles show prominent subunits, but their precise geometrical arrangement is not clear. The particles are rounded and are not penetrated by stain.

Physicochemical and Physical Properties

Virion buoyant density in CsCl is 1.375 g/cm^3 .

Nucleic Acid

Viral nucleic acid comprises two dsRNA segments, which are about 2.1 and 2.25 kbp in size.

Proteins

Not characterized.

Lipids

None reported.

Carbohydrates

None reported.

Antigenic Properties

Some viruses in the genus are serologically related ; none are related to viruses in the genus *Alphacryptovirus*.

List of Species in the Genus

The viruses and their assigned abbreviations ( ) are:

Species in the Genus

carrot temperate virus 2	(CTeV-2)
hop trefoil cryptic virus 2	(HTCV-2)
red clover cryptic virus 2	(RCCV-2)
white clover cryptic virus 2	(WCCV-2)

Tentative Species in the Genus

alfalfa cryptic virus 2	(ACV-2)

List of Unassigned Viruses in the Family

None reported.

Similarity with Other Taxa

None reported.

Derivation of Names

partitus: from Latin *partitius*, 'divided'
crypto: from Greek *crypto*, ' hidden, covered, or secret'

References

Accotto GP, Marzachi C, Luisoni E, Milne RG (1990) Molecular characterization of alfalfa cryptic virus 1. J Gen Virol 71: 433-437

Antoniw JF, White RF, Xie WS (1990) Cryptic viruses of beet and other plants In: Frasser RSS (ed) Recognition and Response in Plant-Virus Interactions. Springer-Verlag, Heidelberg, pp 273-285

Barton RJ, Hollings M (1979) Purification and some properties of two viruses infecting the cultivated mushroom Agaricus bisporus. J Gen Virol 42: 231-241

Boccardo G, Milne RG, Luisoni E, Lisa V, Accotto GP (1985) Three seedborne cryptic viruses containing double-stranded RNA isolated from white clover. Virology 147: 29-40

Bozarth RF (1979) The physicochemical properties of mycoviruses In: Lemke, PA (ed) Viruses and Plasmids in Fungi. Marcel Dekker, New York pp 43-91

Bozarth RF, Wood HA, Goenaga A (1972) Virus-like particles from a culture of *Diplocarpon rosae*. Phytopathology 62: 493

Bozarth RF, Wood HA, Mandelbrot A (1971) The Penicillium stoloniferum virus complex: two similar double stranded RNA virus-like particles in a single cell. Virology 45: 516-523

Buck KW (1979) Replication of double-stranded RNA mycoviruses In: Lemke, PA (ed) Viruses and Plasmids in Fungi. Marcel Dekker, New York pp 94-151

Buck KW, Almond MR, McFadden JJP, Romanos MA, Rawlinson CJ (1981) Properties of thirteen viruses and virus variants obtained from eight isolates of the wheat take-all fungus, *Gaeumannomyces graminis var. tritici*. J Gen Virol 53: 235-245

Buck KW, Grivan RF (1977) Comparison of the biophysical and biochemical properties of Penicillium cyaneo-fulvum virus and Penicillium chrysogenum virus. J Gen Virol 34: 145-154

Buck KW, Kempson-Jones GF (1973) Biophysical properties of Penicillium stoloniferum virus S. J Gen Virol 18: 223-235

Buck KW, McGinty RM, Rawlinson CJ (1981) Two serologically unrelated viruses isolated from a *Phialophora* sp. J Gen Virol 55: 235-239

Edmondson SP, Lang D, Gray DM (1984) Evidence for sequence heterogeneity among double-stranded RNA segments of Penicillium chrysogenum mycovirus. J Gen Virol 65: 1591-1599

Finkler A, Ben-Zavi B, Koltin Y, Barash I (1988) Transcription and in vitro translation of the dsRNA virus isolated from *Rhizoctonia solani*. Virus Genes 1: 206-219

Luisoni E, Milne RG, Accotto GP, Boccardo G (1987) Cryptic viruses in hop trefoil (*Medicago lupulina*) and their relationships to other cryptic viruses in legumes. Intervirology 28: 144-156

Kim JW, Bozarth RF (1985) Intergeneric occurrence of related fungal viruses: the *Aspergillus ochraceous* virus complex and its relationship to the Penicillium stoloniferum virus S. J Gen Virol 66: 1991-2002

McFadden JJP, Buck KW, Rawlinson CJ (1983) Infrequent transmission of double-stranded RNA virus particles but absence of DNA provirus in single ascospore cultures of *Gaeumannomyces graminis*. J Gen Virol 64: 927-937

Natsuaki T, Muroi Y, Okuda S, Teranaka M (1990) Cryptoviruses and RNA-RNA hybridization among their double-stranded RNA segments. Ann Phytopathol Soc Japan 56: 354-358

Natsuaki T, Natsuaki KT, Okuda S, Teranaka M, Milne RG, Boccardo G, Luisoni E (1986) Relationships between the cryptic and temperate viruses of alfalfa, beet and white clover. Intervirology 25: 69-75

Sanderlin RS, Ghabrial SA (1979) Physicochemical properties of two distinct types of virus-like particles from *Helminthosporium victoriae*. Virology 87: 142-151

Tavantzis SM, Bandy BP (1988) Properties of a mycovirus from *Rhizoctonia solani* and its virion-associated RNA polymerase. J Gen Virol 69: 1465-1477

Wood HA, Bozarth RF (1972) Properties of virus-like particles of *Penicillium chrysogenum*: one double-stranded RNA molecule per particle. Virology 47: 604-609

Wood HA, Bozarth RF, Mislivec PB (1971) Virus-like particles associated with an isolate of *Penicillium brevi-compactum*. Virology 44: 592-598

Xie WS, Antoniw JF, White RF, Woods RD (1993) A third cryptic virus in beet (*Beta vulgaris*). Plant Pathology 42: 464-470

Contributed By

Ghabrial SA, Bozarth RF, Buck KW, Yamashita S, Martelli GP, Milne RG

FAMILY *HYPOVIRIDAE*

TAXONOMIC STRUCTURE OF THE FAMILY

Family	***Hypoviridae***
Genus	*Hypovirus*

GENUS *HYPOVIRUS*

Type Species Cryphonectria hypovirus 1-EP713 (CHV1-EP713)

VIRION PROPERTIES

MORPHOLOGY

No true virions are associated with members of this family. Pleomorphic vesicles 50-80 nm in diameter, devoid of any detectable viral structural proteins but containing dsRNA and polymerase activity are the only virus-associated particles that can be isolated from infected fungal tissue.

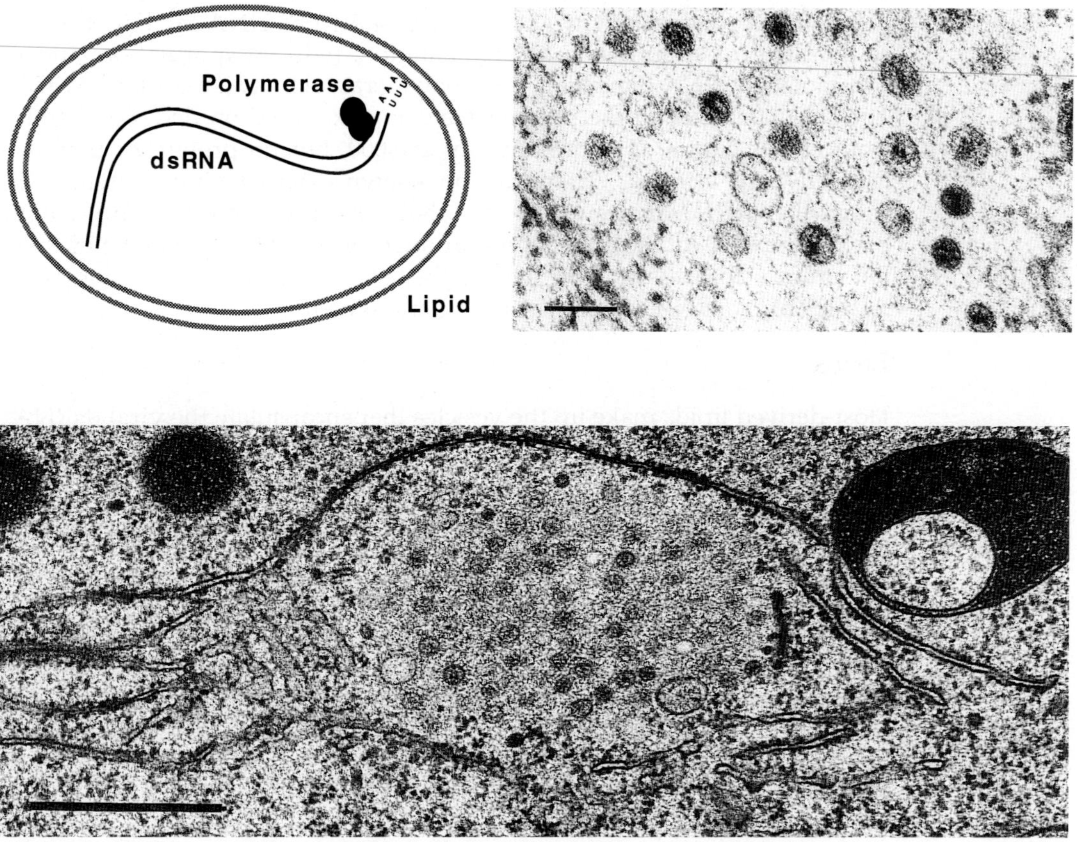

Figure 1: (upper left) Schematic diagram of a vesicle of a member of the family *Hypoviridae*; (upper right) thin section showing vesicles in fungal tissue; (lower) thin section showing vesicle aggregate in fungal tissue surrounded by rough ER (from Newhouse *et al.*, 1983). The bar represents 100 nm.

PHYSICOCHEMICAL AND PHYSICAL PROPERTIES

Mr of vesicles is unknown. They have a buoyant density in CsCl of approximately 1.27-1.3 g/cm^3 and sediment through sucrose as a broad component of approximately 200S. Their pH stability is unknown. The vesicles can be purified in pH 5.0 buffer and resuspended in pH 7.0 buffer. pH optimum for polymerase activity *in vitro* is 8.0; the optimum Mg^{++} for polymerase activity is 5 mM. Activity decreases dramatically at pH less than 7.0 or more than 9.0. The vesicles are unstable when heated, or in lipid solvents. Optimal temperature for polymerase activity is 30° C; temperatures over 40° C inactivate polymerase activity. Deoxycholate at concentrations of more than 0.5% inactivates polymerase activity.

Nucleic Acid

Vesicles contain linear dsRNA, approximately 10-13 kbp in size. The genome of the type species, CHV1-EP713, is 12,712 bp in size. Apparently only one strand is employed in transcription. The coding strand contains a short 3'-poly (A) tail, which is 20-30 residues in length when analyzed as a component of the dsRNA. Apparently one full-length dsRNA molecule is required for virus replication. The presence of shorter-than-full-length dsRNA molecules is common among some members, and satellite-like dsRNAs are present in others. No function has been ascribed to any ancillary dsRNA. The 5' terminus of the positive strand of dsRNA from CHV1-713 blocked, but the blocking group is unknown. The 5' terminus of the negative strand is unblocked. Both 5' termini of dsRNA from CHV3 GH2, are unblocked.

Proteins

No structural proteins have been described for members of this family. No function has been assigned to nonstructural proteins found in all projected members of family. EP713 dsRNA encodes p29, a presumptive NS protein identified *in vitro* and *in vivo*. P29 has papain-like protease activity and has been shown by DNA-mediated transformation to be responsible for suppression of pigmentation, reduced sporulation, and reduced laccase accumulation. RNA-dependent RNA polymerase activity is associated with isolated vesicles. The calculated size of the polymerase complex, based on deduced amino acid sequence from cDNA clones, is approximately 250,000, but no protein of that size has yet been isolated from vesicles. There are no known external viral proteins. The polymerase transcribes ssRNA molecules *in vitro* that correspond in size to full-length dsRNA. Approximately 90% of the polymerase products *in vitro* are of positive polarity. A sequence of ten amino acid residues representing the C-terminal cleavage site for p48, beginning with Ala-419 of L-dsRNA ORF B, has been determined.

Lipids

Host-derived lipids make up the vesicles that encapsulate the viral dsRNA.

Carbohydrates

Carbohydrates similar to those involved in fungal cell wall synthesis are associated with vesicles.

Genome Organization and Replication

The positive (coding) strand is polyadenylated at the 3'-terminus, with an average tail length of approximately 20-24 residues when analyzed as a component of dsRNA. The 5'-terminus of the positive strand appears to be blocked, although the blocking group is unknown. A 5'-leader of approximately 500 nucleotide residues, including several AUG triplets, precedes the AUG codon that initiates the first long ORF, ORF A. The ORF A product may or may not be autocatalytically cleaved, depending on the virus. The UAA termination sequence at the end of ORF A is part of the pentanucleotide UAAUG in all members investigated to date, with the AUG of the UAAUG pentanucleotide initiating the other long ORF, ORF B. In members investigated to date, the N-terminal product of ORF B is a papain-like cysteine protease that autocatalytically releases from the growing polypeptide chain. No further processing *in vitro* has been demonstrated for the remaining 300 kDa polypeptide from this ORF. Phylogenetic relatedness to members of the positive-sense, ssRNA genus *Potyvirus* has been demonstrated, based on protease, polymerase, and helicase domains, although these domains are positioned differently in the two genomes.

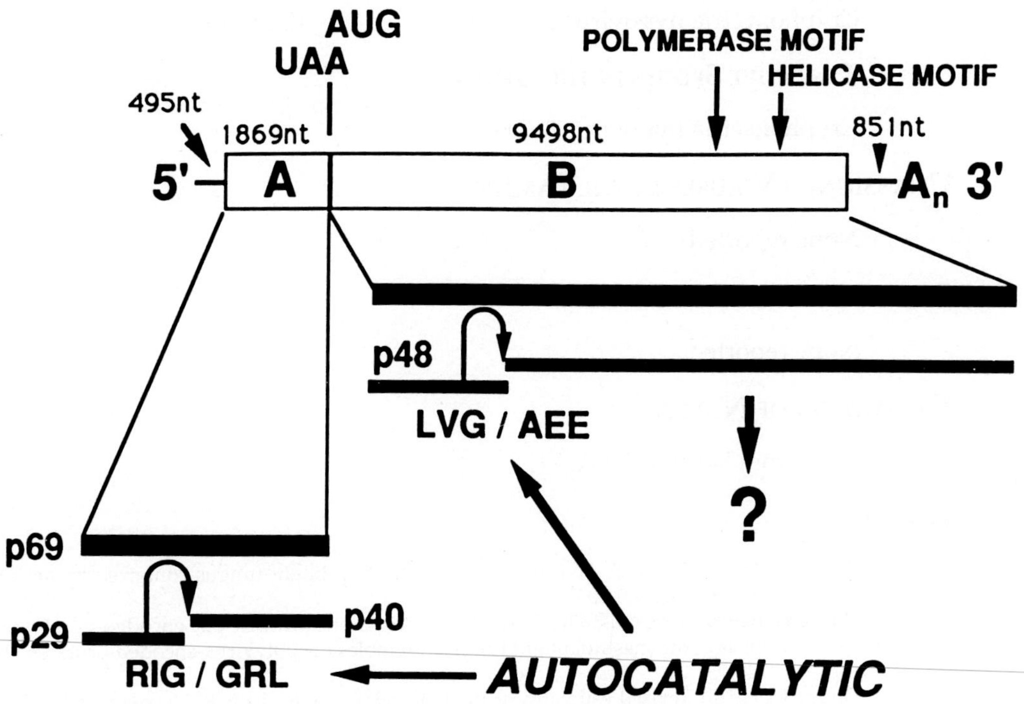

Figure 2: *Hypovirus* genome organization (From Shapira *et al.*, 1991).

ANTIGENIC PROPERTIES

No antibody has ever been raised from virus particle preparations. Anti-dsRNA antibodies have been used to confirm the genomic constituent. Chimeric β-galactosidase/EP713 ORF A fusion proteins have successfully been used to raise antiserum that is immunoreactive with a virus-specific protein in the infected fungal host, but the location of the protein in the cell is unknown.

BIOLOGICAL PROPERTIES

The viruses infect the chestnut blight fungus, *Cryphonectria parasitica*. Confirmed member viruses cause a disease referred to as "hypovirulence" in *C. parasitica*, characterized by reduced virulence of the fungus on its tree host and altered fungal morphology in culture, but many possible family members have little or no discernible effect on their fungal host. Some possible members infect other filamentous fungi, e.g., *Sclerotinia sclerotiorum*. Infection of fungal mycelium is known only through fusion, or anastomosis, between infected and uninfected hyphae. The Transmission rate through asexual spores (conidia) varies from a few to close to 100 percent. Transmission through sexual spores (ascospores) is not known to occur. Transmission via cell-free extracts has not been demonstrated. Confirmed members have been identified throughout chestnut growing areas of Europe and North America. dsRNA-containing vesicles have been associated with abnormal Golgi apparatus in freeze-substituted thin sections. No nuclear or mitochondrial associations, nor virus-associated inclusions, have been noted.

LIST OF SPECIES IN THE GENUS

The viruses, their alternative names ( ), genomic sequence accession numbers [ ] and assigned abbreviations ( ) are:

SPECIES IN THE GENUS

Cryphonectria hypovirus 1-EP713 (hypovirulence-associated virus)	[M57938]	(CHV1-EP713)

Cryphonectria hypovirus 1-EP747		(CHV1-EP747)
Cryphonectria hypovirus 2-NB58	[L29010]	(CHV2-NB58)

Tentative Species in the Genus

Cryphonectria hypovirus 3-GH2	(CHV3-GH2)

Unassigned Viruses in the Family

None reported.

Similarity with Other Taxa

None reported.

Derivation of Names

hypo: from *hypo*virulence

References

Choi GH, Nuss DL (1992) Hypovirulence of chestnut blight fungus conferred by an infectious viral cDNA. Science 257: 800-803

Choi GH, Pawlyk DM, Nuss DL (1991) The autocatalytic protease p29 encoded by a hypovirulence-associated virus of the chestnut blight fungus resembles the potyvirus-encoded protease HC-Pro. Virology 183: 747-752

Dodds JA (1980) Revised estimates of the molecular weights of dsRNA segments in hypovirulent strains of *Endothia parasitica*. Phytopathology 70: 1217-1220

Hansen DR, van Alfen NK, Gillies K, Powell WA (1985) Naked dsRNA associated with hypovirulence of *Endothia parasitica* is packaged in fungal vesicles. J Gen Virol 66: 2605-2614

Hillman BI, Tian Y, Bedker PJ, Brown MP (1992) A North American hypovirulent isolate of the chestnut blight fungus with European isolate-related double-stranded RNA. J Gen Virol 73: 681-686

Hiremath S, L'Hostis B, Ghabrial SA, Rhoads RE (1986) Terminal structure of hypovirulence-associated dsRNAs in the chestnut blight fungus *Endothia parasitica*. Nucl Acid Res 14: 9877-9896

Newhouse JR, Hoch HC, MacDonald WL (1983) The ultrastructure of *Endothia parasitica*. Comparison of a virulent with a hypovirulent isolate. Can Jour Bot 61: 389-399

Newhouse JR, MacDonald WL, Hoch HC (1990) Virus-like particles in hyphae and conidia of European hypovirulent (dsRNA-containing) strains of *Cryphonectria parasitica*. Can Jour Bot 68: 90-101

Shapira R, Choi GH, Nuss DL (1991) Virus-like genetic organization and expression strategy for a double-stranded RNA genetic element associated with biological control of chestnut blight. EMBO J 10: 731-739

Shapira R, Nuss DL (1991) Gene expression by a hypovirulence-associated virus of the chestnut blight fungus involves two papain-like protease activities. J Biol Chem 266: 19419-19425

Tartaglia J, Paul CP, Fulbright DW, Nuss DL (1986) Structural properties of double-stranded RNAs associated with biological control of chestnut blight fungus. Proc Natl Acad Sci USA 83: 9109-9113

van Alfen NK (1986) Hypovirulence of *Endothia (Cryphonectria) parasitica* and *Rhizoctonia solani*. In: Buck KW (ed) Fungal Virology, CRC Press, Boca Raton FL, pp 143-162

Contributed By

Hillman BI, Fulbright DW, Nuss DL, van Alfen NK

ORDER *MONONEGAVIRALES*

TAXONOMIC STRUCTURE OF THE ORDER

Order	***Mononegavirales***
Family	***Paramyxoviridae***
Subfamily	***Paramyxovirinae***
Genus	*Paramyxovirus*
Genus	*Morbillivirus*
Genus	*Rubulavirus*
Subfamily	***Pneumovirinae***
Genus	*Pneumovirus*
Family	***Rhabdoviridae***
Genus	*Vesiculovirus*
Genus	*Lyssavirus*
Genus	*Ephemerovirus*
Genus	*Cytorhabdovirus*
Genus	*Nucleorhabdovirus*
Family	***Filoviridae***
Genus	*Filovirus*

VIRION PROPERTIES

GENERAL

The order comprises the three families of viruses possessing linear, non-segmented, negative sense ssRNA genomes, i.e. the *Paramyxoviridae, Rhabdoviridae* and *Filoviridae.* Common features include negative sense RNA, helical nucleocapsid, the initiation of primary transcription by a virion-associated RNA-dependent RNA polymerase, a similar gene order (3' non-translated region - core protein genes - envelope protein genes - polymerase gene - 5' non-translated region), and a single 3' promoter. Maturation is by budding, predominantly from the plasma membrane, rarely from internal membranes (rabies virus), or the inner nuclear membrane (many plant rhadboviruses). Viruses mature at cytoplasmic locations, except for some plant rhabdoviruses.

MORPHOLOGY

The virions are large, enveloped structures generally with a prominent fringe of spikes that are 5-10 nm long and spaced 7-10 nm apart. The morphologies of the particles are variable but distinguish the three families: simple, branched, U-shaped, 6-shaped, or circular filaments of uniform diameter (about 80 nm) extending up to 14,000 nm are characteristic of the member viruses of the family *Filoviridae,* although purified virions are bacilliform and of uniform length (e.g. 790 nm in the case of Marburg virus); filamentous, pleomorphic, or spherical structures of variable diameter are characteristic of the member viruses of the family *Paramyxoviridae;* and regular bullet-shaped, or bacilliform particles are characteristic of the member viruses of the family *Rhabdoviridae.* The helical ribonucleoprotein core has a diameter of 13-20 nm which in filoviruses and rhabdoviruses is organized into a helical nucleocapsid of about 50 nm diameter. The nucleocapsid of VSV is infectious.

PHYSICOCHEMICAL AND PHYSICAL PROPERTIES

Virion Mr is 300-1000 x 10^6. S_{20w} is 550-1,045 (plant rhabdoviruses have larger S_{20W} values. Virus buoyant density in CsCl is 1.18-1.20 g/cm^3. Virus infectivity is rapidly inactivated by heat treatment, or following UV- or X-irradiation, or exposure to lipid solvents.

NUCLEIC ACID

Virions contain one molecule of linear, non-infectious, negative sense, ssRNA, 11-16 kb in size, Mr of 3.5-5 x 10^6 of which comprises about 0.5-2.0% of the particle weight. The viral

RNA lacks a capped 5' terminus, or a covalently associated protein. The 3' end lacks a poly (A) tract. The genome comprises a linear sequence of non-overlapping genes with short terminal untranscribed regions and intergenic regions ranging from 2 to several hundred nucleotides. Exceptions are a short overlap of some genes (e.g., the 9th and 10th genes of respiratory syncytial virus), and the encoding of genetic information in all three reading frames in the P genes of paramyxoviruses and morbilliviruses.

PROTEINS

There are a limited number of proteins in relation to the large particle size. The 5-7 structural proteins comprise envelope glycoprotein(s), a matrix protein, a major RNA-binding protein, other nucleocapsid-associated protein(s) and a large molecular weight polymerase protein, plus, in some viruses, several non-structural proteins which may be phosphorylated. Enzymes associated with virions may include transcriptase, polyadenylate transferase, mRNA methyl transferase, neuraminidase.

LIPIDS

Virions are composed of about 15-25% lipids, their composition reflecting the host cell membrane where virions bud. Generally, phospholipids represent about 55-60% and sterols and glycolipids about 35-40% of the total lipids. Glycoproteins may have a co-valently associated fatty acid proximal to the lipid envelope.

CARBOHYDRATES

Virions are composed of about 3% carbohydrate by weight. The carbohydrates are present as N- or O-linked glycan chains on surface proteins and as glycolipids. When made in mammalian cells the oligosaccharide chains are generally of the complex type; in insect cells they are of the non-complex types.

GENOME ORGANIZATION AND REPLICATION

Discrete messenger RNAs are transcribed by sequential interrupted synthesis. Generally, genes do not overlap. The P genes of paramyxoviruses and morbilliviruses are exceptional in that all 3 ORFs may be utilized via alternative non-AUG start codons, and mRNA editing by insertion of non-templated nucleotides to change the reading frame for the expression of P gene products. Replication occurs by synthesis of a complete positive sense anti-genome RNA. Maturation of the independently assembled helical nucleocapsids occurs by budding through host membranes and investment by a host-derived lipid envelope containing transmembrane virus proteins.

ANTIGENIC PROPERTIES

Membrane glycoproteins are involved in antibody induced neutralization. Virus serotypes are defined by the surface antigens. Filoviruses are an exception in that they are not neutralized *in vitro*.

BIOLOGICAL PROPERTIES

The host ranges vary from restricted to unrestricted. Filoviruses have only been isolated from primates. Paramyxoviruses occur only in vertebrates and no vectors are known. Rhabdoviruses infect invertebrates, vertebrates and plants. Some rhabdoviruses multiply in both invertebrates and vertebrates, some in invertebrates and plants, but none in all three hosts. The pathology associated with virus infections varies. In human hosts the patho-genic potential tends to be characteristic of the family, i.e., hemorrhagic fever (*Filoviridae*); respiratory and neurological diseases (*Paramyxoviridae*); mild febrile to fatal neurological diseases (*Rhabdoviridae*).

SIMILARITY WITH OTHER TAXA

None reported.

Derivation of Names

cyto: from Greek *kytos*, "cell"
ephemero: from Greek *ephemeros*, "short-lived"
filo: from Latin filo, "thread-like"
lyssa: from Greek *lyssa* "rage, fury, canine madness"
mono: from Greek *monos*, "single"
morbilli: from Latin *morbillus*, diminutive of *morbus*, "disease"
nega: from *nega*tive sense RNA
nucleo: from Latin *nux*, *nucis*, "nut"
paramyxo: from Greek *para*, "by the side of", and *myxa* "mucus"
pneumo: from Greek *pneuma*, "breath
rhabdo: from Greek *rhabdos*, "rod"
rubula: from Latin *ruber*, "red", Rubula inflans - old name for mumps
vesiculo: from Latin *vesicula*, diminutive of *vesica*, "bladder, blister"
virales: from Latin *virales*, "viruses"

References

Kiley MP, Bowen ETA, Eddy GA, Isaacson M, Johnson KM, McCormick JB, Murphy FA, Pattyn SR, Peters D, Prozesky OW, Regnery RL, Simpson DIH, Slenczka W, Sureau P, van der Groen G, Webb PA, Wulff H (1982) *Filoviridae*, a taxonomic home for Marburg and Ebola viruses. Intervirology 18: 24-32

Kingsbury DW (ed) (1991) The Paramyxoviruses. Plenum Press, New York

Pringle CR (1991) The order *Mononegavirales*. Arch Virol 117: 137-140

Wagner RR (ed) (1987) The Rhabdoviruses. Plenum Press, New York

Contributed By

Bishop DHL, Pringle CR

FAMILY PARAMYXOVIRIDAE

TAXONOMIC STRUCTURE OF THE FAMILY

Family	***Paramyxoviridae***
Subfamily	***Paramyxovirinae***
Genus	*Paramyxovirus*
Genus	*Morbillivirus*
Genus	*Rubulavirus*
Subfamily	***Pneumovirinae***
Genus	*Pneumovirus*

VIRION PROPERTIES

MORPHOLOGY

Virions are 150 nm or more in diameter, pleomorphic, but usually spherical in shape, although filamentous and other forms are common. Virions consist of a lipid envelope surrounding a nucleocapsid. The envelope is derived from lipids of the host cell plasma membrane and contains 2 or 3 transmembrane glycoproteins. These are present as homooligomers and form spike-like projections, 8-12 nm in length, spaced 7-10 nm apart (depending on the genus). One or two non-glycosylated membrane proteins are associated with the inner face of the envelope. The viral nucleocapsid consists of a single species of viral RNA and associated proteins. It has helical symmetry and is 13-18 nm in diameter with a 5.5-7 nm pitch (depending on the subfamily); its length can be up to 1,000 nm in some genera. Occasionally, multiploid virions are found.

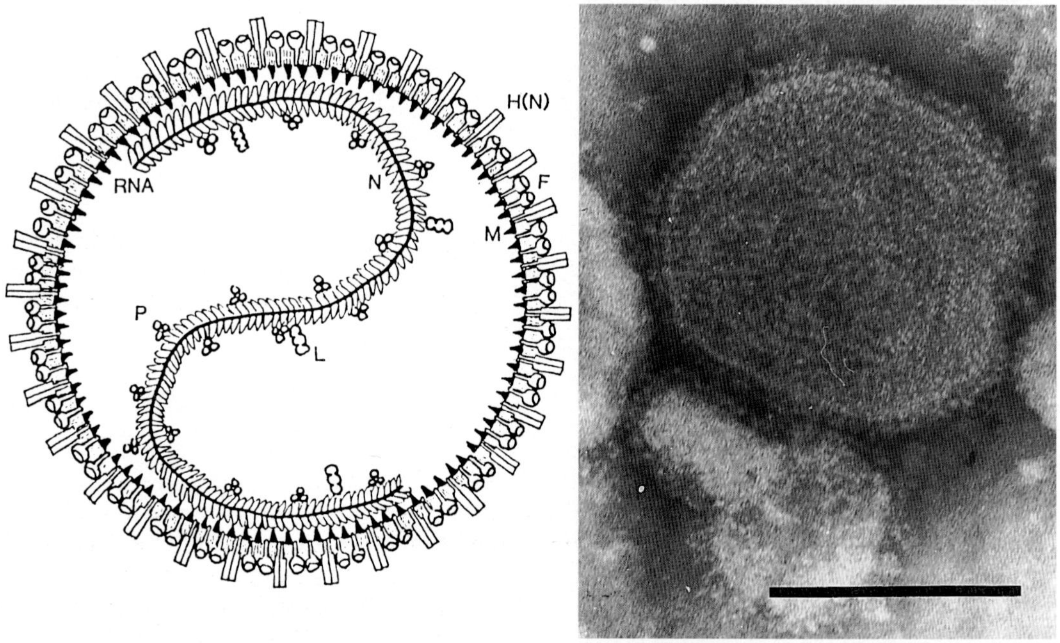

Figure 1: (left) Diagram of virion of a member virus of the subfamily *Paramyxovirinae* in section (N: nucleocapsid; P: phosphoprotein, L: large protein, M: matrix protein, H(N) hemagglutinin (neuraminidase) protein, F: fusion protein; (right) negative contrast electron micrograph of mumps virus (*Rubulavirus*). The bar represents 100 nm.

PHYSICOCHEMICAL AND PHYSICAL PROPERTIES

Virion Mr is about 500 x 10^6, and much greater for multiploid virions. Virion buoyant density in sucrose is 1.18-1.20 g/cm^3. Virion S_{20w} is at least 1000. Virions are very sensitive to heat, lipid solvents, non-ionic detergents, formaldehyde and oxidizing agents.

Nucleic Acid

Virions contain a single molecule of linear, non-infectious, negative sense, ssRNA. The RNA genome size is fairly uniform: 15,156 b for Newcastle disease virus, 15,222 b for human respiratory syncytial virus, 15,244 b for simian virus 5, 15,285 b for Sendai virus, 15,384 b for mumps virus, 15,463 b for human parainfluenza virus 3, 15,646 b for human parainfluenza virus 2 , and 15,892 b for measles virus. Some virions may contain positive sense RNA. Thus, partial self-annealing of extracted RNA may occur. The Mr of the genome is 5-7 x 10^6 and this constitutes about 0.5% of the virion by weight. Intracellularly, or in virions, genome size RNA is found exclusively as nucleocapsids.

Proteins

Members of the subfamily *Paramyxovirinae* contain 6-7 transcriptional elements that encode 10-12 proteins (Mr 5-250 x 10^3) of which 4 or 5 (or more) are derived from the 2-3 overlapping ORFs in the P locus (Fig. 2). Pneumoviruses have 10 ORFs encoding 10 proteins of Mr 4.8-250 x 10^3. Virion proteins common to all genera include: three nucleocapsid-associated proteins, i.e., an RNA-binding protein (N or NP), a phosphoprotein (P), and a large putative polymerase protein (L); three membrane associated proteins, i.e., an unglycosylated envelope protein (M), and two glycosylated envelope proteins, comprising a fusion protein (F) and an attachment protein (G, or H, or HN). The F protein is synthesized within an infected cell as a precursor (F_0) which is activated following cleavage by cellular protease(s) to produce the virion disulfide-linked F_1 and F_2 subunits (order: amino F_2-S-S-F_1 carboxyl). Variable proteins include non-structural proteins (C, IC or NS1, and IB or NS2), a cysterine-rich protein (V) a small integral membrane protein (SH or 1A), and a second inner envelope unglycosylated protein (M2 or 22 kDa protein). Virion enzyme activities (variously represented among the genera) include a transcriptase, an adenylate transferase, mRNA guanylyl and methyl transferases, protein kinase and a neuraminidase.

Lipids

Virions are composed of 20-25% lipid by weight. The lipids are derived from the host cell plasma membrane.

Carbohydrates

Virions are composed of 6% carbohydrate by weight; composition is dependent on the host cell. Fusion and attachment proteins are glycosylated by N-linked carbohydrate side chains. In the subfamily *Pneumovirinae* the attachment protein (G) is heavily glycosylated by O-linked carbohydrate side chains. The SH protein of respiratory syncytial virus contains polylactosamine.

Genome Organization and Replication

The genome organization is illustrated in Fig. 2 for viruses representing the 4 genera of the family. After attachment to cell receptors, virus entry is achieved by fusion of the virus envelope with the cell surface membrane. This can occur at neutral pH. Virus replication occurs in the cell cytoplasm and is thought to be independent of host nuclear functions. The genome is transcribed processively from the 3' end by virion-associated enzymes into 6-10 separate, subgenomic, viral-complementary mRNAs. The mRNAs are capped and possess a 3' poly (A) tract. The intergenic regions may vary in size and sequence between genera and member viruses.

Nucleocapsids assemble independently in the cytoplasm. They are enveloped on the cell surface at sites containing virus envelope proteins. Members of the subfamily *Paramyxovirinae* contains 6-7 transcriptional elements that encode 10-12 proteins. Most proteins are encoded by unique mRNAs. Notable exceptions are the P, C and V mRNAs. These are synthesized by a mechanism involving site-specific stuttering ("editing") on the template. This results in the insertion of one or more non-templated nucleotides and shifts the reading frame to access an alternative ORF. The derived mRNAs synthesize two proteins, P and V, which have identical amino-terminal domains but different sequence in the rest of each protein. Other truncated, or chimeric, proteins can be produced by shifting into the third reading frame. The C ORF present in some viruses overlaps the P ORF and can initiate at a non-AUG codon that is accessed by ribosomal choice. Additional truncated P proteins can be generated by specific internal translation initiation.

Members of the subfamily *Pneumovirinae* have 10 transcriptional elements (mRNAs) each of which encodes a major protein. However, there is overlap between the M2 and L transcriptional elements in some pneumoviruses (Fig. 2).

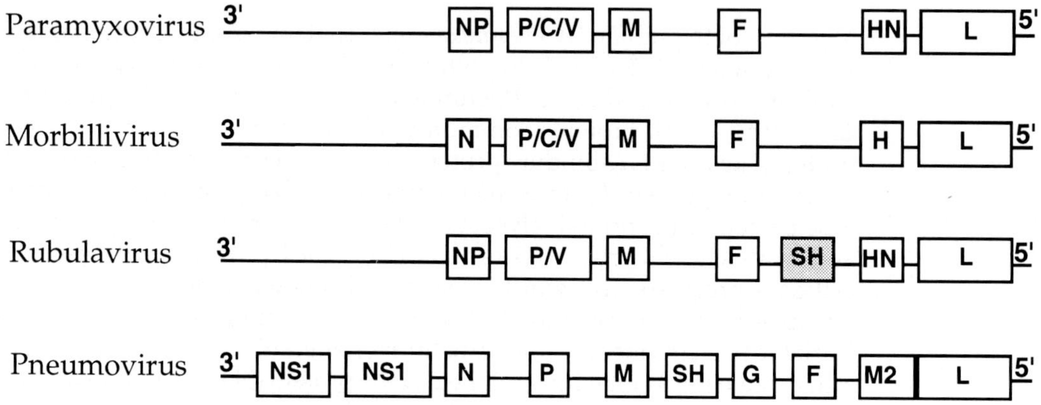

Figure 2: Maps of genomic RNAs (3'-to-5') of the four genera of the family *Paramyxoviridae*. Each box represents a separately encoded mRNA. Boxes identify ORFs; multiple distinct ORFs within a single sequence are indicated by slashes. The lengths of the boxes and intervening or preceding spaces (lines) are not to scale, the spacing only emphasizes the common proteins between genera. The D ORF present in some viruses is not shown. In some viruses (notably in the genus *Paramyxovirus* the V ORF might be a non-expressed relic. In the genus *Rubulavirus* some species lack the SH gene (shaded box). In the genus *Pneumovirus*, respiratory syncytial virus has a transcriptional overlap at M2 and L (black box) although pneumonia virus of mice (PVM) does not. TRTV is also distinct in having a different gene order at the 5' end, i.e., (3') F-M2-SH-G-L (5'). There are conserved trinucleotides that serve as intergenic sequences for the paramyxoviruses and morbilliviruses. For rubulaviruses and pneumoviruses the intergenic sequences are variable (1-31, or 1-57 nucleotides long, respectively).

ANTIGENIC PROPERTIES

The attachment (HN, or H, or G) and fusion (F) proteins are of primary importance in inducing virus-neutralizing antibodies and immunity against reinfection. Antibodies to N and, variably, to other viral proteins also are induced by infection. Various proteins have been reported to serve as antigens for cytotoxic or helper T cells.

BIOLOGICAL PROPERTIES

Paramyxoviruses have only been conclusively identified in vertebrates and almost exclusively in mammals and birds. Most viruses have a narrow specific host range in nature, but in cultured cells they display a broad host range. Transmission is horizontal, mainly through airborne routes; no vectors are known. Temperate and persistent infections are common in cultured cells. Primary replication is mainly in the respiratory tract. Generally, infection is cytolytic, but temperate and persistent infections are common. Other features of infection include the formation of inclusion bodies and syncytia. Cell surface molecules reported to serve as receptors for paramyxovirus attachment include sialoglycoproteins and glyco-lipids. Nucleocapsids associate with viral membrane proteins at the plasma membrane and are enveloped by budding.

SUBFAMILY *PARAMYXOVIRINAE*

TAXONOMIC STRUCTURE OF THE SUBFAMILY

Subfamily	***Paramyxovirinae***
Genus	*Paramyxovirus*
Genus	*Morbillivirus*
Genus	*Rubulavirus*

DISTINGUISHING FEATURES

Member species of the subfamily *Paramyxovirinae* have 6-7 transcriptional elements in contrast to the 10 transcriptional elements in member viruses of the subfamily *Pneumovirinae*. Members of different genera in the subfamily *Paramyxovirinae* exhibit some sequence relatedness between corresponding proteins. Their nucleocapsids have diameters of 18 nm and a pitch of 5.5 nm, the length of the surface spikes is 8 nm.

GENUS *PARAMYXOVIRUS*

Type Species human parainfluenza virus 1 (HPIV-1)

DISTINGUISHING FEATURES

Member viruses of the genus *Paramyxovirus* possess a neuraminidase, in contrast to members of the genus *Morbillivirus*. These viruses have six transcriptional elements. All members encode a C protein. Unedited P mRNA encodes P and C, whereas insertion of a G nucleotide in P mRNA transcripts accesses the V ORF. Corresponding proteins of members of the genus *Paramyxovirus* are highly related. They exhibit intermediate levels of sequence relatedness with the corresponding proteins of mobilliviruses and low levels with those of the rubulaviruses.

LIST OF SPECIES IN THE GENUS

The viruses, their genomic sequence accession numbers [ ] and assigned abbreviations ( ) are:

SPECIES IN THE GENUS

bovine parainfluenza virus 3	[Y00114, Y00115]	(BPIV-3)
human parainfluenza virus 1	[M22347, M31228, M80818]	(HPIV-1)
human parainfluenza virus 3	[Z11575]	(HPIV-3)
Sendai virus	[K01146, M19661, M30202-4]	
(murine parainfluenza virus 1)		
simian parainfluenza virus 10		(SPIV-10)

TENTATIVE SPECIES IN THE GENUS

None reported.

GENUS *MORBILLIVIRUS*

Type Species measles virus (MeV)

DISTINGUISHING FEATURES

All species of the genus *Morbillivirus* lack neuraminidase. Member viruses exhibit intermediate levels of protein sequence relatedness. They have an identical gene order, number of transcriptional elements and size of intergenic sequences with members of the genus *Paramyxovirus* (Fig. 2). All morbilliviruses produce both intracyto-plasmic and intranuclear inclusion bodies which contain viral ribonucleocapsids. Viruses cross-react in serological tests. Hemagglutinin is present in some species of the genus, but absent in most others.

LIST OF SPECIES IN THE GENUS

The viruses, their genomic sequence accession numbers [ ] and assigned abbreviations ( ) are:

SPECIES IN THE GENUS

canine distemper virus	[M12669, M21849, M32418]	(CDV)
dolphin distemper virus		(DMV)
measles (Edmonston) virus	[K01711, X16565]	(MeV)
peste-des-petits-ruminants virus		(PPRV)
phocine (seal) distemper virus	[D10371, X65512, X68311]	(PDV)
porpoise distemper virus		
rinderpest virus	[M17434, M20870, M34018]	(RPV)

TENTATIVE SPECIES IN THE GENUS

None reported.

GENUS *RUBULAVIRUS*

Type Species mumps virus

DISTINGUISHING FEATURES

All species of the genus *Rubulavirus* have hemagglutinin and neuraminidase activities. They show low to intermediate levels of homology in their respective protein sequences. Some members contain an extra gene (SH) between the F and HN loci (Fig. 2). In some members the unedited mRNA from the P locus encodes P in others NS1 (V). The intergenic sequences are of variable length. All members lack a C protein ORF.

LIST OF SPECIES IN THE GENUS

The viruses, their alternative names ( ), genomic sequence accession numbers [ ] and assigned abbreviations ( ) are:

SPECIES IN THE GENUS

avian paramyxovirus 2 (Yucaipa)		(APMV-2)
avian paramyxovirus 3		(APMV-3)
avian paramyxovirus 4		(APMV-4)
avian paramyxovirus 5 (Kunitachi)		(APMV-5)
avian paramyxovirus 6		(APMV-6)
avian paramyxovirus 7		(APMV-7)
avian paramyxovirus 8		(APMV-8)
avian paramyxovirus 9		(APMV-9)
human parainfluenza virus 2	[M37751, X57559]	(HPIV-2)
human parainfluenza virus 4a	[M32982, M55975, D10241]	(HPIV-4a)
human parainfluenza virus 4b	[M32983, M55976, D10242]	(HPIV-4b)
mumps virus	[D00663, D10575, M24731 X57997]	
Newcastle disease virus (avian paramyxovirus 1)	[M11204, X04719, X05399 X60599]	(NDV) (APMV-1)
porcine rubulavirus (La-Piedad-Michoacan-Mexico virus)		
simian parainfluenza virus 5	[J03142, M81442, M81721]	(SV-5)
simian parainfluenza virus 41	[M62733, D90338]	(SV-41)

TENTATIVE SPECIES IN THE GENUS

None reported.

SUBFAMILY *PNEUMOVIRINAE*

TAXONOMIC STRUCTURE OF THE SUBFAMILY

Subfamily	***Pneumovirinae***
Genus	*Pneumovirus*

DISTINGUISHING FEATURES

A single genus in the subfamily is recognized. Member species differ from those of the subfamily *Paramyxovirinae* in several features: (a) possession of 10 separate genes; (b) smaller average gene size; (c) possession of one additional unglycosylated membrane-associated protein (M2 or 22 kDa); (d) extensive O-linked glycosylation of the G protein; (e) the P locus that encodes a single protein; (f) the nucleocapsid diameter (13-14 nm compared with 18 nm in the subfamily *Paramyxovirinae*); (g) nucleocapsid pitch (7 nm); (h) length of glycoprotein spikes (10-12 nm). Species in the subfamily *Pneumovirinae*, genus *Pneumovirus* also lack neuraminidase; hemagglutinin is absent in bovine and human respiratory syncytial viruses, but is present in pneumonia virus of mice. In turkey rhinotracheitis virus, the relative placements of SH-G versus F-M2 in the gene order are reversed. The G protein is structurally unrelated to the HN or H proteins of the other genera of the family *Paramyxoviridae* and exhibit a high level of interstrain diversity (up to 47% non-identity).

GENUS *PNEUMOVIRUS*

Type Species human respiratory syncytial virus (HRSV)

LIST OF SPECIES IN THE GENUS

The viruses, their genomic sequence accession numbers [ ] and assigned abbreviations ( ) are:

SPECIES IN THE GENUS

bovine respiratory syncytial virus	[M58350, M82816]	(BRSV)
human respiratory syncytial virus (A2/18537)	[D00386-397, M17245]	(HRSV)
pneumonia virus of mice	[D01100, D10331]	(PVM)
turkey rhinotracheitis virus	[D00850, X58639]	(TRTV)

TENTATIVE SPECIES IN THE GENUS

None reported.

UNASSIGNED VIRUSES IN THE FAMILY

In addition to three recognized viruses, namely Fer-de-Lance virus of reptiles (FDLV), the chiropteran Mapuera virus (MPRV), and the rodent Nariva virus (NARV), several viruses from penguins are known which are distinct from avian paramyxoviruses 1-9.

SIMILARITY WITH OTHER TAXA

The member viruses of the family *Paramyxoviridae* have a similar strategy of gene expression and replication and gene order to those of other families in the order *Mononegavirales*, that is the families *Rhabdoviridae* and *Filoviridae*.

DERIVATION OF NAMES

paramyxo: from Greek *para*, "by the side of ", and *myxa* 'mucus'

morbilli: from Latin *morbillus*, diminutive of *morbus*, "disease"
pneumo: from Greek *pneuma*, "breath"
rubula: Rubula inflans - old name for mumps

References

Alexander DJ (1986) The classification, host range and distribution of avian paramyxoviruses In: McFerran JB, McNulty MS (eds) Acute Virus Infections in Poultry. Martinus Nijhoff, Dordrecht, pp 52-66

Bishop DHL, Compans RW (eds) (1984) Non-segmented negative strand viruses; paramyxoviruses and rhabdoviruses. Academic Press, Orlando FL

Choppin PW, Compans RW (1975) Reproduction of paramyxoviruses. In: Fraenkel-Conrat H, Wagner RR (eds) Comprehensive Virology Vol 4 Plenum Press, New York, pp 95-178

Kingsbury DW (ed) (1991) The paramyxoviruses. Plenum Press, New York

Kingsbury DW (1990) *Paramyxoviridae* and their replication. In: Fields BN, Knipe JC (eds) Virology, 2nd edn Raven Press, New York, pp 945-962

Morrison TG (1988) Structure, function and intracellular processing of paramyxovirus membrane proteins. Virus Res 10: 113-136

Örvell C, Norrby E (1985) Antigenic structure of paramyxoviruses. In: van Regenmortel MHV, Neurath AR (eds) Immunochemistry of viruses, the basis for serodiagnosis and vaccines. Elsevier Medical Press, Amsterdam, pp 241-264

Pringle CR (1987) Paramyxoviruses and disease. In: Russell WC, Almond JW (eds) SGM Symposium 40, Molecular basis of virus disease. Cambridge University Press, Cambridge, pp 51-90

Scott EJ, Taylor G (1984) Respiratory syncytial virus; brief review. Arch Virol 84: 1-52

Contributed By

Rima B, Alexander DJ, Billeter MA, Collins PL, Kingsbury DW, Lipkind MA, Nagai Y, Örvell C, Pringle CR, ter Meulen V

FAMILY *RHABDOVIRIDAE*

TAXONOMIC STRUCTURE OF THE FAMILY

Family	***Rhabdoviridae***
Genus	*Vesiculovirus*
Genus	*Lyssavirus*
Genus	*Ephemerovirus*
Genus	*Cytorhabdovirus*
Genus	*Nucleorhabdovirus*

VIRION PROPERTIES

MORPHOLOGY

Virions are 100-430 nm long and 45-100 nm in diameter. Defective virus particles are proportionately shorter. Viruses infecting vertebrates are bullet-shaped or cone-shaped; viruses infecting plants mostly appear bacilliform when fixed prior to negative staining; in unfixed preparations they may appear bullet-shaped or pleomorphic. Some putative plant rhabdoviruses lack envelopes. The outer surface of virions (except for the quasiplanar end of bullet-shaped viruses) is covered with projections (peplomers) 5-10 nm long and about 3 nm in diameter. They consist of trimers of the virus glycoprotein. A honeycomb pattern of peplomers is observed on the surface of some viruses. Internally, the nucleocapsid, about 30-70 nm in diameter, exhibits helical symmetry and can be seen as cross-striations (spacing 4.5-5 nm) in negatively stained and thin-sectioned virus particles. The nucleocapsid consists of an RNA and N protein complex together with L and NS (M1) proteins and is surrounded by a lipid envelope containing M (M2) protein. The nucleocapsid contains transcriptase activity and is infectious. Uncoiled it is filamentous, about 700 nm long and 20 nm in diameter.

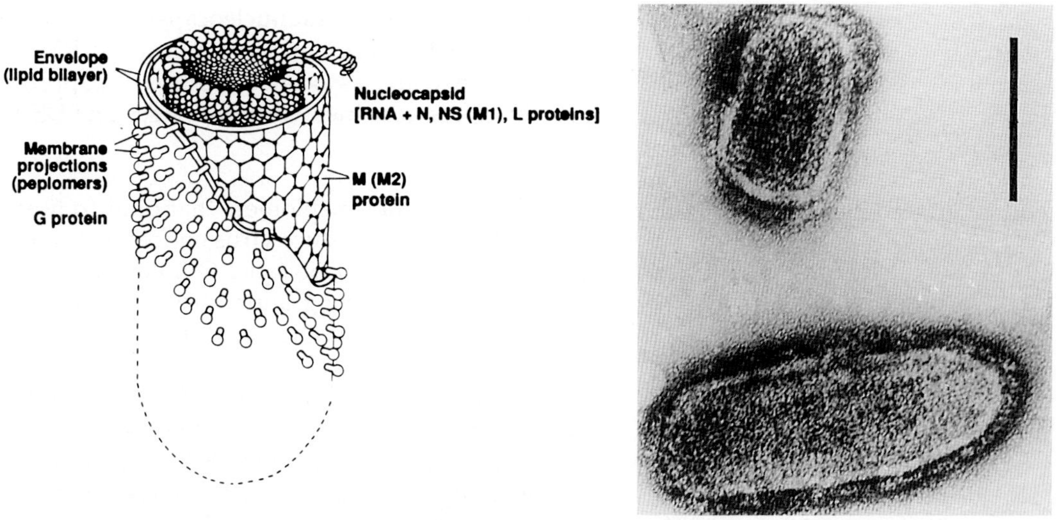

Figure 1: (left) Diagram of virion surface and virion in section (after Francki RIB and Randles JW, 1980); (right) negative contrast electron micrograph of VSIV (courtesy of Nichol ST and Holland JJ). The bar represents 100 nm.

PHYSICOCHEMICAL AND PHYSICAL PROPERTIES

Virion Mr is 300-1,000 x 10^6 and S_{20w} is 550-1,045 (plant rhabdoviruses have larger S_{20w} values). Virus buoyant density in CsCl is 1.19-1.20 g/cm^3, in sucrose it is 1.17-1.19 g/cm^3. Virus infectivity is stable in the range pH 5-10, but is rapidly inactivated at 56° C, or following UV- or X-ray irradiation, or exposure to lipid solvents.

NUCLEIC ACID

Viruses contain a single molecule of linear, negative-sense ssRNA (Mr 4.2-4.6 x 10^6, about 11-15 kb in size). The RNA represents about 1-2% of particle weight. The RNA has a 5' terminal triphosphate and is not polyadenylated. The ends have inverted complementary sequences. Defective RNAs, usually significantly shorter than full-length RNA (less than half size), may be identified in RNA recovered from virus populations. They are usually negative sense, however, hairpin RNA forms are also found. Defectives only replicate in the presence of homologous and, occasionally, certain heterologous helper rhabdoviruses. They may contain functional genes. Full-length positive-strand RNA may constitute up to 5% of a viral RNA population.

PROTEINS

Viruses generally have 5 polypeptides (VSIV: designated L, G, N, NS and M, see Table 1 for summary of their location, sizes and functions). In recognition of its phosphorylated state, NS is sometimes termed P. The presence and functions of other structural proteins (including additional glycoproteins) of certain rhabdoviruses are not known. The structural proteins represent 65-75% of the virus dry weight. For rabies and certain other viruses the NS is designated M1 and M is designated M2. For VSIV the numbers of molecules per infectious virus particle is estimated as: L: 20-50; G: 500-1,500; N: 1,000-2,000; NS: 100-300; and M: 1,500-4,000. The enzymes identified in virions include the RNA transcriptase (L and NS/M1 proteins), a 5' capping enzyme, guanylyl and methyl transferases, a protein kinase (viral-, possibly host-coded), a nucleoside triphosphatase and a nucleoside diphosphate kinase. These activities may be functions of L.

Table: Location, size (kDa) and functions of rhabdovirus structural proteins

Protein	Location, size and function
L	A component of the viral nucleocapsid. Functions (about 220-240 kDa) include transcription and replication. RNA-dependent RNA polymerase with associated mRNA 5'-capping, 3'-poly[A] and protein kinase activities. Observed sizes on SDS-PAGE are 150-190 kDa.
G	Forms virus surface peplomers that bind to host cell (about 65-90 kDa) receptors and induce virus endocytosis and fusion. G is variously N-glycoslyated, it lacks O-linked glycans. G induces and binds virus-neutralizing antibodies and elicits cell-mediated immune responses. G has hemagglutinin activity.
N	N is a major component of the viral nucleocapsid. It(about 47-62 kDa) associates with full-length negative- and positive sense RNAs, or defective RNAs, but not mRNAs. Newly synthesized N modulates genome transcription, promoting replication and read-through of transcription termination and poly[A] signals. N elicits cell-mediated immune responses and humoral antibodies.
NS (or P, or M1)	A component of the viral polymerase (hence, P, (about 20-30 kDa) polymerase associated). It is variously phosphorylated and migrates on SDS-PAGE as a 40-50 kDa protein. The NS of the nucleorhabdoviruses migrates faster. It is required for transcription. A soluble form is present in the cytoplasm of infected cells. May prevent self-aggregation of N protein and aid in N encapsidation of RNA species. NS elicits cell-mediated immune responses.

M (or M2)	A basic protein that is an inner component of the (about 20-30 kDa) virion. It is believed to regulate genome RNA transcription. M binds to nucleocapsids and the cytoplasmic domain of G, thereby facilitating the process of budding. Sometimes M is phosphorylated. M is found in the nucleus and inhibits host cell transcription.

LIPIDS

Virions are composed of about 15-25% lipids; their composition reflecting the host cell membrane where virions bud. Generally phospholipids represent about 55-60%, sterols and glycolipids about 35-40%, of the total lipids. G protein has a covalently associated fatty acid proximal to the lipid envelope.

CARBOHYDRATES

Virions are composed of about 3% carbohydrate by weight. The carbohydrates are present as N-linked glycan chains on G protein and as glycolipids. In mammalian cells, the oligosaccharide chains are generally of the complex type, in insect cells they are of the non-complex types.

GENOME ORGANIZATION AND REPLICATION

The virus codes for at least 5 ORFs in the negative-sense genome in the order 3'-N-NS-M-G-L-5' (e.g., for VSIV), or the equivalent. For certain viruses additional genes are interposed. Genes are transcribed processively (from the 3' to 5' of the template virus RNA and in decreasing molar abundances) as 5' capped, 3' polyadenylated and generally monocistronic mRNAs (Fig. 2). Polycistronic mRNAs have been identified for some species. A short uncapped, unpolyadenylated and untranslated "leader" RNA, corresponding to the complement of the 3' terminus of the viral RNA (i.e., preceding the N mRNA), is also transcribed. Unlike mRNA species, it has a 5' triphosphate terminus (Fig. 2). Leader RNA has been identified in the nucleus of infected cells. For individual viruses and for different viruses, the mRNAs generally have common 5' terminal sequences (generally m7Gppp(m)AmA(m)CA..). Intergenic sequences are generally short. In certain cases the 5' end of an mRNA overlaps the 3' end of the preceding gene. The untranslated region following the L gene is longer than the sequence that preceeds N at the other end of the genome.

Virus adsorption is mediated by G protein attachment to cell surface receptors and penetration of the cell is by endocytosis via coated pits. The identities of the receptors are not known. After penetration, the viral envelope is removed by lysosomal activity leading to deposition of the transcriptionally-active nucleocapsid (RNA, N, L, NS) into the cytoplasm. Virus RNA is repetitively transcribed (primary transcription) by the virion transcriptase into capped and polyadenylated mRNAs that, apart from G mRNA, are translated on cytoplasmic polysomes. G mRNA translation occurs on membrane-bound polysomes. Transcription occurs in the presence of protein synthesis inhibitors indicating that it does not depend on *de novo* host protein synthesis. Following translation, RNA replication occurs in the cytoplasm (full-length positive and then full-length negative RNA synthesis) and depends on the prior translation of the viral mRNA species. Certain plant viruses may replicate RNA in the cell nucleus. Replication requires the newly synthesized N, NS (M1) and L protein species and involves the formation of replicative intermediate nucleocapsids. It may require host factors. It has been proposed that binding of N protein to the 5' proximal (encapsidation) sequences of nascent positive- or negative-sense viral RNA species prevents transcription and, by progressive addition of N, promotes replication, including read-through of transcription termination signals. Following replication, further rounds of transcription (secondary transcription), translation and replication ensue.

Post-translational trafficking and modification of G protein involves transportation across the membrane of the endoplasmic reticulum, removal of the amino-proximal signal sequence and step-wise glycosylation in compartments of the Golgi apparatus. Depending on the cell, the G protein may move to the plasma membrane, in particular, to the basolateral surfaces of polarized cells.

Viral nucleocapsid structures are assembled in association with M (M2) and lipid envelopes containing viral G protein. The site of formation of particles depends on the virus and host cell. For vesiculoviruses, lyssaviruses and ephemeroviruses, nucleocapsids are synthesized in the cytoplasm and viruses bud from the plasma membrane in most, but not all cells. Some lyssaviruses bud predominantly from intracytoplasmic membranes and in some cases prominent virus-specific cytoplasmic inclusion bodies containing N protein in infected cells (rabies inclusion bodies are called Negri bodies). Cytorhabdoviruses bud from intracytoplasmic membranes associated with viroplasms. None has been observed to bud from plasma membranes. Nucleorhabdoviruses bud from the inner nuclear membrane and accumulate in the perinuclear space.

Vesiculoviruses can replicate in enucleated cells, indicating that newly synthesized host gene products are not required. Depending on the virus and host cell type, virus infections may inhibit cellular macromolecular syntheses. The mechanism is not known.

Generally, 5 complementation groups of mutants have been defined by using temperature-sensitive mutants. Host range and temperature-sensitive mutants with altered polymerase

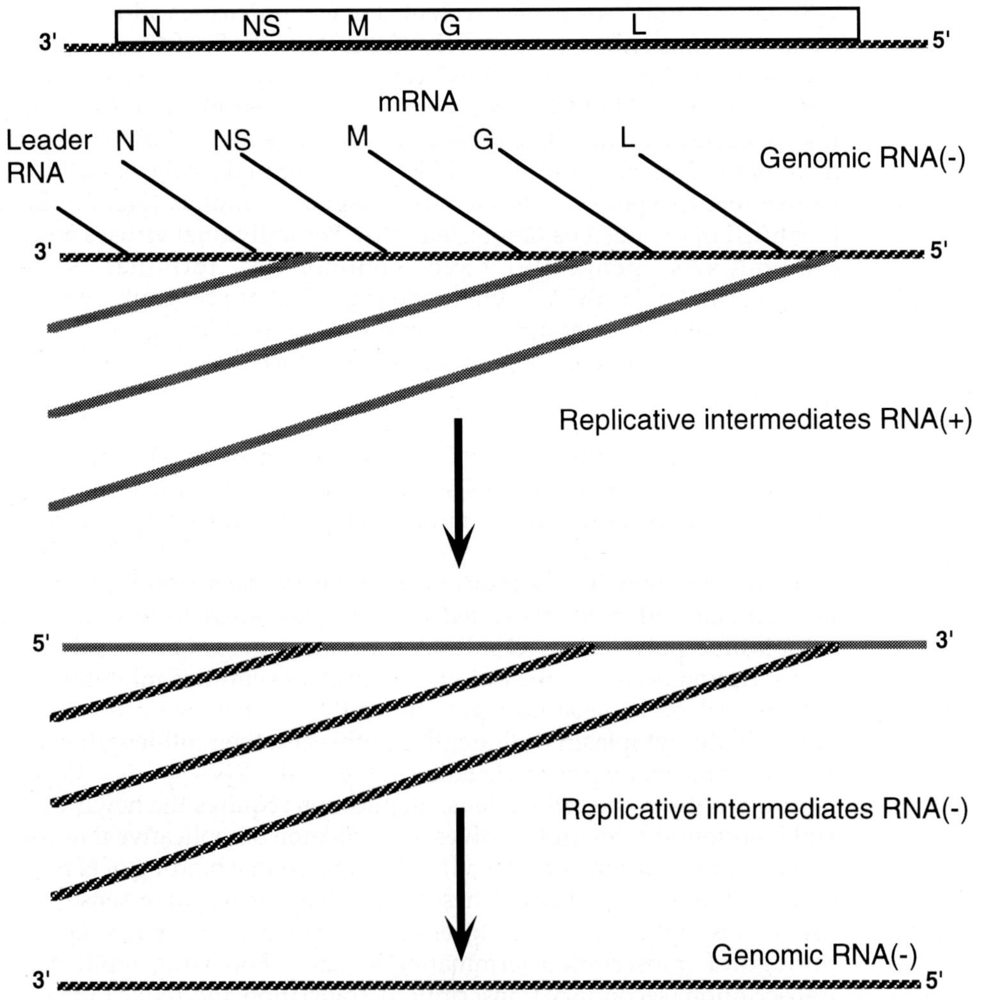

Figure 2: In (upper) is shown the gene order for VSIV; in (lower), which represents the replication cycle, thick lines are replicative intermediate (or genome) RNA-N protein complexes, thin lines are leader RNA, or mRNAs.

functions have also been described. Complementation may occur between related viruses (e.g., between vesiculoviruses), but not between viruses representing distinct genera. Complementation is also reported to occur involving re-utilization of the structural components of UV-irradiated virus (VSIV). Recombination of genes between different virus isolates has not been demonstrated although recombination will occur during the formation of defective RNAs. Phenotypic mixing occurs between some animal rhabdoviruses and other enveloped animal viruses (e.g., paramyxoviruses, orthomyxoviruses, retroviruses, herpesviruses).

ANTIGENIC PROPERTIES

G protein is involved in virus neutralization and defines the virus serotype. N protein is a cross-reacting, complement-fixing (CF) antigen. Weak serological cross-reactions may occur between viruses in different genera. Protection follows vaccination with attenuated viruses, killed viruses, subunits consisting of G protein alone or G protein together with the ribonucleoprotein complex, and expression vectors (e.g., vaccinia virus) that synthesize G and/or N.

BIOLOGICAL PROPERTIES

Some member viruses multiply only in mammals, or fish, or arthropods, or other invertebrates, others have both arthropod and vertebrate hosts (arboviruses), while some members infect plants and certain plant-feeding arthropods. Some of the viruses of vertebrates have a wide experimental host range. A diverse range of vertebrate and invertebrate cells are susceptible to vertebrate rhabdoviruses *in vitro*. The viruses of plant usually have a narrow host range among higher plants; some replicate in insect vectors and grow in insect cell cultures.

Sigma virus was recognized first as a congenital infection of *Drosophila*. No rhabdovirus is transmitted vertically in vertebrates, or plants. Some viruses are transmitted mechanically between plants. Vector transmission may involve mosquitoes, sandflies, mites, culicoides, aphids, lacewings, leafhoppers, or planthoppers (etc.). Some viruses are transmitted mechanically in sap or from the body fluids of infected hosts. Mechanical transmission of viruses infecting vertebrates may be by contact, aerosol, bite, or venereal.

GENUS *VESICULOVIRUS*

Type Species vesicular stomatitis Indiana virus (VSIV)

DISTINGUISHING FEATURES

Vesiculoviruses have 5 major polypeptides (designated L, G, N, NS and M). The 11.2 kb genome includes about 50 nt leader sequence that preceeds N, about 60 nt untranslated region that follows L and intergenic dinucleotides. There is a common (3') AUACUUUUUUU sequence preceeding each intergenic region, and UUGUCNNUAG sequences at the beginning of each gene and following the intergenic sequences that templates the 5' end of the next mRNA species (generally, m7Gppp(m)Am-A(m)CAGNNAUC...). Some viruses (e.g., MEBV, KWAV) are distinctly larger than the type species.

BIOLOGICAL PROPERTIES

Vesiculoviruses have been obtained from a variety of animals, including mammals, fish and invertebrates (insects). Vesicular stomatitis of horses, cattle and swine is one of the oldest known infectious diseases of livestock. It was first recognized as distinct from foot-and-mouth disease early in the nineteenth century. Epidemics of disease occur periodically throughout the Western hemisphere. The disease signs include debilitating lameness in horses and swine and loss of milk production in cattle. VSIV infection of humans (influenza-like symptoms) is common in rural areas where there is animal disease. Certain other vesiculoviruses are recognized as the etiologic agents of disease, including those of human (laboratory infections of VSIV, Piry virus, possibly natural human infections involving Chandipura virus). Several vesiculoviruses infect fish and are responsible for epidemics of disease. Some may be vectored by fish ectoparasites.

TAXONOMIC STRUCTURE OF THE GENUS

The viruses in the *Vesiculovirus* genus exhibit various degrees of cross-neutralization. They cross-react in CF and immunofluorescence tests. Genomic sequence analyses indicate sequence similarities. Higher homologies are observed between the N genes by comparison to the G genes. Apart from the vesicular stomatitis viruses, no serogroups have been established within the genus.

LIST OF SPECIES IN THE GENUS

The viruses, their alternative names ( ), genomic sequence accession numbers [ ] and assigned abbreviations ( ) are:

SPECIES IN THE GENUS

Chandipura virus	[M16608]	(CHPV)
Cocal virus		(COCV)
Isfahan virus		(ISFV)
Maraba virus		(MARAV)
Piry virus	[M14719, M14714, V01208]	(PIRYV)
vesicular stomatitis Alagoas virus		(VSAV)
vesicular stomatitis Indiana virus	[J02428, J02430-2, J02434-8, K00519-20, K01068-70, K01638-9]	(VSIV)
vesicular stomatitis New Jersey virus	[K02379, M35062]	(VSNJV)

TENTATIVE SPECIES IN THE GENUS

BeAn 157575 virus		(BeAnV 157575)
Boteke virus		(BTKV)
Calchaqui virus		(CQIV)
Carajas virus		(CJSV)
eel virus American		(EVA)
Gray Lodge virus		(GLOV)
Jurona virus		(JURV)
Klamath virus		(KLAV)
Kwatta virus		(KWAV)
La Joya virus		(LJV)
Malpais Spring virus		(MSPV)
Mount Elgon bat virus		(MEBV)
Perinet virus		(PERV)
Pike fry rhabdovirus (grass carp rhabdovirus)		(PFRV)
Porton virus		(PORV)
Radi virus		(RADIV)
spring viremia of carp virus	[M35836, K02123]	(SVCV)
Tupaia virus		(TUPV)
ulcerative disease rhabdovirus		(UDRV)
Yug Bogdanovac virus		(YBV)

GENUS *LYSSAVIRUS*

Type Species rabies virus (RABV)

DISTINGUISHING FEATURES

Lyssaviruses such as rabies virus have 5 major polypeptides, designated L (190 kDa), G (65-80 kDa), N (58-62 kDa), M1 (35-40 kDa) and M2 (22-25 kDa). The G protein of rabies virus may be glycosylated at only one or two of the available 3 sites for attachment of N-linked glycans. N and M1 are phosphoproteins, phosphorylation of N may involve a host protein kinase, phosphorylation of M1 probably involves a viral protein kinase (L). The 11.9 kb rabies virus genome includes about 60 nt 3' end sequence that preceeds N, about 70 nt untranslated region that follows L and intergenic di- or pentanucleotides, or a 423 nt spacer (between G and L of the PV rabies virus strain). The lyssaviruses that have been analyzed have intergenic regions that are similar to those identified in vesiculoviruses. Rabies virus characteristically induces the formation of Negri bodies in infected neurons.

BIOLOGICAL PROPERTIES

Rabies is the oldest known disease caused by a rhabdovirus; it is among the most lethal of all infectious diseases. Rabies is enzootic in all regions of the world except Australia and Antarctica. Several island countries (United Kingdom, Ireland, Japan) have remained rabies-free once infected animals were eliminated and strict quarantine and importation regulations were established. Natural animal reservoirs of rabies include many bat species and the skunk, mongoose, raccoon, fox, wolf, jackal, dog (etc.). These animals transmit the disease to other species including livestock, domestic animals and wild-life. Transmission from dogs to human is a major problem in some regions. Transmission usually involves infectious saliva, although other (artificial) forms of transmission have occurred (cornea transplants).

Rabies virus is neurotropic. It multiplies in neurons and myotubes of vertebrates as well as other tissues (e.g., salivary gland). The growth cycle is slow both *in vivo* and *in vitro*. Rabies virus infection does not inhibit cellular macromolecular synthesis.

TAXONOMIC STRUCTURE OF THE GENUS

At present, broadly cross-reacting antigenic sites on the N protein, as recognized by immunofluorescence and complement fixation, determine placement within the *Lyssavirus* genus. More specific antigenic sites on the G protein, as recognized in neutralization tests, determine the placement of a virus isolate as rabies or rabies-related. Cross-neutralization by rabies virus antisera may be moderate (EBV-1, EBV-2, DUVV), to very low (LBV, MOKV), to none (KOTV, OBOV, RBUV). Only one serogroup within the genus has been established. However, the taxonomic significance of the antigenic data is not known. BEFV, which has previously been linked to the lyssaviruses by such data, exhibit greater sequence similarities (albeit distant) to vesiculoviruses than to lyssaviruses. In view of the BEFV results, the postulated assignments of some viruses (KOTV, OBOV, RBUV) to the genus remains to be confirmed. Sequence data are only available for a few lyssaviruses.

LIST OF SPECIES IN THE GENUS

The viruses, their genomic sequence accession numbers [ ] and assigned abbreviations ( ) are:

SPECIES IN THE GENUS

Duvenhage virus	(DUVV)
European bat virus 1	(EBV-1)
European bat virus 2	(EBV-2)
Lagos bat virus	(LBV)

Mokola virus	[D00491, D00492]	(MOKV)
rabies virus	[D10499, D10482, J02293, K02858-69, M12771, M13215, M22013, M31046, M32751, M38452, M61047, M81058-60, X03673, X13357, X55727-29]	(RABV)

TENTATIVE SPECIES IN THE GENUS

Kotonkan virus	(KOTV)
Obodhiang virus	(OBOV)
Rochambeau virus	(RBUV)

GENUS *EPHEMEROVIRUS*

Type Species bovine ephemeral fever virus (BEFV)

DISTINGUISHING FEATURES

BEFV contains at least five structural proteins, designated: L, 180 kDa; G, 81 kDa; N, 52 kDa; M1, 43 kDa; and M2, 29 kDa. The G protein is a virus membrane-associated glycoprotein which contains 5 potential sites for attachment of N-linked glycans and 5 virus-neutralizing antigenic sites. The N protein is phosphorylated. The M2 protein is also phosphorylated in virions. In addition to these proteins, a 90 kDa, non-virion glycoprotein (G_{NS}) has been identified in BEFV-infected mammalian cells. G_{NS} is highly glycosylated (8 potential sites for N-linked glycans). The G and G_{NS} proteins, although not identical, exhibit homologies with each other and to lesser extents with the G proteins of other animal rhabdoviruses. The 14.8 kb negative sense viral RNA genome includes 10 genes in the order (3') N-M1-M2-G-G_{NS}-α_1-α_2 -β-γ-L- (5') and intergenic regions of between 26 and 53 nt. The γ and L genes overlap by 21 nt. Each gene, except $\alpha_{1,}$ is initiated from a UUGUCC sequence (mRNA: 5' cap-AACAGG...) and terminates at a putative polyadenylation site: GNAC(U_{6-7}) 3'. The functions of the α_1, α_2, β, and γ gene products have not been established, at least 2 may be virion components.

The 14.6 kb genome of Adelaide River virus (ARV) contains 9 genes in the negative sense genome in the order (3')-N-M1-M2-G-G_{NS}-α_1-α_2-β-L- (5') and intergenic regions of 1-4 nt. The β and L genes overlap by 22 nt. Each gene is initiated from a viral 3' UUGUC sequence (mRNA: 5' cap-AACAG...), however the putative polyadenylation signals are more variable than those of BEFV and may account for the synthesis of polycistronic mRNAs. The G and G_{NS} genes each encode glycoproteins which share significant amino acid homology with each other and with other rhabdovirus G proteins. The ARV G protein (Mr = 90 x 10^3) contains 6 potential sites for N-linked glycans, the G_{NS} protein 9. Proteins encoded in the ARV $\alpha_{1,}$ α_2 and β genes share homology with the corresponding BEFV proteins, however ARV lacks a γ gene comparable to that of BEFV. Analyses of the amino acid sequences of BEFV and ARV proteins indicate highly significant sequence homologies between most of the corresponding proteins with the higher homologies in the L and N proteins than the G proteins.

Two glycoproteins have been identified in mammalian cells infected with BRMV.

BIOLOGICAL PROPERTIES

Bovine ephemeral fever is an economically important enzootic disease of cattle and water buffalo in most tropical and sub-tropical regions of Africa, Australia, the Middle East and Asia. BEFV infection causes a sudden onset of fever and other clinical signs including lameness, anorexia and ruminal stasis, followed by a sustained drop in milk production. Although the mortality rate is low (1-2%), it is highest in well-conditioned beef cattle and high producing dairy cattle. The virus is transmitted by hematophagous arthropods and has been isolated from both culicoides and mosquitoes.

Other viruses in the genus are not recognized as animal pathogens, but are known to infect cattle and have been isolated from healthy sentinel cattle (ARV, BRMV) or from insects (KIMV, MALV, PUCV).

TAXONOMIC STRUCTURE OF THE GENUS

Viruses in the *Ephemerovirus* genus exhibit low to no cross-neutralization. They cross-react strongly in CF or indirect immunofluorescence tests and may show low level cross-reactions by indirect immuno-fluorescence with viruses of the genus *Lyssavirus*. No serogroups within the genus have been established. Sequence comparisons with other rhabdoviruses indicate that in evolutionary terms the ephemeroviruses are closer to members of the genus *Vesiculovirus* than to those of other defined genera in the family.

LIST OF SPECIES IN THE GENUS

The viruses, their genomic sequence accession numbers [ ] and assigned abbreviations ( ) are:

SPECIES IN THE GENUS

Adelaide River virus	[L09206, L09208]	(ARV)
Berrimah virus		(BRMV)
bovine ephemeral fever virus	[M94266]	(BEFV)

TENTATIVE SPECIES IN THE GENUS

Kimberley virus	(KIMV)
Malakal virus	(MALV)
Puchong virus	(PUCV)

GENUS *CYTORHABDOVIRUS*

Type Species lettuce necrotic yellows virus (LNYV)

In addition to unassigned viruses, two genera of plant rhabdoviruses have been established. The viruses are primarily distinguished on the basis of the sites of virus maturation (cytoplasm: *Cytorhabdovirus*; nucleus: *Nucleorhabdovirus*). However, exceptions exist and the significance of this property is not known. The interrelationships of the different plant viruses within or between the two genera or with the unassigned plant viruses have yet to be established at the genetic level. A wide variety of plants are susceptible to plant rhabdoviruses although each virus usually has a restricted host range. Most of the plant rhabdoviruses are transmitted by leafhoppers, planthoppers, or aphids, although mite- and lacebug-transmitted viruses (one each) have also been identified. Some viruses are transmitted in contaminated sap. In all carefully examined cases, viruses have been shown to replicate in the insect vector as well as in the plant host.

DISTINGUISHING FEATURES

Cytorhabdoviruses replicate in the cytoplasm of infected cells in association with masses of thread-like structures (viroplasms). Virus morphogenesis occurs in association with vesicles of the endoplasmic reticulum. A nuclear phase has been suggested but not proven in the replication of some cytorhabdoviruses, e.g., LNYV. Evidence of the nuclear involvement in the replication of others is lacking (e.g. BYSMV). Information on the genome structure of the cytorhabdoviruses is limited (see nucleorhabdoviruses).

TAXONOMIC STRUCTURE OF THE GENUS

The viruses have not been assigned to groups.

LIST OF SPECIES IN THE GENUS

The viruses, their vector { }, CMI/AAB description # ( ) and assigned abbreviations ( ) are:

SPECIES IN THE GENUS

barley yellow striate mosaic virus {leafhopper} (312)	(BYSMV)
broccoli necrotic yellows virus {aphid} (85)	(BNYV)
Festuca leaf streak virus	(FLSV)
lettuce necrotic yellows virus {aphid} (26, 243)	(LNYV)
Northern cereal mosaic virus {leafhopper} (322)	(NCMV)
Sonchus virus	(SonV)
strawberry crinkle virus {aphid} (163)	(SCV)
wheat American striate mosaic virus {leafhopper} (99)	(WASMV)

TENTATIVE SPECIES IN THE GENUS

None reported.

GENUS *NUCLEORHABDOVIRUS*

Type Species potato yellow dwarf virus (PYDV)

DISTINGUISHING FEATURES

Nucleorhabdoviruses multiply in the nucleus of plants forming large granular inclusions that are thought to be sites of virus replication. Viral proteins are synthesized from discrete polyadenylated mRNAs and accumulate in the nucleus. Virus morphogenesis occurs at the inner nuclear envelope and enveloped virus particles accumulate in perinuclear spaces. In protoplasts treated with tunicamycin, morphogenesis is interrupted and nucleocapsids accumulate in the nucleoplasm. The genome of SYNV virus is about 13.7 kb. Preceded by a non-coding 144 nt leader sequence, the gene order is (3') N-M2-SC4-M1-G-L (5'). N represents the 54 kDa viral nucleocapsid, M2 is probably a 38 kDa phosphoprotein, SC4 is probably a non-structural protein, M1 a 32 kDa matrix protein, G a 70 kDa glycoprotein (unglycosylated form) and L the 241 kDa polymerase. The intergenic regions are similar in length and have sequence relatedness to those of other rhabdoviruses. The 5' non-coding (trailer) region is 162 nt long with extensive complementarity to the leader sequence.

TAXONOMIC STRUCTURE OF THE GENUS

The viruses have not been assigned to serogroups or other taxonomic groupings.

LIST OF SPECIES IN THE GENUS

The viruses, their alternative names ( ), vector { }, genomic sequence accession numbers [ ], CMI/AAB description # ( ) and assigned abbreviations ( ) are:

SPECIES IN THE GENUS

datura yellow vein virus		(DYVV)
eggplant mottled dwarf virus (115)		(EMDV)
(Pittosporum vein yellowingvirus)		(PVYV)
(tomato vein yellowing virus)		(TVYV)
maize mosaic virus {leafhopper} (94)		(MMV)
potato yellow dwarf virus {leafhopper} (35)		(PYDV)
Sonchus yellow net virus {aphid} (205)	[M13950, M17210, M23023, M35689, M73626, M87829]	(SYNV)
sowthistle yellow vein virus {aphid} (62)		(SYVV)

TENTATIVE SPECIES IN THE GENUS

None reported.

LIST OF UNASSIGNED SPECIES IN THE FAMILY (OTHER THAN PLANT VIRUSES)

There are at least six serogroups of rhabdoviruses that infect animals that have not been assigned to an existing genus and there are a number of ungrouped viruses. IHN disease of salmonids and VHS disease of trout cause epidemics involving high mortalities in young fish in North America, Europe, and Japan. IHNV contains a large G-L intergenic region. IHNV also encodes a 6th protein, designated NV (non-viral). Its function is not known. Sigma virus is transmitted vertically through the germinal cells of *Drosophila* species and confers CO_2-sensitivity to infected insects. Both host and viral genes contribute to the maintenance of the virus in the host. Sigma encodes a 6th gene located between the P and M genes. The function of this gene is not known. The intergenic regions of the virus are variable (up to 36 nt in length) and one gene (M) overlaps that of the following gene (G). For most of the other listed viruses, no biochemical characterization has been reported. Their assignment to the family relies on the distinctive morphology of rhabdoviruses.

The groups and viruses, their alternative names ( ), vector { }, genomic sequence accession numbers [ ] and assigned abbreviations ( ) are:

1-Bahia Grande group:	
Bahia Grande virus	(BGV)
Muir Springs virus	(MSV)
Reed Ranch virus	(RRV)
2-Hart Park group:	
Flanders virus	(FLAV)
Hart Park virus	(HPV)
Kamese virus	(KAMV)
Mosqueiro virus	(MQOV)
Mossuril virus	(MOSV)
3-Kern Canyon group:	
Barur virus	(BARV)
Fukuoka virus	(FUKAV)
Kern Canyon virus	(KCV)
Nkolbisson virus	(NKOV)
4-Le Dantec group:	
Le Dantec virus	(LDV)
Keuraliba virus	(KEUV)
5-Sawgrass group:	
Connecticut virus	(CNTV)
New Minto virus	(NMV)
Sawgrass virus	(SAWV)
6-Timbo group:	
Chaco virus	(CHOV)
Sena Madureira virus	(SMV)
Timbo virus	(TIMV)

LIST OF UNASSIGNED VERTEBRATE RHABDOVIRUSES

Almpiwar virus	(ALMV)
Aruac virus	(ARUV)
Bangoran virus	(BGNV)
Bimbo virus	(BBOV)
Bivens Arm virus	(BAV)
blue crab virus	(BCV)
Charleville virus	(CHVV)

Coastal Plains virus		(CPV)
DakArK 7292 virus		
eel virus B12		(EV-B12)
Entamoeba virus		(ENTV)
Garba virus		(GARV)
Gossas virus		(GOSV)
Hirame rhabdovirus		(HIRRV)
Humpty Doo virus		(HDOOV)
infectious hematopoietic necrosis virus	[J04321, M16023]	(IHNV)
Joinjakaka virus		(JOIV)
Kannamangalam virus		(KANV)
Kolongo virus		(KOLV)
Koolpinyah virus		(KOOLV)
Landjia virus		(LJAV)
Manitoba virus		(MNTBV)
Marco virus		(MCOV)
Navarro virus		(NAVV)
Nasoule virus		(NASV)
Ngaingan virus		(NGAV)
Oak-Vale virus		(OVRV)
Oita virus		(OITAV)
Ouango virus		(OUAV)
Parry Creek virus		(PCRV)
Rio Grande cichlid virus		(RGRCV)
Sandjimba virus		(SJAV)
Sigma virus	[X06171]	(SIGMAV)
snakehead rhabdovirus		(SHRV)
Sripur virus		(SRIV)
Sweetwater Branch virus		(SWBV)
Tibrogargan virus		(TIBV)
viral hemorrhagic septicemia virus	[D00687, X59241]	(VHSV)
(Egtved virus)		
(Atlantic cod ulcus syndrome virus)		
(salmonis virus)		
Xiburema virus		(XIBV)
Yata virus		(YATAV)

LIST OF UNASSIGNED PLANT RHABDOVIRUSES

There are many plant rhabdoviruses that have not been assigned to a genus. Their assignment to the family relies on the distinctive morphology of rhabdoviruses. Some have been transmitted experimentally. However, none has been characterized physicochemically.

The viruses, their alternative names ( ), vector { }, CMI/AAB description # ( ), and assigned abbreviations ( ) are:

Atropa belladonna virus	(AtBV)
beet leaf curl virus {lacewing} (268)	(BLCV)
Callistephus chinensis chlorosis virus	(CCCV)
carnation bacilliform virus	(CBV)
carrot latent virus {aphid}	(CLV)
cassava symptomless virus	(CasSV)
cereal chlorotic mottle virus {leafhopper} (251)	(CCMV)
chrysanthemum frutescens virus	(CFV)
chrysanthemum vein chlorosis virus	(CVCV)
clover enation virus	(CLOEV)

coffee ringspot virus {mite}	(CoRSV)
colocasia bobone disease virus {leafhopper}	(CBDV)
coriander feathery red vein virus {aphid}	(CFRVV)
cow parsnip mosaic virus	(CPMV)
Cynara virus	(CyV)
Digitaria striate virus {leafhopper}	(DSV)
Euonymus fasciation virus	(EFV)
finger millet mosaic virus {leafhopper}	(FMMV)
gerbera symptomless virus	(GRBSV)
Gomphrena virus	(GoV)
Holcus lanatus yellowing virus	(HLYV)
Iris germanica leaf stripe virus	(IGLSV)
ivy vein clearing virus	(IVCV)
Laelia red leafspot virus	(LRLV)
Launea arborescens stunt virus	(LASV)
lemon scented thyme leaf chlorosis virus	(LSTCV)
Lolium ryegrass virus	(LoRV)
lotus stem necrosis	(LoSNV)
lucerne enation virus {aphid}	(LEV)
lupin yellow vein virus	(LYVV)
Malva silvestris virus	(MaSV)
maize sterile stunt virus {leafhopper}	(MSSV)
Melilotus latent virus	(MeLV)
melon variegation virus	(MVV)
oat striate mosaic virus {leafhopper}	(OSMV)
orchid fleck virus	(OFV)
parsley virus	(PaV)
pelargonium vein clearing virus	(PVCV)
pigeon pea proliferation virus	(PPPV)
pineapple chlorotic leaf streak virus	(PCLSV)
Pisum virus	(PiV)
plantain mottle virus	(PlMV)
Ranunculus repens symptomless virus	(RaRSV)
Raphanus virus	(RaV)
raspberry vein chlorosis virus {aphid} (174)	(RVCV)
red clover mosaic virus	(RClMV)
rice transitory yellowing virus {leafhopper} (100)	(RTYV)
Sainpaulia leaf necrosis virus	(SLNV)
Sambucus vein clearing virus	(SVCV)
Sarracenia purpurea virus	(SPV)
sorghum virus {leafhopper}	(SSV)
soursop yellow blotch virus	(SYBV)
Triticum aestivum chlorotic spot virus	(TACSV)
Vigna sinensis mosaic virus	(VSMV)
winter wheat Russian mosaic virus {leafhopper}	(WWMV)
wheat chlorotic streak virus {leafhopper}	WCSV)
wheat rosette stunt virus {leafhopper}	(WRSV)
Zea mays virus	(ZMV)

Non-enveloped particles considered as possible members of the family are:

citrus leprosis virus	(CiLV)
Dendrobium leaf streak virus	(DLSV)
Phalaenopsis chlorotic spot virus	(PCSV)

SIMILARITY WITH OTHER TAXA

Rhabdoviruses share several features with viruses of the *Filoviridae* and *Paramyxoviridae* families. Features they have in common include the non-segmeted negative-sense, single-strand, non-infectious RNA genome, the helical nucleocapsid, the initiation of primary transcription by a virion-associated RNA-dependent RNA polymerase, similar gene order, and single 3' promoter with short terminal untranscribed regions and intergenic regions. The virions are large enveloped structures with a prominant fringe of spikes. they replicate in the cytoplasm and mature by budding, predominantly from the plasma membrane with the exception of rabies virus which buds occasionally from internal membranes and plant rhabdoviruses of the *Nucleorhabdovirus* genus which bud from the inner nuclear membrane. They transcribe discrete unprocessed messenger RNAs.

DERIVATION OF NAMES

cyto: from Greek *kytos*, "cell"
ephemero: from Greek ephemeros, "short-lived"
lyssa: from Greek *lyssa* "rage, fury, canine madness"
nucleo: from Latin *nux*, *nucis*, "nut"
rhabdo: from Greek *rhabdos*, "rod"
vesiculo: from Latin *vesicula*, diminutive of *vesica*, "bladder, blister"

REFERENCES

Banerjee AK, Barik S (1992) Gene expression of vesicular stomatitis virus genome. Virology 188: 417-429

Bilsel PA, Nichol ST (1990) Polymerase errors accumulating during natural evolution of the glycoprotein genes of vesicular stomatitis virus Indiana serotype isolates. J Virol 64: 4873-4883

Calisher CH, Karabatsos N, Zeller H, Digoutte J-P, Tesh RB, Shope RE, Travassos da Rosa APA, St George TD (1989) Antigenic relationships among rhabdoviruses from vertebrates and hematophagous arthropods. Intervirology 30: 241-257

Choi T-J, Kuwata S, Koonin EV, Heaton LA, Jackson AO (1992) Structure of the L (polymerase) protein gene of Sonchus yellow net virus. Virology 189: 31-39

Dietzgen RG, Hunter BG, Francki RIB, Jackson AO (1989) Cloning of lettuce necrotic yellows virus RNA and identification of virus-specific polyadenylated RNAs in infected *Nicotiana glutinosa* leaves. J Gen Virol 70: 2299-2307

Francki RIB, Milne RG, Hatta T (eds) (1985) An Atlas of Plant Viruses Vol 1. CRC Press Inc, Boca Raton FL, pp 73-100

Frerichs GN (1989) Rhabdoviruses of fishes. In: Ahne W, Kurstak E (eds) Viruses of Lower Vertebrates. Springer-Verlag, New York, pp 316-332

Goldberg K-B, Modrell B, Hillman BI, Heaton LA, Choi T-J, Jackson AO (1991) Structure of the glycoprotein gene of Sonchus yellow net virus, a plant rhabdovirus. Virology 185: 32-38

Heaton LA, Hillman BI, Hunter BG, Zuidema D, Jackson AO (1989) Physical map of the genome of sonchus yellow net virus, a plant rhabdovirus with six genes and conserved junction sequences. Proc Natl Acad Sci USA 86: 8665-8668

Kurath G, Ahern KG, Pearson GC, Leong JC (1985) Molecular cloning of the six mRNA species of infectious hematopoietic necrosis virus, a fish rhabdovirus, and gene order determination by R-loop mapping. J Virol 53: 469-476

Tordo N, Poch O, Ermine A, Keith G, Rougeon F (1988) Completion of the rabies virus genome sequence determination: highly conserved domains among the L (polymerase) proteins of unsegmented negative-strand RNA viruses. Virology 165: 565-576

Wagner RR (ed) (1987) The Rhabdoviruses. Plenum Press, New York

Walker PJ, Byrne KA, Cybinski DH, Doolan DL, Young W (1991) Proteins of bovine ephemeral fever virus. J Gen Virol 72: 67-74

Walker PJ, Byrne KA, Riding GA, Cowley JA, Wang Y, McWilliams S (1992) The genome of bovine ephemeral fever rhabdovirus contains two related glycoprotein genes. Virology 191: 49-61

Wunner WH (1990) The chemical composition and molecular structure of rabies viruses. In: Baer GE (ed) Natural History of Rabies. CRC Press Inc, Boca Raton FL, pp 31-67

CONTRIBUTED BY

Wunner WH, Calisher CH, Dietzgen RG, Jackson AO, Kitajima EW, Lafon M, Leong JC, Nichol S, Peters D, Smith JS, Walker PJ

FAMILY *FILOVIRIDAE*

TAXONOMIC STRUCTURE OF THE FAMILY

Family *Filoviridae*
Genus *Filovirus*

GENUS *FILOVIRUS*

Type Species Marburg virus (MBGV)

VIRION PROPERTIES

MORPHOLOGY

Viruses are enveloped and pleomorphic, appearing bacilliform, or filamentous (sometimes with extensive branching), or U-shaped, 6-shaped, or circular. Particles vary greatly in length (up to 14,000 nm), but have a uniform diameter, about 80 nm. There are surface projections, about 7 nm in length, spaced at 10 nm intervals. Virions recovered by gradient centrifugation are infectious, generally uniformly bacilliform, (Ebola virus: about 1000 nm; Marburg virus: about 800 nm). Inside the envelope is the virus nucleocapsid. The nucleocapsid has a central axis (about 20 nm in diameter) surrounded by a helical nucleocapsid (about 50 nm in diameter) with cross-striations exhibiting a periodicity of about 5 nm.

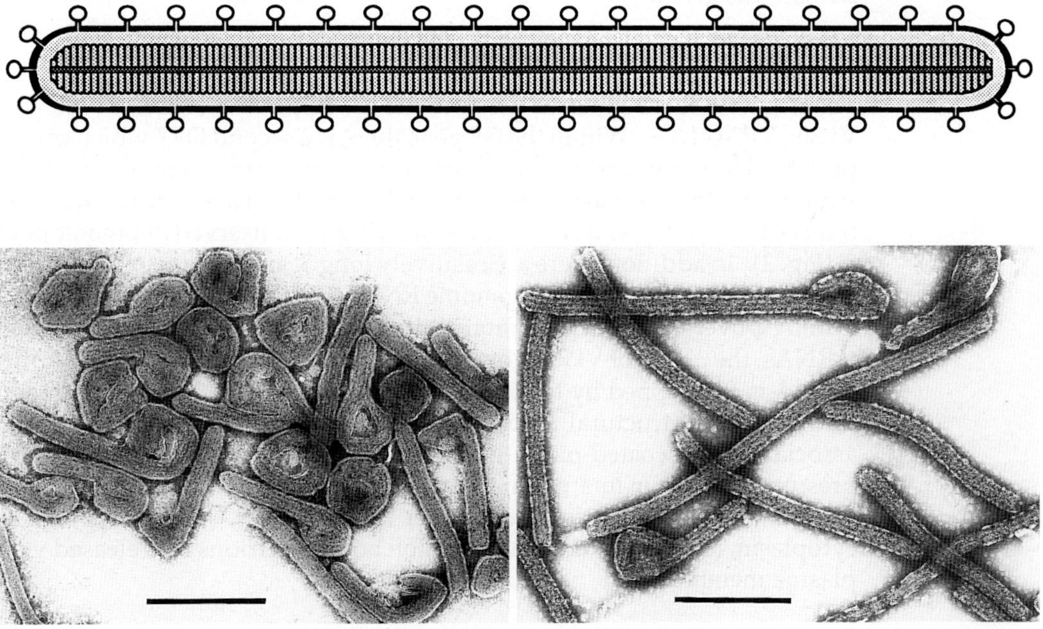

Figure 1: (upper) Schematic of virion in cross-section (not to scale); (lower left) negative contrast electron micrograph of torus and 6-shaped Marburg virus stained with 1% phosphotungstate; (lower right) filamentous forms of Ebola (Reston) virus. The bar represents 500 nm.

PHYSICOCHEMICAL AND PHYSICAL PROPERTIES

Virion Mr is 4.2 x 10^6. The S_{20w} of bacilliform particles is 1,400, for long particles it is very high. The buoyant density is about 1.14 g/cm^3 in potassium tartrate. In CsCl the nucleocapsid has a buoyant density of about 1.32 g/cm^3. Virus infectivity is stable at less than 20° C but not at 60° C. Infectivity is sensitive to lipid solvents, β-propiolactone, formaldehyde, hypochlorite, quarternary ammonium and phenolic disinfectants, or ultraviolet, or gamma irradiation.

Nucleic Acid

Virions contain a non-segmented, negative stranded ssRNA 19.1 kb in size, with complementary end sequences. RNA represents about 1% of the particle mass.

Proteins

Virions contain seven proteins. The sizes estimated from cloned genes (and observed on SDS-PAGE) and functions for Marburg virus are: 267 kDa (180 kDa) L protein that is an RNA transcriptase-polymerase; the 75 kDa (170 kDa) surface glycoprotein (GP) that exists in the form of trimers; the 78 kDa (96 kDa) nucleoprotein (NP); the 32 kDa (38 kDa) matrix or membrane-associated VP-40 protein; the 31 kDa (32 kDa) VP-35 P protein that may be a transcriptase-polymerase component; the 32 kDa (28 kDa) minor nucleoprotein VP-30; and the 29 kDa (24 kDa) second matrix or membrane-associated VP-24 protein. The sizes of the Ebola virus proteins are generally comparable. The nucleocapsid is composed of RNA, L, NP, VP35 and VP30.

Lipids

The viral envelope is derived from host cell membranes and is considered to have a lipid composition similar to that of the plasma membrane.

Carbohydrates

The glycoprotein has N-linked glycans of the complex, hybrid and oligomannosidic type. In addition there are O-linked glycans of the neutral mucin type. The glycans constitute about 50% of the GP mass. Marburg virus glycans lack sialic acids. These are, however, present on Ebola virus glycans.

Genome Organization and Replication

The negative sense filovirus genome has 7 ORFs in the order: 3' -NP - VP35 - VP40 - GP - VP30 - VP24 - L - 5'. Within the GP gene there is a second ORF which could code for a 15 kDa protein. However, besides the seven known structural proteins no other structural or non-structural proteins have been detected. At the gene boundaries there are conserved transcriptional stop and start signals and a highly conserved intergenic pentamer 3' UAAUU 5' (Fig. 2). In addition, there are relatively long 3' and 5' non-coding regions for the mRNAs and the end sequences of the genome RNA. The non-coding 3' end of Marburg virus VP30 mRNA is overlapped by the 5' non-coding end of the VP24 mRNA. Similarly for Ebola virus mRNAs, the 3' end of VP35 is overlapped by the 5' of the VP40 mRNA, the 3' end of the GP mRNA is overlapped by the 5' of VP30 and the 3' of VP24 is overlapped by the 5' of the L mRNA. Ultrastructural studies suggest that at the initiation of infection, virions are associated with coated pits suggesting that they enter cells by endocytosis. Uncoating is presumed to occur in a manner analogous to that of other negative sense RNA viruses. Virus-induced mRNA is abundant in infected cells. Nucleocapsids accumulate in the cytoplasm, forming prominent inclusion bodies. Virions are released via budding through plasma membranes.

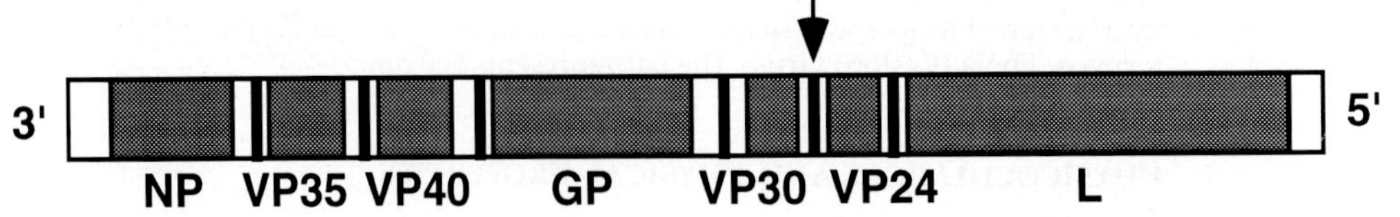

Figure 2: Schematic of the gene organization of the negative sense 19.1 kb Marburg virus RNA. Genes are indicated as shaded boxes, non-coding regions as unshaded areas, conserved intergenic sequences are indicated as heavy vertical lines. The position of the mRNA overlap between VP30 and VP24 is indicated by an arrow.

ANTIGENIC PROPERTIES

Virus infectivity is poorly neutralized *in vitro*. Neutralizing antibody can only be detected in a virus dilution:constant serum format, using serum diluted <1:10. Using this kind of test, there is little antigenic cross-reaction between Marburg and Ebola viruses. Three Ebola virus serotypes, Zaire, Sudan and Reston, can be differentiated antigenically. GP protein epitopes is believed to define the virus serotype.

BIOLOGICAL PROPERTIES

Both Marburg and Ebola viruses are associated with parts of Africa. Some strains cause severe hemorrhagic fever in humans. Marburg virus was first isolated from hemorrhagic fever patients in West Germany and Yugoslavia in 1967 infected by contact with tissues and blood from infected, but apparently healthy, monkeys (*Ceriopithecus aethiops*) imported from Uganda. A second small outbreak of Marburg hemorrhagic fever occurred in South Africa in 1975, and isolated episodes have occurred subsequently in Africa in 1980 and 1987. Overall, Marburg virus mortality rates in humans are reported to be about 30-35%. Ebola virus was first isolated from two separate outbreaks in northern Zaire and southern Sudan in 1976. The estimated case:fatality rates were 88% in Zaire and 53% in the Sudan, with few identified subclinical infections. More recently, Ebola-Reston virus was isolated from cynomolgus monkeys imported from the Philippines into the United States in 1989-1990, and from monkeys at an export facility located in the Philippines. Further isolates have been made from exported Asian monkeys in 1992. While associated with high lethality for naturally and experimentally infected monkeys, Ebola-Reston virus may be less virulent for humans, having infected four animal caretakers without producing serious disease.

The natural reservoir and natural history of filoviruses are unknown. In the laboratory, monkeys, mice, guinea pigs and hamsters have been infected experimentally. The usual pattern seen with large outbreaks of disease in man begins with a focus of infection that disseminates to a number of patients. Secondary and subsequent episodes of disease occur following close contact with patients; such infections usually occur in family members or medical personnel. The major route of interhuman transmission of the virus requires direct contact with blood or body fluids, although droplet and aerosol infections may occur. Transmission of Ebola-Reston virus in colonized monkeys is thought to be similar. Filoviruses have a tropism for cells of the reticulendothelial system, fibroblasts, and interstitial tissues, especially the liver parenchyma. The viruses become distributed in all tissues of the body with high concentrations in the liver, kidney, spleen, and lung. Activation of the clotting cascade with hemorrhagic diathesis and fibrinolysis occurs to varying degrees depending on the virus strain.

LIST OF SPECIES IN THE GENUS

The viruses, their genomic sequence accession numbers [ ] and assigned abbreviations ( ) are:

SPECIES IN THE GENUS

Ebola virus Reston		(EBOV-R)
Ebola virus Sudan		(EBOV-S)
Ebola virus Zaire	[J04337]	(EBOV-Z)
Marburg virus (strain Musoke)	[Z12132]	(MBGV)

TENTATIVE SPECIES IN THE GENUS

None reported.

UNASSIGNED VIRUSES IN THE FAMILY

None reported.

Similarity with Other Taxa

Comparison of filovirus genomes with other non-segmented, negative stranded viruses suggest comparable mechanisms of transcription and translation and a common evolutionary lineage. Sequence analysis of single genes indicate that filoviruses are phylogenetically quite distinct from other families of the order *Mononegavirales*. They are most closely related to the paramyxoviruses, particularly human respiratory syncytial virus.

Derivation of Names

filo: from Latin *filo*, "thread-like", to represent the morphology of virus particles

References

Elliott LH, Kiley MP, McCormick JB (1985) Descriptive analysis of Ebola virus proteins. Virology 147: 169-176

Feldmann H, Muhlberger E, Randolf A, Will C, Kiley MP, Sanchez A, Klenk H-D (1992) Marburg virus, a filovirus: messenger RNAs, gene order, and regulatory elements of the replication cycle. Virus Res 24: 1-19

Geyer H, Will C, Feldmann H, Klenk H-D, Geyer R (1992) Carbohydrate structure of Marburg glycoprotein. Glycobiol 2: 299-312

Jahrling PB, Geisbert TW, Dalgard DW, Johnson TG, Ksiazek TG, Hall WC, Peters CJ (1990) Preliminary report: isolation of Ebola virus from monkeys imported to the USA. Lancet 335: 502-505

Kiley MP, Cox NJ, Elliott LH, Sanchez A, DeFries R, Buchmeier MJ, Richman DD, McCormick JB (1988) Physicochemical properties of Marburg virus: evidence for three distinct virus strains and their relationship to Ebola. J Gen Virol 69: 1957-1967

Muhlberger E, Sanchez A, Randolf A, Will C, Kiley MP, Klenk H-D, Feldmann H (1992) The nucleotide sequence of the L gene of Marburg virus, a filovirus: homologies with paramyxoviruses and rhabdoviruses. Virology 187: 534-547

Richman DD, Cleveland PH, McCormick JB, Johnson KM (1983) Antigenic analysis of strains of Ebola virus: identification of two Ebola virus serotypes. J Infect Dis 147: 268-271

Sanchez A, Kiley MP (1987) Identification and analysis of Ebola virus messenger RNA. Virology 157: 414-420

Sanchez A, Kiley MP, Klenk H-D, Feldmann H (1992) Sequence analysis of the Marburg virus nucleoprotein gene: comparison to Ebola virus and other non-segmented negative strand RNA viruses. J Gen Virol 73: 347-357

Contributed By

Jahrling PB, Kiley MP, Klenk H-D, Peters CJ, Sanchez A, Swanepoel R

FAMILY *ORTHOMYXOVIRIDAE*

TAXONOMIC STRUCTURE OF THE FAMILY

Family	***Orthomyxoviridae***
Genus	*Influenzavirus A,B*
Genus	*Influenzavirus C*
Genus	"Thogoto-like viruses"

VIRION PROPERTIES

MORPHOLOGY

Virions are spherical or pleomorphic, and 80-120 nm in diameter. Filamentous forms several micrometers in length also occur. The virion envelope is derived from cell membrane lipids, incorporating variable numbers of virus glycoproteins (1-3) and non-glycosylated proteins (1-2). Virion surface glycoprotein projections are 10-14 nm in length and 4-6 nm in diameter. The viral nucleocapsid is segmented, has helical symmetry and consists of different size classes, 150-130 nm in length, with loops at one end.

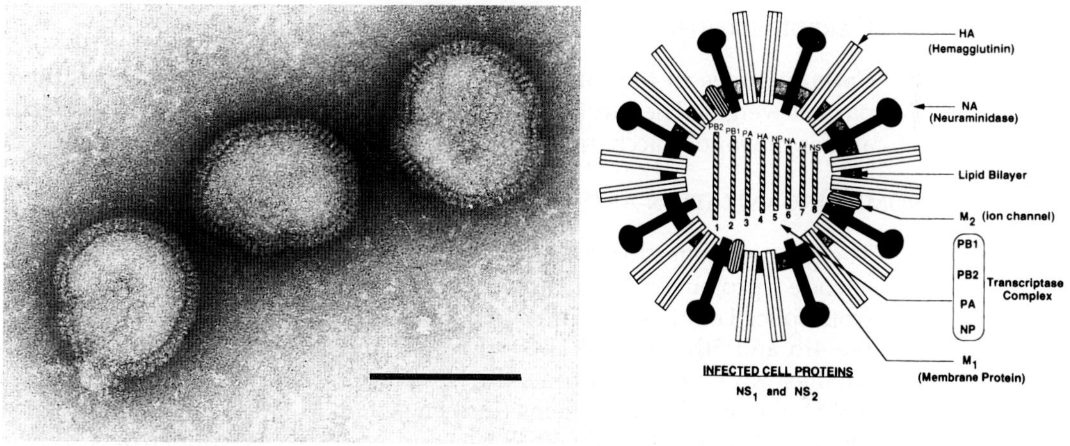

Figure 1: (left) Negative contrast electron micrograph of 3 influenza virus particles; the bar represents 100 nm. (right) Diagram of an influenza A virion in section. The indicated glycoproteins embedded in the lipid membrane are the trimeric hemagglutinin (HA) and the tetrameric neuraminidase (NA). HA predominates. A small number of the membrane ion channel protein M2 is also present in the envelope. The internal components are the M_1 membrane (matrix) protein and the viral ribonucleoprotein (RNP) consisting of RNA segments, associated nucleocapsid protein, NP, and the PA, PB_1 and PB_2 polymerase proteins.

PHYSICOCHEMICAL AND PHYSICAL PROPERTIES

Virion Mr is 250 x 10^6. Virion buoyant density in aqueous sucrose is 1.19 g/cm^3. S_{20w} of non-filamentous particles is 700-800 g/cm^3. Virions are very sensitive to heat, lipid solvents, non-ionic detergents, formaldehyde, irradiation or oxidizing agents.

NUCLEIC ACID

Depending on the virus (genus), virions contain a precise number of segments of linear, negative sense ssRNA (8 segments: influenza A and B viruses; 7 segments: influenza C virus; 6 segments: Thogoto virus; possibly 7 segments: Dhori virus). Segment lengths range from 900 to 2350 nt. The size of the genome ranges from 10.0 - 13.6 kb. Defective (shorter, occasionally chimeric) viral RNAs may occur. Depending on the genus, RNAs possess conserved and partially complementary 5' and 3' end sequences.

PROTEINS

Structural proteins common to all genera include: three polymerase proteins (P, e.g., PA, PB1, PB2 in influenza A), a nucleocapsid protein (NP, a group-specific protein that is phosphorylated and is associated with each genome ssRNA segment in the form of a ribonucleoprotein), a hemagglutinin (HA, HEF, or GP, that is an integral, type I membrane glycoprotein and is involved in virus attachment and envelope fusion), and a non-glycosylated membrane or matrix protein (M_1 or M). The HA of influenza A is acylated at the membrane-spanning region and has N-linked glycans at a number of sites. In addition to its hemagglutinating and fusion properties the HEF protein of influenza C viruses has esterase activity that functions as a receptor destroying enzyme. Depending on the virus (genus) other virion proteins may include an integral, type II envelope glycoprotein (neuraminidase, NA), and an integral, type III membrane protein (M_2 or NB, that may be glycosylated, and may function as an ion channel). In addition to the structural proteins, and depending on the virus (genus), viruses may code for 2 nonstructural proteins (NS_1, NS_2). Virion enzymes (variously represented and reported among genera) include a transcriptase (PB1 in influenza A), an endonuclease (PB2 in influenza A), and a receptor-destroying enzyme (neuraminidase or 9-0-acetyl-neuraminyl esterase in the case of the influenza C HEF protein).

Lipids

Lipids in the virion envelope constitute about 18-37% of the particle weight. They resemble lipids of the host cell plasma membrane.

Carbohydrates

Carbohydrates in the form of glycoproteins and glycolipids constitute about 5% of the particle weight. They are present as N-glycosidic side chains of glycoproteins, as glycolipids, and as mucopolysaccharides. Their composition is host- and virus-dependent.

Genome Organization and Replication

The genome codes for up to 10 proteins (Mr 14-76 x 10^3). The 5 largest genome segments encode 1 protein each, whereas some of the smaller segments code for additional proteins from spliced or bicistronic mRNAs. Generally the three largest RNAs encode the P proteins, the 4th and 5th the viral HA (HEF, GP) and NP proteins. Depending on the virus, the smallest RNA species encode: the NA protein (influenza A NA, influenza B NA, NB: 6th RNA), the membrane proteins (influenza A, B M_1, M_2: 7th RNA; influenza C M: 6th RNA; Dhori (?Thogoto) M_1, M_2: 6th RNA) and NS proteins (influenza A, B NS_1, NS_2: 8th RNA; influenza C NS_1, NS_2: 7th RNA; putative Dhori 7th RNA: unknown). Virus entry involves the virus HA (HEF, GP) and occurs by receptor-mediated endocytosis. The receptor determinant of the influenza viruses is sialic acid bound to glycoproteins or glycolipids. In endosomes low pH-dependent fusion occurs between viral and cell membranes. For influenza viruses, fusion depends on a cleaved virion HA (influenza A, B: HA_1, HA_2; influenza C: HEF_1, HEF_2). No requirement for glycoprotein cleavage has been demonstrated for the GP species of Thogoto virus.

Viral nucleocapsids are transported to the cell nucleus where the virion transcriptase complex synthesizes mRNA species. mRNA synthesis is primed by capped RNA fragments about 8-15 nucleotides in length that are generated from host heterogenous nuclear RNA species by the viral endonuclease activity that is associated with one of the P proteins. Virus-specific mRNA synthesis is inhibited by actinomycin D or α-amanitin due to inhibition of host DNA-dependent RNA transcription and a (presumed) lack of newly synthesized substrates that allow the viral endonuclease to generate the required primers. Virus-specific mRNA species are polyadenylated at the 3'-termini, and lack sequences corresponding to the 5'-terminal (about 16) nucleotides of the viral RNA segment. Certain mRNAs are spliced to provide alternative products (e.g., M_2 of influenza A derives from a spliced mRNA that otherwise is translated to form M_1; likewise NS_2 derives from a spliced mRNA that otherwise encodes NS_1). Other mRNAs may be bicistronic (e.g., the influenza B virus NA and NB are encoded in overlapping ORFs that are translated from segment 6 mRNA; by contrast, influenza B M2 is derived from a second ORF that immediately follows the M1 ORF, i.e., a coupled stop-start). Protein synthesis occurs in the cytoplasm. However,

NP, M_1, and NS_1 proteins accumulate in the cell nucleus during the first few hours of replication, then migrate to the cytoplasm. Nuclear inclusions of NS_1 may be formed.

Complementary RNA molecules which act as templates for new viral RNA synthesis are full-length transcripts and are neither capped nor polyadenylated. These RNAs exist as nucleocapsids in the nucleus of infected cells.

Integral membrane proteins migrate through the Golgi apparatus to localized regions of the plasma membrane. In addition to the activity of signal peptidases, the HA of the influenza viruses must undergo post-translational cleavage by cellular proteases to acquire fusion activity. Cleavability depends, among other factors, on the number of basic amino acids at the cleavage site. It produces a hydrophobic amino terminal HA_2 molecule. New virions form by budding, thereby incorporating matrix protein and the viral nucleocapsids which align below regions of the plasma membrane containing viral envelope proteins. Budding is from the apical surface in polarized cells. Gene reassortment occurs during mixed infections involving virus of the same species, but not between viruses of different types (e.g., influenza A and influenza B) or those of different genera.

ANTIGENIC PROPERTIES

The best studied antigens are the NP, M_1, HA and NA proteins of the influenza A and B viruses. NP and M_1 are species specific for the influenza A and B strains. Considerable variation occurs among the influenza A HA and NA antigens, less for influenza B or the HEF surface antigens of influenza C viruses. Thogoto and Dhori viruses do not cross-react in standard serologic tests. Antibody to HA (HEF, GP) neutralizes virus infectivity.

Erythrocytes of many species are agglutinated by influenza viruses. Agglutination may be blocked by serotype-specific antibodies. Sialic acid-containing virus receptors of erythrocytes may be destroyed by the NA of attached influenza virions, resulting in elution of virus. Hemolysis of erythrocytes may be produced at acid pH.

Thogoto virus exhibits limited hemagglutination by comparison to the influenza viruses and only with certain erythrocyte species.

BIOLOGICAL PROPERTIES

Certain influenza A viruses naturally infect humans causing respiratory disease. Particular influenza A viruses infect other mammalian species and a variety of avian species. Some interspecies transmission is believed to occur. Influenza B strains appear to naturally infect only humans. Influenza B virus causes epidemics every few years. Influenza A and B virus strains grow in the amniotic cavity of embryonated hen's eggs, and after adaption they grow in the allantoic cavity. Primary kidney cells from monkeys, humans, calves, pigs, and chickens support replication of many influenza A and B virus strains. The host range of these viruses may be extended by addition of trypsin to growth medium, so that multiple cycle replication can also be obtained in some continuous cell lines. Clinical specimens from influenza-infected hosts sometimes contain sub-populations of virus with minor sequence differences in at least their HA protein. These subpopulations may differ in their receptor specificity or their propensity for growth in different host cells.

Natural transmission of the influenza viruses is by aerosol (human and most non-aquatic hosts) or is water-borne (ducks).

Thogoto and Dhori viruses are transmitted by ticks and replicate in both ticks and a variety of tissues and organs in mammalian species as well as in mammalian cell cultures. In some laboratory species (e.g., hamsters for Thogoto virus) these infections have a fatal outcome. Unlike influenza viruses, these viruses do not cause respiratory disease and do not replicate in embryonated hen eggs.

Genus *Influenzavirus A, B*

Type Species influenza A virus (A/PR/8/34(H1N1)) (FLUA)

Distinguishing Features

Member viruses of the genus *Influenzavirus A, B* all have 8 genome segments. Hemagglutinin and the neuraminidase receptor destroying enzyme are different glycoproteins. The conserved end sequences of the viral RNAs of the influenza A viruses are 5' AGUAGAAACAAGG..., and 3' UCG(U/C)UUUCGUCC... For influenza B viruses they are 5' AGUAG(A/U)AACAA... and 3' UCGUCUUCGC... The exact order of electrophoretic migration of the RNA segments varies with strain and electrophoretic conditions. On the basis of the gene sequences, for influenza A the segments 1-3 encoded PB1, PB2 and PA proteins are estimated to have a size of about 87 kDa (observed: about 96 kDa), 84 kDa (observed: 87 kDa) and 83 kDa (observed: 85 kDa), respectively. The segment 4 encoded (unglycosylated) HA is about 63 kDa (glycosylated HA_1 is about 48 kDa, HA_2 is about 29 kDa). The segment 5 encoded NP is about 56 kDa (observed: 50-60 kDa). The segment 6 encoded NA is about 50 kDa (observed: 48-63 kDa). The segment 7 encoded M1 and M2 proteins are about 28 kDa (observed: 25 kDa) and 11 kDa (observed: 15 kDa), respectively. The segment 8 encoded NS_1 and NS_2 are 27 kDa (observed: 25 kDa) and 14 kDa (observed: 12 kDa), respectively. Generally the influenza B virus proteins have similar sizes. NB, the second product of influenza B segment 6, is 11 kDa (glycosylated 18 kDa).

Antigenic Properties

Antigenic variation occurring within the HA and NA antigens of influenza A and B viruses has been analyzed in detail. Fourteen subgroups of HA and nine subgroups of NA are recognized for influenza A viruses, with minimal serological cross-reaction between subgroups. Additional variation occurs within subgroups. By convention, new virus types are designated by their serotype / host species / site of origin / month and year of origin and (HA [H] and NA [N] subtype), e.g., A/Tern/South Africa/1/61 (H5N3). Continual evolution of new strains occurs and older strains apparently disappear from circulation. HA and NA antigens of influenza B viruses exhibit less antigenic variation than those of influenza A and no subgroups are defined. Antibody to HA neutralizes infectivity. If NA antibody is present during multicycle replication it inhibits virus release and thus reduces virus yield. Antibody to the amino terminus of M2 greatly reduces virus yield in tissue culture.

Biological Properties

Epidemics of respiratory disease in humans have been caused by influenza A viruses having the antigenic composition H1N1, H2N2, H3N2, and possibly H3N8. Influenza A viruses of subtype H7N7 and H3N8 (previously designated equine 1 and equine 2 viruses) cause outbreaks of respiratory disease in horses. Influenza A (H1N1) viruses, and A (H3N2) viruses have been isolated frequently from swine. The H1N1 viruses isolated from swine in recent years appear to be of three general categories: those closely related to classical "swine influenza" and which cause occasional human cases, those first recognized in avian specimens, but which have caused outbreaks among swine in Germany and France, and those resembling viruses isolated from epidemics in humans since 1977. H3N2 viruses from swine all appear to contain HA and NA genes closely related to those from human epidemic strains. Influenza A (H7N7 and H4N5) viruses have caused outbreaks in seals, with virus spread to non-respiratory tissues in this host. One of these viruses has accidentally caused infection of the conjunctiva of one laboratory worker. Pacific Ocean whales have reportedly been infected with type A (H1N1) virus. Other influenza subtypes have also been isolated from lungs of Atlantic Ocean whales in North America. Influenza A (H10N4) virus has caused outbreaks in mink. All subtypes of HA and NA, in many different combinations, have been identified in isolates from avian species, particularly chickens, turkeys, and ducks. Pathology in avian species varies from inapparent infection (often involving replication in, and probable transmission via, the intestinal tract), to virulent infections (only observed with subtypes H5 and H7) with spread to many tissues and high mortality rates. The structure of the HA protein, in particular the specificity of its receptor binding site and its cleavability by naturally occurring tissue protease(s), appears to be critical in determining the host range and organ tropisms of viruses. In addition, interactions between gene products determine the outcome of infection. Interspecies transmission apparently occurs in some instances without genetic reassortment (e.g., H1N1 virus from swine to humans and *vice versa* or H3N2 virus from humans to swine). In other cases interspecies transmission may involve RNA segment reassortment in hosts infected with more than one strain of virus each with distinct host ranges, or epidemic properties (e.g., 1968 isolates of H3N2 viruses (probably) were derived by reassortment of human H2N2 viruses and an unknown H3-containing virus; seal H7N7 virus probably was derived by reassortment of two or more avian influenza viruses; and reassortment of human H1N1 and H3N2 viruses in 1978 led to outbreaks of virus with H1N1 surface proteins but 4 or 5 other genes of H3N2 origin). Laboratory animals that may be infected with influenza A viruses include ferrets, mice, hamsters, and guinea pigs as well as some small primates such as squirrel monkeys.

TAXONOMIC STRUCTURE OF THE GENUS

A number of subtypes of influenza A virus are recognized on the basis of antigenic differences of their HA and NA proteins. No subtypes of influenza B virus have been described.

LIST OF SPECIES IN THE GENUS

The viruses, their genomic sequence accession numbers [ ] and assigned abbreviations ( ) are:

SPECIES IN THE GENUS

influenza A virus (A/PR/8/34(H1N1))	[V00603, J02151, V01106, J02144, J02148, J02146, V01099, V01104]	(FLUA)
influenza B virus (B/Lee/40)	[M20170, M20168, M20172, K00423, K01395, J02095, J02094, J02096]	(FLUB)

TENTATIVE SPECIES IN THE GENUS

None reported.

GENUS *INFLUENZAVIRUS C*

Type Species influenza C virus (C/California/78) (FLUC)

DISTINGUISHING FEATURES

Member viruses of the genus *Influenzavirus* C naturally infect humans. Viruses have 7 genome segments. They lack neuraminidase. The hemagglutinin (HEF) protein also has the function of a fusion protein and of a receptor-destroying enzyme which is a 9-0-acetylneuraminyl esterase. The conserved end sequences of the viral RNAs of the influenza C viruses are 5' AGCAGUAGCAA..., and 3' UCGU(U/C)UUCGUCC... RNA segments 1-3 encode the P proteins (Mr 87.8 x 10^3, 86.0 x 10^3, 81.9 x 10^3). Segment 4 encodes HEF (unglycosylated: Mr 72.1 x 10^3), segment 5 NP (Mr 63.5 x 10^3), segment 6 M (Mr 27.0 x 10^3) and segment 7 NS_1 (Mr 28.5 x 10^3) and NS_2 (Mr 14.0 x 10^3).

ANTIGENIC PROPERTIES

Antigenic variation among influenza C viruses has not been identified. Viruses exhibit no cross-reactivity with influenza A and B viruses, although homologies of HEF to influenza A and B HA can be identified near the amino and carboxy termini and to several of the cysteines in the co-aligned sequences. Antibody to HEF neutralizes infectivity.

BIOLOGICAL PROPERTIES

Infection in humans is common in childhood. Occasional outbreaks, but not epidemics, have been detected. Swine in China have been reported to be infected by viruses similar to human influenza C strains.

TAXONOMIC STRUCTURE OF THE GENUS

No virus subtypes have been described.

LIST OF SPECIES IN THE GENUS

The viruses, their genomic sequence accession numbers [ ] and assigned abbreviations ( ) are:

SPECIES IN THE GENUS

influenza C virus (C/California/78)	[K01689, M10087, M17700]	(FLUC)

TENTATIVE SPECIES IN THE GENUS

None reported.

GENUS "THOGOTO-LIKE VIRUSES"

Type Species Thogoto virus (THOV)

DISTINGUISHING FEATURES

Morphology and morphogenesis of these viruses show similarities with the influenza viruses. Virions are reported to contain 6 (THOV) or 7 (DHOV) segments of linear, negative sense ssRNA. Total genomic size is about 10,000 kb. Sequences of the ends of vRNA are partially complementary and resemble those of influenza viruses. The conserved end sequences of THOV viral RNAs are 5' AGAGA(U/A)AUCAAAGC... and 3' UCGUUUUUGUUC...; for DHOV they are 5' AGUAGACAUCAA... and 3' UCGUU(A/U)UUGUUCG... The gene encoded by segment 1 is not known. The 2nd [DHOV] and 3rd [THOV] largest RNAs encode proteins (Mr 81 x 10^3, 69 x 10^3, respectively) that exhibit homology to influenza P proteins. The single glycoprotein (GP, DHOV: Mr 65 x 10^3; THOV: Mr 75 x 10^3), is encoded by the 4th segment. It is unrelated to any influenza protein but shows amino acid similarity with the glycoprotein (gp64) of baculoviruses. The DHOV 5th segment encodes a protein (NP, Mr 54 x 10^3) related to influenza NP. The 6th segment (DHOV) encodes the 30 kDa M_1 protein, and another M_2 (15 kDa) of unknown function. The coding of the DHOV putative 7th segment is not known.

ANTIGENIC PROPERTIES

Antigenic relationships between THOV and DHOV viruses are not apparent and none of the virus proteins is related antigenically to the influenza viruses.

BIOLOGICAL PROPERTIES

Thogoto and Dhori viruses are transmitted between vertebrates by ticks. Comparatively low levels of hemagglutination occur at acidic pH and not at physiological pH. No receptor destroying enzyme has been observed. Fusion of infected cells occurs at acidic pH and is inhibited by neutralizing monoclonal antibodies directed against GP, indicating that cell entry is via the endocytotic pathway as for the influenza viruses. Replication is inhibited by actinomycin D. Nucleo-protein accumulates early in replication within the nucleus. GP is synthesized in the cytoplasm and accumulated at the cell surface. Reassortment between

THOV temperature sensitive mutants has been demonstrated experimentally in dually infected ticks and in vertebrates.

TAXONOMIC STRUCTURE OF THE GENUS

THOV and DHOV are unrelated to each other. For each virus several isolates have been made; however, the relationships of these isolates to the prototype viruses are not known.

LIST OF SPECIES IN THE GENUS

The viruses, their genomic sequence accession numbers [ ] and assigned abbreviations ( ) are:

SPECIES IN THE GENUS

Dhori virus	[M65866, M34002, M17435, M95567]	(DHOV)
Thogoto virus	[D00540, M77280]	(THOV)

TENTATIVE SPECIES IN THE GENUS

None reported.

LIST OF UNASSIGNED VIRUSES IN THE FAMILY

None reported.

SIMILARITY WITH OTHER TAXA

None reported.

DERIVATION OF NAMES

ortho: from Greek orthos, "straight"
myxo: from Greek *myxa*, "mucus"
influenza: Italian form of Latin *influentia*, "epidemic", originally used because epidemics were thought to be due to astrological or other occult "influences"

REFERENCES

Kingsbury DW (1990) *Orthomyxoviridae* and their replication In: Fields BN, Knipe DM (eds), Virology 2nd edn Raven Press, New York, pp 1075-1089

Krug RM (ed) (1990) The influenza viruses, Plenum Press, New York

Luong G, Palese P (1992) Genetic analysis of influenza virus. Curr Opinion Gen Develop 2: 77-81

Morse MA, Marriott AC, Nuttall PA (1992) The glycoprotein of Thogoto virus (a tick-borne orthomyxo-like virus) is related to the baculovirus glycoprotein gp64. Virology 186: 640-646

Nuttall PA, Morse MA, Jones LD, Portela A (1992) Adaption of members of the *Orthomyxoviridae* family to transmission by ticks. In: Gibbs AJ, Calisher CH (eds) Molecular Evolution of Viruses, Cambridge University Press (in press)

CONTRIBUTED BY

Klenk H-D, Cox NJ, Lamb RA, Mahy BWJ, Nakamura K, Nuttall PA, Palese P, Rott R

Family *Bunyaviridae*

Taxonomic Structure of the Family

Family	***Bunyaviridae***
Genus	*Bunyavirus*
Genus	*Hantavirus*
Genus	*Nairovirus*
Genus	*Phlebovirus*
Genus	*Tospovirus*

Virion Properties

Morphology

Morphological properties vary among the five genera; however, virions generally are spherical or pleomorphic, 80-120 nm in diameter, and display surface glycoprotein projections 5-10 nm in length which are embedded in a lipid bilayered envelope approximately 5 nm thick. Virion envelopes are usually derived from cellular Golgi membranes or, on occasion, from cell surface membranes. Viral ribonucleocapsids are 2-2.5 nm in diameter, 200-3,000 nm in length, and display helical symmetry.

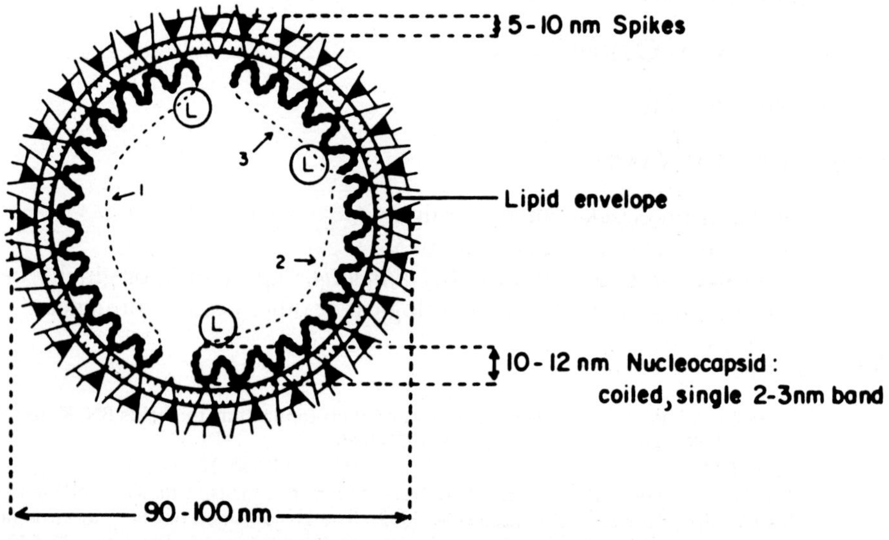

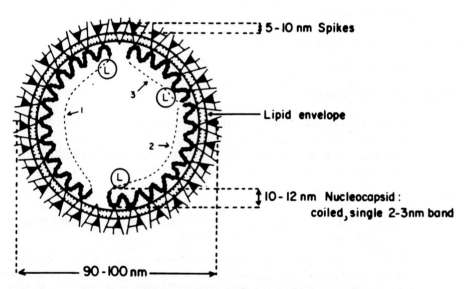

Figure 1: Diagram of virion in section. The surface spikes consist of the viral G1 and G2 proteins. The 3 helical nucleocapsids are circular and consist of non-covalently closed, circular, ssRNA (L, M, or S), plus N and L proteins (courtesy of Bishop DHL).

Physicochemical and Physical Properties

Virion Mr is 300-400 x 10^6; S_{20w} is 350-500. Virion buoyant densities in sucrose and CsCl are 1.16-1.18 and 1.20-1.21 g/cm^3, respectively. Virions are sensitive to heat, detergents and formaldehyde.

Nucleic Acid

Virions contain 3 molecules of negative or ambisense ssRNA. The genome sizes are 11-20 kb (Table 1). Terminal nucleotides of each viral RNA species are base-paired forming non-covalently closed, circular RNAs (and ribonucleocapsids). Terminal sequences of gene segments are conserved among viruses in each genus but are different from those of other genera. The Mr of the genomes range from 4.8-8 x 10^6 and constitute 1-2% of the virion weight. Viral mRNAs are not polyadenylated. By comparison to viral RNAs, they are truncated at the 3' termini. mRNAs have 5' methylated caps and 12-15 non-templated nucleotides derived from host mRNAs.

Table 1: Deduced nucleotide lengths of selected genomic RNAs (ND: not determined)

Genus Virus	RNA segment L	M	S
Bunyavirus			
Aino	ND	ND	850
Bunyamwera	6875	4458	961
Germiston	ND	4534	980
La Crosse	ND	4526	981
Maguari	ND	ND	945
snowshoe hare	ND	4527	982
Hantavirus			
Hantaan (76-118)	6533	3616	1696
Prospect Hill (MP-40)	ND	3707	1675
Puumala (CG 1820)	6550	3682	1784
Puumala (Sotkamo)	ND	3682	1830
Seoul (SR-11)	ND	3651	1769
Seoul (HR80-39)	6530	3651	1769
Nairovirus			
Crimean-Congo hemorrhagic fever	ND	ND	1672
Dugbe	ND	4888	1712
Hazara	ND	ND	1677
Phlebovirus			
Punto Toro	ND	4330	1904
Rift Valley fever	6606	3884	1690
sandfly fever (Sicilian)	ND	ND	1747
Toscana	ND	ND	1869
Uukuniemi	6423	3229	1720
Tospovirus			
impatiens necrotic spot	ND	4972	ND
tomato spotted wilt	8897	4821	2916

Proteins

All viruses have four structural proteins, two external glycoproteins (G1, G2), a nucleocapsid protein (N), and a large transcriptase protein (L). Sizes of the structural proteins and non-structural species (NS) are listed in Table 2.

Table 2: Deduced protein sizes (kDa)

RNA Protein	Genus *Bunyavirus*	*Hantavirus*	*Nairovirus*	*Phlebovirus*	*Tospovirus*
L segment					
L	259	246	>200	241	331
M segment					
G1	108-120	68-76	72-84	55-75	78
G2	29-41	52-58	30-45	50-70	52-58
NS_M	15-18	none	70-110	none or 78	34
S segment					
N	19-25	50-54	48-54	24-30	29
NS_S	10-13	none	none	29-31	52

Lipids

Virions are composed of 20-30% lipid by weight. Lipids are derived from the membranes where viruses mature and include phospholipids, sterols, fatty acids and glycolipids.

Carbohydrates

Virions are composed of 2-7% carbohydrate by weight. N-linked glycans on the G1 and G2 proteins are largely of the high mannose type.

Genome Organization and Replication

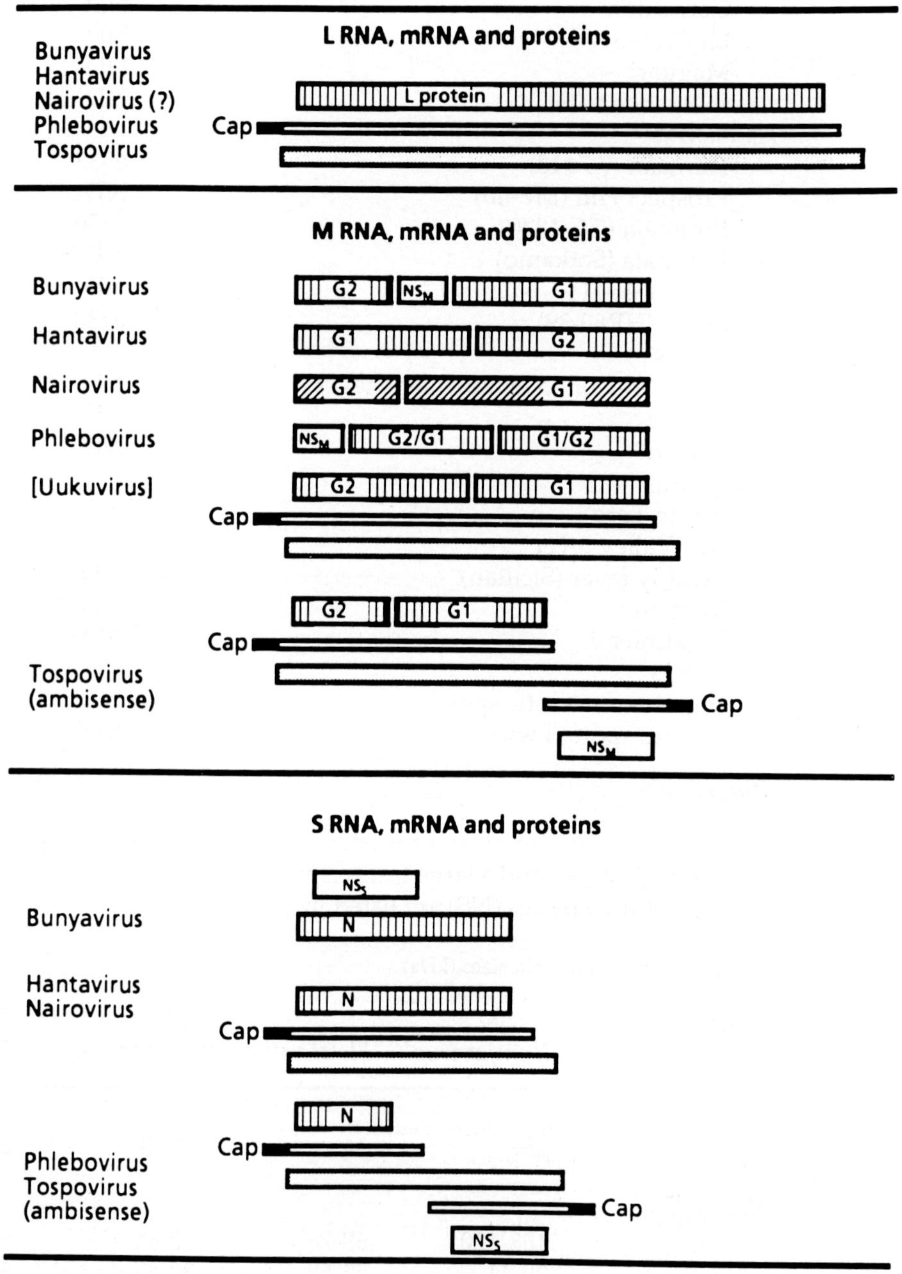

Figure 2: Genome organization of viruses in different genera (not to scale). Although uukuviruses are assigned to the genus *Phlebovirus* they lack an NS_M protein. Stippled boxes are virion RNA species with the 3' terminus on the left. mRNAs are shown with 5' capped primers (in black). Structural proteins are shown as boxes with vertical lines, non-structural proteins as open boxes, gene orders are with respect to the viral-complementary, or viral-sense mRNAs. For nairoviruses the relationships of the M-coded structural to non-structural proteins (Diagonal stripes) and the L coding strategy are unknown.

Bunyaviruses encode a non-structural protein (NS_S) in an ORF that overlaps N in the 5' half of the S mRNA. Phleboviruses and tospoviruses have an ambisense S RNA. They encode The genome organization of different genera is shown in Fig. 2. The virus-complementary L mRNA encodes the viral transcriptase-replicase (L protein), the M mRNA encodes the envelope glycoproteins (G1 and G2), and the S mRNA the nucleocapsid protein (N). NS_S proteins in an ORF in the 5' half of virion S RNA. Hantaviruses and nairoviruses encode no additional proteins in their S genome segments. For all viruses a continuous ORF in the M mRNA encodes the glycoproteins. Other than in nairoviruses, this precursor is cleaved co-translationally to the eventual gene products. Nairoviruses synthesize at least two non-structural proteins which are precursors of the glycoproteins. Bunyaviruses, nairoviruses and phleboviruses (other than Uukuniemi virus) also encode one or more NS_M proteins in the viral-complementary M mRNA. Hantaviruses and Uukuniemi virus (*Phlebovirus*) encode no additional proteins in their M genome segments. Tospoviruses encode an NS_M in an ORF at the 5' end of the ambisense viral M RNA.

All stages of replication occur in the cytoplasm. The principal stages are:
(1) attachment, mediated by an interaction of one or both of the integral viral envelope proteins and host receptors; (2) entry and uncoating, involving endocytosis of virions and fusion of viral membranes with endosomal membranes; (3) transcription involving the synthesis of viral-complementary mRNA species from genome templates and host cell-derived primers by the virion-associated polymerase; (4) translation of primary S and L mRNA transcripts by free ribosomes; translation of primary M segment mRNAs by membrane-bound ribosomes and glycosylation of nascent envelope proteins; co-translational cleavage of precursors to yield G1 and G2, and for some viruses, NS_M; (5) synthesis and encapsidation by N protein of full-length viral complementary RNA to serve as templates for genomic RNA or, in some cases, subgenomic viral-sense mRNA synthesis for RNAs with an ambisense coding strategy; (6) genome RNA replication; (7) secondary transcription involving the amplified synthesis of mRNA species; (8) morphogenesis including accumulation of G1 and G2 in the Golgi, terminal glycosylation, acquisition of modified host membranes and budding generally into Golgi cisternae, also budding at the cell surface in certain cells and tissues; (9) fusion of cytoplasmic vesicles with the plasma membrane and release of mature virions.

Antigenic Properties

One or both of the envelope glycoproteins display hemagglutinating and neutralizing antigenic determinates. Complement fixing antigenic determinants are principally associated with the nucleocapsid protein.

Biological Properties

Viruses in all genera except the genus *Hantavirus* are capable of alternately replicating in vertebrates and arthropods. Viruses are generally cytolytic in their vertebrate hosts, but cause little or no cytopathogenicity in their invertebrate hosts. Various viruses are transmitted by mosquitoes, ticks, phlebotomine flies, thrips, and other arthropod vectors. Some viruses display a very narrow host range, especially in their arthropod vectors. Transovarial and venereal transmission have been demonstrated for some mosquito-borne viruses. Aerosol infection occurs in certain situations, and is the principal means of transmission for some viruses. Hantavirus transmission does not involve arthropods; rather, these viruses are transmitted via rodent host feces, urine and saliva. Some viruses cause a reduction in host-cell protein synthesis in vertebrate cells. Hantaviruses cause no detectable reduction in host macromolecular synthesis and routinely establish persistent, non-cytolytic infections in susceptible mammalian host cells, a finding consistent with their non-pathogenic persistence in rodent hosts. Certain viruses induce cell fusion at low pH. Some viruses exhibit pH-dependent hemagglutinating activities. Genetic reassortment between closely related viruses has been demonstrated for some viruses both *in vitro* and *in vivo*.

GENUS BUNYAVIRUS

Type Species Bunyamwera virus (BUNV)

DISTINGUISHING FEATURES

The morphology of a typical bunyavirus is shown in Fig. 3. Bunyaviruses cross-react serologically to various degrees. They exhibit no antigenic relationship to members of other genera. Generally, the 3' terminal nucleotide sequences of the L, M and S viral RNA segments are: UCAUCACAUGA..., the 5' terminal sequences are: AGUAGUGUGCU... The viral proteins of different bunyaviruses are comparable in terms of size and function and, to varying degrees for those that have been analyzed, by sequence. The proteins exhibit no obvious sequence similarities to proteins of viruses representing other genera. Both G1 and G2 glycoproteins, and a 15-18 kDa NS_M protein, are translated from the M mRNA. The N and NS_S proteins are encoded in overlapping reading frames by the S mRNA. The L protein is translated from the L mRNA. Most bunyaviruses are transmitted by mosquitoes; some (Tete group) are transmitted by ticks. Occasionally, alternate arthropods, e.g. ceratopogonids in the genus *Culicoides*, or phlebotomines, may transmit bunyaviruses. Some viruses are transmitted transovarially in arthropods. Genetic reassortment has been demonstrated among antigenically similar viruses.

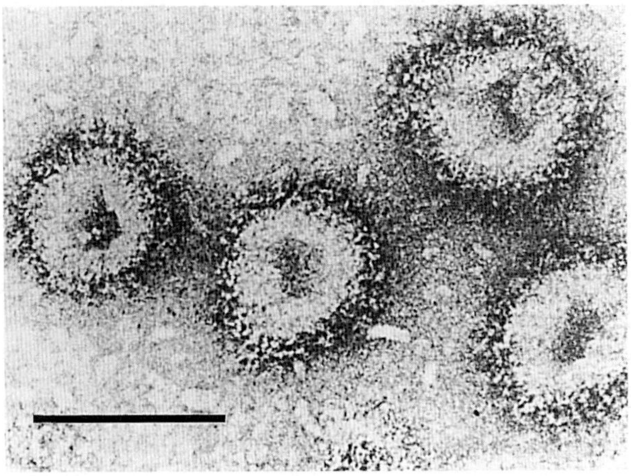

Figure 3: Negative contrast electron micrograph of preparation of La Crosse virus, the bar represents 100 nm. (courtesy of Murphy FA).

TAXONOMIC STRUCTURE OF THE GENUS

There are 18 antigenic groups of the genus *Bunyavirus* (at least 161 viruses) and 4 ungrouped viruses.

LIST OF SPECIES IN THE GENUS

The groups, viruses, their genomic sequence accession numbers [ ] and assigned abbreviations ( ) are:

SPECIES IN THE GENUS

1-Anopheles A virus Group:

Anopheles A virus (reference strain Original)	(ANAV)
CoAr-1071 virus	(CA1071V)
CoAr-3624 virus	(CA3624V)
CoAr-3627 virus	(CA3627V)
ColAn-57389 virus	(CA57389V)
H32580 virus	(H32580V)
Las Maloyas virus	(LMV)
Lukuni virus	(LUKV)

SPAr-2317 virus		(SPAV)
Tacaiuma virus		(TCMV)
Trombetas virus		(TRMV)
Virgin River virus		(VRV)
2-Anopheles B virus Group:		
Anopheles B virus (reference strain Original)		(ANBV)
Boraceia virus		(BORV)
3-Bakau virus Group:		
Bakau virus (reference strain MM-2325)		(BAKV)
Ketapang virus		(KETV)
Nola virus		(NOLAV)
Tanjong Rabok virus		(TRV)
Telok Forest virus		(TFV)
4-Bunyamwera virus Group:		
AG83-1746 virus		(AG1746V)
Anhembi virus		(AMBV)
Batai virus	[S:X73464]	(BATV)
BeAr 328208 virus		(BAV)
Birao virus		(BIRV)
Bozo virus		(BOZOV)
Bunyamwera virus (reference strain Original)	[L: X14383, M: M11852, S: D00353]	(BUNV)
Cache Valley virus	[S:X73465]	(CVV)
CbaAr 426 virus		(CAV)
Fort Sherman virus		(FSV)
Germiston virus	[M: M21951, S: M19420]	(GERV)
Guaroa virus	[S:X73466]	(GROV)
Iaco virus		(IACOV)
Ilesha virus		(ILEV)
Kairi virus	[S:X73467]	(KRIV)
Lokern virus		(LOKV)
Macaua virus		(MCAV)
Maguari virus	[S: D00354]	(MAGV)
Main Drain virus	[S:X73469]	(MDV)
Mboke virus		(MBOV)
Ngari virus		(NRIV)
Northway virus	[S:X73470]	(NORV)
Playas virus		(PLAV)
Potosi virus		(POTV)
Santa Rosa virus		(SARV)
Shokwe virus		(SHOV)
Sororoca virus		(SORV)
Taiassui virus		(TAIAV)
Tensaw virus		(TENV)
Tlacotalpan virus		(TLAV)
Tucunduba virus		(TUCV)
Wyeomyia virus		(WYOV)
Xingu virus		(XINV)
5-Bwamba virus Group:		
Bwamba virus (reference strain M 459)		(BWAV)
Pongola virus		(PGAV)
6-Group C virus Group:		
Apeu virus (reference strain BeAn 848)		(APEUV)
Bruconha virus		(BRUV)
Caraparu virus		(CARV)
Gumbo Limbo virus		(GLV)

Itaqui virus		(ITQV)
Madrid virus		(MADV)
Marituba virus		(MTBV)
Murutucu virus		(MURV)
Nepuyo virus		(NEPV)
Oriboca virus		(ORIV)
Ossa virus		(OSSAV)
Restan virus		(RESV)
Vinces virus		(VINV)
63U-11 virus		(63UV)
7-California encephalitis virus Group:		
AG83-497 virus		(AG497V)
California encephalitis virus (reference strain BFS-283)		(CEV)
Inkoo virus		(INKV)
Jamestown Canyon virus		(JCV)
Keystone virus		(KEYV)
La Crosse virus	[M: D00202, S: K00610]	(LACV)
Melao virus		(MELV)
San Angelo virus		(SAV)
Serra do Navio virus		(SDNV)
snowshoe hare virus	[M: K02539, S: J02390]	(SSHV)
South River virus		(SORV)
Tahyna virus		(TAHV)
trivittatus virus		(TVTV)
8-Capim virus Group:		
Acara virus		(ACAV)
Benevides virus		(BVSV)
Benfica virus		(BENV)
Bushbush virus		(BSBV)
Capim virus (reference strain BeAn 8582)		(CAPV)
Guajara virus		(GJAV)
GU71U-344 virus		(GU344V)
GU71U-350 virus		(GU350V)
Juan Diaz virus		(JDV)
Moriche virus		(MORV)
9-Gamboa virus Group:		
Alajuela virus		(ALJV)
Brus Laguna virus		(BLAV)
Gamboa virus (reference strain MARU 10962)		(GAMV)
Pueblo Viejo virus		(PV)
San Juan virus		(SJV)
75V-2374 virus		(V2374V)
75V-2621 virus		(V2621V)
78V-2441 virus		(V2441V)
10-Guama virus Group:		
Ananindeua virus		(ANUV)
Bertioga virus		(BERV)
Bimiti virus		(BIMV)
Cananeia virus		(CNAV)
Catu virus		(CATUV)
Guama virus (reference strain BeAn 277)		(GMAV)
Guaratuba virus		(GTBV)
Itimirim virus		(ITIV)
Mahogany Hammock virus		(MHV)
Mirim virus		(MIRV)
Moju virus		(MOJUV)
Timboteua virus		(TBTV)

11-Koongol virus Group:		
Koongol virus (reference strain MRM31)		(KOOV)
Wongal virus		(WONV)
12-Minatitlan virus Group:		
Minatitlan virus (reference strain M67U5)		(MNTV)
Palestina virus		(PLSV)
13-Nyando virus Group:		
Eret-147 virus		(E147V)
Nyando virus (reference strain MP 401)		(NDV)
14-Olifantsvlei virus Group:		
Bobia virus		(BIAV)
Botambi virus		(BOTV)
Dabakala virus		(DABV)
Olifantsvlei virus (reference strain SAAr 5133)		(OLIV)
Oubi virus		(OUBIV)
15-Patois virus Group:		
Abras virus		(ABRV)
Babahoya virus		(BABV)
Estero Real virus		(ERV)
Pahayokee virus		(PAHV)
Patois virus (reference strain BT 4971)		(PATV)
Shark River virus		(SRV)
Zegla virus		(ZEGV)
16-Simbu virus Group:		
Aino virus	[S: M22011]	(AINOV)
Akabane virus		(AKAV)
Buttonwillow virus		(BUTV)
Douglas virus		(DOUV)
Facey's Paddock virus		(FPV)
Ingwavuma virus		(INGV)
Inini virus		(INIV)
Kaikalur virus		(KAIV)
Manzanilla virus		(MANV)
Mermet virus		(MERV)
Oropouche virus		(OROV)
Para virus		(PARAV)
Peaton virus		(PEAV)
Sabo virus		(SABOV)
Sango virus		(SANV)
Sathuperi virus		(SATV)
Shamonda virus		(SHAV)
Shuni virus		(SHUV)
Simbu virus (reference strain SAAr 53)		(SIMV)
Thimiri virus		(THIV)
Tinaroo virus		(TINV)
Utinga virus		(UTIV)
Utive virus		(UV)
Yaba-7 virus		(Y7V)
17-Tete virus Group:		
Bahig virus		(BAHV)
Batama virus		(BMAV)
Matruh virus		(MTRV)
Tete virus (reference strain SAAn 3518)		(TETEV)
Tsuruse virus		(TSUV)
Weldona virus		(WELV)
18-Turlock virus Group:		
Lednice virus		(LEDV)

Turlock virus (reference strain S 1954-847-32)	(TURV)
Umbre virus	(UMBV)
Yaba-1 virus	(Y1V)

Tentative Species in the Genus

Kaeng Khoi virus	(KKV)
Leanyer virus	(LEAV)
Mojui dos Campos virus	(MDCV)
Termeil virus	(TERV)

Genus *Hantavirus*

Type Species Hantaan virus (HTNV)

Distinguishing Features

The morphology of a typical hantavirus is shown in Fig. 4. Hantaviruses are serologically related. They exhibit no antigenic relationship with members of other genera. Generally, the terminal 3' nt sequences of the L, M and S viral RNA species are: AUCAUCAUCUG..., 5' nt sequences are: UAGUAGUA... The viral proteins of different hantaviruses are similar in size, function and sequence. The proteins exhibit no obvious sequence similarities to proteins of viruses representing other genera. Hantaviruses lack L-, M-, or S-coded non-structural proteins. Certain hantaviruses are the etiologic agents of hemorrhagic fever with renal syndrome. In contrast to viruses in other genera, hantaviruses are not transmitted by arthropods. The reservoir hosts of hantaviruses are specific rodents, on occasion they infect humans. Hantaviruses cause no detectable cytopathology in vertebrate cell cultures and produce persistent, non-pathogenic infections in rodents.

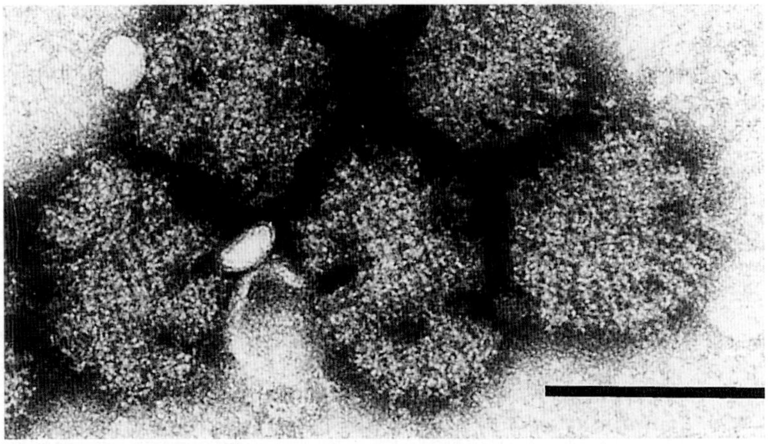

Figure 4: Grid-like surface structure on glutaraldehyde-fixed, negative contrast electron microscopy of Hantaan virus (courtesy of White J). The bar represents 100 nm.

Taxonomic Structure of the Genus

There is 1 recognized group within the genus *Hantavirus* (at least 6 viruses), plus a large number of isolates not yet assigned to an antigenic complex.

List of Species in the Genus

The groups, viruses, their genomic sequence accession numbers [ ] and assigned abbreviations ( ) are:

Species in the Genus

1-Hantaan virus Group:

Dobrava-Belgrade virus		(DOBV)
Hantaan virus (reference strain 76-118)	[L:X55901, M:M14627, S:M14626]	(HTNV)
Prospect Hill virus	[M:X55129, S:X55128]	(PHV)
Puumala virus	[M:X61034, S:X61035, L:M63194]	(PUUV)
Seoul Virus	[L:X56492, M:X56493]	(SEOV)
Thailand virus	[M:L08756]	(THAIV)
Thottapalayam virus		(TPMV)

TENTATIVE SPECIES IN THE GENUS

CG18-20 virus (originally reported as Hällnäs B1 virus)	[L:M63194, M:M29979, S:M32750]	(CG1820V)
HoJo virus	[M:D00376]	(HOJOV)
HV-114 virus	[M:L08753]	(HV114V)
K27 virus	[M:L08754]	(K27V)
Lee virus	[M:D00377]	(LEEV)
P360 virus	[M:L08755]	(P360V)
SR-11 virus	[M: M34882, S: M34881]	(SR11V)

GENUS *NAIROVIRUS*

Type Species Nairobi sheep disease virus (NSDV)

DISTINGUISHING FEATURES

The morphology of a typical nairovirus is shown in Fig. 5. Nairoviruses cross-react serologically to various degrees. Morphologically they are similar, although on fixation some are pleomorphic. Nairoviruses exhibit no antigenic relationship to members of other genera. Generally, the terminal 3' nucleotide sequences of the L, M and S viral RNA species are AGAGUUUCU..., the 5' nucleotide sequences are UCUCAAAGA... The structural proteins of different nairoviruses are similar in terms of size. There are only limited data available concerning the relationships of the observed M-coded non-structural proteins to each other, or to the structural glycoproteins. The S segment does not encode a non-structural protein. No data are available concerning the L gene products. The L RNA is considerably larger than those of other members of the family. From the limited available data, the nairovirus proteins exhibit no obvious sequence similarities to proteins of viruses representing other genera. Most nairoviruses are transmitted by ticks, members of the

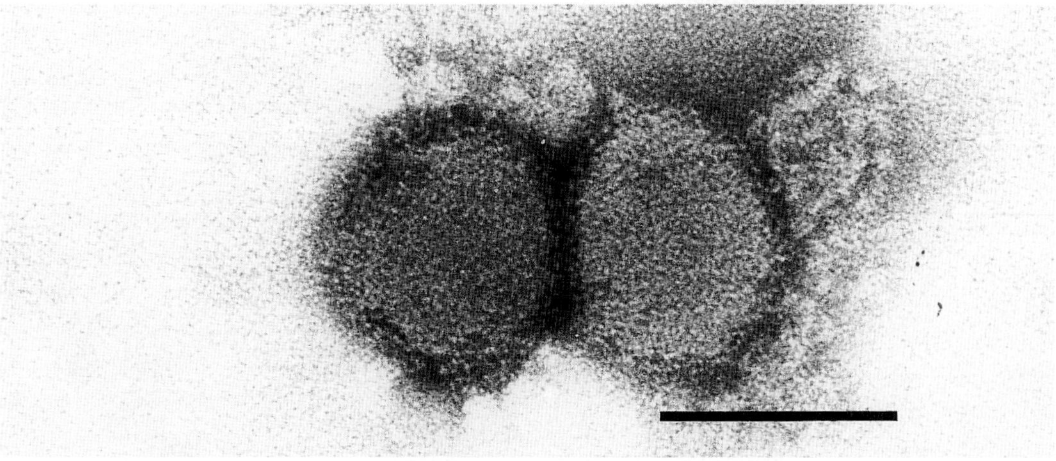

Figure 5: Negative contrast electron micrograph of CCHF virus, the bar represents 100 nm (courtesy of Drier T).

CHFV, NSDV, and SAKV groups mainly by ixodid ticks and DGKV, HUGV and QYBV groups mainly by argasid ticks. Some viruses are transmitted transovarially in arthropods.

Taxonomic Structure of the Genus

There are 7 antigenic groups of the genus *Nairovirus* (at least 33 viruses).

List of Species in the Genus

The groups, viruses, their genomic sequence accession numbers [ ] and assigned abbreviations ( ) are:

Species in the Genus

1-Crimean-Congo hemorrhagic fever virus Group:		
Crimean-Congo hemorrhagic fever virus (reference strain Kodzha)	[S:M86625]	(C-CHFV)
Hazara virus	[S:M86624]	(HAZV)
Khasan virus		(KHAV)
2-Dera Ghazi Khan virus Group:		
Abu Hammad virus		(AHV)
Abu Mina virus		(ABMV)
Dera Ghazi Khan virus (reference strain JD 254)		(DGKV)
Kao Shuan virus		(KSV)
Pathum Thani virus		(PTHV)
Pretoria virus		(PREV)
3-Hughes virus Group:		
Dry Tortugas virus		(DTV)
Farallon virus		(FARV)
Fraser Point virus		(FPV)
Great Saltee virus		(GRSV)
Hughes virus (reference strain Original)		(HUGV)
Puffin Island virus		(PIV)
Punta Salinas virus		(PSV)
Raza virus		(RAZAV)
Sapphire II virus		(SAPV)
Soldado virus		(SOLV)
Zirqa virus		(ZIRV)
4-Nairobi sheep disease virus Group:		
Dugbe virus	[M:M94133, S:M25150]	(DUGV)
Nairobi sheep disease virus (reference strain Original)		(NSDV)
5-Qalyub virus Group:		
Bandia virus		(BDAV)
Omo virus		(OMOV)
Qalyub virus (reference strain Ar 370)		(QYBV)
6-Sakhalin virus Group:		
Avalon virus		(AVAV)
Clo Mor virus		(CMV)
Kachemak Bay virus		(KBV)
Paramushir virus		(PMRV)
Sakhalin virus (reference strain LEIV-71C)		(SAKV)
Taggert virus		(TAGV)
Tillamook virus		(TILLV)
7-Thiafora virus Group:		
Erve virus		(ERVEV)
Thiafora virus (reference strain AnD 11411)		(TFAV)

TENTATIVE SPECIES IN THE GENUS

None reported.

GENUS *PHLEBOVIRUS*

Type Species sandfly fever Sicilian virus (SFSV)

DISTINGUISHING FEATURES

Phleboviruses include the sandfly fever viruses and the tick-transmitted uukuviruses that were previously recognized as a separate genus. However, weak antigenic relationships and significant protein sequence homologies have been demonstrated between uukuviruses and phleboviruses, but none between these viruses and those of members of other genera. For these reasons and the common overall coding and transcriptional strategies of the viruses they are placed in the genus *Phlebovirus*. The morphologies of a typical phlebovirus and Uukuniemi virus are shown in Fig. 6.

Phleboviruses cross-react serologically to different degrees. They are antigenically unrelated to members of other genera. Generally, the 3' terminal nucleotide sequences of the L, M and S viral RNA species segments are: UGUGUUUC..., the 5' terminal sequences are: ACACAAAG... The S RNA has an ambisense coding strategy, i.e., it is transcribed by the virion RNA polymerase to a subgenomic, virus-complementary mRNA that encodes the N protein and, from a full-length viral-complementary S RNA, to a subgenomic, virus-sense mRNA that encodes a non-structural (NS_S) protein. The viral proteins of different phleboviruses are comparable in terms of size and function and, to varying degrees for those that have been analyzed, by sequence. The proteins exhibit no obvious sequence similarities to proteins of viruses representing other genera. Viruses of the sandfly fever virus group, but not of the Uukuniemi virus group, have a pre-glycoprotein coding region that codes for non-structural protein(s) (NS_M). The similar sizes of the G1 and G2 proteins account for the different G1:G2 order in the M gene for different viruses.

Sandfly fever group viruses have been isolated from various vertebrate species and from phlebotomines and occasionally alternative arthropods, e.g., mosquitoes, or ceratopogonids in the genus *Culicoides*. Uukuniemi serogroup viruses have been isolated from various vertebrate species and from ticks.

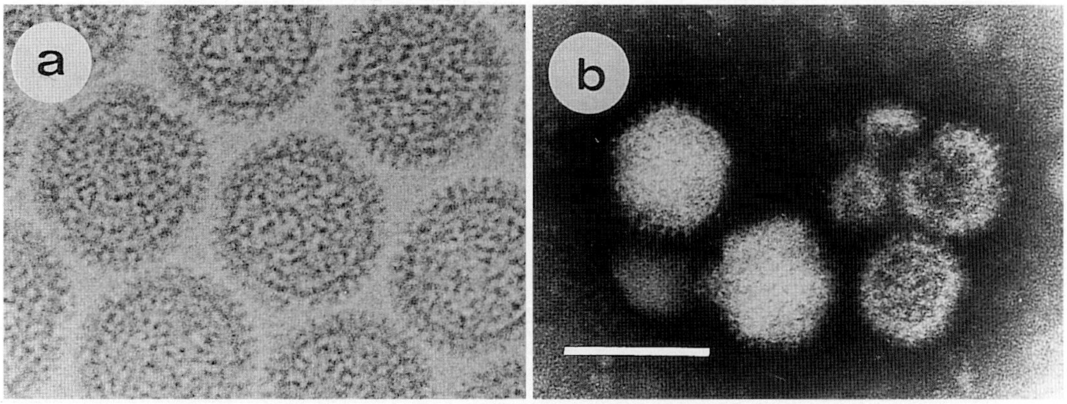

Figure 6: (a) Cryoelectron micrograph of Uukuniemi virus; (b) glutaraldehyde-fixed, negative contrast electron micrograph of Rift valley fever virus, the bar represents 100 nm (courtesy of von Bonsdorff C-H).

TAXONOMIC STRUCTURE OF THE GENUS

There are 8 antigenic complexes (at least 23 viruses) within the sandfly fever group; 16 viruses related to sandfly fever Sicilian virus have not been assigned to an antigenic complex. Uukuniemi group viruses belong to a single serogroup (12 viruses).

The groups and antigenic complexes are:

List of Species in the Genus

The groups, complexes, viruses, their genomic sequence accession numbers [ ] and assigned abbreviations ( ) are:

Species in the Genus

1-sandfly fever virus group
- Bujaru complex:
  - Bujaru virus (reference strain BeAn 47693) (BUJV)
  - Munguba virus (MUNV)
- Candiru complex:
  - Alenquer virus (ALEV)
  - Candiru virus (reference strain BeH 22511) (CDUV)
  - Itaituba virus (ITAV)
  - Nique virus (NIQV)
  - Oriximina virus (ORXV)
  - Turuna virus (TUAV)
- Chilibre complex:
  - Cacao virus (CACV)
  - Chilibre virus (reference strain VP-118D) (CHIV)
- Frijoles complex:
  - Frijoles virus (reference strain VP-161A) (FRIV)
  - Joa virus (JOAV)
- Punta Toro complex:
  - Buenaventura virus (BUEV)
  - Punta Toro virus (reference strain D-4021A) [M:M11156, S:K02736] (PTV)
- Rift Valley fever complex:
  - Arbia virus (ARBV)
  - Belterra virus (BELTV)
  - Icoaraci virus (ICOV)
  - Karimabad virus (KARV)
  - Rift Valley fever virus (reference strain Original) [L:X56464, M:M11157, S:X53771] (RVFV)
- Salehabad complex:
  - Salehabad virus (reference strain 1-81) (SALV)
  - sandfly fever Naples virus (SFNV)
  - Tehran virus (TEHV)
  - Toscana virus [L:X68414, S:X53794] (TOSV)
- No complex assigned in sandfly fever group:
  - Aguacate virus (AGUV)
  - Anhanga virus (ANHV)
  - Arboledas virus (ADSV)
  - Arumowot virus (AMTV)
  - Caimito virus (CAIV)
  - Chagres virus (CHGV)
  - Corfu virus (CFUV)
  - Gabek Forest virus (GFV)
  - Gordil virus (GORV)
  - Itaporanga virus (ITPV)
  - Odrenisrou virus (ODRV)
  - Pacui virus (PACV)
  - Rio Grande virus (RGV)
  - Saint-Floris virus (SAFV)
  - sandfly fever Sicilian virus (reference strain Sabin) [S:J04418] (SFSV)
  - Urucuri virus (URUV)

2-Uukuniemi virus Group:

EgAn 1825-61 virus		(EGAV)
Fin V-707 virus		(FINV)
Grand Arbaud virus		(GAV)
Manawa virus		(MWAV)
Murre virus		(MURRV)
Oceanside virus		(OCV)
Ponteves virus		(PTVV)
Precarious Point virus		(PPV)
St Abbs Head virus		(SAHV)
RML 105355 virus		(RMLV)
Uukuniemi virus (reference strain S 23)	[L:D10759, M:M17417, S:M33551]	(UUKV)
Zaliv Terpeniya virus		(ZTV)

TENTATIVE SPECIES IN THE GENUS

None reported.

GENUS *TOSPOVIRUS*

Type Species tomato spotted wilt virus (TSWV)

DISTINGUISHING FEATURES

Virus morphogenesis occurs in clusters in the cisternae of the endoplasmic reticulum of host cells. Nucleocapsid material may accumulate in the cytoplasm in dense masses. However, these masses may be composed of defective particles. The morphology of a tospovirus is shown in Fig. 7.

The S and M RNAs of tospoviruses exhibit an ambisense coding strategy, and encode non-structural proteins in the virus-sense RNA sequence. Both glycoproteins are encoded in the virus-complementary RNA of the M segment. The S segment encodes the nucleocapsid protein in the virus-complementary mRNA. At least 9 species of thrips have been reported to transmit tospoviruses. Transmission involves the sap of infected plants. More than 360 plant species belonging to 50 families are known to be susceptible to infection with tospoviruses.

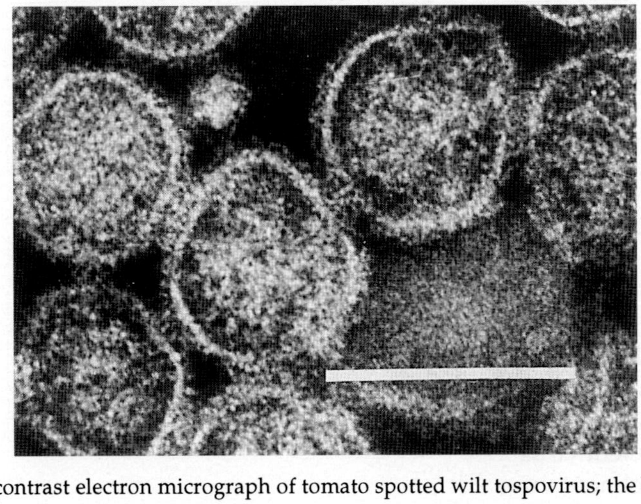

Figure 7: Negative contrast electron micrograph of tomato spotted wilt tospovirus; the bar represents 100nm (courtesy of Peters R).

List of Species in the Genus

The viruses, their genomic sequence accession numbers [ ] and assigned abbreviations ( ) are:

Species in the Genus

impatiens necrotic spot virus	[M: M74904]	(INSV)
tomato spotted wilt virus (reference strain Original)	[L: D10066, M:S48091, S: D00645]	(TSWV)

Tentative Species in the Genus

None reported.

List of Unassigned Viruses in the Family

There are at least 7 groups (19 viruses) and 22 ungrouped viruses which have not been shown to be antigenically related to members of defined genera of the family *Bunyaviridae* . For most, no biochemical characterization of the virus has been reported to confirm their family or genus status.
The groups, viruses and their assigned abbreviations ( ) are:

1-Group 1:	
Bhanja virus	(BHAV)
Forecariah virus	(FORV)
Kismayo virus	(KISV)
2-Group 2:	
Kaisodi virus	(KSOV)
Lanjan virus	(LJNV)
Silverwater virus	(SILV)
3-Group 3:	
Gan Gan virus	(GGV)
Mapputta virus	(MAPV)
Maprik virus	(MPKV)
Trubanaman virus	(TRUV)
4-Group 4:	
Okola virus	(OKOV)
Tanga virus	(TANV)
5-Group 5:	
Antequera virus	(ANTV)
Barranqueras virus	(BQSV)
Resistencia virus	(RTAV)
6-Group 6:	
Aransas Bay virus	(ABV)
Upolu virus	(UPOV)
7-Group 7:	
Kasokero virus	(KASOV)
Yogue virus	(YOGV)

The ungrouped viruses are:

Bangui virus	(BGIV)
Batken virus	(BKNV)
Belem virus	(BLMV)
Belmont virus	(BELV)
Bobaya virus	(BOBV)
Caddo Canyon virus	(CACAV)
Chim virus	(CHIMV)
Enseada virus	(ENSV)

Issyk-Kul virus	(IKV)
Keterah virus	(KTRV)
Kowanyama virus	(KOWV)
Lone Star virus	(LSV)
Pacora virus	(PCAV)
Razdan virus	(RAZV)
Salanga virus	(SGAV)
Santarem virus	(STMV)
Sunday Canyon virus	(SCAV)
Tai virus	(TAIV)
Tamdy virus	(TDYV)
Tataguine virus	(TATV)
Wanowrie virus	(WANV)
Witwatersrand virus	(WITV)
Yacaaba virus	(YACV)

Similarity with Other Taxa

None reported.

Derivation of Names

bunya: from *Bunya*mwera; place in Uganda, where type virus was isolated
nairo: from *Nairo*bi sheep disease; first reported disease caused by a member virus
phlebo: from Greek *phlebos*, "vein", refers to *phlebo*tomine vectors of many of the sandfly fever group viruses
hanta: from *Hanta*an virus; river in South Korea near where the type virus was isolated
tospo: sigla from *to*mato *spo*tted wilt virus

References

Bishop DHL, Shope RE (1979) *Bunyaviridae* In: Fraenkel-Conrat H, Wagner RR (eds) Comprehensive Virology, Vol 14 Plenum Press, New York, pp 1-156
Bishop DHL (1990) *Bunyaviridae* and their replication, In: Fields BN, Knipe DM (eds), Virology, 2nd edn Raven Press, New York, pp 1155-1173
Bouloy M (1991) *Bunyaviridae*: Genome organization and replication strategies. Adv Virus Res 40: 235-266
Elliott RM (1990) Molecular biology of the *Bunyaviridae*. J Gen Virol 781: 501-522
Elliott RM, Schmaljohn CS, Collett MS (1991) Bunyaviridae genome structure and gene expression. Curr Top Micro Immunol, Springer-Verlag, Berlin, pp 91-142
Gonzalez-Scarano F, Nathanson N (1990) Bunyaviruses. In: Fields BN, Knipe DM (eds), Virology, 2nd edn Raven Press, New York, pp 1195-1228
Karabatsos N (ed) (1985) International catalogue of arboviruses including certain other viruses of vertebrates. Amer Soc Trop Med Hyg, San Antonio, Texas, USA
Law MD, Speck J, Moyer JW (1992) The M RNA of impatiens necrotic spot tosposvirus (*Bunyaviridae*) has an ambisense genomic organization. Virology 188: 732-741
Schmaljohn CS, Patterson JL (1991) *Bunyaviridae* and their replication. In: Fields BN, Knipe DM, (eds) Fundamental Virology, 2nd ed Raven Press, New York, pp 545-566

Contributed By

Schmaljohn CS, Beaty BJ, Calisher CH, the late Dalrymple JM, Elliott RM, Karabatsos N, Kolakofsky D, Lee HW, Lvov DK, Marriott AC, Nuttall PA, Peters D, Pettersson RF, Shope RE

Genus *Tenuivirus*

Type Species rice stripe virus (RSV)

Virion Properties

Morphology

Virions have a thin filamentous shape; they consist of nucleocapsids, 3-10 nm in diameter, with lengths proportional to the size of their RNA. The filamentous particles may appear to be spiral, branched or circular (Fig. 1). No envelope has been observed.

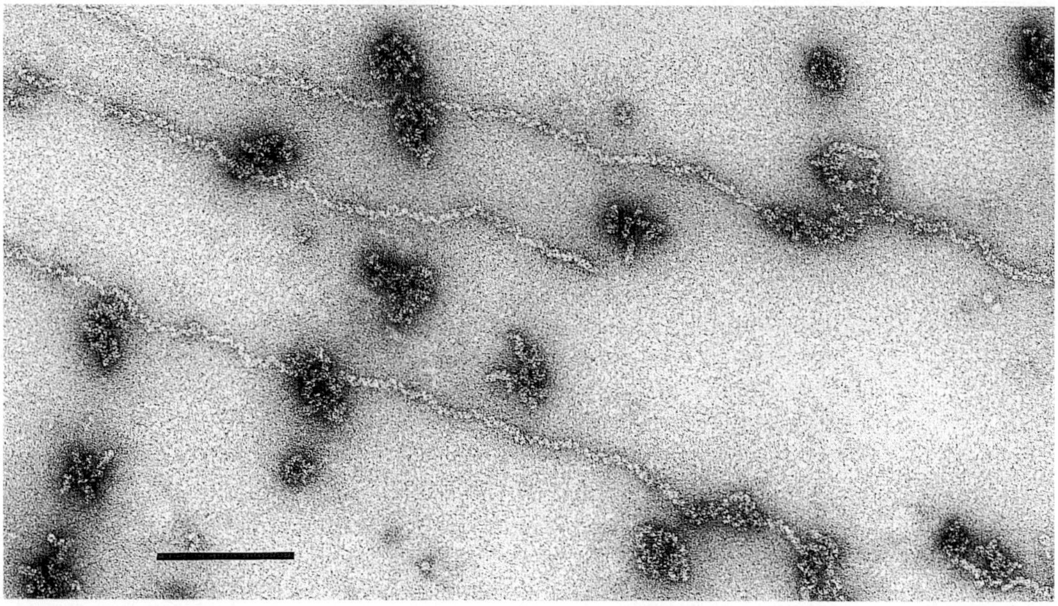

Figure 1: Negative contrast electron micrograph of virions of rice stripe virus. The bar represents 200 nm.

Physicochemical and Physical Properties

Virus preparations are separated into 4 or 5 components by sucrose density gradient centrifugation, but form one component with a buoyant density 1.282-1.288 g/cm^3 when centrifuged to equilibrium in CsCl. The heaviest component is essential for infectivity.

Nucleic Acid

Virions contain ssRNA which is segmented; there are 4 different species, with sizes of 10 kb, 3.4-3.6 kb, 2.3-2.5 kb and 2.0-2.2 kb. Maize stripe virus contains a 5th species of RNA, with a size of 1.3 kb. Virions also contain dsRNA (replicative intermediates). Nucleic acid sequence data for two RNA species of two isolates of rice stripe virus, and maize stripe virus are available.

Proteins

The proteins in nucleocapsid structures has an Mr of 31-34 x 10^3. Two species of coat protein have been detected in rice grassy stunt virus. Non-structural proteins of Mr about 20 x 10^3 have been detected in both plants and viruliferous planthoppers infected with rice stripe virus. A protein of Mr 165 x 10^3 has been found in plants infected with maize stripe virus. Another minor protein, Mr 230 x 10^3 has been detected in rice stripe virus and rice grassy stunt virus. This is a candidate RNA dependent RNA polymerase, the activity of which is associated with filamentous nucleoprotein particles.

Lipids

None reported.

Carbohydrates

None reported.

Genome Organization and Replication

The 3'- and 5'-terminal sequences of each ssRNA are almost complementary for about 20 bases. Either RNA3 or RNA4 of rice stripe virus and maize stripe virus encodes two proteins in an ambisense arrangement. The nucleocapsid protein is encoded by the 5'-proximal region of the negative sense strand of RNA3. A non-structural protein is encoded in the viral sense sequence in the 5'-proximal region of RNA4. The intergenic non-coding region between two ORFs can form a base pair stem configuration (Fig. 2).

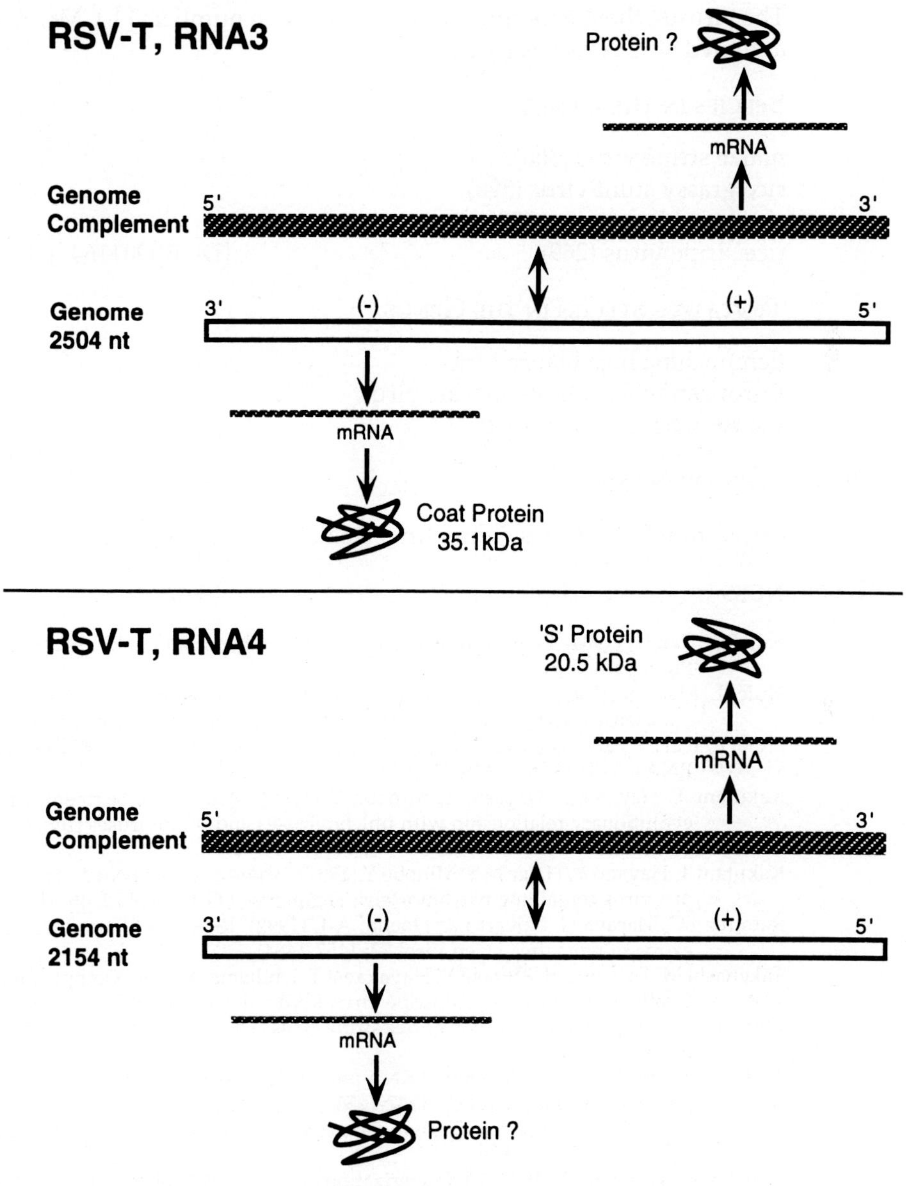

Figure 2: *Tenuivirus* genome organization and strategy of replication of RNA3 and RNA4 of RSV-T isolate.

Antigenic Properties

Rice stripe virus is related serologically to maize stripe virus and distantly related to rice grassy stunt virus. No serological relation has been detected between rice hoja blanca virus, and rice stripe virus or maize stripe virus.

Biological Properties

Host Range

Tenuiviruses are restricted to the host family *Gramineae*.

Transmission

Viruses are transmitted by planthoppers in a persistent manner; in some cases there is transovarial transmission by viruliferous females to progeny. Experimental sap transmission is difficult.

List of Species in the Genus

The viruses, their, genomic sequence accession numbers [ ], CMI/AAB description # ( ), and assigned abbreviations ( ) are:

Species in the Genus

maize stripe virus (300)		(MSpV)
rice grassy stunt virus (320)		(RGSV)
rice hoja blanca virus (299)		(RHBV)
rice stripe virus (269)	[DDBJD01164, DDBJX53563]	(RSV)

Tentative Species in the Genus

Echinochloa hoja blanca virus	(EHBV)
European wheat striate mosaic virus	(EWSMV)
winter wheat mosaic virus	(WWMV)

Derivation of Names

tenui: from Latin *tenuis*, "thin, fine, weak"

References

Falk BW, Tsai JH (1984) Identification of single- and double-stranded RNAs associated with maize stripe virus. Phytopathology 74: 909-915

Huiet L, Klaassen V, Tsai JH, Falk BW (1991) Nucleotide sequence and RNA hybridization analyses reveal an ambisense coding strategy for maize stripe virus RNA3. Virology 182: 47-53

Huiet L, Tsai JH, Falk BW (1992) Complete sequence of maize stripe virus RNA4 and mapping of its subgenomic RNAs. J Gen Virol 73: 1603-1607

Kakutani T, Hayano Y, Hayashi T, Minobe Y (1990) Ambisense segment 4 of rice stripe virus: possible evolutionary relationship with phleboviruses and uukuviruses (*Bunyaviridae*). J Gen Virol 71: 1427-1432

Kakutani T, Hayano Y, Hayashi T, Minobe Y (1991) Ambisense segment 3 of rice stripe virus: the first instance of a virus containing two ambisense segments. J Gen Virol 72: 465-468

Ramirez BC, Macaya G, Calvert LA, Haenni A-L (1992) Rice hoja blanca virus genome characterization and expression *in vitro*. J Gen Virol 73: 1457-1464

Takahashi M, Toriyama S, Kikuchi Y, Hayakawa T, Ishihama A (1990) Complementarity between the 5'- and 3'-terminal sequences of rice stripe virus RNAs. J Gen Virol 71: 2817-2821

Toriyama S (1982) Characterization of rice stripe virus: a heavy component carrying infectivity. J Gen Virol 61: 187-195

Toriyama S (1986) An RNA-dependent RNA polymerase associated with the filamentous nucleoproteins of rice stripe virus. J Gen Virol 67: 1247-1255

Toriyama S (1987) Ribonucleic acid polymerase activity in filamentous nucleoproteins of rice grassy stunt virus. J Gen Virol 68: 925-929

Toriyama S, Watanabe Y (1989) Characterization of single- and double-stranded RNAs in particles of rice stripe virus. J Gen Virol 70: 505-511

Zhu Y, Hayakawa T, Toriyama S, Takahashi M (1991) Complete nucleotide sequence of RNA 3 of rice stripe virus: an ambisense coding strategy. J Gen Virol 72: 763-767

Zhu Y, Hayakawa T, Toriyama S (1992) Complete nucleotide sequence of RNA 4 of rice stripe virus isolate T, and comparison among other isolates and maize stripe virus. J Gen Virol 73: 1309-1312

Contributed By

Toriyama S, Tomaru K

FAMILY *ARENAVIRIDAE*

TAXONOMIC STRUCTURE OF THE FAMILY

Family	***Arenaviridae***
Genus	*Arenavirus*

GENUS *ARENAVIRUS*

Type Species lymphocytic choriomeningitis virus (LCMV)

VIRION PROPERTIES

MORPHOLOGY

Virions are spherical to pleomorphic, 50-300 nm in diameter (mean 110-130 nm), with a dense lipid envelope and a surface layer covered by club-shaped projections, 8-10 nm in length. A variable number of electron dense, 20-25 nm ribosomes are generally present within virus particles. Isolated nucleocapsids, free of contaminating host ribosomes, are organized in closed circles of varying length (450-1300 nm) and display a linear array of nucleosomal subunits.

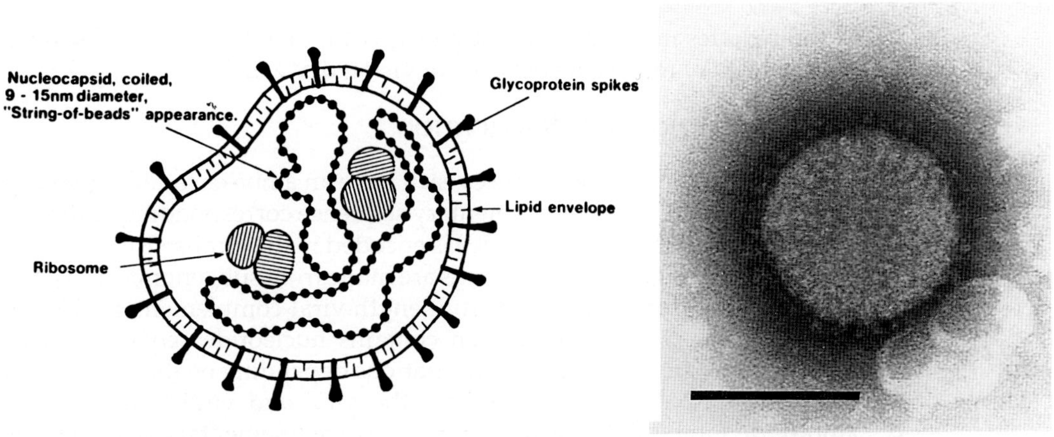

Figure 1: (left) Schematic representation of a section through an arenavirus particle, showing the presence of ribosomes (courtesy of Bishop DHL). The arrangement of the nucleocapsids, ribosomes and surface spikes are hypothetical; (right) negative contrast electron micrograph of Lassa virus, the bar represents 100 nm (courtesy of Lloyd G, Dowsett B).

PHYSICOCHEMICAL AND PHYSICAL PROPERTIES

Virion Mr has not been determined. The S_{20w} is 325-500. The buoyant density in sucrose is about 1.17-1.18 g/cm^3, in CsCl it is about 1.19-1.20 g/cm^3, in amidotrizoate compounds it is about 1.14 g/cm^3. Virions are relatively unstable *in vitro*, and are rapidly inactivated below pH 5.5 and above pH 8.5. Virus infectivity is inactivated at 56° C, or by treatment with organic solvents, or exposure to UV- and gamma- irradiation.

NUCLEIC ACID

RNA constitutes about 2% of the dry weight of virions. The genome consists of two ssRNA molecules, L and S (Mr about 2.2-2.8 x 10^6 and 1.1 x 10^6). The 3' terminal sequences (19-30 nt) are similar between the two RNAs and between different arenaviruses. Overall, they are largely complementary to the 5' end sequences. Although the RNA genomic species may be present in virions in the form of circular nucleocapsids, the genomic RNA is not covalently closed. Variable amounts of full-length viral-complementary RNAs (predominantly S) and viral subgenomic mRNA species can be isolated from virus preparations. Preparations of purified virus may also contain RNAs of cellular origin with sedimentation coefficients of 28S, 18S and 4-6S. These include ribosomal RNAs. The viral mRNA species are presumably associated with encapsidated ribosomes. The proportions of the S to L RNA species are not equimolar apparently due to the packaging of multiple RNA species per virion.

PROTEINS

Proteins constitute about 70% of the dry weight of virions. The most abundant structural protein is a non-glycosylated polypeptide (N or NP, Mr 63-72 x 10^3) found tightly associated with the genomic RNA in the form of a ribonucleoprotein complex. A minor component is the L protein, an RNA polymerase (Mr 200 x 10^3). Two glycosylated proteins (GP-1, GP-2; Mr 34-44 x 10^3) are found in all members of the family and are derived by posttranslational cleavage from an intracellular precursor, GPC; Mr about 75-76 x 10^3. A putative zinc binding protein (Z or p11; Mr 10-14 x 10^3) is apparently an internal structural component of the virus. Other minor proteins and enzymatic activities have been described associated with virions including poly (U) and poly (A) polymerases and a protein kinase that can phosphorylate N. Whether these represent virally encoded enzymes or not is unclear.

LIPIDS

Lipids represent about 20% of virion dry weight and are similar in composition to those of the host plasma membrane.

CARBOHYDRATES

Carbohydrates in the form of complex glycans on GP-1 (5 or 6 sites in LCMV) and GP-2 (2 sites in LCMV) represent about 8% of virion dry weight.

GENOME ORGANIZATION AND REPLICATION

The L and S RNAs of arenaviruses each have an ambisense coding arrangement (Fig. 2). N is encoded in the viral-complementary sequence corresponding to the 3' half of S, while the viral glycoprotein precursor (GPC) is encoded in the viral-sense sequence corresponding to the 5' half of S (Fig. 2). The 2 proteins are made from subgenomic mRNA species transcribed from the viral (for N mRNA) or full-length viral-complementary S RNA species (for GPC mRNA). The S intergenic region contains nucleotide sequences with the potential of forming one or more hairpin configurations depending on the virus. These may function to terminate mRNA transcription from the viral and viral-complementary S RNAs. The ambisense viral L RNA encodes in its viral-complementary sequence the L protein and in the viral-sense 5' end sequence the Z protein. The Z mRNA is small (0.5 kb). The mRNAs are capped and contain 1-5 non-templated nucleotides of heterogeneous sequence at their 5' ends. The mRNAs are not polyadenylated. The transcription mechanism is not fully elucidated. Initiation of transcription may involve cap-snatching. The 3' termini of the mRNAs have been mapped to locations in the intergenic regions. No specific termination sequence can be identified, but characteristic GC-rich, strongly base-paired stem-loop structures in these regions may cause termination.

The process of infection involves attachment to cell receptors (undefined), entry via the endosomal route, uncoating and mRNA transcription in the cytoplasm of infected cells. In view of the ambisense coding arrangement, only N and L mRNAs can be synthesized from the genomic RNAs by the virion polymerase prior to translation. The products of these mRNAs are presumed to be involved in the synthesis of full-length viral complementary species which serve as templates for the synthesis of GPC and Z mRNA and the synthesis of full-length viral RNAs. The process of RNA replication which may involve a slippage mechanism during initiation, and read-through of transcription termination signals, has not been fully elucidated. However, the presence of full-length viral-complementary genomic RNAs and viral subgenomic mRNA species in virus preparations may affect this perceived temporal order of RNA and protein synthesis.

The viral envelope glycoproteins are synthesized in cells as a single mannose-rich precursor molecule which is proteolytically cleaved and processed to contain complex glycans during transport to the plasma membrane. Virions mature by budding at sites on the surface of cells. Ribosomes are also observed at such sites.

Arenavirus strains have the ability to form intrastrain reassortant progeny, including diploid (or multiploid) species with respect to the genomic RNA segments. Some evidence for interspecies reassortment between Lassa and Mopeia viruses has been obtained.

The replication *in vitro* of a number of arenaviruses is inhibited by a variety of antibiotics, including amantadine, alpha-amanitin, glucosamine, and thiosemicarbazones. Ribavirin inhibits the replication of several arenaviruses *in vitro* and is effective in the therapy of humans and primates infected with LASV.

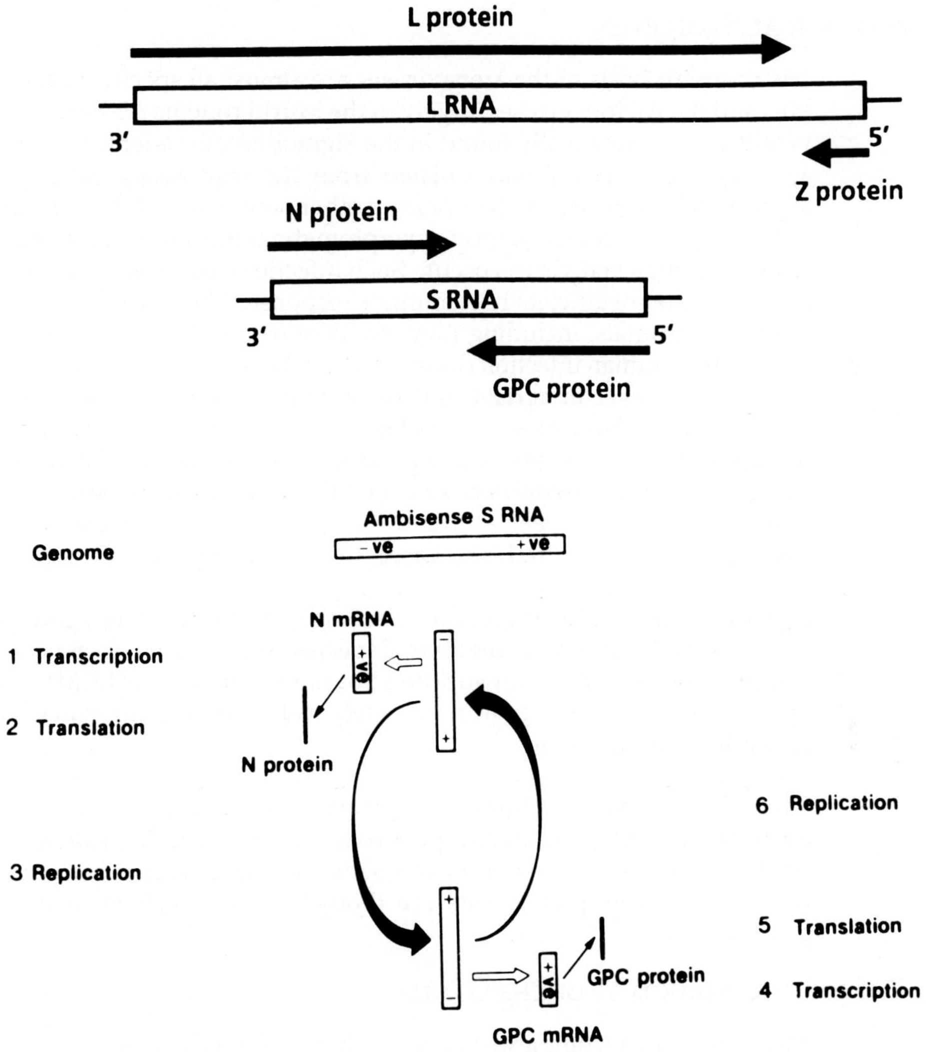

Figure 2: (upper) Organizations of the arenavirus L and S RNAs; (lower) the replication strategy of the ambisense S RNA of arenaviruses (the L RNA is comparable) (courtesy of Bishop DHL).

ANTIGENIC PROPERTIES

Viruses possess a number of distinct antigenic determinants (more than 3) as shown by monoclonal and polyclonal antibody analyses. Antigens on the 44 kDa G1 of LCMV are involved in virus neutralization. These are type-specific, although cross-neutralization tests have demonstrated partially shared antigens between Tacaribe virus and Junin virus and cross-protection against Junin virus following prior infection by Tacaribe virus, or against Lassa virus following infection by Mopeia virus. Major CF antigens are associated with the viral nucleoproteins. CF antigens have been used to define the Tacaribe complex of arenaviruses. Monoclonal antibodies react with common epitopes on the nucleocapsid proteins of all arenaviruses and a single highly conserved epitope has also been described in the transmembrane GP-2 glycoprotein.

By monoclonal and polyclonal antibody analyses, the African arenaviruses (IPPYV, LASV, MOBV, MOPV,) are distinguishable from the New World arenaviruses (TACV complex viruses). Fluorescent antibody studies show that antisera against all TACV complex viruses, as well as those against LASV complex viruses, react with LCMV. Cytotoxic T-lymphocyte epitopes exist on the nucleoprotein and glycoprotein of LCMV. The number and location of epitopes varies depending on the virus strain and host MHC class I molecules. No hemagglutinin has been identified.

BIOLOGICAL PROPERTIES

The reservoir hosts of the arenaviruses are almost all specific rodents. LCMV is found in *Mus* and the African viruses largely in the Murid rodents *Mastomys* and *Praomys*. The New World viruses are mostly found in the Sigmodontine rodents *Calomys*, *Neacomys*, *Orzomys* and *Sigmodon*. TACV was isolated from the fruit-eating bat *Artibeus*, but subsequent attempts to recover it from bats or from other potential hosts have failed. Most of the viruses induce a persistent, frequently asymptomatic infection in their reservoir hosts, in which chronic viremia and viruria occur. Such infections are known or suspected to be caused by a slow and/or insufficient host immune response. The natural spread of many arenaviruses to other mammals, including humans, is unusual. However, Lassa virus is the cause of widespread human infection (Lassa fever) in West Africa, and Junin virus causes Argentine hemorrhagic fever in agricultural workers in an increasingly large area of that country. Machupo virus has caused isolated outbreaks of similar disease in Bolivia, and a recently identified member of the family, Guanarito virus, is associated with human disease in Venezuela. Human infection with LCMV occurs in some urban areas with high rodent populations, and has been acquired from pet hamsters. Severe laboratory-acquired infections have occurred with LCM, Lassa, Junin, Machupo and Flexal viruses.

Experimental infection in laboratory animals (mouse, hamster, guinea pig, rhesus monkey, marmoset, rat) varies with the animal species and the virus. In general, viruses of the TACV complex are pathogenic for suckling but not weaned mice; LCMV and LASV produce the opposite effect. Viruses grow moderately well in many mammalian cells. LCMV can grow in murine T-lymphocytes.

Vertical, venereal and horizontal transmission occurs in the natural hosts, including transuterine, transovarian and post-partum, and by milk-, saliva- or urine-borne routes. Horizontal transmission within and between species occurs by contamination and aerosol routes. No arthropod vectors are thought to be involved in the normal transmission process.

TAXONOMIC STRUCTURE OF THE GENUS

Two serogroups (complexes) of arenaviruses are recognized. These are the LCMV-LASV complex, or Old World arenaviruses, and the TACV complex, or New World arenaviruses. Phylogenetic analysis of currently available amino acid sequences of viral proteins are consistent with this division and provide further data on the relationships between arenaviruses. Such relationships are the same when either N, or G1 (GP-1) or G2 (GP-2) sequences are considered. They show that two strains of Lassa virus (Josiah and GA391, from Nigeria and Sierra Leone) are quite closely related and that another African virus, Mopeia virus, is rather more divergent. The two strains of LCMV are closely related and both are distantly related to the African viruses. Of the New World viruses, Pichinde virus appears to diverge quite extensively from the other three viruses for which sequence data are available (Tacaribe, Machupo, Junin). These 3 viruses are rather closely related to each other.

LIST OF SPECIES IN THE GENUS

The groups, viruses, their genomic sequence accession numbers [ ] and assigned abbreviations ( ) are:

SPECIES IN THE GENUS

1-LCMV-LASV complex (Old World arenaviruses):		
Ippy virus		(IPPYV)
Lassa virus	[LAS-GA391 S:X52400 LAS-Josiah S:J04324]	(LASV)
lymphocytic choriomeningitis virus	[LCM-ARM L:J04331, M27693, S:M20869, LCM-WE S:M22138]	(LCMV)
Mobala virus		(MOBV)
Mopeia virus	[MOP-800150 S:M33879]	(MOPV)
SPH 114202 virus (Brazil)		
2-Tacaribe complex (New world arenaviruses):		
Amapari virus		(AMAV)
Flexal virus		(FLEV)
Guanarito virus		(GUAV)
Junin virus	[JUN-MC2 S:D10072]	(JUNV)
Machupo virus	[AA288-77 S:X62616]	(MACV)
Parana virus		(PARV)
Pichinde virus	[PIC 3739 S:K02734]	(PICV)
Tacaribe virus	[TAC-TRVLII 573 L:J04340 M33513, S:M20304]	(TACV)
Tamiami virus		(TAMV)

TENTATIVE SPECIES IN THE GENUS

Sabio virus

LIST OF UNASSIGNED VIRUSES IN THE FAMILY

None reported.

SIMILARITY WITH OTHER TAXA

None reported.

DERIVATION OF NAMES

arena: from Latin *arena*, "sand" in recognition of the sandy-like ribosomal contents of particles in thin section

REFERENCES

Bishop DHL (1990) *Arenaviridae* and their replication. In: Fields BN, Knipe DM (eds) Virology 2nd edn Raven Press, New York, pp 1231-1243

Buchmeier MJ, Parekh BS (1987) Protein structure and expression among arenaviruses. Curr Topics Microb Immunol 133: 41-57

Salvato M (ed) The *Arenaviridae*. Plenum Press, New York, in press

Garcin D, Kolakofsky D (1992) Tacaribe arenavirus RNA synthesis in vitro is primer dependent and suggests an unusual model for the initiation of genome replication. J Virol 66: 1370-1376

Whitton JL (1990) Lymphocytic choriomeningitis virus CTL. Semin Virol 1: 257-262

CONTRIBUTED BY

Buchmeier MJ, Clegg JCS, Franze-Fernandez MT, Kolakofsky D, Peters CJ, Southern PJ

FAMILY LEVIVIRIDAE

TAXONOMIC STRUCTURE OF THE FAMILY

Family	***Leviviridae***
Genus	*Levivirus*
Genus	*Allolevivirus*

VIRION PROPERTIES

MORPHOLOGY

Virions are spherical and exhibit icosahedral symmetry (T=3); they have a diameter of about 26 nm. There is no envelope.

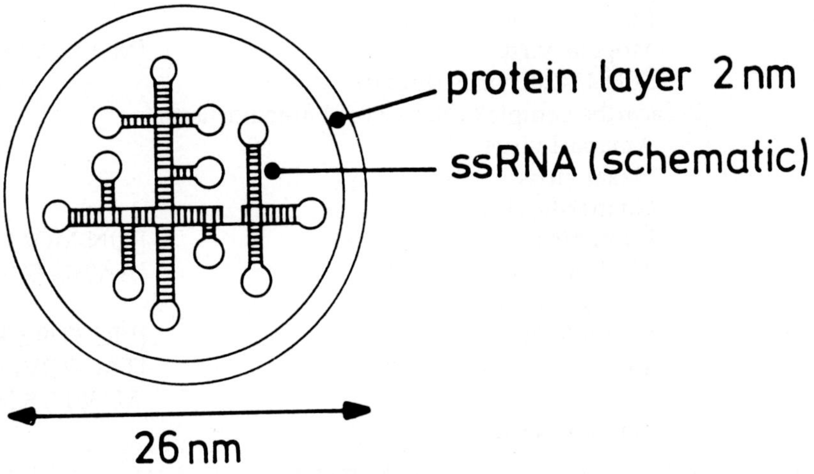

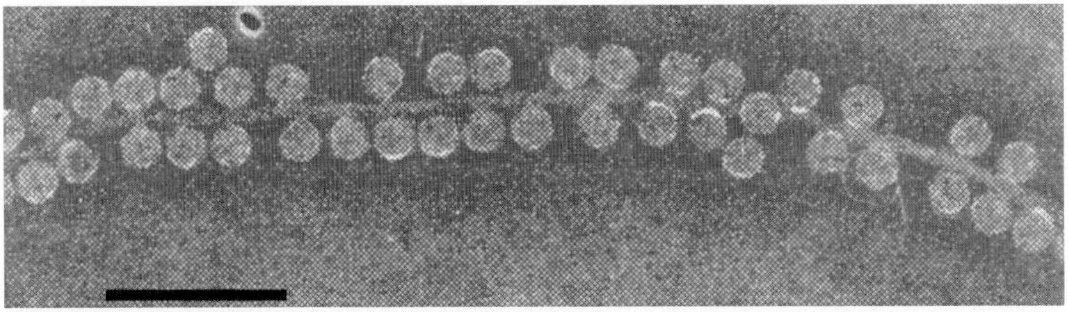

Figure 1: (upper) Diagram of a enterobacteria phage R17 virion in section; (lower) negative contrast electron micrograph of enterobacteria phage MS2. The bar represents 100 nm.

PHYSICOCHEMICAL AND PHYSICAL PROPERTIES

Virion Mr varies between 3.6 and 4.2 x 10^6 depending on the genus. The range in S_{20w} value is from 80 to 84; buoyant density in CsCl is 1.46 g/cm^3. Infectivity is ether and chloroform resistant but sensitive to detergents. Inactivation by UV light and chemicals is comparable to that of other icosahedral viruses containing ssRNA.

NUCLEIC ACID

Virions contain one molecule of positive sense ssRNA ranging in size from 3,466 to 4,276 nt; size and gene arrangement vary with genus. The RNA makes up 30% of the virion weight in almost equimolar amounts of each of the four bases.

Proteins

The capsid contains 180 copies of the coat protein (Mr 14 x 10^3), arranged in 60 identical triangular units which are related by the symmetry elements of an icosahedron. The structure of the protein shell of MS2 has been resolved by X-ray crystallography. The coat protein has no structural similarity to that of other icosahedral RNA viruses. The capsid contains one copy of the A protein (Mr 35-44 x 10^3), which is required for maturation of the virion and for pilus attachment.

Lipids

None reported.

Carbohydrates

None reported.

Genome Organization and Replication

Phages infect by adsorption to the sides of pili. The specificity of this adsorption is determined by a wide variety of different plasmids. The coliphages attach to F pili, which leads to cleavage of the A protein and release of the RNA from the virion. The infecting RNA encodes a replicase, which assembles with four host proteins (ribosomal protein S1, EF-Tu, EF-Ts and a 'host factor') to form the active replicase holoenzyme. This enzyme synthesizes a free negative strand which is the template for positive strand synthesis. Late in infection the coat protein acts as a translational repressor of the replicase gene. Capsids assemble in the cytoplasm around phage RNA. Infection usually results in cell lysis releasing some thousand phages per cell.

Biological Properties

Host Range

The viruses infect enterobacteria, species of the genera *Caulobacter* and *Pseudomonas* and possibly many other gram-negative bacteria, provided that they express appropriate pili on their surface.

Genus *Levivirus*

Type Species enterobacteria phage MS2 (MS2)

Distinguishing Features

Viruses contain the short version of the genome, and have a separate gene for cell lysis, which partly overlaps the replicase coding region in the +1 mode. Overlap with the coat protein gene is variable. Synthesis of the lysis protein is dependent on translation of the coat protein gene. Genome size ranges from 3,466 (GA) to 3,569 nt (MS2), depending on the subgroup.

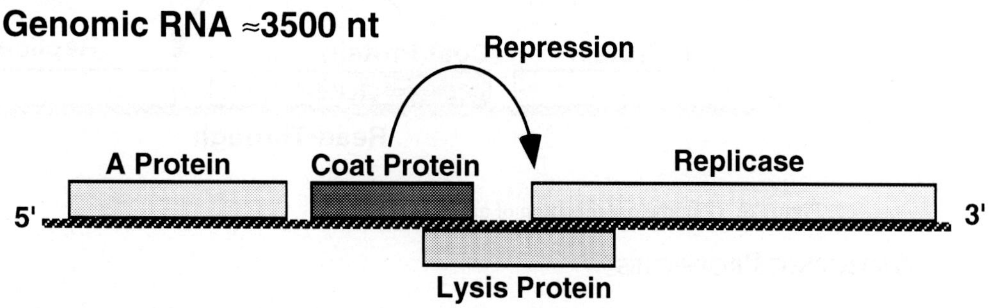

Figure 2: Genome organization of a levivirus.

ANTIGENIC PROPERTIES

Antigenic specificity is distinct from that of members of the genus *Allolevivirus*.

LIST OF SPECIES IN THE GENUS

The groups, viruses and their assigned abbreviations ( ) are:

SPECIES IN THE GENUS

1-Subgroup I:	
enterobacteria phage f2	(f2)
enterobacteria phage fr	(fr)
enterobacteria phage JP501	(JP501)
enterobacteria phage M12	(M12)
enterobacteria phage MS2	(MS2)
enterobacteria phage R17	(R17)
2-Subgroup II:	
enterobacteria phage BZ13	(BZ13)
enterobacteria phage GA	(GA)
enterobacteria phage JP34	(JP34)
enterobacteria phage KU1	(KU1)
enterobacteria phage TH1	(TH1)

TENTATIVE SPECIES IN THE GENUS

None reported.

LIST OF UNASSIGNED VIRUSES IN THE GENUS

Caulobacter phage PP7 (PP7)

GENUS *ALLOLEVIVIRUS*

Type species enterobacteria phage Qβ (Qβ)

DISTINGUISHING FEATURES

Viruses contain the longer version of the genome. The extra RNA encodes a C terminal extension of the coat protein arising by occasional suppression of the coat gene termination codon. The read-through protein is a minor constituent of the capsid and is necessary for infection. There is no separate lysis gene. Cell lysis is ascribed to the A protein. Genome length varies between 4,217 (Qβ) and 4,276 nt (SP), depending on subgroup.

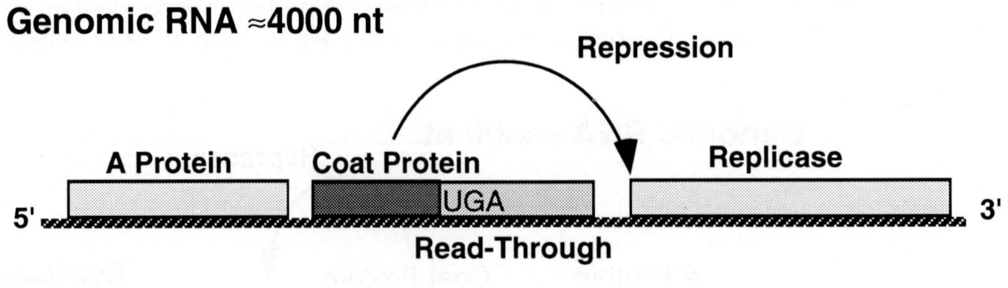

Figure 3: Genome organization of an allolevivirus.

ANTIGENIC PROPERTIES

Antigenic specificity is distinct from that of members of the genus *Levivirus*.

List of Species in the Genus

The groups, viruses and their assigned abbreviations are ( ) are :

Species in the Genus

1-Subgroup III:	
enterobacteria phage M11	(M11)
enterobacteria phage Qβ	(Qβ)
enterobacteria phage ST	(ST)
enterobacteria phage TW18	(TW18)
enterobacteria phage VK	(VK)
2-Subgroup IV:	
enterobacteria phage FI	(FI)
enterobacteria phage ID2	(ID2)
enterobacteria phage NL95	(NL95)
enterobacteria phage SP	(SP)
enterobacteria phage TW28	(TW28)

Tentative Species in the Genus

None reported.

Other Members of the Family

Not yet allocated to genus:

1-Caulobacter:	
Caulobacter phage øCb2	(øCb2)
Caulobacter phage øCb4	(øCb4)
Caulobacter phage øCb5	(øCb5)
Caulobacter phage øCb8r	(øCb8r)
Caulobacter phage øCb9	(øCb9)
Caulobacter phage øCb12r	(øCB12r)
Caulobacter phage øCb23r	(øCb23r)
Caulobacter phage øCP2	(øCP2)
Caulobacter phage øCP18	(øCP18)
Caulobacter phage øCr14	(øCr14)
Caulobacter phage øCr28	(øCr28)
2-Enterobacteria:	
enterobacteria phage B6	(B6)
enterobacteria phage B7	(B7)
enterobacteria phage C-1	(C-1)
enterobacteria phage C2	(C2)
enterobacteria phage fcan	(fcan)
enterobacteria phage Folac	(Folac)
enterobacteria phage Iα	(Iα)
enterobacteria phage M	(M)
enterobacteria phage pilHα	(pilHα)
enterobacteria phage R23	(R23)
enterobacteria phage R34	(R34)
enterobacteria phage ZG/1	(ZG/1)
enterobacteria phage ZIK/1	(ZIK/1)
enterobacteria phage ZJ/1	(ZJ/1)
enterobacteria phage ZL/3	(ZL/3)
enterobacteria phage ZS/3	(ZS/3)
enterobacteria phage α15	(α15)
enterobacteria phage β	(β)
enterobacteria phage μ2	(μ2)

enterobacteria phage τ (τ)
other enterobacteria phages, with many plasmid specificities, have been reported.

3-Pseudomonas:
Pseudomonas phage 7s (7s)
Pseudomonas phage PRR1 (PRR1)

Derivation of Names

levi: from Latin *levis*, 'light'

References

Ackermann H-W, DuBow MS (eds) (1987) Viruses of Prokaryotes, Vol II. CRC Press, Boca Raton FL, pp 171-218

Fiers W (1979) Structure and function of RNA bacteriophages. In: Fraenkel-Conrat H, Wagner RR (eds) Comprehensive Virology, Vol 13. Plenum Press, New York, pp 69-204

Furuse K (1987) Distribution of the coliphages in the environment. In: Goyal SM, Gerber CP, Bitton G (eds) Phage Ecology. John Wiley & Sons, NewYork, pp 87-124

van Duin J (1988) Single-stranded RNA bacteriophages. In: Calendar R (ed) The Bacteriophages. Plenum Press, New York, pp 117-167

Valegaard K, Liljas L, Fridborg K, Unge T (1990) The three-dimensional structure of the bacterial virus MS2. Nature 345: 36-41

Zinder ND (ed) (1975) RNA phages. Cold Spring Harbor Laboratory, Monograph Series, Cold Spring Harbor, N.Y.

Contributed By

van Duin J

FAMILY *PICORNAVIRIDAE*

TAXONOMIC STRUCTURE OF THE FAMILY

Family ***Picornaviridae***
Genus *Enterovirus*
Genus *Rhinovirus*
Genus *Hepatovirus*
Genus *Cardiovirus*
Genus *Aphthovirus*

VIRION PROPERTIES

MORPHOLOGY

Virions are icosahedral (T=1, pseudo T=3) with no envelope; the core consists of ssRNA and a small protein ($3B^{VPg}$) covalently linked to its 5'-end. Electron micrographs reveal no projections, the surface being almost featureless (Fig. 1). Hydrated native particles are 30 nm in diameter but vary from 22-30 nm in micrographs due to drying and flattening during preparation. They sometimes form long ribonucleoprotein strands upon heating at slightly alkaline pH. The capsid is composed of 60 protein subunits (protomers, P1 gene products, Fig. 2), each consisting of four proteins (three of Mr 24-41 x 10^3 e.g., poliovirus VP2, VP3, VP1, and one of Mr 5.5-13.5 x 10^3, e.g., poliovirus VP4). Protomers vary from 80 kDa for aphthovirus to 97 kDa for polioviruses and some may be incompletely cleaved (e.g., the P1 derived poliovirus VP0 precursor to VP4 and VP2). The atomic structures of representative viruses of four of the five picornavirus genera have been solved and are very similar to each other and to certain T=3 icosahedral plant viruses.

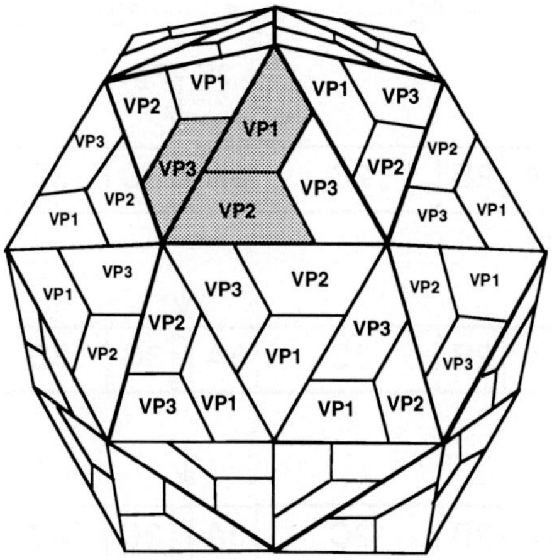

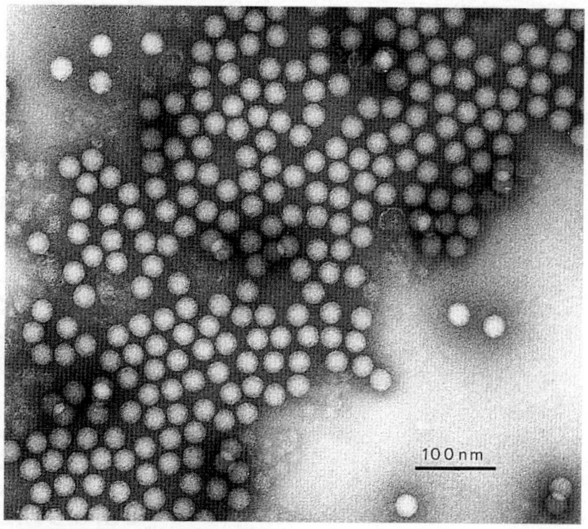

Figure 1: (left) Diagram of poliovirus virion surface showing proteins VP1, VP2 and VP3. The fourth capsid protein, VP4, is located about the internal surface of the pentameric apex of the icosahedron. (right) Negative contrast electron micrograph of poliovirus, the bar represents 100 nm.

PHYSICOCHEMICAL AND PHYSICAL PROPERTIES

Virion Mr is 8-9 x 10^6, S_{20w} is 140-165. Buoyant density in CsCl is 1.33-1.45 g/cm^3, depending on the genus. Some species are unstable below pH 7; many are less stable at low ionic strength than at high ionic strengths. Virions are insensitive to ether, chloroform, or non-ionic detergents. Viruses are inactivated by light when grown with, or in the presence of photodynamic dyes such as neutral red or proflavin. Virions are stabilized by divalent cations. Thermal stability varies with the genus.

NUCLEIC ACID

Virions contain one molecule of infectious, positive sense, ssRNA, 7-8.5 kb in size. A poly (A) tract, heterogenous in length, is located after the 3'-terminal heteropolymeric sequence. A small protein, VPg (Mr about 24 x 10^3), is linked covalently to the 5' terminus. The 5' non-coding region of the genome is believed to possess extensive secondary structure essential to its function. Some viruses have poly (C) tracts in that region (Fig. 2). The sequence identity between viruses of different genera is typically less than 40%.

PROTEINS

Virion proteins include 60 copies each of the four capsid proteins (P1 gene products IA, IB, IC, ID such as poliovirus VP4, VP2, VP3, VP1, respectively, (Fig. 2) and a single copy of the genome linked protein $3B^{VPg}$. In lieu of one or more of the copies of VP4 and VP2 a precursor VP0 protein is commonly identified in virions.

LIPIDS

Virions lack lipids. Some strains of poliovirus may carry 60 molecules each of a sphingosine-like molecule. Polypeptide 1A (VP4), located on the inner surface of the capsid, has a molecule of myristic acid covalently attached to the amino terminal glycine.

CARBOHYDRATES

None of the viral proteins is glycosylated.

GENOME ORGANIZATION AND REPLICATION

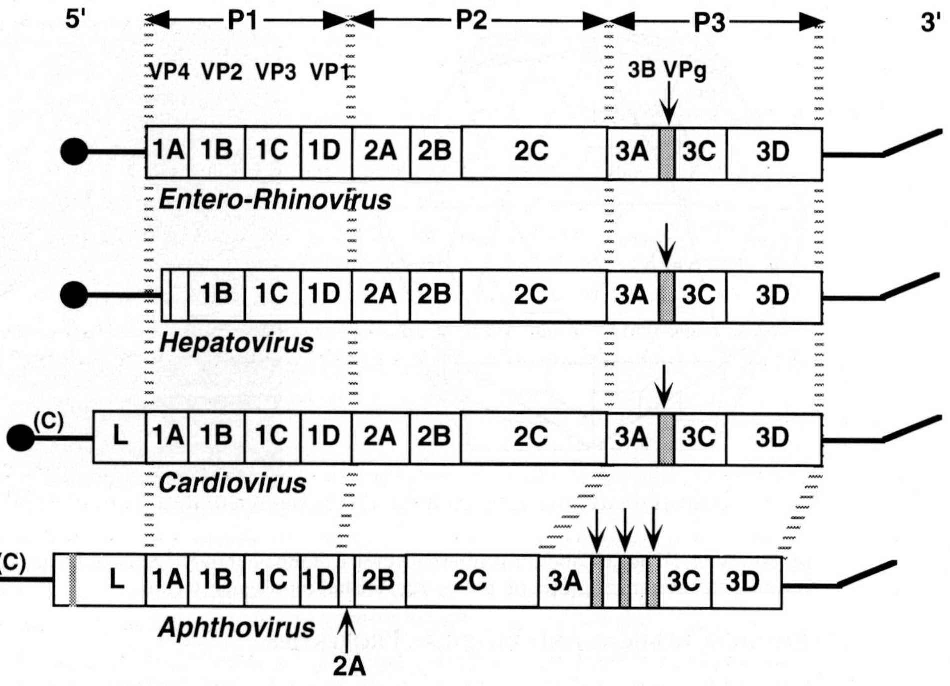

Figure 2: Genome structure and gene organization of picornaviruses. The filled circle at the 5' end is the genome-linked protein VPg (also referred to as the 3B gene product), followed by the 5' non-translated region (5' NTR; line). Letters (C) above the 5' NTR depict poly (C) tracks that are present in some viruses. The 1A gene products are myristylated at the amino terminal glycine, although the situation for hepatoviruses is not known. The open boxes depict the long ORF encoding the polyprotein that is followed by the 3' non-translated region (line) and a poly (A) track (angled line). The eventual cleavage products of the polyprotein are indicated by vertical lines in the boxes, the nomenclature of the polypeptides follows an L:4:3:4 scheme corresponding to the genes (numbers) encoded in the L, P1, P2, P3 regions (Rueckert and Wimmer, 1984). The P1 region encodes the

structural proteins 1A, 1B, 1C and 1D, usually referred to as VP4, VP2, VP3, and VP1, respectively. VP0, not shown here, is the intermediate precursor for VP4 and VP2. In all viruses 3C is a protease, in enteroviruses and rhinoviruses 2A is a protease, while in all viruses 3D is considered to be a component of the RNA replicase. Only aphthoviruses encode 3 VPg proteins that map in tandem.

The genome consists of a ssRNA with a 5' untranslated sequence of variable length followed by an ORF encoding the polyprotein precursor (Mr 240-250 x 10^3) to the structural proteins (P1) and the predominantly nonstructural proteins (P2, P3), followed by a short non-coding sequence and a poly (A) tract of variable length. In some viruses the structural proteins are preceded by a leader protein (L) (Fig. 2). The polyprotein is processed to functional proteins by proteases. One or two of the nonstructural proteins have proteolytic activity (e.g., depending on the virus: L^{pro}, $2A^{pro}$, $3C^{pro}$, some of which, such as the $2A^{pro}$ of cardioviruses and aphthoviruses, are believed to act only in *cis*), other nonstructural proteins include a polymerase ($3D^{pol}$), an ATPase (2C), as well as proteins of unknown function (2B, 3A). The leader protein of aphthoviruses has proteolytic activity (L^{pro}) while that of cardioviruses does not. Intermediates in the polyprotein cleavage process may exhibit functions (e.g., proteolytic activities associated with the poliovirus 3CD intermediate).

Virus entry into cells is believed to involve specific cellular receptors. Initiation of protein synthesis involves recognition sites in the long 5' non-coding region (600-1500 nt in length) which has extensive secondary structure which is believed to be essential to its function as an internal ribosome entry site. Protein synthesis is often accompanied by inhibition of cap-dependent translation of certain cellular mRNAs.

Replication of viral RNA occurs in complexes associated with cytoplasmic membranes. Many compounds that specifically inhibit replication have been described. Mutants resistant to, or dependent on drugs have been reported. Genetic recombination, complementation and phenotypic mixing occur. Defective interfering (DI) particles have been produced experimentally but have not been observed in natural virus populations. They appear only under extreme selection pressure. Infection is generally cytolytic, but persistent infections are common with some species and reported with others.

ANTIGENIC PROPERTIES

Native virions are antigenically type specific (designated "N" or "D" for poliovirus), but after gentle heating are converted to group specificity (designated "H" or "C" for poliovirus). Neutralization by antibody follows first-order inactivation kinetics. Species (equivalent to serotypes) are classified by cross-protection neutralization of infectivity, complement-fixation, specific ELISA using a capture format, or immunodiffusion. Some species can be identified by hemagglutination. Antigenic sites, defined by mutations that confer resistance to neutralization by monoclonal antibodies, typically number 3 or 4 per protomer.

BIOLOGICAL PROPERTIES

Most picornaviruses are specific for one, or a very few host species. Exceptions are the encephalomyocarditis viruses which have been isolated from over 30 host species including mammals, birds and insects, and aphthoviruses which may infect a least 200 species of mammals. Most species can be grown in cell culture. Resistant host cells (e.g., mouse cells in the case of the primate-specific polioviruses) can often be infected (single round) by transfection with naked, infectious RNA. Rhinoviruses and many enteroviruses grow poorly, or not at all, in laboratory animals. Transmission is horizontal, mainly by fecal-oral, fomite or airborne routes. Transmission by arthropod vectors is not known, although EMCV has been isolated from three species of mosquitoes and two species of ticks.

GENUS ENTEROVIRUS

Type Species poliovirus 1 (PV-1)

DISTINGUISHING FEATURES

Virions are stable at acid pH. Buoyant density in CsCl is 1.30-1.34 g/cm^3. Empty capsids are often observed in virus preparations. Sometimes a small proportion (about 1% of the population) of heavy particles (density: 1.43 g/cm^3) are observed. Genomes encode a single VPg and no L protein. Sequence identities for different enteroviruses, or between enteroviruses and rhinoviruses are more than 50% over the genome as a whole. Strains within a species have more than 75% sequence identity over the genome as a whole. Viruses grouped by biological criteria, e.g., the polioviruses, or Coxsackie B viruses, are generally closely related in terms of overall nucleotide sequence identity over the genome as a whole. Viruses primarily multiply in the gastrointestinal tract, but they can also multiply in other tissues, e.g., nerve, muscle, etc. Many different cell surface molecules, most of them unknown, serve as viral receptors. Infection may frequently be asymptomatic. Clinical manifestations include mild meningitis, encephalitis, myelitis, myocarditis and, conjunctivitis.

LIST OF SPECIES IN THE GENUS

Swine vesicular disease virus [D00435] is very similar to human coxsackievirus B5. Certain virus isolates initially reported as novel echoviruses were later shown to have been misidentified. Thus E8 was E1, E10 was a reovirus, E28 was rhinovirus type A1. Similarly coxsackievirus A23 was echovirus 9. Echovirus 22 is distinctive in its genome sequence (exhibiting little or no identity to any other picornavirus) and to some degree in its *in vitro* growth properties. However, its biophysical properties, clinical presentation and occurrence currently support its classification as an atypical enterovirus.

The viruses, serotypes (numbers), their genomic sequence accession numbers [ ] and assigned abbreviations ( ) are:

SPECIES IN THE GENUS

bovine enterovirus 1	[D00214]	(BEV-1)
bovine enterovirus 2		(BEV-2)
human coxsackievirus A 1 to 22	[D00538]	(CAV-1 to 22)
human coxsackievirus A 24		(CAV-24)
human coxsackievirus B 1 to 6	[M33854]	(CBV-1 to 6)
human echovirus 1 to 7		(EV-1 to 7)
human echovirus 9		(EV-9)
human echovirus 11 to 27		(EV-11 to 27)
human echovirus 29 to 33		(EV-29 to 33)
human enterovirus 68 to 71		(HEV68 to 71)
human poliovirus 1	[V01150]	(HPV-1)
human poliovirus 2		(HPV-2)
human poliovirus 3		(HPV-3)
porcine enterovirus 1 to 11		(PEV-1 to 11)
simian enterovirus 1 to 18		(SEV-1 to 18)
Vilyuisk virus		

TENTATIVE SPECIES IN THE GENUS

None reported.

Genus *Rhinovirus*

Type Species human rhinovirus 1A (HRV-1A)

Distinguishing Features

Virions are unstable below pH 5-6. They exhibit buoyant densities in CsCl of 1.38-1.42 g/cm^3. The nucleotide sequence identity over the entire genome for different species of *Rhinovirus*, or between enteroviruses and rhinoviruses is more than 50%, although it may be greater or less than this for particular genomic regions. Human rhinoviruses can be divided into major and minor receptor group viruses; the receptor for the major group is ICAM-1. Others are not defined. Clinical manifestations include the common cold and other upper and lower respiratory tract illnesses of human.

List of Species in the Genus

The viruses, serotypes (numbers), their genomic sequence accession numbers [ ] and assigned abbreviations ( ) are:

Species in the Genus

bovine rhinovirus 1		(BRV-1)
bovine rhinovirus 2		(BRV-2)
bovine rhinovirus 3		(BRV-3)
human rhinovirus 1A	[K02121, K02021]	(HRV-1A)
human rhinovirus 1 to 100		(HRV-1 to 100)

Tentative Species in the Genus

None reported.

Genus *Hepatovirus*

Type Species hepatitis A virus (HAV)

Distinguishing Features

Viruses are very stable, resistant to acid pH and elevated temperatures (60° C for 10 min.). Buoyant density in CsCl is 1.32-1.34 g/cm^3. The viruses infect liver cells, causing disease in those tissues, and are found in feces at high titre shortly before clinical signs of hepatitis develop. Viruses are strongly conserved in their antigenic properties and generally establish persistent virus infections *in vitro*. The VP4 protein (1A gene product), if it exists at all, is small. There is little similarity between the genome sequences of hepatoviruses and those of enteroviruses, or rhinoviruses. Nucleotide sequence identity between different hepatitis A strains, as determined by amplification of limited regions of the genomes of viruses from unpassaged material, is greater than 80%. Clinical manifestations are hepatitis and gastroenteritis.

List of Species in the Genus

The viruses, their genomic sequence accession numbers [ ] and assigned abbreviations ( ) are:

Species in the Genus

hepatitis A virus	[M14707]	(HAV)
simian hepatitis A virus		(SHAV)

Tentative Species in the Genus

None reported.

Genus *Cardiovirus*

Type Species encephalomyocarditis virus (EMCV)

Distinguishing Features

Virion buoyant density in CsCl is 1.33-1.34 g/cm^3. The viruses have a poly (C) tract of variable length (usually 80-250 bases) about 150 bases from the 5' terminus of the viral RNA. Empty capsids are only seen rarely, if ever. The viral genome encodes an L protein. Clinical manifestations include encephalitis and myocarditis in mice and certain other animals. The nucleotide sequence identity over the entire genome for different species of cardiovirus is more than 50%.

List of Species in the Genus

The viruses, their alternative names ( ), genomic sequence accession numbers [ ] and assigned abbreviations ( ) are:

Species in the Genus

encephalomyocarditis virus	[M81861]	(EMCV)
(Columbia SK virus)		
(mengovirus)		
(mouse Elberfield virus)		
Theiler's murine encephalomyelitis virus	[M20562]	(TMEV)
(murine poliovirus)		

Mengovirus, Columbia SK virus and mouse Elberfield virus are best regarded as strains of EMCV, based on serological cross-reaction and sequence identity. Theiler's encephalomyelitis virus, also known as murine poliovirus, lacks a poly (C) tract but has 54% nucleotide sequence identity with EMCV and less than 40% with other picornavirus groups. The location and nature of its antigenic sites are comparable to those of the other cardioviruses.

Tentative Species in the Genus

None reported.

Genus *Aphthovirus*

Type Species foot-and-mouth disease virus O (FMDV-O)

Distinguishing Features

Virions are unstable below pH 6.5. Virion buoyant density in CsCl is 1.43-1.45 g/cm^3. Poly (C) tracts of variable length (100-250 bases) occur about 360 bases from the 5' terminus of RNA. The genome encodes 3 species of VPg. Translation starts at two alternative in-frame initiation sites, resulting in two forms of the L protein (Lab and Lb). The nucleotide sequence identity over the entire genome for different species of aphthoviruses is more than 50%. Clinical manifestations include foot-and-mouth disease of cloven hoofed animals and myocarditis in young animals.

List of Species in the Genus

The viruses, their genomic sequence accession numbers [ ] and assigned abbreviations ( ) are:

Species in the Genus

foot-and-mouth disease virus A	[M10975, M32257]	(FMDV-A)
foot-and-mouth disease virus ASIA 1		(FMDV-ASIA1)
foot-and-mouth disease virus C		(FMDV-C)

foot-and-mouth disease virus O	(FMDV-O)
foot-and-mouth disease virus SAT 1	(FMDV-SAT1)
foot-and-mouth disease virus SAT 2	(FMDV-SAT2)
foot-and-mouth disease virus SAT 3	(FMDV-SAT3)

TENTATIVE SPECIES IN THE GENUS

None reported.

LIST OF UNASSIGNED VIRUSES IN THE FAMILY

Unassigned viruses that are considered possible members of the family are:

cricket paralysis virus	(CrPV)
Drosophila C virus	(DCV)
equine rhinovirus 1	(ERV-1)
equine rhinovirus 2	(ERV-2)
equine rhinovirus 3	(ERV-3)
Gonometa virus	

The significance of the reported serological cross-reaction between CrPV and EMCV is not presently understood.

There are a number of small RNA viruses that have been described for which the taxonomic status is not known. These include the following:

1-three acid stable viruses of horses, two of which belong to a single serotype. Their properties are similar to equine rhinoviruses, which themselves vary in acid liability.

2-several diseases of domesticated birds caused by small RNA viruses which have often been referred to as 'enteroviruses'. They include avian encephalomyelitis (AEV), duck hepatitis virus I and duck hepatitis virus III (type II is an astrovirus), avian nephrites virus (ANV) and a number of poorly characterized isolates.

3-at least 25 small RNA viruses from various insect species. These are described in the literature as picornaviruses, or picornavirus-like viruses. The position of all these viruses within the family *Picornaviridae* is currently under review. They include agents such as bee acute paralysis, bee slow paralysis virus, bee virus X, Drosophila P and Drosophila A virus, sacbrood virus, Queensland fruitfly virus, Triatoma virus and aphid lethal paralysis virus.

4-viruses morphologically resembling picornaviruses isolated from harbor seals and sea bass.

5-Members of the family *Sequiviridae* have many properties in common with picornaviruses.

SIMILARITY WITH OTHER TAXA

None reported.

DERIVATION OF NAMES

picorna: from the prefix "pico" (= 'micro-micro'), and RNA, the sigla for ribonucleic acid
entero: from Greek *enteron*, "intestine"
rhino: from Greek *rhis, rhinos*, "nose"
hepato: from Greek *hepatos*, "liver"
cardio: from Greek *kardia* "heart"
aphtho: from Greek *aphtha*, "vesicles in the mouth"; English: aphtho, "thrush"; French: fievre aphtheuse

References

Acharya R, Fry KE, Stuart D, Fow G, Rowlands D, Brown F (1989) The three dimensional structure of foot-and-mouth disease virus at 2.9° resolution. Nature 337: 709-716

Adair BM, Kennedy S, McKillop ER, McNulty MS, McFerran JB (1987) Bovine, porcine and ovine picornaviruses: identification of viruses with properties similar to human coxsackieviruses. Arch Virol 97: 49-60

Grant RA, Filman DJ, Fujinami RS, Icenogle JP, Hogle JM (1992) Three dimensional structure of Theiler virus. Proc Natl Acad Sci USA 89: 2061-2065

Hamparian VV, Colonno RJ, Dick EC, Gwaltney JM, Hughes JH, Jordan WS, Kapikian AZ, Mogabgab WJ, Mores A, Phillips CA, Rueckert RR, Scheble JH, Stott EJ, Tyrrell DAJ (1987) A collaborative report: rhinoviruses - extension of the numbering system from 89 to 100. Virology 159: 191-192

Hyypia T, Horsnell C, Maaronen M, Khan M, Kalkinnen N, Auvinen P, Kinnuren L, Stanway G (1992) A novel picornavirus group identified by sequence analysis. Proc Natl Acad Sci USA 89: 8847-8851

Knowles N, Barnett ITR (1985) A serological classification of bovine enteroviruses. Arch Virol 83: 141-155

McFerran JB, McNulty MS (1986) Recent advances in enterovirus infections of birds. In: McFerran JB, McNulty MS (eds) Acute Virus Infections of Poultry. CEC Agriculture Research Programme Seminar, Martinus Nijhoff, Dordrecht, pp195-202

Rueckert RR (1990) Picornavirus and their multiplication. In: Fields BN, Knipe DM (eds) Virology, 2nd edn. Raven Press, New York, pp 507-548

Rueckert RR, Wimmer E (1984) Systematic nomenclature of picornavirus proteins. J Virol 50: 957-959

Stanway G (1990) Structure, function and evolution of picornaviruses. J Gen Virol 71: 2483-2501

Williamson C, Rybicki E, Kasdorf GCF, von Wechmar MB (1988) Characterisation of a new picorna-like virus isolated from aphids. J Gen Virol 69: 787-795

Contributed By

Minor PD, Brown F, Domingo E, Hoey E, King A, Knowles N, Lemon S, Palmenberg A, Rueckert RR, Stanway G, Wimmer E, Yin-Murphy M

FAMILY SEQUIVIRIDAE

TAXONOMIC STRUCTURE OF THE FAMILY

Family	***Sequiviridae***
Genus	*Sequivirus*
Genus	*Waikavirus*

VIRION PROPERTIES

MORPHOLOGY

Particles are isometric, about 30 nm in diameter.

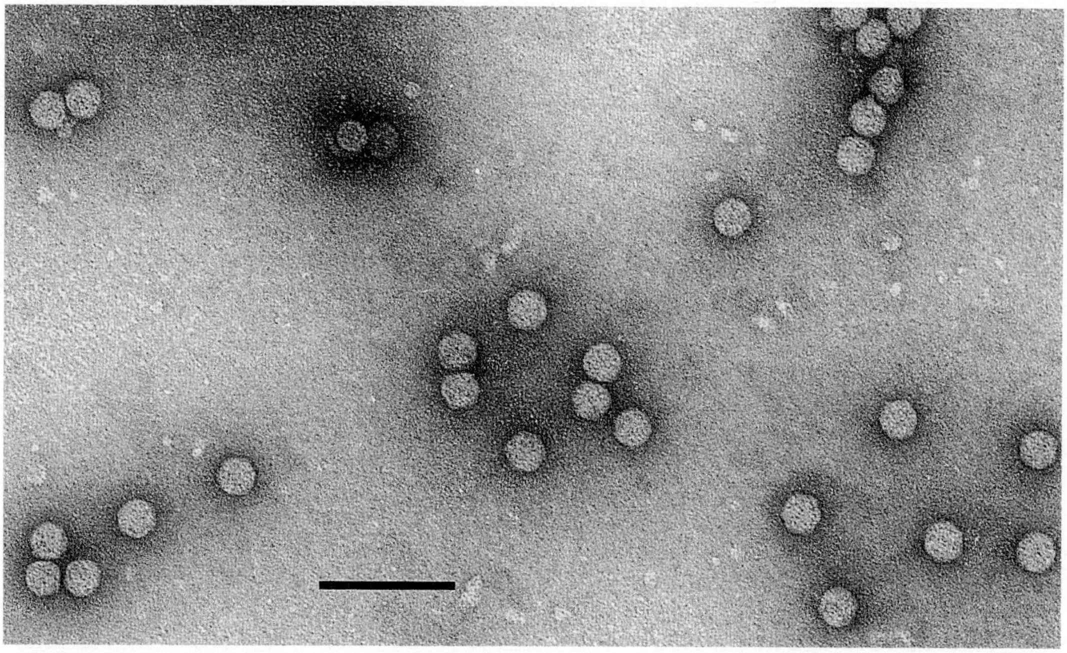

Figure 1: Negative contrast electron micrograph of parsnip yellow fleck virus stained in 1% uranyl acetate. The bar represents 100 nm.

PHYSICOCHEMICAL AND PHYSICAL PROPERTIES

The main virion component sediments at 150-190 *S*, contains about 40% RNA and has a correspondingly high equilibrium density in caesium salts. Some preparations also contain a slower sedimenting (about 60 *S*), less dense particle.

NUCLEIC ACID

The main virion component contains one molecule of infective, positive sense ssRNA, 9-12 kb in size. Sequivirus RNA is not poly-adenylated but waikavirus RNA is. Infectivity is protease-sensitive and a 5'-linked VPg molecule is probably present. There are some reports of an about 1 kb RNA being present in the 60*S* particles.

PROTEINS

Virions contain three major species with Mr of about 32×10^3 (CP1), 26×10^3 (CP2) and 23×10^3 (CP3). Particles of some waikaviruses are thought to contain other proteins which may be derived from one of the three major proteins. Virion and non-structural proteins arise by proteolytic cleavage of a polyprotein.

LIPIDS

None reported.

CARBOHYDRATES

None reported.

GENOME ORGANIZATION AND REPLICATION

The virus genome consists of a single infective ssRNA containing one major ORF which encodes a polyprotein of about 3,000 to 3,500 amino acids. The structural proteins are in the N-terminal half of the polyprotein but are separated from the N-terminus by polypeptide(s) of Mr 40-60 x 10^3. Sequences downstream of the structural proteins contain domains characteristic of proteins with nucleoside triphosphate binding, protease and RNA polymerase activities. RTSV, but not PYFV, has small 3'-co-terminal sub-genomic RNA which correspond to small ORFs downstream of the major large ORF.

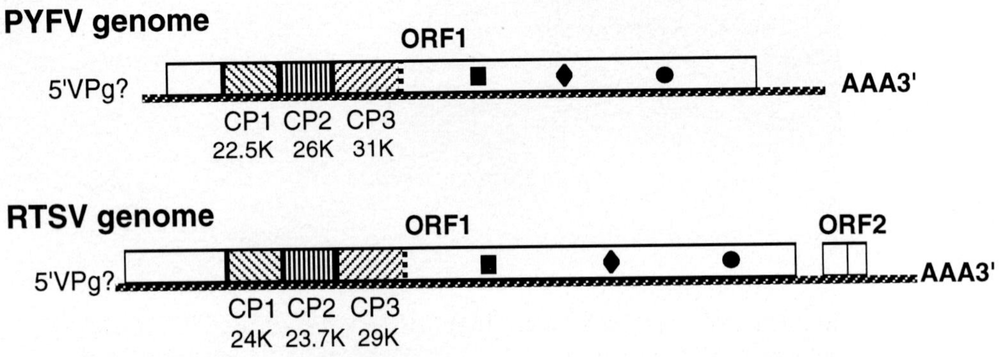

Figure 2: Genome structures of parsnip yellow fleck sequivirus and rice tungro spherical waikavirus. The boxes represent the polyproteins encoded by the large ORFs. The vertical solid lines show where cleavages are known to occur in the polyproteins and the dashed lines show where cleavages are presumed to occur. The approximate positions of protease (filled square), polymerase (filled diamond) and NTP-binding (filled circle) are shown.

ANTIGENIC PROPERTIES

Polyclonal sera contain antibodies to all virion proteins.

BIOLOGICAL PROPERTIES

Natural host ranges are restricted. Transmission is semi-persistent by aphids or, for most waikavirus species, by leafhoppers. A helper protein is needed which may be self-encoded (*Waikavirus*) or encoded by a helper virus (*Sequivirus*).

GENUS *SEQUIVIRUS*

Type species parsnip yellow fleck virus (parsnip serotype) (PYFV)

DISTINGUISHING FEATURES

The RNA is about 10 kb. PYFV RNA is not polyadenylated and lacks small ORF near the 3'-end. There are about 400 amino acids upstream of the structural proteins in the large polyprotein. Transmission of PYFV depends on the presence of a helper protein encoded by anthriscus yellows waikavirus.

LIST OF SPECIES IN THE GENUS

The viruses, their alternative names ( ), CMI/AAB description # ( ) and assigned abbreviations ( ) are:

SPECIES IN THE GENUS

dandelion yellow mosaic virus (DYMV)
parsnip yellow fleck virus (129) (PYFV)
(parsnip serotype)

parsnip yellow fleck virus A421 (129) (PYFV-A421)
(*Anthriscus* serotype)

Tentative Species in the Genus

None reported.

Genus *Waikavirus*

Type species rice tungro spherical virus (RTSV)

Distinguishing Features

The RNA is longer than 11 kb and has a poly (A) tail. RTSV RNA contains a small ORF near the 3'-end and has about 600 amino acids upstream of the structural proteins in the large polyprotein. Transmission depends on a self-encoded helper protein. The helper protein of some members can assist transmission of other unrelated viruses.

List Of Species In The Genus

The viruses, their CMI/AAB description # ( ) and assigned abbreviations ( ) are:

Species in the Genus

Anthriscus yellows virus (AYV)
maize chlorotic dwarf virus (194) (MCDV)
rice tungro spherical virus (67) (RTSV)

Tentative Species in the Genus

None reported.

Derivation Of Names

sequi: from Latin *sequi*, to follow, accompany, attend (in reference to the dependent aphid transmission of PYFV
waika: from Japanese describing the symptoms induced in rice by infection with RTSV alone (i.e. without rice tungro bacilliform badnavirus being present)

Similarity with Other Taxa

The amino acid sequences in the conserved NTP-binding and RNA polymerase domains of the polyproteins resemble those in the polyproteins encoded by RNA of viruses in the families *Comoviridae* and *Picornaviridae*. The number and sizes of the coat proteins resemble those of the *Picornaviridae* although the size of the protein(s) upstream of the coat proteins is larger than the L protein of aphthoviruses. The properties of the particles and the genomes of these viruses have prompted their description as 'plant picornaviruses'. There is insufficient information available to make comparisons with picornaviruses or picorna-like viruses that infect insects.

References

Bos L, Huijberts N, Huttinga H, Maat DZ (1983) Further characterization of dandelion yellow mosaic virus from lettuce and dandelion. Neth J Pl Path 89: 207-222

Elnagar S, Murant AF (1976) The role of the helper virus, anthriscus yellows, in the transmission of parsnip yellow fleck virus by the aphid Cavariella aegopodii. Ann Appl Biol 84: 169-181

Ge X, Gordon DT, Gingery RE (1989) Characterization of maize chlorotic dwarf virus (MCDV) RNA. Phytopathology 79: 1157

Ge X, Gordon DT, Gingery RE (1989) Occurrence of a small RNA in maize chlorotic dwarf virus-like particles. Phytopathology 79:1195

Gingery RE (1988) Maize chlorotic dwarf and related viruses. In: Koenig R (ed) The plant viruses; polyhedral virions with monopartite RNA Vol 3, Plenum Press, New York, pp 259-272

Hemida SK, Murant AF, Duncan GH (1989) Purification and some particle properties of Anthriscus yellows virus, a phloem-limited, semi-persistent, aphid-borne virus. Ann Appl Biol 114: 71-86
Hemida SK, Murant AF (1989) Particle properties of parsnip yellow fleck virus. Ann Appl Biol 114: 87-100
Hunt RE, Nault LR, Gingery RE (1988) Evidence for infectivity of maize chlorotic dwarf virus and for a helper component in its leafhopper transmission. Phytopathology 78: 499-504
Maroon CM, Gordon DT, Gingery RE (1989) Serological relationships of the capsid proteins of the type isolate of maize chlorotic dwarf virus (MCDV-T). Phytopatho-logy 79: 1157
Murant AF (1988) Parsnip yellow fleck virus, type member of a proposed new plant virus group, and a possible second member, dandelion yellow mosaic virus. In: Koenig R (ed), The plant viruses, polyhedral virions with monopartite genomes, Vol 3, Plenum Press, New York, pp 273-288
Murant AF, Goold RA (1968) Purification, properties and transmission of parsnip yellow fleck, a semi-persistent, aphid-borne virus. Ann Appl Biol 62: 123-137
Reavy B, Mayo MA, Turnbull-Ross AD, Murant AF (1993) Parsnip yellow fleck and rice tungro spherical viruses resemble picornaviruses and represent two genera in a proposed new plant picornavirus family (*Sequiviridae*). Arch Virol 131: 441-446
Shen P, Kaniewska MB, Smith C, Beachy RN (1993) Nucleotide sequence and genomic organization of rice tungro spherical virus. Virology 193: 621-630
Turnbull-Ross AD, Reavy B, Mayo MA, Murant AF (1992) The nucleotide sequence of parsnip yellow fleck virus: a plant picorna-like virus. J Gen Virol 73: 3203-3211
Turnbull-Ross AD, Mayo MA, Reavy B, Murant AF (1993) Sequence analysis of the parsnip yellow fleck virus polyprotein: evidence of affinities with picornaviruses. J Gen Virol 74: 555-561
Zhang S, Jones MC, Barker P, Davies JW, Hull R (1993) Molecular cloning and sequencing of coat protein-encoding cDNA of rice tungro spherical virus - a plant picornavirus. Virus Genes 7: 121-132

CONTRIBUTED BY

Mayo MA, Murant AF, Turnbull-Ross AD, Reavy B, Hamilton RI, Gingery RE

FAMILY *COMOVIRIDAE*

TAXONOMIC STRUCTURE OF THE FAMILY

Family	***Comoviridae***
Genus	*Comovirus*
Genus	*Fabavirus*
Genus	*Nepovirus*

VIRION PROPERTIES

MORPHOLOGY

Virions are non-enveloped 28-30 nm in diameter and exhibit icosahedral symmetry (T=1). The core consists of two positive sense RNA molecules, each having a small protein (VPg) (not known for fabaviruses) at their 5'-end. Virus preparations contain three sedimenting components, T (empty particles), M (particles usually containing a single molecule of RNA2) and B (particles containing a single molecule of RNA1).

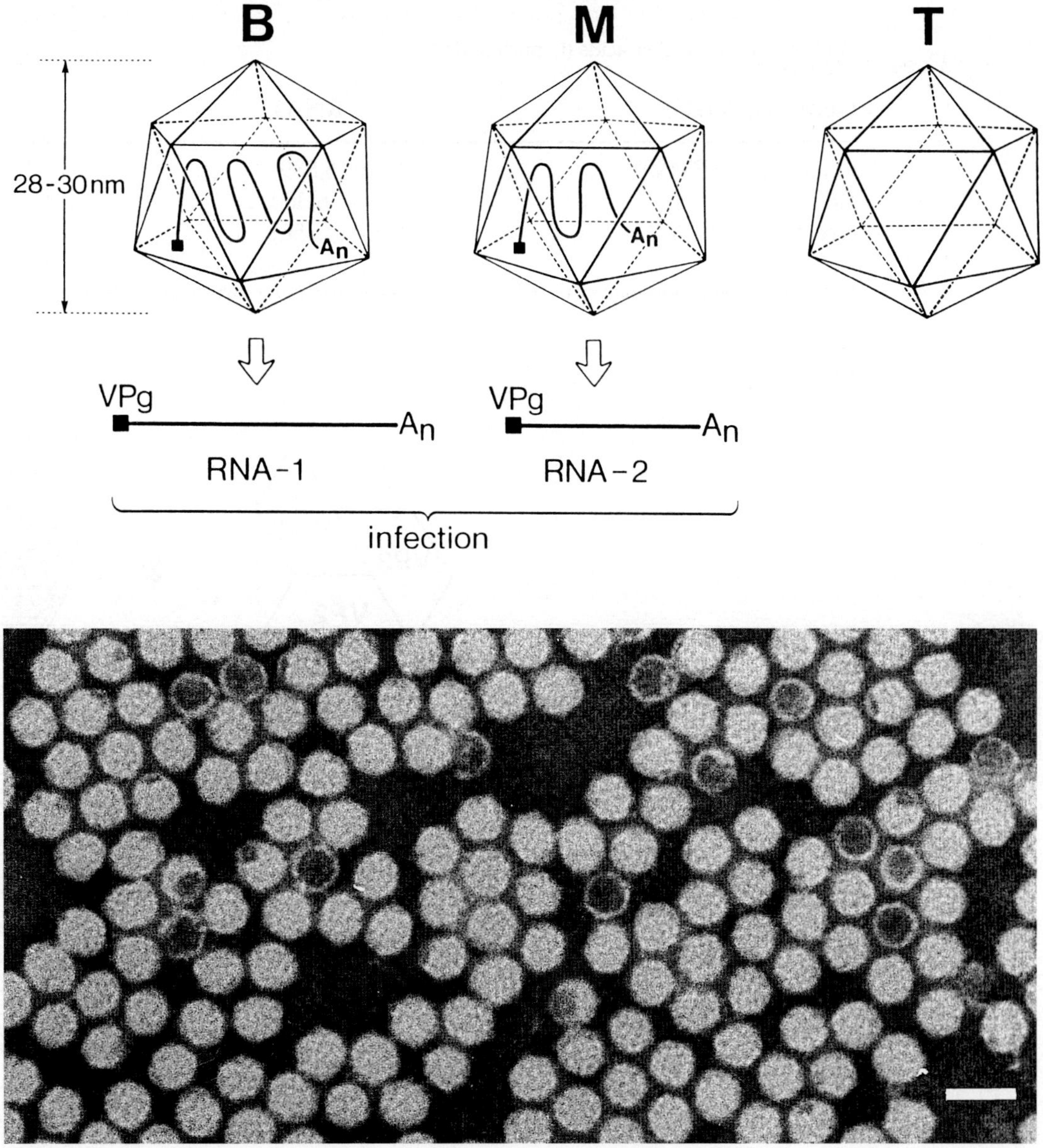

Figure 1: (upper) Diagram of the three different particles; (lower) negative contrast electron micrograph of cowpea mosaic virus (genus *Comovirus*). The bar represents 50 nm.

PHYSICOCHEMICAL AND PHYSICAL PROPERTIES

Virions are heat stable (thermal inactivation is usually above 60° C), and most are insensitive to organic solvents. Particles sediment as three components, T, M and B, with S_{20w} values of 49-63, 86-128 and 113-134, respectively, (values vary within each genus). Mr of particles is 3.2-3.8 x 10^6 (T), 4.6-5.8 x 10^6 (M) and 6.0-6.2 x 10^6 (B). Buoyant densities in CsCl are 128-130 (T), 141-148 (M) and 144-153 (B) g/cm^3 (density values refer only to genera *Comovirus* and *Nepovirus*).

NUCLEIC ACID

The genome consists of two species of linear positive sense ssRNA. Both RNAs are necessary for systemic infection. Sizes of RNAs differ among genera; nepovirus RNA1 (7.2-8.4 kb) and RNA2 (3.9-7.2 kb) are larger than fabavirus and comovirus RNA1 (5.9-7.2 kb) and RNA2 (3.5-4.5 kb). For the genera *Comovirus* and *Nepovirus* the genomic RNAs have been shown to contain a 3'-terminal poly (A) tract of variable length, and a protein VPg (Mr 4-6 x 10^3) at the 5'-end. For some species, complete nucleotide sequences are available in the EMBL database. For genus *Fabavirus*, information about RNA termini and nucleotide sequences is not yet available.

Table 1: Sizes of nucleic acids (in nucleotides)

Genus (species)	RNA1	RNA2
Comovirus (CPMV)*	5,900-7,200 (5,889)	3,500-4,500 (3,481)
Fabavirus	6,300	4,500
Nepovirus (TBRV)*	7,200-8,400 (7,356)	3,900-7,200 (4,662)

* values for cowpea mosaic virus (CPMV) and tomato black ring virus (TBRV) refer to sizes exclusive poly (A) tract.

PROTEINS

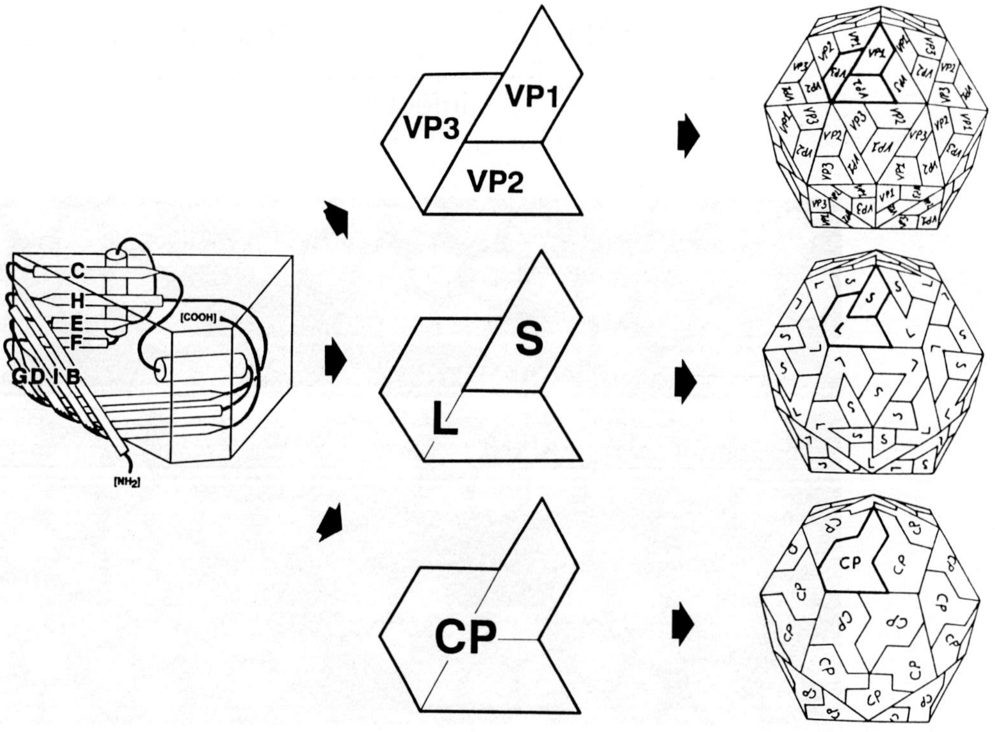

Figure 2: Architecture of the capsids of a picornavirus (top), comovirus (middle) and nepovirus (bottom). (With permission from Le Gall *et al.* 1992).

Como- and fabaviruses have two coat polypeptides (Mr 40-43 x 10^3 and 22-27 x 10^3); nepoviruses normally have a single coat polypeptide species (Mr 55-60 x 10^3). Virions probably have 60 copies per protein species per particle. For two comoviruses (CPMV, BPMV) the atomic structure has been solved and found to be very similar (pseudo T = 3) to that of the *Picornaviridae*. Como- and nepoviruses (fabaviruses not known) produce polyproteins from which the structural and nonstructural proteins are generated by proteolytic cleavages. Nonstructural proteins of como- and nepoviruses include a (putative) cell-to-cell movement protein (encoded by RNA2), an NTP-binding motif-containing protein, a VPg, a proteinase, and a polymerase (all coded for by RNA1).

LIPIDS

None reported.

CARBOHYDRATES

None reported for faba- and nepoviruses; coat proteins of comoviruses possibly are glycosylated.

GENOME ORGANIZATION AND REPLICATION

Unfractionated RNA is highly infective but neither RNA species alone can infect plants. Cytoplasm of infected cells contains conspicuous inclusions consisting primarily of membranous elements and electron dense material which may be the site of viral genome replication and expression. Virions assemble and accumulate in the cytoplasm, often in crystalline or paracrystalline arrays. They are also found within tubules, which penetrate through cell walls, and which may be implicated in cell-to-cell transport. The following information only refers to como- and nepoviruses (fabaviruses have not been studied): RNA1 can replicate in protoplasts but in the absence of RNA2 (encoding the coat proteins) no virus particles are produced. RNA1 carries all information for RNA replication, including the polymerase. Both RNA species are translated into polyproteins that are cleaved to give the functional proteins.

ANTIGENIC PROPERTIES

The viruses serve as good immunogens. Species belonging to the same genus are serologically interrelated, often distantly.

BIOLOGICAL PROPERTIES

HOST RANGE AND SYMPTOMS

Comoviruses have narrow host ranges; nepo- and fabaviruses have wide host ranges. Symptoms vary widely within each genus.

TRANSMISSION

Member viruses of the family *Comoviridae* all have biological vectors, comoviruses being transmitted by beetles (especially members of the family *Chrysomelidae*), fabaviruses by aphids and (most) nepoviruses by nematodes. All are readily transmissible experimentally by mechanical inoculation. Seed transmission is very common among nepoviruses, but is rare or does not occur with como- and fabaviruses.

GENUS *COMOVIRUS*

Type Species cowpea mosaic virus (CPMV)

DISTINGUISHING FEATURES

Capsids are constructed from two polypeptide species (Large and Small). Comoviruses have narrow host ranges, 11 of the 15 species being restricted to a few species of the family

Leguminosae. Mosaic and mottle symptoms are characteristic, not ringspots. Transmission in nature is exclusively by beetles, especially members of the family *Chrysomelidae*. Beetles retain their ability to transmit virus for days or weeks.

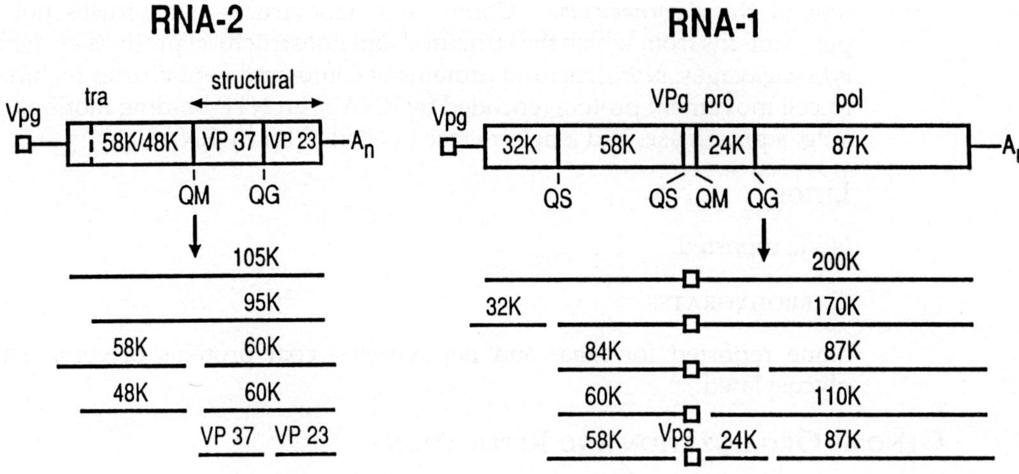

Figure 3: Organization and expression of the CPMV genome (genus *Comovirus*). Proteolytic cleavage sites are indicated below the ORFs in both RNAs. Tra, transport protein; pro, proteinase; pol, polymerase.

List of Species in the Genus

The viruses, their genomic sequence accession numbers [ ], CMI/AAB description # ( ) and assigned abbreviations ( ) are:

Species in the Genus

Andean potato mottle virus (203)		(APMoV)
bean pod mottle virus (108)	[M62738]	(BPMV)
bean rugose mosaic virus (246)		(BRMV)
broad bean stain virus (126)		(BBSV)
broad bean true mosaic virus (20)		(BBTMV)
cowpea mosaic virus (47, 197)	[X00206, X00729]	(CPMV)
cowpea severe mosaic virus (209)	[M83830, M83309]	(CPSMV)
Glycine mosaic virus		(GMV)
pea green mottle virus		(PGMV)
pea mild mosaic virus		(PMiMV)
quail pea mosaic virus (238)		(QPMV)
radish mosaic virus (121)		(RaMV)
red clover mottle virus (74)	[M14193]	(RCMV)
squash mosaic virus (43)		(SqMV)
Ullucus virus C (277)		(UVC)

Tentative Species in the Genus

None reported.

Genus *Fabavirus*

Type Species broad bean wilt virus 1 (BBWV-1)

Distinguishing Features

Fabaviruses have wide host ranges among dicotyledons and some families of monocotyledons. Symptoms are ringspots, mottle, mosaic, distortion, wilting and apical necrosis. In nature fabaviruses are transmitted nonpersistently by aphids. In other respects, fabaviruses are similar to comoviruses.

LIST OF SPECIES IN THE GENUS

The viruses, their CMI/AAB description # ( ) and assigned abbreviations ( ) are:

SPECIES IN THE GENUS

broad bean wilt virus 1 (81)	(BBWV-1)
broad bean wilt virus 2	(BBWV-2)
Lamium mild mosaic virus	(LMMV)

TENTATIVE SPECIES IN THE GENUS

None reported.

GENUS *NEPOVIRUS*

Type Species tobacco ringspot virus (TRSV)

DISTINGUISHING FEATURES

Capsids are composed of a single polypeptide species (Mr 55-60 x 10^3), whereas the capsids of most unassigned viruses yield, upon degradation, two or three smaller polypeptides (Mr 21-44 x 10^3). Genome organization and expression are similar to those of comoviruses, except that RNA2 specifies a single primary translation product (Mr 105-165 x 10^3) which is processed into three, rather than four mature proteins. Nepoviruses are widely distributed in temperate regions. Natural host ranges vary from wide to restricted to a single plant species, depending on the virus. Ringspot symptoms are characteristic, but mottling and spotting are equally frequent. Linear or circular satellite RNAs, which sometimes modulate symptoms, are found associated with several viruses. Eleven species are acquired and transmitted persistently by longidorid nematodes (*Xiphinema* or *Longidorus*), three are transmitted by pollen, and the others have no known vector.

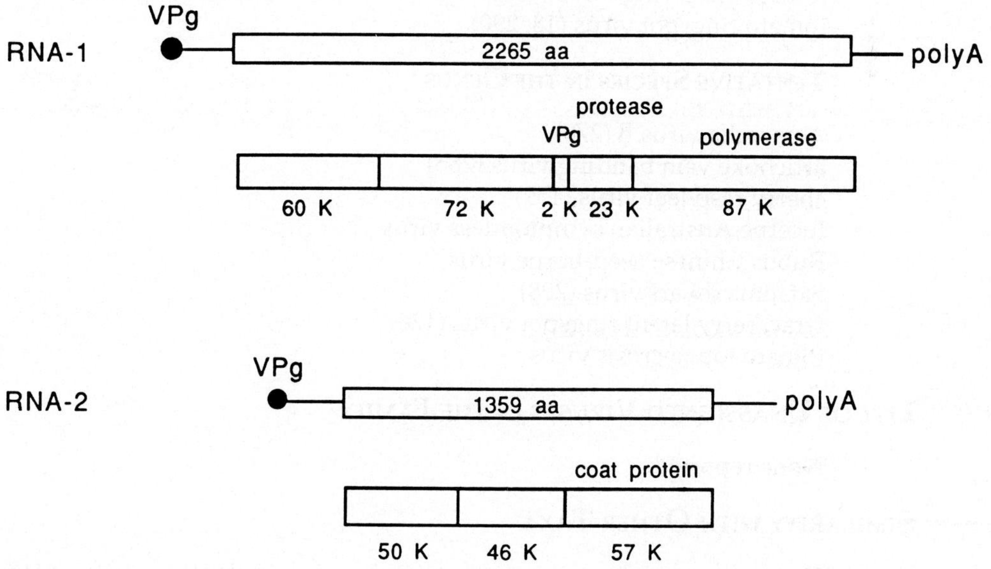

Figure 4: Organization and expression of the TBRV genome (*Nepovirus*). Positions and sizes of the mature proteins are indicated in the primary translation products of RNA1 and RNA2.

LIST OF SPECIES IN THE GENUS

The viruses, their genomic sequence accession numbers [ ], CMI/AAB description # ( ) and assigned abbreviations ( ) are:

SPECIES IN THE GENUS

Arabis mosaic virus (16)		(ArMV)
Arracacha virus A (216)		(AVA)
artichoke Italian latent virus (176)		(AILV)
artichoke yellow ringspot virus (271)		(AYRSV)
blueberry leaf mottle virus (267)		(BLMV)
cassava American latent virus		(CsALV)
cassava green mottle virus		(CGMV)
cherry leaf roll virus (80, 306)		(CLRV)
chicory yellow mottle virus (132)		(ChYMV)
cacao necrosis virus (173)		(CNV)
crimson clover latent virus		(CCLV)
Cycas necrotic stunt virus		(CNSV)
grapevine Bulgarian latent virus (186)		(GBLV)
grapevine chrome mosaic virus (103)	[X15346, X15163]	(GCMV)
grapevine fanleaf virus (28)	[X16907]	(GFLV)
grapevine Tunisian ringspot virus		(GTRSV)
hibiscus latent ringspot virus (233)		(HLRSV)
lucerne Australian latent virus (225)		(LALV)
mulberry ringspot virus (142)		(MRSV)
myrobalan latent ringspot virus (160)		(MLRSV)
olive latent ringspot virus (301)		(OLRSV)
peach rosette mosaic virus (150)		(PRMV)
potato black ringspot virus (206)		(PBRSV)
potato virus U		(PVU)
raspberry ringspot virus (6, 198)		(RpRSV)
tobacco ringspot virus (17, 309)		(TRSV)
tomato black ring virus (138)	[D00322, X04062]	(TBRV)
tomato ringspot virus (18, 290)		(ToRSV)

TENTATIVE SPECIES IN THE GENUS

Arracacha virus B (270)	(AVB)
artichoke vein banding virus (285)	(AVBV)
cherry rasp leaf virus (159)	(CRLV)
lucerne Australian symptomless virus	(LASV)
Rubus Chinese seed-borne virus	(RCSV)
Satsuma dwarf virus (208)	(SDV)
strawberry latent ringspot virus (126)	(SLRSV)
tomato top necrosis virus	(ToTNV)

LIST OF UNASSIGNED VIRUSES IN THE FAMILY

None reported.

SIMILARITY WITH OTHER TAXA

Comoviruses and nepoviruses have properties similar to members of the families *Potyviridae* and *Picornaviridae*; e.g. genome organization, VPg at 5'-end and poly (A) tract at 3'-end of genomes, post-translational processing of polyproteins and sequence similarities among nonstructural proteins. Moreover, como-, nepo- and picornaviruses have very similar capsid morphology.

DERIVATION OF NAMES

como: sigla from *co*wpea *mo*saic
faba: Latin *Faba*, bean; also *Vicia faba*, broad bean
nepo: sigla from *ne*matode, *po*lyhedral to distinguish these viruses from the tobraviruses

REFERENCES

Brault V, Hibrand L, Candresse T, Le Gall O, Dunez J (1989) Nucleotide sequence and genetic organization of Hungarian grapevine chrome mosaic nepovirus RNA 2. Nucl Acids Res 17: 7809-7819

Chen Z, Stauffacher C, Li Y, Schmidt T, Bomu W, Kamer G, Shanks M, Lomonossoff GP, Johnson JE (1989) Protein-RNA interactions in an icosahedral virus at 3.0 Å resolution. Science 245: 154-159

Eggen R, van Kammen A (1988) RNA replication in comoviruses. In: Domingo E, Holland JJ, Ahlquist P (eds) RNA genetics Vol I. CRC Press, Boca Raton FL, pp 49-69

Francki RIB, Milne RG, Hatta T (eds) (1985) Comovirus group. In: Atlas of plant viruses Vol II. CRC Press, Boca Raton FL, pp 1-22

Franssen H, Leunissen J, Goldbach R, Lomonossoff GP, Zimmern D (1984) Homologous sequences in non-structural proteins from cowpea mosaic virus and picornaviruses. EMBO J 3: 855-861

Fulton JP, Scott HA (1979) A serogrouping concept for legume comoviruses. Phytopathology 69: 305-306

Goldbach R (1987) Genome similarities between plant and animal RNA viruses. Microbiol Sci 4: 197-202

Goldbach R, van Kammen A (1985) Structure, replication, and expression of the bipartite genome of cowpea mosaic virus. In: Davies JW (ed) Molecular Plant Virology Vol II. CRC Press, Boca Raton FL, pp 83-120

Greif C, Hemmer O, Fritsch C (1988) Nucleotide sequence of tomato black ring virus RNA-1. J Gen Virol 69: 1517-1529

Le Gall O, Candresse T, Brault V, Dunez J (1989) Nucleotide sequence of Hungarian grapevine chrome mosaic nepovirus RNA 1. Nucl Acids Res 17: 7795-7807

Le Gall O, Lanneau M, Candresse T, Dunez J (1992) Sequence comparison of two strains of tomato black ring virus transmitted by two different nematode species. 5th International Plant Virus Epidemiology Symposium, Valenzano (Bari), Italy 27-31 July

Meyer M, Hemmer O, Mayo MA, Fritsch C (1986) The nucleotide sequence of tomato black ring virus RNA-2. J Gen Virol 67: 1257-1271

Ritzenthaler C, Viry M, Pinck M, Margis R, Fuchs M, Pinck L (1991) Complete nucleotide sequence and genetic organization of grapevine fanleaf nepovirus. J Gen Virol 72: 2357-2365

Stace-Smith R, Ramsdell DC (1987) Nepoviruses of the Americas. In: Harris KF (ed) Current Topics in Vector Research Vol 5. Springer, Wien, New York, pp 131-166

van Lent JWM, Wellink J, Goldbach R (1990) Evidence for the involvement of the 58K and 48K proteins in the intercellular movement of cowpea mosaic virus. J Gen Virol 71: 219-223

Wellink J, van Lent JWM, Goldbach R (1988) Detection of viral proteins in cytopathic structures in cowpea protoplasts infected with cowpea mosaic virus. J Gen Virol 69: 751-755

Xu ZG, Cockbain AJ, Woods RD, Govier DA (1988) The serological relationships and some other properties of isolates of broad bean wilt virus from faba bean and pea in China. Ann Appl Biol 113: 287-296

CONTRIBUTED BY

Goldbach R, Martelli GP, Milne RG

Family *Potyviridae*

Taxonomic Structure of the Family

Family	***Potyviridae***
Genus	*Potyvirus*
Genus	*Rymovirus*
Genus	*Bymovirus*

Virion Properties

Morphology

Virions are flexuous filaments with no envelope and are 11-15 nm in diameter, with a helical pitch of about 3.4 nm. Particle lengths of members of the three genera differ. Members of the genera *Potyvirus* and *Rymovirus* and the unassigned viruses are monopartite with particle modal lengths of 650-900 nm; members of the genus *Bymovirus* are bipartite with particles of two modal lengths of 250-300 and 500-600 nm.

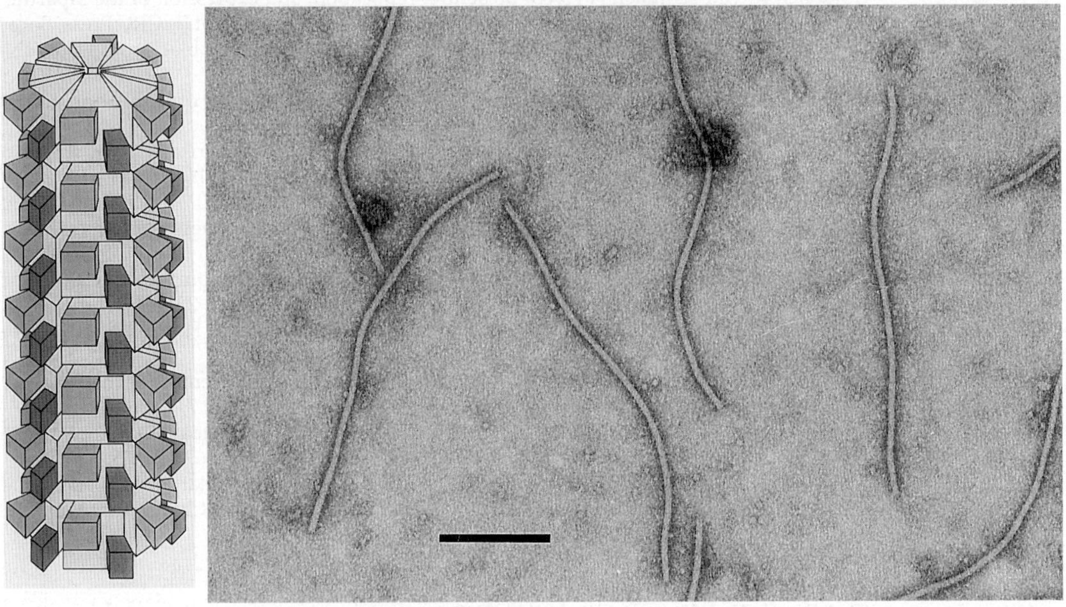

Figure 1: (left) Schematic diagram of potyvirus particle. The N-terminal ~30 amino acids (large rectangle) and C-terminal ~19 amino acids (small rectangle) of the coat protein molecules are exposed on the surface of the intact virus particle (from Shukla and Ward, 1989). (right) Virions of plum pox potyvirus stained with 1% PTA, pH 6.0, the bar represents 200 nm (from Scottish Crop Research Institute).

Physicochemical and Physical Properties

Member viruses of the genera *Potyvirus* and *Rymovirus* have a density in CsCl of about 1.31 g/cm^3 and an S_{20w} of 150-160. Member viruses of the genus *Bymovirus* have a density in $CsCl_2$ of about 1.29 g/cm^3.

Nucleic Acid

Member viruses of the genera *Potyvirus* and *Rymovirus* have a single molecule of positive sense, ssRNA, 8.5 - 10.0 kb in size (Mr 3.0 x 10^6). Virions are infectious. A protein (VPg about 24 kDa) is covalently linked to the 5' terminal nucleotide. A polyadenylate tract (20 to 160 adenosines) is present at the 3' terminus. The complete nucleotide sequence is known for at least 10 members of the genus *Potyvirus*. Member viruses of the bipartite genus *Bymovirus* have two positive sense, ssRNA molecules; RNA1 is 7.90 kb in size (Mr 2.6 x 10^6) and RNA2 is 4.56 kb in size (Mr 1.5 x 10^6). Both RNAs have 3' terminal polyadenylate tracts but it is not known if a VPg is present at the 5' terminus. The complete nucleotide sequence of BaYMV RNAs has been determined, about 70% of WSMV has been sequenced.

Proteins

The genome-derived polyprotein is cleaved into several proteins, some of which form inclusion bodies in the cell (see genus descriptions). Virions contain one coat protein, Mr of 30-47 x 10^3. N- and C- terminal residues are positioned on the exterior of the virion. Mild trypsin treatment removes N- and C-terminal segments, leaving a trypsin resistant core of about 24 kDa. Plant proteases may degrade the coat protein *in vivo* similar to the *in vitro* degradation which occurs during purification with some procedures or hosts. All potyviruses display significant amino acid sequence homology in the trypsin resistant core, but little homology in their N and C-terminal segments.

Lipids

None reported.

Carbohydrates

None reported.

Genome Organization and Replication

Genetic information encoded by the RNA genome is organized as a single ORF. Genetic maps for TEV, a member of the genus *Potyvirus*, and BaYMV, a member of the genus *Bymovirus* are presented in genera descriptions. For members of the genus *Potyvirus*, the genome is expressed initially as a polyprotein which undergoes co- and post-translational proteolytic processing by three viral-encoded proteinases to form individual gene products. Little information is available on the replication of RNA.

Antigenic Properties

The viral proteins are moderately immunogenic; there are serological relationships between members. A conserved internal trypsin-resistant core coat protein epitope has been identified, which is, shared by most members of the family.

Biological Properties

Inclusion Body Formation

All members of the family *Potyviridae* form cytoplasmic cylindrical inclusion (CI) bodies during infection. The CI is an aggregate of about 70 kDa viral protein which possesses ATPase and helicase activities. The viruses encode and express the following proteins, but inclusion bodies comprised of these proteins are not formed in all instances (some potyviruses induce nuclear inclusion bodies which are co-crystals of two viral-encoded proteins present in equimolar amounts): The small nuclear inclusion (NIa) protein (49 kDa) is a polyprotein consisting of the VPg and proteinase. The large nuclear inclusion (NIb) protein (58 kDa) has amino acid motifs of RNA-dependent RNA-polymerases. Amorphous inclusion bodies are also evident in the cytoplasm during certain potyvirus infections and represent aggregations of 52 kDa protein. This protein, referred to as HC-PRO, has a helper component activity and a proteolytic activity associated with it. Bymoviruses may not encode a protein analogous to the helper component in length, but a 28 kDa protein from RNA2 of BayMV has amino acid domains with sequence homologies to the potyvirus helper component protease.

Host Range

Some members have a narrow host range, most members infect an intermediate number of plants, and a few members infect species in up to 30 families. Transmission is readily accomplished by mechanical inoculation. Many viruses are widely distributed. Distribution may be aided by seed transmission.

TRANSMISSION

Potyviruses are vectored by a variety of organisms. Members of the genus *Potyvirus* are vectored by aphids in a non-persistent, non-circulative manner. A helper component and a particular coat protein amino acid triplet (i.e., DAG for some potyviruses) are required for aphid transmission. Rymoviruses are transmitted by mites. Bymoviruses are transmitted by a fungal vector. Two of the unassigned viruses, sweetpotato mild mottle and sweetpotato yellow dwarf viruses, may be transmitted by whiteflies.

GENUS *POTYVIRUS*

Type Species potato virus Y (PVY)

DISTINGUISHING FEATURES

VIRION PROPERTIES

MORPHOLOGY

Virions are flexuous filaments, 680-900 nm long and 11-13 nm wide, with helical symmetry and a pitch of about 3.4 nm. Particles of some viruses are longer in the presence of divalent cations than in the presence of EDTA.

PHYSICOCHEMICAL AND PHYSICAL PROPERTIES

Virion S_{20w} is 150-160; density in CsCl is 1.31 g/cm^3; $E^{0.1\%}_{1\,cm,\,260\,nm}$ = 2.4-2.7.

NUCLEIC ACID

Virions contain a single molecule of linear, positive sense ssRNA, about 9.7 kb in size (Mr 3.0-3.5 x 10^6); virions contain 5% RNA by weight. RNA molecules have poly (A) tracts at their 3' ends. A genome-linked protein of about 24 kDa is covalently linked at or near the 5' terminus.

PROTEINS

Virions contain a single coat protein, Mr 30-47 in size. The coat protein of the type species, PVY, contains 267 amino acids.

GENOME ORGANIZATION AND REPLICATION

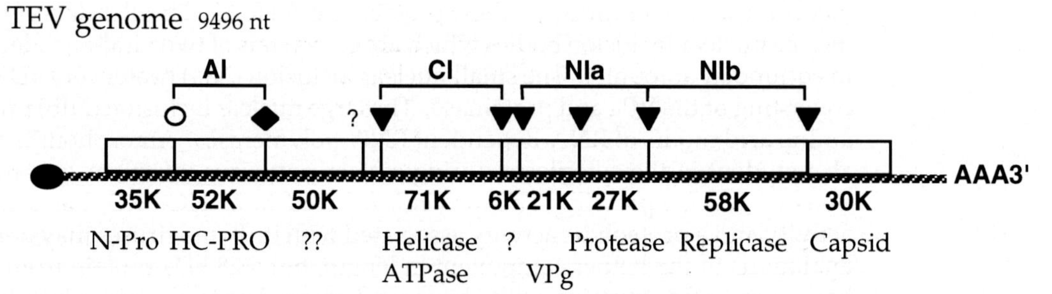

Figure 2: Genomic map of TEV, a member of genus *Potyvirus*. The RNA genome is represented by thin lines and an open box which represent untranslated and translated segments of the ssRNA, respectively. The filled box at the 5' end represents a VPg molecule. The Mr (x 10^3) of the individual gene products are shown below the box. Activities associated with these products are shown beneath the molecular masses, as follows: N-Pro, a protein with a proteolytic activity responsible for cleavage at Phe-Ser (o); HC-PRO, a protein with helper component activity and proteolytic activity responsible for cleavage at a Gly-Gly (◆); VPg, genome-linked viral protein covalently attached to the 5' terminal nucleotide (represented by the filled black circle); Pro, serine-like proteolytic activity responsible for cleavage at the Gln-(Ser/Gly) (▼). Some of these proteins of particular member viruses of the family *Potyviridae* aggregate to form inclusion bodies during infection. The protein involved and the particular type of inclusion body is shown above the genetic map; AI, amorphous inclusion; CI, cylindrical-shaped inclusion body found in the cytoplasm; NIa and NIb, small and large nuclear inclusion proteins which aggregate in the nucleus to form a nuclear inclusion body.

ANTIGENIC PROPERTIES

Virions are moderately immunogenic; there are serological relationships among many members. One monoclonal antibody reacts with most aphid transmitted potyviruses. The coat protein amino acid sequence homology among aphid transmitted viruses is 40-70%. Some species are serologically related to species in the genera *Rymovirus* and *Bymovirus*.

BIOLOGICAL PROPERTIES

Many individual viruses have a narrow host range, but a few infect species in up to 30 host families. The viruses are transmitted by aphids in a non-persistent manner and are transmissible experimentally by mechanical inoculation. Some isolates are inefficiently transmitted by aphids and others are not transmissible by aphids at all. This is apparently due to mutations within the helper component and/or coat protein cistrons. Some viruses are seed transmitted.

LIST OF SPECIES IN THE GENUS

The viruses, their alternative names ( ), genomic sequence accession numbers [ ], CMI/AAB description # ( ) and assigned abbreviations ( ) are:

SPECIES IN THE GENUS

Alstroemeria mosaic virus	(AlMV)
Amaranthus leaf mottle virus	(AmLMV)
Araujia mosaic virus	(ArjMV)
artichoke latent virus	(ArLV)
asparagus virus 1	(AV-1)
bean common mosaic virus (73, 337)	(BCMV)
(blackeye cowpea mosaic virus) (305)	
(Azuki bean mosaic virus)	
(peanut stripe virus)	
(peanut mild mottle virus)	
(peanut chlorotic ring mottle virus)	
(sesame yellow mosaic virus)	
bean common mosaic necrosis virus	(BCMNV)
(serotype A of BCMV)	
bean yellow mosaic virus	(BYMV)
(Crocus tomasinianus virus)	
(white lupinmosaic virus)	
(pea mosaic virus) (40)	
beet mosaic virus (53)	(BtMV)
bidens mottle virus (209)	(BiMoV)
cardamom mosaic virus	(CdMV)
carnation vein mottle virus (78)	(CVMV)
carrot thin leaf virus (218)	(CTLV)
celery mosaic virus (50)	(CeMV)
chilli veinal mottle virus	(ChiVMV)
clover yellow vein virus (131)	(ClYVV)
(pea necrosis virus)	
(statice virus Y)	
cocksfoot streak virus (59)	(CSV)
Colombian datura virus	(CDV)
Commelina mosaic virus	(ComMV)
cowpea aphid-borne mosaic virus (134)	(CABMV)
(South African passiflora virus)	
cowpea green vein banding virus	(CGVBV)
dasheen mosaic virus (191)	(DsMV)
datura shoestring virus	(DSTV)

Dendrobium mosaic virus		(DeMV)
Gloriosa stripe mosaic virus		(GSMV)
groundnut eyespot virus		(GEV)
guinea grass mosaic virus (190)		(GGMV)
Helenium virus Y		(HVY)
henbane mosaic virus (95)		(HMV)
Hippeastrum mosaic virus (117)		(HiMV)
Iris fulva mosaic virus (310)		(IFMV)
iris mild mosaic virus (116, 324)		(IMMV)
iris severe mosaic virus (147, 338)		(ISMV)
(bearded iris mosaic virus) (147, 338)		
Johnsongrass mosaic virus	[Z26920]	(JGMV)
konjac mosaic virus		(KMV)
leek yellow stripe virus (240)		(LYSV)
lettuce mosaic virus (9)		(LMV)
maize dwarf mosaic virus		(MDMV)
narcissus degeneration virus		(NDV)
narcissus yellow stripe virus (76)		(NYSV)
Nothoscordum mosaic virus		(NoMV)
onion yellow dwarf virus (158)		(OYDV)
Ornithogalum mosaic virus		(OrMV)
papaya ringspot virus	[X67673]	(PRSV)
(watermelon mosaic virus 1) (63, 84, 292)		
parsnip mosaic virus (91)		(ParMV)
passion fruit woodiness virus (122)		(PWV)
pea seed-borne mosaic virus (146)	[D10930, D01152]	(PSbMV)
peanut mottle virus (141)		(PeMoV)
pepper mottle virus (253)	[M96425]	(PepMoV)
pepper severe mosaic virus		(PeSMV)
pepper veinal mottle virus (104)		(PVMV)
Peru tomato mosaic virus (255)		(PTV)
plum pox virus (70)	[D00424, M92280, X16415, D13751]	(PPV)
pokeweed mosaic virus (97)		(PkMV)
potato virus A (54)		(PVA)
potato virus V (316)		(PVV)
potato virus Y (37, 242)	[D00441, M95491]	(PVY)
Rembrandt tulip breaking virus		(ReTBV)
sorghum mosaic virus		(SrMV)
soybean mosaic virus (93)	[S42280]	(SMV)
sugarcane mosaic virus (88, 341)		(SCMV)
sweet potato feathery mottle virus		(SPFMV)
(sweet potato russet crack virus)		
(sweet potato A virus)		
(sweet potato chlorotic leafspot virus)		
(sweet potato internal cork virus)		
tamarillo mosaic virus		(TamMV)
Telfairia mosaic virus		(TeMV)
tobacco etch virus (55, 258)	[M15239]	(TEV)
tobacco vein mottling virus (325)	[X04083]	(TVMV)
tulip band breaking virus		(TBBV)
(lily mottle virus)		
tulip breaking virus (71)		(TBV)
tulip chlorotic blotch virus		(TCBV)
turnip mosaic virus (8)	[D10927]	(TuMV)
(tulip top breaking virus)		
watermelon mosaic virus 2 (63,293)		(WMV-2)

(vanilla necrosis virus)	
Wisteria vein mosaic virus	(WVMV)
yam mosaic virus (314)	(YMV)
(Dioscorea green banding virus)	
zucchini yellow fleck virus	(ZYFV)
zucchini yellow mosaic virus (282)	(ZYMV)

Tentative Species in the Genus

Aphid-borne (*aphid transmission not confirmed; +name inadequate but denotes plant species with a report of a potyvirus infection)

Alstroemeria streak virus	(AlSV)
Amazon lily mosaic virus	(ALiMV)
Aneilema virus+	(AneV)
Anthoxanthum mosaic virus*	(AntMV)
Aquilegia virus*+	(AqV)
Arracacha virus Y	(AVY)
Asystasia gangetica mottle virus*	(AGMoV)
bidens mosaic virus	(BiMV)
bramble yellow mosaic virus	(BrmYMV)
brandle yellow mosaic virus	(BrnYMV)
Bryonia mottle virus	(BryMV)
canary reed mosaic virus	(CRMV)
Canavalia maritima mosaic virus	(CnMMV)
carrot mosaic virus	(CtMV)
Cassia yellow spot virus	(CasYSV)
celery yellow mosaic virus	(CeYMV)
chickpea bushy dwarf virus	(CpBDV)
chickpea filiform virus	(CpFV)
Clitoria yellow mosaic virus	(CtYMV)
cowpea rugose mosaic virus	(CPRMV)
Crinum mosaic virus*	(CriMV)
Croatian clover virus+	(CroCV)
Cypripedium calceolus virus*	(CypCV)
daphne virus Y	(DVY)
datura virus 437	(DV-437)
datura distortion mosaic virus	(DDMV)
datura mosaic virus*	(DTMV)
datura necrosis virus	(DNV)
Desmodium mosaic virus	(DesMV)
Dioscorea alata ring mottle virus	(DARMV)
Dioscorea trifida virus+	(DTV)
Dipladenia mosaic virus	(DipMV)
dock mottling mosaic virus	(DMMV)
eggplant green mosaic virus	(EGMV)
eggplant severe mottle virus	(ESMV)
Euphorbia ringspot virus	(EuRV)
Ficus carica virus+	(FicCV)
freesia mosaic virus	(FreMV)
garlic yellow streak virus	(GYSV)
guar symptomless virus*	(GSLV)
Habenaria mosaic virus	(HaMV)
Holcus streak virus*	(HSV)
Hungarian datura innoxia virus*	(HDIV)
hyacinth mosaic virus*	(HyaMV)
Indian pepper mottle virus	(IPMV)
isachne mosaic virus*	(IsaMV)

Kennedya virus Y	(KVY)
lily mild mottle virus	(LiMMV)
Malva vein clearing virus	(MVCV)
marigold mottle virus	(MaMoV)
Melilotus mosaic virus	(MeMV)
melon vein-banding mosaic virus	(MVBMV)
Moroccan watermelon mosaic virus	(MWMV)
mungbean mosaic virus*	(MbMV)
mungbean mottle virus	(MMTV)
Narcissus late season yellows virus (jonquil mild mosaic virus)	(NLSYV)
nasturtium mosaic virus	(NasMV)
Nerine virus*+	(NV)
palm mosaic virus*	(PalMV)
papaya leaf distortion mosaic virus	(PLDMV)
passion fruit mottle virus	(PFMV)
passion fruit ringspot virus	(PFRSV)
patchouli mottle virus	(PatMV)
peanut green mottle virus	(PeGMV)
peanut mosaic virus	(PeMsV)
Pecteilis mosaic virus	(PcMV)
pepper mild mosaic virus	(PMMV)
Perilla mottle virus	(PerMV)
plantain virus 7	(PlV-7)
Pleioblastus mosaic virus	(PleMV)
Populus virus*	(PV)
primula mosaic virus	(PrMV)
primula mottle virus	(PrMoV)
ranunculus mottle virus	(RanMV)
Sri Lankan passionfruit mottle virus	(SLPMV)
sunflower mosaic virus*	(SuMV)
sweet potato latent virus	(SwPLV)
sweet potato vein mosaic virus	(SPVMV)
sword bean distortion mosaic virus	(SBDMV)
teasel mosaic virus	(TeaMV)
tobacco vein banding mosaic virus	(TVBMV)
tobacco wilt virus	(TWV)
Tongan vanilla virus	(TVV)
Tradescantia/Zebrina virus+	(TZV)
Trichosanthes mottle virus	(TrMV)
Tropaeolum virus 1	(TV-1)
Tropaeolum virus 2	(TV-2)
Ullucus mosaic virus	(UMV)
Vallota mosaic virus	(ValMV)
vanilla mosaic virus	(VanMV)
white bryony virus	(WBV)
wild potato mosaic virus	(WPMV)
Zoysia mosaic virus	(ZMV)

GENUS *RYMOVIRUS*

Type Species ryegrass mosaic virus (RGMV)

DISTINGUISHING FEATURES

VIRION PROPERTIES

Virions are flexuous filaments 690-720 x 11-15 nm in size. Virion density in CsCl is 1.33 g/ cm^3 (for RGMV). Virion S_{20w} is 165-166 for most members. Virions contain a single molecule of linear positive sense ssRNA with a 3' poly (A) terminus. Virion RNA is about 8.2 kb in size (Mr 2.7 x 10^6). WSMV RNA is about 8.5 kb in size (Mr 2.8 x 10^6). Sequences of CI, NIb, and CP are known for WSMV. Rymoviruses have a single capsid protein 29.2 kDa in size (RGMV). WSMV has capsid protein species 42 kDa, 36 kDa and 32 kDa in size; the two smaller proteins are subsets of the 42 kDa protein.

GENOME ORGANIZATION AND REPLICATION

The WSMV capsid protein sequence shows limited (22-25%) homology with capsid protein sequences of some aphid-transmitted potyviruses. Likewise, WSMV shows significant amino acid sequence homology with aphid-transmitted potyviruses in the potyviral cylindrical inclusion protein and portions of the potyviral nuclear inclusion protein. There is an *in vitro* translation product that is precipitated with antiserum to HC-PRO helper component of a potyvirus. The 3'-terminal non-coding region sequences of five WSMV isolates are greater than 90% identical to each other; these isolates were not similar to the 3'-terminal sequence of hordeum mosaic virus. Characteristic cytoplasmic cylindrical ("pinwheel") inclusions composed of a 66 kDa protein are present in infected cells. The WSMV capsid protein gene has been mapped to the 3'-terminal region of the genome. WSMV RNA has been translated *in vitro* into several large proteins immunoprecipitable with WSMV capsid protein antiserum, suggesting that WSMV uses a proteolytic processing strategy to express functional proteins such as the capsid protein. Antiserum to tobacco etch potyvirus 58 kDa nuclear inclusion protein also reacts with WSMV *in vitro* translation products.

ANTIGENIC PROPERTIES

Most rymoviruses are moderately immunogenic. No serological relationships among member viruses have been found except for a weak reaction between WSMV and ONMV.

BIOLOGICAL PROPERTIES

HOST RANGE

Most rymoviruses have limited but widespread host ranges within the family *Graminae* but some have relatively narrow host ranges.

TRANSMISSION

Transmission by eriophyid mites and mechanical transmission have been reported for most members.

LIST OF SPECIES IN THE GENUS

The viruses, their CMI/AAB description # ( ) and assigned abbreviations ( ) are:

SPECIES IN THE GENUS

Agropyron mosaic virus (118)	(AgMV)
Hordeum mosaic virus	(HoMV)
oat necrotic mottle virus (169)	(ONMV)
ryegrass mosaic virus (86)	(RGMV)
wheat streak mosaic virus (48)	(WSMV)

TENTATIVE SPECIES IN THE GENUS

brome streak virus (BStV)
Spartina mottle virus (SpMV)

GENUS *BYMOVIRUS*

Type Species barley yellow mosaic virus (BaYMV)

DISTINGUISHING FEATURES

VIRION PROPERTIES

MORPHOLOGY

Virions are flexuous filaments of two modal lengths, 250-300 and 500-600 nm; both are 13 nm in width.

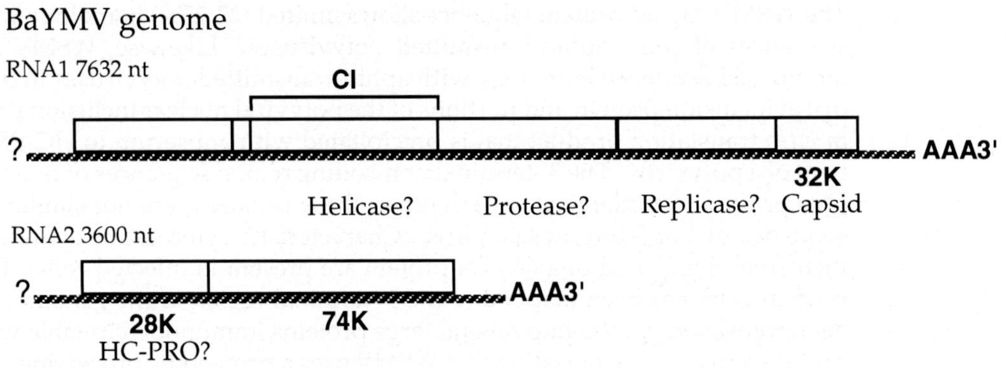

Figure 3: Genomic map of the barley yellow mosaic virus (BaYMV) bipartite genome. The same conventions as for TEV are employed. The boundaries of possible gene products are represented by vertical lines. Activities of the gene products are postulated by analogy with genus *Potyvirus*.

PHYSICOCHEMICAL AND PHYSICAL PROPERTIES

Virion buoyant density in CsCl is 1.28-1.30 g/cm^3.

NUCLEIC ACID

Virions contain two molecules of linear positive sense, ssRNA. RNA1 is 7.9 kb (Mr 2.6 x 10^6) and RNA2 is 4.56 kb (Mr 1.5 x 10^6) in size; RNA makes up 5% by weight of particles. Both RNA molecules have 3'-terminal poly (A) tracts. There is little base sequence homology between the two RNAs except in the 5' noncoding regions. The coat protein gene is located in the 3'-proximal region of RNA1.

PROTEINS

Virions have a single coat protein 28.5-33 kDa in size. The coat protein of the type species, BaYMV contains 297 amino acids.

GENOME ORGANIZATION AND REPLICATION

The two RNA molecules appear to be translated initially into precursor polypeptides from which functional proteins are derived by proteolytic processing.

ANTIGENIC PROPERTIES

The viral proteins are moderately immunogenic; serological relationships exist among members except barley mild mosaic virus (BaMMV). The coat protein amino acid sequence homology between BaYMV and BaMMV is 35-38%.

BIOLOGICAL PROPERTIES

INCLUSION BODY FORMATION

There are characteristic pinwheel-like inclusions and membranous network structures are formed in the cytoplasm of infected plant cells. No nuclear inclusions are found.

HOST RANGE

The host range of member viruses is narrow, restricted to the host family *Graminae*.

TRANSMISSION

The viruses are transmitted by the plasmodiophoraceous fungus *Polymyxa graminis*; transmissible experimentally by mechanical inoculation.

LIST OF SPECIES IN THE GENUS

The viruses, their alternative names ( ), genomic sequence accession numbers [ ], CMI/AAB description # ( ) and assigned abbreviations ( ) are:

SPECIES IN THE GENUS

barley mild mosaic virus		(BaMMV)
barley yellow mosaic virus (143)	[D01091, D01092]	(BaYMV)
oat mosaic virus (145)		(OMV)
rice necrosis mosaic virus (172)		(RNMV)
wheat spindle streak mosaic virus (167)		(WSSMV)
(wheat yellow mosaic virus)		

TENTATIVE SPECIES IN THE GENUS

None reported.

LIST OF UNASSIGNED VIRUSES IN THE FAMILY

1-whitefly transmitted:	
sweet potato mild mottle virus (162)	(SPMMV)
sweet potato yellow dwarf virus	(SPYDV)
2-aphid transmitted:	
Maclura mosaic virus	(MacMV)
narcissus latent virus	(NLV)

SIMILARITY WITH OTHER TAXA

Viruses of the family *Potyviridae* are similar to members of the families *Comoviridae*, *Picornaviridae*, and *Hypoviridae*. Genomes of member viruses of these taxa are single-stranded, positive sense RNAs. Most have a VPg at their 5' termini and a poly (A) tract at their 3' termini. Their genomes are expressed initially as high molecular weight polyprotein precursors which are processed by viral-encoded proteases. Gene products involved in replication are conserved in gene order and gene sequence.

DERIVATION OF NAMES

poty: siglum from *pot*ato *Y*
rymo: siglum from *ry*egrass *mo*saic
bymo: siglum from *b*arley *y*ellow *mo*saic

REFERENCES

Barnett OW (ed) (1992) Potyvirus Taxonomy. Arch Virol, Suppl 5. Springer-Verlag, Wien New York
Davidson AD, Prols M, Schell J, Steinbiss A-H (1991) The nucleotide sequence of RNA 2 of barley yellow mosaic virus. J Gen Virol 72: 989-993

Dougherty WG, Carrington JC (1988) Expression and function of potyviral gene products. Ann Rev Phytopathol 26: 123-143

Dougherty WG, Parks TD (1991) Post-translational processing of the tobacco etch virus 49 kDa small nuclear inclusion polyprotein: identification of an internal cleavage site and delimitation of VPg and proteinase domains. Virology 182: 17-27

Edwardson JR, Christie RG (eds) (1991) The Potyvirus Group. Vol I-IV. Univ Fla Agric Exp Sta Mono 16, Gainesville FL

Francki RIB, Milne RG, Hatta T (eds) (1985) Atlas of Plant Viruses, Vol II. CRC Press, Boca Raton FL

Hollings M, Brunt AA (1981) Potyviruses. In: Kurstak E (ed) Handbook of Plant Virus Infections and Comparative Diagnosis. Elsevier/North Holland Biomedical Press, Amsterdam, pp 731-807

Jordan R, Hammond J (1991) Comparison and differentiation of potyvirus isolates and identification of strain - , virus -, subgroup - specific and potyvirus group - common epitopes using monoclonal antibodies. J Gen Virol 72: 25-36

Kashiwazaki S, Minobe Y, Omura T, Hibino H (1990) Nucleotide sequence of barley yellow mosaic virus RNA 1: a close evolutionary relationship with potyviruses. J Gen Virol 71: 2781-2790

Kashiwazaki S, Minobe Y, Hibino H (1991) Nucleotide sequence of barley yellow mosaic virus RNA 2. J Gen Virol 72: 995-999

Lain S, Martin MT, Riechmann JL, Garcia JA (1991) Novel catalytic activity associated with positive-strand RNA virus infection: nucleic acid-stimulated ATPase activity of the plum pox potyvirus helicase-like protein. J Virol 65: 1-6

Milne RG (ed) (1988) The Plant Viruses, Vol 4. The Filamentous Plant Viruses. Plenum Press, New York

Mowat WP, Dawson S, Duncan GH, Robinson DJ (1991) Narcissus latent, a virus with filamentous particles and a novel combination of properties. Ann Appl Biol 119: 31-46

Niblett CL, Zagula KR, Calvert LA, Kendall TL, Stark DM, Smith CE, Beachy RN, Lommel SA (1991). cDNA cloning and nucleotide sequence of the wheat streak mosaic virus capsid protein gene. J Gen Virol 72: 499-504.

Restrepo MA, Freed DD, Carrington JC (1990) Nuclear transport of plant potyviral proteins. Plant Cell 2: 987-998

Schenk PM, Steinbiss H-H, Muller B, Schmitz K (1993) Association of two barley yellow mosaic virus (RNA2) encoded proteins with cytoplasmic inclusion bodies revealed by immunogold localization. Protoplasma 173: 113-122

Shukla DD, Brunt AA, Ward CW (1994) *Potyviridae*. Descriptions of Plant Viruses N° 245. Assoc Appl Biol, Wellesbourne UK

Shukla DD, Ward CW, Brunt AA, (1994) Potyviruses: Biology, Molecular Structure, and Taxonomy. CAB International, Wallingford, UK (in press)

Shukla DD, Ward CW (1989) Structure of potyvirus coat protein and its application to the taxonomy of the potyvirus group. Adv Virus Res 36: 273-314

Ward CW, Shukla DD (1991) Taxonomy of potyviruses: current problems and some solutions. Intervirology 32: 269-296

Contributed By

Barnett OW, Adam G, Brunt AA, Dijkstra J, Dougherty WG, Edwardson JR, Goldbach R, Hammond J, Hill JH, Jordan RL, Kashiwazaki S, Lommel SA, Makkouk K, Morales FJ, Ohki ST, Purcifull D, Shikata E, Shukla DD, Uyeda I

FAMILY *CALICIVIRIDAE*

TAXONOMIC STRUCTURE OF THE GENUS

Family ***Caliciviridae***
Genus *Calicivirus*

GENUS *CALICIVIRUS*

Type Species vesicular exanthema of swine virus (VESV)

VIRION PROPERTIES

MORPHOLOGY

Virions are 30-38 nm in diameter with 32 cup-shaped surface depressions arranged in T=3 icosahedral symmetry. The capsid is comprised of 180 protein molecules arranged in dimers and forming 90 capsomers.

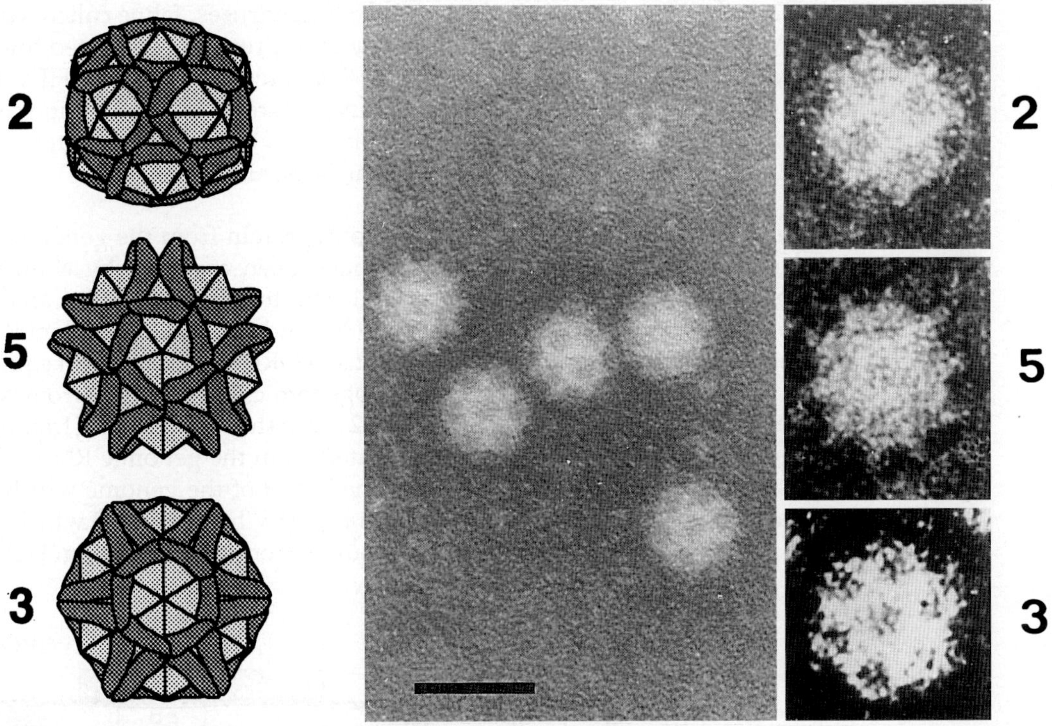

Figure 1: (left) Diagram of virion; (center) negative contrast electron micrograph of canine calicivirus particles (CaCV); (right) negative contrast electron micrograph of human calicivirus (HuCV) particles illustrating the surface appearance of particles orientated along the indicated 2-, 5- and 3-fold axes of symmetry. The bar represents 50 nm.

PHYSICOCHEMICAL AND PHYSICAL PROPERTIES

Virion Mr is about 15 x 10^6. Virion buoyant density is 1.33-1.40 g/cm^3 in CsCl and 1.29 g/cm^3 in glycerol-potassium tartrate gradients. Virion S_{20w} is 170-187. A second peak at 160-170 is believed to consist of defective interfering particles. Virions are insensitive to treatment with ether, chloroform, or mild detergents. Inactivation occurs at pH values between 3 and 5. Thermal inactivation is accelerated in high concentrations of Mg^{++} ions. Some calicivirus strains are inactivated by trypsin, whereas replication of others appears to be enhanced by trypsin. Several strains are readily disrupted by freezing and thawing.

NUCLEIC ACID

Virions contain a single molecule of linear, positive sense, ssRNA 7.4-7.7 kb in size. A protein (VPg, Mr 10-15 x 10^3) is covalently attached to the 5' end of most viruses. Hepatitis

E virus (HEV) lacks this structure and is capped. Subgenomic RNAs (2.2-2.4 kb) are synthesized intracellularly and may also be encapsidated by some members, e.g. rabbit hemorrhagic disease virus (RHDV).

PROTEINS

Virions are constructed from one major species of protein (Mr 59-71 x 10^3), the N-terminus of which is usually blocked. A minor 'soluble' protein (Mr 28-30 x 10^3) has been detected in Norwalk virus, amyelosis chronic stunt virus and porcine enteric calicivirus.

LIPIDS

None reported.

CARBOHYDRATES

None reported.

GENOME ORGANIZATION AND REPLICATION

The genomic organization and ORFs of three caliciviruses, feline calicivirus (FCV), RHDV and HEV are illustrated in (Fig. 2). Non structural proteins are located towards the 5' end, structural proteins towards the 3' end. As indicated, in FCV and HEV these genes are distinct, located in different reading frames and separated by termination codons. Norwalk virus and the Southampton strain of human calicivirus which is antigenically related to Snow Mountain virus have a genomic organization similar to FCV.

Non structural proteins are translated as a polyprotein from the genomic RNA. Putative roles for calicivirus non-structural genes have been assigned by comparison with the functional motifs in picornavirus proteins. The terminology is by analogy with those viruses. A helicase (2C), a cysteine protease (3C) and an RNA-dependent RNA polymerase (3D) are located towards the carboxy terminus of ORF 1; HEV lacks an equivalent to the 3C region. FCV and HEV each synthesize a subgenomic RNA (ORF 2) from which the capsid gene is expressed. In RHDV ORF 1 and ORF 2 are in the same reading frame. In this case the capsid protein (ORF 2) is apparently translated from the genomic RNA. FCV and RHDV also possess a potential ORF 3 at the extreme 3' end of the genome which could specify a small basic protein (Mr 10-12 x 10^3). In contrast, HEV has an ORF 3 which specifies a type-specific antigen that is distinct from the putative products of ORF 3 in FCV and RHDV.

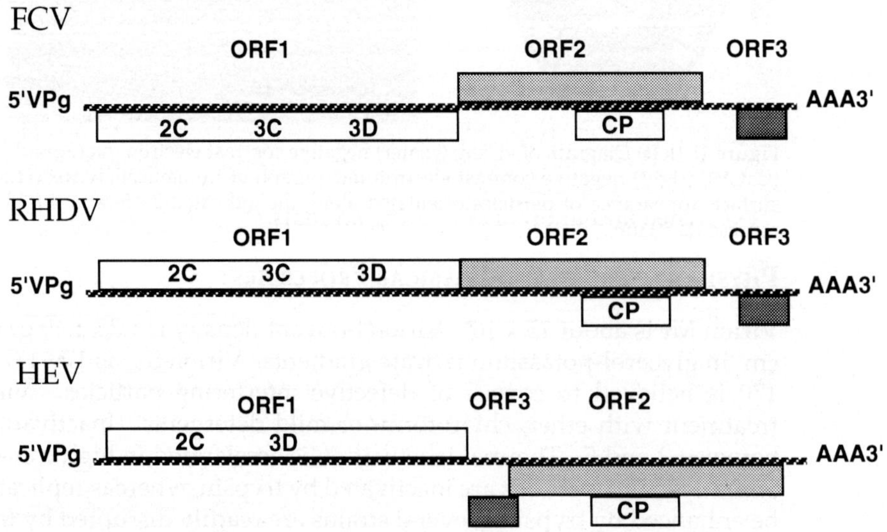

Figure 2: Genomic organization of FCV (7.69 kb), RHDV (7.437 kb) and HEV (7.194 kb). Open boxes are non-structural proteins (2C, 3C, 3D), grey boxes are capsid proteins (CP), dark grey boxes are putative proteins encoded by ORF3.

Two major virus specific ssRNA species are found in infected cells: a genome-sized RNA and a smaller RNA of 2.2-4 kb. Genome RNA serves as the mRNA for the non-structural proteins, and in the case of RHDV, for the capsid protein. Otherwise a subgenomic RNA codes for the capsid protein. Genome is replicated via a negative-sense RNA template. A negative-stranded form of the subgenomic RNA is readily detected in certain caliciviruses (e.g., FCV) but its function has yet to be established. The capsid polypeptide is the major protein product. An uncertain number of additional polypeptides are also synthesized. Precursor-product relationships among these proteins are not fully established. Virions mature in the cytoplasm of infected cells.

ANTIGENIC PROPERTIES

There are multiple distinct serotypes of vesicular exanthema of swine virus (VESV) and San Miguel sealion virus (SMSV). There is considerable cross-reactivity among feline caliciviruses. There is also cross-reactivity between SMSV and feline caliciviruses. By contrast, canine calicivirus, Norwalk virus and RHDV appear to be antigenically distinct.

BIOLOGICAL PROPERTIES

A variety of calicivirus hosts have been identified; e.g., VESV (swine and pinnipeds), SMSV (pinnipeds, fish and swine), FCV (cats and dogs), canine calicivirus (dogs), RHDV (rabbits), and Norwalk virus (human). Experimental hosts are diverse; e.g., VESV (some species of horse, dogs), SMSV (primates, mink), Norwalk virus (possibly chimpanzees). In cell culture a variety of host cells can be infected; e.g., VESV and SMSV (porcine, primate), FCV (feline, dolphin, porcine, primate), and porcine enteric calicivirus (porcine), HEV (human).

Transmission is via contaminated food, water, fomites, and on occasion via aerosolization of faecal material, vomitus or respiratory secretions. (e.g., FCV, canine calicivirus, Norwalk virus, RHDV). No vectors appear to be involved.

VESV produces in swine clinical signs some of which are indistinguishable from foot-and-mouth disease. These include vesicles in the mouth, tongue, lips, snout and between the toes. In addition, the virus may cause encephalitis, myocarditis, fever, diarrhea and failure of infected animals to thrive. Pregnant sows often abort. High mortality is associated with some strains. SMSV is similar to VESV. FCV produces in cats conjunctivitis, rhinitis, pneumonia, mucosal vesiculation, diarrhea and paresis. FCV produces a carrier state with virus latent in the tonsils. High mortality is associated with some strains of FCV. RHDV causes in rabbits haemorrhagic septicaemia, infectious necrosis of the liver and high mortality in adult animals. HEV in human causes acute hepatitis and in some outbreaks has caused high mortality in pregnant women. Fowl calicivirus produces stunting and high mortality in chicks. Amyelosis chronic stunt virus also results in stunting and high mortality in insects. Primate calicivirus produces mucosal vesiculation and persistent infection. Norwalk virus and human calicivirus induce diarrhea, vomiting, fever, nausea, colic and myalgia. Bovine enteric calicivirus and porcine enteric calicivirus infections result in diarrhea and anorexia in young animals.

GEOGRAPHIC DISTRIBUTION

Although most viruses have a worldwide distribution, some have only come from certain regions, e.g., vesicular exanthema of swine virus has come from whales and seals in North America, San Miguel sealion virus has been isolated from pinnipeds and fish in North America, rabbit haemorrhagic disease virus has come from China and Europe, canine calicivirus, primate calicivirus, reptile calicivirus and amyelosis chronic stunt virus have come from the USA, bovine enteric calicivirus (Newbury agents) have come from the UK and USA, fowl calicivirus from the UK, European brown hare syndrome virus has come from Europe, and porcine calicivirus has come from the UK, USA and Japan. In some cases the geographic distribution reflects host distribution; in other cases the distribution may be restricted or incompletely recognized.

List of Species in the Genus

The viruses, their genomic sequence accession numbers [ ] and assigned abbreviations ( ) are:

Species in the Genus

canine calicivirus		(CaCV)
feline calicivirus	[M86379, N32296, M32819, D90357]	(FCV)
hepatitis E virus	[M73218]	(HEV)
human caliciviruses	[M62825, M87661]	(HuCV)
Norwalk virus		(NV)
Hawaii strain		
Taunton strain		
Snow Mountain strain		
Southampton strain	[L07418]	
rabbit hemorrhagic disease virus	[M67473, Z11535]	(RHDV)
San Miguel sealion virus	[M87481, M87482]	(SMSV)
vesicular exanthema of swine virus		(VESV)

Tentative Species in the Genus

None reported.

List of Unassigned Viruses in the Family

amyelosis chronic stunt virus (insects)	(ACSV)
bovine enteric calicivirus	(BoCV)
fowl calicivirus	(FCVV)
European brown hare syndrome virus	(EBHSV)
human calicivirus	(HuCV)
mink calicivirus	(MCV)
primate calicivirus (Pan-1)	(PCV)
porcine enteric calicivirus	(PoCV)
reptile calicivirus (Cro-1)	(RCV)
walrus calicivirus	(WCV)

Similarity with Other Taxa

None reported.

Derivation of Names

calici: from Latin *calix*, "cup" or "goblet", from cup-shaped depressions observed by electron microscopy.

References

Bradley DW (1990) Enterically transmitted non-A, non-B hepatitis. Br Med Bull 46: 442-461

Carter MJ (1989) Feline calicivirus protein synthesis investigated by Western blotting. Arch Virol 108: 69-79

Carter MJ, Milton ID, Meanger J, Bennett MJ, Gaskell RM, Turner PC (1992) The complete nucleotide sequence of a feline calicivirus. Virology 190: 443-448

Cubitt D (1989) Diagnosis, occurrence and clinical significance of the human 'candidate caliciviruses'. Prog Med Virol 36: 103-119

Jiang X, Graham DY, Wang K, Estes MK (1991) Norwalk virus genome cloning and characterization. Science 250: 1580-1583

Meyers G, Wirblich C, Thiel HJ (1991) Rabbit haemorrhagic disease virus - molecular cloning and nucleotide sequencing of a calicivirus genome. Virology 184: 664-676

Parwani AV, Saif LJ, Kang SY (1990) Biochemical characterization of porcine enteric calicivirus: analysis of structural and non-structural viral proteins. Arch Virol 112: 41-53

Schaffer FL (1979) Caliciviruses, In: Fraenkel-Conrat H, Wagner RR (eds) Comprehensive Virology Vol 14. Plenum Press, New York, pp 249-284

Smith AW, Boyt PM (1990) Caliciviruses of ocean origin. J Zoo Wildlife Med 21: 3-23
Studdert MJ (1978) Caliciviruses: a brief review. Arch Virol 58: 157-191
Tam AW, Smith MW, Guerra ME, Huang CC, Bradley DW, Fry KE, Reyes GR (1991) Hepatitis E virus (HEV) molecular cloning and sequencing of the full length viral genome. Virology 185: 120-131

Contributed By

Cubitt D, Bradley DW, Carter MJ, Chiba S, Estes MK, Saif LJ, Schaffer FL, Smith AW, Studdert MJ, Thiel HJ

Family *Astroviridae*

Taxonomic Structure of the Family

Family	***Astroviridae***
Genus	*Astrovirus*

Genus *Astrovirus*

Type Species human astrovirus 1 (HAstV-1)

Virion Properties

Morphology

Virions are 28-30 nm in diameter, spherical in shape and non-enveloped. A distinctive five- or six-pointed star is discernible on the surface of about 10% of virions.

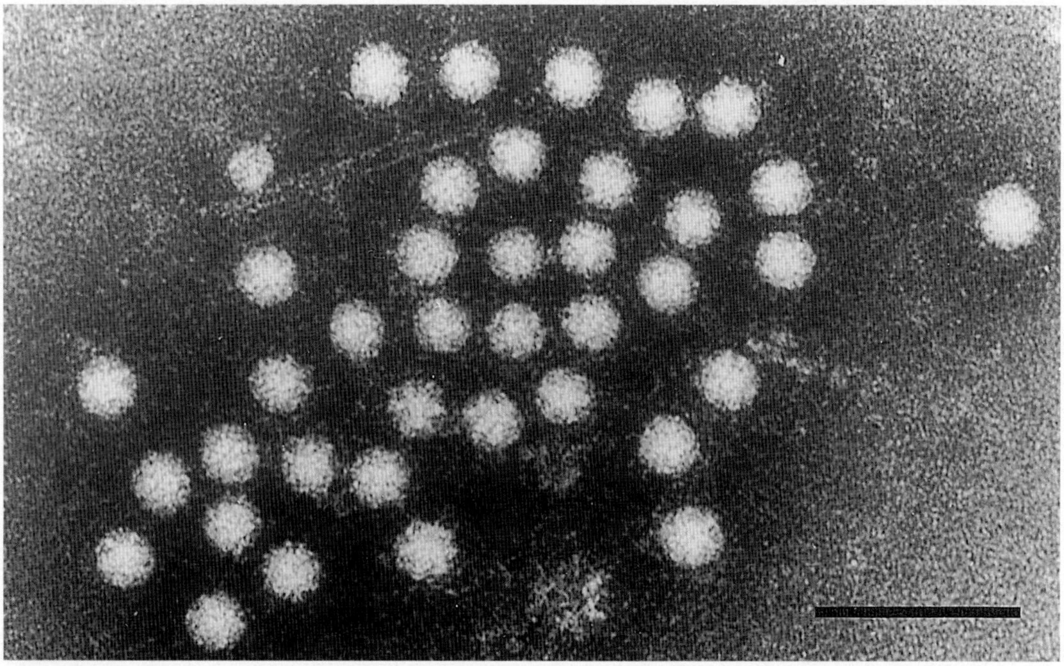

Figure 1: Negative contrast electron micrograph of human astrovirus from stool specimen. The bar represents 100 nm (courtesy of Humphrey C).

Physicochemical and Physical Properties

Virion Mr is about 8 x 10^6. Virion buoyant density in CsCl is 1.36 - 1.39 g/cm^3. Virion S_{20w} is about 160. Virions are resistant to pH 3, 50° C for 1 hr, 60° C for 5 min., chloroform, lipid solvents and non-ionic, anionic and zwitterionic detergents.

Nucleic Acid

Virions contain a single molecule of linear ssRNA. The genome is between 6.8 and 7.9 kb in size, is polyadenylated at the 3' end and presumed to be of positive polarity. The structure of the 5' end of the genome is unknown.

Proteins

Virion protein composition remains unclear; however, all isolates have at least two, possibly 3, major proteins with a Mr between 29 - 39 x 10^3. Several isolates also contain smaller proteins with Mr 13 - 36 x 10^3. Reportedly, a smaller protein is removed from virions following purification in SDS.

Lipids

Virions do not contain a lipid envelope. No information exists concerning fatty acid modification of any capsid protein.

Carbohydrates

No information exists concerning carbohydrate modification of any capsid protein.

Genome Organization and Replication

The genome organization and replication strategy of two human astroviruses have been determined. A polyadenylated, sub-genomic RNA (about 2.8 kb) has been detected in the cytoplasm of infected cells. Viral RNA replication is resistant to actinomycin D. Post-translational processing of viral proteins has not been examined. There is a single report of a presumed capsid precursor protein with Mr of 90 x 10^3 in the cytoplasm of infected cells where viral proteins accumulate. Early in infection proteins have been detected in the cell nucleus. Mature virus is often seen in crystalline arrays in the cytoplasm of infected cells.

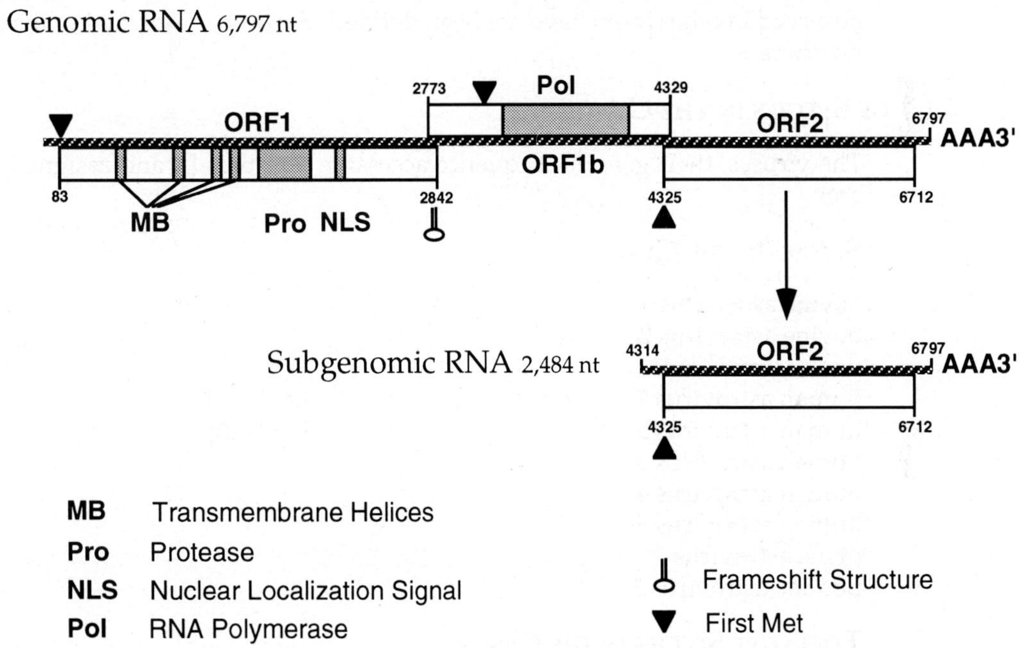

Figure 2: The arrangements of the genome, subgenomic RNA and deduced coding information for human astrovirus are shown. ORF 1b, encoding a putative polymerase is in a different reading frame to that of ORF 1a; translation may involve a ribosomal frameshift.

Antigenic Properties

At least seven serotypes of human astroviruses have been defined by immune electron microscopy and neutralization tests. They share at least one common epitope recognized by monoclonal antibody. At least two distinct serotypes of bovine astroviruses have been described by neutralization.

Biological Properties

Astroviruses appear to be host restricted, and have been detected in stool samples from humans, cats, cattle, deer, dogs, ducks, mice, pigs, sheep and turkeys. Transmission is by the fecal-oral route and no intermediate vectors have been described. Astroviruses are distributed worldwide and have been associated with about 2-8% of acute, non-bacterial gastroenteritis in children. The predominant feature of astrovirus infection in humans and animals is a self-limiting gastroenteritis. In humans, astrovirus has been detected in duodenal biopsies in ephithelial cells located in the lower part of villi. In experimentally infected sheep, astrovirus was found in the small intestine in the apical two-thirds of villi. In calves, astrovirus infection was localized to specialized M cells overlying the Peyer's patches. An often fatal hepatitis has been described in ducklings. The duck astrovirus, (duck hepatitis virus type 2) (types 1 and 3 are considered picornaviruses) is distinct from astrovirus isolates from turkeys and chickens in cross-protection and transmission studies.

Human, bovine, feline and porcine astroviruses have been isolated in primary embryonic kidney cells, but only the human and porcine viruses have been adapted to growth in established cell lines. Tryspin is required in the growth medium for serial propagation of the virus. Duck astrovirus grows in embryonated chicken eggs following blind passage in the amniotic sac. Few infected embryos die in less than 7 days. Infected embryos appeared stunted and have greenish, necrotic livers in which astrovirus-like particles have been identified.

TAXONOMIC STRUCTURE OF THE GENUS

At least 7 serotypes of human astroviruses, two serotypes of bovine astroviruses and one serotype of duck astrovirus are recognized. Their relationships to each other and those observed in other hosts have not been defined. Serotypes assigned to the groups are given numbers.

LIST OF SPECIES IN THE GENUS

The viruses, their genomic sequence accession numbers [ ] and assigned abbreviations ( ) are:

SPECIES IN THE GENUS

bovine astrovirus 1		(BAstV-1)
bovine astrovirus 2		(BAstV-2)
duck astrovirus 1		(DAstV-1)
human astrovirus 1	[Z25771]	(HAstV-1)
human astrovirus 2	[L13745]	(HAstV-2)
human astrovirus 3		(HAstV-3)
human astrovirus 4		(HAstV-4)
human astrovirus 5		(HAstV-5)
ovine astrovirus 1		(OAstV-1)
porcine astrovirus 1		(PAstV-1)

TENTATIVE SPECIES IN THE GENUS

None reported.

LIST OF UNASSIGNED VIRUSES IN THE FAMILY

None reported.

SIMILARITY WITH OTHER TAXA

None reported.

DERIVATION OF NAMES

astro: from Greek *astron*, "star", representing the star-like surface structure on virions

References

Aroonprasert D, Fagerland JA, Kelso NE, Zheng S, Woode GN (1989) Cultivation and partial characterization of bovine astrovirus. Vet Microbiol 19: 113-125

Herring AJ, Gray EW, Snodgrass DR (1981) Purification and characterization of ovine astrovirus. J Gen Virol 53: 47-55

Herrmann JE, Hudson RW, Perron-Henry DM, Kurtz JB, Blacklow NR (1988) Antigenic characterization of cell-cultivated astrovirus serotypes and development of astrovirus-specific monoclonal antibodies. J Infect Dis.158: 182-185

Hudson RW, Herrmann JE, Blacklow NR (1989) Plaque quantitation and virus neutralization assays for human astroviruses. Arch Virol 108: 33-38

Jiang B, Monroe SS, Koonin EV, Stine SE, Glass RI (1993) RNA sequence of astrovirus: Distinctive genomic organization and a putative retrovirus-like ribosomal frameshifting signal that directs the viral replicase synthesis. Proc Natl Acad Sci USA 90: 10539-10543

Kurtz JB, Lee TW (1987) Astroviruses: human and animal. In: Bock G, Whelan J (eds) Ciba Foundation Symposium 128. Novel diarrhoea viruses. John Wiley & Sons, New York, pp 92-107

Lee TW, Kurtz JB (1994) Prevaleance of human astrovirus serotypes in the Oxford region 1976-1992, with evidence for two new serotypes. Epidemiology & Infection 112: 187-193

Lewis TL, Greenberg HB, Herrmann JE, Smith LS, Matsui SM (1994) Analysis of astrovirus serotype 1 RNA, identification of the viral RNA-dependent RNA polymerase motif, and expression of a viral structural protein. J Virol 68: 77-83

Monroe SS, Jiang B, Stine SE, Koopmans M, Glass RI (1993) Subgenomic RNA sequence of human astrovirus supports classification of *Astroviridae* as a new family of RNA viruses. J Virol 67: 3611-3614

Monroe SS, Stine SE, Gorelkin L, Herrmann JE, Blacklow NR, Glass RI (1991) Temporal synthesis of proteins and RNAs during human astrovirus infection of cultured cells. J Virol 65: 641-648

Willcocks MM, Carter MJ, Laidler FR, Madeley CR (1990) Growth and characterization of human faecal astrovirus in a continuous cell line. Arch Virol 113: 73-81

Willcocks MM, Carter MJ, Madeley CR (1992) Astroviruses. Rev Med Virol 2: 97-106

Woolcock PR, Fabricant J (1991) Duck virus hepatitis. In: Calnek BW (ed) Diseases of poultry 9th edn. Marinus Nijhoff Dordrecht, Netherlands, pp 597-608

Woode GN, Gourley NEK, Pohlenz JF, Lieber EM, Mathews SL, Hutchinson MP (1985) Serotypes of bovine astrovirus. J Clin Microbiol 22: 668-670

Contributed By

Monroe SS, Carter MJ, Herrmann JE, Kurtz JB, Matsui SM

FAMILY *NODAVIRIDAE*

TAXONOMIC STRUCTURE OF THE FAMILY

Family ***Nodaviridae***
Genus *Nodavirus*

GENUS *NODAVIRUS*

Type Species Nodamura virus (NoV)

VIRION PROPERTIES

MORPHOLOGY

Virions are unenveloped, roughly spherical in shape, 30 nm in diameter and have icosahedral symmetry (T=3). No distinct surface structure is seen by electron microscopy. Empty shells are rarely, if ever observed in virus preparations.

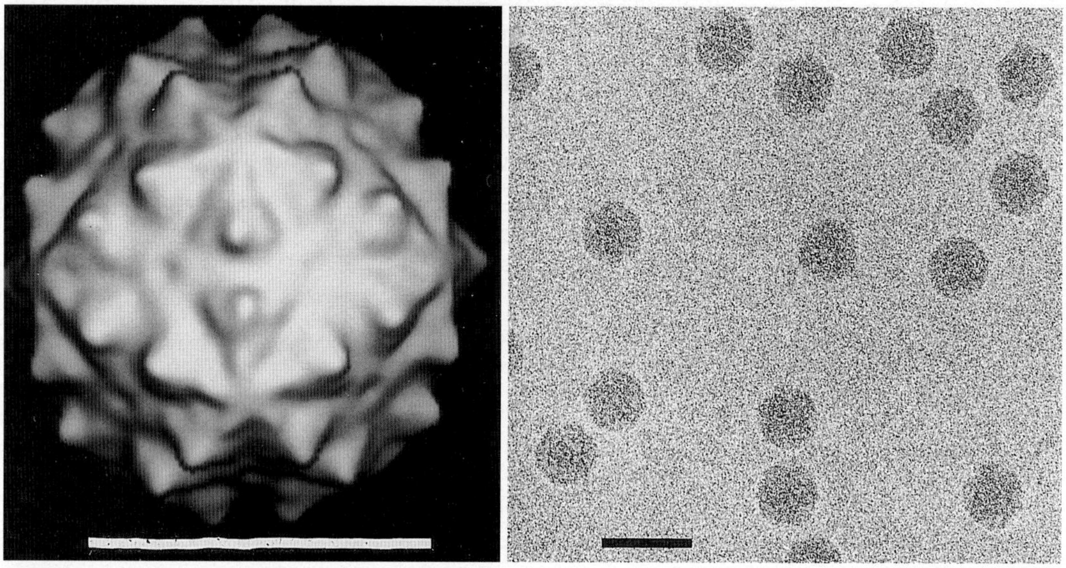

Figure 1: (left) Image reconstruction of flock house virus; the bar represnts 20 nm. (right) Cryo-electron micrograph of flock house virus; the bar represents 50 nm. (Photos courtesy of Norman Olson & Tim Baker, Purdue University).

PHYSICOCHEMICAL AND PHYSICAL PROPERTIES

Virion Mr is about 8 x 10^6; S_{20w} is about 135 to 140. Virion buoyant density in CsCl is 1.30 to 1.34 g/cm^3 (varies with species). Infectivity of aqueous suspensions is stable to extraction with chloroform. Infectivity of Nodamura virus, black beetle virus, and flock house virus is stable at room temperature in 1% sodium dodecyl sulfate but Boolarra virus is inactivated. Virions are stable at acid pH. The RNA content of the virion is about 16%.

NUCLEIC ACID

The genome consists of two molecules of ssRNA molecules, with an Mr of 1.1 x 10^6 and 0.48 x 10^6, respectively. Both molecules are apparently encapsidated in the same particle. Both molecules are capped at their 5'-end and lack a poly (A) tail at their 3'-end. The 3'-ends cannot be chemically derivatized even after treatment with denaturing solvents suggesting they are blocked, possibly with a protein.

PROTEINS

The capsid consists of 180 protein subunits (protomers). Morphogenesis involves formation of a virus-like "provirion" which acquires infectivity by autocatalytic cleavage of the coat protein precursor alpha (Mr 44 x 10^3) to form two smaller proteins, called beta (Mr 40 x 10^3) and gamma (Mr 4 x 10^3). This "maturation" cleavage is often incomplete; Virions typically contain residual uncleaved precursor chains, the proportion varying from 10 to 50%, depending upon virus species and probably also conditions of propagation and purification.

LIPIDS

Virions are not known to carry lipid; however the amino terminus of the coat protein precusor, alpha, is blocked by an unidentified entity.

CARBOHYDRATES

None reported.

GENOME ORGANIZATION AND REPLICATION

The virus replicates in the cytoplasm. RNA synthesis is resistant to actinomycin D. Infected cells contain three ssRNAs: RNA1 (Mr 1 x 10^6); RNA2 (Mr 0.5 x 10^6) and a subgenomic RNA3 (Mr 0.15 x 10^6). RNA3 is not packaged into virions. RNA1 codes for protein A (Mr 112 x 10^3); the latter is probably a component of the viral RNA polymerase. RNA2 codes for the coat protein precursor, alpha (Mr 44 x 10^3). RNA3 encodes protein B (Mr 10 x 10^3) which may play a role in synthesis of positive-strand RNA. Cells infected with isolated RNA1 synthesize RNA1 and RNA3 but not RNA2. Both RNA1 and RNA2 are required for

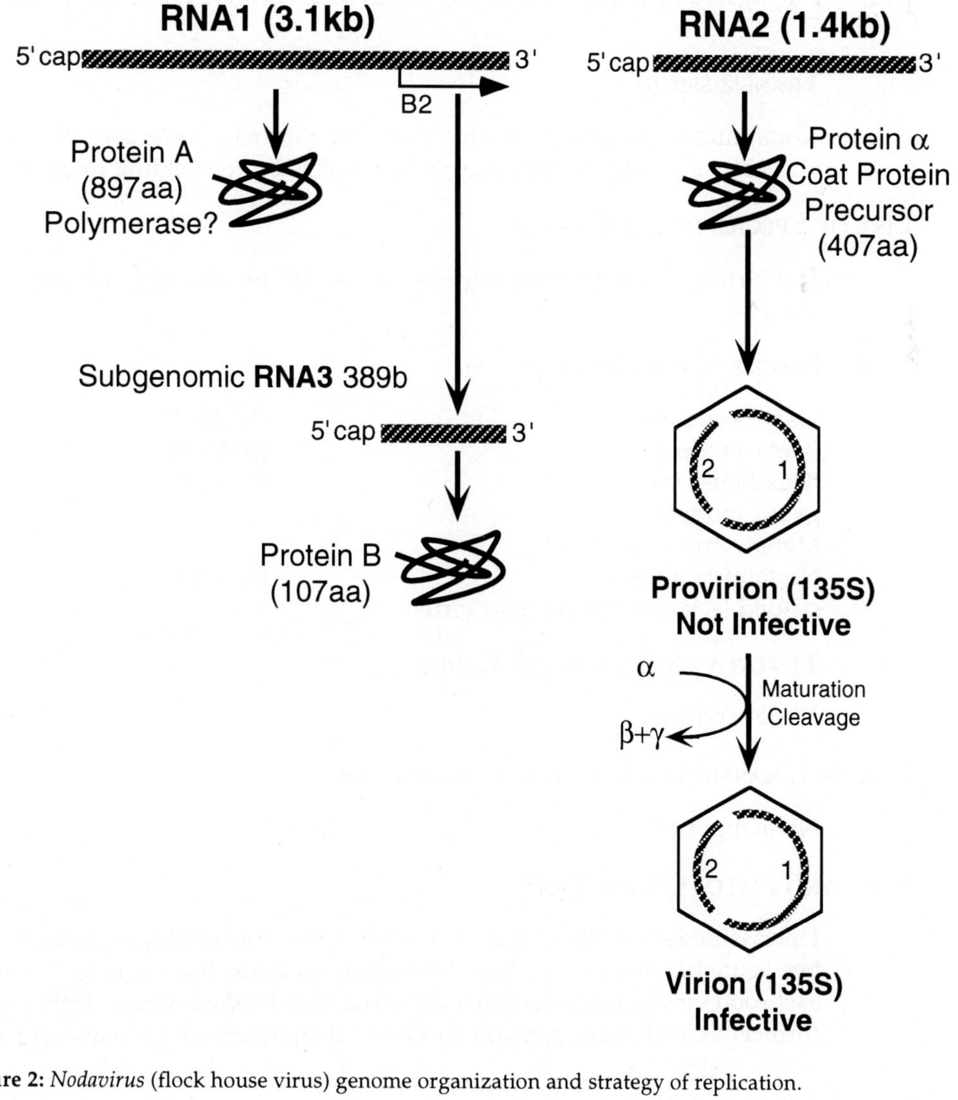

Figure 2: *Nodavirus* (flock house virus) genome organization and strategy of replication.

production of virions. RNA2 strongly inhibits synthesis of RNA3. Messenger activity of the RNAs in infected cells is in relative terms RNA3>RNA2>RNA1. Defective-interfering particles are formed readily if virus is not passaged at low multiplicity of infection. Persistent infection, with subsequent resistance to superinfection, occurs readily in cultured cells.

Antigenic Properties

Nodamura virus, black beetle virus, flock house virus and Boolarra virus are cross-reactive by double-diffusion precipitin tests but all four viruses represent different serotypes (neutralization titer of each antiserum less than 0.5% in heterotypic crosses).

Biological Properties

Host Range

Nature: All species, except striped jack nervous necrosis virus, were isolated from insects. Viruses do not seem to be notably host-specific.
Laboratory: Most, if not all member viruses, can be propagated in larvae of the common wax moth, *Galleria mellonella*. Nodamura virus, isolated from mosquitoes, grows in suckling mice but not in cultured cells of *Drosophila melanogaster*; flock house virus, isolated from larvae of a grass grub *Costelytra zealandica*, multiplies in tobacco plants as well as in cultured *Drosophila* cells. Black beetle virus, flock house virus and Nodamura virus form plaques in cultured *Drosophila* cells. Nodamura virus multiplies poorly in cell culture but can be propagated by transfecting cell cultures with virion RNA at temperatures below about 34° C.

Transmission

Nodamura virus is transmissible to suckling mice by *Aedes aegypti* mosquitoes. Nodamura virus causes paralysis and death when injected into suckling mice or wax moth larvae.

List of Species in the Genus

The viruses, their genomic sequence accession numbers [ ] and assigned abbreviations ( ) are:

Species in the Genus

black beetle virus	[K02560]	(BBV)
Boolarra virus	[X15960]	(BoV)
flock house virus	[X15959]	(FHV)
gypsy moth virus		(GMV)
Manawatu virus		(MwV)
Nodamura virus	[X15961]	(NoV)
striped Jack nervous necrosis virus		(SJNNV)

Tentative Species in the Genus

None reported.

List of Unassigned Viruses in the Family

None reported.

Similarity with Other Taxa

Unclassified small RNA viruses: Viruses with a morphology similar to nodaviruses include bee acute paralysis virus, bee slow paralysis virus, bee virus X, *Drosophila P* and *A* virus, sacbrood virus, Queensland fruitfly virus, and *Triatoma* virus. Aphid lethal paralysis virus, formerly listed here, appears to have 3 major capsid proteins and is likelier related to picornaviruses. Two tetraviruses (NωV & HaSV) contain a bipartite single-stranded genome, but they have larger capsids with T=4 icosahedral symmetry and have capped genomic strands that are twice as long with no 3' terminal blockage.

DERIVATION OF NAMES

Nodamura: formerly a village (now a city, Nodashi), in the vicinity of the site where the virus was isolated in Japan. Other nodaviruses are similarly named after the place of isolation or after the common name of the animal from which the virus was isolated. Striped Jack is a species of fish.

REFERENCES

Ball LA, Amann JM, Garret BK (1992) Replication of Nodamura virus after transfection of viral RNA into mammalian cells in culture. J Virol 66: 2326-2334

Dearing SC, Scotti PD, Wigley PJ, Dhana SD (1980) A small RNA virus isolated from the grass grub, *Costelytra zealandica* (*Coleoptera: Scarabaeidae*). N Z J Zool 7: 267-269

Garzon S, Charpentier G (1992) *Nodaviridae*. In: Adams JR, Bonami JR (eds) Atlas of Invertebrate Viruses. CRC Press, Boca Raton FL, pp 351-370

Hendry DA, (1991) *Nodaviridae* of Invertebrates. In: Kurstak E (ed) Viruses of Invertebrates. Marcel Deker, New York NY, pp 227-276

Mori KI, Nakai T, Muroga K, Arimoto M, Musiake K, Firusawa I (1992) Properties of a new virus belonging to *Nodaviridae* found in larval Striped Jack (*Pseudocaranx dentiex*) with nervous necrosis. Virology 187: 368-371

Reinganum C, Bashirrudin JB, Cross GF (1985) Boolarra virus: a member of the *Nodaviridae* isolated from *Oncopera intricoides* (*Lepidoptera: Hapealidae*). Intervirology 24: 10-17

Schneemann, Zhong W, Gallagher TM, Rueckert RR (1992) Maturation cleavage required for infectivity of a nodavirus. J Virol 66: 6728-6734

Scotti PD, Fredericksen S (1987) Manawatu virus: a nodavirus isolated from *Costelytra zealandica* (White) (*Coleoptera: Scarabaeidae*). Arch Virol 97: 85-92

CONTRIBUTED BY

Hendry DA, Johnson JE, Rueckert RR, Scotti PD

Family *Tetraviridae*

Taxonomic Structure of the Family

Family	***Tetraviridae***
Genus	"Nudaurelia capensis β–like viruses"
Genus	"Nudaurelia capensis ω–like viruses"

Virion Properties

Morphology

Virions are unenveloped, roughly spherical, about 40 nm in diameter and exhibit icosahedral symmetry. Distinct capsomers have been resolved by cryo-electron microscopy and image reconstruction. The genome consists of ssRNA. Member viruses of the unnamed genus comprising the Nudaurelia capensis β–like viruses have a monopartite genome and those of the genus comprising the Nudaurelia capensis ω like-viruses have a bipartite genome.

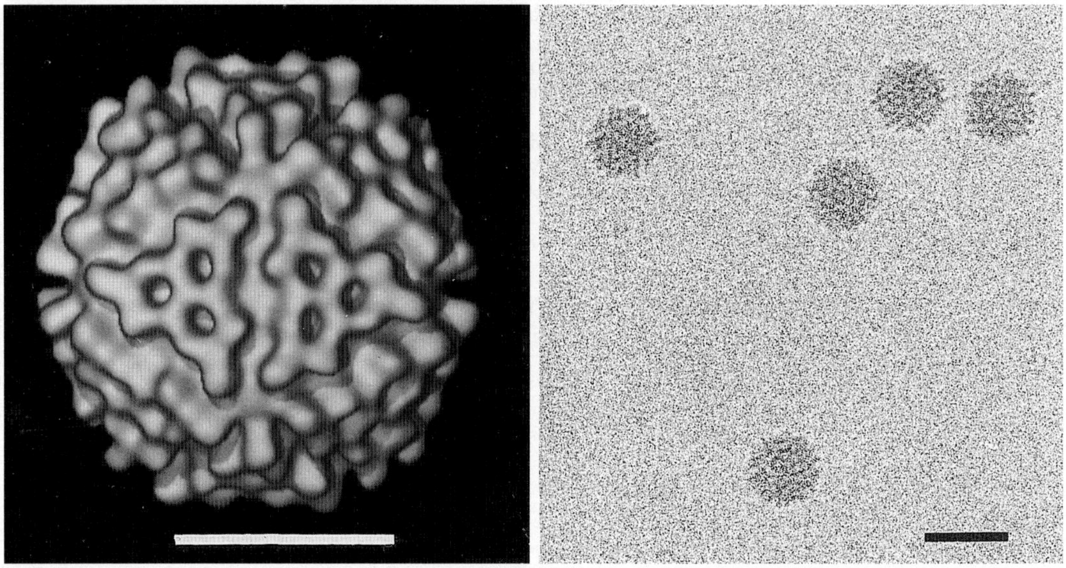

Figure 1: (left) Image reconstruction of Nudaurelia capensis β virus; the bar represents 20 nm. (right) Cryo-electron micrograph of Nudaurelia capensis β virus; the bar represents 50 nm. (Photos courtesy of Holland R, Cheng, Norman Olson & Tim Baker, Purdue University).

Physicochemical and Physical Properties

Virion Mr is about 16 x 10^6. Virion S_{20w} is about 194-210. Virion buoyant density in CsCl is 1.29-1.30 g/cm^3 (varies with species). Virion is stable at pH 3.

Nucleic Acid

The type virus, Nudaurelia capensis β virus, contains a single molecule of RNA, about 5.5 kb in size (Mr 1.8 x 10^6). This represents about 11% of the virion mass. At least two other tetravirus-like agents, Nudaurelia capensis ω virus and Heliothis armigera stunt virus (HaSV) have bipartite genomes. RNA1 is about 5.5 kb in size (Mr 1.8 x 10^6) whereas RNA2 is about 2.5 kb in size (Mr 0.8 x 10^6). Neither RNA is polyadenylated at the 3'-end. It is not known if the two RNA segments are packaged together in the same particle or separately.

Proteins

The capsid consists of 240 protein subunits (protomers). Each protomer consists of one 70 x 10^3 precursor or a pair of cleavage products of 62 and 8 x 10^3, respectively.

LIPIDS

Virions are not known to contain lipid; however the amino terminus of the coat precusor, alpha, is blocked by an unidentified entity.

CARBOHYDRATES

No carbohydrates have been identified.

GENOME ORGANIZATION AND REPLICATION

The viruses replicate primarily in the cytoplasm of gut cells of several *Lepidoptera*. Crystalline arrays of virus particles are often seen within cytoplasmic vesicles. The genome organization of the genus comprising the Nudaurelia capensis β-like viruses is not known; that of the genus comprising the Nudaurelia capensis ω-like viruses is depicted in fig. 2.

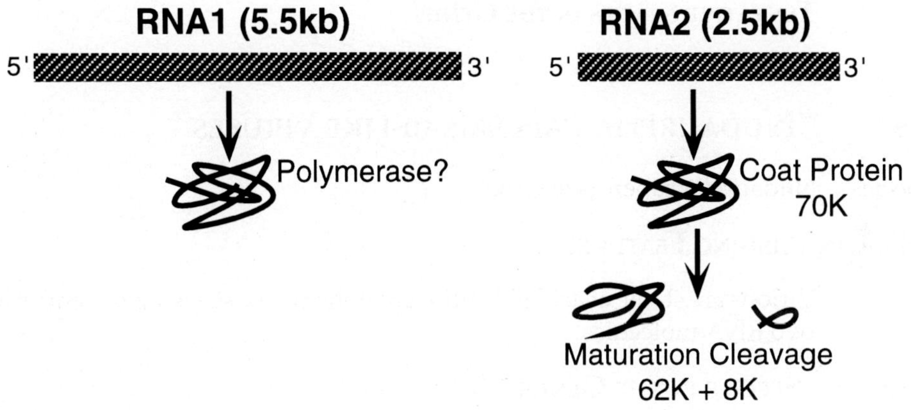

Figure 2: Nudaurelia capensis ω virus genome organization.

ANTIGENIC PROPERTIES

Most of the members of the family are serologically interrelated but distinguishable. The majority of the isolates were identified on the basis of their serological reaction with antiserum raised against Nudaurelia capensis β virus.

BIOLOGICAL PROPERTIES

HOST RANGE

Nature: All species were isolated from species of *Lepidoptera*, principally from Saturniid, Limacodid and Noctuid moths. Individual viruses exhibit a broad range of infection and pathogenicity. Infection leads to rapid death or to growth retardation of larval stages.

Laboratory: No infections by the viruses of the genus comprising the Nudaurelia capensis β-like viruses have yet been achieved in cultured invertebrate cells. A virus of the genus comprising the Nudaurelia capensis ω-like viruses, Heliothis armigera stunt virus, grows slowly and without cytopathic effect in cultured *Drosophila* and *Spodoptera* cells.

TRANSMISSION

Heliothis armigera stunt virus is transmitted orally. Oral transmission can be inferred from reports of tetraviruses being used as sprayed insecticides in Malaysia; e.g. Darma trima virus and the Setora nitens virus

GENUS "NUDAURELIA CAPENSIS β-LIKE VIRUSES"

Type Species Nudaurelia capensis β virus (NβV)

DISTINGUISHING FEATURES

Virions are stable at acid pH; buoyant density in CsCl is 1.29 g/cm^3. Virions appear to contain a single molecule of RNA.

LIST OF SPECIES IN THE GENUS

The viruses and their assigned abbreviations ( ) are:

SPECIES IN THE GENUS

Nudaurelia capensis β virus (NβV)

TENTATIVE SPECIES IN THE GENUS

Trichoplusia ni virus (TnV)

GENUS "NUDAURELIA CAPENSIS ω-LIKE VIRUSES"

Type Species Nudaurelia capensis ω virus (NωV)

DISTINGUISHING FEATURES

Virions are stable at acid pH; buoyant density in CsCl is 1.29 g/cm^3. They appear to contain two RNA molecules.

LIST OF SPECIES IN THE GENUS

The viruses and their assigned abbreviations ( ) are:

SPECIES IN THE GENUS

Nudaurelia capensis ω virus (NωV)

TENTATIVE SPECIES IN THE GENUS

Helicoverpa armigera stunt virus (HaSV)

LIST OF UNASSIGNED VIRUSES IN THE FAMILY

Unassigned viruses that are considered possible members of the family are:

Acherontia atropas virus
Agraulis vanillae virus
Antheraea eucalypti virus (AeV)
Darna trim virus (DtV)
Dasychira pudibunda virus (DpV)
Eucocystis meeki virus
Euploea corea virus
Hyalophora cecropia virus
Hypocrita jacobeae virus
Lymantria ninayi virus
Nudaurelia capensis ε virus (epsilon virus) (NεV)
Philosamia ricini virus (PxV)
Pseudoplusia includens virus (PiV)
Thosea asigna virus (TaV)
Saturnia pavonia virus
Setora nitens virus

Derivation of Names

tetra: from Greek *tettares* 'four' as T=4

References

Agrawal DK, Johnson JE (1992) Sequence and analysis of the capsid protein of Nudaurelia capensis w virus, an insect virus with T=4 icosahedral symmetry. Virology 190: 806-814

du Plessis DH, Mokhosi G, Hendry DA (1991) Cell-free translation and identification of the replicative form of Nudaurelia b virus RNA. J Gen Virol 72: 267-273

Hanzlik TN, Dorrian S, Gordon KHJ, Christian PD (1993) A novel small RNA virus isolated from the cotton bollworm, *Helicoverpa armigera*. J Gen Virol pp 1805-1810

Hendry DA, Agrawal DK (1993) Small RNA Viruses of Insects: *Tetraviridae*. In: Granoff A, Webster RG (eds) Encyclopedia of Virology. Academic Press, London pp 1416-1422

Moore NF (1991) The Nudaurelia b family of insect viruses. In: Kurstak E (ed) Viruses of Invertebrates. Marcel Dekker, New York, pp 277-285

Olson NH, Baker TS, Johnson JE, Hendry DA (1990) The three-dimensional structure of frozen-hydrated Nudaurelia capensis β virus, a T=4 insect virus. J Struct Biol 105: 111-122.

Contributed By

Hendry DA, Johnson JE, Rueckert RR, Scotti PD, Hanzlik TN

Genus *Sobemovirus*

Type Species Southern bean mosaic virus (SBMV)

Virion Properties

Morphology

Virions are about 30 nm in diameter and exhibit icosahedral symmetry (T=3). Virions are composed of 180 subunits. Each protein subunit has two domains. One forms parts of the icosahedral shell about 3.5 nm thick and the other forms a partially ordered 'arm' into the interior of the virus.

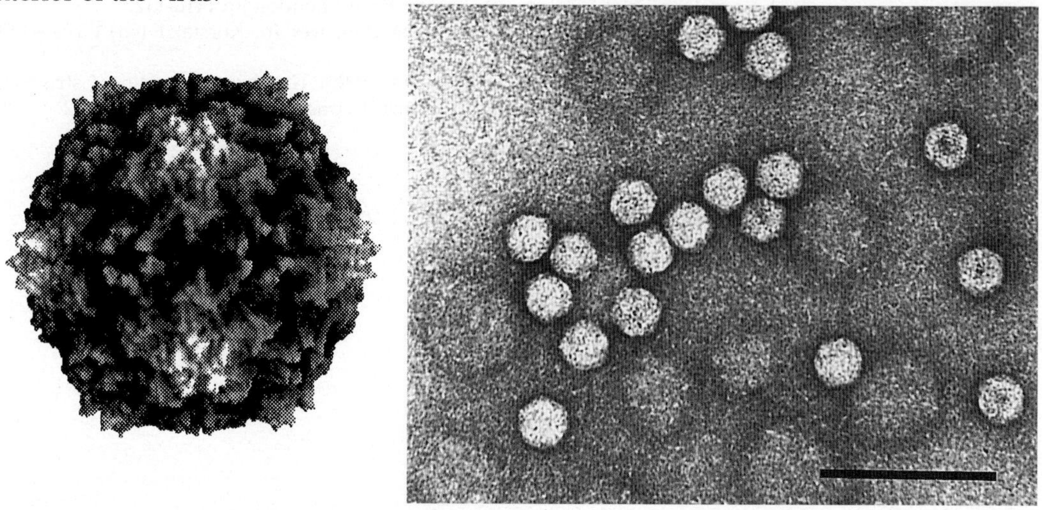

Figure 1: (left) Electronic image of a SBMV particle (T=3), (courtesy of Sgro JY, Wisconsin). (right) Negative contrast electron micrograph of rice yellow mottle virus stained in uranyl acetate. The bar represents 100 nm.

Physicochemical and Physical Properties

Virion Mr is about 6.6×10^6; S_{20w} is about 115; density is about 1.36 g/cm^3 in CsCl (but virus forms two or more bands in Cs_2SO_4); particles swell reversibly in EDTA and higher pH with concomitant changes in capsid conformation and partial loss of stability.

Nucleic Acid

Particles contain a single molecule of positive sense ssRNA, approximately 4.2 kb in size (Mr 1.4×10^6). Vpg, which is probably essential for infectivity, is associated with the 5'-end of the genome. The 3'-end does not contain poly (A) or a tRNA-like structure. A subgenomic, 3'-coterminal RNA (Mr 0.38×10^6) is also found in SBMV. Satellite viroid-like RNAs are associated with some member viruses.

Proteins

There is one coat protein species with an Mr about 30×10^3. No functions have been attributed to products of other ORFs.

Lipids

None reported.

Carbohydrates

None reported.

Genome Organization and Replication

Genomic RNA remains associated with swollen virions during cell-free translation in wheat germ extract. Sequencing of the cowpea strain of SBMV has indicated four possible ORFs, with coding capacity for proteins of Mr 21 x 10^3 (ORF 1; 49-603), 105 x 10^3 (ORF 2; 570-3,437), 18 x 10^3 (ORF 3; 1,895-2,380) and 31 x 10^3 (ORF 4; 3,217-4,053). *In vitro* translation of full-length SBMV genomic RNA in wheat germ, or of turnip rosette virus RNA in rabbit reticulocyte lysate, yields three proteins (P1, 105 x 10^3; P2, 60 x 10^3; P4, 14-25 x 10^3); however, coat protein (P3, 28 x 10^3) is only translated from 0.3-0.4 x 10^6 virion-associated RNA 2, indicating that this is a subgenomic mRNA. It is suggested that ORF 1 encodes P4(s); ORF 2 encodes P1; P2 is derived by proteolysis from P1; ORF 4 encodes P3. No protein or subgenomic mRNA has been associated with ORF 3. Genome homologies suggest similarities to picorna- and potyviruses. Replication is thought to be mediated by an RNA-dependent RNA polymerase via a (-)-strand intermediate.

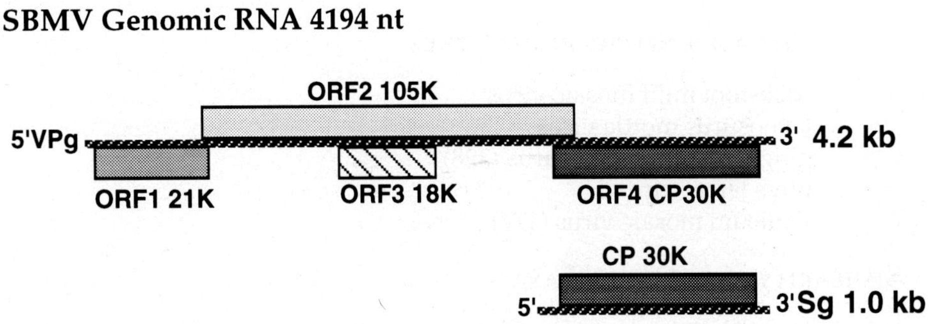

Figure 2: Genome organization of SBMV (cowpea strain). The lines represent the viral RNA genome and the subgenomic RNA. The boxes indicate the ORFs with the size of the corresponding protein; CP = coat protein.

ANTIGENIC PROPERTIES

Viral proteins serve as efficient immunogens. A single precipitin line is formed in gel diffusion tests. There are serological relationships between strains and some members of the genus.

BIOLOGICAL PROPERTIES

HOST RANGE

The natural host range of each virus species is relatively narrow. Disease symptoms are mainly mosaics and mottles. Systemic infections are caused in most natural hosts with most cell types being infected.

TRANSMISSION

Seed transmission occurs in several host plants. The viruses are transmitted by beetles or a myrid in the case of velvet tobacco mottle virus. The viruses are readily transmitted mechanically.

GEOGRAPHIC DISTRIBUTION

Most members have limited distribution but, as a whole, they are found worldwide.

CYTOPATHIC EFFECTS

Virions are found in both the cytoplasm and nuclei, and late in infection occur as large crystalline aggregates in the cytoplasm. Infected cells show extensive cytoplasmic vacuolation

LIST OF SPECIES IN THE GENUS

The viruses, their genomic sequence accession numbers [ ], CMI/AAB description # ( ) and assigned abbreviations ( ) are:

SPECIES IN THE GENUS

blueberry shoestring virus (204)		(BSSV)
cocksfoot mottle virus (23)		(CoMV)
lucerne transient streak virus (224)		(LTSV)
rice yellow mottle virus (149)	[EM_VI: RYVCGEN]	(RYMV)
Solanum nodiflorum mottle virus (318)		(SNMV)
Southern bean mosaic virus (57,274)		(SBMV)
sowbane mosaic virus (64)		(SoMV)
subterranean clover mottle virus (329)		(SCMoV)
turnip rosette virus (125)		(TRoV)
velvet tobacco mottle virus (317)		(VTMoV)

TENTATIVE SPECIES IN THE GENUS

cocksfoot mild mosaic virus	(CMMV)
Cynosurus mottle virus	(CnMoV)
ginger chlorotic fleckvirus (328)	(GCFV)
olive latent virus 1	(OLV-1)
Panicum mosaic virus (177)	(PMV)

SIMILARITY WITH OTHER TAXA

Virions of the members of the family *Tombusviridae* (genera *Tombusvirus* and *Carmovirus*) and of the genus *Necrovirus* are isometric, encapsidating a single genomic RNA species about 4 kb in size. These viruses are generally similar to the member viruses of the genus *Sobemovirus*. These other viruses differ from sobemoviruses in the Mr of their coat proteins and in their genome organizations.

DERIVATION OF NAMES

sobemo: sigla derived from the name of type species *so*uthern *be*an *mo*saic

REFERENCES

Abad-Zapatero C, Abdel-Meguid SS, Johnson JE, Leslie AGW, Rayment I, Rossmann MG, Suck D, Tsukihara T (1980) Structure of southern bean mosaic at 2.8Å resolution. Nature 286: 33-39

Carrington JC, Morris TJ, Stockley PG, Harrison SC (1987) Structure and assembly of turnip crinkle virus. IV. Analysis of the coat protein gene and implications of the subunit primary structure. J Mol Biol 194: 265-276

Francki RIB, Milne RG, Hatta T (eds) (1985) Sobemovirus group, In: Atlas of Plant Viruses Vol I. CRC Press, Boca Raton FL, pp 153-169

Francki RIB, Randles JW, Hatta T, Davies C, Chu PWG, McLean GD (1983) Subterranean clover mottle virus: another virus from Australia with encapsidated viroid-like RNA. Plant Pathol 32: 47-59

Ghosh A, Dasgupta R, Salerno-Rife T, Rutgers T, Kaesberg P (1979) Southern bean mosaic viral RNA has a 5'-linked protein but lacks 3' terminal poly (A). Nucl Acids Res 7: 2137-2146

Goldbach R (1987) Genome similarities between plant and animal RNA viruses. Microbiol Sci 4: 197-202

Gorbalenya AE, Koonin EV, Blinov VM, Donchenko AP (1988) Sobemovirus genome appears to encode a serine protease related to cysteine proteases of picornaviruses. FEBS Lett 236: 287-290

Hull R (1988) The sobemovirus group. In Koenig R (ed), The Plant Viruses, Polyhedral virions with monopartite genomes Vol 3. Plenum Press, New York, pp 113-146

Jones AT, Mayo MA (1984) Satellite nature of the viroid-like RNA-2 of Solanum nodiflorum mottle virus and the ability of other plant viruses to support the replication of viroid-like RNA molecules. J Gen Virol 65: 1713-1721

Rossmann MG, Abad-Zapatero C, Hermodson MA, Erickson JW (1983) Subunit interactions in southern bean mosaic virus. J Mol Biol 166: 37-83

Salerno-Rife T, Rutgers T, Kaesberg P (1980) Translation of southern bean mosaic virus RNA in wheat embryo and rabbit reticulocyte extracts. J Virol 34: 51-58

Wu S, Rinehart CA, Kaesberg P (1987) Sequence and organization of southern bean mosaic virus genomic RNA. Virology 161: 73-80

CONTRIBUTED BY

Hull R

GENUS *LUTEOVIRUS*

Type Species barley yellow dwarf virus (BYDV)

VIRION PROPERTIES

MORPHOLOGY

Virions are 25 to 30 nm in diameter, hexagonal in outline and have no envelope or surface features. They exhibit icosahedral symmetry (T = 3). Particle cores consist of the genomic RNA; a small protein covalently linked to the 5' end of the genomic RNA (VPg) has been reported for PGRV and BYDV-RPV, but it is not yet clear if this is the case for all luteoviruses.

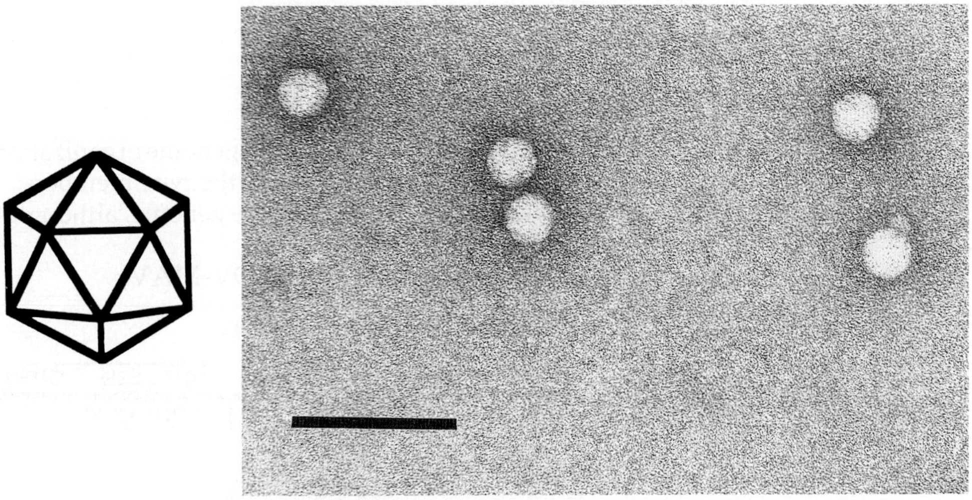

Figure 1: Negative contrast electron micrograph of subterranean clover red leaf virus particles stained with uranyl acetate. The bar represents 100 nm.

PHYSICOCHEMICAL AND PHYSICAL PROPERTIES

Virion Mr is 6.5 x 10^6; buoyant density in CsCl is 1.40 g/cm^3; S_{20w} is 104-127. Virions are moderately stable, and insensitive to freezing, chloroform, and non-ionic detergents.

NUCLEIC ACID

Virions contain a single molecule of infectious linear, positive sense ssRNA. The genome size is fairly uniform among the member viruses; 5,677 nt for the PAV strain of barley yellow dwarf virus, 5,882 nt for potato leafroll virus, 5,600 nt for the RPV strain of BYDV, 5,861 nt for soybean dwarf virus and 5,641 nt for beet western yellows virus. A VPg is linked to the 5' end of the genome of the subgroup II luteoviruses PLRV and BYDV-RPV, however it is not yet known whether subgroup luteoviruses also possess a VPG. There is no 3'-terminal poly (A) tract.

PROTEINS

Table: Proteins of the different ORFs of luteoviruses with their size (kDa) and their possible functions.

ORF	BYDV MAV	BYDV PAV	PLRV	BWYV	BYDV RPV	SDV	Function of protein product
0	-	-	28	29	29	-	Unknown function
1	39	39	70	66	71	40	Contains helicase motifs
2	61	60	69	70	72	59	probable RNA-dependent RNA polymerase
3	22	22	23	23	22	22	Coat protein gene
4	17	17	17	20	17	21	Possibly VPg or movement protein
5	51	43	56	52	50	48	Possible aphid transmission factor
6	4	7	-	-	-	-	Unknown function

Luteoviruses contain 5 or 6 ORFs which encode proteins of between 4 and 72 kDa (Table). Only the coat protein gene has been unequivocally assigned; it resides in ORF 3. ORFs 1 and 2 of the subgroup I are not homologous to the corresponding ORFs of subgroup II. Additionally, ORF 0 is found only in subgroup II, and ORF 6 exists only in subgroup I. ORF 0 overlaps ORF 1 (subgroup II only), which overlaps ORF 2 (both subgroups). ORF 4 is contained completely within ORF 3. Finally, ORF 5 is positioned directly downstream of and contiguous with ORF 3.

LIPIDS

Virions contain no lipids.

CARBOHYDRATES

Virions contain no carbohydrates.

GENOME ORGANIZATION AND REPLICATION

The two luteovirus subgroups possess different genome organization BYDV-PAV (subgroup I) and PLRV (subgroup II) may be considered the type members. The difference between the subgroups is principally in the 5' end of the genome, although BYDV-PAV contains an

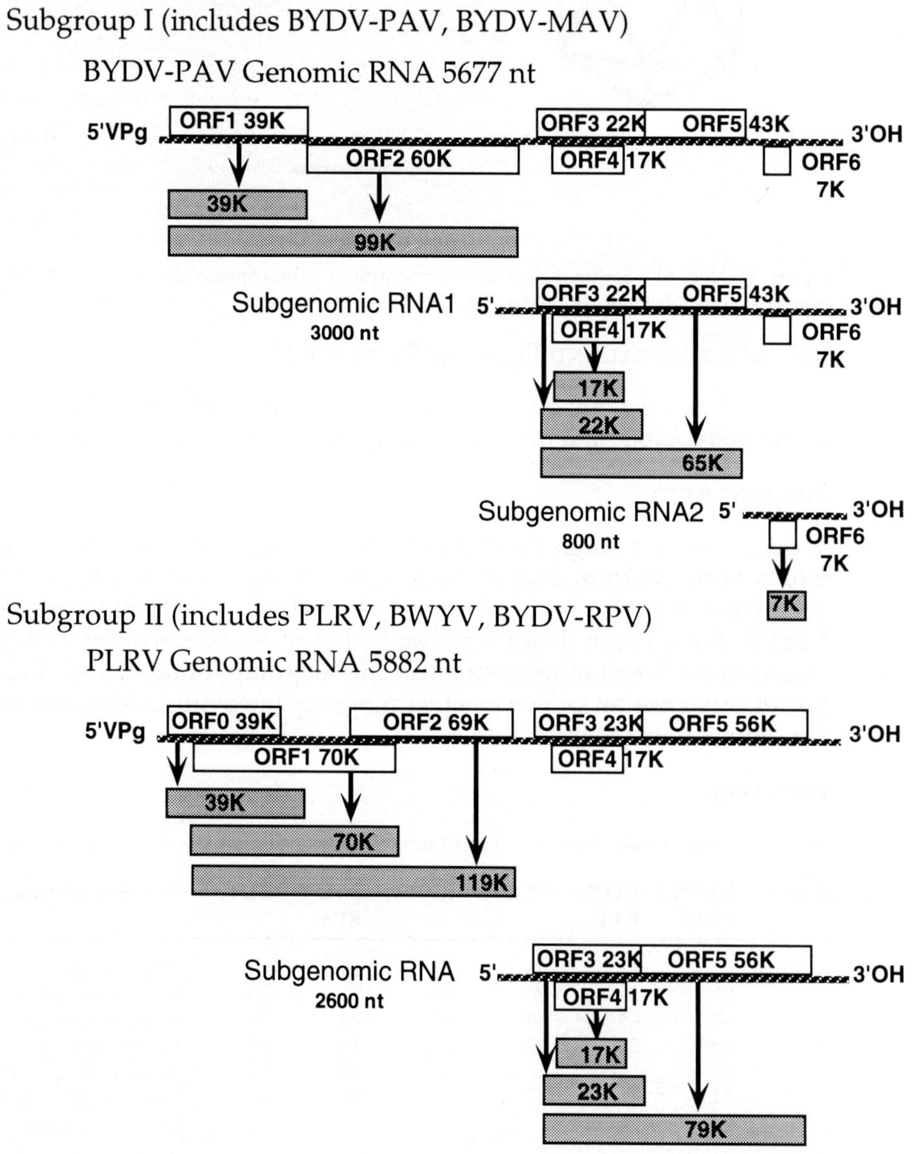

Figure 2: Diagram of the genome organization and map of the translation products of the luteovirus subgroups.

additional ORF (ORF 6) at the 3' end. ORFs 0, 1, and 2 are probably translated from the genomic RNA. It is likely that ORF 2 is translated via a frameshift from ORF 1, and is thus coterminal with the ORF 1 product. ORFs 3, 4, and 5 are expressed from a subgenomic RNA in both genome types. ORF 5 is probably translated via a readthrough following translation of ORF 3. In BYDV-PAV (subgroup I), ORF 6 seems to be expressed from a separate subgenomic RNA. There are no data on post-translational modification events in the luteoviruses. Mature virions have been observed in the phloem tissue of infected plants.

ANTIGENIC PROPERTIES

The viruses are strongly immunogenic. Luteoviruses form a serologic continuum but with some clustering. The clusters are: beet western yellows, beet mild yellowing, malva yellows and turnip mild yellows virus; bean leaf roll, legume yellows and Michigan alfalfa viruses; potato leaf roll, solanum yellows, tomato yellow top, and tobacco necrotic dwarf viruses; soybean dwarf and subterranean clover red leaf viruses; barley yellow dwarf viruses, MAV, PAV and SGV; barley yellow dwarf viruses RPV, RMV and RGV.

BIOLOGICAL PROPERTIES

Most luteoviruses have natural host ranges largely restricted to one plant family. Luteoviruses are transmitted in a circulative non-propagative manner by specific aphid vectors. Virus is acquired by phloem feeding, enters the hemocoel of the aphid via the hind gut, circulates in hemolymph, and probably enters the accessory salivary gland. Inoculation probably results from transport of virus into the salivary duct, and introduction of saliva into the plant during feeding. Luteoviruses occur worldwide, some viruses have restricted distribution. Luteoviruses are tissue-specific and particles are detectable in phloem. Phloem necrosis spreads from inoculated sieve elements, and causes symptoms by inhibiting translocation, slowing plant growth, and inducing loss of chlorophyll.

LIST OF SPECIES IN THE GENUS

The viruses, their alternative names ( ), genomic sequence accession numbers [ ], CMI/AAB description # ( ) and assigned abbreviations ( ) are:

SPECIES IN THE GENUS

1-BYDV subgroup I		
barley yellow dwarf virus - MAV (32)	[D01213]	(BYDV-MAV)
barley yellow dwarf virus - PAV	[D01214]	(BYDV-PAV)
barley yellow dwarf virus - SGV		(BYDV-SGV)
2-BYDV subgroup II		
barley yellow dwarf virus - RGV		(BYDV-RGV)
barley yellow dwarf virus - RMV		(BYDV-RMV)
barley yellow dwarf virus - RPV	[Y07496]	(BYDV-RPV)
bean leafroll virus		(BLRV)
(legume yellows virus)		
(Michigan alfalfa virus)		
(pea leafroll virus) (286)		
beet western yellows virus (89)	[X13062, X13063]	(BWYV)
(beet mild yellowing virus)		
(Malva yellows virus)		
(turnip mild yellows virus)		
carrot red leaf virus (249)		(CtRLV)
groundnut rosette assistor virus		(GRAV)
Indonesian soybean dwarf virus		(ISDV)
potato leafroll virus (291)		(PLRV)
Solanum yellows virus		(SYV)
tomato yellow top virus		(ToYTV)
soybean dwarf virus (179)	[L24049]	(SbDV)

(subterranean clover red leaf virus)
(strawberry mild yellow edge virus)
tobacco necrotic dwarf virus (234) (TNDV)

TENTATIVE SPECIES IN THE GENUS

beet yellow net virus	(BYNV)
celery yellow spot virus	(CeYSV)
chickpea stunt virus	(CpSV)
cotton anthocyanosis virus	(CAV)
filaree red leaf virus	(FLRV)
grapevine ajinashika virus	(GAV)
milk vetch dwarf virus	(MVDV)
millet red leaf virus	(MRLV)
Physalis mild chlorosis virus	(PhyMCV))
Physalis vein blotch virus	(PhyVBV)
raspberry leaf curl virus	(RLCV)
tobacco vein distorting virus	(TVDV)
tobacco yellow net virus	(TYNV)
tobacco yellow vein assistor virus	(TYVAV)

SIMILARITY WITH OTHER TAXA

The organization of RNA1 of pea enation mosaic enamovirus is similar to subgroup II luteoviruses, whereas the organization of RNA2 of PEMV resembles that of the subgroup I luteoviruses. The RNA associated with the S19 strain of BWYV (subgroup II) also show genomic similarities to the subgroup I luteoviruses. The member viruses of the genus *Luteovirus* shows evolutionary relationships to members of the genera *Sobemovirus* and *Carmovirus*.

DERIVATION OF NAMES

luteo: from Latin *luteus*, "yellow"

REFERENCES

Brault V, Miller WA (1992) Translational frameshifting mediated by a viral sequence in plant cells. Proc Natl Acad Sci USA 89: 2262-2266
Chin L-S, Foster JL, Falk BW (1993) The beet western yellows virus ST9-associated RNA shares structural and nucleotide sequence homology with carmo-like viruses. Virology 192: 473-482
Demler SA, de Zoeten GA (1991) The nucleotide sequence and luteovirus-like nature of RNA1 of an aphid non-transmissible strain of pea enation mosaic virus. J Gen Virol 72: 1819-1834
Francki RIB, Milne RG, Hatta T (1985) Luteovirus group, In: Atlas of Plant Viruses Vol I. CRC Press, Boca Raton FL, pp 137-152
Habili N, Symons RH (1989) Evolutionary relationship between luteoviruses and other RNA plant viruses based on sequence motifs in their putative RNA polymerases and nucleic acid helicases. Nucl Acids Res 17: 9543-9555
Johnstone GR, Ashby JW, Gibbs AJ, Duffus JE, Thottapilly G, Fletcher JD (1984) The host ranges, classification and identification of eight persistent aphid-transmitted viruses causing diseases in legumes. Neth J Pl Path 90: 225-245
Martin RR, Keese P, Young MJ, Waterhouse PM, Gerlach WL (1990) Evolution and molecular biology of luteoviruses. Ann Rev Phytopathol 28: 341-363
Martin RR, D'arcy CJ (1990) Relationships among luteoviruses based on nucleic acid hybridization and serological studies. Intervirology 31: 23-30
Mayo MA, Robinson DJ, Jolly CA, Hyman L (1989) Nucleotide sequence of potato leafroll luteovirus RNA. J Gen Virol 70: 1037-1051
Miller WA, Waterhouse PM, Gerlach WL (1988) Sequence and organization of barley yellow dwarf virus genomic RNA. Nucl Acids Res 16: 6097-6111
Prufer D, Tacke E, Schmitz J, Kull B, Kaufmann A, Rohde W (1992) Ribosomal frameshifting in plants: a novel signal directs the -1 frameshift in the synthesis of the putative viral replicase of potato leaf roll luteovirus. EMBO J 11: 1111-1117
Rathjen JP, Karageorgos LE, Habili N, Waterhouse PM, Symons RH (1994) Soybean dwarf luteovirus contains the third variant genome type in the luteovirus group. Virology 198:671-679
Ueng PP, Vincent JR, Kawata EE, Lei C-H, Lister RM, Larkins BA (1992) Nucleotide sequence analysis of the genomes of the MAV-PS1 and P-PAV isolates of barley yellow dwarf virus. J Gen Virol 73: 487-492

Vincent JR, Lister RM, Larkins BA (1991) Nucleotide sequence analysis and genomic organization of the NY-RPV isolate of barley yellow dwarf virus. J Gen Virol 72: 2347-2355
Waterhouse PM, Gildow FE, Johnstone GR (1988) Luteovirus group. CMI/AAB Description of Plant Viruses N° 339, 4pp

Contributed By

Randles JW, Rathjen JP

GENUS *ENAMOVIRUS*

Type Species pea enation mosaic virus (PEMV)

VIRION PROPERTIES

MORPHOLOGY

Virions are polyhedral and are of two distinct sizes, approximately 25 nm and 28 nm for the top (T) and bottom (B) components, respectively. A 180 subunit arrangement in a T=3 icosahedron has been proposed for the B component, and a 150 subunit arrangement lacking quasi-equivalence has been suggested for the T component.

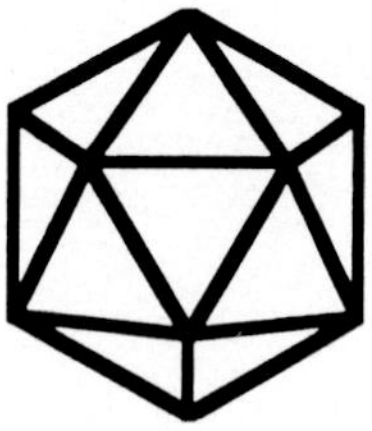

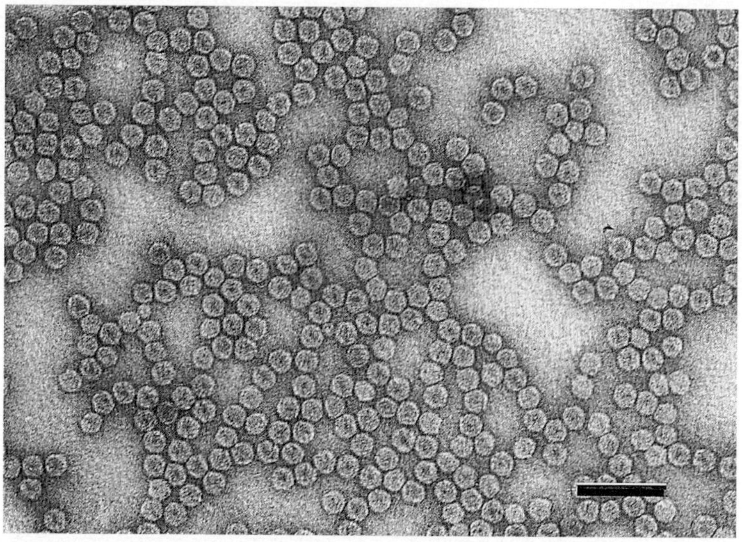

Figure 1: (left) diagramatic representation of pea enation mosaic virus particle (PEMV) (T=3). (right) Negative contrast electron micrograph of PEMV particles isolated by means of sucrose density gradient centrifugation. The bar represents 100 nm.

PHYSICOCHEMICAL AND PHYSICAL PROPERTIES

The Mr of the B component is about 5.6-5.7 x 10^6 and the of the T component is about 4.4-4.6 x 10^6. S_{20w} ranges from 107-122 for the B component to 91-106 for the T component. The buoyant density in CsCl for the B component is approximately 1.42 g/cm^3. The T component is disrupted in CsCl; in Cs_2SO_4 both components have a density of approximately 1.38 g/cm^3. The T component is less stable under high salt conditions than the B component, although both are eventually disrupted under these conditions.

NUCLEIC ACID

Virus preparations contain two species of linear positive sense ssRNA. RNA1 consists of 5,705 nt, RNA2 consists of 4,253 nt. Some strains contain a third RNA component comprising 717 bases. The latter is considered to be a satellite RNA. The RNAs are not polyadenylated and are not aminoacylatable. A genome linked protein (Mr 17.5 x 10^3) is associated with virion RNA. It is not known whether all RNA species carry this protein covalently linked to their 5' ends. The covalently linked protein is not necessary for infectivity of the RNAs. The 3' and 5' termini of the two RNAs (RNA1 and 2) of PEMV are not identical. The only similarity between termini occurs between the 5' and 3' ends of RNA2 and the satellite RNA in which 12 of the first 14 nucleotides and 7 of the final 8 nucleotides are homologous.

PROTEINS

The structural proteins of PEMV are encoded by RNA1. They consist of a major coat protein (Mr 21 x 10^3) and a minor protein (Mr 54 x 10^3). The latter is associated generally with virions of aphid transmissible isolates and represents a fusion of the products of the CP gene (21 kDa) and the 3' terminal gene (33 kDa).

LIPIDS

None reported.

CARBOHYDRATES

None reported.

GENOME ORGANIZATION AND REPLICATION

Sequence analysis of RNA1 indicates 5 ORFs. The 21 kDa and the 54 kDa fusion proteins form structural subunits and are thought to be translated from a subgenomic messenger RNA. The products predicted for the 34 and 84 kDa ORFs have been confirmed by *in vitro* translation studies. A 130 kDa polypeptide postulated to represent the RNA-dependent-RNA polymerase may be generated by a translational frameshift fusion of the 84 and 67 kDa ORF products. RNA1 is capable of autonomous replication in protoplasts, although both RNA1 and RNA2 are necessary for supporting the systemic invasion of RNA1 when introduced by mechanical transmission.

The ORFs of RNA2 also potentially code for 5 polypeptides, although the 3'-terminal 15 kDa ORF is dispensable in infection. A 93 kDa peptide identified in *in vitro* translation studies is thought to represent a second RNA-dependent RNA polymerase, and is composed of a translational frameshift fusion of the 33 and 65 kDa products. RNA 2 is also capable of autonomous replication in pea protoplasts. Unlike RNA1, RNA2 can be transmitted mechanically to plants, resulting in a largely asymptomatic systemic infection.

No translational activity of the satellite RNA has been detected. In protoplasts, the replication of the satellite RNA is solely under the control of RNA2. RNA2 is also responsible for both the replication and the systemic movement of the satellite *in planta*. The encapsidation and aphid transmission of the satellite RNA is under the control of the RNA1 encoded structural proteins.

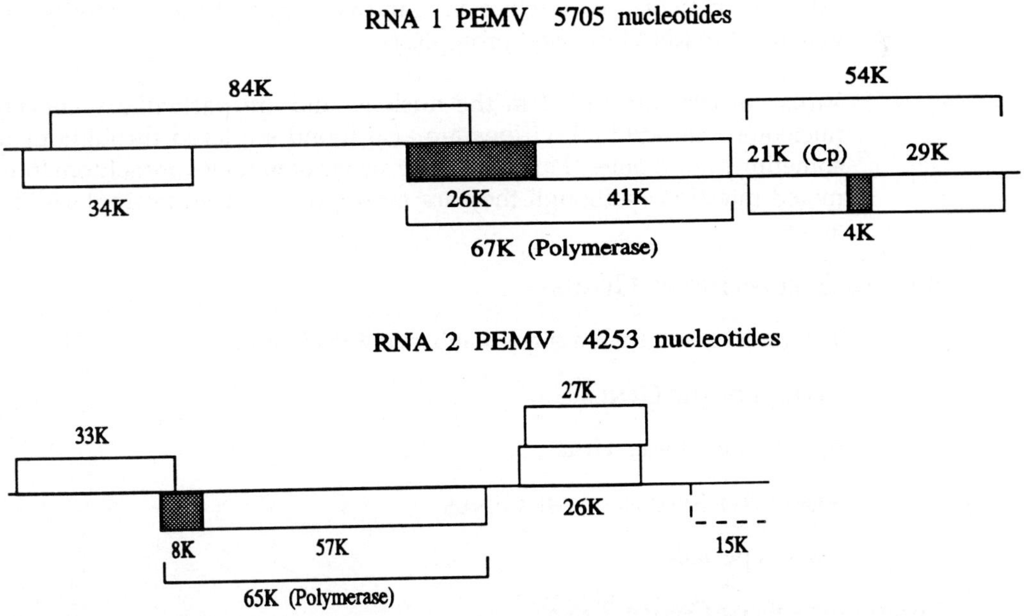

Figure 2: Genomic organization of RNA1 and RNA2 of PEMV. Open boxes depict prominent ORFs. Black boxes represent ORF extensions preceding the first initiation codon of respective ORFs. The dashed line outlining the 15 kDa of RNA 2 signifies the nonessential role of this reading frame in infection.

Antigenic Properties

The virus is moderately antigenic. In gel diffusion assays, aphid transmissible isolates display an additional antigenic determinant absent in aphid non-transmissible isolates.

Biological Properties

Host Range

The virus infects many legumes but only a few species of other host families. *Chenopodium quinoa* seems to be the preferred local lesion host for this virus.

Transmission

The virus is transmitted by aphids in a persistent, non-propagative manner. The virus is readily transmitted mechanically, but loss of aphid transmissibility occurs after increasing numbers of mechanical passages.

Cytopathic effects

The most distinctive characteristic of PEMV replication is its intimate association with the host nucleus. A replication complex is generated from the inner membrane of the nuclear envelope, resulting in the formation of vesicles in the perinuclear space. The vesicles bud from the nucleus into the cytosol surrounded by the outer membrane of the nuclear envelope. These vesicles are found in all cell types, and are particularly prominent within phloem tissue, implicating these structures in the systemic movement of infection. Both isolated nuclei of healthy peas and the replication complex isolated from infected peas can sustain RNA replication when provided with the appropriate energy sources and requisite nucleotides. Protoplasts inoculated solely with RNA1 also demonstrate this cytopathology, thus linking the emergence of this complex with RNA1 replication. Since both viral RNAs are independently capable of replication in pea protoplasts, it is currently unknown whether RNA2 also uses this complex in some capacity in mixed infections.

Plant tissues infected solely with RNA2 display a marked proliferation of the endoplasmic reticulum, with extensive branching and separation of cisternae. These cells also display extensive networks of single-membrane vesicles that are decidedly different from those generated in RNA1 infected protoplasts.

Virus particles are found in the nucleus, and are particularly concentrated within the nucleolus. In addition, virions are also found scattered throughout the cytoplasm and sometimes in vacuoles. Paracrystalline arrays of particles are seldom found in RNA1-RNA2 mixed infections, although they are more prominent in protoplasts infected with RNA1 alone.

List of Species in the Genus

The viruses, and their assigned abbreviations ( ) are:

Species in the Genus

pea enation mosaic virus (PEMV)

Tentative Species in the Genus

None reported.

Similarity with Other Taxa

The taxonomic status of the genus *Enamovirus* is currently in a state of transition. Pea enation mosaic virus can best be characterized as a symbiotic association of two taxonomically distinct viral genomes. RNA2 is a coat protein-deficient viral RNA with a polymerase domain which is closely related to those of member viruses of the family *Tombusviridae* (genera *Tombusvirus* and *Carmovirus*) and the genera *Dianthovirus*, *Necrovirus* and *Luteovirus*. The RNA2 encoded polymerase also has strong sequence homology with carrot mottle virus, the type species of the genus *Umbravirus*. These taxonomic affiliations, the dependence on a luteo-like virus for encapsidation and aphid transmission, and the ability of RNA2 to initiate an autonomous systemic infection would strongly argue that RNA2 should be included within the genus *Umbravirus*.

In contrast, RNA1 of PEMV has many characteristics (aphid transmission, cytopathology, genomic organization) that would indicate a stronger affiliation with the BWYV-PLRV subgroup of the genus *Luteovirus*. At this time, the limitations to this analogy centers on whether RNA1 alone can induce a phloem-limited infection *in planta*. If RNA1 and RNA2 infections are separable at the whole plant level, then PEMV should be considered a true mixed infection of taxonomically distinct viruses. However, if RNA1 retains some form of dependence on RNA2, then the retention of the *Enamovirus* genus would be more appropriate.

References

Adam G, Sander E, Shepherd RJ (1979) Structural differences between pea enation mosaic virus strains affecting transmissibility by *Acyrthosiphon pisum* (Harris). Virology 92: 1-14

Clarke RG, Bath JE (1977) Serological properties of aphid-transmissible and aphid-nontransmissible pea enation mosaic virus isolates. Phytopathology 67: 1035-1040

Demler SA, de Zoeten GA (1989) Characterization of a satellite RNA associated with pea enation mosaic virus. J Gen Virol 70: 1075-1084

Demler SA, de Zoeten GA (1991) The nucleotide sequence and luteovirus-like nature of RNA 1 of an aphid non-transmissible strain of pea enation mosaic virus. J Gen Virol 72: 1819-1834

Demler SA, Rucker DG, de Zoeten GA (1993) The chimeric nature of the genome of pea enation mosaic virus: The independent replication of RNA 2. J Gen Virol 74: 1-14

Demler SA, Borkhsenious ON, Rucker DG, de Zoeten GA (1994) Assessment of the autonomy of replicative and structural functions encoded by the luteo-phase of pea enation mosaic virus. J Gen Virol 75: 997-1007

Demler SA, Rucker DG, Nooruudin L, de Zoeten GA (1994). Replication of the satellite RNA of pea enation mosaic virus is controlled by RNA 2 encoded functions. submitted J Gen Virol

de Zoeten GA, Gaard G, Diez FB (1972) Nuclear vesiculation associated with pea enation mosaic virus-infected plant tissue. Virology 48: 638-647

de Zoeten GA, Powell CA, Gaard G, German TL (1976) *In situ* localization of pea enation mosaic virus double-stranded ribonucleic acid. Virology 70: 459-469

Gabriel CJ, de Zoeten GA (1984) The *in vitro* translation of pea enation mosaic virus. Virology 139: 223-230

German TL, de Zoeten GA (1975) Purification and properties of the replicative forms and replicative intermediates of pea enation mosaic virus. Virology 66: 172-184

German TL, de Zoeten GA, Hall TC (1978) Pea enation mosaic virus genome RNA contains no polyadenylate sequences and cannot be aminoacylated. Intervirology 9: 226-230

Gonsalves D, Shepherd RJ (1972) Biological and physical properties of the two nucleoprotein components of pea enation mosaic virus and their associated nucleic acids. Virology 48: 709-723

Harris KF, Bath JE (1972) The fate of pea enation mosaic virus in its pea aphid vector, *Acyrthosiphon pisum* (Harris). Virology 50: 778-790

Harris KF, Bath JE, Thottapilly G, Hooper GR (1975) Fate of pea enation mosaic virus in PEMV injected pea aphids. Virology 65: 148-162

Hull R, Lane LC (1973) The unusual nature of the components of a strain of pea enation mosaic virus. Virology 55: 1-13

Hull R (1977) Particle differences related to aphid-transmissibility of a plant virus. J Gen Virol 34: 183-187

Powell CA, de Zoeten GA (1977) Replication of pea enation mosaic virus RNA in isolated pea nuclei. Proc Natl Acad Sci USA 74: 2919-2922

Powell CA, de Zoeten GA, Gaard G (1977) The localization of pea enation mosaic virus-induced RNA-dependent RNA polymerase in infected peas. Virology 78: 135-143

Reisman D, de Zoeten GA (1982) A covalently linked protein at the 5'-ends of the genomic RNAs of pea enation mosaic virus. J Gen Virol 62: 187-190

Contributed By

de Zoeten GA, Demler SA

GENUS *UMBRAVIRUS*

Type Species carrot mottle virus (CMoV)

VIRION PROPERTIES

MORPHOLOGY

Approximately 52 nm-diameter enveloped structures occur in vacuoles of CMoV infected cells, and in partially purified preparations from such cells. It is not known whether these are (i) virus particles of a kind unusual among plant viruses but resembling those of some viruses infecting insects or vertebrates, or (ii) cytopathological structures involved in virus replication. Similar structures occur in plants infected with BYVBV, GRV and LSMV, but no information is available for cells infected with other umbraviruses.

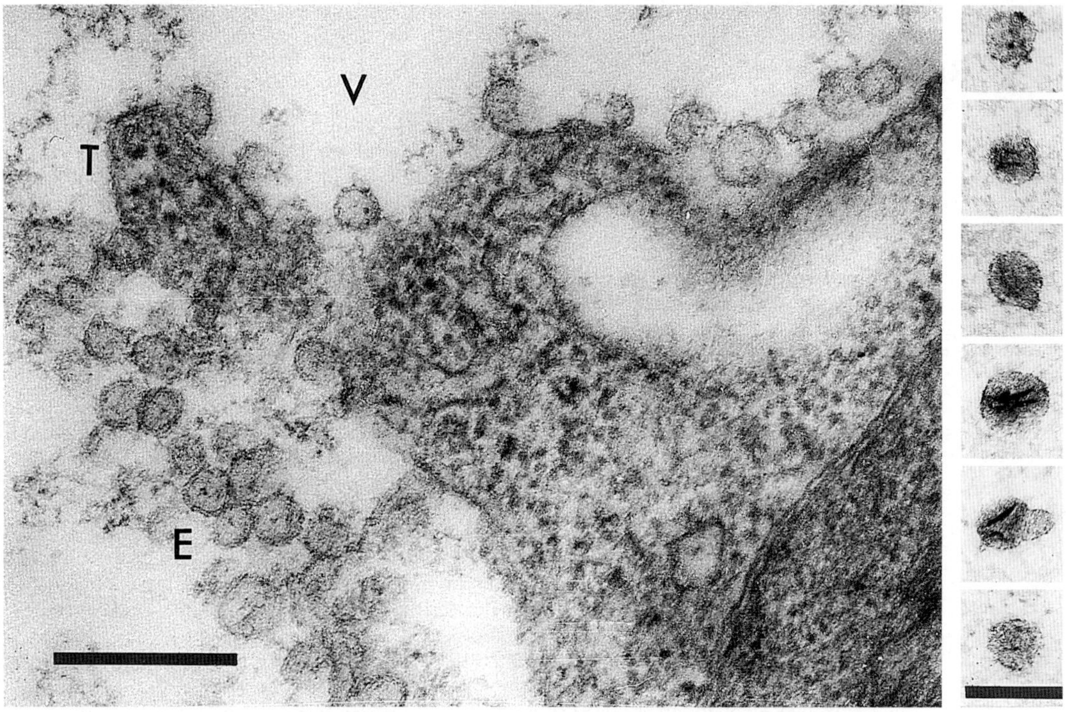

Figure 1: (left) Section of palisade mesophyll cell from a leaf of *Nicotiana clevelandii* systemically infected with CMoV, showing enveloped structures (*E*) about 52 nm in diameter in the cell vacuole (*V*) in association with the tonoplast (*T*). The bar represents 250 nm. (right) Enveloped structures about 52 nm in diameter in a partially purified preparation from CMoV-infected *N. clevelandii*, stained with 2% uranyl acetate; bar represents 100 nm.

PHYSICOCHEMICAL AND PHYSICAL PROPERTIES

Infectivity in leaf extracts is stable for several hours at room temperature or several days at 5° C, but is abolished by treatment with organic solvents. Partially purified preparations of CMoV consist predominantly of cell membrane but contain infective components which, because they sediment at about 270 S_{20W} and have a buoyant density of about 1.15 g/cm^3 in CsCl, are probably the 52 nm-diameter enveloped structures observed in these preparations.

NUCLEIC ACID

Phenol extracts of leaves are often much more infective than buffer extracts. The infective RNA is single-stranded, about 4.5 kb in size, and is probably not polyadenylated.

PROTEINS

None reported.

LIPIDS

The sensitivity to organic solvents of the infective components in partially purified preparations and their low buoyant density suggests the presence of lipid, and also indicates that they probably correspond to the enveloped structures seen in sections of infected leaves.

CARBOHYDRATES

None reported.

GENOME ORGANIZATION AND REPLICATION

Infected leaf tissue contains abundant dsRNA. Two species are common to all members: one (dsRNA1) is about 4.2-4.8 kbp in size and another (dsRNA2) is about 1.1-1.5 kbp in size. cDNA copies of the larger species hybridize with the smaller and it is thought that they represent double-stranded forms of, respectively, the genomic and a sub-genomic ssRNA. The native dsRNA is not infective but becomes so when heat-denatured. Some umbraviruses may have one or more additional dsRNA species, which at least in one instance (GRV) is known to represent a satellite RNA.

Complementary DNA to a central portion of the CMoV genome has been sequenced. This includes an ORF encoding a sequence that contains motifs typical of RNA-dependent RNA polymerases but which is not closely similar to those of viruses in any of the existing taxonomic groups.

ANTIGENIC PROPERTIES

None reported.

BIOLOGICAL PROPERTIES

HOST RANGE

Individual umbraviruses are confined in nature to one or a few host plant species. Their experimental host range is broader but still restricted. The symptoms are mottles or mosaics but, at least with GRV, are greatly influenced by associated satellite RNA.

TRANSMISSION

Umbraviruses are transmissible, sometimes with difficulty, by mechanical inoculation, but in nature each is dependent on a specific helper virus, commonly a luteovirus, for transmission in a persistent (circulative, non-propagative) manner by aphids. The mechanism of this dependence is packaging of the dependent virus RNA in the coat protein of the helper. In GRV the satellite RNA plays an essential role in mediating this luteovirus-dependent aphid-transmission. There is no evidence for multiplication of umbraviruses in the insect vector. Seed transmission has not been reported.

GEOGRAPHICAL DISTRIBUTION

CMoV apparently occurs worldwide, but other umbraviruses have a restricted distribution. Several umbraviruses, notably GRV, occur only in Africa.

CYTOPATHIC EFFECTS

Umbraviruses, even in the absence of their helper viruses, exhibit rapid systemic spread in plants. They infect cells throughout the leaf, though presumably the aphid-transmissible particles, like the luteoviruses that provide their coat protein, occur only in the phloem. In infected mesophyll cells there is extensive development of cell wall outgrowths sheathing elongated plasmodesmatal tubules.

LIST OF SPECIES IN THE GENUS

The viruses, their genomic sequence accession numbers [ ], (CMI/AAB description # ( ) and assigned abbreviations ( ) are:

SPECIES IN THE GENUS

bean yellow vein-banding virus		(BYVBV)
carrot mottle virus (137)		(CMoV)
groundnut rosette virus	[Z29702, Z29711]	(GRV)
lettuce speckles mottle virus		(LSMV)
tobacco mottle virus		(TMoV)

TENTATIVE SPECIES IN THE GENUS

sunflower crinkle virus (sunflower rugose mosaic virus)	(SCV)
sunflower yellow blotch virus (sunflower yellow ringspot virus)	(SYBV)
tobacco bushy top virus	(TBTV)
tobacco yellow vein virus	(TYVV)

SIMILARITY WITH OTHER TAXA

The CMoV RNA-dependent RNA polymerase sequence is distantly related (less than 45% amino acid sequence identity) to those of member viruses of the family *Tombusviridae* (genera *Tombusvirus* and *Carmovirus*) and the genera *Necrovirus*, *Dianthovirus*, *Machlomovirus* and *Luteovirus*. The polymerase has 63% amino acid sequence identity to the polymerase found in RNA2 of pea enation mosaic virus (PEMV) (genus *Enamovirus*) but only 20% identity to the polymerase found in PEMV RNA1.

DERIVATION OF NAMES

umbra: From Latin, a shadow. In English, a shadow, an uninvited guest that comes with an invited one

REFERENCES

Adams AN, Hull R (1972) Tobacco yellow vein, a virus dependent on assistor viruses for its transmission by aphids. Ann Appl Biol 71: 135-140

Cockbain AJ, Jones P, Woods RD (1986) Transmission characteristics and some other properties of bean yellow vein-banding virus, and its association with pea enation mosaic virus. Ann Appl Biol 108: 59-69

Falk BW, Duffus JE, Morris TJ (1979) Transmission, host range, and serological properties of the viruses that cause lettuce speckles disease. Phytopathology 69: 612-617

Falk BW, Morris TJ, Duffus JE (1979) Unstable infectivity and sedimentable ds-RNA associated with lettuce speckles mottle virus. Virology 96: 239-248

Gibbs MJ (1994) The luteovirus supergroup: rampant recombination and persistent partnerships. In: Gibbs AJ, Calisher CH, Garcia-Arenal F (eds) Molecular Basis of Viral Evolution. Cambridge University Press, Cambridge (in press)

Halk EL, Robinson DJ, Murant AF (1979) Molecular weight of the infective RNA from leaves infected with carrot mottle virus. J Gen Virol 45: 383-388

Hull R, Adams AN (1968) Groundnut rosette and its assistor virus. Ann Appl Biol 62: 139-145

Murant AF, Goold RA, Roberts IM, Cathro J (1969) Carrot mottle - a persistent aphid-borne virus with unusual properties and particles. J Gen Virol 4: 329-341

Murant AF, Rajeshwari R, Robinson DJ, Raschke JH (1988) A satellite RNA of groundnut rosette virus that is largely responsible for symptoms of groundnut rosette disease. J Gen Virol 69: 1479-1486

Murant AF, Roberts IM, Goold RA (1973) Cytopathological changes and extractable infectivity in *Nicotiana clevelandii* leaves infected with carrot mottle virus. J Gen Virol 21: 269-283

Murant AF, Waterhouse PM, Raschke JH, Robinson DJ (1985) Carrot red leaf and carrot mottle viruses: observations on the composition of the particles in single and mixed infections. J Gen Virol 66: 1575-1579

Reddy DVR, Murant AF, Raschke JH, Mayo MA, Ansa OA (1985) Properties and partial purification of infective material from plants containing groundnut rosette virus. Ann Appl Biol 107: 65-78

Smith KM (1946) The transmission of a plant virus complex by aphides. Parasitology 37: 131-134

Theuri JM, Bock KR, Woods RD (1987) Distribution, host range and some properties of a virus disease of sunflower in Kenya. Trop Pest Management 33: 202-207
Waterhouse PM, Murant AF (1983) Further evidence on the nature of the dependence of carrot mottle virus on carrot red leaf virus for transmission by aphids. Ann Appl Biol 103: 455-464
Watson M, Serjeant EP, Lennon EA (1964) Carrot motley dwarf and parsnip mottle viruses. Ann Appl Biol 54: 153-166

Contributed By

Murant AF, Robinson DJ, Gibbs MJ

FAMILY TOMBUSVIRIDAE

TAXONOMIC STRUCTURE OF THE FAMILY

Family	***Tombusviridae***
Genus	*Tombusvirus*
Genus	*Carmovirus*

VIRION PROPERTIES

MORPHOLOGY

Virions exhibit icosahedral symmetry (T=3); virions are composed of 180 protein subunits. Virions have a rounded outline, a granular surface, and a diameter of about 30 nm. Each subunit folds in three distinct structural domains: R, the N-terminal internal domain interacting with RNA; S, the shell domain constituting the capsid backbone; and P, the protruding C-terminal domain. P domains are clustered in pairs to form 90 projections. These dimeric contacts are important in the assembly and stabilization of the virion structure. R domain, which contains many positively charged residues, binds RNA. S domain forms a barrel structure made up of β-strands. Two Ca^{++} binding sites stabilize contacts between S domains.

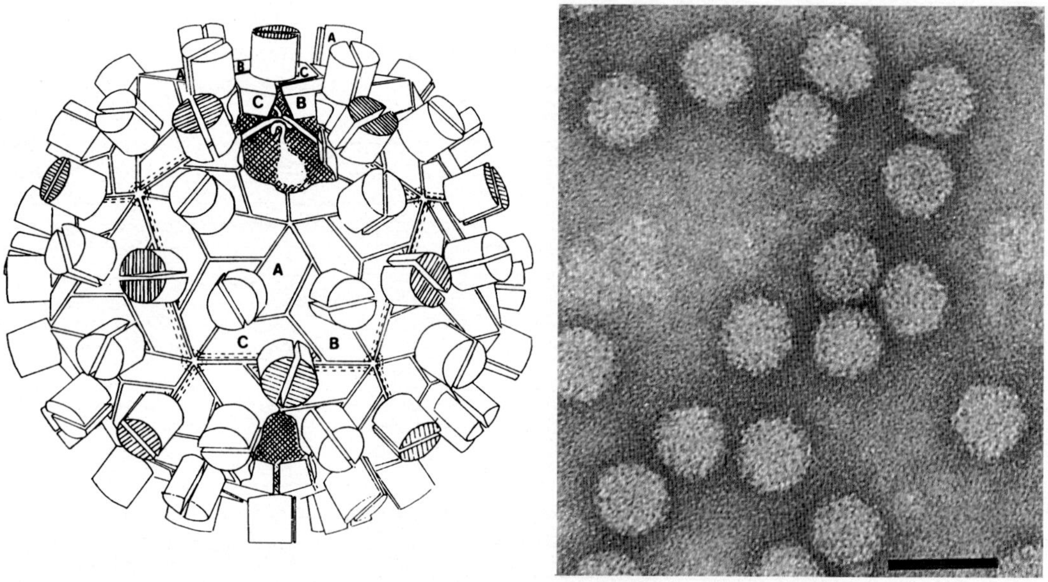

Figure 1: (left) Diagrammatic representation of a TBSV particle (from Hopper et al., 1984, with permission). (right) Negative contrast electron micrograph of TBSV particles. The bar represents 50 nm.

PHYSICOCHEMICAL AND PHYSICAL PROPERTIES

Depending on the genus, the virion Mr is 8.2-8.9 x 10^6, S_{20w} is 118-140, and buoyant density in CsCl is 1.34-1.36 g/cm^3. Virions sediment as a single component in sucrose and CsCl gradients, are stable at acidic pH, but expand above pH 7 and in the presence of EDTA. Lowering pH or adding of Ca^{++} recompacts the particles. Virions are resistant to elevated temperatures (thermal inactivation usually occurs above 80° C) and are insensitive to organic solvents.

NUCLEIC ACID

Virions contain a single molecule of positive sense, linear ssRNA, that constitutes about 17% of the particle weight, and has a size ranging from 4 to 4.7 kb, depending on the genus. The 3' end is not polyadenylated. The 5' terminus is protected but the presence of a cap was demonstrated only in carnation mottle virus, the type species of the *Carmovirus* genus.

Addition of a cap analogue to *in vitro* RNA transcripts enhances infectivity little or not at all. Defective interfering (DI) and satellite RNAs are known to occur.

PROTEINS

Depending on the genus, the single major capsid polypeptide has an Mr of 38-43 x 10^3. Nonstructural proteins include a polypeptide of Mr 28-33 x 10^3 and a readthrough product of Mr 88-92 x 10^3. Readthrough polypeptides contain the GDD motif of RNA polymerases and two motifs of NTP-binding proteins (helicases). Additional nonstructural proteins are polypeptides with Mr of 8 and 9 x 10^3 (carmoviruses) and Mr of 19 and 22 x 10^3 (tombusviruses), for which a cell-to-cell movement function has been established.

LIPIDS

Virions contain no lipids.

CARBOHYDRATES

Virions contain no carbohydrates.

GENOME ORGANIZATION AND REPLICATION

The viral genome contains five ORFs differing in size and relative location in the two genera. Replication occurs in the cytoplasm, possibly in membranous vesicles that may be associated with endoplasmic reticulum, or modified organelles such as peroxisomes, mitochondria and, more rarely, chloroplasts. Products of the 5'-proximal ORFs 1 and 2 are expressed through genome-size RNA translation, whereas translation products of the 3'-proximal ORFs 3, 4 and 5, are expressed through subgenomic RNAs. dsRNAs corresponding in size to virus-related RNAs (genomic and subgenomic) are present in infected tissues. Virions are assembled in the cytoplasm and occasionally in mitochondria and nuclei. Virions accumulate, sometimes in crystalline form, in the cytoplasm and in vacuoles.

ANTIGENIC PROPERTIES

Virions are efficient immunogens. Antisera yield single precipitin lines in immunodiffusion tests. Depending on the genus, serological cross-reactivity among species ranges from nil to near-homologous titers.

BIOLOGICAL PROPERTIES

HOST RANGE

The natural host range of individual virus species is relatively narrow and restricted to dicotyledons. The experimental host range is wide. Infection is often limited to the root system, but when hosts are invaded systemically, viruses enter all tissues. Diseases are characterized by mottling, crinkling and deformation of foliage. Certain virus species infect natural hosts symptomlessly.

TRANSMISSION

All species are readily transmitted by mechanical inoculation and through propagative plant material. Some may be transmitted by contact and through seeds. Viruses are often found in natural environments, i.e. surface waters and soils from which they can be acquired without assistance of vectors. Transmission by the chytrid fungus *Olpidium radicale* and beetles has also been reported.

GEOGRAPHICAL DISTRIBUTION

Geographical distribution of particular species varies from wide to restricted. The majority of the species occur in temperate regions. Legume-infecting carmoviruses have been recorded from tropical areas.

CYTOPATHIC EFFECTS

Distinctive cytopathological features occur in association with exceedingly high accumulations of virus particles in cells and "multivesicular bodies", i.e. cytoplasmic membranous inclusions originated from profoundly modified mitochondria and/or peroxisomes.

GENUS *TOMBUSVIRUS*

Type Species tomato bushy stunt virus (TBSV)

DISTINGUISHING FEATURES

Virion Mr is 8.9 x 10^6 and S_{20w} is 132-140. Genomic RNA has a size of about 4.7 kb and consists of five ORFs. Translation products of genome-length RNA are a 33 kDa protein encoded in ORF 1 and a 92 kDa polypeptide (ORF 1 plus ORF 2) originating from readthrough of the amber terminator of ORF 1. ORF 3 codes for coat protein (41 kDa) and is located internally. Coat protein and the polypeptides of 19 and 22 kDa encoded in ORF 4 and 5, are expressed through subgenomic RNAs of 2.1 and 0.9 kb, respectively. Most species are serologically interrelated, though to a variable extent, and all elicit formation of multivesicular inclusion bodies. Tombusvirus-induced diseases prevail in temperate climates. All species are soil-borne, but only one (CNV) has a recognized fungal vector (*Olpidium radicale*).

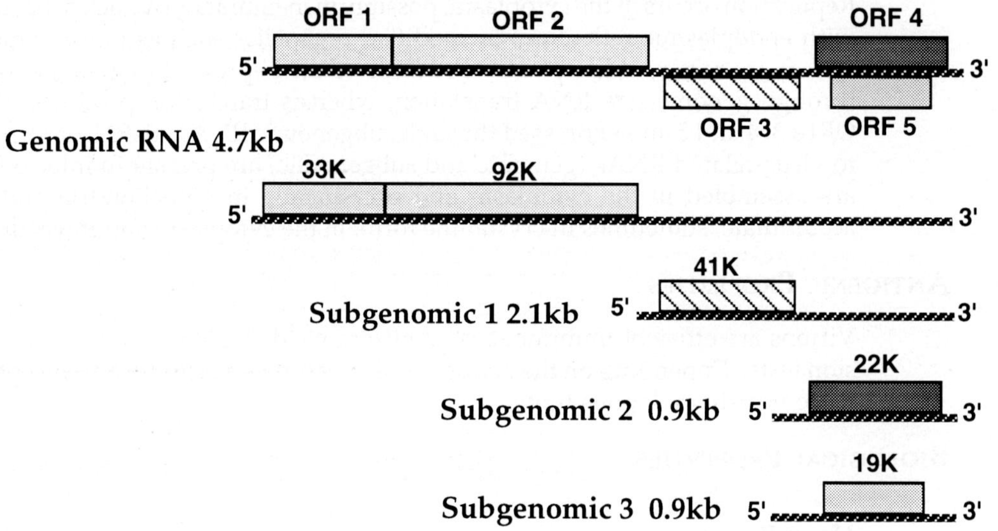

Figure 2: *Tombusvirus* (CymRSV) genome organization and strategy of replication.

LIST OF SPECIES IN THE GENUS

The viruses, their genomic sequence accession numbers [ ], CMI/AAB description # ( ) and assigned abbreviations ( ) are:

SPECIES IN THE GENUS

artichoke mottled crinkle virus (69)		(AMCV)
carnation Italian ringspot virus (69)		(CIRV)
cucumber necrosis virus (178)	[M25270]	(CNV)
Cymbidium ringspot virus (178)	[X15511]	(CymRSV)
eggplant mottled crinkle virus		(EMCV)
grapevine Algerian latent virus		(GALV)
Moroccan pepper virus		(MPV)
Lato river virus		(LRV)
Neckar river virus		(NRV)
pelargonium leaf curl virus (69)		(PLCV)

petunia asteroid mosaic virus (69)		(PAMV)
Sikte water-borne virus		(SWBV)
tomato bushy stunt virus (69)	[M21958]	(TBSV)

TENTATIVE SPECIES IN THE GENUS

None reported.

GENUS *CARMOVIRUS*

Type Species carnation mottle virus (CarMV)

DISTINGUISHING FEATURES

Virion Mr is 8.2 x 10^6 and S_{20w} is 118-130. Some viruses sediment as two density species in cesium sulphate gradients. Genomic RNA is about 4.0 kb on size and consists of five ORFs. Full-size genome translation products are a 28 kDa polypeptide encoded in ORF 1 and a 88 kDa polypeptide (ORF 1 plus ORF 2) originating from readthrough of the amber terminator of ORF 1. ORF 3 and 4 code for two small polypeptides of 7-8 kDa and 8-9 kDa, respectively, depending on the virus. Coat protein is encoded in ORF 5 which is 3' coterminal. Translation products of ORFs 3, 4 and 5 are expressed through subgenomic RNAs with a size of about 1.7 and 1.5 kb, respectively. Viral species are not serologically related. Multivesicular bodies are formed only by some viruses. Most species are found in temperate regions. Those infecting legumes are reported from tropical areas. Several viruses are soil-borne, but only two (CLSV and MNSV) are transmitted by *Olpidium radicale.* Others are transmitted by beetles (CpMoV, BMMV, BMoV, TCV).

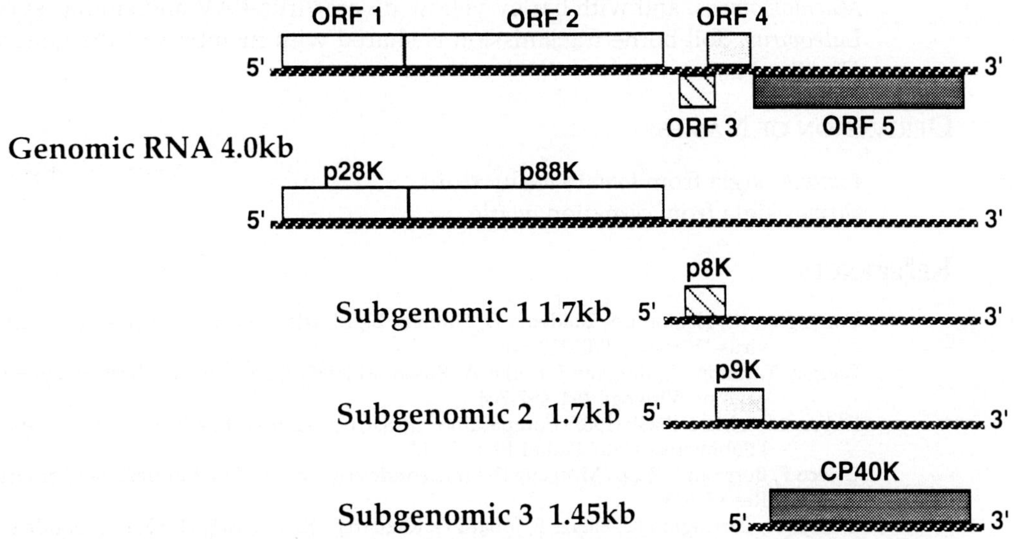

Figure 3: *Carmovirus* (TCV) genome organization and strategy of replication.

LIST OF SPECIES IN THE GENUS

The viruses, their genomic sequence accession numbers [ ], CMI/AAB description # ( ) and assigned abbreviations ( ) are:

SPECIES IN THE GENUS

Ahlum water-borne virus		(AWBV)
bean mild mosaic virus (231)		(BMMV)
carnation mottle virus (7)	[X02986]	(CarMV)
cucumber soil-borne virus		(CSBV)
cucumber leaf spot virus (319)		(CLSV)
Galinsoga mosaic virus (252)		(GaMV)

hibiscus chlorotic ringspot virus (227)		(HCRSV)
melon necrotic spot virus (302)	[M29671]	(MNSV)
pelargonium flower break virus (130)		(PFBV)
saguaro cactus virus (148)		(SCV)
turnip crinkle virus (109)	[M22445]	(TCV)
Weddel water-borne virus		(WWBV)

TENTATIVE SPECIES IN THE GENUS

blackgram mottle virus (237)	(BMoV)
cowpea mottle virus (212)	(CPMoV)
eldeberry latent virus (127)	(ELV)
Glycine mottle virus (166)	(GMoV)
narcissus tip necrosis virus	(NTNV)
plantain virus 6	(PlV-6)
Tephrosia symptomless virus	(TeSV)

LIST OF UNASSIGNED VIRUSES IN THE FAMILY

None reported.

SIMILARITY WITH OTHER TAXA

There are significant structural similarities in the capsid protein with respect to polypeptide folding topology and subunit interactions are shared with member viruses of the genus *Dianthovirus*. Putative nucleic acid helicase and polymerase gene sequences show similarities with comparable regions of member viruses of the genus *Dianthovirus*, *Necrovirus*, *Machlomovirus*, and with barley yellow dwarf virus-PAV and similar species of the genus *Luteovirus*. Soil-borne transmission is shared with members of the genera *Necrovirus* and *Dianthovirus*.

DERIVATION OF NAMES

tombus: sigla from *tom*ato *bus*hy stunt
carmo: sigla from *car*nation *mo*ttle

REFERENCES

Carrington JC, Heaton LA, Zuidema D, Hillman BI, Morris TJ (1989) The genome structure of turnip crinkle virus. Virology 170: 219-226

Dalmay T, Rubino L, Burgyan J, Kollar A, Russo M (1993) Functional analysis of Cymbidium ringspot virus genome. Virology 194: 697-704

Di Franco A, Martelli GP (1987) Comparative ultrastructural investigations on four soil-borne cucurbit viruses. J Submicrosc Cytol Pathol 19: 605-613

Grieco F, Burgyan J, Russo M (1989) The nucleotide sequence of Cymbidium ringspot virus RNA. Nucl Acids Res 17: 6383

Guilley H, Carrington JC, Balazs E, Jonard G, Richards KE, Morris TJ (1985) Nucleotide sequence and genome organization of carnation mottle virus RNA. Nucl Acids Res 13: 6663-6677

Hacker DL, Petty IRD, Wei N, Morris TJ (1992) Turnip crinkle virus genes required for RNA replication and virus movement. Virology 186: 1-8

Hearne PQ, Knorr DA, Hillman BI, Morris TJ (1990) The complete genome structure and synthesis of infectious RNA from clones of tomato bushy stunt virus. Virology 177: 141-151

Koenig R, Gibbs AJ (1986) Serological relationships among tombusviruses. J Gen Virol 67: 75-82

Martelli GP, Gallitelli D, Russo M (1988) Tombusviruses. In: Koenig R (ed) The plant viruses Vol 3. Polyhedral virions with monopartite RNA genome. Plenum Press, New York, pp 13-72

Morris TJ, Carrington JC (1988) Carnation mottle virus and viruses with similar properties. In: Koenig R (ed) The plant viruses Vol 3. Polyhedral virions with monopartite RNA genome. Plenum Press, New York, pp 73-112

Olson AJ, Bricogne G, Harrison SC (1983) Structure of tomato bushy stunt virus. IV. The virus particle at 2.9 A resolution. J Mol Biol 171: 61-93

Riviere CJ, Rochon DM (1990) Nucleotide sequence and genomic organization of melon necrotic spot virus. J Gen Virol 69: 395-400

Rochon DM, Johnston JC, Riviere CJ (1991) Molecular analysis of the cucumber necrosis virus genome. Can J Pl Pathol 13: 162-154.

Rubino L, Burgyan J, Grieco F, Russo M (1989) Sequence analysis of Cymbidium ringspot virus satellite and defective interfering RNAs. J Gen Virol 71: 1655-1660

Russo M, Di Franco A, Martelli GP (1990) Cytopathology in the identification and classification of tombusviruses. Intervirology 28: 134-143

CONTRIBUTED BY

Martelli GP, Russo M

Genus *Necrovirus*

Type Species tobacco necrosis virus (TNV)

Virion Properties

Morphology

Virions exhibit icosahedral symmetry (T = 3) and are approximately 28 nm in diameter. The virion associated satellite virus is 16.8 nm in diameter with T=1 icosahedral symmetry.

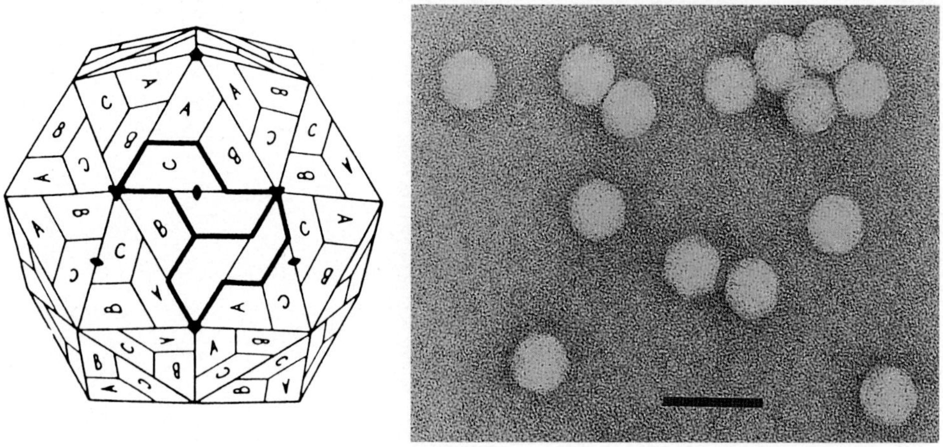

Figure 1: (left) Diagram of (T=3) TNV virion. (right) Negative contrast electron micrograph of TNV virions. The bar represents 50 nm.

Physicochemical and Physical Properties

Virion Mr is 7.6 x 10^6; S_{20w} is 118; buoyant density in CsCl is 1.40 g/cm^3. Mr of the satellite is 1.64 x 10^6. Virions of both the parent and satellite are insensitive to ether, chloroform and non-ionic detergents. The thermal inactivation point of TNV is between 85 and 95° C. Virion isoelectric point is pH 4.5.

Nucleic Acid

Virions contain one molecule of infectious linear positive sense ssRNA. The D strain RNA is 3,759 nt in size. The 5' end of the RNA does not have a covalently linked virion protein and is uncapped possessing a ppA... terminus. The RNA does not contain a 3' terminal poly (A) tract. The satellite virus RNA is 1,239 nt in size with the same lack of terminal structures as the parent virus. The complete nucleic acid sequence of the D strain, nearly complete sequence of the A stain, and the satellite virus are in the EMBL/GenBank databases.

Proteins

The virion is composed of 180 copies of a single capsid protein species. This protein has 268-275 amino acids and has an Mr of 29-30 x 10^3. The satellite virion is composed of 60 copies of a capsid protein species which has 195-197 amino acids and an Mr of 21.8 x 10^3.

Lipids

None reported.

Carbohydrates

None reported.

GENOME ORGANIZATION AND REPLICATION

The genomic RNA contains 5 ORFs. However, the A strain also contains a small 3' proximal ORF 6. ORF 1 is capable of encoding a Mr 23 x 10^3 peptide. Readthrough of the ORF 1 amber termination codon allows translation to continue into ORF 2 for the expression of an Mr 82 x 10^3 polypeptide. The Mr 82 x 10^3 protein is predicted to be the RNA-dependent RNA polymerase found in infected plants. ORF 3 can encode for a Mr 7.9 x 10^3 and ORF 4 a Mr 6.2 x 10^3 polypeptide. ORF 5 encodes the Mr 30 x 10^3 capsid protein. ORF 6 present only in the A strain can encode a Mr 6.7 x 10^3 protein. Two subgenomic RNAs of 1.6 and 1.3 kb are synthesized in infected cells. The smaller subgenomic RNA is the translational template for capsid protein and the larger for the ORF 3 and possibly ORF 4 products. The functions of the ORF 3, ORF 4, and ORF 6 products are not known. The satellite virus is dependent on helper virus for replication. The satellite virus genome contains a single ORF which encodes a capsid protein. Crystalline aggregates of virions are prominent in infected cells. Sometimes patches of electron-dense amorphous material can be seen. The satellite virus readily forms crystalline arrays.

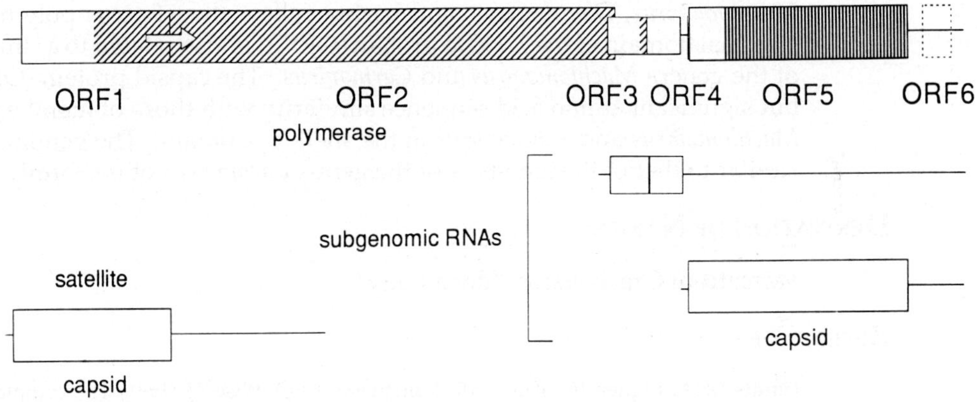

Figure 2: Organization and expression of the genome of TNV and its satellite virus. Arrow identifies translational readthrough of ORF 1 amber termination codon to produce 82 x 10^3 protein. Hatched regions ORF 1/ORF 2 identifies amino acid sequence similarity to member viruses of the family *Tombusviridae* and genera *Dianthovirus*, and *Machlomovirus* polymerases. Shaded area identifies capsid protein shell domain with amino acid sequence similarity to member viruses of the genera *Machlomovirus* and *Sobemovirus* capsid proteins. The two subgenomic RNAs are illustrated below the genomic RNA.

ANTIGENIC PROPERTIES

Member viruses are moderately immunogenic and the associated satellite virus is highly immunogenic. Two major TNV serotypes (A and D) with several strains of each may be distinguished serologically. Antisera yield a single precipitin line in agar gel-diffusion assays.

BIOLOGICAL PROPERTIES

HOST RANGE

Necroviruses have a wide host range among both monocotyledonous and dicotyledonous plant species. Infections are typically restricted to roots in natural infections. Experimental inoculations usually cause necrotic lesions on the inoculated leaves, rarely resulting in systemic infection.

TRANSMISSION

Virions are readily transmitted by mechanical inoculation. Member viruses are transmitted naturally by the chytrid fungus *Olpidium brassicae*.

LIST OF SPECIES IN THE GENUS

The viruses, their genomic sequence accession numbers [ ], CMI/AAB description # ( ) and assigned abbreviations ( ) are:

SPECIES IN THE GENUS

Chenopodium necrosis virus		(ChNV)
tobacco necrosis virus (14)	[D00942, M33002, M64479]	(TNV)

TENTATIVE SPECIES IN THE GENUS

carnation yellow stripe virus	(CYSV)
Lisianthus necrosis virus	(LNV)

SIMILARITY WITH OTHER TAXA

The polymerase (ORF 1, ORF 2) has a high degree of sequence similarity to the member viruses of the family *Tombusviridae* (genera *Tombusvirus* and *Carmovirus*) and the genera *Machlomovirus*, *Dianthovirus*, and barley yellow dwarf virus polymerases. The carboxy-terminal domain of the 7.9 kDa protein (ORF 3) is also related to a similar domain in viruses of the genera *Machlomovirus* and *Carmovirus*. The capsid protein (ORF 5) contains limited but significant amino acid sequence similarity with those of member viruses of the genera *Machlomovirus* and *Sobemovirus* in the shell (S) domain. The genome organization is most similar to that of the members of the genus *Carmovirus* of the family *Tombusviridae*.

DERIVATION OF NAMES

necro: from Greek *nekros*, "dead body"

REFERENCES

Coutts RHA, Rigden JE, Slabas AR, Lomonossoff GP, Wise PJ (1991) The complete nucleotide sequence of tobacco necrosis virus strain D. J Gen Virol 72: 1521-1529

Danthinne X, Seurinck J, van Montagu M, Pleij CWA, van Emmelo J (1991) Structual similarities between the RNAs of two satellites of tobacco necrosis virus. Virology 185: 605-614

Francki RIB, Milne RG, Hatta T (eds) (1985) Tobacco necrosis virus group. In: Atlas of plant viruses Vol I. CRC Press, Boca Raton FL, pp 171-180

Gallitelli D, Castellano MA, Di Franco A, Rana GIL (1979) Properties of carnation yellow stripe virus, a member of the tobacco necrosis virus group. Phytopathol Medit 18: 31-40

Gama MICS, Kitajima EW, Lin MT (1982) Properties of tobacco necrosis virus isolate from *Pogostemum patchuli* in Brazil. Phytopathology 72: 529-532

Iwaki M, Hanada K, Maria ERA, Onogi S (1987) Lisianthus necrosis virus, a new necrovirus from *Eustoma russellianum*. Phytopathology 77: 867-870

Koonin EV (1991) The phylogeny of RNA-dependent RNA polymerases of positive-strand RNA viruses. J Gen Virol 72: 2197-2206

Koonin EV (1991) Phylogeny of capsid proteins of small icosahedral RNA plant viruses. J Gen Virol 72: 1481-1486

Lesnaw JA, Reichmann ME (1969) The structure of tobacco necrosis virus I. The protein subunit and the nature of the nucleic acid. Virology 39: 729-737

Liljas L, Unge T, Jones TA, Fridborg K, Lovgren S, Skoglund U, Strandberg B (1982) Structure of satellite tobacco necrosis virus at 3.0 Å Resolution. J Mol Biol 159: 93-108

Meulewaeter F, Seurinck J, van Emmelo J (1990) Genome structure of tobacco necrosis virus strain A. Virology 177: 699-709

Stussi-Garaud C, Lemius J, Fraenkel-Conrat H (1977) RNA polymerase from tobacco necrosis virus-infected and uninfected tobacco. Virology 81: 224-236

Tomlinson JA, Faithfull EM, Webb MJW, Fraser RSS, Seeley ND (1983) Chenopodium necrosis: a distinctive strain of tobacco necrosis virus isolated from river water. Ann Appl Biol 102: 135-147

Uyemoto JK (1981) Tobacco necrosis and satellite viruses. In: Kurstak E (eds) Handbook of plant virus infections and comparative diagnosis. Elsevier/North Holland, Amsterdam, pp 123-146

CONTRIBUTED BY

Lommel SA

GENUS *DIANTHOVIRUS*

Type Species carnation ringspot virus (CRSV)

VIRION PROPERTIES

MORPHOLOGY

Virions are 32-35 nm in diameter and exhibit icosahedral symmetry (T=3). Virions have a distinctively granular surface. Detailed structure of the virion is not known. However, based on capsid protein sequence similarity, it is predicted that the virion is structurally similar to the T=3 virions of the member viruses of the family *Tombusviridae*.

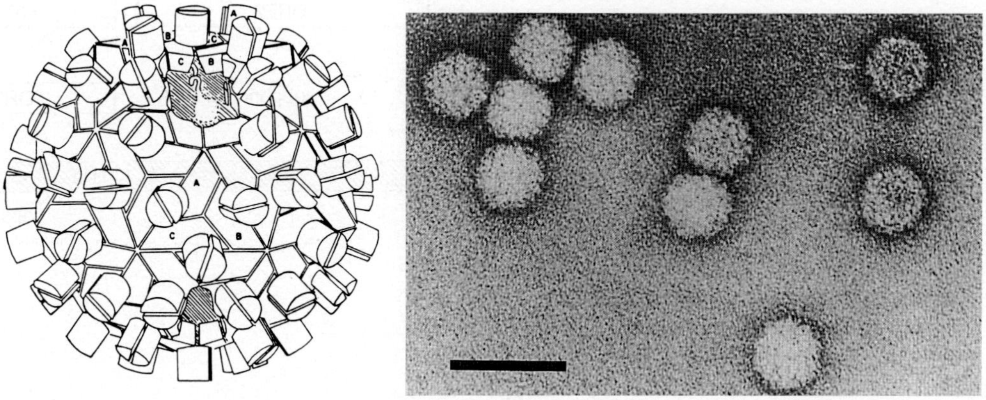

Figure 1: (left) Diagrammatic representation of tomato bushy stunt virus particle (*Tombusvirus*), best representing the structure of dianthoviruses. (right) Negative contrast electron micrograph of RCNMV virions; the bar represents 50 nm.

PHYSICOCHEMICAL AND PHYSICAL PROPERTIES

Virion Mr is 8.6 x 10^6; S_{20w} is 133; buoyant density in CsCl is 1.37 g/cm^3. Virions are insensitive to ether, chloroform and non-ionic detergents. Virions are stable at pH 6 and lower; alkaline conditions (pH 7-8) induce particle swelling. Virions are stabilized by divalent cations.

NUCLEIC ACID

Virions contain two molecules of infectious linear positive sense ssRNA. RNA1 is 3,889 nt and RNA2 is 1,448 nt in size. The 5' end of each RNA is capped with a m^{7}G linked to an A residue. The RNAs do not contain a 3' terminal poly (A) tract.

PROTEINS

Virions are composed of 180 copies of a 339 amino acid capsid protein species (Mr 37 x 10^3).

LIPIDS

None reported.

CARBOHYDRATES

None reported.

GENOME ORGANIZATION AND REPLICATION

Only the 5' terminal 6 nucleotides and 3' terminal 27 nucleotides are identical between RNA1 and RNA2. The 3' 27 nucleotides are predicted to form a stem-loop structure. RNA1 contains three ORFs. ORF 1 is capable of encoding a Mr 27 x 10^3 protein (unknown function). An internal ORF 2 could encode a Mr 57 x 10^3 protein. The ORF 2 gene product has not been observed *in vivo* and the independent production of this protein may be an *in vitro* translation artifact. RNA1 also directs the synthesis of a Mr 88 x 10^3 fusion protein by translational readthrough of ORF 1 into ORF 2 by a ribosomal frameshift mechanism similar to that of retroviruses. This fusion protein is the virus encoded polymerase.

The 3' proximal ORF 3 encodes the Mr 37 x 10^3 capsid protein. Capsid protein is expressed *in vivo* from a 1.4 kb subgenomic RNA. RNA2 contains a single ORF encoding the Mr 35 x 10^3 movement protein. RNA1 replicates in plant protoplasts and produce virions in the absence of RNA2. RNA1 is capable of replication in the absence of both the capsid protein gene and RNA2. The RCNMV capsid protein is not necessary for cell-to-cell movement, but is required for rapid systemic infection through the vascular tissue.

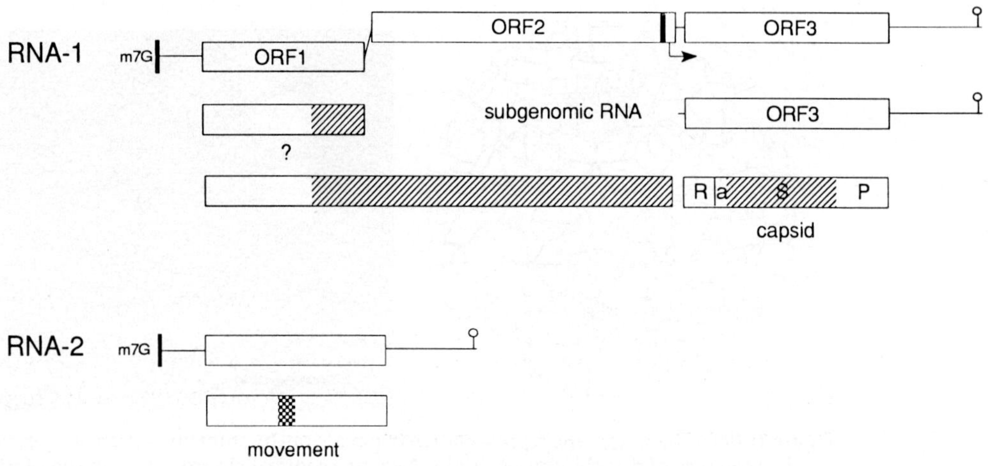

Figure 2: Organization and expression of the RCNMV genome. RCNMV RNA1 and RNA2 are depicted as solid lines. ORFs are identified as open rectangles. Rectangles below the RNAs represent virus encoded polypeptides and shaded areas identify domains with significant amino acid sequence similarity to like proteins in the family *Tombusviridae* and genera *Necrovirus* and *Machlomovirus*. The checkered region in the RNA2 encodes a movement protein motif which is conserved between dianthoviruses and members of the family *Bromoviridae*. The R (RNA binding), a (arm), S (shell), and P (protruding) domains of the RCNMV capsid protein are indicated.

ANTIGENIC PROPERTIES

The viruses are moderately to highly immunogenic. Various serological strains have been identified. Antisera yield a single precipitin line in agar gel-diffusion assays.

BIOLOGICAL PROPERTIES

HOST RANGE

Nature: Dianthoviruses have moderately broad natural host ranges restricted to dicots.
Laboratory: The experimental host range of the dianthoviruses is much broader than that found in nature, including a wide range of herbaceous species in the families *Solanaceae*, *Leguminosae*, *Cucurbitaceae*, and *Compositae*. Members of the group infect an even larger number of plants locally (non-systemically).

TRANSMISSION

The viruses are readily transmitted by mechanical inoculation. The viruses are not known to be seed transmitted. The viruses are not transmitted by insects, nematodes, or soil inhabiting fungi. However, viruses are readily transmitted through the soil without the aid of a biological vector.

GEOGRAPHIC DISTRIBUTION

Dianthoviruses, with the possible exception of FNSV, which appears to be tropical in range, are widespread throughout the temperate regions of the world.

PATHOGENICITY, ASSOCIATION WITH DISEASE

CRSV is a pathogen of carnations, orchard, and vine crops. RCNMV and SCNMV cause a mild disease of forage legumes. In general, dianthovirus infections do not kill host plants; however, necrosis and other symptoms can become quite severe at sustained temperatures between 15 to 20° C.

LIST OF SPECIES IN THE GENUS

The viruses, their genomic sequence accession numbers [ ], CMI/AAB description # ( ) and assigned abbreviations ( ) are:

SPECIES IN THE GENUS

carnation ringspot virus (21, 308)	[M88589]	(CRSV)
red clover necrotic mosaic virus (181)	[J04357, X08021]	(RCNMV)
sweet clover necrotic mosaic virus (321)		(SCNMV)

TENTATIVE SPECIES IN THE GENUS

Furcraea necrotic streak virus	(FNSV)

SIMILARITY WITH OTHER TAXA

The Mr 88 x 10^3 polymerase has a high degree of sequence similarity to those of members of the family *Tombusviridae* and the genera *Necrovirus*, *Machlomovirus* and *Luteovirus*. The movement proteins contains a motif conserved among species of the family *Bromoviridae*. The capsid protein S domain (160 residues) is highly conserved and the P domain moderately conserved among members of the family *Tombusviridae*.

DERIVATION OF NAMES

diantho: from *Dianthus*, the generic name of carnation

REFERENCES

Gould AR, Francki RIB, Hatta T, Hollings M (1981) The bipartite genome of red clover necrotic mosaic virus. Virology 108: 499-506

Hiruki C (1987) The Dianthoviruses: A distinct group of isometric plant viruses with bipartite genome. Adv Virus Res 33: 257-300

Hiruki C, Rao ALN, Furuya Y, Figueiredo G (1984) Serological studies of dianthoviruses using monoclonal and polyclonal antibodies. J Gen Virol 65: 2273-2275

Koonin EV (1991) The phylogeny of RNA-dependent RNA polymerases of positive-strand RNA viruses. J Gen Virol 72: 2197-2206

Koonin EV (1991) Phylogeny of capsid proteins of small icosahedral RNA plant viruses. J Gen Virol 72: 1481-1486

Lommel SA, Weston-Fina M, Xiong Z, Lomonossoff GP (1988) The nucleotide sequence of red clover necrotic mosaic virus RNA-2. Nucl Acids Res 16: 8587-8602

Morales FJ, Castaño M, Calvert LA, Arroyave J (1992) Furcraea necrotic streak virus: an apparent new member of the dianthovirus group. J Phytopathology 134: 247-254

Osman TAM, Buck KW (1987) Replication of red clover necrotic mosaic virus RNA in cowpea: RNA-1 replicates independently of RNA-2. J Gen Virol 68: 289-296

Osman TAM, Buck KW (1990) Double-stranded RNAs isolated from plant tissue infected with red clover necrotic mosaic virus correspond to genomic and subgenomic single-stranded RNAs. J Gen Virol 71: 945-948

Osman TAM, Miller SJ, Marriott AC, Buck KW (1991) Nucleotide sequence of RNA-2 of a Czechoslovakian isolate of red clover necrotic mosaic virus. J Gen Virol 72: 213-216

Tremaine JH, Ronald WP, Valcic A (1976) Aggregation properties of carnation ringspot virus. Phytopathology 66: 34-39

Xiong Z, Lommel SA (1989) The complete nucleotide sequence and genome organization of red clover necrotic mosaic virus. Virology 171: 543-554

CONTRIBUTED BY

Lommel SA

GENUS *MACHLOMOVIRUS*

Type Species maize chlorotic mottle virus (MCMV)

VIRION PROPERTIES

MORPHOLOGY

Virions are approximately 30 nm in diameter and exhibit icosahedral symmetry. Detailed structure of virions is not known. Based on capsid protein sequence similarity, it is predicted that the virion is structurally similar to the T=3 virions of southern bean mosaic virus (Genus *Sobemovirus*).

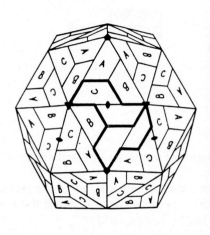

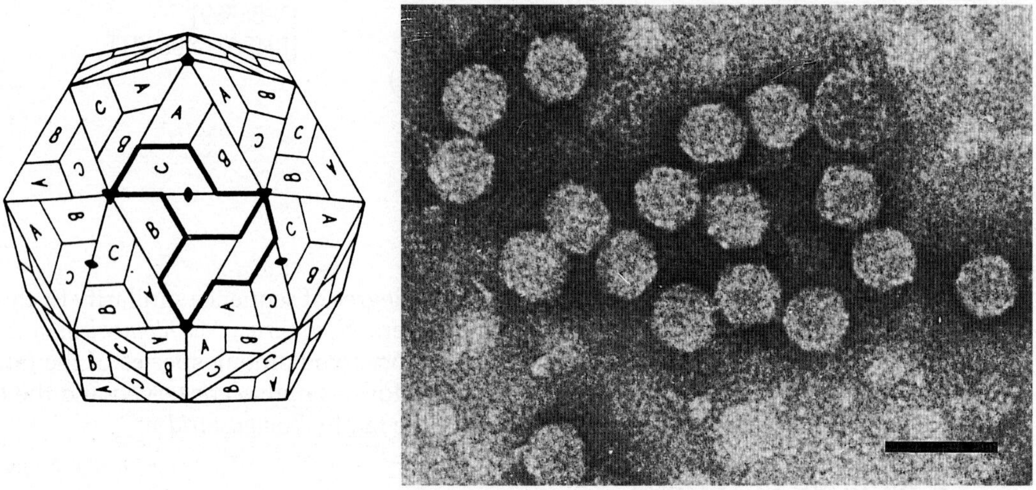

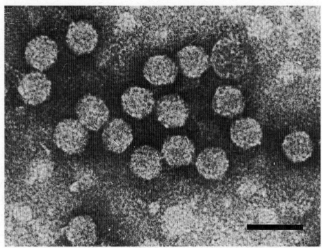

Figure 1: (left) Diagrammatic representation of a machlomovirus particle. (right) Negative contrast electron micrograph of virions. The bar represents 100 nm.

PHYSICOCHEMICAL AND PHYSICAL PROPERTIES

Mr of virions is 6.1 x 10^6; S_{20w} is 109; buoyant density in CsCl is 1.365 g/cm^3. Virions are insensitive to ether, chloroform and non-ionic detergents. Virions are stable *in vitro* for up to 33 days and the thermal inactivation point of virions is between 80-85° C. Virions are stable at pH 6 and lower. Virions are stabilized by divalent cations.

NUCLEIC ACID

Virions contain a single molecule of infectious linear positive sense ssRNA. The RNA is 4,437 nt in length. The 5' end of the RNA is capped with a m^7G linked to an A residue. The RNA does not contain a 3' terminal poly (A) tract. A 1,100 nt subgenomic RNA is also packaged into virions at a very low frequency.

PROTEINS

The virion is probably composed of 180 copies a single capsid protein species made up of 238 amino acids (Mr 25.1 x 10^3).

LIPIDS

None reported.

CARBOHYDRATES

None reported.

GENOME ORGANIZATION AND REPLICATION

The genomic RNA contains 4 ORFs. ORF 1 is capable of encoding a Mr 32 x 10^3 protein. ORF 2 can encode a Mr 50 x 10^3 protein. Readthrough of the ORF 1 amber termination codon allows the expression of a Mr 111 x 10^3 protein. A Mr 111 x 10^3 protein is observed upon translation of virion RNA in an *in vitro* translation system. ORF 3 can encode a Mr 9 x 10^3 protein. Assuming readthrough of the ORF 3 opal termination codon, a Mr 33 x 10^3 protein could be produced. ORF 4 encodes the Mr 25.1 x 10^3 capsid protein. A subgenomic RNA of 1.1 kb synthesized in infected cells is the translational template for capsid protein. The functions of ORF 1 and ORF 3 encoded proteins and the ORF 3 readthrough product are not known. The ORF 2 encoded protein and its readthrough product are thought to be the viral polymerase. Two dsRNAs corresponding to the genomic RNA and capsid protein subgenomic RNA are detected in infected tissue.

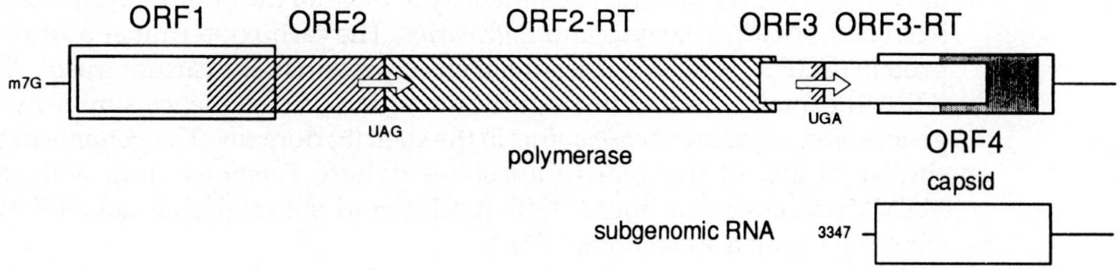

Figure 2: Organization and expression of the MCMV genome. Arrows identify ORF extensions assuming suppression and translational readthrough of the identified termination codon. Hatched area in ORF 2/ORF 2RT identifies amino acid sequence similarity to the family *Tombusviridae* and genera *Necrovirus* and *Machlomovirus* polymerases. Shaded area identifies capsid protein shell domain with amino acid sequence similarity to *Sobemovirus* capsid proteins. Capsid protein subgenomic RNA is illustrated below the genomic RNA.

ANTIGENIC PROPERTIES

The virus is moderately to highly immunogenic. Various serological variants have been identified. Antisera yield a single precipitin line in agar gel-diffusion assays.

BIOLOGICAL PROPERTIES

HOST RANGE

Nature: The virus systemically infects maize (*Zea mays*) varieties.
Laboratory: The virus is restricted to members of the host family *Gramineae*.

TRANSMISSION

The virus is readily transmitted by mechanical inoculation. The virus is seed transmitted. Kansas and Nebraska isolates can be transmitted by six species of chrysomelid beetles in the laboratory. A Hawaiian isolate is transmitted by thrips.

GEOGRAPHIC DISTRIBUTION

The virus has been reported in Argentina, Mexico, Peru, and the United States. Within the United States, the virus is restricted to the Republican river valley of Kansas and Nebraska, and to Kauai, Hawaii.

PATHOGENICITY, ASSOCIATION WITH DISEASE

MCMV causes a mild mosaic on maize in nature. When plants are also infected with one of several gramineae-specific potyviruses, a severe necrotic disease results, termed corn lethal necrosis.

List of Species in the Genus

The viruses, their genomic sequence accession numbers [ ], CMI/AAB description # ( ) and assigned abbreviations ( ) are:

Species in the Genus

maize chlorotic mottle virus (284)	[X14736]	(MCMV)

Tentative Species in the Genus

None reported.

Similarity with Other Taxa

The pre-readthrough and post-readthrough portions of the polymerase (ORF 1, ORF 1 RT) have a high degree of sequence similarity to those of the family *Tombusviridae* and the genera *Necrovirus*, *Macholomovirus* and *Luteovirus*. The carboxy-terminal portion of the Mr 9 x 10^3 protein (ORF 3) is related to a similar sized protein in the carmoviruses. The capsid protein (ORF 4) contains limited but significant amino acid sequence similarity with those of the genera *Necrovirus* and *Sobemovirus* in the shell (S) domain. The genome organization is most similar to that of the genus *Carmovirus* (family *Tombusviridae*), with the exception that MCMV possess an additional ORF (ORF 1) and the small internal ORF 3 appears not to be expressed from a subgenomic RNA.

Derivation of Names

Machlomo: sigla from *ma*ize *chlo*rotic *mo*ttle

References

Goldberg KB, Brakke MK (1987) Concentration of maize chlorotic mottle virus increased in mixed infections with maize dwarf mosaic virus, strain B. Phytopathology 77: 162-167

Gordon DT, Bradfute OE, Gingery RE, Nault LR, Uyemoto JK (1984) Maize chlorotic mottle virus. CMI/AAB Descriptions of plant viruses N° 284, 4pp

Jaing XQ, Wilkinson DR, Berry JA (1990) An outbreak of maize chlorotic mottle virus in Hawaii and possible association with thrips. Phytopathology 80: 1060

Jensen SG (1985) Laboratory transmission of maize chlorotic mottle virus by three species of corn root worm. Plant Dis 69: 864-868

Jensen SG, Wysong DS, Ball EM, Higley PM (1991) Seed transmission of maize chlorotic mottle virus. Plant Dis 75: 497-498

Koonin EV (1991) The phylogeny of RNA-dependent RNA polymerases of positive-strand RNA viruses. J Gen Virol 72: 2197-2206

Koonin EV (1991) Phylogeny of capsid proteins of small icosahedral RNA plant viruses. J Gen Virol 72: 1481-1486

Lommel SA, Kendall TL, Siu NF, Nutter RC (1991) Characterization of maize chlorotic mottle virus. Phytopathology 81: 819-823

Lommel SA, Kendall TL, Xiong Z, Nutter RC (1991) Identification of the maize chlorotic mottle virus capsid protein cistron and characterization of its subgenomic messenger RNA. Virology 181: 382-385

Nutter RC, Sheets K, Panganiban LC, Lommel SA (1989) The complete nucleotide sequence of maize chlorotic mottle virus. Nucl Acids Res 17: 3163-3177

Uyemoto JK, Bockelman DL, Claflin LE (1980) Severe outbreak of corn lethal necrosis disease in Kansas. Plant Dis 64: 99-100

Contributed By

Lommel SA

FAMILY *CORONAVIRIDAE*

TAXONOMIC STRUCTURE OF THE FAMILY

Family	***Coronaviridae***
Genus	*Coronavirus*
Genus	*Torovirus*

VIRION PROPERTIES

MORPHOLOGY

Virions are enveloped, those of coronaviruses being commonly 120-160 nm in diameter, pleomorphic but roughly spherical in shape, those of toroviruses being 120-140 nm in diameter and disc-, kidney-, or rod-shaped. Two to four proteins, some glycosylated, are associated with the envelope. The largest surface projections (S) are glycoproteins and vary in size and appearance, being about 20 nm in length. The viral nucleocapsid is helical (coronavirus), or tubular (torovirus).

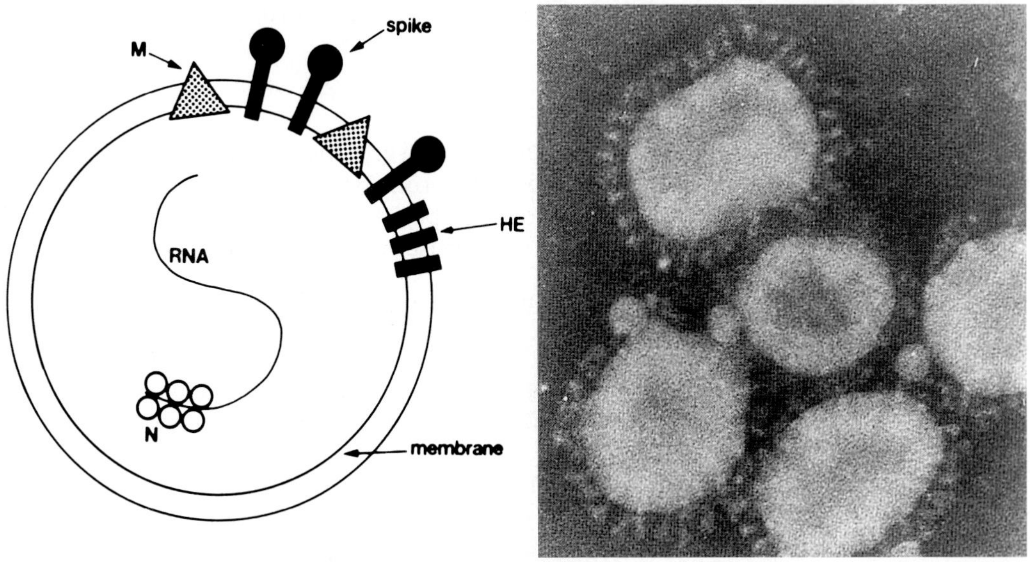

Figure 1: (left) Diagram of a coronavirus virion in section. The HE glycoprotein is only present in a subset of the genus. The location of the sM protein is not clear and is not shown; (right) negative contrast electron micrograph of IBV particles.

PHYSICOCHEMICAL AND PHYSICAL PROPERTIES

Virion Mr has been estimated at 400 x 10^6 for coronaviruses. Virion buoyant density in sucrose is 1.15-1.19 g/cm^3; density in CsCl is 1.23-1.24 g/cm^3 for coronaviruses. Virion S_{20w} is 300-500. Virions are sensitive to heat, lipid solvents, non-ionic detergents, formaldehyde and oxidizing agents. Some viruses in both genera are stable at pH 3.0. Magnesium ions (1 M) reduce heat inactivation of MHV.

NUCLEIC ACID

Virions contain a single molecule of linear, positive sense ssRNA, about 30 kb (coronavirus) or 20 kb (torovirus) in size. Virion RNA has a 5' terminal cap and a 3' terminal poly (A) tract.

PROTEINS

Virions contain a large surface glycoprotein (or spike, S), an integral membrane protein (M) which spans the virus envelope three times with only 10% protruding at the virion surface, and a nucleocapsid protein (N) (Table). The S protein is responsible for attachment to cells, hemagglutination and membrane fusion. It has a carboxy-terminal half with a coiled-coil structure. In addition, a sub-set of coronaviruses contains a hemagglutinin-esterase protein (HE) which forms short surface projections. In BCV this has receptor binding, hemagglutination and receptor destroying activities. The HE protein has identity with part of the hemagglutinin-esterase protein of influenza C virus; the nature of the presumed gene acquisition is uncertain. A small (approximately 100 amino acid) protein, tentatively named sM (small membrane), has been detected in virions of IBV and TGEV.

Table: Size of virion associated proteins (kDa) (NK: presence not known)

Protein	*Coronavirus*	*Torovirus*
S	180-220	200
M	30-35	27
N	50-60	19
HE	65	NK
sM	10-12	NK

LIPIDS

Virions have lipid-containing envelopes. The S protein of coronaviruses is acylated (MHV, BCV).

CARBOHYDRATES

The S and HE proteins contain N-linked glycans, the S protein is heavily glycosylated (about 20-35 glycans). The M proteins of coronaviruses contain a small number of either N- or O-linked glycans, depending on the species. These side chains are located near the amino-terminus. The M protein of toroviruses is not glycosylated.

GENOME ORGANIZATION AND REPLICATION

The genomic RNA is considered to be the mRNA for the RNA polymerase (*Pol*). When translated, the *Pol* products are responsible for amplification of the viral genome, the formation of full-length viral-complementary and viral-sense RNA species and the production of subgenomic mRNAs. The *Pol* is derived from the 5' proximal gene. This encodes two overlapping ORFs termed *Pol* 1a and *Pol* 1b. For coronaviruses, *Pol* 1a is about 440-500 kDa, *Pol* 1b is about 300-308 kDa in size. The sizes of the torovirus *Pol* products are not known. In addition to *Pol* and the structural protein genes (Fig. 2), the viral genomes contain several additional ORFs (not indicated in Fig. 2).

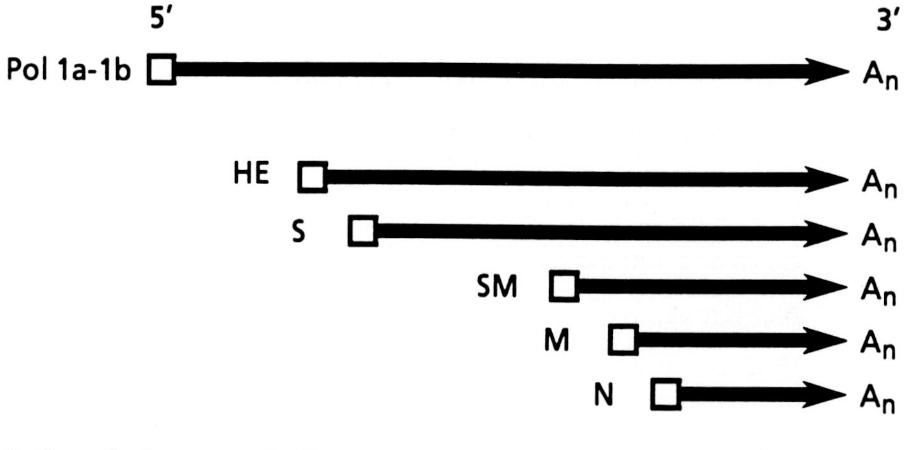

Figure 2: Generalized genome and mRNA organization of the member viruses of the family *Coronaviridae*. A leader sequence (open box) corresponding to the viral 5' terminus initiates each coronavirus mRNA. HE is present only in a subgroup of the coronaviruses. So far, neither an sM protein nor a leader sequence have been demonstrated in toroviruses. Genes (mRNA) with the potential to encode non structural proteins (other than Pol) are not shown.

One species of genome length negative stranded RNA is believed to act as template for the synthesis of a 3'-coterminal nested set of subgenomic mRNAs that are capped (5') and polyadenylated (3' A_n). Synthesis of coronavirus mRNA species from this template involves a process of discontinuous transcription, probably by a leader-priming mechanism (open boxes, Fig. 2). Coronavirus mRNAs may serve as templates for their own replication since negative stranded subgenomic RNAs of mRNA length are also found in infected cells. It is also possible that the negative stranded subgenomic RNAs may arise by discontinuous transcription from the genome template. The number of major subgenomic mRNAs varies from 5-7 depending on the virus. Only the 5' unique regions of the mRNAs, i.e., those absent from the next smaller mRNA, are thought to be translationally active. Translation of *Pol* 1b ORF involves ribosomal frame-shifting. Virions mature in the cytoplasm by budding through the endoplasmic reticulum and Golgi membranes. Viruses are not thought to mature at the plasma membrane. A high frequency of recombination has been demonstrated for mouse hepatitis virus with circumstantial evidence for other coronaviruses.

ANTIGENIC PROPERTIES

There are 3 or 4 major antigens corresponding to each of the major virion proteins. Spike and HE are the predominant antigens involved in virus neutralization. Neutralization with anti-M antibodies involves complement (coronaviruses). Anti-N and anti-M antibodies, in addition to those against S, give some protection in vivo.

BIOLOGICAL PROPERTIES

Coronaviruses are known to infect many mammals, including humans. They cause respiratory, gastrointestinal organs and neurological infections. Biological vectors are not known. Respiratory, fecal-oral and mechanical transmission are common. Toroviruses infect ungulates and humans, probably also carnivores (mustellids). Torovirus transmission is probably by the fecal-oral route.

GENUS *CORONAVIRUS*

Type Species avian infectious bronchitis virus (IBV)

DISTINGUISHING FEATURES

The N protein is much larger than that of toroviruses (Table); the M protein is glycosylated and an HE glycoprotein is present in some species. There is little sequence similarity between coronavirus and torovirus proteins. Coronavirus mRNAs have been shown to contain a 5' leader sequence.

LIST OF SPECIES IN THE GENUS

The viruses, their genomic sequence accession numbers [ ] and assigned abbreviations ( ) are:

SPECIES IN THE GENUS

avian infectious bronchitis virus	[M95169]	(IBV)
bovine coronavirus		(BCV)
canine coronavirus		(CCV)
feline infectious peritonitis virus		(FIPV)
human coronavirus 229E		(HCV-229E)
human coronavirus OC43		(HCV-OC43)
murine hepatitis virus		(MHV)
porcine epidemic diarrhea virus		(PEDV)
porcine hemagglutinating encephalomyelitis virus		(HEV)
porcine transmissible gastroenteritis virus		(TGEV)
rat coronavirus		(RCV)
turkey coronavirus		(TCV)

Tentative Species in the Genus

rabbit coronavirus (RbCV)

Genus *Torovirus*

Type Species Berne virus (BEV)

Distinguishing Features

The nucleocapsid has a tubular appearance and virions are disc-, kidney- or rod-shaped (Fig. 3). Data are based mostly on one virus, Berne virus. The N protein is small and M is not glycosylated. The viral genome contains an ORF potentially encoding a 142 amino acid proteins with 30-35% identity to the much larger HE protein of coronaviruses (Table). So far, an RNA leader sequence has not been identified on the mRNAs.

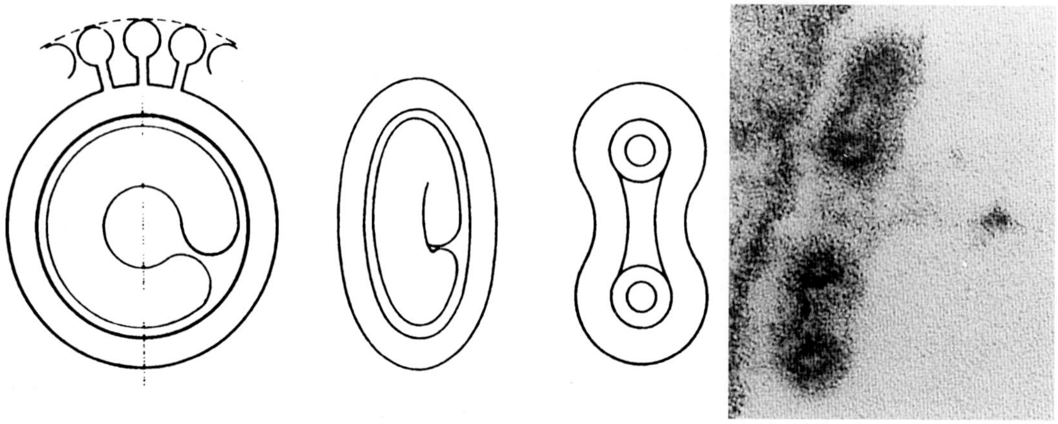

Figure 3: (left) Schematic representation of a torovirus virion in three projections; (right) thin section showing two virions of BEV.

List of Species in the Genus

The viruses, their genomic sequence accession numbers [ ] and assigned abbreviations ( ) are:

Species in the Genus

Berne virus	[X52374, X52505, X52506]	(BEV)
Breda virus		(BRV)

Tentative Species in the Genus

None reported.

List of Unassigned Viruses in the Family

None reported.

Similarity with Other Taxa

None reported.

Derivation of Names

corona: from Latin corona for "crown", representing the appearance of surface projections in negatively stained electron micrographs of members of the Coronavirus genus
toro: from Latin torus, "lowest convex moulding in the base of a column"

REFERENCES

Cavanagh D, Brian DA, Brinton MA, Enjuanes L, Holmes KV, Horzinek MC, Lai MMC, Laude H, Plagemann PGW, Siddell SG, Span W, Taguchi F and Talbot PJ (1994) Revision of the taxonomy of the Coronavirus, Torovirus and Arterivirus genera. Arch Virol 135: 227-237

Cavanagh D, Brown TDK (eds) (1990) Coronaviruses and their diseases. Advances in Experimental Medicine and Biology Vol 276. Plenum Press, New York

Cavanagh D, MacNaughton MR (1994) Coronaviruses and toroviruses. In: Zuckerman AJ, Pattison JR, Banatvala JE (eds) Principle and Practice of Clinical Virology 3rd edn. John Wiley & Sons, New York

Cavanagh D, Brian DA, Enjuanes L, Holmes KV, Lai MMC, Laude H, Siddell SG, Spaan WJM, Taguchi F, Talbot PJ (1990) Recommendations of the coronavirus study group for the nomenclature of the structural proteins, mRNAs and genes of coronaviruses. Virology 176: 306-307

Chirnside ED (1992) Equine arteritis virus: an overview. Br Vet J 148: 181-198

den Boon JA, Snijder EJ, Chirnside ED, de Vries AAF, Horzinek MC, Spaan WJM (1991) Equine arteritis virus is not a togavirus but belongs to the coronavirus-like superfamily. J Virol 65: 2910-2920

Kuo L, Chen Z, Rowland RRR, Faaberg KS, Plagemann PGW (1992) Lactate dehydrogenase-elevating virus (LDV): subgenomic mRNAs, mRNA leader and comparison of 3'-terminal sequences of two LDV isolates. Vir Res 23: 55-72

MacNaughton MR, Davies HA (1986) Coronaviridae. In: Nermut MV, Steven AC (eds) Animal Virus Structure. Elsevier Biomedical Press, Amsterdam, pp 173-183

Snijder EJ, den Boon JA, Horzinek MC, Spaan WJM (1991) Comparison of the genome organization of toro- and coronaviruses; evidence for two nonhomologous RNA recombination events during Berne virus evolution. Virology 180: 448-452

Spaan WJM, Cavanagh D, Horzinek MC (1988) Coronaviruses: structure and genome expression. J Gen Virol 69: 2939-2952

Spaan WJM, Cavanagh D, Horzinek MC (1990) Coronaviruses. In: van Regenmortel MHV, Neurath AR (eds) Immunochemistry of Viruses Vol II. Elsevier/North Holland, Amsterdam, pp 359-379

Wege H, Siddell SG, ter Meulen V (1982) The biology and pathogenesis of coronaviruses. Curr Top Microbiol Immunol 99: 165-200

Weiss M, Horzinek MC (1987) The proposed family *Toroviridae*: agents of enteric infections. Arch Virol 92: 1-15

CONTRIBUTED BY

Cavanagh D, Brain DA, Brinton MA, Enjuanes L, Holmes KV, Horzinek MC, Lai MMC, Laude H, Plagemann PGW, Siddell SG, Spaan WJM, Taguchi F, Talbot PJ

Genus *Arterivirus*

Type Species equine arteritis virus (EAV)

Virion Properties

Morphology

Virions are 60 nm in diameter and consist of an isometric nucleocapsid of about 35 nm in diameter, surrounded by a lipid envelope possessing 12-15 nm ring-like surface structures (Fig. 1).

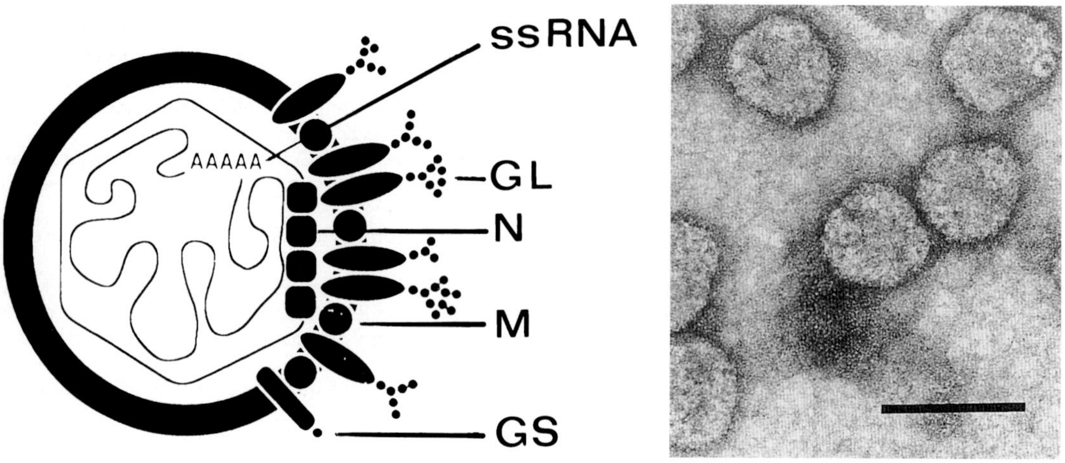

Figure 1: (left) Diagram of an arterivirus virion (EAV); (right) negative contrast electron micrograph of arterivirus virions. The bar represents 100 nm.

Physicochemical and Physical Properties

Virion buoyant density is about 1.13-1.17 g/cm^3 in sucrose and 1.17-1.20 g/cm^3 in CsCl. Virion S_{20w} is 200-230.

Nucleic Acid

Virions contain a single molecule of linear, positive-stranded RNA of about 13 kb in size. Virion RNA has a 5'-terminal cap (SHFV) and a 3'-terminal poly (A) tract.

Proteins

Virions are composed of a nucleocapsid protein (N), about 12 kDa in size; a non-glycosylated triple-membrane spanning integral membrane protein (M), about 16 kDa in size and at least two N-glycosylated surface proteins. The latter are associated with small (G_S) and large (G_L) glycopolypeptides of 25 kDa and 30-42 kDa, respectively, G_L being heterogeneously glycosylated (EAV; de Vries AFF, Horzinek MC and Rottier, unpublished data).

Lipids

The virions have lipid-containing envelopes.

Carbohydrates

The S but not the M protein have N-linked glycans.

Genome Organization and Replication

Genome organization is similar to that of member viruses of the family *Coronaviridae*. A leader is present on the mRNAs.

Antigenic Properties

No antigenic relationship between EAV, LDV and SIRSV has been found.

Biological Properties

Arteriviruses infect horses (EAV), mice (LDV), monkeys (SHFV) and swine (SIRSV). EAV causes abortion. EAV causes necrosis in muscle cells of small arteries. Primary host cells are macrophages. Persistent infections are established frequently. Spread is horizontal (respiratory, biting) and, for EAV, by semen.

List of Species in the Genus

The viruses, their alternative names ( ), genomic sequence accession numbers [ ] and assigned abbreviations ( ) are:

Species in the Genus

equine arteritis virus	[X53459]	(EAV)
lactate dehydrogenase-elevating virus	[L13298]	(LDV)
swine infertility and respiratory syndrome virus	[M96262]	(SIRSV)
(porcine respiratory and reproductive syndrome)		
simian hemorrhagic fever virus		(SHFV)

Tentative Species in the Genus

None reported.

Similarity with Other Taxa

Member viruses of the genus *Arterivirus* have a genome organization and replication strategy similar to that of the viruses of the family *Coronaviridae*. However, there are major differences. The arterivirus genome (13 kb) and virions are only about half the size of those of members of the family *Coronaviridae*. The nucleocapsid of arterivirus is isometric and their surface projections are relatively small and indistinct. The structure of arterivirus surface projection proteins does not include a coiled-coil structure and they are considerably smaller. The M and N polypeptides are also smaller than those of members of the family *Coronaviridae*.

Derivation of Names

arteri: from equine *arteri*tis, the disease caused by the reference virus

References

Cavanagh D, Brian DA, Brinton MA, Enjuanes L, Holmes KV, Horzinek MC, Lai MMC, Laude H, Plagemann PGW, Siddell SG, Spaan WJM, Taguchi F, Talbot PJ (1994) Revision of the taxonomy of the Coronavirus, Torovirus and Arterivirus genera. Arch Virol 135: 227-237

Cavanagh D, Brown TDK (eds) (1990) Coronaviruses and their diseases. Advances in Experimental Medicine and Biology Vol 276. Plenum Press, New York

Cavanagh D, Macnaughton MR (1994) Coronaviruses and toroviruses. In: Zuckerman AJ, Pattison JR, Banatvala JE (eds) Principle and Practice of Clinical Virology Chapter 9, 3rd edn. John Wiley & Sons, New York

Cavanagh D, Brian DA, Enjuanes L, Holmes KV, Lai MMC, Laude H, Siddell SG, Spaan WJM, Taguchi F, Talbot PJ (1990) Recommendations of the coronavirus study group for the nomenclature of the structural proteins, mRNAs and genes of coronaviruses. Virology 176: 306-307

Chirnside ED (1992) Equine arteritis virus: an overview. Brit Vet J 148: 181-198

den Boon JA, Snijder EJ, Chirnside ED, de Vries AAF, Horzinek MC, Spaan WJM (1991) Equine arteritis virus is not a togavirus but belongs to the coronavirus-like superfamily. J Virol 65: 2910-2920

Kuo L, Chen Z, Rowland RRR, Faaberg KS, Plagemann PGW (1992) Lactate dehydrogenase-elevating virus (LDV): subgenomic mRNAs, mRNA leader and comparison of 3'-terminal sequences of two LDV isolates. Vir Res 23: 55-72

Macnaughton MR, Davies HA (1986) Coronaviridae. In: Nermut MV, Steven AC (eds) Animal Virus Structure. Elsevier Biomedical Press, Amsterdam, pp 173-183

Snijder EJ, den Boon JA, Horzinek MC, Spaan WJM (1991) Comparison of the genome organization of toro- and coronaviruses; evidence for two nonhomologous RNA recombination events during Berne virus evolution. Virology 180: 448-452

Spaan WJM, Cavanagh D, Horzinek MC (1988) Coronaviruses: structure and genome expression. J Gen Virol 69: 2939-2952

Spaan WJM, Cavanagh D, Horzinek MC (1990) Coronaviruses. In: Van Regenmortel MHV, Neurath AR (eds) Immunochemistry of Viruses Vol II. Elsevier/North Holland, Amsterdam, pp 359-379

Wege H, Siddell SG, ter Meulen V (1982) The biology and pathogenesis of coronaviruses. Curr Top Microbiol Immunol 99: 165-200

Weiss M, Horzinek MC (1987) The proposed family *Toroviridae*: agents of enteric infections. Arch Virol 92: 1-15

Contributed By

Cavanagh D, Brain DA, Brinton MA, Enjuanes L, Holmes KV, Horzinek MC, Lai MMC, Laude H, Plagemann PGW, Siddell SG, Spaan WJM, Taguchi F, Talbot PJ

Family *Flaviviridae*

Taxonomic Structure of the Family

Family	***Flaviviridae***
Genus	*Flavivirus*
Genus	*Pestivirus*
Genus	"Hepatitis C like-viruses"

Virion Properties

Morphology

Virions are 40-60 nm in diameter, spherical in shape and contain a lipid envelope (Fig. 1). The protein spikes on the virion surface do not show a characteristic structure detectable by currently available methodology. The viral core is spherical. Detailed structural properties, such as triangulation numbers, are not yet known. Hepatitis C virus has not been visualized. The behavior of hepatitis C virus during filtration, and its sensitivity to chemical and physical treatments, suggest the virus is structurally similar to the flaviviruses and pestiviruses.

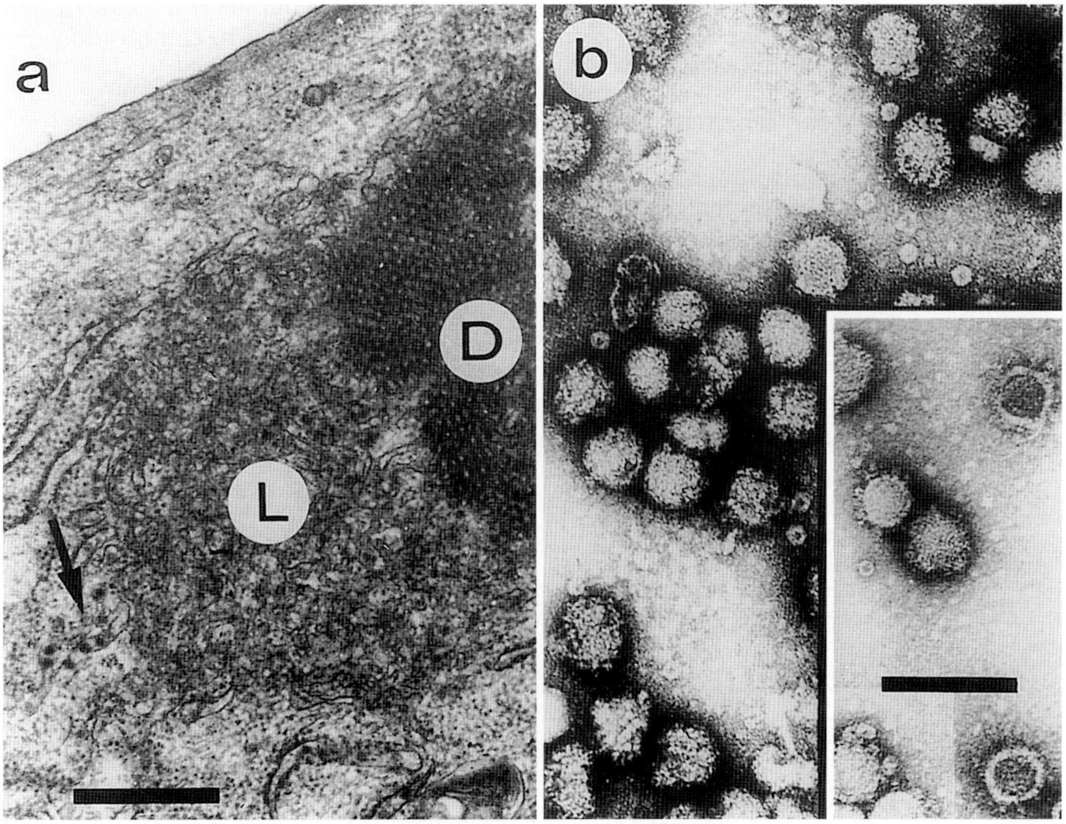

Figure 1: (a) West Nile virus infection of BHK-21 cell showing virus particles (arrow) and loose (L) and dense (D) proliferation of host cell membranes; the bar represents 1 µm. (b) Negative contrast electron micrograph of West Nile virus; insert shows particles where stain has penetrated; the bar represents 100 nm.

Physicochemical and Physical Properties

Virion Mr has not been determined precisely; Mr is estimated from virus composition to be about 60 x 10^6. Virion buoyant density in sucrose is 1.1-1.23 g/cm^3 and their S_{20W} is 140-200. Virions are sensitive to heat, organic solvents and detergents.

NUCLEIC ACID

Virions contain a single molecule of linear positive sense ssRNA. The genome sizes of flaviviruses, pestiviruses and hepatitis C virus are about 10.7 kb, 12.5 kb, and 9.5 kb, respectively. The 5' end structure of the viral RNA has not been characterized for members of all genera. Except for a few strains of the tick-borne encephalitis complex of flaviviruses, the genome RNA does not contain a poly (A) tract at the 3'-end.

PROTEINS

Virions contain two or three membrane-associated proteins and a core protein. The analogous structural proteins of flaviviruses, pestiviruses and hepatitis C virus show no detectable sequence similarities. By contrast, several amino acid sequence motifs in the non-structural proteins indicate that there are specific functional activities that have been conserved among the three genera.

LIPIDS

Virions are composed of about 15-20% lipid by weight; lipids are host cell derived.

CARBOHYDRATES

Virions contain carbohydrates in the form of glycolipids and usually glycoproteins. Some viruses do not contain glycosylated surface proteins. The composition and structure of the carbohydrates are host cell dependent.

GENOME ORGANIZATION AND REPLICATION

The only viral messenger RNA is the genome. A single long ORF codes for a polyprotein which is proteolytically cleaved into all the virus-encoded proteins. The structural proteins are located at the 5' end, non-structural proteins including proteases, helicases and polymerases, are encoded at the 3' end. The viruses multiply in the cytoplasm of infected cells in association with membranes and mature in cytoplasmic vesicles. Replication commonly is accompanied by a characteristic proliferation of intracellular membranes (Fig. 1).

ANTIGENIC PROPERTIES

Members of each genus are serologically related to each other, but not to members of other genera.

BIOLOGICAL PROPERTIES

The biological properties vary widely between the different genera. See the corresponding sections of the genus descriptions for details.

GENUS *FLAVIVIRUS*

Type Species yellow fever virus (YFV)

VIRION PROPERTIES

MORPHOLOGY

Virions are 40-50 nm in diameter and spherical in shape (Fig. 1). The virion envelope contains a dense layer of surface projections about 6 nm in length which are constructed from two viral proteins: E and preM in the case of cell-associated virus particles, or E and M in the case of extracellular particles. The viral core is spherical with a diameter of about 30 nm. Its symmetry is unknown.

PHYSICOCHEMICAL AND PHYSICAL PROPERTIES

Virion Mr has not been precisely determined but can be estimated from the virus composition to be about 60 x 10^6. Buoyant density in CsCl is 1.22-1.24 g/cm^3; S_{20W} is 170-210. Viruses are stable at slightly alkaline pH (8.0) and unstable at temperatures above 40° C. Solvents and detergents rapidly inactivate the viruses.

NUCLEIC ACID

Virions contain a single molecule of linear positive sense ssRNA (Fig. 2). The genome length varies between 10,976 nt (Japanese encephalitis virus) and 10,488 nt (tick-borne encephalitis virus). The genome is capped at the 5' end and, except for some strains of tick-borne encephalitis virus, no poly (A) track is present at the 3'-end. The gene order is 5'-C-preM-E-NS1-NS2A-NS2B-NS3-NS4A-NS4B-NS5 3'.

PROTEINS

Since flaviviruses mature into cytoplasmic vesicles, two types of virus particles can be distinguished: cell-associated virus and extracellular virus. Extracellular virus contains the two envelope proteins E and M and an internal, RNA-associated protein, C. Instead of the M protein, cell-associated virus particles contain a larger precursor protein, termed preM, which is cleaved during or shortly after release of virus from infected cells. Only the carboxy terminal part of preM remains associated with the extracellular virus particle as the M protein. The E membrane protein (50 kDa) is usually glycosylated. It contains twelve conserved cysteine residues which form six disulfide bridges. The M membrane protein (8 kDa) is singly glycosylated and contains six disulfide bridges. The C core protein (13 kDa) is rich in arginine and lysine residues.

LIPIDS

Virions contain about 17% lipid by weight; lipids are derived from host cell membranes.

CARBOHYDRATES

Virions contain about 9% carbohydrate by weight (glycolipids, glycoproteins); their composition and structure are dependent on the host cell (vertebrate or arthropod).

GENOME ORGANIZATION AND REPLICATION

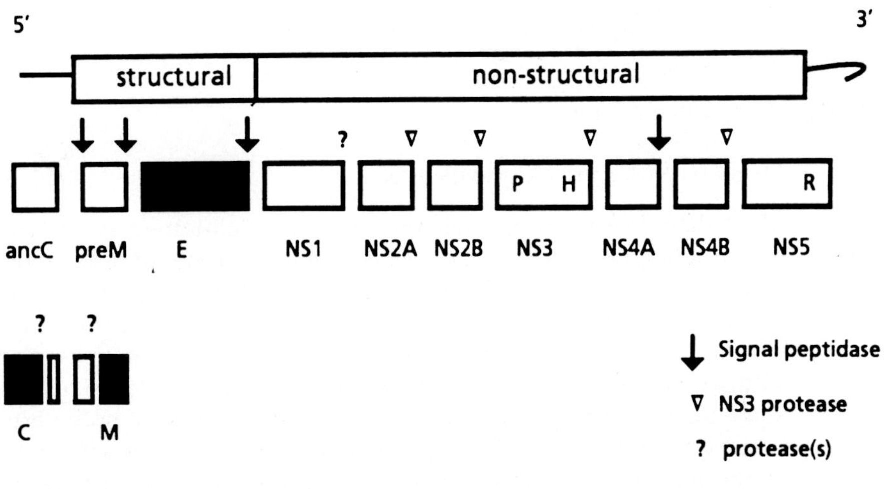

Figure 2: Flavivirus genome organization (not to scale). The total RNA of YFV contains 10,862 nt. The 5' non-coding region contains 118 , the 3' 511, and the ORF 10,233 nt. The loop at the 3' end of the RNA indicates the existence of a stem and loop structure which is present at the 3' terminus of almost all flavivirus RNAs. P, H, R indicate the location of the NS3 protease, NS3 helicase and NS5 RNA replicase, respectively. The viral structural proteins are in black. The proteases (where known) and proteolytic steps involved in the generation of the individual proteins are indicated.

The genome RNA is the sole viral mRNA molecule (Fig. 2). It contains a single long ORF which is translated on membrane-bound polyribosomes. The corresponding polyprotein is co-translationally and post-translationally cleaved into the individual viral structural and non-structural proteins. Virus attachment is mediated by the viral E protein, the availability of receptors for E is believed to determine tissue and cell tropisms (hence to some extent the host range). After endocytosis and uncoating the virus RNA is translated, the products processed and RNA replication ensues. Replication occurs in the cytoplasm, and is associated with proliferation of rough and smooth endoplasmic reticulum. Nucleocapsids have not been visualized in cells. Virus particles accumulate within lamellae and vesicles. RNA replication occurs in foci in the perinuclear region through a negative strand intermediate.

Polyprotein processing has been difficult to observe in infected cells but has been studied in cell-free translation systems. Signal peptidase is believed to effect the three cleavages required to produce the structural proteins (Fig. 2). The 13 kDa C and 8 kDa M proteins are derived from precursor polypeptides called anchored C and the preM, respectively, which are cleaved during virus maturation to their final forms. PreM is present as part of an E-preM heterodimer. The non-structural proteins following the structural 50 kDa E protein (in order) are: NS1 (a 50 kDa glycoprotein found on the cell surface and in the culture medium); NS2A (a 21 kDa integral membrane protein); NS2B (a 15 kDa integral membrane protein that cooperates during proteolysis with NS3); NS3 (a 70 kDa peripheral membrane protein with an amino terminal portion that is a serine protease with the amino acid H-D-S catalytic triad and a carboxy portion that has a ssRNA-stimulated triphosphatase-RNA helicase); NS4A (a 15 kDa integral membrane protein); NS4B (a 29 kDa integral membrane protein); and NS5 (a 100 kDa peripheral membrane protein that is a component of the RNA-dependent RNA polymerase). At least four of the cleavages to separate these proteins from the nascent polyprotein are made by the NS3 protease. Signal peptidase makes at least one of the two other cleavages required to separate the non-structural proteins (Fig. 2). Both NS3 and NS5 are believed to be components of the RNA replicase. In vertebrate cells, the latent period is 12-16 h and virus production continues over 3-4 days. Host cell RNA and protein synthesis continue throughout infection.

ANTIGENIC PROPERTIES

An hypothetical structural model of protein E assigns antigenic domains and epitopes to distinct sequence elements and protein domains. These antigenic domains induce antibodies with type, or subtype, complex, or group reactivities as determined by ELISA tests, RIA, immunofluorescence, virus neutralization, or enhancement of infectivity assays. Antibodies to E neutralize virus infectivity. In some cases it has been shown that antibodies to NS1 (a soluble complement-fixing antigen also found on infected cell surfaces) can prevent lethal infection.

BIOLOGICAL PROPERTIES

HOST RANGE AND TRANSMISSION

Most flaviviruses are arboviruses and are maintained in nature by transmission from hematophagous arthropod vectors (either mosquitoes or ticks, depending on the species) to vertebrate hosts (mammals, or marsupials, or birds) when the arthropod takes a blood meal. For certain isolates (predominantly bat isolates) no arthropod host has been identified. Viruses may also be passed trans-ovarially (mosquitoes, ticks) and trans-stadially (ticks). Transplacental and horizontal transmission between vertebrates has been demonstrated for some viruses. Viruses replicate in susceptible species of both vertebrates and arthropods. Some viruses have a limited vertebrate host range (e.g. only primates), for others host range can cover a wide variety of species (birds, mammals, etc.). Transmission usually derives from the presence of a viremia in the vertebrate host, and virus in the arthropod salivary gland secretions. The non-arbovirus members of the genus have been isolated either from arthropods, or from vertebrates, but not from both.

Pathogenicity

Essentially no pathogenicity has been demonstrated in arthropods. In vertebrate species, pathogenicity is highly variable. Some 30 flaviviruses cause disease in humans, varying from febrile illness with or without rash, to life-threatening conditions, such as hemorrhagic fever, encephalitis, or hepatitis. Some 8 to 10 flaviviruses cause severe and economically important diseases in domestic animals.

Experimental Isolation and Adaptation

Initial virus isolation is usually undertaken in newborn mice by intracranial inoculation. Tissue culture can also be employed. In certain inbred mouse strains, a single dominant gene determines resistance specific for flaviviruses. Genetic resistance is often associated with the generation of DI genomes and virions. Arthropods can be infected by feeding on infected animals, by capillary feeding or by inoculation.

Cell Cultures

Many vertebrate and arthropod cells support flavivirus replication. Some viruses induce cytopathic changes (plaques), others do not (depending on the virus and cell). Syncytium formation occurs upon infection of certain cell systems. Persistent infection is common.

Hemagglutination

Red blood cells from adult geese, or 1-2 day-old chicks, are agglutinated optimally at slightly acid pH.

Taxonomic Structure of the Genus

The taxonomic structure of the genus is generally based on cross-neutralization tests with polyclonal hyper-immune mouse ascitic fluids prepared against each of the viruses, except where indicated otherwise. Nine serologically defined groups are recognized. Unassigned viruses denote those which gave no significant cross-neutralization in such experiments. They are designated flaviviruses on the basis of some serological cross-reaction with at least one accepted member of the genus. Available nucleotide and amino acid sequence analyses have demonstrated conservation of sequences both within a subgroup and between serogroups. The extent of sequence conservation varies depending on the viruses and the particular genes under consideration.

List of Species in the Genus

The groups, and viruses, their alternative names ( ), genomic sequence accession numbers [ ] and assigned abbreviations ( ) are:

Species in the Genus

1-yellow fever virus group (mosquito-borne):		
yellow fever virus	[X03700, X15065]	(YFV)
2-tick-borne encephalitis virus group (known tick-borne viruses):		
tick-borne encephalitis virus		(TBEV)
(a) European subtypes		
Hanzalova virus		(HANV)
Hypr virus	[M76660]	(HYPRV)
Kumlinge virus	[M27157]	(KUMV)
Neudoerfl virus	[M27157, M33668]	(NEUV)
(b) Far eastern subtypes		
(Russian spring summer encephalitis virus)		(RSSEV)
Absettarov virus		(ABSV)
Karshi virus		(KSIV)
Kyasanur forest disease virus		(KFDV)
Langat virus	[M73835, M86650]	(LGTV)
louping ill virus	[M59376, M94957, X59815]	(LIV)

Negishi virus	[94956]	(NEGV)
Omsk hemorrhagic fever virus	[X66694]	(OMSKV)
Powassan virus		(POWV)
Royal farm virus		(RFV)
Sofyn virus	[X07755]	(SOFV)
(no known vector):		
Carey Island virus		(CIV)
Phnom-Penh bat virus		(PPBV)
3-Rio Bravo virus group (no known vector):		
Apoi virus		(APOIV)
Bukalasa bat virus		(BUBV)
Dakar bat virus		(DBV)
Entebbe bat virus		(ENTV)
Rio Bravo virus		(RBV)
Saboya virus		(SABV)
4-Japanese encephalitis virus group (mosquito-borne):		
Alfuy virus		(ALFV)
Japanese encephalitis virus	[M18370]	(JEV)
Kokobera virus		(KOKV)
Koutango virus		(KOUV)
Kunjin virus	[D00246]	(KUNV)
Murray Valley encephalitis virus	[X03467]	(MVEV)
St. Louis encephalitis virus	[M1661]	(SLEV)
Stratford virus		(STRV)
Usutu virus		(USUV)
West Nile virus	[M12294]	(WNV)
5-Tyuleniy virus group (tick-borne):		
Meaban virus		(MEAV)
Saumarez Reef virus		(SREV)
Tyuleniy virus		(TYUV)
6-Ntaya virus group (mosquito-borne):		
Bagaza virus		(BAGV)
Israel turkey meningoencephalitis virus		(ITV)
Ntaya virus		(NTAV)
Tembusu virus		(TMUV)
Yokase virus		(YOKV)
7-Uganda S virus group (mosquito-borne):		
Banzi virus		(BANV)
Bouboui virus		(BOUV)
Edge Hill virus		(EHV)
Uganda S virus		(UGSV)
8-Dengue virus group (mosquito-borne):		
Dengue virus 1	[M23027]	(DENV-1)
Dengue virus 2	[M19197]	(DENV-2)
Dengue virus 3	[A34774]	(DENV-3)
Dengue virus 4	[M14931]	(DENV-4)
9-Modoc virus group (no known vector):		
Cowbone Ridge virus		(CRV)
Jutiapa virus		(JUTV)
Modoc virus		(MODV)
Sal Vieja virus		(SVV)
San Perlita virus		(SPV)

TENTATIVE SPECIES IN THE GENUS

1-tick-borne viruses:	
Gadget's Gully virus	(GGYV)
Kadam virus	(KADV)

2-mosquito-borne viruses:

Bussuquara virus	(BSQV)
Ilheus virus	(ILHV)
Jugra virus	(JUGV)
Kedougou virus	(KEDV)
Naranjal virus	(NJLV)
Rocio virus	(ROCV)
Sepik virus	(SEPV)
Spondweni virus	(SPOV)
Wesselsbron virus	(WSLV)
Yaounde virus	(YAOV)
Zika virus	(ZIKAV)

3-viruses with no known vector:

Aroa virus	(AROAV)
Cacipacore virus	(CPCV)
Montana myotis leukoencephalitis virus	(MMLV)
Sokoluk virus	(SOKV)
Tamana bat virus	(TABV)

GENUS *PESTIVIRUS*

Type Species bovine diarrhea virus (BDV)

VIRION PROPERTIES

MORPHOLOGY

Virions are 40-60 nm in diameter and spherical in shape (Fig. 3). The virion envelope has 10-12 nm ring-like subunits on its surface. The structure and symmetry of the core have not been characterized.

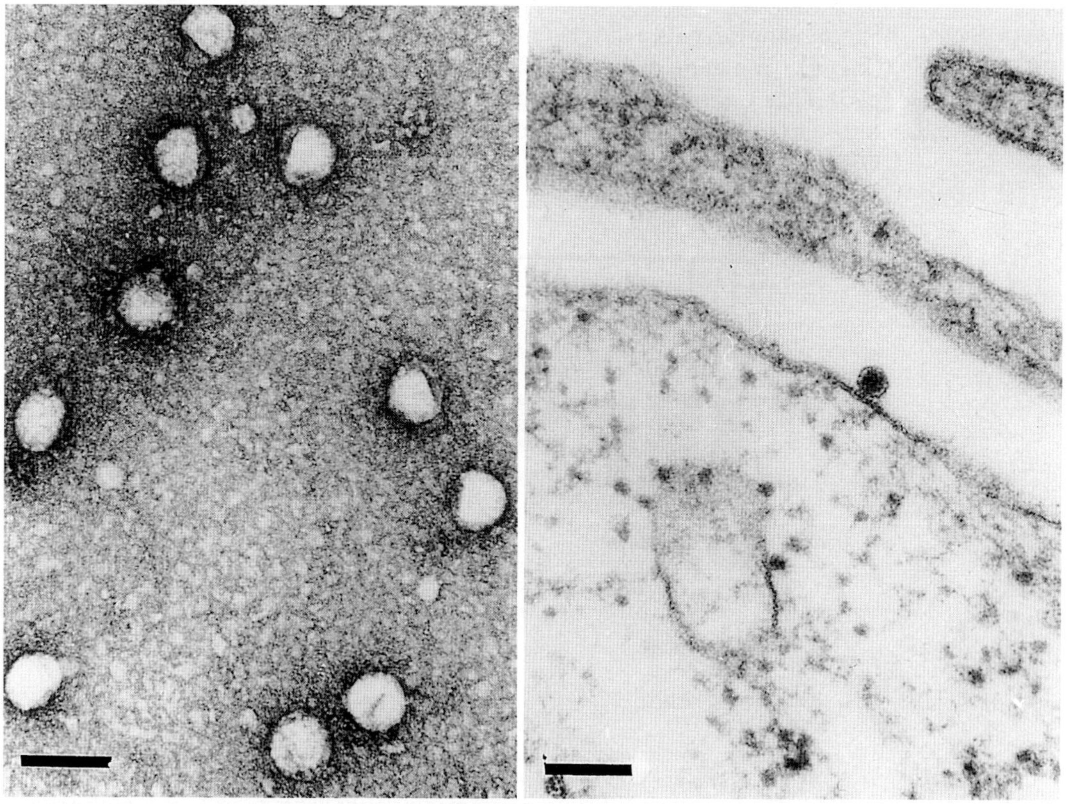

Figure 3: (left) Negative contrast electron micrograph of bovine viral diarrhea virus (BVDV); (right) thin section of BVDV, bars represent 100 nm (courtesy of Weiland F).

Physicochemical and Physical Properties

Virion Mr has not been precisely determined but can be estimated from the virus composition to be about 60 x 10^6. Buoyant density in sucrose is 1.12-1.13 g/cm^3; S_{20W} is 140. Virions are stable at slightly alkaline pH (8.0) and unstable at temperatures above 40° C. Solvents and detergents rapidly inactivate the viruses.

Nucleic Acid

Virions contain one positive sense molecule of ssRNA about 12.5 kb in size. The 5' end has not yet been characterized; no poly (A) tract is present at the 3'-end. For cytopathic biotypes of BVDV, a small and variable segment of host cell nucleic acid may be integrated into one particular region (p54) of the viral genome. This insertion maintains the ORF. Additionally, cytopathic BVDV isolates have been identified in which viral gene duplications involving all or part of the p20 or p80 protein coding regions have occurred, resulting in genomic RNA sizes significantly larger than 12.5 kb.

Proteins

Virion proteins are designated according to the Mr of the proteins of BVDV (NADL strain). However, protein sizes for member viruses vary by up to 25%. Virions are composed of four structural proteins: the nucleocapsid protein, p14, and three envelope glycoproteins, gp48, gp25, and gp53.

Lipids

Although the viruses are enveloped, no reports have described the lipid composition.

Carbohydrates

All virus envelope glycoproteins contain N-linked glycans.

Genome Organization and Replication

Sequencing reveals a single large ORF encoding a polyprotein of about 4,000 amino acids (Fig. 4). The gene order is 5'-p20-p14-gp48-gp25-gp53-p125(p54/p80)-p10-p30-p133(p58/p75)-3', as established by sequence-specific antibody reactivities. All four structural proteins (p14, gp48, gp25, gp53) are encoded within the amino-terminal portion of the large ORF. However, they are preceded by the first polypeptide of the ORF, the non-structural

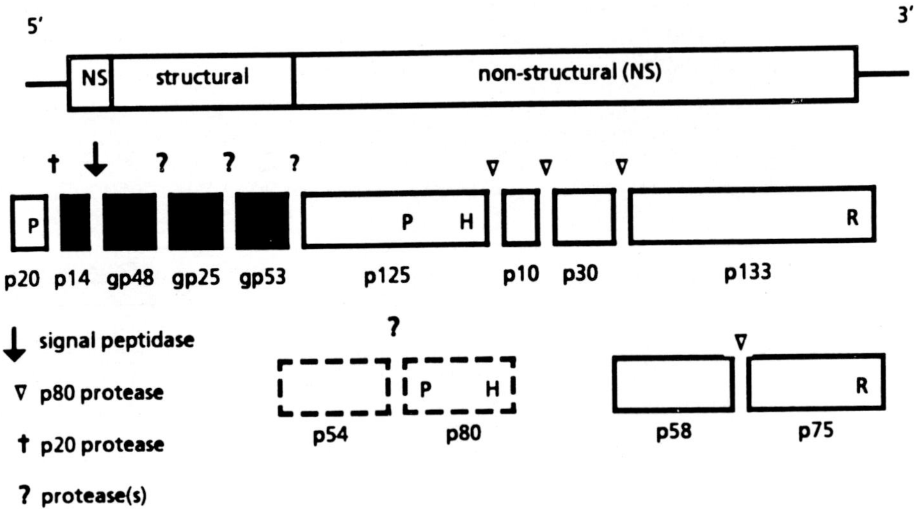

Figure 4: *Pestivirus* genome organization (not to scale). The RNA is about 12.5 kb in size. The 5' non-coding region is about 360-385 nt, the 3' about 230 nt, the ORF about 12 kb (depending on the virus). The proteases (where known) involved in the post-translational modifications are indicated. The structural proteins are shown in black. P, H and R represent the locations of the p80/p125 protease, the predicted p80/p125 RNA helicase and the putative RNA-dependent RNA polymerase, respectively.

protein p20. The p20 protein possesses proteolytic activity. The major nucleocapsid protein is p14. The envelope glycoproteins (gp48, gp25 and gp53 which are targets for virus neutralizing antibodies and are believed to be responsible for virus adsorption, tissue and cell tropisms), form intermolecular disulfide bridges. Following the glycoprotein coding regions are the remaining viral non-structural polypeptides. The p125 non-structural protein has an extremely hydrophobic amino-terminal region (perhaps a membrane-spanning domain) and possesses amino acid sequence motifs indicative of a zinc finger, a serine protease, and an NTPase/RNA helicase (possibly involved in RNA binding and replication). It is believed to be involved in both protein processing and RNA replication. In cytopathic BVDV, but not in non-cytopathic BVDV infected cells, two products encompassing the p125 coding region are observed: p54 and p80. This p54 has a small host cell gene insert (not in non-cytopathic BVDV). The function of p54 is unknown (it may be a membrane protein involved in binding nucleic acids). The p80 protein has been shown to be a serine protease and an RNA-stimulated NTPase with possible roles in RNA replication (RNA helicase) which induces cytopathic effects. No roles for p10 and p30 have been suggested. The p133 protein serves as a precursor for p58 and p75. The p75 protein possesses amino acid sequence motifs characteristic of RNA-dependent RNA polymerases.

Replication occurs in association with intracytoplasmic membranes. Replicative forms of viral RNA have not yet been described. Replication is sensitive to proflavine and acriflavine. No subgenomic mRNAs are found in infected cells. The genomic RNA is translated into a polyprotein that is rapidly processed co-translationally and post-translationally. Translation initiation may occur via ribosome entry at an internal site within the 385 nucleotide 5' non-coding region of the viral RNA. Polyprotein translation from the first AUG of the large ORF leads to the synthesis of the p20 protein which autocatalytically releases itself from the nascent polyprotein. Glycoprotein translocation to the endoplasmic reticulum likely occurs by an internal signal sequence, perhaps within the nucleocapsid protein p14. Glycoprotein processing involves host cell proteases, or signalases. Carboxy-terminal, non-structural protein processing is carried out by the viral p80 serine-type protease or, in the case of non-cytopathic BVDV, is suspected to be carried out by the p125 protein. The p58 and p75 proteins are believed to be components of the RNA-dependent RNA polymerase. Host cell RNA and protein syntheses continue throughout infection.

Antigenic Properties

Monoclonal antibodies reactive with two of the viral envelope glycoproteins, gp48 and gp53, have been obtained that neutralize virus infectivity. Monoclonal antibodies, as well as monospecific antisera, to the non-structural protein p80 fail to neutralize virus. Infected animals mount potent antibody responses to the three viral structural glycoproteins and to the non-structural p80 protein, which likely represents the virus "soluble antigen". Antibody responses to all other virus-encoded polypeptides, including the nucleocapsid protein, p14, are extremely weak or non-existent.

Biological Properties

Host Range

The viruses have limited host ranges (mammals). There are no invertebrate hosts.

Transmission

Transmission occurs by direct and indirect contact (e.g., fecally contaminated food, urine, or nasal secretions, etc.). Transplacental and congenital transmission occur in all target species.

Pathogenicity

Infection with pestiviruses produce inapparent infections, acute or persistent subclinical infections, acute fatal disease (mucosal disease), fetal death or congenital abnormalities, and a wasting disease. In mucosal disease, two natural virus biotypes (cytopathic and non-cytopathic) may collaborate to induce a fatal disease. Pestivirus infections of livestock are economically important worldwide.

EXPERIMENTAL HOSTS

No experimental infection models have been established outside the natural mammalian hosts.

CELL CULTURES

Only cells derived from natural host species (bovine, porcine, ovine) support virus replication. Most virus isolates do not produce cytopathic effects. Many cause persistent infections in cell culture. For BVDV, cytopathic viruses are routinely identified, capable of plaque formation and extensive cytopathology.

HEMAGGLUTINATION

No hemagglutinating activity has been found associated with pestiviruses.

LIST OF SPECIES IN THE GENUS

The viruses, their genomic sequence accession numbers [ ] and assigned abbreviations ( ) are:

SPECIES IN THE GENUS

border disease virus (sheep)		(BDV)
bovine diarrhea virus	[M31182]	(BDV)
hog cholera virus	[M31768, J04358]	(HCV)

TENTATIVE SPECIES IN THE GENUS

None reported.

GENUS "HEPATITIS C - LIKE VIRUSES"

Type Species hepatitis C virus (HCV)

VIRION PROPERTIES

MORPHOLOGY

Hepatitis C virus has not been visualized by electron microscopy. Virion diameter is estimated to be about 40-50 nm by filtration (filtrate assays by chimpanzee titration). Virions are enveloped (inferred from chloroform sensitivity).

NUCLEIC ACID

Virions contain one positive sense molecule of ssRNA about 9.4 kb in size. The genome has a 5' untranslated end (341 nt) and a 3'-end untranslated region (about 50 nt). The ORF encodes a polyprotein of about 3,000 amino acids. The majority of isolates lack a 3' poly (A) tail, although A-rich regions exist.

PHYSICOCHEMICAL AND PHYSICAL PROPERTIES

Virion Mr has not been determined. Buoyant density in sucrose is 1.09-1.11 g/cm^3. The S_{20W} is greater than or equal to 150. The virus is stable in buffer at pH 8.0-8.7. Organic solvents and detergents rapidly inactivate the virus.

Proteins

The nature of structural proteins has not been established by conventional biochemical methods.

Lipids

Lipids have not been demonstrated directly; on the basis of solvent sensitivity, it is presumed that virions are enveloped.

Carbohydrates

Carbohydrates have not been demonstrated directly. The presence of carbohydrates is inferred on the basis that virions are probably enveloped and probably contain glycoproteins.

Genome Organization and Replication

From cDNA analyses the HCV genome appears to be organized in a fashion similar to that of flaviviruses and pestiviruses (Fig. 5). The 5' end of the genome encodes the putative structural proteins. Sequence analysis indicates that the rest of the genome probably includes a non-structural viral protease, an helicase (NS3) and an RNA-dependent RNA polymerase (NS5). Hydrophobicity plots of the indicated HCV polyprotein show similar spacing of the putative hydrophobic NS2 and NS4 regions to those found in both the pestiviruses and flaviviruses.

The indicated protein order of the structural proteins is 5': p22-gp33-gp70. The highly basic p22 is thought to be the virion core (C) protein with gp33 and gp70 being envelope proteins E1 and E2. However, the possibility that E2 may be equivalent to the flavivirus non-structural NS1 protein has not been ruled out. The gp33 and gp70 proteins have been shown to be glycosylated using an *in vitro* translation system employing transcribed RNA and each can be deglycosylated by treatment with endoglycosidase H to yield proteins of 21 kDa and 38 kDa, respectively. The region encompassing these two proteins contains 15 potential N-linked glycosylation sites. The conservation and locations of sequence motifs representing serine proteases (amino-terminal segment of NS3), helicases (carboxy-terminal segment of NS3), and RNA-dependent RNA polymerases (NS5) are the same as those of the pestiviruses and flaviviruses, which suggests a similar genome organization. By analogy to members of those genera, the putative HCV non-structural proteins also include NS2 (33 kDa), NS4 (50 kDa), and NS5 (116 kDa).

No information is available on the replication strategy or evidence of RNA intermediates. dsRNA has been detected in both infected liver tissue and serum. Subgenomic RNAs of defined length have not been reported. It is believed that the large ORF is translated into

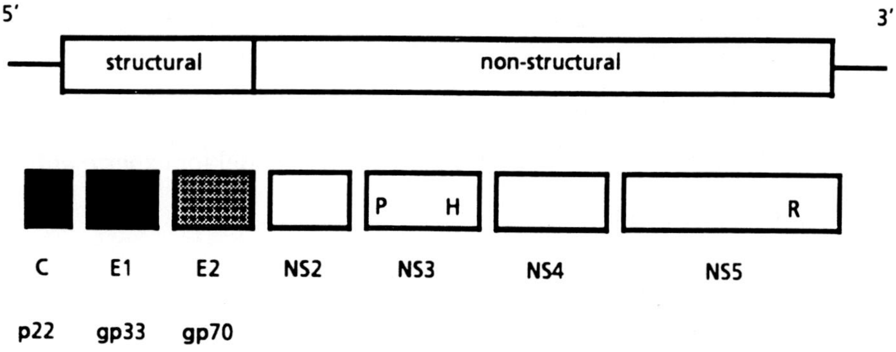

Figure 5: Proposed organization of the HCV genome (not to scale). The total RNA contains about 9.5 kb. The proteases involved in post-translational processing have not been defined. The P, H, and R symbols indicate the locations of the predicted protease, helicase and RNA replicase, respectively. The indicated structural proteins are in black. E2 may be non-structural and equivalent to the NS1 protein of flaviviruses.

One long polyprotein which is processed by a combination of cellular and viral-encoded proteases to yield the mature viral proteins. The 5'-untranslated region possesses structural similarities to those of picornaviruses, including multiple AUG triplets upstream of that which initiates the ORF, and the presence of a consensus sequence indicative of an internal ribosome binding site.

Immunofluorescence data indicates that the viral proteins are accumulated within the cytoplasm of infected cells with NS3/NS4 being the main components. *In situ* hybridization to viral RNA has demonstrated that the cytoplasm is also the site of viral replication. Extensive nucleotide sequence variation exists amongst HCV isolates with the 5' untranslated region and capsid coding sequences being most conserved and E1, E2 the least conserved. Such data have prompted proposals that HCV is a group of at least 4 related genotypes. However, serotypic differences have not been well documented.

ANTIGENIC PROPERTIES

Recombinant-expressed core, NS3, and NS4 proteins have been used successfully to detect virus-specific antibodies in individuals infected with HCV. Amino acids sequence comparisons between numerous HCV isolates have revealed the existence of a variable region within E1 and a hypervariable region within the amino-terminal portion of E2. Such data indicate that these regions may be subject to host immune selection. Assays utilizing recombinant-expressed E1 and E2 have been developed. The role of conformational determinants in the structural proteins in relation to immune responses is unknown. No neutralizing antibodies have been described.

BIOLOGICAL PROPERTIES

HOST RANGE

Humans are the natural host and apparent reservoir of HCV. No other natural host and no invertebrate vectors, have been identified.

TRANSMISSION

Risk factors for acquiring HCV have been largely, but not completely, identified. Approximately 60% of all disease caused by HCV occurs as a result of parenteral exposure (blood contact). In the United States, serologic studies of blood donors for virus-specific antibody suggests that about 0.5-1.5% may be infected with HCV. Epidemiological studies indicate that about 30% of all acute hepatitis in the United States is caused by HCV.

PATHOGENICITY

Virus infections range from inapparent, subclinical infections to fulminant disease, resulting in hepatic failure and death. Persistent infection occurs in about 60-70% of HCV infected individuals. Of these about 20% develop chronic, active hepatitis and/or cirrhosis. Persistent HCV infection has been serologically linked to primary liver cancer, cryptogenic cirrhosis, and some forms of auto-immune disease.

EXPERIMENTAL HOSTS

The chimpanzee remains the only proven model for experimental HCV infection.

CELL CULTURE

The virus has proved difficult to culture *in vitro*.

LIST OF SPECIES IN THE GENUS

The viruses, their genomic sequence accession numbers [ ] and assigned abbreviations ( ) are:

Species in the Genus

hepatitis C virus	[M62321, M58406, M58407, D90208, M58335]	(HCV)

Tentative Species in the Genus

None reported.

Unassigned Viruses in the Family

None reported.

Similarity with Other Taxa

None reported.

Derivation of Names

flavi: from Latin *flavus*, "yellow"
pesti: from Latin *pestis*, "plague"
hepat: from Greek *hepar*, *hepatos*, "liver"

References

Alter HJ, Margolis HS, Krawzcynski K, Judson FN, Mares A, Alexander WJ, Hu PY, Miller JK, Gerber MA, Samplinar RE, Meeks EL, Beach MJ (1992) The natural history of community-acquired hepatitis C in the United States. N Engl J Med 327: 1899-1905

Calisher CH, Karabatsos N, Dalrymple JM, Shope RE, Porterfield JS, Westaway EG, Brandt WE (1989) Antigenic relationships between flaviviruses as determined by cross-neutralization tests with polyclonal antisera. J Gen Virol 70: 37-43

Chamberlain RW (1980) Epidemiology of arthropod-borne viruses: the role of arthropods as hosts and vectors and of vertebrate hosts in natural transmission cycles. In: Schlesinger RW (ed) The Togaviruses. Academic Press, New York, pp 175-227

Chambers TJ, Hahn CS, Galler R, Rice CM (1990) Flavivirus genome organization, expression, and replication. Ann Rev Microbiol 44: 649-688

Choo QL, Richman KH, Han JH, Berger K, Lee C, Dong C, Gallegos C, Coit D, Medina-Selby A, Barr PJ, Weiner AJ, Bradley DW, Kuo G, Houghton M (1991) Genetic organization and diversity of the hepatitis C virus. Proc Natl Acad Sci USA 88: 2451-2455

Collett MS, Moennig V, Horzinek MC (1989) Recent advances in pestivirus research. J Gen Virol 70: 253-266

Collett MS (1992) Molecular genetics of pestiviruses. Comp Immunol Micro infect Dis 15: 145-154

Donis RO, Dubovi EJ (1987) Molecular specificity of the antibody responses of cattle naturally and experimentally infected with cytopathic and noncytophatic bovine viral diarrhea virus biotypes. Am J Vet Res 48: 1549-1554

Karabatsos N (ed) (1985) International catalogue of arboviruses 3rd edn. Amer Soc Trop Med Hyg, San Antonio

Mandl CW, Guirakhoo F, Holzmann H, Heinz FX, Kunz C (1989) Antigenic structure of the flavivirus envelope protein E at the molecular level, using tick-borne encephalitis virus as a model. J Virol 63: 564-571

Meyers G, Tautz N, Dubovi EJ, Thiel HJ (1991) Viral cytopathogenicity correlated with integration of ubiquitin-coding sequences. Virology 180: 602-616

Monath TP (ed) (1988) The Arboviruses: Epidemiology and Ecology Vol 5. CRC Press, Boca Raton FL

Murphy FA (1980) Togavirus morphology and morphogenesis. In: Schlesinger RW (ed) The Togaviruses. Academic Press, New York, pp 241-316

Okamoto H, Kurai K, Okada SI, Yamamoto K, Lizuka H, Tanaka T, Fukada S, Tsuda F, Mishiro S (1992) Full length sequence of a hepatitis C virus genome having poor homology to reported isolates: comparative study of four distinct genotypes. Virology 188: 331-341

Plagemann PGW (1991) Hepatitis C virus. Arch Virol 120: 165-180

Schlesinger S, Schlesinger MJ (eds) (1986). The *Togaviridae* and *Flaviviridae*. Plenum Press, New York

Strauss JH (ed) (1990) Viral proteinases. Sem Virol 1: 307-384

Thiel HJ, Stark R, Weiland E, Rumenapf T, Meyers G (1991) Hog cholera virus: molecular composition of virions from a pestivirus. J Virol 65: 4705-4712

Contributed By

Wengler G, Bradley DW, Collett MS, Heinz FX, Schlesinger RW, Strauss JH

Family *Togaviridae*

Taxonomic Structure of The Family

Family	***Togaviridae***
Genus	*Alphavirus*
Genus	*Rubivirus*

Virion Properties

Morphology

Virions are 70 nm in diameter, spherical, with a lipid envelope containing glycoprotein spikes composed of two virus glycoproteins forming heterodimers. At least for alphaviruses, the heterodimers are organized in a T=4 icosahedral lattice consisting of 80 trimers (Fig. 1). The envelope is tightly organized around an icosahedral nucleocapsid that is 40 nm in diameter. The nucleocapsid is composed of the capsid protein, organized in a T=4 icosahedral symmetry, and the viral RNA. The one-to-one relation between glycoprotein heterodimers and nucleocapsid proteins is believed to be important in virus assembly. Virions of rubella virus are pleomorphic.

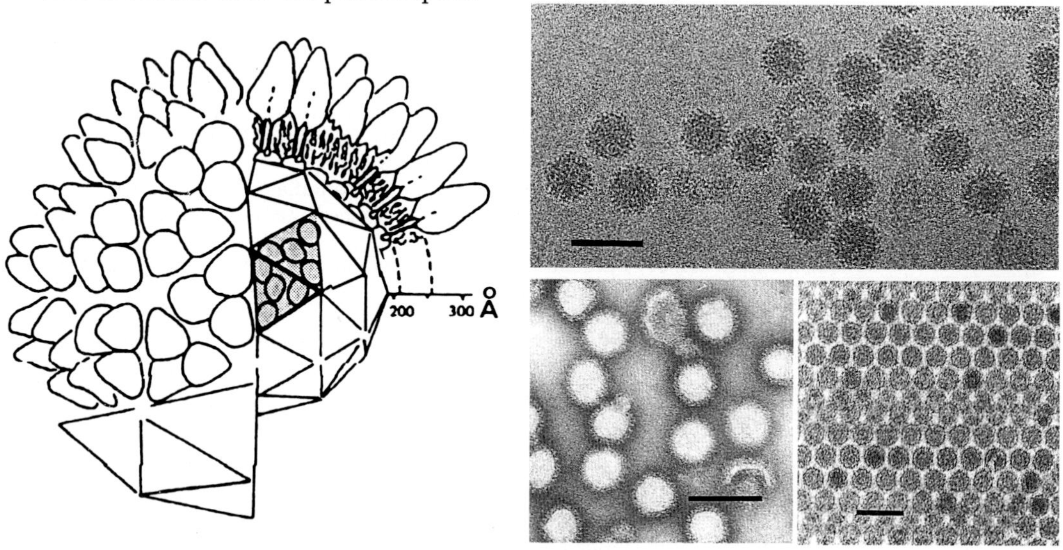

Figure 1: (left) Diagrammatic representation of Sindbis virus. On the left, the exterior of the particle is shown, on the right, the nucleocapsid is revealed. The knobs on the surface represent the external portions of the E1 + E2 heterodimers. The heterodimers associate to form trimers. The 240 heterodimers and 240 copies of the Sindbis capsid protein are arranged in an icosahedral lattice with a T=4 symmetry (modified from Harrison, 1990); (right) upper panel: cryoelectron micrograph of Sindbis viruses (courtesy of Prasad BVV); lower right: negative contrast electron micrograph of Semliki Forest virus (SFV) (courtesy of von Bonsdorff C-H); lower left: thin section of pelleted SFV (courtesy of von Bonsdorff C-H), the bars represent 100 nm.

Physicochemical and Physical Properties

Virion Mr is about 52 x 10^6. Alphaviruses have a buoyant density in sucrose of 1.22 g/cm^3 and an S_{20w} of 280. Rubella has a buoyant density of 1.18-1.19 g/cm^3 and a similar S value. Alphaviruses are stable between pH 7 and 8, but are rapidly inactivated by acidic pH. Virions have a half-life at 37° C of about 7 hrs. in culture medium. Most alphaviruses are rapidly inactivated at 58° C with a half-life measured in minutes. Rubella virions are less stable than alphaviruses, with a half-life at 37° C of 1 to 2 hr. and a half-life at 58° C of 5-20 min. Generally, the viruses are sensitive to organic solvents and detergents which solubilize their lipoprotein envelopes. Sensitivity to irradiation is proportional to the size of the viral genome.

NUCLEIC ACID

The genome consists of a linear, positive sense, ssRNA molecule 9.7-11.8 kb in size. The viral RNA is capped at the 5' terminus and polyadenylated at the 3' end.

PROTEINS

The structural proteins of togaviruses consist of a basic capsid protein (C, Mr 30-33 x 10^3) and two envelope glycoproteins (E1 and E2, Mr 45-58 x 10^3). Some alphaviruses may have a third envelope protein, E3 (Mr 10 x 10^3).

LIPIDS

Lipids comprise about 30% of the dry weight of virions. They are derived from the host-cell plasma membrane. Their composition depends upon the cells in which the virus was grown. Phospholipids (including phosphatidyl ethanolamine, phosphatidyl choline, phosphatidyl serine, and sphingomyelin) and cholesterol are present in a molar ratio of about 2:1 for alphaviruses, 4:1 for rubella, presumably because the latter matures primarily at intracellular membranes.

CARBOHYDRATES

Both high mannose and complex N-linked glycans are found on the envelope glycoproteins. In addition, rubella virus E2 protein contains O-linked glycans.

GENOME ORGANIZATION AND REPLICATION

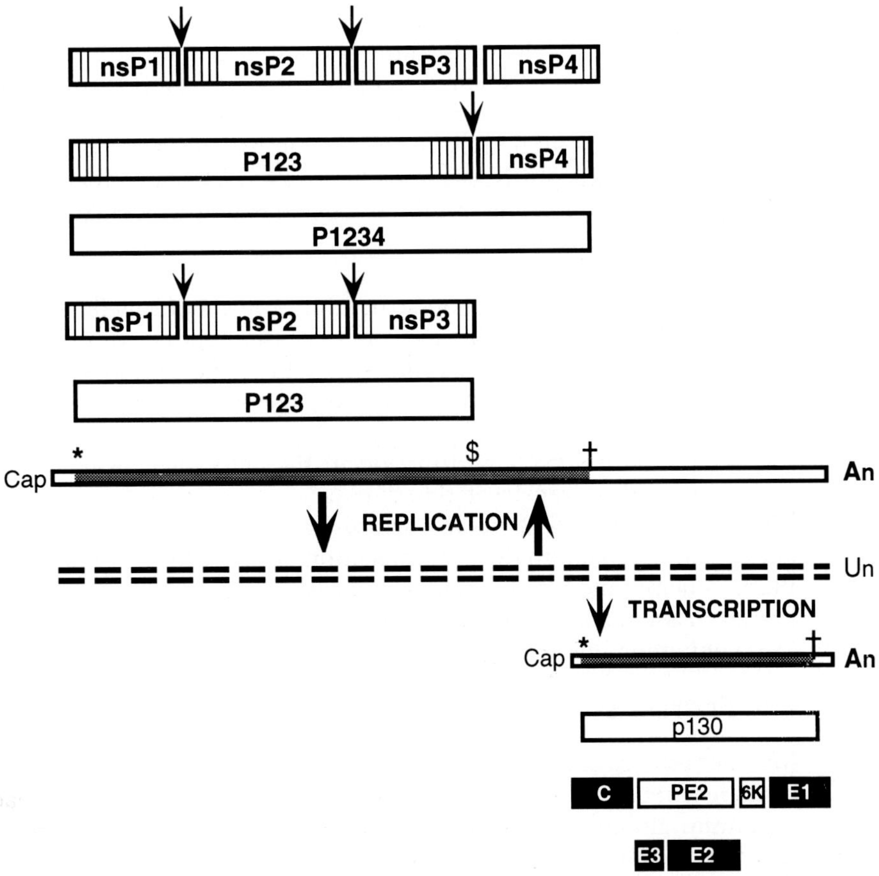

Figure 2: Genome organization, translation, transcription and replication strategies of Sindbis alphavirus (nearly to scale). The regions of the 11.7 kb 49S viral RNA and 26S subgenomic mRNA (lines) that code for the non-structural (striped boxes) and structural proteins (black boxes) are shown. Replication and transcription are indicated by thick arrows. The dashed line is the replicative intermediate that is also the template for 26S mRNA. E3 is a structural protein in some alphaviruses (not present in rubella). Initiation codons are indicated by *, termination codons by † and $ (the latter is read-through to produce P1234, hence nsP4). Thin arrows represent

nsP2 protease activity; see text for proteases that cleave the structural proteins. (This diagram is adapted from Strauss and Strauss, 1983).

The genomic RNA serves as the mRNA for the non-structural proteins of the virus. The polyprotein precursor is cleaved by a viral-encoded proteinase in nsP2 to produce four final products, nsP1, nsP2, nsP3 and nsP4. In four of six alphaviruses sequenced, there is a termination codon (UGA) between nsP3 and nsP4 which is read-through with moderate efficiency (20%), whereas in the two other alphaviruses this codon has been replaced by a codon for arginine (CGA). Polyproteins containing nsP2 are enzymes and function primarily in *trans* to produce the cleaved non-structural proteins.

The nonstructural proteins, as individual entities and as polyproteins, are required to replicate viral RNA and probably act in association with cellular proteins. The alphavirus nsP1 protein is thought to be involved in capping of viral RNAs and in initiation of negative-strand RNA synthesis. The nsP2 functions as a protease to process the non-structural proteins and is believed to be a helicase required for RNA replication. Protein nsP4 is believed to be the viral RNA polymerase. Protein nsP3 is also required for RNA replication; P123 and nsP4 form the replicase complex for minus strand synthesis whereas efficient plus-strand synthesis requires cleavage of P123. For replication, a negative-strand copy is produced that is used as template in the synthesis of both genome-sized RNA as well as a subgenomic 26S mRNA that corresponds to the 3' third of the viral genome and encodes the viral structural proteins. This mRNA is capped and polyadenylated. It is translated as a polyprotein which is processed in alphaviruses by a combination of an autoprotease activity present in the capsid protein and cellular organelle-bound proteases to produce the viral structural proteins.

Cis-acting regulatory elements in the 5' non-translated region and in the 3' non-translated region of the genomic RNA are required to produce alphavirus minus strands and to copy the minus strand into plus strands. There are believed to be other cis-acting regulatory elements within the viral RNA as well. For alphaviruses, the promoter for the production of the 26S subgenomic mRNA is a stretch of 24 nucleotides that span the start point of the subgenomic mRNA. This minimal 24 nucleotide sequence element is upregulated by upstream sequences.

Details of the processing of non-structural proteins of rubella are not known. The rubella polyprotein has motifs indicative of replicase, helicase, and protease functions that are shared with alphaviruses, as well as a motif found in alphavirus nsP3. However, these motifs are present in a different order to those present in the alphavirus genome.

The non-structural proteins function in the cytoplasm of infected cells, although some alphavirus nsP2 is translocated to the nucleus. The capsid protein assembles with the viral RNA to form the viral nucleocapsids in the cytosol. Glycoproteins inserted into the endoplasmic reticulum during translation are translocated via the Golgi apparatus to the plasma membrane for alphaviruses; for rubella they are also found at intracellular membranes. Assembled nucleocapsids bud through these membranes acquiring a lipid envelope containing the two integral membrane glycoproteins.

ANTIGENIC PROPERTIES

Member viruses of the genus *Alphavirus* were originally defined on the basis of serological cross-reactions. Thus, all alphaviruses are antigenically related to each other. They share a minimum amino acid sequence identity of about 40% in the more divergent structural proteins and about 60% in the non-structural proteins. Rubella virus is serologically distinct from alphaviruses and no amino acid sequence similarity can be detected between the structural proteins of rubella virus and those of alphaviruses.

BIOLOGICAL PROPERTIES

Alphaviruses are transmitted between vertebrates by mosquitoes and certain other hematophagous arthropods. Alphaviruses have a wide host range and worldwide distribution. The infection of cells of vertebrate origin by alphaviruses is cytolytic and involves the shutdown of host-cell macromolecular synthesis. In mosquito cells, alphaviruses usually establish a non-cytolytic infection in which the cells survive and become persistently infected. The assembly of virions in mosquito cells appears to differ from that for vertebrate cells in that most, perhaps all, virus assembly occurs in association with intracellular membranes rather than by budding through the plasma membrane. The details may differ in different types of cells. In contrast, humans are the only known host for rubella virus.

GENUS *ALPHAVIRUS*

Type Species Sindbis virus (SINV)

DISTINGUISHING FEATURES

Genomes are 11-12 kb in size (exclusive of the 3' terminal poly (A) tract: SINV, 11,703 nt; ONNV, 11,835 nt; RRV, 11,851 nt; VEEV, 11,444 nt; SFV 11,442 nt; S_{20w} about 49). The order of the genes for the non-structural proteins in the genomic RNA is (Fig. 2) nsP1, nsP2, nsP3, nsP4. These are made as polyprotein precursors and processed by the nsP2 protease (Fig. 2). The gene order in the 26S mRNA is C-E3-E2-6K-E1. The derived polyprotein is processed by an auto-proteolytic activity in the capsid protein, by cellular signal peptidase, and by an enzyme thought to be a component of the Golgi apparatus (Fig. 2). Glycoprotein E2 is produced as a precursor, PE2 (otherwise called p62), that is cleaved during virus maturation. For some viruses the N-terminal cleavage product of PE2, referred to as E3 (about 10 kDa), remains associated with the virion. Carbohydrates comprise about 14% of the mass of the envelope glycoproteins and about 5% of the mass of the alphavirus virion.

Alphaviruses possess the ability to replicate and pass horizontally in mosquitoes, or transovarially in certain vectors. Each virus usually has a preferred mosquito vector, however as a group the viruses use a wide range of mosquitoes. Isolation of SINV from a mite has also been reported. FMV is transmitted by arthropods of the family *Cimicidae* (Hemiptera-Heteroptera) associated with house sparrows. Most alphaviruses can infect a wide range of vertebrates. Many alphaviruses have different species of birds as their primary vertebrate reservoir host, but most are able to replicate in mammals as well. A number of alphaviruses have various mammals as their primary vertebrate reservoir host. Some of these, such as RRV, replicate poorly in birds. Alphavirus isolations from reptiles and amphibians have been reported. As group, the viruses are found on all continents except Antarctica and on many islands. However, most viruses have a more limited distribution. SINV, the type virus, has been isolated from many regions of Europe, Africa, Asia, the Philippines and Australasia. WEEV is distributed discontinuously from Canada to Argentina. At the other extreme, ONNV has been isolated only from East Africa where it caused an epidemic in the years 1959-60 and subsequently disappeared. Many Old World alphaviruses cause serious, but not life threatening illnesses that are characterized by fever, rash and a painful arthralgia. RRV, MAYV, and the Ockelbo strain of SINV cause epidemic polyarthritis in humans with symptoms (in a minority of cases) that may persist for months, or years. The New World alphaviruses, EEEV and WEEV, regularly cause fatal encephalitis in humans, although the fraction of infections that lead to clinical disease is small. These viruses, together with VEEV, cause encephalitis in horses and are serious veterinary as well as human pathogens.

LIST OF SPECIES IN THE GENUS

The viruses, their alternative names ( ), genomic sequence accession numbers [ ] and assigned abbreviations ( ) are:

SPECIES IN THE GENUS

Aura virus		(AURAV)
Babanki virus		(BBKV)
Barmah Forest virus		(BFV)
bebaru virus		(BEBV)
Buggy Creek virus		
chikungunya virus		(CHIKV)
Eastern equine encephalitis virus	[D00145]	(EEEV)
Everglades virus		(EVEV)
Fort Morgan virus		(FMV)
getah virus		(GETV)
Highlands J virus	[J02206]	(HJV)
Kyzylagach virus		(KYZV)
Mayaro virus		(MAYV)
Middelburg virus	[J02246]	(MIDV)
Mucambo virus		(MUCV)
Ndumu virus		(NDUV)
Ockelbo virus	[M69205]	(OCKV)
o'nyong-nyong virus		(ONNV)
Pixuna virus		(PIXV)
Ross River virus	[M20162]	(RRV)
Sagiyama virus		(SAGV)
Semliki Forest virus	[X04129]	(SFV)
Sindbis virus	[V00073]	(SINV)
Una virus		(UNAV)
Venezuelan equine encephalitis virus	[X04368]	(VEEV)
Western equine encephalitis virus	[J03854]	(WEEV)
Whataroa virus		(WHAV)

TENTATIVE SPECIES IN THE GENUS

None reported.

GENUS *RUBIVIRUS*

Type Species rubella virus (RUBV)

DISTINGUISHING FEATURES

The genome is 9,757 nt in size exclusive of the 3' terminal poly (A) tract. The virus has a capsid protein (33 kDa) and two envelope glycoproteins (E1, 58 kDa; E2, 44.5 kDa), but no equivalent of E3 or the 6K protein of the alphaviruses. The order of the RUBV proteins in the polyprotein precursor of the structural proteins is C-E2-E1-COOH. The two cleavages that separate these three structural proteins are effected by signal peptidase. The nsP2 and nsP4 motifs of RUBV are similar to those of alphaviruses. Carbohydrates make up 10% of the mass of E1 and 30-40% of the mass of E2. E2 is heterogeneous in size due to differential processing of glycans (N- and O-linked). RUBV is transmitted primarily as an aerosol but congenital transmission can occur. It causes a trivial illness under normal circumstances but is teratogenic and often leads to fetal abnormalities when infection occurs in the first trimester of pregnancy.

LIST OF SPECIES IN THE GENUS

The viruses, their genomic sequence accession numbers [ ] and assigned abbreviations ( ) are:

rubella virus	[M15240, M18901, M32735]	(RUBV)

Tentative Species in the Genus

None reported.

List of Unassigned Viruses in the Family

None reported.

Similarity with Other Taxa

Alphavirus nonstructural proteins nsP1, nsP2, and nsP4 share some sequence homology with the nonstructural proteins of several groups of plant viruses, including tobamoviruses, bromoviruses and tobraviruses, suggesting a common origin for the replicases of these viruses.

Derivation of Names

toga: from Latin *toga* "cloak"
alpha: from Greek letter α
rubi: from Latin *rubeus* "reddish"

References

Baron MD, Ebel T, Suomalainen M (1992) Intracellular transport of rubella virus structural proteins expressed from cloned cDNA. J Gen Virol 73: 1073-1086

Choi H-K, Tong L, Minor W, Dumas P, Boege U, Rossmann MG, Wengler G (1991) Structure of Sindbis virus core protein reveals a chymotryspin-like serine proteinase and the organization of the virion. Nature 354: 37-43

Dominguez GD, Wang C-Y, Frey TK (1990) Sequence of the genome RNA of rubella virus: evidence for genetic rearrangement during togavirus evolution. Virology 177: 225-238

Lemm AJ, Rumenapt T, Strauss EG, Strauss JH, Rice CM (1994) Polypeptide requirements for assembly of functional Sindbis virus replication complexes: A model for the temporal regulation of minus and plus-strand RNA synthesis. EMBO J (in press)

Monath TP (ed) (1988) The Arboviruses: epidemiology and ecology, 5 Vols. CRC Press, Boca Raton FL

Schlesinger S, Schlesinger MJ (eds) (1986) The *Togaviridae* and *Flaviviridae*. Plenum Press, New York

Shirako Y, Strauss JH (1994) Regulation of Sindbis virus RNA replication: Uncleaved P123 and nsP4 function in minus strand RNA synthesis whereas cleaved products from P123 are required for efficient plus-strand synthesis. J Virol 68: 1874-1885

Strauss JH (ed) (1990) Viral proteinases. Sem Virol 1: 307-384

Strauss JH, Strauss EG (1988) Evolution of RNA viruses. Ann Rev Microbiol 42: 657-683

Wolinsky JS (1990) Rubella. In: Fields BN, Knipe DM (eds), Virology 2nd edn. Raven Press, New York, pp 815-838

Contributed By

Strauss JH, Calisher CH, Dalgarno L, the late Dalrymple JM, Frey TK, Pettersson RF, Rice CM, Spaan WJM

Genus *Tobamovirus*

Type Species tobacco mosaic virus (TMV)

Virion Properties

Morphology

Virions are elongated rigid cylinders, about 18 nm in diameter and 300 nm long, with helical symmetry (pitch 2.3 nm).

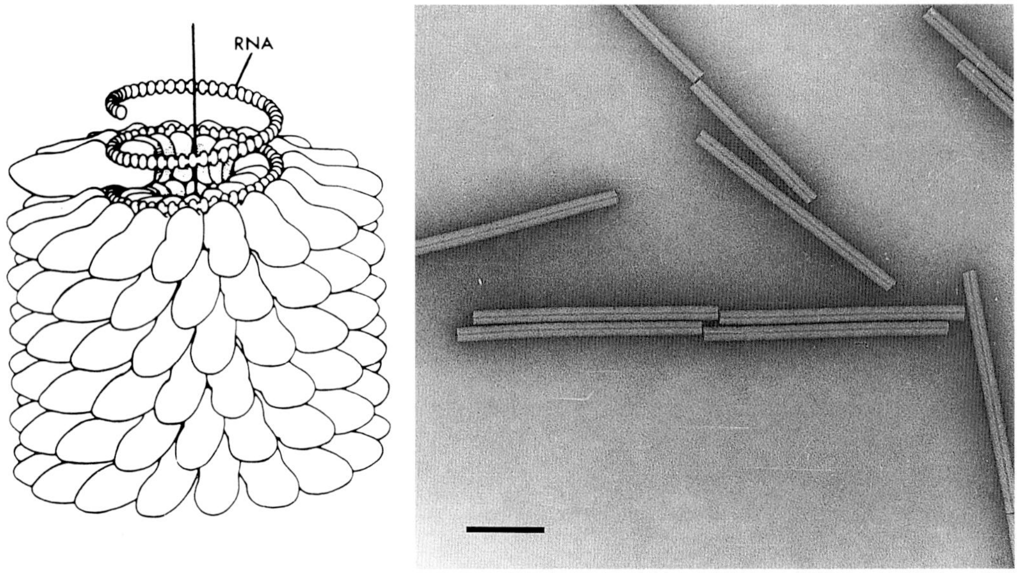

Figure 1: (left) Schematic diagram of TMV particle showing about one-twentieth of its total length. (right) Negative contrast electron micrograph of TMV particles stained with uranyl formate, (Courtesy of Dr. Finch JT). The bar represents 100 nm.

Physicochemical and Physical Properties

Virion Mr is 40 x 10^6. Buoyant density in CsCl is 1.325 g/cm^3. S_{20w} is 194. Virions are very stable.

Nucleic Acid

Virions contain a single molecule of positive sense linear ssRNA; 6.4 kb in size. Virion RNA has a Mr of approximately 2 x 10^6. Subgenomic RNAs having the origin for assembly are found in virions. A cap structure is found at the 5' terminus, followed by an approximately 70 nt long 5' non-translated sequence, containing many AAC repeats and few or no G residues. The 0.2 kb non-translated region at the 3' terminus can be folded into a tRNA-like, amino-acid-accepting structure, with consecutive pseudoknot structures.

Proteins

Virions contain one coat polypeptide, with an Mr of 17-18 x 10^3 (CP). There are three nonstructural proteins with Mr of 126-129 x 10^3, 183-187 x 10^3 and 28-31 x 10^3 respectively. The 183-187 x 10^3 kDa polypeptide is produced by readthrough of the termination codon of the gene V coding for the 126-129 kDa polypeptide. These proteins are involved in replication (replicase or its components), are found in cytoplasm and show sequence similarity with replicative proteins of alpha-like supergroup RNA viruses. The N- and C-terminal halves of the 126-129 kDa polypeptide show similarity to methyltransferase/ guanylyl transferase and RNA helicase (including an NTP-binding motif), respectively. The C-terminal one-third of the 183-187 kDa polypeptide has a motif common to RNA-dependent RNA polymerases. The 28-31 kDa polypeptide (movement protein, MP), the least conserved among the encoded proteins, is involved in cell-to-cell movement. It is found in plasmodesmata and can bind *in vitro* single stranded nucleic acids.

GENOME ORGANIZATION AND REPLICATION

The genome encodes at least 4 proteins with Mr of: 126 x 10^3, 183 x 10^3 (replicase or its components), 30 x 10^3 (movement protein) and 17 x 10^3 (capsid protein) in the 5' to 3' order. The positive sense genomic RNA is copied into a negative-sense RNA which is used to produce the positive sense genomic and subgenomic RNAs. The 183 kDa polypeptide is synthesized by readthrough of the leaking termination codon of the gene for the 126 kDa polypeptide. These 2 polypeptides are translated from the genomic RNA. Movement and capsid proteins are synthesized from their 3' co-terminal respective mRNAs.

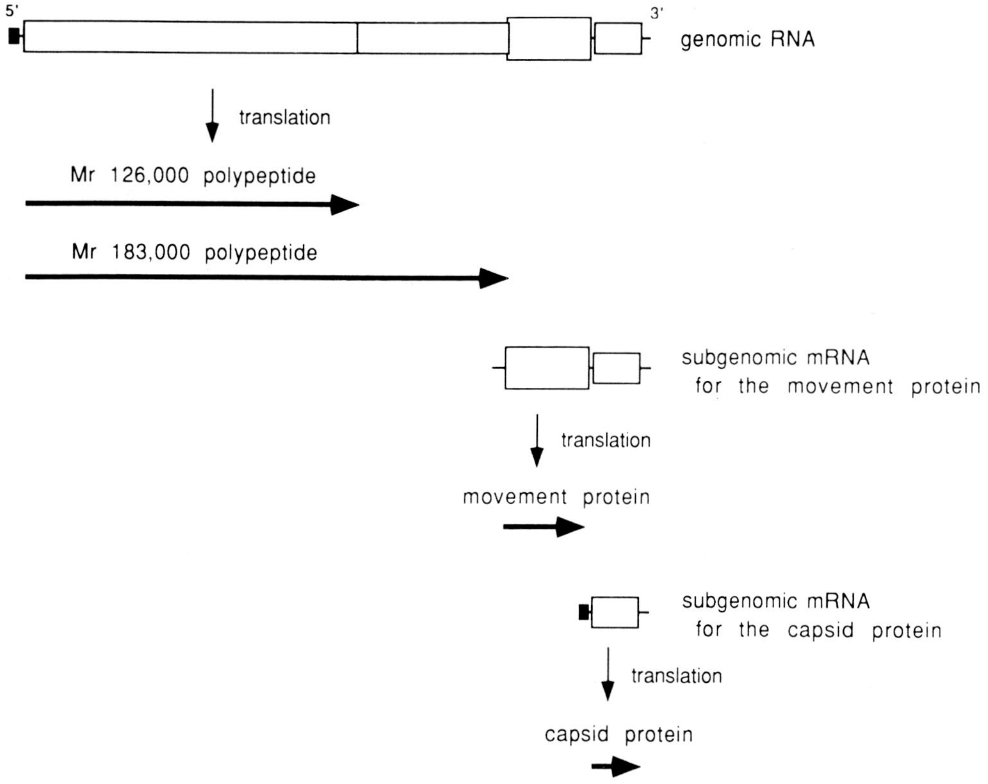

Figure 2: Genome organization and replication strategy of TMV.

ANTIGENIC PROPERTIES

The viruses act as strong immunogens. Different species can be identified by intragel cross-absorption immunodiffusion tests using polyclonal antiserum or by ELISA using monoclonal antibodies. Antigenic distances between individual species expressed as serological differentiation indices are correlated with the degree of sequence difference in their coat proteins.

BIOLOGICAL PROPERTIES

Most species have moderate to wide host ranges; they are transmitted in nature without the help of vectors by contact between plants and sometimes by seed. Geographic distribution is world-wide. The viruses are found in all parts of host plants. Virions often form large crystalline arrays visible by light microscopy.

List of Species in the Genus

The viruses, their genomic sequence accession numbers [ ], CMI/AAB description # ( ) and assigned abbreviations ( ) are:

Species in the Genus

cucumber green mottle mosaic virus (SH strain) (154)		(CGMMV)
	[D12505, D01188]	
frangipani mosaic virus (196)		(FrMV)
kyuri green mottle mosaic virus		(KGMMV)
Odontoglossum ringspot virus (155)		(ORSV)
paprika mild mottle virus		(PaMMV)
pepper mild mottle virus (S strain) (330)		(PMMoV)
	[S76816, M81413]	
ribgrass mosaic virus (152)		(RMV)
Sammons' Opuntia virus		(SOV)
sunn-hemp mosaic virus (153)		(SHMV)
tobacco mild green mosaic virus (U2 strain) (351)		(TMGMV)
	[M34077, M22483]	
tobacco mosaic virus (151)		(TMV)
(vulgare strain; ssp. NC82 strain)	[J02415, X68110]	
tomato mosaic virus (L strain) (156)	[X02144]	(ToMV)
Ullucus mild mottle virus		(UMMV)

Tentative Species in the Genus

Chara corallina virus	(ChaCV)
Maracuja mosaic virus	(MarMV)

Derivation of Names

tobamo: siglum from *toba*cco *mo*saic virus

References

Alonso E, Garcia-Luque I, de la Cruz A, Wicke B, Avila-Roncon MJ, Serra MT, Castresana C, Diaz-Ruiz JR (1991) Nucleotide sequence of the genomic RNA of pepper mild mottle virus, a resistance-breaking tobamovirus in pepper. J Gen Virol 72: 2875-2884

Dekker EL, Dore I, Porta C, van Regenmortel MHV (1987) Conformational specificity of monoclonal antibodies used in the diagnosis of tomato mosaic virus. Arch Virol 94: 191-203

Dawson WO, Lehto KM (1990) Regulation of Tobamovirus Gene Expression. Adv Virus Res 38: 307-342

Dubs MC, van Regenmortel MHV (1990) Odontoglossum ringspot virus coat protein: sequence and antigenic comparisons with other tobamoviruses. Arch Virol 115: 239-249

Fraile A, Garcia-Arenal F (1990) A classification of the tobamoviruses based on comparisons among their 126K proteins. J Gen Virol 71: 2223-2228

Francki RIB, Hu J, Palukaitis P (1986) Taxonomy of cucurbit-infecting tobamoviruses as determined by serological and molecular hybridization analyses. Intervirology 26: 156-163

Garcia-Luque I, Ferrero ML, Rodriguez JM, Alonso E, de la Cruz A, Sanz AI, Vaquero C, Serra MT, Diaz-Ruiz JR (1993) The nucleotide sequence of the coat protein genes and 3' non-coding regions of two resistance-breaking tobamoviruses in pepper shows that they are different viruses. Arch Virol 131: 75-88

Gibbs AJ (1977) Tobamovirus group. CMI/AAB Descriptions of Plant Viruses N° 184, 4pp

Goelet P, Lomonossoff GP, Butler PJG, Akam ME, Gait MJ, Karn J (1982) Nucleotide sequence of tobacco mosaic virus RNA. Proc Natl Acad Sci USA 79: 5818-5822

Hills GJ, Plaskitt KA, Young ND, Dunigan DD, Watts JW, Wilson TMA, Zaitlin M (1987) Immunological localization of the intra-cellular sites of structural and nonstructural tobacco mosaic virus proteins. Virology 161: 488-496

Meshi T, Watanabe Y, Okada Y (1992) Molecular pathology of tobacco mosaic virus revealed by biologically active cDNAs. In: Wilson TMA, Davies JW (eds) Genetic Engineering with Plant Viruses. CRC Press, Boca Raton FL, pp 149-186

Ohno T, Aoyagi M, Yamanashi Y, Saito H, Ikawa S, Meshi T, Okada Y (1984) Nucleotide sequence of the tobacco mosaic virus (tomato strain) genome and comparison with the common strain genome. J Biochem 96: 1915-1923

Solis I, Garcia-Arenal F (1990) The complete nucleotide sequence of the genomic RNA of the tobamovirus tobacco mild green mosaic virus. Virology 177: 553-558

Ugaki M, Tomiyama M, Kakutani T, Hidaka S, Kiguchi T, Nagata R, Sato T, Motoyoshi F, Nishiguchi M (1991) The complete nucleotide sequence of cucumber green mottle mosaic virus (SH strain) genomic RNA. J Gen Virol 72: 1487-1495

van Regenmortel MHV (1975) Antigenic relationships between strains of tobacco mosaic virus. Virology 64: 415-420

van Regenmortel MHV (1981) Tobamoviruses. In: Kurstak E (ed) Handbook of Plant Virus Infection. Elsevier/ North-Holland Publication, Amsterdam, pp 541-564

van Regenmortel MHV, Fraenkel-Conrat H (eds) (1986) The Plant Viruses. The Rod-Shaped Plant Viruses. Plenum Press, New York

Wittmann-Liebold B, Wittmann HG (1967) Coat proteins of strains of two RNA viruses: comparison of their amino acid sequences. Mol Gen Genet 100: 358-363

Contributed By

van Regenmortel MHV, Meshi T

Genus *Tobravirus*

Type Species tobacco rattle virus (TRV)

Virion Properties

Morphology

Virions are tubular with no envelope. They are of two predominant lengths, (L) 180-215 nm and (S) ranging from 46 to 115 nm, depending on the isolate. Many strains produce in addition small amounts of shorter particles. The particle diameter is 21.3-23.1 nm by electron microscopy and 20.5-22.5 nm by X-ray diffraction, and there is a central canal 4-5 nm in diameter. Virions have helical symmetry with a pitch of 2.5 nm; the number of subunits per turn has been variously estimated as 25 or 32.

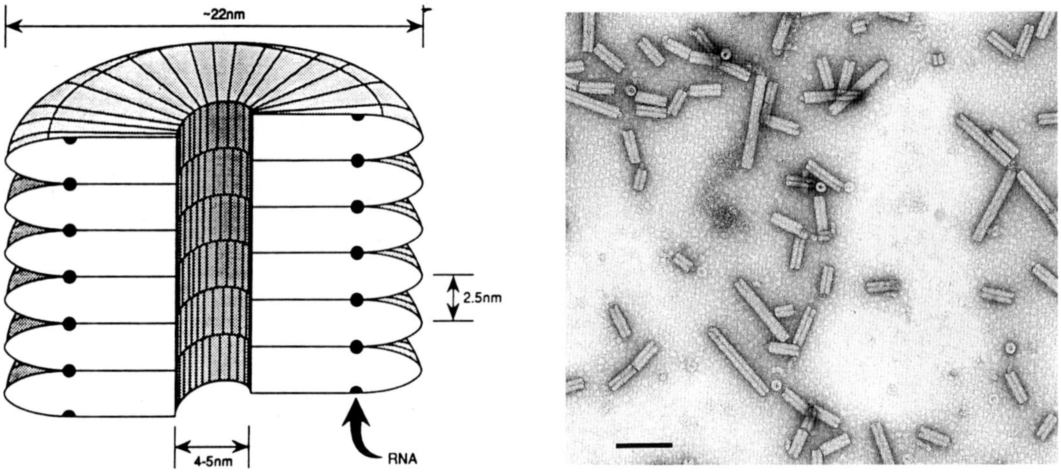

Figure 1: (left) Diagram of TRV virion in section; (right) negative contrast electron micrograph of particles of TRV, the bar represents 100 nm.

Physicochemical and Physical Properties

Virion Mr is 48-50 x 10^6 (L particles) and 11-29 x 10^6 (S particles). Buoyant density in CsCl is 1.306-1.324 g/cm^3. S_{20w} is 286-306 (L particles) and 155-245 (S particles). Virions are stable over a wide range of pH and ionic conditions, and are resistant to many organic solvents, but are sensitive to treatment with EDTA.

Nucleic Acid

The genome consists of two molecules of linear positive sense ssRNA; RNA1 is about 6.8 kb in size and RNA2 ranges from 1.8 kb to about 4.5 kb in size (varying in different isolates). The 5' terminus is capped with the structure $m^7G^{5'}ppp^{5'}Ap$... There is no genome-linked protein or poly (A) tract.

Proteins

Virions contain a single structural protein (Mr 22-24 x 10^3). RNA1 of tobacco rattle virus (TRV) codes for four nonstructural proteins: a 134 kDa protein terminated by an opal stop codon and a 194 kDa protein produced by readthrough of this stop codon, both of which are probably involved in RNA replication; a 29 kDa protein, probably involved in intracellular transport of the virus; and a 16 kDa protein of unknown function. The sizes of the analogous proteins in pea early-browning virus (PEBV) are 141 kDa, 201 kDa, 30 kDa and 12 kDa, respectively. In addition to the virion structural protein, RNA2 of PEBV and of some strains of TRV codes for a nonstructural protein of 29-30 kDa, of unknown function.

LIPIDS

Virions contain no lipids.

CARBOHYDRATES

Virions contain no carbohydrates.

GENOME ORGANIZATION AND REPLICATION

RNA1 is capable of independent replication and systemic spread in plants. The 134/141 kDa and 194/201 kDa proteins are translated directly from it, whereas the 29/30 kDa and 16/12 kDa proteins are translated from subgenomic RNA species 1a and 1b, respectively. RNA2 does not itself have messenger activity; the particle protein is translated from subgenomic RNA2a, and the additional nonstructural protein, when present, from subgenomic RNA2b. There is sequence homology between RNA1 and RNA2 at both ends, but the extent of the homology varies between strains. In some strains, the homologous region at the 3' end is large enough to include some or all of the 16/12 kDa and 29/30 kDa genes of RNA1, but it is not known whether these genes are expressed from RNA2. Accumulation of virus particles is sensitive to cycloheximide but not to chloramphenicol, suggesting that cytoplasmic ribosomes are involved in viral protein synthesis. Virions accumulate in the cytoplasm. L particles of pepper ringspot virus become radially arranged around mitochondria, which are often distorted, and in cells infected with some other isolates, 'X-bodies' largely composed of abnormal mitochondria and containing small aggregates of virus particles may be produced.

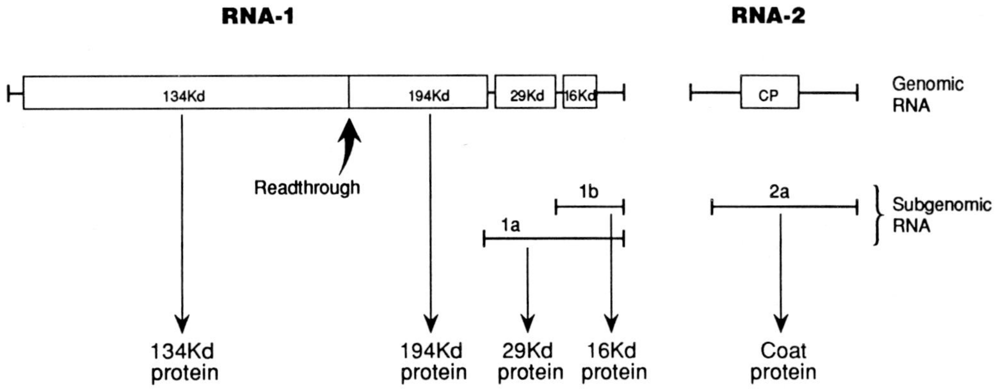

Figure 2: Genome organization and strategy of expression of tobacco rattle virus (TRV).

ANTIGENIC PROPERTIES

Viruses are moderately immunogenic. There is little or no serological relationship between members of the genus, and considerable antigenic heterogeneity among different isolates of the same virus.

BIOLOGICAL PROPERTIES

The host ranges are wide, including members of more than 50 monocotyledonous and dicotyledonous plant families. The natural vectors are nematodes in the genera *Trichodorus* and *Paratrichodorus* (*Trichodoridae*), different species being specific for particular virus strains. Adults and juveniles can transmit, but virus is probably not retained through the moult. Virus can be retained for many months by non-feeding nematodes. Virus particles become attached to the esophageal wall of the nematodes and are thought to be egested with saliva into root cells when the nematodes feed. There is no evidence for multiplication of virus in the vector and it is probably not transmitted through nematode eggs. The viruses are transmitted through seed of at least some host species. Tobacco rattle virus occurs in Europe (including Russia), Japan, New Zealand and North America; pea early-browning virus occurs in Europe and North Africa, and pepper ringspot virus occurs in South

America. Tobacco rattle virus causes diseases in a wide variety of crop plants as well as weeds and other wild plants, including spraing (corky ringspot) and stem mottle in potato, rattle in tobacco, streaky mottle in narcissus and tulip, ringspot in aster, notched leaf in gladiolus, malaria in hyacinth and yellow blotch in sugar beet. Pea early-browning virus is the cause of diseases in several legumes, including broad bean yellow band, distorting mosaic of *Phaseolus* bean and pea early-browning. Pepper ringspot virus causes diseases in artichoke, pepper and tomato.

Most tissues of systemically invaded plants can become infected, but in many species virus remains localized at the initial infection site. In some virus-host combinations, notably tobacco rattle virus in potato, limited systemic invasion occurs, and virus may not be passed on to all the vegetative progeny of infected mother plants.

Normal particle-producing isolates (called M-type) are readily transmitted by inoculation with sap and by nematodes. Other isolates (called NM-type) have only RNA1, do not produce particles, are transmitted with difficulty by inoculation with sap, and are probably not transmitted by nematodes. NM-type isolates are obtained from M-type isolates by using inocula containing only L particles, and are also found in naturally infected plants. They often cause more necrosis in plants than do their parent M-type cultures.

LIST OF SPECIES IN THE GENUS

The viruses, their genomic sequence accession numbers [ ], (CMI/AAB description # ( ) and assigned abbreviations ( ) are:

SPECIES IN THE GENUS

pea early-browning virus (120)	[M90705, X14006, X51828]	(PEBV)
pepper ringspot virus (347)	[L23972, X03241]	(PepRSV)
tobacco rattle virus (12, 346)	[X06172, D00155, X03955, J04347, X03685, X03686]	(TRV)

TENTATIVE SPECIES IN THE GENUS

None reported.

SIMILARITY WITH OTHER TAXA

The 134/141 kDa and 194/201 kDa nonstructural proteins contain conserved sequence motifs common to RNA-dependent RNA polymerases of many viruses, but are most closely related to the analogous proteins of tobacco mosaic virus. The 29/30 kDa protein encoded by RNA1 also shares sequence similarities with the analogous 30 kDa protein of tobacco mosaic virus and, to a lesser extent, with nonstructural proteins of some other plant viruses.

DERIVATION OF NAMES

tobra: sigla from *to*bacco *ra*ttle

REFERENCES

Harrison BD (1973) Pea early-browning virus. CMI/AAB Description of Plant Viruses N° 120, 4pp
Harrison BD, Robinson DJ (1978) The tobraviruses. Adv Virus Res 23: 25-77
Harrison BD, Robinson DJ (1981) Tobraviruses. In: Kurstak E (ed) Handbook of plant virus infections and comparative diagnosis. Elsevier/North-Holland, Amsterdam, pp 515-540
Harrison BD, Robinson DJ (1986) Tobraviruses. In: van Regenmortel MHV, Fraenkel-Conrat H (eds) The plant viruses Vol 2. Plenum Press, New York, pp 339-369
Robinson DJ, Harrison BD (1985) Unequal variation in the two genome parts of tobraviruses and evidence for the existence of three separate viruses. J Gen Virol 66: 171-176
Robinson DJ, Harrison BD (1985) Evidence that broad bean yellow band virus is a new serotype of pea early-browning virus. J Gen Virol 66: 2003-2009
Robinson DJ, Harrison BD (1989) Tobacco rattle virus. CMI/AAB Description of Plant Viruses. N° 346, 6pp
Robinson DJ, Harrison BD (1989) Pepper ringspot virus. CMI/AAB Description of Plant Viruses N° 347, 4pp

CONTRIBUTED BY

Robinson DJ

GENUS *HORDEIVIRUS*

Type Species barley stripe mosaic virus (BSMV)

VIRION PROPERTIES

MORPHOLOGY

Virions are non-enveloped, elongated and rigid, about 20 x 110-150 nm in size; they are helically symmetrical with a pitch of 2.5 nm.

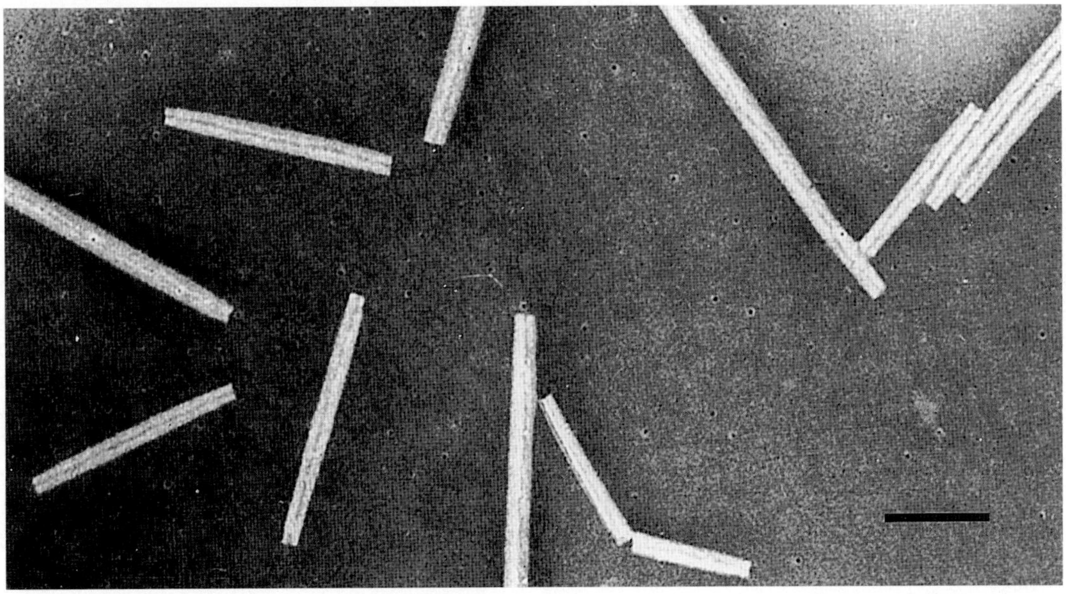

Figure 1: Electron micrograph of purified BSMV particles (Jackson and Brakke, 1973) stained with 2% uranyl acetate. The particles are approximately 20 nm wide and have a length that varies depending on the size of the encapsidated RNA. The particles in the top left, bottom center and upper left side of the micrograph are end to end aggregates that occur during purification. The field was selected to represent monomers, but a range of heterodisperse end to end aggregates up to one µ in length may predominate in various purified preparations. The bar represents 150 nm.

PHYSICOCHEMICAL AND PHYSICAL PROPERTIES

Virions occur as a major sedimenting species with an S_{20w} of about 182-193; other species have an S_{20w} of about 165-200, depending on the virus. Virion isoelectric point is pH 4.5. Anionic detergents, added in purification procedures, increase virus yield by preventing particle aggregation. Thermal inactivation of infectivity occurs at 63-70° C. Virions are rather stable; survival in sap ranges from a few days to several weeks.

NUCLEIC ACID

Virions contain three molecules of positive sense ssRNA, 3.8 kb (RNA α), 3.3 kb (RNA β) and 3.2 or 2.8 kb (RNA γ) in size. A fourth RNA, 2.5 kb in size arising from a deletion, is present in the Argentine mild strain of BSMV. Other RNAs of varying length (800-2900 nt) are found, depending on the strain, and may represent subgenomic or defective RNAs. There is no extensive sequence similarity in the coding regions of BSMV RNAs α, β and γ, but there is a putative helicase region in βb that has amino acid sequence relatedness to the αa helicase motif. No extensive hybridization can be detected between RNAs of BSMV, poa semilatent virus (PSLV) and lychnis ring spot virus (LRSV). The nt sequence similarity of anthoxanthum latent bleaching virus (ALBV), a recently discovered hordeivirus, has not been established. Each RNA has m^7GpppGUA at its 5'-end and a poly(A) tract of 8-40 nt followed by a highly conserved 236-238 nt rRNA-like structure at its 3'-end which accepts tyrosine. A close sequence similarity between the first 70 nucleotides of RNA α and RNA γ of one strain of BSMV suggests that RNA recombination may have a significant role in the evolution of BSMV strains.

PROTEINS

The virion capsid is constructed from protein subunits of a single protein (Mr 22 x 10^3).

LIPIDS

None reported.

CARBOHYDRATES

The virion capsid protein is reported to be glycosylated, but independent confirmation has not been reported. Glycosylation sites are not present in the deduced protein sequence.

GENOME ORGANIZATION AND REPLICATION

The BSMV genome encodes seven proteins: αa (130 kDa) is possibly the viral replicase; βa (22 kDa) is the capsid protein; βb (60 kDa), βc (17 kDa) and βd (14 kDa) are associated with virus movement *in situ*; γa (87 kDa) is a putative polymerase; and γb (17 kDa) is apparently involved in regulating expression of genes encoded in RNA β.

RNA α has a single ORF from which the putative replicase (130 kDa) is translated *in vitro*. RNA β encodes the capsid protein (βa) near the 5' end; further downstream, separated by a 147 nt intergenic region, a triple block sequence codes for three nonstructural proteins (βb, βc and βd) in which βd overlaps the other two genes; the block sequences may be involved with viral movement *in situ*. BSMV RNA γ is bicistronic and encodes a polymerase protein (γa) with putative replicase motifs and a 3' nonstructural gene product that is expressed by a subgenomic RNA. RNA γ is unusually variable in size and number, depending on the strain, especially the Argentine mild strain of BSMV. Downstream from the coding region of each genomic RNA there is a poly (A) sequence separating a 238 nt 3′ terminal tRNA-like structure that can be aminoacylated with tyrosine.

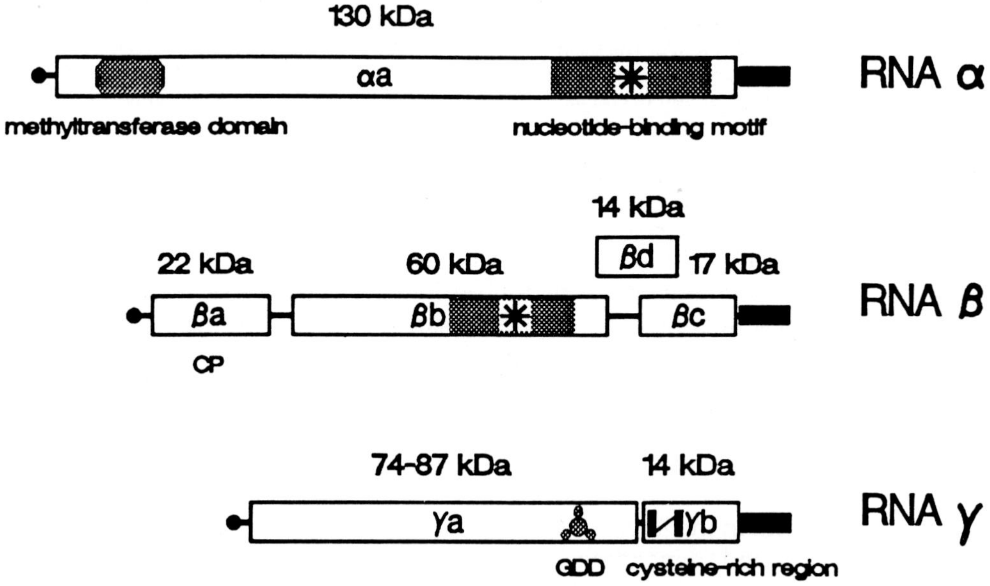

Figure 2: Genome organization of the BSMV genome (Jackson *et al.*, 1991). The filled circle, open rectangles and solid rectangles represent the 5' cap structure, the ORFs and the 3' terminal rRNA-like structure. RNA α encodes a single protein, αa with a putative methyltransferase domain near the amino terminus and a nucleotide binding motif near the carboxy-terminus. RNA β encodes four proteins: βa, the coat protein; βb ORF by 173 nt and terminates with a UAA to initiate the short poly (A) tract that precedes the 238 nt tRNA-like terminus; and βd, a 14 kDa polypeptide which overlaps the βb and the βc ORFs. RNAγ, which varies in size among BSMV strains due to a tandemly duplicated region near the 5' terminus, encodes two polypeptides. The γa polypeptide contains the GDD domain that is present in other viral proteins involved in RNA replication. The 17 kDa γ protein, which is translated from a subgenomic RNA, contains a cysteine-rich region and can affect the expression of genes encoded by RNAβ.

All three BSMV genomic RNAs are required for systemic invasion of plants, but only RNAs α and γ are required for replication in protoplasts. ORFs in RNA β (b,c,d) are required for systemic invasion of plants, but the capsid protein gene (βa) is dispensable and the γb gene is not required in some genetic backgrounds. A mutation in the 5' leader sequence of the γa ORF prevented systemic infection of *Nicotiana benthamiana* , suggesting that modulation of γa expression is involved in movement. RF RNAs corresponding to all viral genomic ssRNAs can be isolated from infected plants. Virus particles accumulate predominantly in the cytoplasm and also in nuclei. Virus particles and dsRNAs are associated with peripheral vesicles in proplastids and chloroplasts in infected barley suggesting that replication and/ or assembly of virions occurs in such organelles.

ANTIGENIC PROPERTIES

The viruses are efficient immunogens. Member species are very distantly related serologically.

BIOLOGICAL PROPERTIES

HOST RANGE

The natural hosts of three species (ALBV, BSMV, PSLV) are grasses (family *Gramineae*); strains of LRSV occur naturally in dicotyledonous plants of the families *Caryophyllaceae* and *Labiatae*.

TRANSMISSION

BSMV is efficiently transmitted by the seed of barley, to some extent by pollen and field spread is by direct leaf contact. There are no known vectors.

GEOGRAPHIC DISTRIBUTION

ALBV has been reported only from Wales; BSMV occurs world-wide wherever barley is grown; LRSV (mentha strain) has only been isolated in Hungary, but the type strain which is highly seed-transmissible in the family *Caryophyllaceae*, was initially discovered in California from seed of *Lychnis divaricata* introduced from Europe.

LIST OF SPECIES IN THE GENUS

The viruses, their genomic sequence accession numbers [ ], CMI/AAB description # ( ) and assigned abbreviations ( ) are:

SPECIES IN THE GENUS

Anthoxanthum latent blanching virus		(ALBV)
barley stripe mosaic virus (68, 344)	[X03854, X52774]	(BSMV)
lychnis ringspot virus		(LRSV)
Poa semilatent virus		(PSLV)

TENTATIVE SPECIES IN THE GENUS

None reported.

DERIVATION OF NAMES

None reported.

REFERENCES

Beczner L, Hamilton RI, Rochon DM (1992) Properties of the mentha strain of lychnis ringspot virus. Intervirology 33: 49-56

Carroll TW (1986) Hordeiviruses: biology and pathology. In: van Regenmortel MHV, Fraenkel-Conrat H (eds) The Plant Viruses, The Rod-shaped Plant Viruses, Vol 2. Plenum Press, New York, pp 373-395

Edwards MC, Petty ITD, Jackson AO (1992) RNA recombination in the genome of barley stripe mosaic virus. Virology 189: 389-392

Edwards ML, Kelley SE, Arnold MK, Cooper JI (1989) Properties of a hordeivirus from *Anthoxanthum odoratum*. Plant Pathol 38: 209-21

Jackson AO, Brakke MK (1973) Multicomponent properties of barley stripe mosaic virus ribonucleic acid. Virology 55: 483-494

Jackson AO, Hunter BG, Gustafson GD (1989) Hordeivirus relationships and genome organization. Ann Rev Phytopathol 27: 95-121

Jackson AO, Petty ITD, Jones RW, Edwards MC, French R (1991) Analysis of barley stripe mosaic virus pathogenicity. Semin Virol 2: 107-119

Jackson AO, Petty ITD, Jones RW, Edwards MC, French R (1991) Molecular genetic analysis of barley stripe mosaic virus pathogenicity determinants. Can J Plant Pathol 13: 163-177

Na-Sheng L, Langenberg WG (1985) Peripheral vesicles in proplastids of barley stripe mosaic virus-infected wheat cells contain double-stranded RNA. Virology 142: 291-298

Partridge JE, Shannon LM, Gumpf DJ, Colbaugh P (1974) Glycoprotein in the capsid of plant viruses as a possible determinant of seed transmissibility. Nature 247: 491-492

Petty ITD, Jackson AO (1990) Mutational analysis of barley stripe mosaic virus RNA b. Virology 179: 712-718

Petty ITD, Edwards MC, Jackson AO (1990) Systemic movement of an RNA plant virus determined by a point substitution in a 5'leader sequence. Proc Natl Acad Sci USA 87: 8894-8897

Petty ITD, French R, Jones RW, Jackson AO (1990) Identification of barley stripe mosaic virus involved in viral RNA replication and systemic movement. EMBO J 9: 3453-3457

Contributed By

Hamilton RI, Jackson AO

GENUS *FUROVIRUS*

Type Species soil-borne wheat mosaic virus (SBWMV)

VIRION PROPERTIES

MORPHOLOGY

Virions are rod-shaped, about 20 nm in diameter, with predominant lengths of 92-160 nm and 250-300 nm; two unassigned species also have particles 380-390 nm in length. The viral capsid has helical symmetry; that of the unassigned beet necrotic yellow vein virus (BNYVV) has a pitch of 2.6 nm, with 12.25 subunits per turn of the right-handed helix.

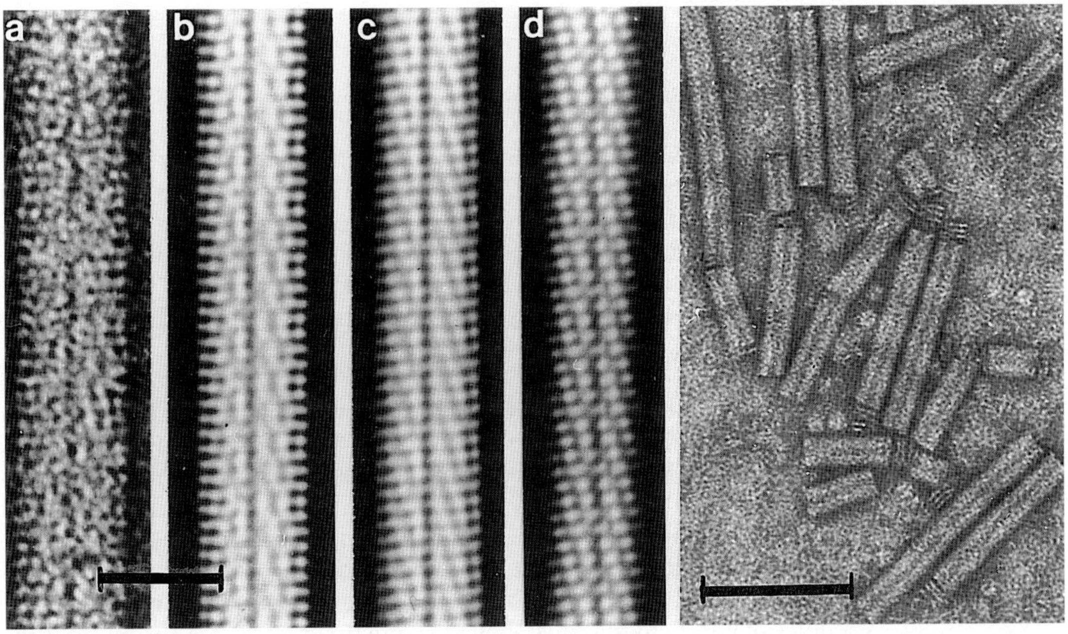

Figure 1: From left (a) negative contrast electron micrograph of beet necrotic yellow vein virus (BNYVV) particle; bar represents 20 nm; (b, c, d) computer-filtered micrographs of BNYVV particles (courtesy of Steven AC); (right) negative contrast electron micrograph of potato mop-top virus; bar represents 250 nm (courtesy of Woods RD).

PHYSICOCHEMICAL AND PHYSICAL PROPERTIES

Virions sediment as two or more components, the number dependent on the virus; those of soil-borne wheat mosaic virus have a S_{20w} of 220-230 (longer particles), 170-225 (shorter particles) and 126-177 (deletion mutants). Virions have a buoyant density in CsCl of about 1.32 g/cm^3.

NUCLEIC ACID

Virions contain two molecules of linear positive sense ssRNA (RNA3 of potato mop-top virus is possibly a delated form of RNA2). Both RNAs of the type virus are capped at their 5' termini; their 3' termini are not polyadenylated, but each has a tRNA-like structure. The RNAs of less well-studied members are reported to lack a 5' cap structure or VPg but, like the type member, are not 3' polyadenylated. RNA1, present in longer particles, is 5.9-7.1 kb in size (Mr 1.83-2.49 x 10^6); RNA2, present in shorter particles, is either 3.5-4.3 kb in size (Mr 1.23-1.83 x 10^6) or, in deleted molecules, 2.1-2.4 kb in size (Mr 0.74-0.84 x 10^6). RNA1 and RNA2 of the wild type isolate of soil-borne wheat mosaic virus contain 7,099 and 3,593 nt, respectively. The complete sequence of both has been determined, and the relevant data are deposited at the GenBank. Beet necrotic yellow vein virus, an unassigned species, differs in usually having a quadripartite ssRNA genome (RNAs 1-4 6.75, 4.61, 1.77 and 1.47 kb in size, respectively, excluding poly (A) tails). Some Japanese isolates also contain RNA5 (1.4 kb) and some European isolates a subgenomic RNA (0.55 kb) of RNA3. All are 3'-polyadenylated

(65-140 residues) and have 5'-terminal caps (m^7 GpppA); RNAs 3 and 4 also have unusually long (445 and 379 nt, respectively) 5' -non-coding regions.

Proteins

Virions are composed of a single protein (Mr 19.7-23.0 x 10^3). That of most species is about 20 x 10^3; however the coat protein subunits of potato mop-top virus are readily degraded by plant proteases and undegraded polypeptides are estimated to be 23.9 x 10^3.

Lipids

None reported.

Carbohydrates

None reported.

Genome Organization and Replication

RNA1 of SBWMV encodes a 150 kDa protein, a readthrough product of 209 kDa and a 37 kDa protein. The 150 and 209 kDa proteins contain NTB-binding helicase and RNA polymerase motifs of a putative replication complex, and the 37 kDa protein is possibly a cell-to-cell transport protein. RNA2 encodes the capsid protein (19 kDa), 84 and 19 kDa readthrough proteins and a 28 kDa protein. Potato mop-top virus-infected plants contain three dsRNAs (6.5, 3.2 and 2.4 kbp in size) corresponding to the three viral ssRNAs, 6.5, 3.2 and 2.5 kb in size.

RNA1, RNA3 and RNA4 of beet necrotic yellow vein virus each contain a single ORF which encodes proteins, respectively, with Mr of 200 (probably the viral replicase), 25 and 31 x 10^3. RNA2 has six ORFs encoding polypeptides, respectively, with Mr of 21 (capsid protein), 75, 42 , 13 , 15 and 14 x 10^3. RNA4, together with the 75 kDa readthrough protein of RNA2, is probably essential for the efficient transmission of the virus by its fungal vector and RNA3 may facilitate virus movement in roots and development of rhizomania symptoms. The function of RNA5 is not yet known. The virus particles usually occur in the cytoplasm and vacuoles of parenchyma cells; they are sometimes scattered throughout the cytoplasm but,

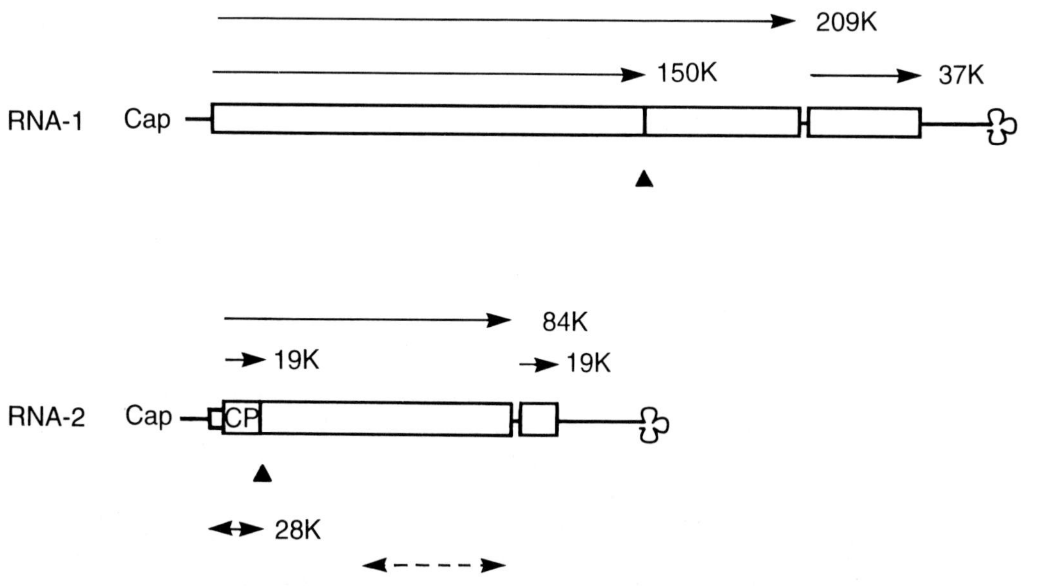

Figure 2: Genomic organization of SBWMV RNAs. ORFs are indicated by rectangles and corresponding translation production by arrows. CP, coat protein; ▲, suppressible termination codons. Broken line beneath RNA2 indicates approximate location and extent of deletions of "lab" isolates. (From Shirako Y & Wilson TMA, 1993).

especially in older cells, occur more frequently in aggregates. Some species also induce cytoplasmic inclusions consisting of interwoven masses of tubules, ribosomes and virus particles.

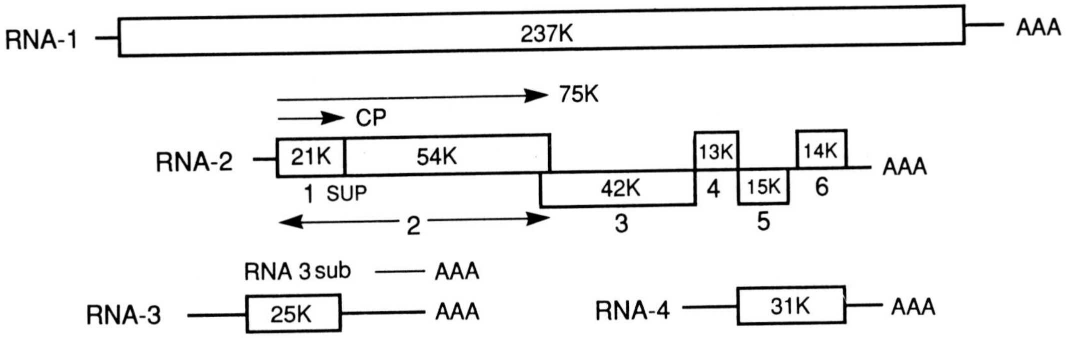

Figure 3: Genomic organization of beet necrotic yellow vein virus RNAs 1-4. ORFs are indicated by hollow rectangles, coat protein (CP), the 75 kDa readthrough translation product by arrows, and the position of the suppressible termination codon by "sup". (From Brunt AA & Richards KE; 1989).

Antigenic Properties

Most species are fairly good immunogens. The type species is serologically distantly related to potato mop-top, broadbean necrosis, oat golden stripe and sorghum chlorotic spot viruses.

Biological Properties

Host Range

The natural host range of individual species is very narrow, but the experimental host range of some is moderately wide.

Transmission

The viruses are transmitted naturally by plasmodiophorid fungi (*Polymyxa graminis*, *P. betae* or *Spongospora subterranea*); virions are carried internally within motile zoospores of the vector fungus, and can be retained for many years within resting spores. Peanut clump virus is also seedborne. All the viruses are mechanically transmissible.

Geographical Distribution

With the notable exceptions of soil-borne wheat mosaic virus and Hypochoeris mosaic virus, most species and tentative species of the genus have restricted geographical distributions. Most of the viruses occur in temperate countries, but peanut clump and rice stripe necrosis virus infect tropical crops.

Cytopathic Effects

Virions can be detected in the cytoplasm and less commonly in vacuoles of comparatively few host cells; the particles are scattered throughout the cytoplasm or occur in parallel arrays to form aggregates or, rarely, paracrystals. The arrays of virions are sometimes found in layers which alternate at about 45° to form angled layer aggregates. Some viruses also induce the formation of intracellular inclusions which are readily detectable by light microscopy and consist of masses of microtubules alone or interwoven masses of tubules, ribosomes and virus particles. Virus-like particles have been detected also within viruliferous zoospores of the vectors.

List of Species in the Genus

The viruses, their genomic sequence accession numbers [ ], CMI/AAB description # ( ) and assigned abbreviations ( ) are:

Species in the Genus

oat golden stripe virus		(OGSV)
peanut clump virus (235)	[L07269]	(PCV)
potato mop-top virus (138)		(PMTV)
soil-borne wheat mosaic virus (77)	[L07937, L07938]	(SBWMV)
sorghum chlorotic spot virus		(SgCSV)

Tentative Species in the Genus

beet necrotic yellow vein virus (144)	[X05147, D00115, X04197]	(BNYVV)
beet soil-borne virus		(BSBV)
broad bean necrosis virus (223)		(BBNV)
Hypochoeris mosaic virus (273)		(HyMV)
rice stripe necrosis virus		(RSNV)

Nicotiana velutina mosaic virus, previously included as a tentative species of the genus, also has a bipartite genome (8 kb and 3 kb); the sizes of its two RNAs, however, differ from those of furoviruses, and it is now probably best excluded from the genus.

Similarity with Other Taxa

The type and two other species are reported to be serologically related to one or more tobamoviruses; the relationship of the type species to tobacco mosaic virus is moderately close. Comparative amino acid analysis of capsid proteins suggests that beet necrotic yellow vein virus also has a distant relationship to tobamoviruses. Similarities in the amino acid sequence of their RNA replicase genes indicate that furoviruses are more closely related to tobamo-, tobra- and hordeiviruses than to beet necrotic yellow vein virus. Similarities in genome organization and RNA sequences indicate that the type member is a member of the "Sindbis-like" superfamily of RNA-containing viruses. Beet necrotic yellow vein virus RNA2 shares some sequence homology with RNAs of barley stripe mosaic hordeivirus, potexviruses and carlaviruses. Thus, the 42K and 13K polypeptides encoded respectively by ORFs 3 and 4 have sequence homology with polypeptides encoded by two contiguous ORFs in barley stripe mosaic hordeivirus RNA2, and in the RNA of potexviruses and carlaviruses.

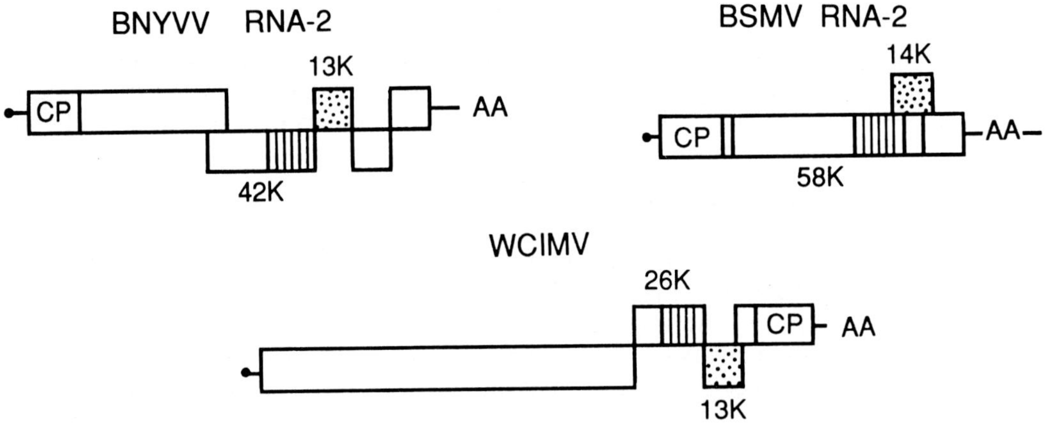

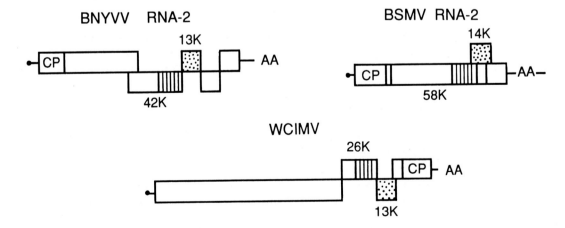

Figure 4: Regions of sequence homologies between beet necrotic yellow vein virus RNA2, barley stripe mosaic virus RNA2 and white clover mosaic virus RNA indicated by hatched and stippled areas within ORFs. (From Brunt AA & Richards KE; 1989).

Derivation of Names

furo: siglum from *fu*ngus-borne, *ro*d-shaped virus

References

Adams MJ (1991) Transmission of plant viruses by fungi. Ann Appl Biol 118: 479-492

Brunt AA, Richards KE (1989) Biology and molecular biology of furoviruses. Adv Virus Res 36: 1-32

Gilmer D, Bouzoubaa S, Hehn A, Guilley H, Richards KE, Jonard G (1992) Efficient cell-to-cell movement of beet necrotic yellow vein virus requires 3' proximal genes located on RNA2. Virology 189: 40-47

Jupin I, Tamada T, Richards KE (1991) Pathogenesis of beet necrotic yellow vein virus. Semin Virol 2: 121-129

Jupin I, Richards KE, Jonard G, Guilley H, Pleij CWA (1990) Mapping sequences required for productive replication of beet necrotic yellow vein virus RNA 3. Virology 178: 273-280

Jupin I, Bouzoubaa S, Richards KE, Jonard G, Guilley H (1990) Multiplication of beet necrotic yellow vein virus RNA 3 lacking a 3′ poly(A) tail is accompanied by reappearance of the poly(A) tail and a short U-rich tract preceeding it. Virology 178: 281-284

Kallender H, Buck KW, Brunt AA (1990) Association of three RNA molecules with potato mop-top virus. Neth J Pl Path 96: 47-50

Kiguchi T, Saito M, Harada T, Tamada T (1989) Nucleotide sequence of beet necrotic yellow vein virus RNA-5. Ann Phytopath Soc Jpn 55: 108

Koenig R, Commandeur U, Lesemann D-E, Burgermeister W, Torrance L, Grassi G, Alric M, Kallerhoff J, Schots A (1990) Antigenic analysis of the coat protein of beet necrotic yellow vein virus by means of monoclonal antibodies. J Gen Virol 71: 2229-2232

Lesemann D-E, Koenig R, Torrance L, Buxton G, Boonekamp PM, Peters D, Schots A (1990) Electron microscopical demonstration of different binding sites for monoclonal antibodies on particles of beet necrotic yellow vein virus. J Gen Virol 71: 731-733

Niesbach-Klösgen U, Guilley H, Jonard G, Richards KE (1990) Immunodetection of beet necrotic yellow vein virus encoded proteins *in vitro*. Virology 178: 52-61

Randles JW, Rohde W (1990) Nicotiana velutina mosaic virus: evidence for a bipartite genome comprising 3 kb and 8 kb RNAs. J Gen Virol 71: 1019-1027

Schmitt C, Balmori E, Jonard G, Richards KE, Guilley H (1992) *In vitro* mutagenesis of biologically active transcripts of beet necrotic yellow vein RNA2: evidence that a domain of the 75 kDa read-through protein is important for efficient virus assembly. Proc Natl Acad Sci USA 89: 5715-5719

Shirako Y, Wilson TMA (1992) Furoviruses. In: Webster RG, Granoff A (eds) Encyclopedia of Virology. Academic Press, London, pp 508-516

Shirako Y, Wilson TMA (1992) Complete nucleotide sequence and organization of the bipartite RNA genome of soil-borne wheat mosaic virus. Virology 195: 16-32

Tamada T, Kusume T (1991) Evidence that the 75 K readthrough protein of beet necrotic yellow vein virus RNA-2 is essential for transmission by the fungus Polymyxa betae. J Gen Virol 72: 1497-1504

Torrance L, Cowan GH, Pereira LG (1993) Monoclonal antibodies specific for potato mop-top virus, and some properties of the coat protein. Ann Appl Biol 122: 311-322

Contributed By

Brunt AA

FAMILY BROMOVIRIDAE

TAXONOMIC STRUCTURE OF THE FAMILY

Family	***Bromoviridae***
Genus	*Alfamovirus*
Genus	*Ilarvirus*
Genus	*Bromovirus*
Genus	*Cucumovirus*

VIRION PROPERTIES

MORPHOLOGY

Virions of members of the genera *Bromovirus*, *Cucumovirus* and *Ilarvirus* are 26-35 nm in diameter, spherical and exhibit icosahedral symmetry (T=3). Virions contain three genomic and one subgenomic ssRNA molecules: RNA1 and RNA2 are contained in separate particles while RNA3 and RNA4 (subgenomic) are contained in one particle. Surface details

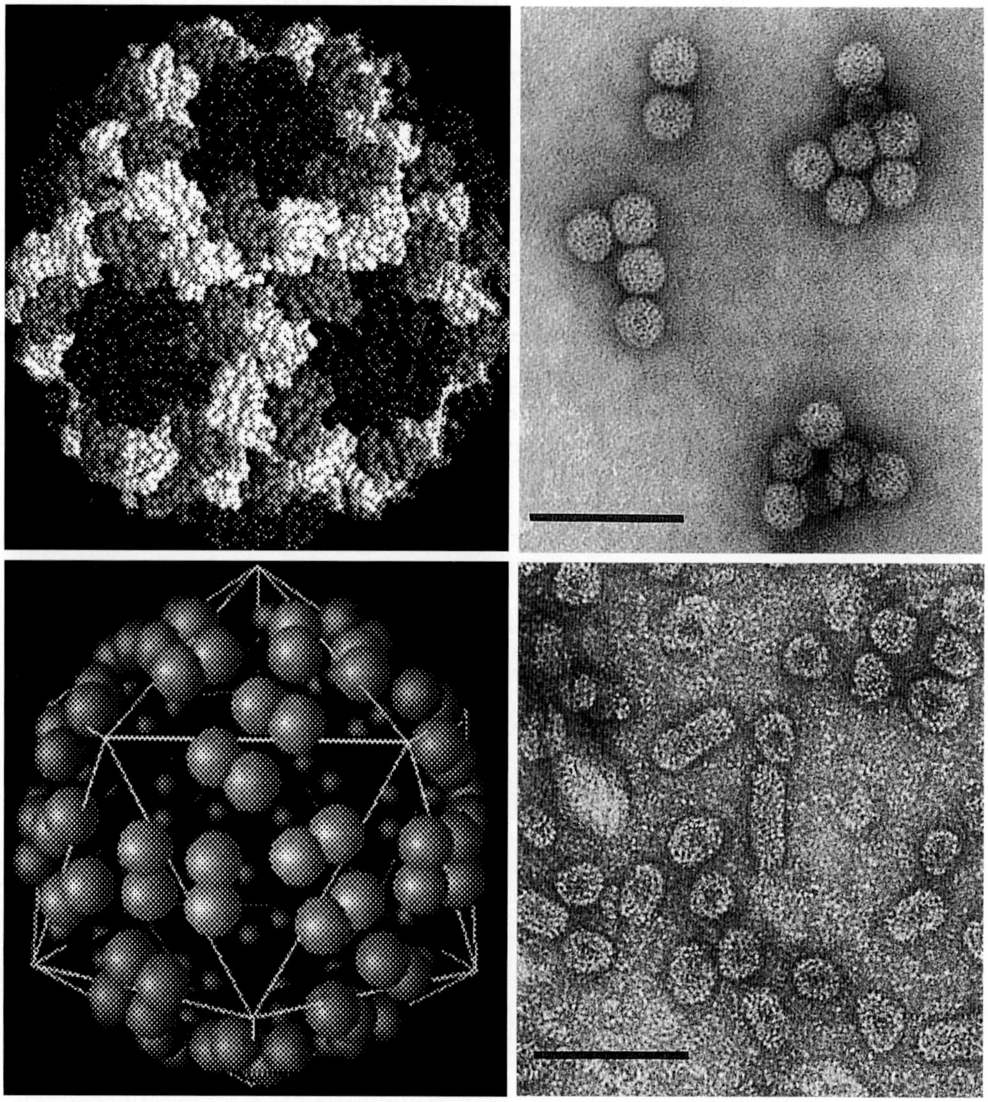

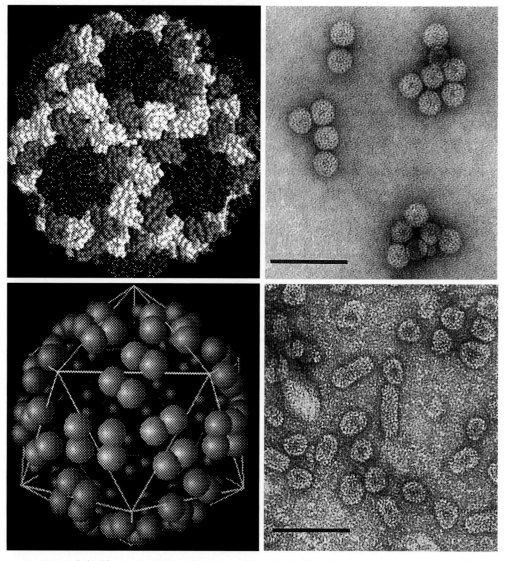

Figure 1: (upper left) Electronic image of cowpea chlorotic mottle virus showing pentamer and hexamer clustering in a T = 3 quasi-icosahedron, (courtesy of Sgro JY); (lower left) diagram of alfalfa mosaic virus Ta particle showing T = 1 structure, (courtesy of Sgro JY) (upper right) negative contrast electron micrograph of cucumber mosaic virus particles, (courtesy of Kasdorf G (lower right); negative contrast electron micrograph of prune dwarf mosaic virus, (courtesy of Kasdorf G). The bar represents 100 nm.

(pentamer and hexamer rings) are visible on virions. Virions of members of the genus *Alfamovirus* (and sometimes of members of the genus *Ilarvirus*) are mostly bacilliform, with different lengths (30 - 57 nm) but a constant diameter of 18 nm. There are four particle sizes, three containing single copies of each of RNAs 1 (B), 2 (M) and 3 (Tb), and the fourth containing two copies of RNA4 (Ta).

PHYSICOCHEMICAL AND PHYSICAL PROPERTIES

Virion Mr varies according to nucleic acid content and coat protein. RNA1, RNA2 and RNA3 & RNA4-containing particles have an Mr of 4.6 - 6.0 x 10^6; alfamovirus virions have an Mr ranging from 3.5 to 6.9 x 10^6. Buoyant densities of aldehyde-fixed virions in CsCl are 1.35 - 1.37 g/cm^3; particles are readily disrupted by neutral chloride salts and SDS, and nucleic acid is RNAse-susceptible *in situ*, at neutral pH. S_{20w} of virions is 78 - 99, and 73 and 63 for alfamovirus Tb and Ta particles, respectively. Virion RNA content ranges from 14 - 25%.

NUCLEIC ACID

Table: Sizes of genome segments

RNA Species	BMV (*Bromovirus*)	CMV (*Cucumovirus*)	TSV (*Ilarvirus*)	AMV (*Alfamovirus*)
RNA1	3,234[a]	3,410	2,940	3,644
RNA2	2,865	3,035	2,770	2,593
RNA3	2,114	2,193	2,205	2,037
RNA4	876	1,027	850	881
5' end	m^7 Gppp	m^7 Gppp	?	m^7 Gppp
3' end	tRNA-like[b]	tRNA-like	complex[c]	complex

a=size in bases
b=aminocylatable, pseudoknot folding
c=coat protein-binding, complex secondary structure

The genome consists of three molecules of linear positive sense ssRNA, 3,200-3,644 nt (RNA1), 2,600-3,050 nt (RNA2), and 2,100-2,216 nt (RNA3) in size. A subgenomic coat protein mRNA, derived from RNA3, 800 - 1000 nt in size is also encapsidated. 5'-termini of all RNAs are capped ($m^7G^{5'}ppp^{5'}Gp$...); 3'-termini of all RNAs of most viruses contain long (150-200 nt) regions of strong sequence and predicted structural similarity, and are not polyadenylated. 3'-termini of cucumo- and bromoviruses can be aminoacylated with tyrosine; alfamo- and ilarvirus RNA 3'-termini cannot be aminoacylated. The 3'-termini are presumed to be telomeric. Short regions at the 5'-termini of genomic RNAs of any one virus bear limited similarity to one another. The total genomes of representatives of each genus except the ilarviruses have been sequenced, and infectious clones are available for a number of viruses.

PROTEINS

Viruses have a single coat polypeptide, Mr 20-26 x 10^3. Virions are constructed from 180 subunits, apparently arranged in pentamer and hexamer clusters. Alfamovirus (and some ilarvirus) bacilliform particles of different lengths apparently have different hexamer net expansions from a basic T=1 icosahedral structure. Proteins have highly basic N-termini (+/-25 residues) which may be degraded *in vivo* and *in vitro*.

LIPIDS

Virions contain no lipid.

Carbohydrates

Virion capsid proteins are not glycosylated.

Genome Organization and Replication

RNA1 and RNA2 each encode single polypeptides of Mr 110 - 126 x 10^3 (P1) and 90-95 x 10^3 (P2), respectively; RNA3 is dicistronic, and encodes polypeptides of Mr 30-35 x 10^3 (P3 or P3a) and 20-26 x 10^3 (coat protein, CP). P1 and P2 are implicated in viral RNA synthesis; P3 is implicated in cell-to-cell spread of the genome. Genomic RNAs replicate via full-length negative sense RNAs in cytoplasmic membrane-associated structures containing P1 and P2 and cellular components. CP is translated *in vivo* and *in vitro* only from subgenomic RNA4: this is derived from RNA3 negative strand template by recognition of a subgenomic promoter by the virus replicase in the P3 - CP intergenic region. Recombination can occur during replication. Bromo- and cucumoviruses require intact RNA 3'-termini, and alfamo- and ilarvirus RNAs require coat protein specifically associated with 3'-terminal sequences for replicase recognition. Particles assemble and accumulate in the cytoplasm, and are found occasionally in nuclei and vacuoles. Inclusion bodies, if present, may be granular or crystalline in appearance.

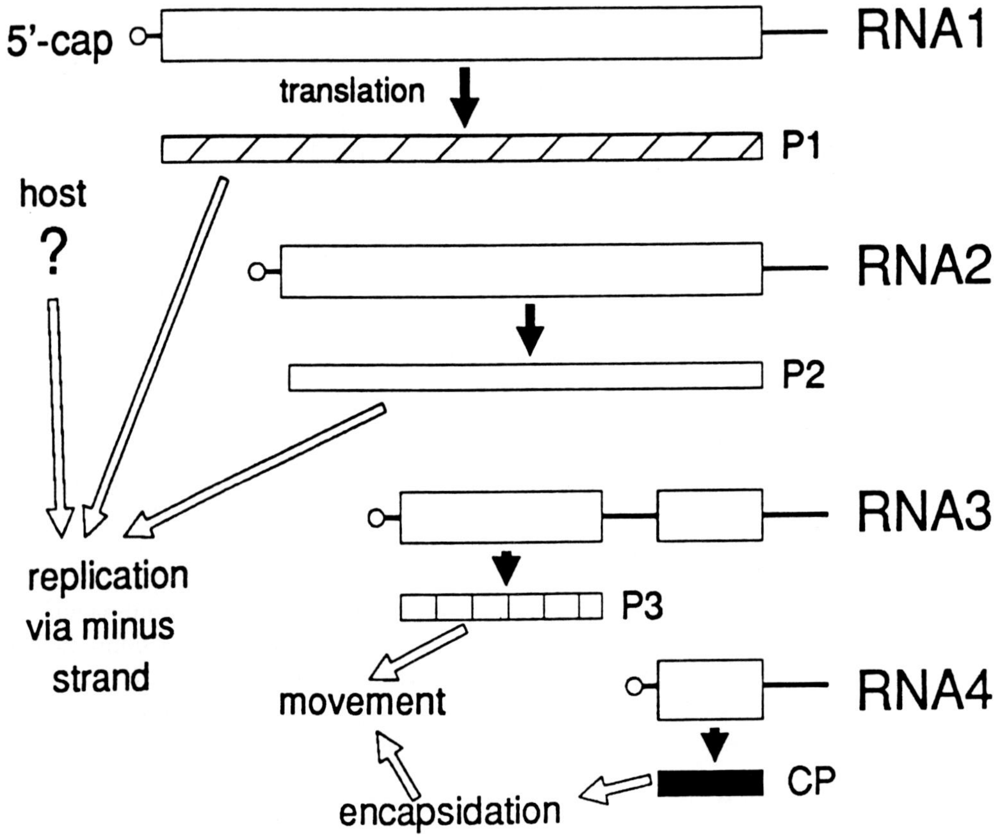

Figure 2: Virus genome organization and replication strategy of members of the family *Bromoviridae*.

Antigenic Properties

Native virions are typically moderate to poor immunogens, and serological reactions are often complicated by sensitivity of particles to salts. Virions are usually satisfactorily stabilized for use as antigens or immunogens by fixation with aldehydes. There are no serological relationships between genera.

Biological Properties

Host Range

Representative viruses of all of the genera of the family *Bromoviridae* have a cosmopolitan distribution, and several are important pathogens of crop and horticultural species. Different viruses in different genera have a variety of host ranges. Individual member viruses of the genus *Bromovirus* typically have narrow host ranges, while the range of the genus includes a variety of species in the families *Gramineae* and *Leguminoseae*. Cucumoviruses as a whole have narrow host ranges in the families *Leguminoseae* and *Solanaceae*, but cucumber mosaic virus has a very wide host range (more than 1000 species). Most ilarviruses infect only woody hosts, but the host range is wide. Alfamoviruses infect over 300 species, including many legumes.

Transmission

All of the viruses are readily transmissible by mechanical inoculation; otherwise, cucumo- and alfamoviruses are non-persistently transmitted by a wide variety of aphids, and certain of these and some ilarviruses are seed-transmitted in some host species. Some ilarviruses are transmitted via pollen, and some bromoviruses are purported to be beetle-transmitted.

Genus *Alfamovirus*

Type Species alfalfa mosaic virus (AMV)

Distinguishing Features

Virions are bacilliform; there is activation of replication by coat protein binding (and reciprocal cross-activation of ilarvirus replication). Viruses are non-persistently transmitted by aphids and have a very wide host range, often causing yellowing symptoms in the field. There is a close serological relationship among all members. There is a weak sequence similarity between P3 proteins of AMV and tobacco streak virus, though not between the coat proteins. Sequence similarities between AMV and other member viruses of the family *Bromoviridae* are only apparent at the level of P1 and P2 proteins, indicating a more distant relationship with these than to ilarviruses.

List of Species in the Genus

The viruses, their genomic sequence accession numbers [ ], CMI/AAB description # ( ) and assigned abbreviations ( ) are:

Species in the Genus

alfalfa mosaic virus (46, 229) [X01572, J02002, K02702] (AMV)

Tentative Species in the Genus

None reported.

Genus *Ilarvirus*

Type Species tobacco streak virus (TSV)

Distinguishing Features

Virions are quasi-isometric or occasionally bacilliform, and are about 30 nm in diameter. There is coat protein activation of replication (and cross-activation of alfamoviruses). There is a short homologous region at RNA 3' ends. The viruses infect mainly woody plants. Viruses in each subgroup are all serologically related, and there are some serological cross-reactions between certain subgroups; however, there are no cross-reactions between subgroup 1 viruses and any other viruses.

LIST OF SPECIES IN THE GENUS

The viruses, their alternative names ( ), genomic sequences accession numbers [ ], CMI/AAB description # ( ) and assigned abbreviations ( ) are:

SPECIES IN THE GENUS

The basic criteria used to subdivide the genus have been serology and host relations; Hamilton (1991) proposed 10 subgroups as follows:

-Subgroup 1:		
tobacco streak virus (44)	[X00435, V00600, J02416, J02417]	(TSV)
-Subgroup 2:		
asparagus virus 2 (288)		(AV-2)
blueberry shock virus		(BlShV)
citrus leaf rugose virus (164)		(CiLRV)
(citrus crinkly leaf virus)		
citrus variegation virus (164)		(CVV)
elm mottle virus (139)		(EMoV)
Tulare apple mosaic virus (42)		(TAMV)
-Subgroup 3:		
apple mosaic virus (83)	[L03726, U03857]	(ApMV)
(some isolates of rose mosaic virus)		(RMV)
Prunus necrotic ringspot virus (5)		(PNRSV)
(some isolates of rose mosaic virus)		(RMV)
-Subgroup 4:		
prune dwarf virus (19)	[L28145]	(PDV)
-Subgroup 5:		
American plum line pattern virus (280)		(APLPV)
-Subgroup 6:		
spinach latent virus (281)		(SPLV)
-Subgroup 7:		
lilac ring mottle virus (201)		(LRMV)
-Subgroup 8:		
hydrangea mosaic virus		(HdMV)
-Subgroup 9:		
Humulus japonicus virus	[X65990]	(HJV)
-Subgroup 10:		
Parietaria mottle virus		(PMoV)

TENTATIVE SPECIES IN THE GENUS

None reported.

GENUS *BROMOVIRUS*

Type Species brome mosaic virus (BMV)

DISTINGUISHING FEATURES

Virions are polyhedral, and all the same size. Virions prepared below pH 6.0 have S_{20w} of 88, a diameter of 27 nm, are stable to high salt and low detergent concentrations, and are nuclease- and protease-resistant. At pH 7.0 and above virions swell to a diameter of 31 nm, S_{20w} decreases to 78, salt and detergent stability decreases dramatically, and protein and RNA are susceptible to hydrolytic enzymes. This swelling is accompanied by conformational changes of the capsid which are detectable by physical and serological means. Coat protein Mr is 20 x 10^3, unlike the 24-26 x 10^3 of other member viruses of the family *Bromoviridae*. RNA 3'-termini are tRNA-like, are very similar in all viruses sequenced so far, and can be aminoacylated with tyrosine. All members are serologically related, although species differences are large. All species are supposedly beetle-transmitted, though BMV is inefficiently transmitted by aphids in a non-persistent manner. Coat proteins of bromovirus species share sequence similarities with one another, and more distantly with cucumoviruses, but not with ilar viruses or alfamoviruses. The same is true of P3 proteins, though distant sequence similarities are apparent between bromoviruses, cucumoviruses and alfamoviruses at the level of P1 and P2 proteins. These relationships indicate a closer relationship between bromoviruses and cucumoviruses than between either of these and viruses of the other two genera.

LIST OF SPECIES IN THE GENUS

The viruses, their genomic sequence accession numbers [ ], CMI/AAB description # ( ) and assigned abbreviations ( ) are:

SPECIES IN THE GENUS

broad bean mottle virus (101)	[K01776, K01777, K01778, M64713, M65138, M60291]	(BBMV)
brome mosaic virus (3, 180)	[V00099, J02042, J02043 K02706, K02707, X01678 X02380, M25172]	(BMV)
Cassia yellow blotch virus		(CYBV)
cowpea chlorotic mottle virus (49)	[M28817, M28818, J02052 K01779, K01780, M65139, M18658, M65155]	(CCMV)
Melandrium yellow fleck virus (236)		(MYFV)
spring beauty latent virus		(SBLV)

TENTATIVE SPECIES IN THE GENUS

None reported.

GENUS *CUCUMOVIRUS*

Type Species cucumber mosaic virus (CMV)

DISTINGUISHING FEATURES

Virions are polyhedral, all the same size, and appear doughnut-shaped in by negative contrast electron microscopy (similar to bromoviruses). Virions are generally labile and sensitive to neutral salts and anionic detergents. RNA 3'-termini (200 nt) are tRNA-like, aminoacylatable with tyrosine, and very similar in all members. All cucumoviruses are serologically related to one another, though species relationships are distant, and all are aphid-transmissible in a non-persistent manner. Cucumber mosaic virus has a very wide host range; others are more limited. Satellite RNAs (330-390 nt; eg. CARNA5, PARNA5) are often associated with cucumoviruses: these typically depend on the virus genome for encapsidation and replication, and may exacerbate or ameliorate symptoms in the host plant.

LIST OF SPECIES IN THE GENUS

The viruses, their alternative names ( ), genomic sequence accession numbers [ ], CMI/AAB description # ( ) and assigned abbreviations ( ) are:

SPECIES IN THE GENUS

cucumber mosaic virus (1, 213)	[D10538, X00985, D10539, D00356, D00385 D10209, D00355, J02059]	(CMV)

peanut stunt virus (91)	[X56544, D11126, D11127 D01123, D01124, D00668]	(PSV)
(robinia mosaic virus) (65)		
tomato aspermy virus (79)	[L15335, D01102, D01015 D01015, M10345, M10346 M10344, M10342, D10044]	(TAV)

TENTATIVE SPECIES IN THE GENUS

None reported.

LIST OF UNASSIGNED VIRUSES IN THE FAMILY

None reported.

SIMILARITY WITH OTHER TAXA

The viruses are members of the "alpha-like supergroup": proteins P3 of member viruses of the family *Bromoviridae* and the 35 kDa protein of the members of the genus *Dianthovirus* (RCNMV) form a distinct "family" of movement-associated proteins. Putative replication-associated proteins P1 and P2 share extensive sequence similarities with proteins of certain rod-shaped viruses (tobra-, hordei- and tobamoviruses), filamentous viruses (potex- and carlaviruses), and spherical viruses (genus *Tymovirus*) of plant and animal alphaviruses (family *Togaviridae*). P1 proteins contain methyl transferase-related and helicase-related domains, while P2 proteins contain motifs characteristic of polymerases. No easily discernible "superfamily" can be defined on the basis of sequence similarities of the entire genomes. Raspberry bushy dwarf virus and olive latent virus 2 (Unassigned Viruses) have similarities in genome organization and in sequence of certain genes with the *Bromoviridae,* but insufficient data is available to satisfactorily define their taxonomic status as yet.

DERIVATION OF NAMES

alfamo: sigla derived from *alf*alfa *mo*saic virus
ilar: sigla from *i*sometric *la*bile *r*ingspot
cucumo: sigla derived from *cucu*mber *mo*saic virus
bromo: sigla derived from *bro*me *mo*saic, also, from Bromus (host of brome mosaic virus bromovirus)

REFERENCES

Bernal JJ, Moriones E, Garcia-Arenal F (1991) Evolutionary relationships in the Cucumoviruses: nucleotide sequence of tomato aspermy virus RNA 1. J Gen Virol 72: 2191-2195

Bruenn JA (1991) Relationships among the positive strand and double-strand RNA viruses as viewed through their RNA-dependent RNA polymerases. Nucl Acids Res 19: 217-226

Dzianott AM, Bujarski JJ (1991) The nucleotide sequence and genome organisation of the RNA-1 segment in two Bromoviruses: broad bean mottle virus and cowpea chlorotic mottle virus. Virology 185: 553-562

Francki RIB (1985) The viruses and their taxonomy. In: Francki RIB (ed) The Plant Viruses I: polyhedral virions with tripartite genomes. Plenum Press, New York, pp 1-18

Goldbach RW (1986) Molecular evolution of plant RNA viruses. Ann Rev Phytopathol 24: 289-310

Johnson JE, Argos P (1985) Virus particle stability and structure. In: Francki RIB (ed) The Plant Viruses I: polyhedral virions with tripartite genomes. Plenum Press, New York, pp 19-56

Karasawa A, Nakaho K, Kakutani T, Minobe Y, Ehara Y (1991) Nucleotide sequence of RNA 3 of peanut stunt Cucumovirus. Virology 185: 464-467

Koonin EV (1991) The phylogeny of RNA-dependent RNA polymerases of positive-strand RNA viruses. J Gen Virol 72: 2197-2206

Koonin EV, Mushegian AR, Ryabov EV, Dolja VV (1992) Diverse groups of plant RNA and DNA viruses share related movement proteins that may possess chaperone-like activity. J Gen Virol 72: 2895-2903

Martelli GP, Russo M (1985) Virus-host relationships: symptoma-tological and ultrastructural aspects. In: Francki RIB (ed) The Plant Viruses Vol I: Polyhedral virions with tripartite genomes. Plenum Press, New York, pp 163-206

Moriones E, Roossinck M, Garcia-Arenal F (1991) Nucleotide sequence of tomato aspermy virus RNA 2. J Gen Virol 72: 779-783

O'Reilly D, Thomas CJR, Coutts RHA (1991) Tomato aspermy virus has an evolutionary relationship with other tripartite RNA plant viruses. J Gen Virol 72: 1-7

Rybicki EP, von Wechmar MB (1985) Serology and immunochemistry. In: Francki RIB (ed) The Plant Viruses Vol I. Polyhedral virions with tripartite genomes. Plenum Press, New York, pp 207-244
Valverde RA, Glascock CB (1991) Further examination of the RNA and coat protein of spring beauty latent virus. Phytopathology 81: 401-404

CONTRIBUTED BY

Rybicki EP

Genus *Idaeovirus*

Type Species raspberry bushy dwarf virus (RBDV)

Virion Properties

Morphology

Virions are isometric, about 33 nm in diameter and are not enveloped. They appear flattened in electron micrographs of preparations negatively stained with uranyl salts.

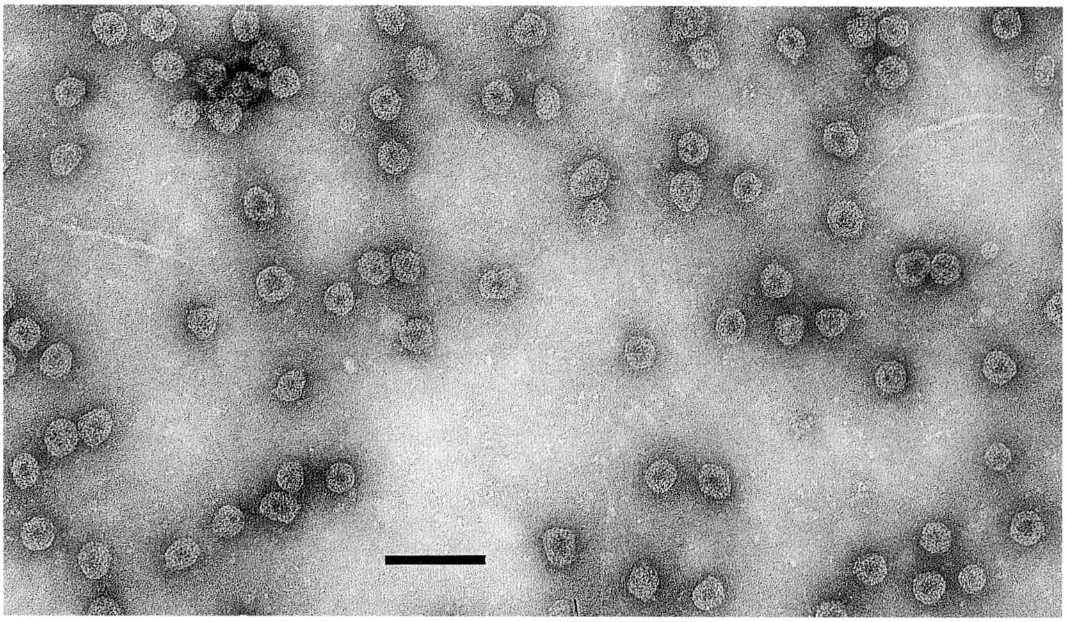

Figure 1: Negative contrast electron micrograph of raspberry bushy dwarf virus stained with uranyl formate/ sodium hydroxide. The bar represents 100 nm.

Physicochemical and Physical Properties

Virion Mr is about 7.5 x 10^6 (calculated from the S_{20w} of 115). The buoyant density of aldehyde-fixed particles in CsCl is 1.37 g/cm^3. Particles are readily disrupted in neutral chloride salts and by sodium dodecyl sulphate.

Nucleic Acid

Virion preparations contain three species of linear, positive sense, ssRNA, 5.4 kb (RNA1), 2.2 kb (RNA2) and 1 kb (RNA3) in size. These RNA molecules are not polyadenylated.

Proteins

Virions possess one major coat protein species (Mr 30 x 10^3). Sequence data indicate that there are two non-structural proteins with Mr of 188 x 10^3 and 39 x 10^3.

Lipids

None reported.

Carbohydrates

None reported.

Genome Organization and Replication

The genome is bipartite. RNA1 has one major ORF encoding a Mr 188 x 10^3 protein which contains sequence motifs characteristic of helicases and polymerases. RNA2 has two in-

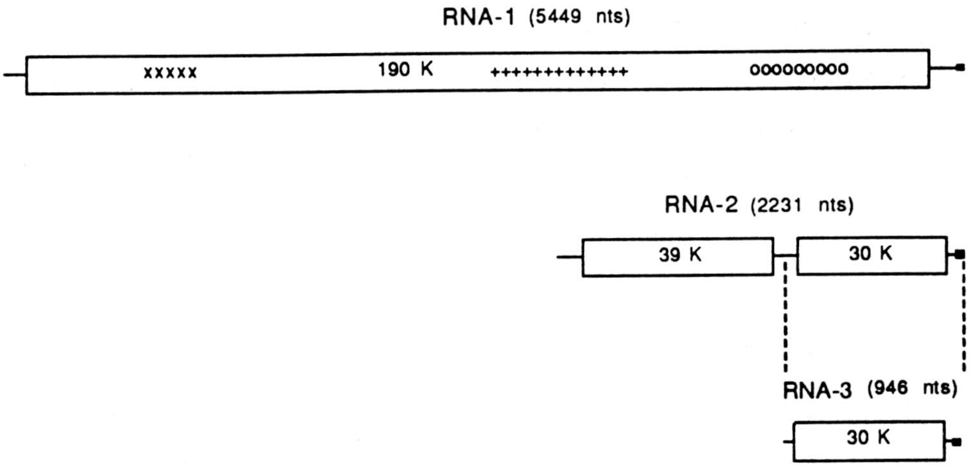

Figure 2: Scale diagram of RNA species found in particles of raspberry bushy dwarf virus. The open boxes represent the ORFs. The dashed lines indicate the derivation of RNA3 from the 3'-end of RNA2. 'xxx' indicates the position of the methyl transferase motif, '+++' indicates the position of the helicase motif and 'ooo' indicates the position of the RNA polymerase motif.

frame ORFs: that in the 5'-terminal half encodes a Mr 39 x 10^3 protein which has some slight sequence similarities with proteins of other viruses that are thought to have roles in virus transport; that in the 3'-terminal half encodes the coat protein. RNA2 is probably a template for the production of RNA3 which comprises the 3'-most 946 nucleotides of RNA2 and is a subgenomic mRNA for coat protein. The 3'-terminal non-coding 18 nt of RNA1 and RNA2 (and hence of RNA3) are the same and the 3'-terminal 70 nt can be arranged in similar extensively base-paired structures. Infected leaves contain dsRNA corresponding in size to double-stranded forms of RNA1 and RNA2. *In vitro* translation yields three major proteins, Mr 190 x 10^3, 44 x 10^3 and 31 x 10^3 (coat protein), which are products, respectively, of RNA1, RNA2 and RNA3.

ANTIGENIC PROPERTIES

Particles are moderate immunogens.

BIOLOGICAL PROPERTIES

In nature the host range is confined to *Rubus* species, all but one in the subgenus *Idaeobatus*; the experimental host range is fairly wide. The virus occurs in all tissues of the plant, including seed and pollen, and RBDV is transmitted in association with pollen, both vertically to the seed and horizontally to the pollinated plant. This is the only known method of natural spread, but experimentally, the virus can be transmitted by mechanical inoculation. The virus occurs throughout the world wherever raspberry is grown. Infection of raspberry is often symptomless but in some cultivars may be associated with 'yellows' or 'crumbly fruit'. Confusingly, RBDV does not seem to be the cause of raspberry bushy dwarf disease of Lloyd George raspberry, though it might contribute to it in association with black raspberry necrosis virus.

LIST OF SPECIES IN THE GENUS

The viruses, their genomic sequence accession numbers [ ], CMI/AAB description # ( ) and assigned abbreviations ( ) are:

SPECIES IN THE GENUS

raspberry bushy dwarf virus (165)	[S51557, S55890, D01052]	(RBDV)

Tentative Species in the Genus

None reported.

Similarity with Other Taxa

RBDV resembles viruses of the genus *Ilarvirus,* family *Bromoviridae,* in having easily deformable particles that are transmitted in association with pollen. RNA2 resembles RNA3 of viruses in family *Bromoviridae* in the arrangement and sizes of its encoded gene products, the generation of a 3'-terminal subgenomic RNA and in the structured nature of the 3' ends of the molecules. The sequence of the translation product of RBDV RNA1 resembles, in different parts, sequences in the translation products of viruses in the family *Bromoviridae* and to a lesser extent the sequence of the helicase + polymerase protein (Mr 183 x 10^3) of tobamoviruses. Idaeoviruses, therefore, belong to the 'Sindbis-like' supergroup.

Derivation of Names

idaeo: from *idaeus*, specific name of raspberry, *Rubus idaeus*

References

Barnett OW, Murant AF (1970) Host range, properties and purification of raspberry bushy dwarf virus. Ann Appl Biol 65: 435-449

Mayo MA, Jolly CA, Murant AF, Raschke JH (1991) Nucleotide sequence of raspberry bushy dwarf virus RNA-3. J Gen Virol 72: 469-472

Murant AF (1975) Some properties of the particles of raspberry bushy dwarf virus. Proc Am Phytopath Soc 2: 116-117

Murant AF (1987) Raspberry bushy dwarf. In: Converse RH (ed) Virus Diseases of Small Fruits. USDA Agriculture Handbook N° 631, pp 229-234

Murant AF, Chambers J, Jones AT (1974) Spread of raspberry bushy dwarf virus by pollination, its association with crumbly fruit, and problems of control. Ann Appl Biol 77: 271-281

Murant AF, Mayo MA, Raschke JH (1986) Some biochemical properties of raspberry bushy dwarf virus. Acta Hort 186: 23-30

Natsuaki T, Mayo MA, Jolly CA, Murant AF (1991) Nucleotide sequence of raspberry bushy dwarf virus RNA-2: a bicistronic component of a bipartite genome. J Gen Virol 72: 2183-2189

Ziegler A, Natsuaki T, Mayo MA, Jolly CA, Murant AF (1992) Nucleotide sequence of raspberry bushy dwarf virus RNA-1. J Gen Virol 73: 3213-3218

Contributed By

Murant AF, Mayo MA

GENUS *CLOSTEROVIRUS*

Type Species beet yellows virus (BYV)

VIRION PROPERTIES

MORPHOLOGY

Virions are very flexuous filaments, 1200-2200 nm long and about 12 nm wide. Virions have helical symmetry, and exhibit distinct cross-banding with a pitch of 3.4-3.8 nm. There are about 10 protein subunits per turn of the helix.

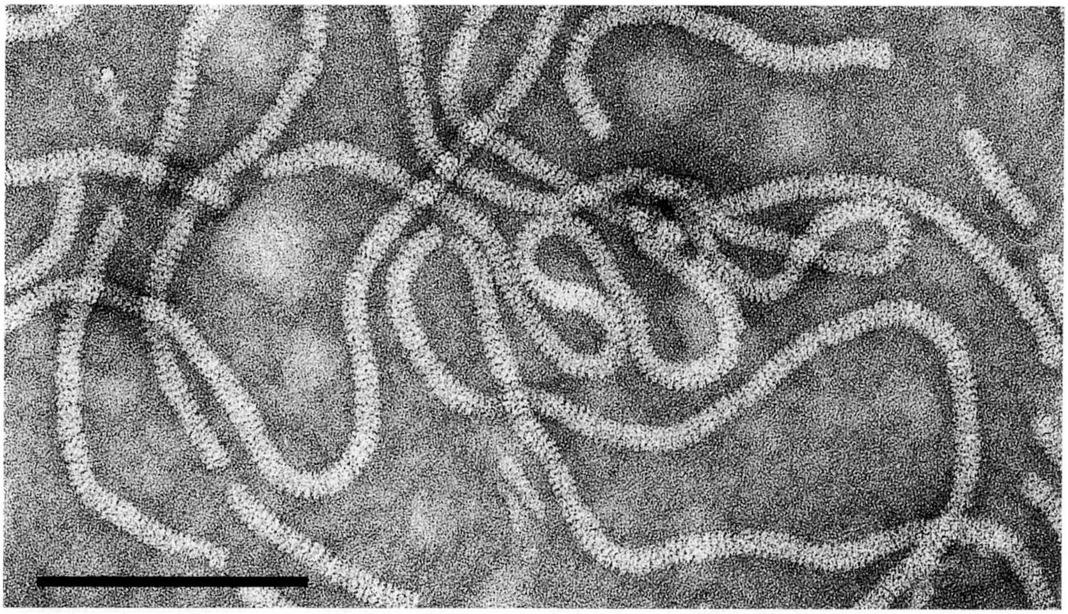

Figure 1: Negative contrast electron micrograph of virions of citrus tristeza virus, the bar represents 100 nm. (Courtesy of Milne RG).

PHYSICOCHEMICAL AND PHYSICAL PROPERTIES

Virions usually sediment as a single band in sucrose or Cs_2SO_4 gradients. S_{20w} ranges from 96 to 140, buoyant density in CsCl is 1.30-1.34 g/cm^3, in Cs_2SO_4 is 1.24-1.27 g/cm^3. Virions of most species are degraded by CsCl and are unstable in high salt concentration. Virions resist moderately high temperatures (thermal inactivation is around 45-55° C) and organic solvents, but are sensitive to RNase and chelation.

NUCLEIC ACID

Virions contain a single molecule of linear, positive sense, ssRNA, constituting 5-6% of the virion weight. The genome of beet yellows virus (BYV), the type species of the genus, is about 15.5 kb in size. The genome size of other viral species is related to particle length, with a maximum size of about 20 kb. The 3' end is not polyadenylated and may not possess a tRNA-like structure. The genomic RNA of BYV has been completely sequenced, whereas that of the tentative species citrus tristeza (CTV) and cucumber chlorotic spot (CCSV) viruses has been sequenced in part.

PROTEINS

Virions are composed of a single major protein Mr 23-28 kDa. CTV is reported to have major and minor coat protein subunits with Mr 27-28 x 10^3 and 26 x 10^3, respectively. The major virion protein of several of the grapevine leaf-roll associated viruses have an Mr ranging from 35 to 43 x 10^3. Structural proteins of some of the species (BYV, carnation necrotic fleck and lilac chlorotic leafspot viruses) lack tryptophan, which is reflected in the high A_{260}/A_{280}

ratio (1.4-1.8) of the viruses. The BYV genome expresses eight nonstructural proteins, the largest of which (295) contains cysteine protease, methyltransferase, aspartyl protease (putative), and helicase signatures. For BYV and CTV, but not CCSV, one of these nonstructural proteins (24 kDa) is closely related to CP. It is a diverged duplicate of the CP gene which is not part of the virion. The above three viruses encode one (CCSV) or two (BYV and CTV) polypeptides which may have a transport function and show sequence homology with heat shock-related proteins (HSP). These genes, together with those coding for ordinary and diverged duplicate CPs make a characteristic four-gene module (HSP/CP) the organization of which is conserved in BYV and CTV.

LIPIDS

None reported.

CARBOHYDRATES

None reported.

GENOME ORGANIZATION AND REPLICATION

BYV genome contains nine ORFs, two of which are located downstream of the coat protein (CP) gene. The organization of the 3' region of BYV differs from that of CTV and CCSV, which have four ORFs downstream of the CP gene. The strategy of expression of the BYV genome is complex, being based on proteolytic processing, frameshifting, and subgenomic RNA production. Analysis of dsRNA patterns of other viral species suggests that some of their ORFs may also be expressed via subgenomic messenger RNAs. Replication occurs in the cytoplasm, possibly in association with membranous vesicles and vesiculated mitochondria.

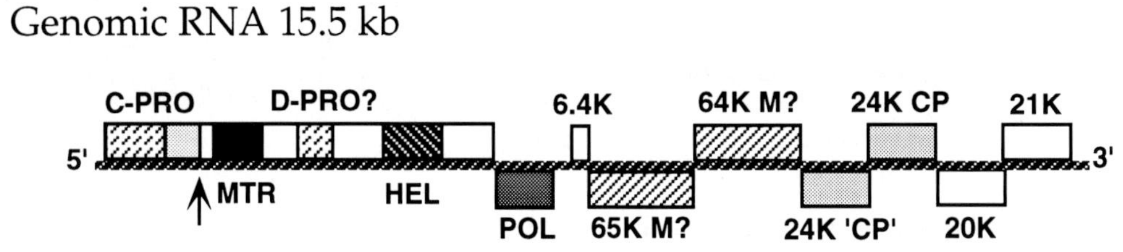

Figure 2: Genetic map of BYV showing the relative position of the ORFs and their products. C-PRO, cysteine protease and its cleavage site (arrow); MTR, methyltranferase; D-PRO, putative aspartyl protease; HEL, helicase; POL, RNA polymerase; 65K-M and 64K-M, polypetides showing homology with heat shock-related proteins, possibly representing movement proteins; 24K 'CP', coat protein; CP, polypeptide related to coat protein. The function of 6.4 K, 20K and 21K polypeptides is unknown (courtesy of Agranowsky AA).

ANTIGENIC PROPERTIES

Virion proteins are moderately antigenic. Most of the species are serologically unrelated to one another.

BIOLOGICAL PROPERTIES

HOST RANGE

The natural and experimental host ranges of individual virus species are restricted. Disease symptoms are of the yellowing type (i.e. rolling, yellowing or reddening of the leaves), or pitting and/or grooving of the woody cylinder. Infection is systemic, but usually limited to the phloem, which may necrotize to a varying extent.

Transmission

Few species are transmissible with difficulty by mechanical inoculation. In vegetatively propagated crops, virus dissemination is primarily through infected propagating material. Transmission through seeds is very rare. Natural vectors are aphids, which transmit in a semi-persistent manner, whiteflies (*Bemisia, Trialeurodes*), and pseudococcid mealybugs (*Pseudococcus, Planococcus*).

Geographical Distribution

Geographical distribution varies from restricted to widespread, depending on the species, most of which occur in temperate regions.

Cytopathic Effects

Virions are usually in the phloem where they accumulate in bundles or conspicuous fibrous masses intermingled with single or clustered membranous vesicles. These may derive either from the endoplasmic reticulum, or from peripheral vesiculation of mitochondria.

List of Species in the Genus

Molecular investigations still in progress show viral species that have been sequenced in toto (BYV) or in part (CTV and CCSV) to possess genomes with a different number and distribution of ORFs. Moreover, certain species (e.g. lettuce infectious yellows virus) may have a divided genome. This may call for a re-classification of species now included among the closteroviruses.

The viruses, their genomic sequence accession numbers [ ], CMI/AAB description # ( ) and assigned abbreviations ( ), are:

Species in the Genus

beet yellow stunt virus (207)		(BYSV)
beet yellows virus (13)	[X73476]	(BYV)
burdock yellows virus		(BuYV)
carnation necrotic fleck virus (136)		(CNFV)
carrot yellow leaf virus		(CYLV)
wheat yellow leaf virus (157)		(WYLV)

Tentative Species in the Genus

1-Aphid-transmitted:		
citrus tristeza virus (33, 353)	[L12175, M76485]	(CTV)
Dendrobium vein necrosis virus		(DVNV)
Heracleum virus 6		(HV-6)
2-Mealybug-transmitted:		
grapevine leafroll-associated virus 3		(GLRaV-3)
pineapple mealybug wilt-associated virus		(PMWaV)
sugarcane mild mosaic virus		(SMMV)
3-Vector unknown:		
alligatorweed stunting virus		(AWSV)
Festuca necrosis virus		(FNV)
grapevine corky bark-associated virus		(GCBaV)
grapevine leafroll-associated virus 1		(GLRaV-1)
grapevine leafroll-associated virus 2		(GLRaV-2)
grapevine leafroll-associated virus 4		(GLRaV-4)
grapevine leafroll-associated virus 5		(GLRaV-5)
4-Whitefly-transmitted:		
beet pseudoyellows virus		(BPYV)
cucumber chlorotic spot virus		(CCSV)

cucumber yellows virus	(CuYV)
Diodia vein chlorosis virus	(DVCV)
lettuce infectious yellows virus	(LIYV)
muskmelon yellows virus	(MYV)

Unassigned Species

The whitefly-transmitted sweet potato sunken vein virus (SPSVV) has virions with the same general structure of those of closteroviruses, but they are shorter (about 850 nm).

Similarity with Other Taxa

Virions of capilloviruses and trichoviruses have the same typical flexuous particle morphology as those of closteroviruses. However, the sequence of the coat protein of BYV has little homology with that of coat proteins of capillo- and trichoviruses, and major differences exist in genome organization and strategy of expression. BYV replication-associated proteins (polymerase, methyltransferase and helicase) resemble those of member viruses of the family *Bromoviridae* and the genera *Tobravirus* and *Tobamovirus* .

Derivation of Names

clostero: from Greek *kloster*, 'spindle, thread', from the appearance of the very long thread-like particles

References

Agranowsky AA, Boyko VP, Karasev AV, Lunina NA, Koonin EV, Dolja VV (1991) Nucleotide sequence of the 3' terminal half of beet yellows closterovirus RNA genome: unique arrangement of eight virus genes. J Gen Virol 72: 15-23

Agranowsky AA, Koonin EV, Boyko VP, Maiss E, Frotschl R, Lunina NA, Atabekov JG (1994) Beet yellows closteroviruses: complete genome structure and identification of a leader papain-like thiol protease. Virology, (in press)

Candresse T (1993) Closteroviruses and clostero-like elongated plant viruses. In: Webster RG, Granoff A (eds) Encyclopedia of Virology. Academic Press, New York (in press)

Dodds JA, Bar-Joseph M (1983) Double stranded RNA from plants infected by closteroviruses. Phytopathology 73: 419-423

Faoro F, Tornaghi R, Cinquanta S, Belli G (1992) Cytopathology of grapevine leafroll-associated virus III (GLRaV III). Riv Patol Veg SV 2: 67-83

Francki RIB, Milne RG, Hatta T (eds) (1985) Closteroviruses. Atlas of Plant Viruses Vol 2. CRC Press, Boca Raton FL, pp 219-234

Lister RM, Bar-Joseph M (1981) Closteroviruses. In: Kurstak E (ed) Handbook of Plant Virus Infections and Comparative Diagnosis. Elsevier/North Holland Biomedical Press, Amsterdam, pp 809-844

Milne RG (ed) (1989) The Plant Viruses, The filamentous plant viruses Vol 4. Plenum Press, New York London

Pappu HR, Karasev AV, Anderson EJ, Pappu SS, Hilf NE, Febres VJ, Eckloff RMG, McCaffery M, Boyko VP, Gowda S, Dolja VV, Koonin EV, Gumpf DJ, Cline KC, Garnsey SN, Dawson WO, Lee RF, Niblett CL (1994) Nucleotide sequence and organization of eight 3' open reading frames of the citrus tristeza closterovirus genome. Virology 199: 35-46

Sekya M, Lawrence SD, McCaffery M, Cline KC (1991) Molecular cloning and nucleotide sequencing of the coat protein gene of citrus tristeza virus. J Gen Virol 72: 1013-1020

Woudt LP, de Rover AP, de Haan PT, van Grisven MQJM (1993) Sequence analysis of the RNA genome of cucumber chlorotic spot virus (CCSV), a whitefly transmitted closterovirus. IXth Internatl Congr Virology, Glasgow pp 60-26

Contributed By

Candresse T, Martelli GP

Genus *Capillovirus*

Type Species apple stem grooving virus (ASGV)

Virion Properties

Morphology

Virions are flexuous filaments, 640 x 12 nm, constructed from helically arranged protein subunits in a primary helix with a pitch of 3.4 nm and between 9 and 10 subunits per turn.

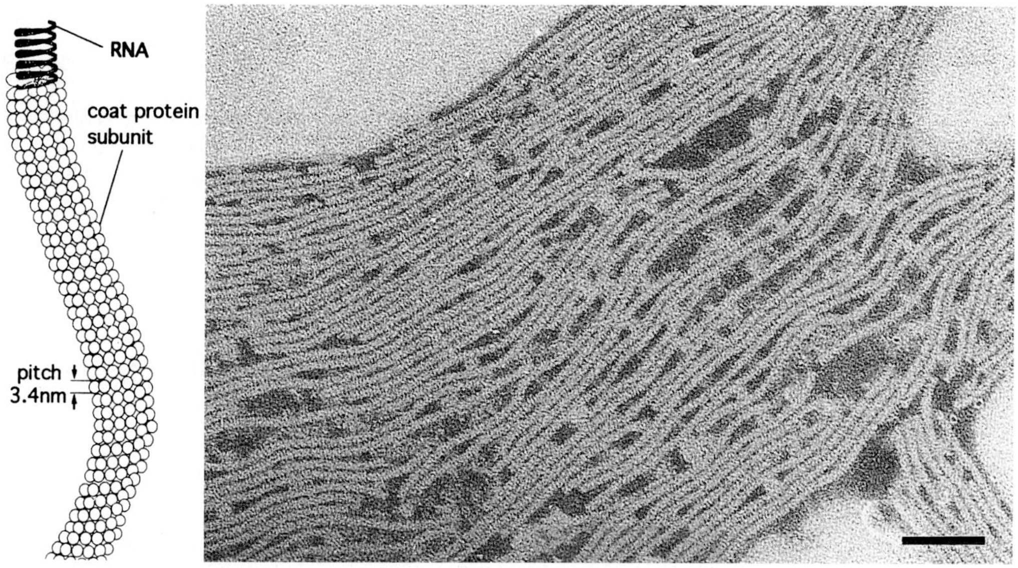

Figure 1: (left) Schematic representation of a portion of a capillovirus. (right) Negative contrast electron micrograph of citrus tatter leaf virus. The bar represents 100 nm.

Physicochemical and Physical Properties

S_{20w} is 112. Isoelectric point is about pH 4.3 at ionic strength 0.1M. Electrophoretic mobility is 10.3 and 6.5 x 10^{-5} cm^2/sec/volt respectively at pH 7.0 and 6.0 (ionic strength 0.1M).

Nucleic Acid

Virions contain linear positive sense ssRNA, 6.5 kb in size, constituting about 5% by weight of virions. The RNA is polyadenylated at its 3'-end. The complete nucleotide sequence of ASGV and citrus tatter leaf virus (CTLV) genomic RNA was determined.

Proteins

Virions are composed of a single protein (Mr about 27 x 10^3). Nonstructural proteins include a 36 kDa protein with sequence homology with supposed viral movement proteins, and proteins of undetermined size with conserved NTP-binding helicase and RNA polymerase motifs.

Lipids

None reported.

Carbohydrates

None reported.

GENOME ORGANIZATION AND REPLICATION

The genomic RNA of ASGV contains two ORFs. ORF 1 encodes a putative 240 kDa protein (about 2,100 amino acids) followed by an untranslated region of 142 nt upstream of the 3' poly (A) tail. ORF 2 is nested within ORF 1 near its 3'-end, and encodes a protein with Mr of 36 x 10^3 (about 320 amino acids). ORF 1-encoded product has homologies with putative polymerase proteins of the "alpha-like" supergroup of RNA viruses. The coat protein cistron is located in the C-terminal end of ORF 1 and is translated as part of the 240 kDa polyprotein. Presumably, replication occurs in the cytoplasm, in which virus particles accumulate in discrete bundles.

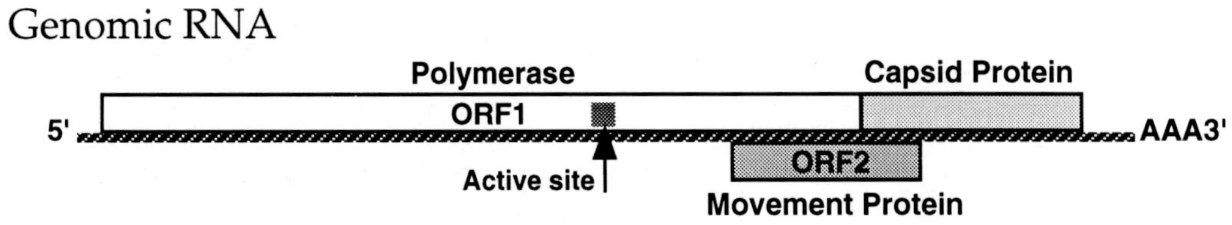

Figure 2: *Capillovirus* (ASGV and CTLV) genome organization.

ANTIGENIC PROPERTIES

Virions are moderately antigenic. CTLV is serologically related to ASGV.

BIOLOGICAL PROPERTIES

HOST RANGE

Most species exhibit narrow host specificity. CTLV has been isolated from citrus and lily. Several species induce destructive diseases, e.g. ASGV and CTLV elicit stock/scion incompatibility in apple (top-working disease) and citrus (budunion crease syndrome), respectively.

TRANSMISSION

No vectors are known. ASGV and CTLV have been transmitted through seed to progeny seedlings of *Chenopodium quinoa*, and lily (CTLV). CTLV, ASGV and NSPV have been transmitted by grafting. NSPV has not been transmitted by sap inoculation, but by grafting and by slashing stems with a partially purified preparation.

GEOGRAPHIC DISTRIBUTION

Geographical distribution ranges from wide to restricted according to the virus. ASGV has been reported wherever apples are cultivated. CTLV occurs in China, Japan, United States, Australia, and South Africa. LCLV occurs in England, The Netherlands, and possibly in Europe and the United States. NSPV is found only in the United States.

CYTOPATHOLOGY

No distinct cytological alterations have been observed in infected cells. Virus particles occur in bundles in mesophyll and phloem parenchyma cells, but not in epidermis and sieve elements.

LIST OF SPECIES IN THE GENUS

The viruses, their CMI/AAB description # ( ), genomic sequence accession numbers and assigned abbreviations ( ) are:

SPECIES IN THE GENUS

apple stem grooving virus (31)		(ASGV)
citrus tatter leaf virus	[D16681]	(CTLV)
lilac chlorotic leafspot virus (202)		(LCLV)

TENTATIVE SPECIES IN THE GENUS

Nandina stem pitting virus	(NSPV)

SIMILARITY WITH OTHER TAXA

Member viruses of the genus *Capillovirus* have the same morphology as members of the genera *Closterovirus* and *Trichovirus*. Similarities exist between members of the genera *Capillovirus* and *Trichovirus* in amino acid sequences around conserved helicase and polymerase motifs, in their respective 36 kDa and 50 kDa polypeptides, and in their coat proteins. The genome organization and replication strategy, however, are different.

DERIVATION OF NAMES

capillo: from Latin *capillus*, a hair

REFERENCES

Ahmed NA, Christie SR, Zettler FW (1983) Identification and partial characterization of a closterovirus infecting *Nandina domestica*. Phytopathology 73: 470-475

De Sequeira OA, Lister RM (1969) Purification and relationships of some filamentous viruses from apple. Phytopathology 59: 1740-1749

Inouye N, Maeda T, Mitsuhata K (1989) Citrus tatter leaf virus isolated from lily. Ann Phytopath Soc Japan 45: 712-720

Nishio T, Kawai A, Takahashi T, Namba S, Yamashita S (1989) Purification and properties of citrus tatter leaf virus. Ann Phytopath Soc Japan 55: 254-258

Ohira K, Ito T, Kawai A, Namba S, Kusumi T, Tsuchizaki T (1994) Nucleotide sequence of the 3'-terminal region of citrus tatter leaf virus RNA. Virus Genes 8: 169-172

Ohki ST, Yoshikawa N, Inouye N, Inouye T (1989) Comparative electron microscopy of *Chenopodium quinoa* leaves infected with apple chlorotic leaf spot, apple stem grooving, or citrus tatter leaf virus. Ann Phytopath Soc Japan 55: 245-249

Semancik JS, Weathers LG (1965) Partial Purification of a mechanically transmissible virus associated with tatter leaf of citrus. Phytopathology 55: 1354-1358

Yoshikawa N, Takahashi T (1988) Properties of RNAs and proteins of apple stem grooving and apple chlorotic leaf spot viruses. J Gen Virol 69: 241-245

Yoshikawa N, Takahashi T (1992) Evidence for genomic translation of apple stem grooving capillovirus RNA. J Gen Virol 73: 1313-1315

Yoshikawa N, Sasaki E, Kato M, Takahashi T (1992) The nucleotide sequence of apple stem grooving capillovirus genome. Virology 191: 98-105

CONTRIBUTED BY

Namba S

Genus *Trichovirus*

Type species apple chlorotic leaf spot virus (ACLSV)

Virion Properties

Morphology

Virions are very flexuous filaments, 640-800 x 12 nm in size, helically constructed with a pitch of 3.3- 3.5 nm and about 10 subunits per turn of the helix. Virions may show cross banding, criss-cross or rope-like features according to the negative contrast material used.

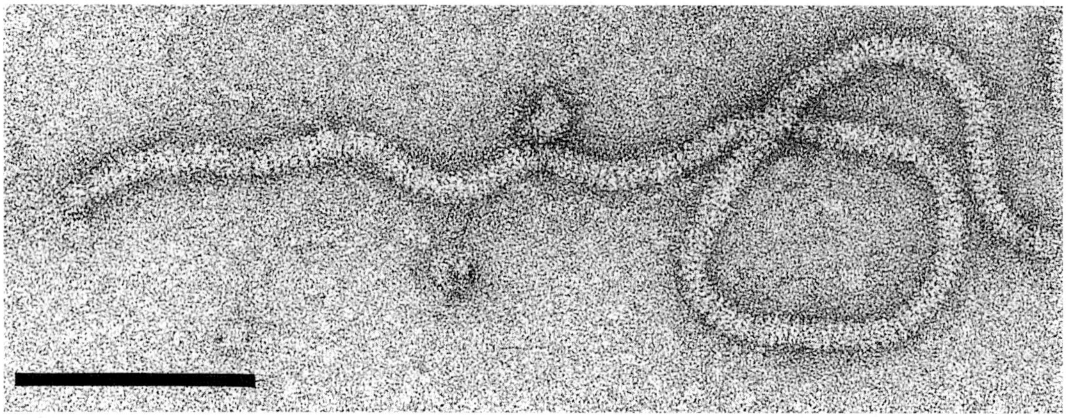

Figure 1: Negative contrast electron micrograph of grapevine virus A particles, the bar represents 100 nm (courtesy of Milne RG).

Physicochemical and physical properties

Virions sediment as single or as two very close bands with an S_{20w} of 92-99. Virions of apple chlorotic leaf spot (ACLSV) and heracleum latent (HLV) viruses are sensitive to ribonucleases. Virions of all species resist moderately high temperatures (thermal inactivation is around 55-60° C) and organic solvents.

Nucleic acid

Virions contain a single molecule of linear, positive sense, ssRNA, 6.3-7.6 kb in size (Mr 2.2-2.5 x 10^6). The RNA has a polyadenylated 3' terminus. Indirect evidence suggests that the genomic RNA of ACLSV is capped at its 5' end. RNA accounts for about 5% of the particle weight. The complete nucleotide sequences are available for some members.

Proteins

Virions of all species are composed of a single major polypeptide (Mr 22-27 x 10^3). Non structural proteins of ACLSV and PVT are: (i) a protein of about 180-220 kDa containing RNA-dependent RNA polymerase (GDD), nucleotide binding (helicase) and methyltransferase signature sequences, all typical of replication-associated proteins of the "alpha-like" supergroup of ssRNA viruses; (ii) a polypeptide of 40-50 kDa with weak homologies to some plant virus movement proteins. GVA and GVB may encode an additional non structural polypeptide of 10-13 kDa with weak homologies to proteins with RNA-binding properties.

Lipids

None reported.

Carbohydrates

None reported.

Genome Organization and Replication

The genome of ACLSV and PVT contains three slightly overlapping ORFs. The large 5' ORF is directly expressed from genomic RNA, whereas the two smaller downstream ORFs that code, respectively, for the putative movement protein and coat protein, are probably expressed from subgenomic messenger RNAs. ACLSV-infected tissues contain 5 dsRNA species, three of which are 5' coterminal with genomic RNA, and two of which are dsRNA forms of the respective subgenomic RNAs. The most abundant dsRNA species, the functions of which are unknown, are 5' coterminal with genomic RNA, and have a size of 6.5 and 5.5 kbp, respectively. The tentative species GVA and GVB have an additional small ORF downstream of the coat protein cistron and produce at least four subgenomic RNAs. Replication is presumed to be cytoplasmic and to involve the product of ORF 1.

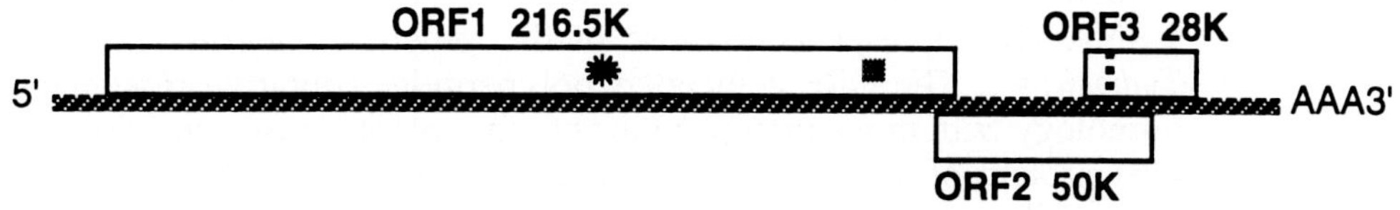

Figure 2: Genome organization of ACLSV, showing position and translation products of the three ORFs. The asterisk and square indicate the position of helicase and polymerase motifs, respectively (from German *et al.*, 1990).

Antigenic Properties

The viruses serve as moderate to poor antigens. Species are not serologically interrelated.

Biological Properties

Host range

The natural host range of individual species is narrow (ACLSV), or restricted to a single host (PVT, GVA, GVB). Infections induce little or no symptoms (PVT, HLV, ACLSV in certain hosts), or mottling, rings and line patterns (ACLSV), or pitting and grooving of the wood (GVA and GVB).

Transmission

The viruses are transmitted by mechanical inoculation, some (GVA) with difficulty, by grafting (ACLSV, GVA, GVB) and through propagating material. PVT is seed-transmitted in several hosts, including *Solanum* spp. GVA and GVB are transmitted by pseudococcid mealybugs (*Pseudococcus, Planococcus*), and HLV is transmitted in a semipersistent manner by aphids in association with a helper virus. No natural vectors of ACLSV and PVT are known.

Geographical distribution

Geographical distribution varies from wide to restricted, according to the virus species. PVT reported only from the Andean region of South America.

Cytopathic effects

Infected cells are damaged to a varying extent. GVA, GVB and HLV elicit the formation of vesicular evaginations of the tonoplast containing finely fibrillar material, possibly representing replicating forms of viral RNA. Virions are found in phloem and parechyma cells of leaves and roots and accumulate in the cytoplasm in bundles or paracrystalline aggregates.

List of Species in the Genus

The viruses, their genomic sequence accession numbers [ ], CMI/AAB description # ( ) and assigned abbreviations ( ) are:

Species in the Genus

apple chlorotic leaf spot virus (30)	[M13714]	(ACLSV)
potato virus T (187)	[D10172]	(PVT)

Tentative species in the genus

grapevine virus A	[X75433]	(GVA)
grapevine virus B	[X75448]	(GVB)
Heracleum latent virus (228)		(HLV)

Similarity with Other Taxa

Virions resemble, somewhat, those of member viruses of the genera *Closterovirus* and *Capillovirus*. The ORF 1-encoded polypeptide (putative polymerase) contains signature sequences homologous to those found in other members of the "alpha-like" supergroup of ssRNA viruses, especially those of the genera *Carlavirus*, *Capillovirus*, *Potexvirus*, and *Tymovirus* . The ORF 2-encoded polypeptide (putative movement protein) has weak homology with movement proteins of other plant viruses, the closest relative being the 36 kDa protein of apple stem grooving capillovirus (ASGV). The 10-13 kDa polypeptide potentially encoded by the putative 3' ORF of GVA and GVB has weak homologies with the 12-15 kDa product of carlaviruses, which has RNA-binding properties. Coat proteins of ACLSV, PVT, GVA, and GVB share distinct homology with that of ASGV, but not with coat proteins of beet yellows and citrus tristeza closteroviruses.

Derivation of Names

Tricho: from Greek "thrix", hair

References

Bem F, Murant AF (1979) Comparison of particle properties of heracleum latent and apple chlorotic leaf spot virus. J Gen Virol 44: 817-826

Boscia D, Savino V, Minafra A, Namba S, Elicio V, Castellano MA, Gonsalves D, Martelli GP (1993) Properties of a filamentous virus isolated from grapevine affected by corky bark. Arch Virol 130: 109-120

Candresse T (1993) Closteroviruses and clostero-like elongated plant viruses. In : Webster RG, Granoff A (eds) Encyclopedia of Virology. Academic Press, London 1: 242-248

Castrovilli S, Gallitelli D (1985) A comparison of two isolates of grapevine virus A. Phytopathol Medit 24: 219-220

Conti M, Milne RG, Luisoni E, Boccardo G (1980) A closterovirus from a stem-pitting diseased grapevine. Phytopathology 70: 394-399

de Sequeira OA, Lister RM (1969) Purification and relationships of some filamentous viruses of apple. Phytopathology 59: 1740-1749

German S, Candresse T, Lanneau M, Pernollet JC, Dunez J (1990) Nucleotide sequence and genomic organization of apple chlorotic leaf spot closterovirus. Virology 179: 104-112

German S, Candresse T, Le Gall O, Lanneau M, Dunez J (1992) Analysis of dsRNAs of apple chlorotic leaf spot virus. J Gen Virol 73: 767-773

Ochi M, Kashiwazaki S, Hiratsuka K, Namba S, Tsuchizaki T (1992). Nucleotide sequence of the 3'-terminal region of potato virus T RNA. Ann Phytopath Soc Japan 58: 416-425

Ohki ST, Yoshikawa N, Inouye N, Inouye T (1989) Comparative electron microscopy of *Chenopodium quinoa* leaves infected with apple chlorotic leaf spot, apple stem groving or citrus tatter leaf virus. Ann Phytopath Soc Japan 55: 245-249

Rosciglione B, Castellano MA, Martelli GP, Savino V, Cannizzaro G (1983) Mealybug transmission of grapevine virus A. Vitis 22: 331-347

Salazar LF, Harrison BD (1978) Host range, purification and properties of potato virus T. Ann Appl Biol 89: 223-235

Sato K, Yoshikawa N, Takahashi T (1993) Complete nucleotide sequence of the genome of an apple isolate of apple chlorotic leaf spot virus. J Gen Virol 74: 1927-1931

Tollin P, Wilson HR, Roberts IM, Murant AF (1992) Diffraction studies of the particles of two closteroviruses: heracleum latent virus and heracleum virus 6. J Gen Virol 73: 3045-3048

Contributed By

Candresse T, Namba S, Martelli GP

Genus *Tymovirus*

Type Species turnip yellow mosaic virus (TYMV)

Virion Properties

Morphology

Virions exhibit icosahedral symmetry (T = 3); they are non-enveloped, and have a diameter of about 30 nm. Morphological subunits formed by the 20 hexamers and 12 pentamers of the coat protein subunits are clearly visible. Virions and 'empty particles' are readily distinguished.

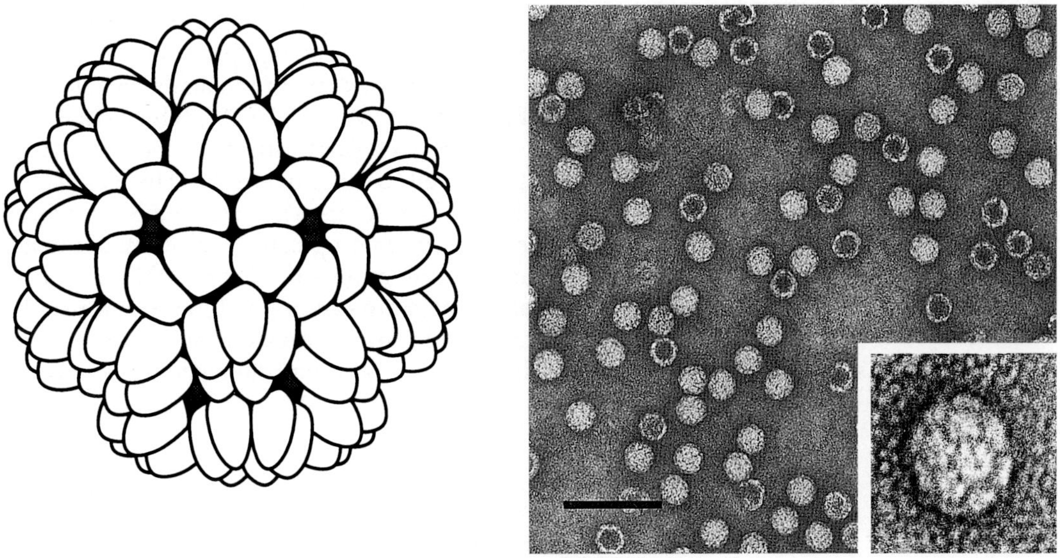

Figure 1: (left) Diagram of virion with coat protein clusters in hexa- and pentamers; (right) negative contrast electron micrograph of belladonna mottle virus virions and 'empty particles', inset shows intact virion. The bar represents 100 nm.

Physicochemical and Physical Properties

The two major classes of stable particles (B and T) have an Mr of 5.6 and 3.6 x 10^6, a buoyant density of 1.42 and 1.29 g/cm^3 and a S_{20w} of about 115 and 55, respectively. Only the B component containing the genomic RNA is infectious. Several minor nucleoproteins have densities intermediate between those of the two major particle types and in the case of turnip yellow mosaic virus (TYMV) they contain the subgenomic coat protein messenger RNA or less than full-length pieces of the genomic RNA. Virions are stable at neutral pH. The isoelectric point of TYMV is 3.75, those of other species cover a wide range. The structure of the particles is stabilized by protein-protein interactions which are mainly hydrophobic. The thermal inactivation points range from 65 to 95° C for different species. The overall structure of virions is stable to ether, chloroform and butanol, but the RNA and a few coat protein subunits may be released. Virions are readily disrupted by sodium dodecylsulphate.

Nucleic Acid

B particles contain one molecule of infectious linear positive sense ssRNA of about 6.3 kb which is capped on the 5' end and has a tRNA-like structure on the 3' end which accepts valine in the case of TYMV. Tymovirus RNAs are characterized by a high cytidine content and in several species they are apparently neutralized in the particles by several hundred molecules of polyamines (spermine, spermidine).

PROTEINS

Virions contain 180 copies of a single 20 kD coat protein species.

LIPIDS

None reported.

CARBOHYDRATES

None reported.

GENOME ORGANIZATION AND REPLICATION

The genomic RNA contains 3 ORFs. ORF 1 encodes a 206 kD protein which contains sequence motifs characteristic for nucleotide binding and RNA polymerase functions. In *in vitro* translation experiments, this protein is at least in part proteolysed in cis to give a larger N-coterminal (Mr 150 x 10^3) and a smaller C-terminal product (Mr 70 x 10^3). The protease activity apparently resides in a domain between amino acids 555 and 1051 of the 206 kDa protein. ORF 2 encodes a 69 kDa protein (Mr 75-80 x 10^3) which can be detected in *in vitro* translation experiments and also *in vivo* early during infection. It is dispensable for replication, but is required for viral cell to cell movement. The 20 kDa viral coat protein is expressed from a subgenomic RNA. Tymoviruses induce double-membrane bound vesicles which invaginate in the periphery of the chloroplasts. They contain membrane-bound viral RNA polymerase and are probably the main site of viral RNA replication. Presumably hexa- and pentamers of the coat protein are synthesized in the cytoplasm, become inserted in the outer chloroplast membrane in an orientated fashion and encapsidate the RNA strands which emerge from the vesicles. Empty protein shells accumulate in nuclei.

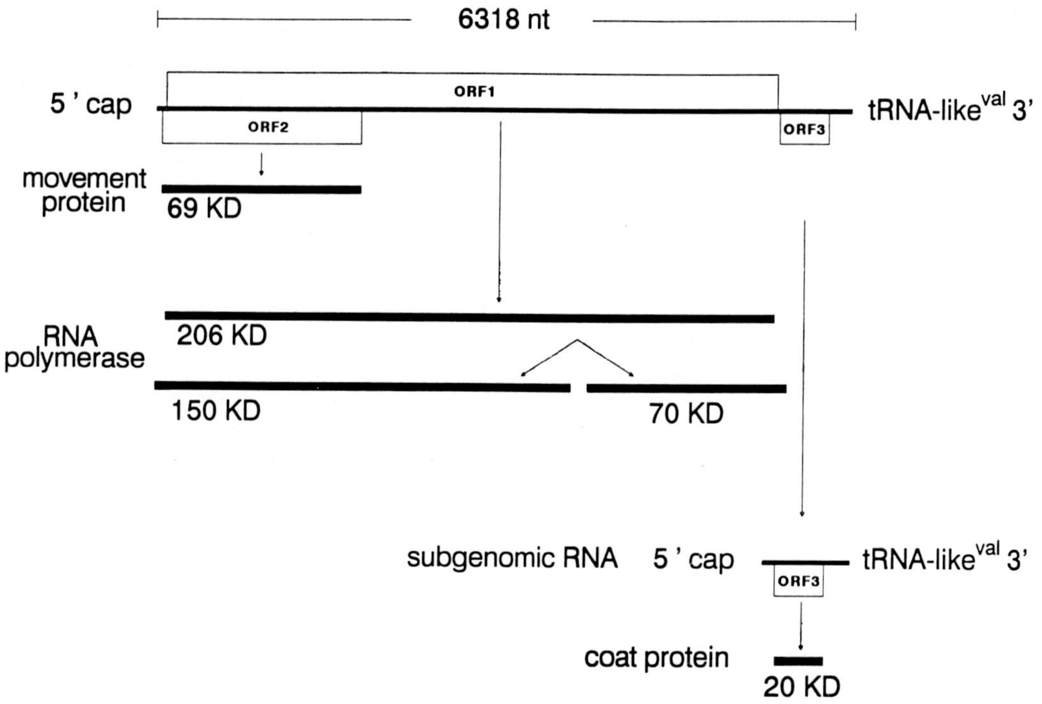

Figure 2: Organization and expression of the TYMV genome.

ANTIGENIC PROPERTIES

Virions are moderately to highly antigenic and form single precipitin lines in agar gel double diffusion tests. Serological relationships between different species range from very close, to distant, to not detectable.

BIOLOGICAL PROPERTIES

Tymoviruses are possibly restricted to dicotyledonous hosts. They have been reported from most parts of the world. Restricted host ranges and lack of vector insects are probably the main reasons for the limited distribution of individual tymoviruses. The viruses are transmitted mechanically and by beetles of the families *Chrysomelidae* and *Curculionidae*. They invade all main tissues of their host plants and cause bright yellow mosaic symptoms or mottling.

LIST OF SPECIES IN THE GENUS

The viruses, their genomic sequence accession numbers [ ], CMI/AAB description # ( ) and assigned abbreviations ( ) are:

SPECIES IN THE GENUS

Belladonna mottle virus (52)	[X54529]	(BeMV)
cacao yellow mosaic virus (11)		(CYMV)
Clitoria yellow vein virus (171)	[M15963]	(CYVV)
Desmodium yellow mottle virus (168)		(DYMV)
Dulcamara mottle virus (124)		(DuMV)
eggplant mosaic virus		(EMV)
(Andean potato latent virus) (124)	[M15284, M58313]	
Erysimum latent virus (222)		(ErLV)
Kennedya yellow mosaic virus (193)	[D00637]	(KYMV)
okra mosaic virus (128)		(OkMV)
passion fruit yellow mosaic virus		(PaYMV)
peanut yellow mosaic virus		(PeYMV)
Physalis mosaic virus		(PhyMV)
Plantago mottle virus		(PlMoV)
Scrophularia mottle virus (113)		(ScrMV)
(Anagyris vein yellowing virus)		
(Ononis yellow mosaic virus)	[J04375]	
turnip yellow mosaic virus (2; 230)	[J04373, X16378, X07441]	(TYMV)
Voandzeia necrotic mosaic virus (279)		(VNMV)
wild cucumber mosaic virus (105)		(WCMV)

TENTATIVE SPECIES IN THE GENUS

poinsettia mosaic virus (311)	(PnMV)

SIMILARITY WITH OTHER TAXA

Tymoviruses are morphologically similar to marafiviruses. The latter, however, have two coat protein species, are not transmitted mechanically but only by leafhoppers and do not induce double-membrane bound vesicles in chloroplasts. The derived amino acid sequences for the putative RNA polymerases of tymoviruses have the closest relationships to those of potexviruses, but no relationships are found between the coat proteins of potex- and tymoviruses.

DERIVATION OF NAMES

tymo: sigla from *t*urnip *y*ellow *mo*saic virus

REFERENCES

Bozarth CS, Weiland JJ, Dreher TW (1992) Expression of ORF-69 of turnip yellow mosaic virus is necessary for viral spread in plants. Virology 187: 124-130

Bransom KL, Weiland JJ, Dreher TW (1991) Proteolytic maturation of the 206-kDa nonstructural protein encoded by turnip yellow mosaic virus RNA. Virology 184: 351-358

Candresse T, Morch M-D, Dunez J (1990) Multiple alignment and hierarchical clustering of conserved amino acid sequences in the replication-associated proteins of plant RNA viruses. Res Virol 141: 315-329

Crestani OA, Kitajima EW, Lin MT, Marinho VLA (1986) Passionfruit yellow mosaic virus, a new tymovirus found in Brazil. Phytopathology 76: 951-955

Ding S-W, Keese P, Gibbs AJ (1989) Nucleotide sequence of the ononis yellow mosaic tymovirus genome. Virology 172: 555-563

Finch JT, Klug A (1966) Arrangement of protein subunits and the distribution of nucleic acid in turnip yellow mosaic virus. II. Electron microscopic studies. J Mol Biol 15: 344-364

Francki RIB, Milne RG, Hatta T (eds) (1985) Atlas of Plant Viruses Vol I. CRC Press, Boca Raton FL

Hirth L, Givord L (1985) Tymoviruses. In: Koenig R (ed) The Plant Viruses Vol 3. Polyhedral Virions with Monopartite RNA Genomes. Plenum Press, New York London, pp 163-212

Kadaré G, Drugeon G, Savithri HS, Haenni A-L (1992) Comparison of the strategies of expression of five tymovirus RNAs by *in vitro* translation studies. J Gen Virol 73: 493-498

Koenig R (1976) A loop structure in the serological classification system of tymoviruses. Virology 72: 1-5

Koenig R, Lesemann D-E (1979) Tymovirus group. CMI/AAB Descriptions of Plant Viruses N° 214, 4pp

Koenig R, Lesemann D-E (1981) Tymoviruses. In: Kurstak E (ed) Handbook of Plant Virus Infections. Comparative Diagnosis. Elsevier/North Holland Biomedical Press, Amsterdam New York Oxford, pp 33-60

Lesemann D-E (1977) Virus group-specific and virus-specific cytological alterations induced by members of the tymovirus group. Phytopathology Z 90: 315-336

Matthews REF (1991) (ed) Plant Virology 3 edn. Academic Press, San Diego

Morch M-D, Boyer J-C, Haenni A-L (1988) Overlapping open reading frames revealed by complete nucleotide sequencing of turnip yellow mosaic virus genomic RNA. Nucl Acids Res 16: 6157-6173

Morch M-D, Drugeon G, Szafranski P, Haenni A-L (1989) Proteolytic origin of the 150 Kilodalton protein encoded by turnip yellow mosaic virus genomic RNA. J Virol 63: 5153-5158

Rana GIL, Castellano MA, Koenig R (1988) Characterization of a tymovirus isolated from *Anagyris foetida* as a strain of scrophularia mottle virus. J Phytopathology 121: 239-249

CONTRIBUTED BY

Koenig R, Lesemann D-E, Commandeur U

GENUS CARLAVIRUS

Type Species carnation latent virus (CLV)

VIRION PROPERTIES

MORPHOLOGY

Virions are slightly flexuous filaments, 610-700 nm in length and 12-15 nm in diameter. Virions exhibit helical symmetry with a pitch of about 3.4 nm.

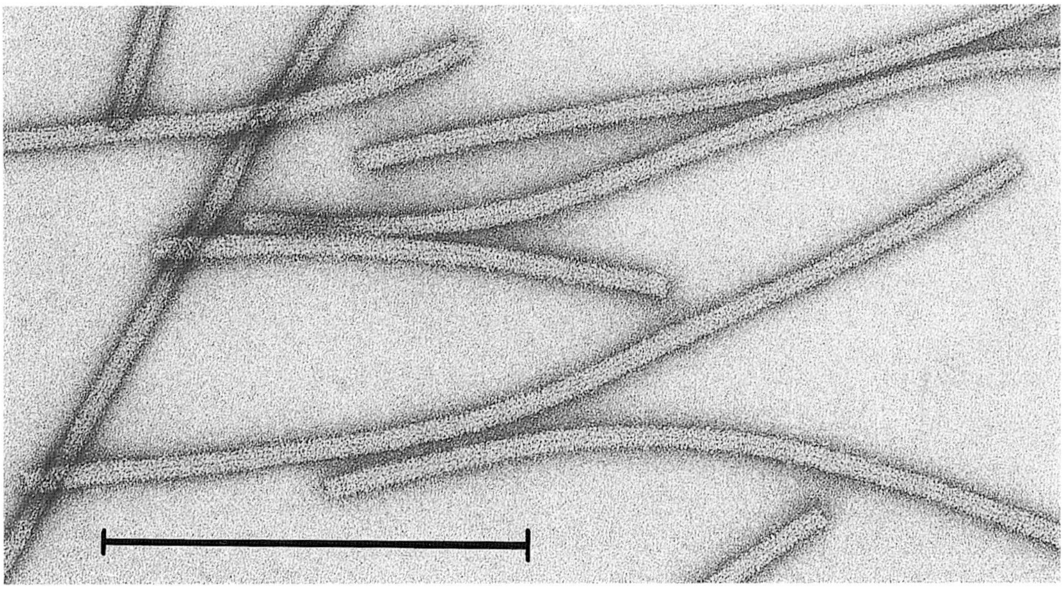

Figure 1: Filamentous particles of carnation latent virus, the bar represents 100 nm (courtesy of Milne RG).

PHYSICOCHEMICAL AND PHYSICAL PROPERTIES

Virion Mr is about 60 x 10^6. Virion S_{20w} is 147-176, and buoyant density in CsCl is 1.3 g/cm^3.

NUCLEIC ACID

Virions contain a single molecule of linear ssRNA, 7.4-7.7 kb in size (although potato virus M is 8.53 kb in size). Some species also have two subgenomic RNAs (2.1-3.3 kb and 1.3-1.6 kb) which are possibly encapsidated in shorter particles. The genomic RNAs have a 3' poly (A) tract, and some have a 5' VPg or a cap structure without a VPg. The RNAs contain six ORFs; the one located at the 3' terminus, which codes for a polypeptide of 10-15 kDa, is apparently similar to that of carlaviruses. The nucleotide sequences of partial sequences of eight carlaviruses have been determined.

PROTEINS

Virions are composed of a single polypeptide (Mr 31-36 x 10^3).

LIPIDS

None reported.

CARBOHYDRATES

None reported.

GENOME ORGANIZATION AND REPLICATION

The genomic RNA of potato virus M contains six large ORFs and non-coding sequences of 75 nt at the 5' terminus, 70 nt followed by a poly (A) tail at the 3' terminus and 38 and 21 nt between the three large blocks of coding sequences. The ORFs code for polypeptides of 5'-223 kDa, 25 kDa, 12 kDa, 7 kDa, 34 kDa and 11 kDa-3'. The gene arrangement of five other

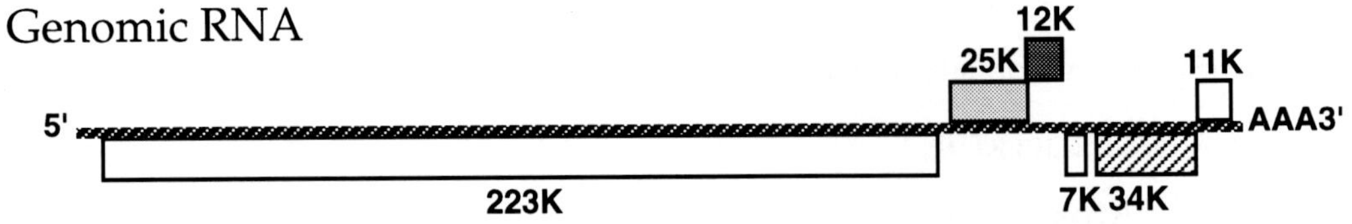

Figure 2: Genome organization of potato M carlavirus (from Zavriev *et al.*, 1991).

incompletely sequenced carlaviruses is similar. The 223 kDa polypeptide is probably the viral RNA replicase. The proteins encoded by the triple gene block (25 kDa, 12 kDa and 7 kDa) may facilitate cell-to-cell movement of virus. The 34 kDa polypeptide is the capsid protein. The function of the 11 kDa polypeptide has yet to be determined, but its ability to bind nucleic acid indicates that it possibly facilitates aphid transmission or is involved in host gene transcription and/or viral RNA replication.

ANTIGENIC PROPERTIES

The viruses are good immunogens. Some members of the group are serologically interrelated, but others are apparently distinct.

BIOLOGICAL PROPERTIES

HOST RANGE

Individual viruses have restricted natural host ranges, but some can infect a wide range of experimental hosts.

TRANSMISSION

Member viruses are transmitted naturally by aphids in a non-persistent manner; two possible member viruses are transmitted by whiteflies. Three of the viruses naturally occurring in leguminous species are seedborne. All the viruses are mechanically transmissible.

GEOGRAPHICAL DISTRIBUTION

The geographic distribution of many species is restricted, but those infecting vegetatively-propagated crops are usually more widely distributed. Most species commonly occur in temperate climates.

CYTOPATHIC EFFECTS

Virions are scattered throughout cytoplasm or occur in membrane-associated bundle-like or plate-like aggregates. Many species also induce the formation of ovoid or irregularly shaped inclusions which are seen by light microscopy as vacuolate bodies; these consist of aggregates of virus particles, mitochondria, endoplasmic reticulum and lipid globules.

LIST OF SPECIES IN THE GENUS

The viruses, their alternative names ( ), genomic sequence accession numbers [ ], CMI/AAB description # ( ) and assigned abbreviations ( ) are:

SPECIES IN THE GENUS

American hop latent virus (262)		(AHLV)
blueberry scorch virus		(BlSV)
cactus virus 2		(CV-2)
caper latent virus		(CapLV)
carnation latent virus (61)	[X55331, X55897]	(CLV)
chrysanthemum virus B (110)	[S60150]	(CVB)
dandelion latent virus		(DaLV)

elderberry virus (263)		(EV)
(elderberry virus A)		
garlic common latent virus		(GCLV)
Helenium virus S (265)	[D10454]	(HVS)
honeysuckle latent virus (289)		(HnLV)
hop latent virus (261)		(HpLV)
hop mosaic virus (241)		(HpMV)
hydrangea latent virus		(HdLV)
kalanchoe latent virus		(KLV)
lilac mottle virus		(LiMV)
lily symptomless virus (96)	[X15343]	(LSV)
(Alstroemeria virus)		
mulberry latent virus		(MLV)
muskmelon vein necrosis virus		(MuVNV)
Nerine latent virus		(NeLV)
(Hippeastrum latent virus)		
Passiflora latent virus		(PLV)
pea streak virus (112)		(PeSV)
(alfalfa latent virus) (211)		
poplar mosaic virus (75)	[X65102, D13364]	(PopMV)
potato virus M (87)	[X53062, X57440, D144449]	(PVM)
potato virus S (60)	[D00461, S45593]	(PVS)
(pepino latent virus)		
red clover vein mosaic virus (22)		(RCVMV)
shallot latent virus (250)		(SLV)
Sint-Jem's onion latent virus		(SJOLV)
strawberry pseudo mild yellow edge virus		(SPMYEV)

Tentative Species in the Genus

1-Aphid-borne:

Anthriscus virus	(AntV)
Arracacha latent virus	(ALV)
artichoke latent virus M	(ArLVM)
artichoke latent virus S	(ArLVS)
butterbur mosaic virus	(ButMV)
caraway latent virus	(CawLV)
Cardamine latent virus	(CaLV)
Cassia mild mosaic virus	(CasMMV)
chicory yellow blotch virus	(ChYNMV)
Chinese yam necrotic mosaic virus	(ChYNMV)
cole latent virus	(CoLV)
Cynodon mosaic virus	(CynMV)
daphne virus S	(DVS)
Dulcamara virus A	(DuVA)
Dulcamara virus B	(DuVB)
eggplant mild mottle virus	(EMMV)
(eggplant virus)	
Euonymus mosaic virus	(EuoMV)
fig virus S	(FVS)
fuchsia latent virus	(FLV)
garlic mosaic virus	(GarMV)
Gentiana virus	(GenV)
Gynura latent virus (strain of Chrysanthemum B?)	(GyLV)
Helleborus mosaic virus	(HeMV)
impatiens latent virus	(ILV)
lilac ringspot virus	(LacRSV)
plantain virus 8	(PlV-8)

Prunus virus S (PruVS)
Southern potato latent virus (SoPLV)
white bryony mosaic virus (WBMV)

2-Whitefly-borne:
cassava brown streak-associated virus (CBSaV)
cowpea mild mottle virus (140) (CPMMV)
(Psophocarpus necrotic mosaic virus)
(groundnut crinkle virus)
(tomato pale chlorosis virus)
(Voandzeia mosaic virus)

SIMILARITY WITH OTHER TAXA

The putative viral replicase gene of carlaviruses shows some sequence similarity with those of alphaviruses, tobamoviruses, tobraviruses and furoviruses, but shows closer homology with those of potexviruses, tymoviruses and closteroviruses. The 25 kDa polypeptide of carlaviruses has some similarity with the 42 kDa and 58 kDa polypeptides of, respectively, beet necrotic yellow vein furovirus and barley stripe mosaic hordeivirus RNA2. The 12 kDa and 7 kDa polypeptides of carlaviruses is similar to comparable polypeptides of potexviruses.

Narcissus latent virus virions are filamentous and about 650 nm long. It was previously considered to be a carlavirus. However, it differs from carlaviruses in inducing the formation of intracellular inclusions ("pinwheels") and having a capsid protein of 46 kDa; it is thus now probably better placed in a separate possible genus of the family *Potyviridae* with maclura mosaic virus to which it is serologically related.

DERIVATION OF NAMES

carla: sigla from *car*nation *la*tent

REFERENCES

Foster GD (1992) The structure and expression of the genome of carlaviruses. Res Virol 143: 103-112

Foster GD, Mills PR (1991) Nucleotide sequence of the 7K gene of carnation latent virus. Pl Molec Biol 15: 937-939

Foster GD, Mills PR (1990) Investigation of the 5' terminal structures of genomic and subgenomic RNAs of potato virus S. Virus Genes 4: 359-366

Foster GD, Mills PR (1990) Evidence for subgenomic RNAs in leaves infected with an Andean strain of potato virus S. Acta virol 35: 260-267

Foster GD, Mills PR (1991) Cell-free translation of American hop latent virus RNA. Virus Genes 5: 327-334

Foster GD, Mills PR (1991) Occurrence of chloroplast ribosome recognition sites within conserved elements of the RNA genomes of carlaviruses. FEBS Lett 280: 341-343

Foster GD, Mills PR (1992) Translation of potato virus S RNA in vitro: evidence of protein processing. Virus Genes 6: 47-52

Gramstat A, Courtpozanis A, Rohde W (1990) The 35 kDa protein of potato virus M displays properties of a nucleic acid-binding regulatory protein. FEBS Lett 276: 34-38

Haylor MTM, Brunt AA, Coutts RHA (1990) Conservation of the 3' terminal nucleotide sequence in five carlaviruses. Nucl Acids Res 18: 6127

Henderson J, Gibbs MJ, Edwards ML, Clark VA, Gardner KA, Cooper JI (1992) Partial nucleotide sequence of poplar mosaic virus RNA confirms classification as a carlavirus. J Gen Virol 73: 1887-1890

Meehan BM, Mills PR (1991) Nucleotide sequence of the 3'-terminal region of carnation latent virus. Intervirology 32: 262-267

Memelink J, van der Vlugt CIM, Linthorst HJM, Derks AFLM, Asjes CJ, Bol JF (1990) Homologies between the genomes of a carlavirus (lily symptomless virus) and a potexvirus (lily virus X) from lily plants. J Gen Virol 71: 917-924

Mowat WP, Dawson S, Duncan GH, Robinson DJ (1991) Narcissus latent, a virus with filamentous particles and a novel combination of properties. Ann Appl Biol 119: 31-46

Turner RL, Mills PR, Foster GD (1993) Nucleotide sequence of the 7kDa gene of Helenium virus S. Acta Virol 37: 523-528

Zavriev SK, Kanyuka KV, Leavy KE (1991) The genome organisation of potato virus M RNA. J Gen Virol 72: 9-14

CONTRIBUTED BY

Brunt AA

Genus *Potexvirus*

Virion Properties

Morphology

Virions are flexuous helical rods; 470-580 nm in length and 13 nm in diameter. The pitch of the helix is between 3.3 and 3.7 Å. A central axial hole (canal) has been seen only occasionally (about 3 nm in diameter). The number of protein subunits per turn of the primary helix is slightly less than 9.0. The RNA backbone is at a radial position of 3.3 nm.

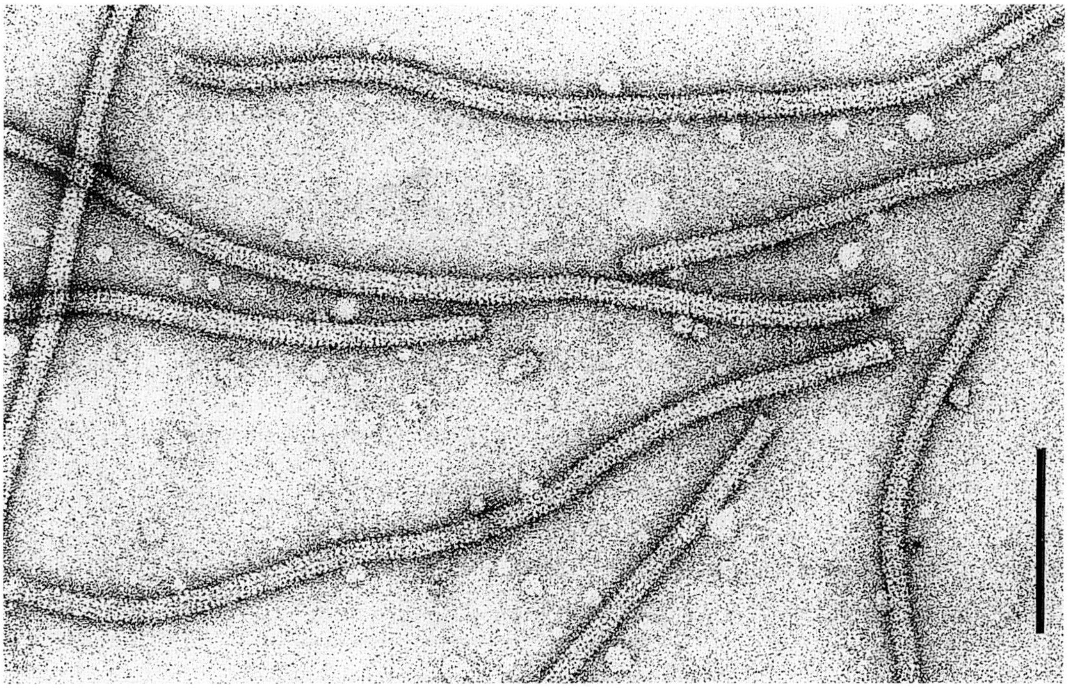

Figure 1: Negative contrast electron micrograph of potato virus X particles. The bar represents 100 nm.

Physicochemical and Physical Properties

Virion Mr is about 3.5×10^6; S_{20w} is 115-130; buoyant density in CsCl is 1.31 g/cm^3.

Nucleic Acid

The genome is a single linear molecule of positive sense ssRNA; Mr of genomic RNA is 2.1-2.3×10^6 (about 6% by weight of the virion). The RNA is capped and 3' polyadenylated. The size of the genomic RNA of potato virus X (the type species of the genus) is 6,435 bases, white clover mosaic virus is 5,845 bases, of clover yellow mosaic virus is 7,015 bases, of papaya mosaic virus is 6,656 bases, of narcissus mosaic virus is 6,955 bases. All these RNAs have been sequenced.

Proteins

Virion nucleocapsids consist of 1,000-1,500 protein subunits of a single type; (Mr 18-27 x 10^3). Partial proteolytic cleavage of coat protein (CP) molecules can occur during storage of purified virus. Four non-structural proteins are coded by the PVX genome including an RNA polymerase (165 kDa) and three proteins (25 kDa, 12 kDa and 8 kDa) involved in cell-to cell spread of infection (Fig. 2).

Lipids

None reported.

CARBOHYDRATES

None reported.

GENOME ORGANIZATION AND REPLICATION

Virions of PVX contain only genomic RNA; however some potexviruses may also encapsidate the subgenomic RNA for the CP. Genomic RNA is translated as functionally monocistronic: only the 5'-proximal RNA-polymerase gene is translated directly by ribosomes, producing the 150-181 kDa protein (RNA polymerase).

The 5'-untranslated leader sequence of PVX RNA (αβ -leader) consists of 83 nts (apart from cap-structure) and has been shown to act as an efficient translational enhancer.

The CP gene (ORF 5) is located at the 3'-proximal position of PVX RNA and between ORF 1 and ORF 5 a block of three overlapping ORFs is present. The products of the triple gene block (25 kDa, 12 kDa, 8 kDa) are involved in cell-to-cell movement of viral genetic material. The 25 kDa protein (as well as the 165 kDa replicase) contain an NTPase-helicase domain, however the 25 kDa protein is not involved in RNA replication. The 12 kDa and 8 kDa contain large blocks of uncharged amino acids and are membrane-bound. A similar triple gene block has been revealed in genomic RNAs of furo-, hordei- and carlaviruses. In all these cases the products of the triple gene block are responsible for the movement function.

All the 5' -distal genes (ORFs 2 to 5) are expressed via the production (and subsequent translation) of appropriate subgenomic RNAs (sgRNAs). From two to three 3'-coterminal sgRNAs can be isolated from plants infected with potexviruses (2.1; 1.2 and 1.0 kb). And the double-stranded counterparts of these sgRNAs have been also revealed. It is probable that the medium-size sgRNA (1.2 kb) is functionally bicistronic, producing the 12 kDa and 8 kDa proteins upon translation.

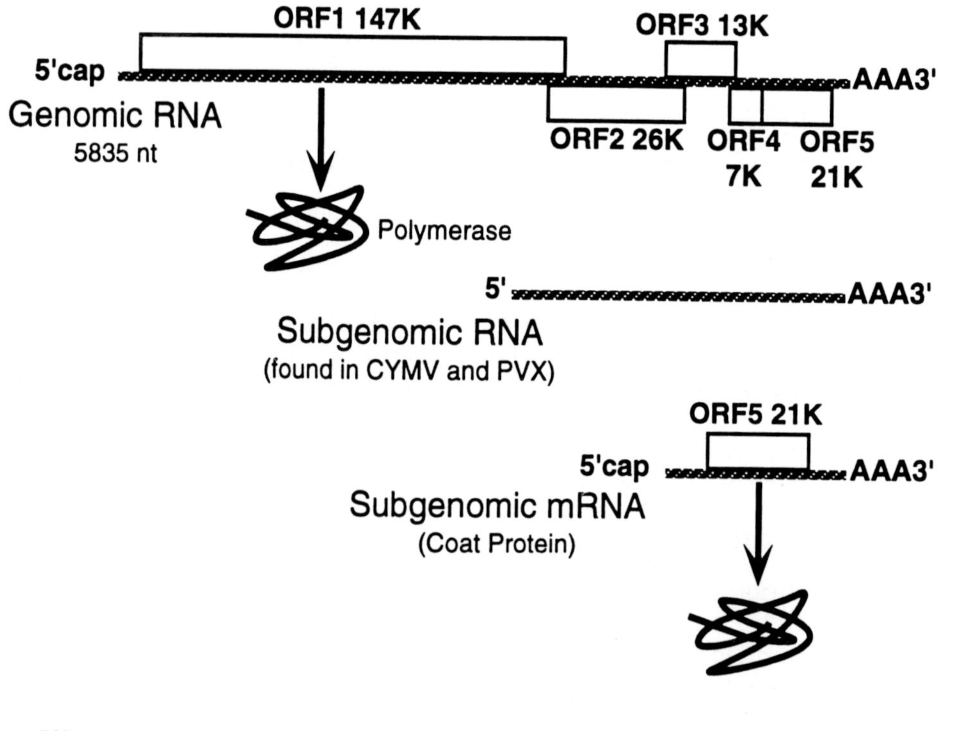

Figure 2: PVX genome structure and expression.

ANTIGENIC PROPERTIES

Virions are highly immunogenic; some members are antigenically related.

Biological Properties

The viruses are usually moderately pathogenic, causing mosaic or ringspot symptoms in a wide range of mono- and dicotyledonous plants. The host range of individual members is limited. The viruses are readily transmissible by manual inoculation; no vectors are known. The viruses are transmitted in nature by mechanical contacts and have world-wide distribution.

List of Species in the Genus

The viruses, their genomic sequence accession numbers [ ], CMI/AAB description # ( ) and assigned abbreviations ( ) are:

Species in the Genus

asparagus virus 3		(AV-3)
cactus virus X		(CVX)
cassava virus X		(CsVX)
clover yellow mosaic virus	[M63511, M63512, M63513 M63514, D00485]	(ClYMV)
Commelina virus X		(ComVX)
Cymbidium mosaic virus	[X62663, X62664, X62133]	(CymMV)
foxtail mosaic virus	[M62730]	(FoMV)
hydrangea ringspot virus		(HRSV)
lily virus X		(LVX)
narcissus mosaic virus		(NMV)
Nerine virus X		(NVX)
papaya mosaic virus		(PapMV)
pepino mosaic virus		(PepMV)
Plantago severe mottle virus		(PlSMV)
plantain virus X		(PlVX)
potato aucuba mosaic virus		(PAMV)
potato virus X		(PVX)
tulip virus X		(TVX)
viola mottle virus		(VMV)
white clover mosaic virus		(WClMV)

Tentative Species in the Genus

artichoke curly dwarf virus		(ACDV)
bamboo mosaic virus		(BaMV)
barley virus B1		(BarV-B1)
Boletus virus		(BolV)
cassava common mosaic virus (90)		(CsCMV)
Centrosema mosaic virus		(CenMV)
daphne virus X (195)		(DVX)
Dioscorea latent virus		(DLV)
lychnis virus		
Malva veinal necrosis virus		(MVNV)
Nandina mosaic virus		(NaMV)
negro coffee mosaic virus		(NeCMV)
parsley virus 5		(PaV-5)
parsnip virus 3		(ParV-3)
parsnip virus 5		(ParV-5)
rhododendron necrotic ringspot virus		(RoNRSV)
rhubarb virus 1		(RV-1)
Smithiantha virus		(SmiV)
strawberry mild yellow edge-associated virus	[D12517, D12515, D01227, D00866]	(SMYEaV)

wineberry latent virus	(WLV)
Zygocactus virus	(ZV)

References

Sit TL, Abou Haidar MG, Holy S (1989) Nucleotide sequence of papaya mosaic virus RNA. J Gen Virol 70: 2325-2331

Dolja VV, Grama DP, Morozov SY, Atabekov JG (1987) Potato virus X-related ss and ds RNAs. FEBS Lett 214: 308-312

Forster RLS, Bevan MV, Harrison SA, Gardner RC (1988) The complete nucleotide sequence of the potexvirus white clover mosaic virus. Nucl Acids Res. 16: 291-303

Huisman MJ, Linthorst HJM, Bol JF, Cornellisen BJC (1988) The complete nucleotide sequence of potato virus X and its homologies at the amino acid level with various plus-stranded RNA Viruses. J Gen Virol 69: 1789-1798

Jelkman W, Martin RR, Lesemann D-E, Velten HJ, Skelton F (1990) A new potexvirus associated with strawberry mild yellow edge disease. J Gen Virol 71: 1251-1258

Koenig R, Lesemann D-E (1978) Potexvirus group. In CMI/ABB Description of plant viruses N° 200, 4pp

Morozov SY, Miroshnichenko NA, Solovyev AG, Fedorkin ON, Zelenina DA, Lukasheva LI, Karasev AV, Dolja VV, Atabekov JG (1991). Expression strategy of the potato virus X tripple gene block. J Gen Virol 72: 2039-2042

Purcifull D, Edwardson JR, (1981) Potexviruses In: Kurstak E (ed), Plant Virus Infections: Comparative Diagnosis Elsevier/North Holland, Amsterdam pp 627-624

Sit TL, White KA, Holy S, Padmanabhan U, Eweida M, Hiebert M, Mackie GA, Abou Haidar MG (1990) Complete sequence of clover yellow mosaic virus RNA. J Gen Virol 71: 1913-1920

Skryabin KG, Morozov SY, Kraev AS, Rozanov MV, Chernov BK, Lukasheva LI, Atabekov JG (1988) Conserved and variable elements in RNA genomes of potexviruses. FEBS Lett 240: 33-40

Smirnyagina EV, Morozov SY, Rodionova NP, Miroshnichenko NA, Solovyev AG, Fedorkin ON, Atabekov JG (1991). Translational efficiency and competitive ability of mRNAs with 5' -untranslated -leader of potato virus X RNA. Biochimie 73: 587-598

Tollin P, Wilson MR (1988) Particle structure. In: Milne RG (ed) The plant viruses Vol 4. Plenum Press, New York London, pp 51-83

Tomashevskaya OL, Solovyev AG, Karpova OV, Fedorkin ON, Rodionova NP, Morozov SY, Atabekov JG (1993) Effects of sequence elements in the potato virus X RNA non-translated ab-leader on its tramslation enhancing activity. J Gen Virol 74: 2717-2724

Wodnar-Filipowicz A, Skizeczkowski LJ, Filipowicz W (1990) Translation of potato virus X RNA into high molecular weight proteins. FEBS Lett 109: 151-155

Zuidema D, Linthorst HJM, Huisman MJ, Asjes CJ, Bol JF (1989) Nucleotide sequence of narcissus mosaic virus RNA. J Gen Virol 70: 267-276

Contributed By

Atabekov JG

FAMILY *BARNAVIRIDAE*

TAXONOMIC STRUCTURE OF THE FAMILY

Family *Barnaviridae*
Genus *Barnavirus*

GENUS *BARNAVIRUS*

Type Species mushroom bacilliform virus (MBV)

VIRION PROPERTIES

MORPHOLOGY

Virions are bacilliform, nonenveloped and lack prominent surface projections. Typically, virions are 19 x 50 nm, but range between 18-20 nm in width and 48-53 nm in length. Optical diffraction patterns of the virions resemble those of alfalfa mosaic virus, suggesting a morphological subunit diameter of about 10 nm and a T = 1 icosahedral symmetry.

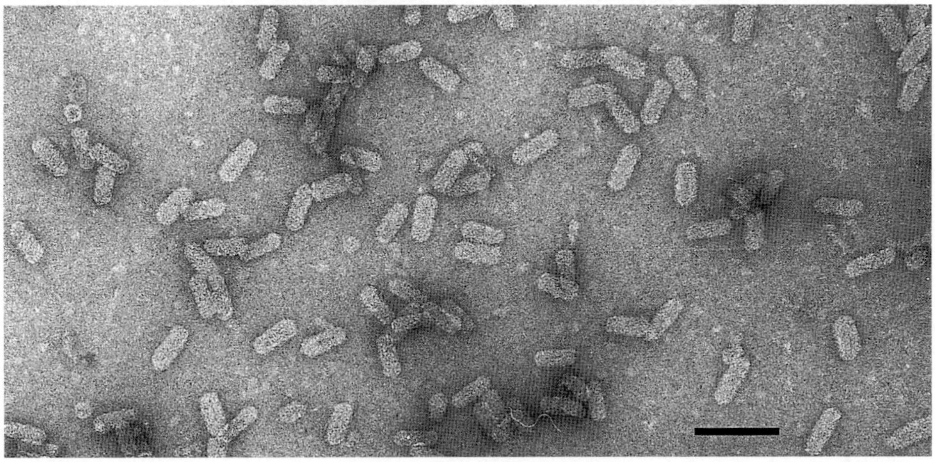

Figure 1: Negative contrast electron micrograph of mushroom bacilliform virus (MBV). The bar represents 100 nm.

PHYSICOCHEMICAL AND PHYSICAL PROPERTIES

Virion Mr is 7.1×10^6, buoyant density in $CsSO_4$ is 1.32 g/cm^3. Virions are stable between pH 6 and 8 and ionic strength of 0.01 to 0.1 M phosphate, and are insensitive to chloroform.

NUCLEIC ACID

Virions contain a single molecule of a positive sense ssRNA, 4.4 kb in size. The RNA has a Mr of 1.4×10^6 and constitutes about 20% of virion weight.

PROTEINS

Virions are composed of a single major capsid protein (Mr 24.4×10^3). There are probably 240 molecules forming the capsid. No RNA-dependent RNA polymerase activity has been found associated with purified virions.

LIPIDS

None reported.

CARBOHYDRATES

None reported.

Genome Organization and Replication

In a cell-free system, the genomic RNA directs the synthesis of a major 77 kDa polypeptide and possibly four minor translation products of 37 kDa, 28 kDa, 24 kDa, and 21 kDa. The full-length genomic RNA and a 1.8 kb RNA, probably a subgenomic RNA, are found in infected cells. Virions accumulate singly or as aggregates in the cytoplasm of infected cells.

Antigenic Properties

Mushroom bacilliform virus (MBV) is highly immunogenic.

Biological Properties

The virus is restricted to the common cultivated mushroom *Agaricus bisporus*. However, bacilliform particles, which are morphologically-identical to MBV, have been observed in the field mushroom *Agaricus campestris*. Transmission is horizontal via mycelium and probably basidiospores. Distribution of MBV coincides with that of the commercial cultivation of mushrooms (*A. bisporus*); the virus has been reported to occur in most major mushroom-growing countries. MBV occurs as a single infection, but more commonly as a mixed infection with a dsRNA virus (La France isometric virus, LIV) in mushrooms affected with La France disease. MBV is not involved in all episodes of the disease, suggesting it does not have an obligatory role in pathogenesis. MBV RNA and LIV dsRNAs do not share extensive sequence homology.

List of Species in the Genus

The viruses, and their assigned abbreviations ( ) are:

Species in the Genus

mushroom bacilliform virus	(MBV)

Tentative Species in the Genus

None reported.

Similarity with Other Taxa

None reported.

Derivation of Name

barna: from *ba*cilliform-shaped *RNA* viruses

References

Buck KW (1986) Fungal virology- an overview. In: Buck KW (ed) Fungal virology. CRC Press, Boca Raton FL, pp 1-84

Ghabrial SA (1994) New developments in Fungal Virology, Adv Virus Res. 43: 303-388

Goodin MM, Schlagnhaufer B, Romaine CP (1992) Encapsidation of the La France disease specific double-stranded RNAs in 36-nm isometric virus-like particles. Phytopathology 82: 285-290

Moyer JW, Smith SH (1976) Partial purification and antiserum production to the 19 x 50 nm mushroom virus particle. Phytopathology 66: 1260-1261

Moyer JW, Smith SH (1977) Purification and serological detection of mushroom virus-like particles. Phytopathology 67: 1207-1210

Romaine CP, Schlagnhaufer B (1991) Hybridization analysis of the single-stranded RNA bacilliform virus associated with La France disease of *Agaricus bisporus*. Phytopathology 81: 1336-1340

Tavantzis SM, Romaine CP, Smith SH (1980) Purification and partial characterization of a bacilliform virus from *Agaricus bisporus*: a single-stranded RNA mycovirus. Virology 105: 94-102

Tavantzis SM, Romaine CP, Smith SH (1983) Mechanism of genome expression in a single-stranded RNA virus from the cultivated mushroom *Agaricus bisporus*. Phytopathology 106: 45-50

van Zaayen AM (1979) Mushroom viruses. In: Lemke PA (ed) Viruses and plasmids in fungi. Marcel Dekker, New York, pp 239-324

Contributed By

Romaine CP

Genus *Marafivirus*

Type Species maize rayado fino virus (MRFV)

Virion Properties

Morphology

Virions exhibit icosahedral symmetry, are 28-32 nm in diameter, and do not have an envelope. Capsomer arrangement is readily seen in electron micrographs (Fig. 1).

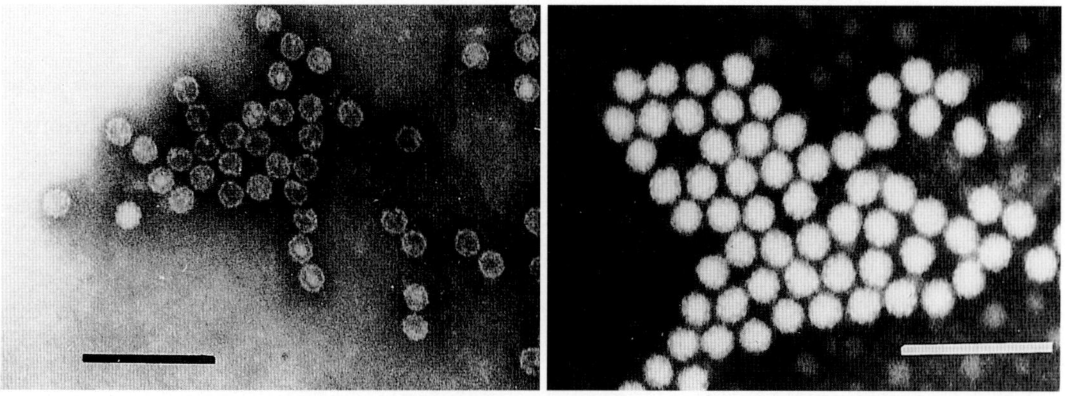

Figure 1: Negative contrast electron micrographs of particles of maize rayado fino virus, (left) top component, (right) bottom component. Bars represent 100 nm.

Physicochemical and Physical Properties

Purified virus sediments as two components: top component (no RNA) (T) and bottom component (B). S_{20w} are 52 (47-57) (T) and 120 (118-124) (B). Buoyant densities in CsCl are 1.26-1.28 g/cm^3 (T) and 1.42-1.46 g/cm^3 (B) and in Cs_2SO_4 are 1.24 g/cm^3 (T) and 1.37 g/cm^3 (B).

Nucleic Acid

Virions contain one molecule of linear positive sense ssRNA, Mr 2.0-2.4 x 10^6. RNA constitutes 25-30% of B particles by weight.

Proteins

Virions are composed of a single major capsid protein (Mr 27 x 10^3) (Bermuda grass etched-line virus) or a major protein (Mr 22 x 10^3) and a sequence related minor protein (Mr 28 x 10^3) (some isolates of maize rayado fino virus).

Lipids

None reported.

Carbohydrates

None reported.

Genome Organization and Replication

Virion RNA is translated to yield polypeptides ranging in size from Mr 15-165 x 10^3. However, no viral coat protein has been detected in *in vitro* translation products.

Antigenic Properties

Virions are moderately immunogenic. No serological relationship exists between maize rayado fino virus and oat blue dwarf virus. Bermuda grass etched-line virus is serologically related to both maize rayado fino virus and oat blue dwarf virus.

Biological Properties

The viruses generally have narrow host ranges restricted to the family *Gramineae*. One member, oat blue dwarf virus, has a wide host range including dicotyledonous plants. The viruses are transmitted by leafhoppers; manual transmission is difficult. Replication of marafiviruses in their vectors is suggested by serial passage experiments and an increase of virus structural proteins in vectors with time after infection.

List of Species in the Genus

The viruses, their CMI/AAB description # ( ) and assigned abbreviations ( ) are:

Species in the Genus

Bermuda grass etched-line virus	(BELV)
maize rayado fino virus (220)	(MRFV)
oat blue dwarf virus (123)	(OBDV)

Tentative Species in the Genus

None reported.

Derivation of Names

marafi: sigla from *ma*ize *ra*yado *fi*no

References

Banttari EE, Moore MB (1962) Virus cause of blue dwarf of oats and its transmission to barley and flax. Phytopathology 52: 897-902

Banttari EE, Zeyen RJ (1969) Chromatographic purification of the oat blue dwarf virus. Phytopathology 59: 183-186

Espinoza AM, Ramirez P, Leon P (1988) Cell-free translation of maize rayado fino virus genomic RNA. J Gen Virol 69: 757-762

Falk BW, Tsai JH (1986) The two capsid proteins of maize rayado fino virus contain common peptide sequences. Intervirology 25: 111-116

Gamez RA (1973) Transmission of rayado fino virus of maize (*Zea mays*) by *Dalbulus maidis*. Ann Appl Biol 73: 285-292

Gingery RE, Gordon DT, Nault LR(1982) Purification and properties of an isolate of maize rayado fino virus from the United States. Phytopathology 72: 1313-1318

Leon P, Gamez RA (1986) Biologia molecular del virus del rayado fino del maiz. Rev Biol Trop 34: 111-114

Leon P, Gamez RA (1981) Some physicochemical properties of maize rayado fino virus. J Gen Virol 56: 67-75

Lockhart BEL, Khaless N, Lennon AM, El Maatauoi M (1985)Properties of Bermuda grass etched-line virus, a new leafopper-transmitted virus related to maize rayado fino ad oat blue dwarf viruses. Phytopathology 75: 1258-1262

Pring DR, Zeyen RJ, Banttari EE (1973) Isolation and characterization of oat blue dwarf virus ribonucleic acid. Phytopathology 63: 393-396

Rivera C, Gamez RA (1986) Multiplication of maize rayado fino virus in the leafhopper vector *Dalbulus maidis*. Intervirology 25: 76-80

Contributed By

Tomaru K, Toriyama S, Gingery RE, Gamez RA

SUBVIRAL AGENTS: SATELLITES

DEFINITION

Satellites are sub-viral agents composed of nucleic acid molecules that depend for their productive multiplication on co-infection of a host cell with a helper virus. Satellite nucleic acids have substantially distinct nucleotide sequences from those of the genomes of either their helper virus or host. When a satellite encodes the coat protein in which its nucleic acid is encapsidated it is referred to as a satellite virus.

CATEGORIES OF SATELLITES

dsDNA satellites
ssDNA satellite viruses
dsRNA satellites
ssRNA satellite viruses
ssRNA satellites

DISTINGUISHING FEATURES

Satellites are characterized by their dependence on a helper virus. However, the reasons for their dependency are various. For most satellites, dependence is for genome replication functions, but for others dependence is for encapsidation. Some viruses are defective in other biologically essential properties, such as vector transmission, but these have usually been classified along with similar viruses with intact genomes rather than as satellites. Some satellites multiply poorly or only in rare circumstances in the absence of their helper virus, but most are absolutely dependent on the helper virus being present. Thus, the boundary between satellites and viruses is not always clear cut.

Satellites are genetically distinct from their helper virus by virtue of having a substantially nucleotide sequence different from that of their helper virus. However, some satellites have short sequences, often at termini that are the same as those of the helper. This is presumably because nucleic acids of both satellite and helper depend on the same viral enzymes for replication. Satellites are thus distinct from defective interfering particles or RNAs because these are wholly derived from their 'helper' virus genomes.

Satellites do not constitute a homogeneous taxonomic group. Some are related to viruses in particular families or genera; the dsDNA satellite P4 is classified in the family *Myoviridae*, the ssDNA adeno-associated viruses are classified in the family *Parvoviridae* and the ssRNA hepatitis delta virus is classified in the genus *Deltavirus*. However, others are not classified among the viruses. The descriptions in this section are meant only to provide a classification framework and nomenclature to assist in the description and identification of satellites. The arrangement adopted is based largely on features of the genetic material of the satellites. The nature of the helper virus and of the helper virus host are important secondary characters.

There appears to be no taxonomic correlation between the viruses that are associated with satellites; satellitism would appear to have arisen many times during virus evolution. A further complication is that some viruses are associated with more than one satellite. Satellites can even depend on both a second satellite and a helper virus for multiplication.

Most known satellites are ssRNA satellites, with ssRNA plant viruses as helpers. It can be very difficult to distinguish between satellite and genome RNA (e.g., in the case of the dsRNA satellites of fungus viruses) and it is very likely that other satellites, some with novel combinations of characters, remain to be discovered.

dsDNA SATELLITES

DISTINGUISHING FEATURES

The only example in this category is the satellite bacteriophage P4 in the family *Myoviridae*. The helper viruses are bacteriophage P2 and related phages. P4 contains 10-15 genes and depends on P2 for late gene functions. No P4-specific antigens are present in particles containing P4 DNA but the P4 particles have smaller heads than P2 particles. P4 DNA can infect its enterobacterial host, replicate and cause lysogeny without P2 being present.

LIST OF SPECIES

P4

REFERENCES

Bertani LE, Six EW (1988) The P2-like phages and their parasite, P4. Annu Rev Genetics 24: 465-490
Christie GE, Calendar R (1990) Interactions between satellite bacteriophage P4 and its helper. In Calendar R (ed) The Bacteriophages 2: 73-143. Plenum press, New York

ssDNA SATELLITE VIRUSES

DISTINGUISHING FEATURES

This category comprises the satellites with ssDNA encapsidated in satellite-encoded protein structures. The only examples are members of the genus *Dependovirus* in the family *Parvoviridae*. In some cultured cells, dependoviruses can replicate without a helper virus being present. Normally, it is infection by a helper adenovirus or herpesvirus which renders the intracellular milieu permissive for dependovirus replication. Under non-permissive conditions the dependovirus genome integrates in the host genome to establish a latent infection.

LIST OF SPECIES

adeno-associated virus 1	(AAV-1)
adeno-associated virus 2	(AAV-2)
adeno-associated virus 3	(AAV-3)
adeno-associated virus 4	(AAV-4)
adeno-associated virus 5	(AAV-5)
avian adeno-associated virus	(AAAV)
bovine adeno-associated virus	(BAAV)
canine adeno-associated virus	(CAAV)
equine adeno-associated virus	(EAAV)
ovine adeno-associated virus	(OAAV)

REFERENCES

Berns KI (1990) Parvovirus replication. Microbiol Rev 54: 316-329

dsRNA SATELLITES

DISTINGUISHING FEATURES

The only examples in this category are satellites found in association with viruses of the family *Totiviridae*. The 1 to 1.8 kbp dsRNA genomes encode 'killer' proteins and are encapsidated in helper virus coat protein; these particles often also contain a positive sense single-stranded copy of the dsRNA

LIST OF SPECIES

M satellites of yeast

LIST OF TENTATIVE SPECIES

M satellites of Ustilago maydis killer virus

REFERENCES

Wickner RB (1992) Double-stranded and single-stranded RNA viruses of *Saccharomyces cerevisiae*. Annu Rev Microbiol 46: 347-375

Shelbourn SL, Day PR, Buck KW (1988) Relationships and functions of virus double-stranded RNA in a P4 killer strain of *Ustlilago maydis*. J Gen Virol 69: 975-982

ssRNA SATELLITE VIRUSES

DISTINGUISHING FEATURES

This category comprises the satellites with ssRNA genomes encapsidated in satellite-encoded protein structures. Several types are known. In all cases the satellite virus particles are antigenically, and usually morphologically, distinct from those of the helper virus.

Two different subgroups of satellite viruses are distinguished: chronic bee-paralysis virus associated satellite and tobacco necrosis virus satellite.

SUBGROUP 1: CHRONIC BEE-PARALYSIS VIRUS ASSOCIATED SATELLITE

DISTINGUISHING FEATURES

Satellite particles are found in bees infected with the helper, chronic bee-paralysis virus (CPV). Particles are about 17 nm in diameter and serologically unrelated to those of CPV. Satellite RNA is also found encapsidated in CPV coat protein. The RNA consists of 3 species, about 1 kb in size, which are distinct from CPV RNA but some T1 oligonucleotides appear to be common to CPV RNA and to satellite RNA. The satellite interferes with CPV replication.

LIST OF SPECIES

chronic bee-paralysis virus associate satellite

REFERENCES

Overton HA, Buck KW, Bailey L, Ball BV (1982) Relationships between the RNA components of chronic bee-paralysis virus and those of chronic bee-paralysis virus associate. J Gen Virol 63: 171-179

SUBGROUP 2: TOBACCO NECROSIS VIRUS SATELLITE

DISTINGUISHING FEATURES

Satellite particles are found in plant hosts in association with taxonomically diverse helper viruses. Particles are isometric, about 17 nm in diameter, and comprise 60 copies of a single protein (Mr 17×10^3 to 24×10^3). Some satellite RNAs contain a second ORF.

LIST OF SPECIES

maize white line mosaic virus satellite
Panicum mosaic virus satellite
tobacco mosaic virus satellite
tobacco necrosis virus satellite

REFERENCES

Masuta C, Zuidema D, Hunter BG, Heaton LA, Sopher DS, Jackson AO (1987) Analysis of the satellite panicum mosaic virus. Virology 159: 329-338

Mirkov TE, Mathews DM, du Plessis DH, Dodds JA (1989) Nucleotide sequence and translation of satellite tobacco mosaic virus RNA Virology 170: 139-146

Ysebaert M, van Emmelo J, Fiers W (1980) Total nucleotide sequence of a nearly full-size DNA copy of satellite tobacco necrosis virus RNA. J Mol Biol 143: 273-287
Zhang L, Zitter TA, Palukaitis P (1991) Helper virus-dependent replication, nucleotide sequence and genome organization of the satellite virus of maize white line mosaic virus. Virology 180: 467-473

ssRNA SATELLITES

DISTINGUISHING FEATURES

This category comprises the satellites with ssRNA genomes which do not encode a capsid protein. Particles containing satellite RNA are antigenically identical to those of the helper virus and can sometimes be distinguished by physical features such as sedimentation rates. Four different subgroups of virus satellites are distinguished: genus deltavirus, B type mRNA satellites, C type linear satellites, D type circular satellites.

SUBGROUP 1: GENUS DELTAVIRUS

DISTINGUISHING FEATURES

The only described example of this category is hepatitis delta virus. It is described more fully under the genus *Deltavirus.* The RNA is circular, 1.7 kb in size and encodes proteins used during its replication. The natural helper virus is hepatitis B virus; woodchuck hepatitis virus can act as a surrogate helper virus.

LIST OF SPECIES

hepatitis delta virus (HDV)

REFERENCES

Taylor JM (1992) The structure and replication of hepatitis delta virus. Annu Rev Microbiol 46: 253-276

SUBGROUP 2: B TYPE mRNA SATELLITES

DISTINGUISHING FEATURES

This category comprises satellites with genomes that are 0.8 to 1.5 kb in size and encode a non-structural protein which, at least in some cases, is essential for satellite RNA multiplication. Little sequence homology exists between satellite and helper, some satellites can be exchanged among different helper viruses. These satellites rarely modify the disease induced in host plants by the helper virus.

LIST OF SPECIES

Arabis mosaic virus large satellite
bamboo mosaic virus satellite
chicory yellow mottle virus large satellite
grapevine Bulgarian latent virus satellite
grapevine fanleaf virus satellite
myrobalan latent ringspot virus satellite
pea enation mosaic virus satellite
strawberry latent ringspot virus satellite
tomato black ring virus satellite

LIST OF TENTATIVE SPECIES

beet western yellows virus satellite
groundnut rosette virus satellite

REFERENCES

Demler SA, de Zoeten GA (1989) Characterisation of a satellite RNA associated with pea enation mosaic virus. J Gen Virol 70: 1075-1084

Falk BW, Duffus JE (1984) Identification of small single- and double-stranded RNAs associated with severe symptoms in beet western yellows virus-infected *Capsella bursa-pastoris*. Phytopathology 74: 1224-1229
Fritsch C, Mayo MA, Hemmer O (1993) Properties of satellite RNA of nepoviruses. Biochimie 75: 561-567
Fuchs M, Pinck M, Serghini MA, Ravelonandro M, Walter B, Pinck L (1989) The nucleotide sequence of satellite RNA in grapevine fanleaf virus strain F13. J Gen Virol 70: 955-962
Hemmer O, Meyer M, Greif C, Fritsch C (1987) Comparison of the nucleotide sequences of five tomato black ring virus satellite RNAs. J Gen Virol 68: 1823-1833
Kreiah S, Cooper JI, Strunk G (1993) The nucleotide sequence of a satellite RNA associated with strawberry latent ringspot virus. J Gen Virol 74: 1163-1165
Liu YY, Helen CUT, Cooper JI, Bertioli DJ, Coates D, Bauer G (1990) The nucleotide sequence of a satellite RNA associated with arabis mosaic nepovirus. J Gen Virol 71: 1259-1263
Rubino L, Tousignant ME, Steger G, Kaper JM (1990) Nucleotide sequence and structural analysis of two satellite RNAs associated with chicory yellow mottle virus. J Gen Virol 71: 1897-1903

SUBGROUP 3: C TYPE LINEAR RNA SATELLITES

DISTINGUISHING FEATURES

This category comprises the satellites with genomes less than 0.7 kb that do not encode functional proteins. No circular molecules are present in infected cells. Satellites can substantially modify the symptoms of helper virus infection.

LIST OF SPECIES

cucumber mosaic virus satellite (several types)
Panicum mosaic virus small satellite
peanut stunt virus satellite
turnip crinkle virus satellite

LIST OF TENTATIVE SPECIES

Cymbidium ringspot virus satellite
tobacco necrosis virus small satellite
tomato bushy stunt virus satellite

REFERENCES

Collmer C, Howell S (1992) Role of satellite RNA in the expression of symptoms caused by plant viruses. Annu Rev Phytopath 30: 419-442
Gallitelli D, Hull R (1985) Characterization of satellite RNAs associated with tomato bushy stunt virus and five other definitive tombusviruses. J Gen Virol 66: 1533-1543
Masuta C, Zaidema D, Hunter BG, Heath LA, Sopher DS, Jackson AO (1987) Analysis of the genome of panicum mosaic virus. Virology 159: 321-338
Naidu RA, Collins GB, Ghabrial SA (1991) Symptom-modulating properties of peanut stunt virus satellite RNA sequence variants. Mol Plant Microbe Interact 4: 268-275
Rubino L, Burgyan J, Grieco F, Russo M (1990) Sequence analysis of cymbidium ringspot virus satellite and defective interfering RNAs. J Gen Virol 71: 1655-1660
Simon AE, Howell SH (1986) The virulent satellite RNA of turnip crinkle virus has a major domain homologous to the 3' end of the helper virus genome. EMBO J 5: 3423-3428

SUBGROUP 4: D TYPE CIRCULAR RNA SATELLITES

DISTINGUISHING FEATURES

This category comprises the satellites with genomes that are about 350 nucleotides long and occur as circular as well as linear molecules. Replication of some has been shown to involve self-cleavage of circular progeny molecules by an RNA-catalyzed reaction.

LIST OF SPECIES

Arabis mosaic virus small satellite
barley yellow dwarf virus satellite
lucerne transient streak virus satellite
Solanum nodiflorum mottle virus satellite

subterranean clover mottle virus satellite (2 types)
tobacco ringspot virus satellite
velvet tobacco mottle virus satellite

REFERENCES

Abou Haidar MG, Paliwal YC (1988) Comparison of the nucleotide sequences of viroid-like satellite RNAs of the Canadian and Australian strains of lucerne transient streak virus. J Gen Virol 69: 2369-2373

Buzayan JM, Gerlach WL, Bruening G, Keese P, Gould AR (1986) Nucleotide sequence of satellite tobacco ringspot virus RNA and its relationship to multimeric forms. Virology 151: 186-199

Davies C, Haseloff J, Symons RH (1990) Structure self-cleavage and replication of two viroid-like satellite RNAs (virusoids) of subterranean clover mottle virus. Virology 177: 216-224

Kaper JM, Tousignant ME, Steger MT (1988) Nucleotide sequence predicts circula-rity and self-cleavage of 300-ribonucleotide satellite of arabis mosaic virus. Biochem Biophys Res Commun 154: 318-325

Miller WA, Hercus T, Waterhouse PM, Gerlach WL (1991) A satellite of barley yellow dwarf virus contains a novel hammerhead structure in the self-cleavage domain. Virology 183: 711-720

Rubino L, Tousignant ME, Steger G, Kaper JM (1990) Nucleotide sequence and structural analysis of two satellite RNAs associated with chicory yellow mottle virus. J Gen Virol 71: 1897-1903

CONTRIBUTED BY

Mayo MA, Berns KI, Fritsch C, Kaper JM, Jackson AO, Leibowitz MJ, Taylor JM

SIMILARITY WITH OTHER TAXA

The involvement of reverse transcription in the replication of the hepatitis B and delta viruses is similar to that of retroviruses and cauliflower mosaic virus.

The genome structure and catalytic activities of HDV closely resemble those of viroids and satellite viruses found in certain plants and animals. The translation of HDV RNA and its helper-dependency on other hepadnaviruses for the formation of new particles distinguishes it from plant associated viroids.

DERIVATION OF NAMES

delta: from Greek letter Δ, "D"

REFERENCES

Lai MMC, Chao Y-C, Chang M-F, Lin J-H, Gust ID (1991) Functional studies of hepatitis delta antigen and delta virus RNA. In: Gerin JL, Purcell RH, Rizzeto M (eds) The hepatitis delta virus. Alan R. Liss, New York, pp 283-292

Taylor JM (1991) Human hepatitis delta virus. Curr Top Micro Immunol 168: 141-166

Taylor JM (1991) Structure and replication of hepatitis delta virus. In: Hollinger FB, Lemon SM, Margolis HS (eds), Viral hepatitis and liver disease. Williams and Wilkins, Baltimore, pp 460-463

CONTRIBUTED BY

Howard CR, Burrell CJ, Gerin JL, Gerlich WH, Gust ID, Koike K, Marion PL, Mason WS, Neurath AR, Newbold J, Robinson W, Schaller H, Tiollais P, Wen Y-M, Will H

GENUS *DELTAVIRUS*

Type Species hepatitis delta virus (HDV)

DISTINGUISHING FEATURES

Hepatitis delta virus is defective and requires certain helper functions for replication; such functions can be supplied by hepatitis B virus or woodchuck hepatitis virus. Virions are spherical, about 34 nm in diameter with no surface projections. The envelope is acquired from the helper virus (HBsAg, when the helper is hepatitis B virus); within is a stable ribonucleoprotein complex forming a spherical core structure 18 nm in diameter. The genome consists of a single molecule of circular, negative sense, ssRNA, about 1,700 nt in size; it exists as an unbranched, rod shaped structure formed by intramolecular base-pairing. Genome replication involves RNA-directed RNA synthesis via a rolling circle mechanism that generates complementary oligomeric forms and involves site-specific autocatalytic cleavage and ligation to generate monomers. The complementary intermediate form is referred to as the antigenome. Only one hepatitis delta virus mRNA is found in infected liver; it directs the synthesis of the single virus protein, hepatitis delta antigen (HDAg, Mr 22-27 x 10^3); this protein exists in two forms which differ by a 19-amino acid carboxy-terminal extension. The smaller form is needed for genome replication, the larger for particle assembly. The genome structure and catalytic activities of hepatitis delta virus closely resemble those of some viroids and satellite viruses found in certain plants. The translation of hepatitis delta antigen and the dependency on hepadnavirus replication distinguish hepatitis delta virus from plant associated agents.

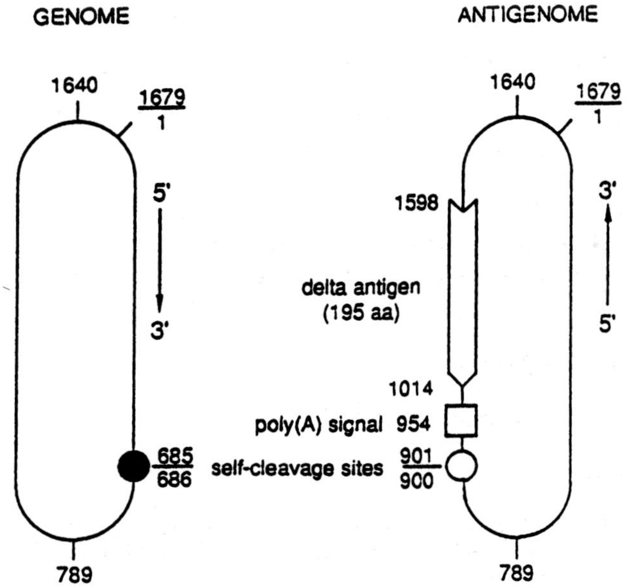

Figure 1: Organization of the HDV genome and antigenome (from Taylor *et al.*, 1991).

LIST OF SPECIES IN THE GENUS

The viruses, their genomic sequence accession numbers [ ] and assigned abbreviations ( ) are:

SPECIES IN THE GENUS

hepatitis delta virus [M21012, X04451, X60193] (HDV)

TENTATIVE SPECIES IN THE GENUS

None reported.

Similarity with Other Taxa

The involvement of reverse transcription in the replication of the hepatitis B and delta viruses is similar to that of retroviruses and cauliflower mosaic virus.

The genome structure and catalytic activities of HDV closely resemble those of viroids and satellite viruses found in certain plants and animals. The translation of HDV RNA and its helper-dependency on other hepadnaviruses for the formation of new particles distinguishes it from plant associated viroids.

Derivation of Names

delta: from Greek letter Δ, "D"

References

Lai MMC, Chao Y-C, Chang M-F, Lin J-H, Gust ID (1991) Functional studies of hepatitis delta antigen and delta virus RNA. In: Gerin JL, Purcell RH, Rizzeto M (eds) The hepatitis delta virus. Alan R. Liss, New York, pp 283-292

Taylor JM (1991) Human hepatitis delta virus. Curr Top Micro Immunol 168: 141-166

Taylor JM (1991) Structure and replication of hepatitis delta virus. In: Hollinger FB, Lemon SM, Margolis HS (eds), Viral hepatitis and liver disease. Williams and Wilkins, Baltimore, pp 460-463

Contributed By

Howard CR, Burrell CJ, Gerin JL, Gerlich WH, Gust ID, Koike K, Marion PL, Mason WS, Neurath AR, Newbold J, Robinson W, Schaller H, Tiollais P, Wen Y-M, Will H

SUBVIRAL AGENTS: VIROIDS

Type species potato spindle tuber viroid (PSTVd)

DEFINITION

Viroids are unencapsidated, small, circular, single-stranded RNAs which replicate autonomously when inoculated into host plants. Some are pathogenic, others replicate without eliciting symptoms.

PHYSICOCHEMICAL AND PHYSICAL PROPERTIES

Viroid molecules display extensive internal base pairing to give, in most cases, rod-like secondary structures 50 nm long. These structures denature by cooperative melting (Tm in 10 mM Na^+ = 50° C) to single-stranded circles of 100 nm contour length. Metastable conformations with hairpins may be physiologically important. MW is 80-125 x 10^3.

Sequences vary from 246 to 375 nt in length and are rich in G+C (53-60%) with the only exception of ASBVd (38%). All except ASBVd and PLMVd share a model of five structural-functional domains. The central domain contains a conserved region. The upper strand of the central conserved region can form either a hairpin or, in oligomers, a palindromic structure possibly relevant in replication. CCCVd is unusual in occurring as RNAs of different sizes, the larger ones having sequence repetitions of the smallest one. CLVd, GYSVd-2, AGVd, CBLVd, PBCVd and CVd-IV appear to have emerged from RNA recombination events since they seem to consist of a mosaic of sequences present in other viroids. There is no evidence that viroids encode protein.

ANTIGENIC PROPERTIES

No antigenicity demonstrated.

REPLICATION

Viroids differ fundamentally from viruses in that whereas virus replication parasitizes host translation, viroid replication parasitizes host transcription, possibly by using RNA polymerase II and/or other cellular RNA polymerases. Multimers isolated from infected tissues may be replicative intermediates produced by a rolling circle mechanism with two variants (symmetric and asymmetric) and three steps (RNA polymerization, cleavage and ligation). ASBVd and PLMVd multimers self-cleave *in vitro* and very probably *in vivo* to produce unit length strands but others do not, and may rely on host factors for cleavage. PSTVd accumulates mostly in nucleoli, ASBVd accumulates mostly in chloroplasts.

BIOLOGICAL PROPERTIES

HOST RANGE

Some viroids have wide host ranges in the angiosperms but others have narrow host ranges. CCCVd and CTiVd infect monocotyledons, the remainder infect dicotyledons. Grapevine and *Citrus* can harbor at least five different viroids.

TRANSMISSION

Viroids are transmitted mainly by vegetative propagation. Some are transmissible in seed or mechanically. Only TPMVd is known to be efficiently transmitted by aphids.

CROSS PROTECTION

Interactions at the level of symptom expression and viroid accumulation have been detected in plants co-infected by two strains of a viroid or by two different viroids sharing extensive sequence similarities.

CLASSIFICATION

Two criteria based on the sequence of the central conserved region or on a consensus phylogenetic tree have been proposed. Both lead essentially to the same grouping (Table 1). ASBVd and PLMVd form a special group of viroids with self-cleaving RNAs. A tentative nomenclature based on the phylogenetic analysis has been offered. Variation occurs within each viroid species and an arbitrary level of 90% sequence similarity currently separates variants from species.

LIST OF SPECIES SEQUENCED

The viroids, their genomic sequence accession numbers and assigned abbreviations are:

Table: Groups of viroids which have been sequenced

Viroid	Abbreviation	Accession #	Size (nt)	CCR*	MG*
potato spindle tuber (66)	PSTVd	V01465	356, 359-360	PSTVd	PSTVd
citrus exocortis (226)[†]	CEVd	M34917	370-375	PSTVd	PSTVd
chrysanthemum stunt	CSVd	V01107	354, 356	PSTVd	PSTVd
tomato apical stunt	TASVd	K00818	360, 363	PSTVd	PSTVd
tomato planta macho	TPMVd	K00817	360	PSTVd	PSTVd
Columnea latent[‡]	CLVd	X15663	370, 372	PSTVd	PSTVd
hop stunt (326)[§]	HSVd	X00009	297-303	PSTVd	HSVd
coconut cadang-cadang(287)	CCCVd	J02049	246-247	PTSVd	CCCVd
coconut tinangaja	CTiVd	M20731	254	PTSVd	CCCVd
hop latent	HLVd	X07397	256	PTSVd	CCCVd
citrus IV	CVd-IV	X14638	284	PTSVd	CCCVd
apple scar skin (349)[¶]	ASSVd	M36646	329-330	ASSVd	ASSVd
grapevine yellow speckle 1	GYSVd-1	X06904	366-368	ASSVd	ASSVd
grapevine yellow speckle 2	GYSVd-2	J04348	363	ASSVd	ASSVd
Australian grapevine	AGVd	X17101	369	ASSVd	ASSVd
citrus bent leaf	CBLVd	M74065	318	ASSVd	ASSVd
pear blister canker	PBCVd	S46812	315	ASSVd	ASSVd
Coleus blumei 1	CbVd-1	X52960	248	CbVd-1	CbVd-1
avocado sunblotch (254)	ASBVd	J02020	246-250	-	-
peach latent mosaic	PLMVd	M83545	336-337		

*CCR refers to central conserved region and MG to monophyletic group.
[†]Agent also of Indian tomato bunchy top and isolated from grapevine.
[‡]Isolated also from *Nematanthus wettsteinii*.
[§]Agent also of cucumber pale fruit, plum dapple, peach dapple, *Citrus cachexia* and isolated from grapevine, pear, apricot, banana, raspberry, *Hibiscus* and croton *(Codiaeum)*.
[¶]Agent also of dapple apple and pear rusty skin.

LIST OF SPECIES NOT YET SEQUENCED

burdock stunt viroid	(BSVd)
citrus viroids	(CVds)
Coleus blumei viroid 2	(CbVd-2)
Coleus blumei viroid 3	(CbVd-3)
Nicotiana glutinosa stunt viroid	(NgSVd)
pigeon pea mosaic mottle viroid	(PMMVd)
tomato bunchy top viroid	(TBTVd)

LIST OF TENTATIVE SPECIES

carnation stunt associated viroid-like RNA	(CarSAVd)
chrysanthemum chlorotic mottle viroid-like RNA	(CChMVd)

DERIVATION OF NAMES

viroid: from the name given to the sub-viral RNA agent of potato spindle tuber disease

REFERENCES

Branch AD, Robertson HD (1984) A replication cycle for viroids and other small infectious RNAs. Science 223: 450-455

Diener TO (1971) Potato spindle tuber "virus" IV. A replicating, low molecular weight RNA. Virology 45: 411-428

Diener TO (1991) The frontiers of life: the viroids and viroid-like satellite RNAs, In: Maramorosch K (ed) Viroids and Satellites: Molecular Parasites at the Frontier of Life. CRC Press, Boca Raton FL, pp 1-20

Durán-Vila N, Roistacher CN Rivera-Bustamante R, Semancik JS, (1988) A definition of citrus viroid groups and their relationship to the citrus exocortis disease. J Gen Virol 69: 3069-3080

Elena SF, Dopazo J, Flores R, Diener TO, Moya A (1991) Phylogeny of viroids, viroid like satellite RNAs, and the viroidlike domain of hepatitis virus RNA. Proc Natl Acad Sci USA 88: 5631-5634

Garnsey SM, Randles JW (1987) Biological interactions and agricultural implications of viroids; In: Semancik JS (ed) Viroids and Viroidlike Pathogens. CRC Press, Boca Raton FL, pp 127-160

Gross HJ, Domdey H, Lossow C, Jank P, Raba M, Alberty H, Sänger HL (1978) Nucleotide sequence and secondary structure of potato spindle tuber viroid. Nature 273: 203-208

Harders J, Lukács N, Robert-Nicoud M, Jovin JM, Riesner D (1989) Imaging viroids in nuclei from tomato leaf tissue by *in situ*- hybridization and confocal laser scanning microscopy. EMBO J 8: 3941-3949

Hernández C, Flores R (1992) Plus and minus RNAs of peach latent mosaic viroid self-cleave *in vitro* via hammerhead structures. Proc Natl Acad Sci USA 89: 3711-3715

Hutchins CJ, Rathjen PD, Forster AC, Symons RH (1986) Self-cleavage of plus and minus RNA transcripts of avocado sunblotch viroid. Nucl Acids Res 14: 3627-3640

Keese P, Symons RH (1987) Physical-chemical properties: molecular structure (primary and secondary); In: Diener TO (ed) The Viroids (The Viruses). Plenum Press, New York pp 37-62

Koltunow AM, Rezaian MA (1989) A scheme for viroid classification. Intervirology 30: 194-201

Rezaian MA (1990) Australian grapevine viroid - evidence for extensive recombination between viroids. Nucl Acids Res 18: 1813-1818

Riesner D (1991) Viroids: from thermodynamics to cellular structure and function. Mol Plant-Microbe Interact 4: 122-131

Sänger HL (1987) Viroid function: viroid replication; In: Diener TO (ed) The Viroids (The Viruses), Plenum Press, New York pp 117-166

Sano T, Candresse T, Hammond R, Diener TO, Owens RA (1992) Identification of multiple structural domains regulating viroid pathogenicity. Proc Natl Acad Sci USA 89: 10104-10108

CONTRIBUTED BY

Flores R

Subviral Agents: Agents of Spongiform Encephalopathies (Prions)

Prions are small, proteinaceous infectious particles that resist inactivation by procedures which affect nucleic acids. To date, no detectable nucleic acids of any kind and no virus-like particles have been associated with prions. Prions cause scrapie and other spongiform encephalopathies of animals and humans (Table 1).

Table 1: The spongiform encephalopathies.

Disease abbreviation	Natural host	Prion	Abnormal PrP Term	Alternate PrP Term
scrapie	sheep & goats	Scrapie	$ShePrP^{Sc}$	$ShePrP^{Sc}$
transmissible mink encephalopathy (TME)	mink	TME prion	$MkPrP^{Sc}$	$MkPrP^{TME}$
chronic wasting disease (CWD)	mule deer & elk	CWD prion	$MDePrP^{Sc}$	$MDePrP^{CWD}$
bovine spongiform encephalopathy (BSE)	cattle	BSE prion	$BovPrP^{Sc}$	$BovPrP^{BSE}$
feline spongiform encephalopathy (FSE)	cats	FSE prion	$FePrP^{Sc}$	$FePrP^{FSE}$
exotic ungulate encephalopathy (EUE)	nyala & greater kudu	EUE prion	$NyaPrP^{Sc}$	$NyaPrP^{EUE}$
kuru	humans	Kuru prion	$HuPrP^{Sc}$	$HuPrP^{Ku}$
Creutzfeldt-Jakob disease (CJD)	humans	CJD prion	$HuPrP^{Sc}$	$HuPrP^{CJD}$
Gerstmann-Straussler-Scheinker syndrome (GSS)	humans	GSS prion	$HuPrP^{Sc}$	$HuPrP^{GSS}$
fatal familial insomnia (FFI)	humans	FFI prion	$HuPrP^{Sc}$	$HuPrP^{FFI}$

Prions are composed largely, if not entirely, of a protein designated as the scrapie isoform of the prion protein, PrP^{Sc} (see Table 2 for glossary). A post-translational process, as yet undefined, generates PrP^{Sc} from the normal cellular isoform of the protein, designated PrP^{C}. Both PrP^{Sc} and PrP^{C} are encoded by a single copy chromosomal gene. Although the inoculated prion initiates the production of PrP^{Sc}, its synthesis originates from the host PrP gene.

Several features distinguish prions from viruses. First, prions can exist in multiple molecular forms, whereas viruses exist in a single form with distinct ultrastructural morphology. Second, prions are non-immunogenic, in contrast to viruses, which almost always provoke an immune response. Third, there is no evidence for an essential nucleic acid within the infectious prion particle, whereas viruses have a nucleic acid genome which serves as the template for the synthesis of progeny virus. Fourth, the only known component of the prion is PrP^{Sc}, which is encoded by a chromosomal gene, whereas viruses are composed of nucleic acid, proteins, and often other constituents.

PRION PROPERTIES

MORPHOLOGY

Microsomal fractions from infected tissues enriched for prion infectivity contain numerous membrane vesicles (Fig. 1a); detergent extraction and limited proteolysis of brain microsomes generate rod-shaped particles (Fig. 1b). Most are of uniform diameter (11 nm) with mean lengths of 165 nm (range 25-550 nm). The rods are smooth, almost ribbon-like, and infrequently are twisted. The rods resemble purified amyloid, both ultrastructurally and histochemically (Fig. 1b). The rods are not considered the infectious entity since large PrP 27-30 polymers are not required for infectivity (Fig. 1c).

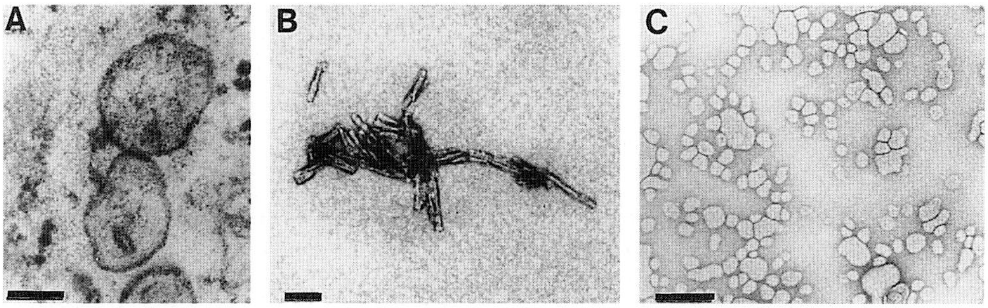

Figure 1: Multiple forms of scrapie prions isolated from infected Syrian hamster brains: (left) microsomal membranes containing submicroscopic, infectious prion particles; (center) purified prion rods representing a polymeric form of the infectious prion particle and generated by limited proteolysis in the presence of detergent; (right) prion liposomes generated by sonication of infectious prion rods isolated from scrapie-infected Syrian hamster brains using limited proteolysis, detergent extraction and sedimentation through a discontinuous sucrose gradient. All three forms contain high levels of prion infectivity (>10^7 ID_{50} units/ml). Bars represent 100 nm.

Table 2: Glossary of prion terminology.

Term	Description
Prion	A small proteinaceous infectious particle which resists inactivation by procedures that affect nucleic acids. Prions are composed largely, if not entirely, of PrP^{Sc} molecules.
PrP^{Sc}	Scrapie isoform of the prion protein. This protein is the only identifiable macromolecule in purified preparations of scrapie prions.
PrP^{C}	Cellular isoform of the prion protein.
PrP 27-30	Digestion of PrP^{Sc} with proteinase K generates PrP 27-30 by hydrolysis of the N-terminal 67 amino acids.
PRNP	PrP gene located on human chromosome 20.
Prn-p	PrP gene located on mouse chromosome 2.
Pid-1	Gene on mouse chromosome 17 which appears to influence experimental CJD and scrapie incubation times.
Prn-i	Gene on mouse chromosome 2 controlling experimental scrapie and CJD incubation times. *Prn-i* and *Prn-p* form the prion gene complex (*Prn*).
Sinc	Gene in mice controlling experimental scrapie incubation times. This genetic locus is probably the same as *Prn-i*.
PrP amyloid	Amyloid plaque composed of PrP in brain of animals and humans with spongiform encephalopathy.
Prion rod	An aggregate of prions composed largely, if not entirely, of PrP 27-30 molecules. Created by detergent extraction and limited proteolysis of PrP^{Sc}.

PHYSICOCHEMICAL AND PHYSICAL PROPERTIES

The Mr of PrP^{Sc} is 33-35 x 10^3. The Mr of PrP^{Sc} dimers or trimers are consistent with an ionizing radiation target size of 55±9 kDa. Prions aggregate into particles of non-uniform size and cannot be solubilized by detergents, except under denaturing conditions where infectivity is lost. However, solubilization of PrP^{Sc} and prions can be achieved with phospholipids (Fig. 1). Prions resist inactivation by nucleases, UV-irradiation at 254 nm, treatment with psoralens, divalent cations, metal ion chelators, acid (between pH 3 and 7), hydroxyl-amine, formalin, boiling, and proteases. Prion infectivity is diminished by prolonged digestion with proteases, or by treatments such as urea, boiling in SDS, alkali (>pH 10), autoclaving at 132° C for more than 2 hr., denaturing organic solvents (e.g., phenol), or chaotropic agents such as guanidine isocyanate.

NUCLEIC ACIDS

No prion-specific nucleic acid has been detected.

PROTEINS

PrP^{Sc} is derived from PrP^{C} by a post-translational process. The molecular events in the conversion are unknown but may involve only a change in the conformation of the protein. PrP^{Sc} may be readily distinguished from PrP^{C} by its different biochemical and biophysical properties. Limited proteolysis of PrP^{Sc} produces a smaller, protease-resistant molecule of about 142 amino acids, designated PrP 27-30. Under the same conditions PrP^{C} is completely hydrolyzed. The amino acid sequence of PrP^{Sc} that has been established by protein sequencing and mass spectrometry is identical to that deduced from the genomic DNA sequence. No proteins other than PrP^{Sc} have been consistently found in fractions enriched for prion infectivity.

LIPIDS

PrP^{Sc} contains a glycosylinositol phospholipid (GPI) attached to amino acid residue 231 (serine) of the Syrian hamster PrP. The lipids of the diradylglycerol moiety of the GPI anchor are not well characterized.

CARBOHYDRATES

In addition to the GPI anchor which contains sialic acid, PrP^{Sc} has two consensus sites where it can undergo N-linked glycosylation (residues 181 and 197 of the Syrian hamster PrP). Bi-, tri- and tetra-antennary structures have been reported for the N-linked, complex type glycans of PrP^{Sc}. Some of these complex-type oligosaccharides have branched fucose residues, some have terminal sialic acid residues. Six different GPI glycans have been found, two of which are sialylated.

ORGANIZATION AND REPLICATION

The entire ORF of all known mammalian and avian PrP genes is contained within a single exon. The two exons of the Syrian hamster PrP gene are separated by a 10 kb intron. Exon-1 of this gene encodes a portion of the 5' untranslated leader sequence while exon-2 encodes the ORF and the 3' untranslated region. The mouse PrP gene is comprised of three exons with exon-3 analogous to exon-2 of the Syrian hamster. The ORF of both the mouse and hamster PrP genes encode proteins of 254 amino acids. The promoters of both the PrP genes of both animals contain 3 or 2 repeats, respectively, of G-C nonamers, but are devoid of TATA boxes. These nonamers represent a motif which may function as a canonical binding site for transcription factor Sp1.

The multiplication of prion infectivity involves the post-translational conversion of PrP^{C}, or another precursor, to PrP^{Sc}. Studies with transgenic mice expressing a Syrian hamster PrP gene argue that prion synthesis involves "propagation", whereby infecting PrP^{Sc} molecules combine with homologous host-encoded PrP^{C} molecules giving rise to new PrP^{Sc} molecules.

Additional evidence to support this proposed model for prion replication comes from studies of transgenic mice expressing a chimeric mouse: Syrian hamster PrP gene, where the prions produced from these transgene products have an artificial host range. In the absence of any candidate post-translational chemical modification that differentiates PrP^C from PrP^{Sc}, it seems likely that these two isoforms may be distinguishable only by their conformation.

Antigenic Properties

PrP^{Sc} is a weak antigen. The immunoreactivity of PrP^{Sc} is significantly enhanced by denaturation. Antibodies raised to denatured PrP 27-30 of Syrian hamsters have been used to neutralize prion infectivity that is dispersed into liposomes.

Biological Properties

The prion diseases are a group of neurodegenerative disorders afflicting mammals (Table 1). The diseases are transmissible under some circumstances but, unlike other transmissible disorders, the prion diseases can also be caused by mutations in the host PrP gene. The mechanism of prion spread among sheep and goats developing natural scrapie is unknown. CWD, TME, BSE, FSE and EUE are all thought to occur after the consumption of prion-infected materials. Similarly, kuru of the New Guinea Fore people is thought to have resulted from the consumption of brains during ritualistic cannibalism. Familial CJD, GSS and FFI are all dominant, inherited prion diseases which have been shown to be genetically linked to mutation in the PrP gene. While iatrogenic CJD cases can be traced to inoculation of prions through human pituitary-derived growth hormone, cornea transplants, dura mater grafts, or cerebral electrode implants, the number of cases recorded to date is small. Most cases of CJD are sporadic, probably the result of somatic mutation of the PrP gene or the spontaneous conversion of PrP^C into PrP^{Sc}. About 10-15% of CJD cases and virtually all cases of GSS and FFI appear to be caused by germline mutations in the PrP gene. Twelve different mutations of the PrP gene have been shown to segregate with the human prion diseases (Table 3).

Table 3: Proposed designation of human PrP gene mutations.

Disease	PrP Gene Mutation
Gerstmann-Straussler-Scheinker syndrome	(PrP P102L)
Gerstmann-Straussler-Scheinker syndrome	(PrP A117V)
familial Creutzfeldt-Jakob disease; fatal familial insomnia	(PrP D178N)
Gerstmann-Straussler-Scheinker syndrome	(PrP F198S)
familial Creiutzfeldt-Jakob disease	(PrP E200K)
Gerstmann-Straussler-Scheinker syndrome	(PrP Q217R)
familial Creutzfeldt-Jakob disease	(PrP octarepeat insert)

Prion isolates

There is good evidence for multiple "strains" or distinct isolates of prions as defined by specific incubation times, distribution of vacuolar lesions and patterns of PrP^{Sc} accumulation. The mechanism by which isolate-specific information is carried by prions is unknown. Two different isolates from mink dying of TME exhibit different sensitivities of PrP^{Sc} to proteolytic digestion, supporting the suggestion that isolate-specific information might be carried by PrP^{Sc}.

Mutant PrP genes

Humans carrying point mutations or inserts in their PrP genes produce mutant PrP^C molecules that are believed to spontaneously convert into PrP^{Sc}. While the initial stochastic event could be inefficient, once it happens the process may become autocatalytic. The proposed mechanism is consistent with individuals harboring germline mutations who develop CNS dysfunction only after decades but then rapidly progress to death.

INCUBATION TIME GENES

Studies of PrP genes (*Prn-p*) in mice with short and long incubation times have demonstrated genetic linkage between a *Prn-p* restriction fragment length polymorphism and a gene modulating incubation times (*Prn-i*). Although it seems likely that the genes for PrP, *Prn-i* and *Sinc* are all congruent, it has not formally been proven.

NOMENCLATURE

A listing of the different animal prions is given in Table 1. Although the prions that cause TME and BSE are referred to as TME prions and BSE prions, this may be unjustified, because both are thought to originate from the oral consumption of scrapie prions in sheep-derived foodstuffs and because many lines of evidence argue that the only difference among the various prions is the sequence of PrP which is dictated by the host and not the prion itself.

The human prions present a similar semantic conundrum. Transmission of human prions to laboratory animals produces prions carrying PrP molecules with sequences dictated by the PrP gene of the host, not that of the inoculum. To simplify the terminology, it has been suggested that the disease-related PrP isoform be designated PrP^{Sc} without regard to the origin of the prion (Table 1). Alternatively, the superscript of the disease-related PrP isoform can be used to signify the host in which the prion disease originated. For added specificity, a variant or mutant PrP can be noted in parentheses (Table 3) [e.g., the prion found in the I/Ln mouse which has a PrP variant with F at codon 108 and V at 189 can be identified as $MoPrP(108F, 189V)^{Sc}$; similarly, the prion found in a Libyan Jewish CJD patient homozygous for the mutation K at codon 200 can be identified as $HuPrP(200K)^{Sc}$]. For heterozygous situations, and where the allele that determines the PrP form is not known, $HuPrP^{Sc}$, or $HuPrP^{CJD}$, can be used as a default.

Distinguishing among CJD, GSS and FFI has grown increasingly difficult with the recognition that familial CJD, GSS and FFI are autosomal dominant diseases that are caused by mutations in the PRNP gene. Initially, it was thought that a specific PrP mutation was associated with a particular clinical / neuropathological presentation. Now, an increasing number of exceptions are being recognized. In a single family with a particular PrP mutation, different clinical / neuropathologic manifestations can be seen. It has been suggested that the disorders be labeled "*inherited prion disease,*" followed by an identification of the mutation. For example, most patients with a PrP mutation at codon 102 present with ataxia and have PrP amyloid plaques; these patients are generally diagnosed as GSS, but some individuals within these families present with dementia characteristic of CJD.

DERIVATION OF NAMES

prion: singla for *pro*teinaceous and *in*fectious particle

REFERENCES

Aiken JM, Marsh FR (1990) The search for scrapie agent nucleic acid. Microbiol Rev 54: 242-246

Brown P, Goldfarb LG, Gajdusek DC (1991) The new biology of spongiform encephalopathy: infectious amyloidoses with a genetic twist. Lancet 337: 1019-1022

Carp RI, Kascsak RJ, Wisniewski HM, Mertz PA, Rubenstein R, Bendheim P,Bolton D (1989) The nature of the unconventional slow infection agents remains a puzzle. Alzheimer Dis Assoc Disord 3: 79-99

Dickinson AG, Outram GW (1988) Genetic aspects of unconventional virus infections: the basis of the virino hypothesis. In: Bock G, Marsh J (eds), Novel Infectious Agents and the Central Nervous System. Ciba Foundation Symposium 135. John Wiley and Sons, Chichester, UK, pp 63-83

Gabizon R, Prusiner SB (1990) Prion liposomes. Biochem J 266: 1-14

Gajdusek DC (1977) Unconventional viruses and the origin and disappearance of kuru. Science 197: 943-960

Gibbs CJ Jr, Gajdusek DC (1978) Subacute spongiform virus encephalopathies: the transmissible virus dementias. In: Katzman R, Terry RD, Bick KL (eds) Alzheimer's Disease: Senile Dementia and Related Disorders, Aging, Vol 7, Raven Press, New York USA, pp 559-577

Prusiner SB (1991) Molecular biology of prion diseases. Science 252: 1515-1522

Prusiner SB, Collinge J, Powell J, Anderton B (eds) (1992) Prion Diseases of Humans and Animals. Ellis Horwood, London UK, pp 583
Rohwer RG (1991) The scrapie agent: "a virus by any other name". Curr Top Microbiol Immunol 172: 195-232
Weissmann C (1991) A "unified theory" of prion propagation. Nature 352: 679-683

CONTRIBUTED BY

Prusiner SB, Baldwin M, Collinge J, DeArmond SJ, Marsh R, Tateishi J, Weissmann C

Unassigned Viruses

Although many of the known viruses have been classified into genera in this Report, a significant number have not yet been assigned to a recognized genus, or sufficiently distinguished from recognized genera so as to form a new genus. Some examples are listed here. These are viruses for which some key characteristics are known but which are as yet unplaced; poorly characterized viruses are excluded. The listing is not exhaustive, rather it contains examples which illustrate that the task of devising a universally applicable virus taxonomy is not yet complete.

Animal viruses

Borna disease virus

Virions are enveloped and contain a negative sense, 8.9 kb ssRNA. The nucleotide sequence (database accession number L27077), which contains 5 substantial ORFs, suggests a string relationship to the family *Rhabdoviridae*. Replication and transcription take place inside nuclei.

Cubitt B, Oldstone C, de la Torre JC (1994) Sequence and genome organization of Borna disease virus. J Virol 68: 1382-1396

Nyamanini virus

Virions are enveloped and contain RNA. The virus has been isolated from cattle egrets and ticks in Africa.

Karabatsos N (ed) (1985) International Catalogue of Arboviruses Including Certain Other Virus of Vertebrates 3rd edn, San Antonio, Texas. Am Soc Trop Med Hyg.

Plant Viruses With ssDNA

Banana bunchy top virus (BBTV)

Particles are isometric, 18 to 20 nm in diameter and comprise a coat protein with an Mr of about 20 x 10^3 and circular ssDNA of about 1 kb. The genome consists of 2 or more DNA molecules. The virus is persistently transmitted by aphids.

Harding RM, Burns TM, Hafner G, Dietzgen RG, Dale JL (1993) Nucleotide sequence of one component of the banana bunchy top virus genome contains a putative replicase gene. J Gen Virol 74: 323-328.

Coconut foliar decay virus (CFDV)

Particles consist of a Mr 25 x 10^3 coat protein and more than one molecule of circular ssDNA 1.29 kb in length. The virus is transmitted by plant hoppers.

Rohde W, Randles JW, Langridge P, Hanold D (1990) Nucleotide sequence of a circular single-stranded DNA associated with coconut foliar decay virus. Virology 176: 648-651

Subterranean clover stunt virus (SCSV)

Particles are 17 to 19 nm in diameter and comprise a Mr 19 x 10^3 coat protein and circular ssDNA molecules 850 to 880 nt in length. There are 7 or more DNA species present. The virus is transmitted by aphids.

Chu PWG, Helms K (1988) Novel virus-like particles containing circular single-stranded DNAs associated with subterranean clover stunt disease. Virology 167: 38-49

PLANT VIRUSES WITH dsDNA

CUCUMBER VEIN YELLOWING VIRUS (CVYV)

Particles are filamentous, about 740-800 nm in length and 15-18 nm in width, sediment at about 220S and comprise a Mr 39 x 10^3 coat protein and dsDNA. The virus is mechanically transmissible and transmitted in nature by the whitefly *Bemisia tabaci* in a semi-persistent manner.

Sela I, Assouline I, Tanne E, Cohen S, Marco S (1980) Isolation and characterization of a rod-shaped, whitefly transmissible, DNA-containing plant virus. Phytopathology 70: 226-228

PLANT VIRUSES WITH dsRNA

TOBACCO STUNT VIRUS (TStV)

Particles are rod-shaped, 18 nm x 300-340 nm and contain dsRNA of about 7 kbp and about 6 kbp. Coat protein has an Mr of 48 x 10^3. Virus is transmitted by the fungus *Olpidium brassicae*. Lettuce big vein virus is similar and may be related.

Kuwata S, Kubo S (1986) Tobacco stunt virus. CMI/AAB Descriptions of Plant Viruses, N° 313, 4pp

PLANT VIRUSES WITH SSRNA

GARLIC VIRUSES A,B,C,D (GarVA,B,C,D)

Particles are filamentous, about 700 nm in length and comprise a Mr 34 x 10^3 coat protein and a ssRNA of about 10 kb with a poly (A) tail. The 3'-terminal sequences resemble those of RNA from carlaviruses except that one ORF is distinctly larger. The coat protein sequence suggests only a distant relationship with carlaviruses and potexviruses.

Sumi S, Tsuneyoshi T, Furutani H (1993) Novel rod-shaped viruses isolated from garlic, *Allium sativum*, possessing a unique genome organization. J Gen Virol 74: 1879-1885.

GRAPEVINE FLECK VIRUS (GFkV)

Particles are isometric, about 30 nm in diameter, and either RNA-free or contain ssRNA of about 7.5 kb. Coat protein has an Mr of about 28 x 10^3. Particles are phloem-limited and not mechanically transmissible. The vector is not known.

Boulila M, Boscia D, Di Terlizzi B, Castellano MA, Minafra A, Savino V, Martelli GP (1990) Some properties of a phloem-limited non-mechanically transmissible grapevine virus. J Phytopathol 129: 151-158.

MAIZE WHITELINE MOSAIC VIRUS (MWLMV)

Particles are isometric about 35 nm in diameter and contain about 4 kb ssRNA. Coat protein has an Mr of about 33 x 10^3. Virus is soil-borne; transmission may be by fungi but is not possible mechanically.

de Zoeten GA, Reddick BB (1984) Maize white line mosaic virus. CMI/AAB Descriptions of Plant Viruses, N° 283, 4pp

OLIVE LATENT VIRUS 2 (OLV-2)

Particles range in shape from quasi-spherical, 26 nm in diameter, to bacilliform, 37, 43, 48 and 55 nm long and 18 nm wide. They consist of four separately encapsidated major species of ssRNA of about 3.3 kb, 2.8 kb, 2.45 kb and 2.1 kb, and a coat protein with an Mr of about 24 x 10^3. The 2.1 kb RNA is part of the 2.45 kb RNA. There are three minor RNA species of 0.5 kb, 0.3 kb and 0.2 kb but infectivity is associated with the four larger species. The vector is not known.

Grieco F, Martelli GP, Savino V, Piazzolla P, (1992) Properties of olive latent virus 2. Rivista di Patologia Vegetale, S.V, 2: 125-136

Ourmia melon virus (OuMV)

Particles are short rods, 18.5 nm in diameter and either 30 nm or 37 nm in length, and have somewhat pointed ends. Particles contain positive sense ssRNA of about 3 kb, 1.1 kb or 1 kb and proteins with Mr of 26.3 x 10^3 and 23.3 x 10^3. No vector is known.

Lisa V, Milne RG, Accotto GP, Boccardo G, Caciagli P, Parvizy R (1988) Ourmia melon virus, a virus from Iran with novel properties. Ann Appl Biol 112: 291-302

Pelargonium zonate spot virus (PZSV)

Particles are quasi-isometric 25-35 nm in diameter, and sediment as three components. Nucleic acid is ssRNA of about 4.4 kb and about 3.3 kb. Coat protein has an Mr of 44 x 10^3. The virus is readily transmitted by sap inoculation. Natural transmission is by thrips.

Gallitelli D, Quacquarelli A, Martelli GP (1983) Pelargonium zonate spot virus. CMI/AAB Descriptions of Plant Viruses, N° 272, 4pp

Fungus Viruses

Agaricus bisporus virus 1

Particles are isometric, about 25 nm in diameter and sediment at 90-100 S. The single Mr 25 x 10^3 coat protein encapsidates two dsRNA species of about 2 kb.

Barton RJ, Hollings M (1979) Purification and some properties of two viruses infecting the cultivated mushroom *Agaricus bisporus*. J Gen Virol 42: 231-240

Allomyces arbuscula virus

Particles are isometric, about 40 nm in diameter and sediment as 67 *S* and 75 S components. Particles consist of proteins, with Mr of 38 x 10^3, 34 x 10^3, 28 x 10^3 and 21 x 10^3, and dsRNA of 3.6 kbp, 2 kbp and 1.6 kbp.

Khandjian EW, Turian G, Eisen H (1977) Characterization of the RNA mycovirus infecting *Allomyces arbuscula*. J Gen Virol 35: 415-424

Aspergillus foetidus virus F

Particles are isometric, 40-42 nm in diameter and sediment as 164 *S* and 145 S components. Particles contain a major Mr 87 x 10^3 protein and minor species of Mr 125 x 10^3 and 100 x 10^3. The ds RNA are 3.8 kbp, 2.7 kbp, 2.5 kbp, 2.1 kbp and 1.8 kbp.

Buck KW, Ratti G (1975) Biophysical and biochemical properties of two viruses isolated from *Aspergillus foetidus*. J Gen Virol 27: 211-224

Colletotrichum lindemuthianum virus

Particles are isometric, 30 nm in diameter and sediment as 110 *S* and 85 *S* components. Particles contain a major Mr 52 x 10^3 protein and a minor species of Mr 45 x 10^3. The ds RNA are 3.6 kbp, 1.6 kbp and 1.5 kbp.

Rawlinson CJ, Carpenter JM, Muthyalu G (1975) Double-stranded RNA virus in *Colletotrichum lindemuthianum*. Trans Brit Mycol Soc 65: 305-341

Gaeumannomyces graminis virus 45/101-C

Particles are isometric, 29 nm in diameter and sediment at 127 *S*. They consist of a Mr 66 x 10^3 protein and a ds RNA of 1.8 kbp.

Buck KW (1984) A new double-stranded RNA virus from *Gaeumannomyces graminis*. J Gen Virol 65: 987-990

HELMINTHOSPORIUM MAYDIS VIRUS

Particles are isometric, 48 nm in diameter and sediment at 283 *S*. Particles consist of a Mr 121 x 10^3 protein and ds RNA of 8.3 kbp.

Bozarth RF (1977) Biophysical and biochemical characterization of virus-like particles containing a high molecular weight dsRNA from *Helminthosporium maydis*. Virology 80: 149-157

LENTINUS EDODES VIRUS

Particles are isometric, 39 nm in diameter and contain 1 dsRNA of 6.5 kbp.

Ushiyama R, Nakai Y (1982) Ultrastructural features of virus-like particles from *Lentinus edodes*. Virology 123: 93-101

LAFRANCE ISOMETRIC VIRUS

Particles are isometric, 36 nm in diameter and contain dsRNA species of 3.6 kbp, 3 kbp, 2.8 kbp, 2.7 kbp, 2.5 kbp, 1.6 kbp, 1.4 kbp, 0.9 kbp, and 0.8 kbp.

Goodin MM, Schlagnhaufer B, Romaine CP (1992) Encapsidation of the LaFrance disease-specific double-stranded RNAs in 36-nm isometric virus like particles. Phytopathol 82: 285-290

PERICONIA CIRCINATA VIRUS

Particles are isometric, 32 nm in diameter and sediment as 150 S and 140 S components. Particles contain ds RNA of 2.5 kbp, 2 kbp, 1.8 kbp, 1.6 kbp, 0.7 kbp and 0.6 kbp.

Dunkle LD (1974) Double-stranded RNA mycovirus in *Perconia circinata*. Physiol Plant Pathol 4: 107-116

INVERTEBRATE VIRUSES

ORYCTES RHINOCEROS VIRUS (OrV)

The Oryctes rhinoceros virus is a pathogenic virus of invertebrates, infecting a number of coleopteran insects in the family *Scarabaeidae*. The mature virion of Oryctes rhinoceros virus consists of an enveloped, rod-shaped nucleocapsid and contains a unique tail-like structure protruding from one end. The mature virion is produced by virus budding from the plasma membrane and contains two unit membranes. The genome is a single supercoiled circular dsDNA of approximately 130 kbp. Although these viruses were previously classified as members of the family *Baculoviridae*, they differ in several respects including virion morphology and the lack of an occlusion body.

Crawford A (1994) Nonoccluded baculoviruses. In: Encyclopedia of Virology Webster RG, Granoff A (eds). Academic Press, New York, pp 133-139

HELIOTHIS ZEA VIRUS 1 (HzV-1)

The Heliothis zea virus 1 virus was isolated as a persistent virus of an insect cell line derived from *Heliothis zea*. Although the virus can infect a number of insect (lepidopteran) cell lines, infection of an insect has not been observed. The virion of Heliothis zea virus 1 is composed of an enveloped rod-shaped nucleocapsid. Virions are released from infected cells by cell lysis. The Heliothis zea virus 1 genome consists of a single molecule of circular dsDNA, approximately 240 kbp in length. The Heliothis zea virus 1 was previously classified as a non-occluded member of the family *Baculoviridae*.

Burand J (1991) Molecular biology of the HzV-1 and Oryctes nonoccluded baculoviruses. In: Viruses of Invertebrates, Kurstak E (ed) Marcel Dekker, Inc. New York, pp 111-126

CONTRIBUTED BY

Mayo MA

Part III: The International Committee on Taxonomy of Viruses

Officers and Members of ICTV 1990-93

Executive committee

President	Murphy, F.A.	USA
Vice-President	Buck, K.W.	UK
Secretary	Pringle, C.R.	UK
Secretary	Fauquet, C.M	France

Subcommittee Chairpersons

Coordination SC	Murphy, F.A.	USA
Bacterial viruses SC	Jarvis, A.W.	New Zealand
Fungal viruses SC	Ghabrial, S.A.	USA
Invertebrate viruses SC	Summers, M.D.	USA
Plant viruses SC	Martelli, G.P.	Italy
Vertebrate viruses SC	Bishop, D.H.L.	UK
Virus Data SC	Gibbs, A.J.	Australia

Elected Members

Member	Ackermann, H.-W.	Canada
Member	Ahlquist, P.	USA
Member	Berthiaume, L.	Canada
Member	Calisher, C.H.	USA
Member	Goldbach, R.	The Netherlands
Member	Maniloff, J.	USA
Member	Mayo, M. A.	UK
Member	Rohrmann, G.F.	USA

Life Members

Life-Member	Brown, F.	UK
Life-Member	Fenner, F.J.	Australia
Life-Member	Lwoff, A.	France
Life-Member	Matthews, R.E.F.	New-Zealand
Life-Member	Maurin, J.	France
Life-Member	Melnick, J.L.	Texas
Life-Member	Pereira, H.G.	Brazil

Coordination Subcommittee

Chair	Murphy, F.A.	USA
Member	Buck, K.W.	UK
Member	Goldbach, R.	The Netherlands
Member	McGeoch, D.	UK
Member	Strauss, J.H.	USA
Member	van Regenmortel, M.	France

Bacterial Virus Subcommittee

Chair	Jarvis, A.W.	New Zealand
Vice-Chair	Ackermann, H.-W.	Canada
Member	Chatterjee, S.N.	India
Member	Dubow, M.S.	Canada
Member	Jones, L.A.	USA
Member	Krylov, V.N.	USSR
Member	Maniloff, J.	USA
Member	Ogata, S.	Japan
Member	Rocourt, J.	France

Member	Safferman, R.S.	USA
Member	Schneider, J.	Germany
Member	Seldin, L.	Brazil
Member	Sozzi, T.	Italy
Member	Stewart, P.R.	Australia
Member	van Duin, J.	The Netherlands
Member	Werquin, M.	France
Member	Wunsche, L.	Germany

ACTINOPHAGES STUDY GROUP

Chair	Jones, L.A.	USA
Member	Korn-Wendisch, F.	Germany
Member	Pigac, J.	Yugoslavia
Member	Schneider, J.	Germany
Member	Sonnen, H.	Germany

BACILLUS PHAGES STUDY GROUP

Chair	Seldin, L.	Brazil
Member	Dean, D.H.	USA
Member	Doskocill, J.	Czechoslovakia
Member	Lovett, P.S.	USA
Member	Nagy, E.	Sweden
Member	Tikhonenko, A.S.	USSR
Member	Trautner, T.A.	Germany
Member	Vary, P.S.	USA

CLOSTRIDIUM PHAGES STUDY GROUP

Chair	Ogata, S.	Japan
Member	Eklund, M.W.	USA
Member	Jones, D.T.	New Zealand
Member	Mahony, D.E.	Canada
Member	Oguma, K.	Japan
Member	Schallehn, G.	Germany

CUBIC, FILAMENTOUS & PLEOMORPHIC PHAGES STUDY GROUP

Chair	van Duin, J.	The Netherlands
Member	Bamford, D.	Finland
Member	Denhardt, D.T.	Canada
Member	Havelaar, A.H.	The Netherlands
Member	Kodaira, K.	Japan
Member	Maniloff, J.	USA

CYANOPHAGES STUDY GROUP

Chair	Safferman, R.S.	USA
Member	Cannon, R.	USA
Member	Desjardins, P.R.	USA
Member	Gromov, B.V.	USSR
Member	Haselkorn, R.	USA
Member	Sherman, L.	USA

LACTOBACILLUS PHAGES STUDY GROUP

Chairperson	Sozzi, T.	Italy
Member	Accolas, J-P	France
Member	Alatossava, T.	Finland

Member	Mata M.	France
Member	Ritzenthaler, P.	France
Member	Sechaud, L.	France
Member	Trevors, K.E.	Canada
Member	Watanabe, K.	Japan

LACTOCOCCAL & STREPTOCOCCAL PHAGES STUDY GROUP

Chair	Jarvis, A.W.	New Zealand
Member	Fitzgerald, G.	Ireland
Member	Mata, M.	France
Member	Mercenier, A.	France
Member	Neve, H.	Germany
Member	Powell, I.B.	Australia
Member	Ronda, C.	Spain
Member	Saxelin, M.-L.	Finland

LISTERIA & CORYNEFORM PHAGES STUDY GROUP

Chair	Rocourt, J.	France
Member	Groman, N.	USA
Member	Ortel, S.	Germany
Member	Rappuoli, R.	Italy
Member	Trautwetter, A.	France

MYCOPLASMA PHAGES STUDY GROUP

Chair	Maniloff, J.	USA
Member	Bove, J.M.	France

RHIZOBIUM PHAGES STUDY GROUP

Chair	Werquin, M.	France
Member	Kowalski, M.	Poland

STAPHYLOCCUS PHAGES STUDY GROUP

Chair	Stewart, P.R.	Australia
Member	Bes, M.	France
Member	Duval-Iflah, Y.	France

TAILED PHAGES OF ENTEROBACTERIA STUDY GROUP

Chair	Ackermann, H.-W.	Canada
Member	Dhillon, T.S.	Hong Kong
Member	Dubow, M.S.	Canada
Member	Gershman, M.M.	USA
Member	Grimont, F.	France
Member	Hausmann, R.	Germany
Member	Karska-Wysocki, B.	Canada
Member	Kasatiya, S.S.	Canada
Member	Mamet-Bratley, M.D.	Canada
Member	McCorquodale, D.J.	USA
Member	Regue, M.	Spain
Member	Stocker, B.A.D.	USA
Member	Zachary, A.	USA

VIBRIO PHAGES STUDY GROUP

Chair	Chatterjee, S.N.	India
Member	Amad, S.A.	India

Member	Ansari, M.Q.	India
Member	Bhattacharya, S.C.	India
Member	Das, J.	India
Member	Kasatiya, S.S.	Bangladesh
Member	Kawata, T.	Japan
Member	Koga, T.	Japan
Member	Maiti, M.	India
Member	Pal, S.C.	India

FUNGAL, ALGAL & PROTOZOAL SUBCOMMITTEE

Chair	Ghabrial, S.A.	USA
Member	Bozarth, R.F.	USA
Member	Bruenn, J.A.	USA
Member	Buck, K.W.	UK
Member	Koltin, Y.	Israel
Member	Romaine, C.P.	USA
Member	van Etten, J.	USA
Member	Wickner, R.B.	USA
Member	Yamashita, S.	Japan

CRYPHONECYRIA PARASITICA VIRUS STUDY GROUP

Chair	Hillman, B.I.	
Member	Fulbright, D.W.	USA
Member	Nuss, D.L.	USA
Member	van Alfen, N.K.	USA

PROTOZOAL VIRUS SUBCOMMITTEE

Chair	Patterson, J.L.	USA
Member	Stuart, K.	USA
Member	Wang, A.L.	USA
Member	Wang, C.C.	USA

ALGAL VIRUS SUBCOMMITTEE

Chair	van Etten, J.L.	USA
Member	Dodds, J.A.	USA
Member	Gibbs, A.J.	Australia

PLANT VIRUS SUBCOMMITTEE

Chair	Martelli, G.P.	Italy
Member	Adam, G.	Germany
Member	Atabekov, J.G.	USSR
Member	Barnett, O.W.	USA
Member	Goldbach, R.	The Netherlands
Member	Koenig, R.	Germany
Member	Lecoq, H.	France
Member	Makkouk, K.	Syria
Member	Milne, R.G.	Italy
Member	Mink, G.I.	USA
Member	Morris, T.J.	USA
Member	Murant, A.F.	UK
Member	Randles, J.W.	Australia
Member	Rybicki, E.	South Africa
Member	Salazar, L.F.	Peru
Member	Tomaru, K.	Japan

POTYVIRIDAE STUDY GROUP

Chair	Barnett, O.W.	USA
Member	Adam, G.	Germany
Member	Brunt, A.A.	UK
Member	Dijkstra, J.	The Netherlands
Member	Dougherty, W.G.	USA
Member	Edwardson, J.R.	USA
Member	Goldbach, R.	The Netherlands
Member	Hammond, J.	USA
Member	Hill, J.H.	USA
Member	Jordan, R.	USA
Member	Makkouk, K.	Syria
Member	Morales, F.	Colombia
Member	Ohki, S.T.	Japan
Member	Purcifull, D.	USA
Member	Shikata, E.	Japan
Member	Shukla, D.D.	Australia
Member	Uyeda, I.	Japan

SATELLITE STUDY GROUP

Chair	Mayo, M.A.	UK
Member	Berns, K.I.	USA
Member	Fritsch, C.	France
Member	Jackson, A.O.	USA
Member	Kaper, J.M.	USA
Member	Leibowitz, M.J.	USA
Member	Taylor, J.M.	USA

INVERTEBRATE VIRUS SUBCOMMITTEE

Chair	Summers, M.D.	USA
Member	Carstens, E.	Canada
Member	Payne, C.	UK
Member	Rueckert, R.R.	USA
Member	Stoltz, D.B.	Canada
Member	Vlak, J.M.	The Netherlands
Member	Volkman, L.	USA
Member	Willis, D.	USA
Member	Wilson, M.	USA

BACULOVIRIDAE STUDY GROUP

Chair	Volkman, L.	USA
Member	Blissard, G.W	USA
Member	Friesen, P.	USA
Member	Keddie, P.	Canada
Member	Possee, R.	UK
Member	Theilmann, D.	Canada

IRIDOVIRIDAE STUDY GROUP

Chair	Willis, D.	USA
Member	Cameron, I.	UK
Member	Devauchelle, G.	France
Member	Kalmakoff, J.	NewZealand
Member	Seligy, V.L.	Canada
Member	van Etten, J.	USA

NODAVIRIDAE & TETRAVIRIDAE STUDY GROUP

Chair	Rueckert, R.R.	USA
Member	Hendry, D.A.	South Africa
Member	Johnson, J.	USA
Member	Scotti, P.	New Zealand

POLYDNAVIRIDAE STUDY GROUP

Chair	Stoltz, D.B.	Canada
Member	Beckage, N.	USA
Member	Blissard, G.W.	USA
Member	Fleming, J.A.	USA
Member	Krell, P.	Canada
Member	Summers, M.D.	USA
Member	Theilmann, D.	Canada
Member	Webb, B.A.	USA

VERTEBRATE VIRUS SUBCOMMITTEE

Chair	Bishop, D.H.L.	UK
Member	Buchmeier, M.J.	USA
Member	Calisher, C.H.	USA
Member	Cavanagh, D.	USA
Member	Coffin, J.M.	USA
Member	Cubitt, D.	UK
Member	Dobos, P.	Canada
Member	Esposito, J.J.	USA
Member	Frisque, R.J.	USA
Member	Holmes, I.H.	Australia
Member	Howard, C.R.	UK
Member	Klenk, H.-D.	Germany
Member	McCormick, J.B.	USA
Member	Minor, P.	UK
Member	Pringle, C.R.	UK
Member	Roizman, B.	USA
Member	Russell, W.C.	UK
Member	Siegl, G.	Switzerland
Member	Strauss, J.H.	USA
Member	Wengler, G.	Germany
Member	Wunner, W.H.	USA

ADENOVIRIDAE STUDY GROUP

Chair	Russell, W.C.	UK
Member	Adrian, T.	Germany
Member	Bartha, A.	Hungary
Member	De Jong, J.C.	The Netherlands
Member	Fujinaga, K.	Japan
Member	Ginsberg, H.	USA
Member	Hierholzer, C.	USA
Member	Li, Q.G.	China
Member	Mautner, V.	UK
Member	Nasz, I.	Hungary
Member	Shenk, T.	USA
Member	Wadell, G.	Sweden

ARENAVIRIDAE STUDY GROUP

Chair	Buchmeier, M.J.	USA
Member	Auperin, D.D.	USA
Member	Franze-Fernandez, M.T.	Argentina
Member	Gonzalez, J-P	Senegal
Member	Howard, C.R.	UK
Member	Lehman-Grube, F.	Germany
Member	McCormick, J.B.	USA
Member	Peters, C.J.	USA
Member	Romanowski, V.	Argentina
Member	Southern, P.J.	USA

BIRNAVIRIDAE STUDY GROUP

Chair	Dobos, P.	Canada
Member	Azad, A.	Australia
Member	Chistie, K.E.	Norway
Member	Kibenge, F.S.B.	Canada
Member	Leong, T.-A.	USA
Member	Muller, H.	Germany
Member	Nicholson, B.	USA

BUNYAVIRIDAE STUDY GROUP

Chair	Calisher, C.H.	USA
Member	Beaty, B.J.	USA
Member	Dalrymple, J.M.	USA
Member	Elliott, R.M.	UK
Member	Karabatsos, N.	USA
Member	Kolakovsky, D.	Switzerland
Member	Lee, H.-W.	Korea
Member	Lvov, D.K.	USSR
Member	Nuttall, P.A.	UK
Member	Peters, D.	The Netherlands
Member	Pettersson, R.	Sweden
Member	Schmaljohn, C.S.	USA
Member	Shope, R.E.	USA

CALICIVIRIDAE STUDY GROUP

Chair	Cubitt, D.	UK
Member	Black, D.	UK
Member	Chiba, S.	Japan
Member	Smith, A.W.	USA
Member	Studdert, M.J.	Australia

CIRCOVIRIDAE STUDY GROUP

Chair	Lukert, P.D.	USA
Member	De Boer, G.F.	Germany
Member	McNulty, D.S.	Ireland
Member	Tischer, I.	Germany

CORONAVIRIDAE STUDY GROUP

Chair	Cavanagh, D.	UK
Member	Brian, D.A.	USA
Member	Enjuanes, L.	Spain
Member	Holmes, K.V.	USA

Member	Lai, M.M.C.	USA
Member	Laude, H.	France
Member	Siddell, S.	Germany
Member	Spaan, W.J.M.	The Netherlands
Member	Taguchi, F.	Japan
Member	Talbot, P.J.	Canada

Filoviridae Study Group

Chair	McCormick, J.B.	USA
Member	Kiley, M.	USA
Member	Kingsbury, D.W.	USA
Member	Klenk, H.-D.	Germany
Member	Wertz, G.W.	USA

Flaviviridae Study Group

Chair	Wengler, G.	Germany
Member	Bradley, D.	USA
Member	Collett, M.	USA
Member	Heinz, F.X.	Austria
Member	Schlesinger, R.W.	USA
Member	Strauss, J.H.	USA

Hepadnaviridae Study Group

Chair	Howard, C.R.	UK
Member	Burrell, C.J.	Australia
Member	Gerin, J.L.	USA
Member	Gerlich, W.H.	Germany
Member	Gust, J.	Australia
Member	Koike, K.	Japan
Member	Marion, P.L.	USA
Member	Mason, W.	USA
Member	Neurath, A.R.	USA
Member	Newbold, J.	USA
Member	Robinson, W.	USA
Member	Schaller, H.	Germany
Member	Tiollais, P.	France
Member	Wen, Y.-M.	China
Member	Will, H.	Germany

Herpesviridae Study Group

Chair	Roizman, B.	USA
Member	Desrosiers, R.C.	USA
Member	Fleckenstein, B.	Germany
Member	Lopez, C.	USA
Member	Minson, A.C.	UK
Member	Studdert, M.J.	Australia

Orthomyxoviridae Study Group

Chair	Klenk, H.-D.	Germany
Member	Cox, N.	USA
Member	Lamb, R.A.	USA
Member	Lwow, D.	USSR
Member	Mahy, B.	USA
Member	Nakamura, K.	Japan
Member	Palese, P.	USA
Member	Rott, R.	Germany

PAPOVAVIRIDAE STUDY GROUP

Chair	Frisque, R.J.	USA
Member	Barbanti-Brodano, G.	Italy
Member	Crawford, L.V.	UK
Member	Gardner, S.D.	UK
Member	Howley, P.M.	USA
Member	Orth, G.	France
Member	Shah, K.V.	USA
Member	van der Noordaa, J.	The Netherlands
Member	zur Hausen, H.	Germany

PARAMYXOVIRIDAE STUDY GROUP

Chair	Rima, B.	UK
Member	Alexander, D.J.	UK
Member	Billeter, M.A.	Switzerland
Member	Collins, P.L.	USA
Member	Kingsbury, D.W.	USA
Member	Lipkind, M.A.	Israel
Member	Nagai, Y.	Japan
Member	Orwell, C.	Sweden
Member	Pringle, C.R.	UK
Member	Rott, R.	Germany
Member	ter Meulen, V.	Germany

PARVOVIRIDAE STUDY GROUP

Chair	Siegl, G.	Switzerland
Member	Berns, K.I.	USA
Member	Bloom, M.	USA
Member	Carter, B.J.	USA
Member	Cotmore, S.	USA
Member	Lenderman, M.	USA
Member	Tal, T.	Israel
Member	Tattersall, P.	USA
Member	Tijssen, P.	Canada

PICORNAVIRIDAE STUDY GROUP

Chair	Minor, P.	UK
Member	Brown, F.	UK
Member	Knowles, N.	UK
Member	Lemon, S.	USA
Member	Palmenberg, A.	USA
Member	Rueckert, R.R.	USA
Member	Stanway, G.	UK
Member	Wimmer, E.	USA
Member	Yin Murphy, M.	Malaysia

POXVIRIDAE STUDY GROUP

Chair	Esposito, J.J.	USA
Member	Baxby, D.	UK
Member	Black, D.	UK
Member	Dales, S.	Canada
Member	Darai, G.	Germany
Member	Dumbell, K.	South Africa
Member	Granados, R.	USA
Member	Joklik, W.K.	USA

Member	McFadden, G.	Canada
Member	Moss, B.	USA
Member	Moyer, R.	USA
Member	Pickup, D.	USA
Member	Robinson, A.	New Zealand
Member	Rouhandeh, H.	USA
Member	Tripathy, D.	USA

REOVIRIDAE STUDY GROUP

Chair	Holmes, I.H.	Australia
Member	Boccardi, G.	Italy
Member	Estes, M.K.	USA
Member	Furuichi, M.K.	Japan
Member	Hoshino, Y.	USA
Member	Joklik, W.K.	USA
Member	Knudson, D.	USA
Member	Lopez-Ferber, M.	UK
Member	McCrae, M.A.	UK
Member	Mertens, P.	UK
Member	Milne, R.G.	Italy
Member	Nuss, D.	USA
Member	Shikata, E.	Japan
Member	Winton, J.R.	USA

RETROVIRIDAE STUDY GROUP

Chair	Coffin, J.M.	USA
Member	Essex, M.	USA
Member	Gallo, R.	USA
Member	Graf, T.M.	Germany
Member	Hinuma, Y.	Japan
Member	Hunter, E.	USA
Member	Jaenisch, R.	USA
Member	Nusse, R.	USA
Member	Oroszlan, S.	USA
Member	Svoboda, J.	Czechoslovakia
Member	Toyoshima, K.	Japan
Member	Varmus, H.	USA

RHABDOVIRIDAE STUDY GROUP

Chair	Wunner, W.H.	USA
Member	Calisher, C.H.	USA
Member	Dietzgen, R.	Australia
Member	Jackson, A.O.	USA
Member	Kitajima, E.W.	Brazil
Member	Lafon, M.	France
Member	Leong, J.C.	USA
Member	Nichol, S.T.	USA
Member	Peters, D.	The Netherlands
Member	Smith, J.S.	USA

TOGAVIRIDAE STUDY GROUP

Chair	Strauss, J.H.	USA
Member	Calisher, C.H.	USA
Member	Dalgarno, L.	Australia
Member	Dalrymple, J.M.	USA

Member	Frey, T.K.	USA
Member	Rice, C.M.	USA
Member	Spaan, W.J.M.	The Netherlands

Torovirus Study Group

Chair	Horzinek, M.C.	The Netherlands
Member	Flewett, T.H.	UK
Member	Saif, L.	USA
Member	Spaan, W.J.M.	The Netherlands
Member	Woode, G.N.	USA

Virus Data Subcommittee

Chairperson	Gibbs, A.J.	Australia
Member	Barnett, O.W.	USA
Member	Berthiaume, L.	Canada
Member	Blaine, L.	USA
Member	Bove, J.M.	France
Member	Brunt, A.A.	UK
Member	Buck, K.W.	UK
Member	Calisher, C.H.	USA
Member	Dallwitz, M.J.	Australia
Member	Dellaporta, T.	Australia
Member	Dobos, P.	Canada
Member	Fauquet, C.M	France
Member	Federici, B.	USA
Member	Jarvis, A.W.	New Zealand
Member	Kingsbury, D.W.	USA
Member	Kitajima, E.W.	Brasil
Member	Kolaskar, A. S.	India
Member	Krichevsky, M.	USA
Member	Lvov, D.K.	USSR
Member	McManus, C.	USA
Member	van Etten, J.	USA
Member	Watson, L.	Australia

National Representatives

Member	Allam, E.	Egypt
Member	Banerjee, K.	India
Member	Barnett, O.W.	USA
Member	Becker, Y.	Israel
Member	Bellamy, A.R.	New Zealand
Member	Bouguermouh, A.	Algeria
Member	Bozakaya, E	Turkey
Member	Brunt, A.A.	UK
Member	Cajal, N.	Romania
Member	Chernesky, M.A.	Canada
Member	Cvetnic, S.	Yugoslavia
Member	Dundarov, S.G.	Bulgaria
Member	French, G.L.	Hong Kong
Member	Gaidamovich, S.Y.	Russia
Member	Haller, O.	Germany
Member	Haukenes, G.	Norway
Member	Hiebert, E.	USA
Member	Holmes, I.H.	Australia
Member	Hovi, T.	Finland
Member	Iroegbu, C.	Nigeria

Member	Jarzabek, Z.	Poland
Member	Korych, B.	Czechoslovakia
Member	La Placa, M.	Italy
Member	Najera, R.	Spain
Member	Nasz, I.	Hungary
Member	Ouf, M.	Egypt
Member	Oya, A.	Japan
Member	Palmenberg, A.C.	USA
Member	Pang, Q.F.	China
Member	Peters, D.	The Netherlands
Member	Pfister, H.	Germany
Member	Rishi, N.	India
Member	Sarker, A.J.	Bangladesh
Member	Suarez, M.	Chile
Member	Tsotsos, A.	Greece
Member	Verwoerd, D.W.	South Africa
Member	Vestergaard, B.F.	Denmark
Member	Wadell, G.	Sweden
Member	Watson, D.H.	UK
Member	Weissenbacher, M.	Argentina

THE STATUTES OF ICTV 1993

ARTICLE 1

OFFICIAL NAME

International Committee on Taxonomy of Viruses (ICTV).

ARTICLE 2

STATUS

The ICTV is a Committee of the Virology Division of the International Union of Microbiology Societies (IUMS).

ARTICLE 3

OBJECTIVES

1. To develop an internationally agreed taxonomy for viruses.
2. To establish internationally agreed names for taxonomic groups of viruses.
3. To communicate the latest results on the classification and nomenclature of viruses to virologists by holding meetings and publishing reports.

ARTICLE 4

MEMBERSHIP

Membership of the ICTV shall be comprised as follows.

A. President and Vice-President
These shall be nominated and seconded by any members of the ICTV and elected at a plenary meeting of the full ICTV membership. They shall be elected for a term of three years and may not serve for more than two consecutive terms of three years.

B. Secretaries
Two permanent secretaries shall be nominated by the Executive Committee and elected at a plenary meeting of the full ICTV membership. The Secretaries shall be elected for a period of six years, with provision for renewal at three year intervals.

C. Members of the Executive Committee (EC)
The President, Vice-President and Secretaries
Chairs of the Subcommittees (SC)
Bacterial Virus SC
Co-ordination Virus SC (The President ex officio)
Fungal Virus SC
Invertebrate Virus SC
Plant Virus SC
Vertebrate Virus SC
Virus Data SC
Eight elected members.

The Chairs of the Subcommittees shall be elected by the Executive Committee at its mid-term meeting preceding the next plenary meeting of the full ICTV membership for a term of three years and may not serve more than two consecutive terms of three years each.

The eight elected members shall be nominated and seconded by any ICTV member and elected at a plenary meeting of the ICTV for a term of three years and may not serve for more than two consecutive terms of three years each. Generally four of the elected members shall be replaced every three years.

D. National Members
National members shall be nominated by Member Societies of the Virology Division of the IUMS. Societies belonging to the IUMS are considered to be Member Societies of the Division if they have members actively interested in virology. Wherever practicable, each country shall be represented by at least one National Member and no country by more than five National Members. Nominated National Members shall not require further approval by the ICTV.

E. Life Members
Life members shall be nominated by the Executive Committee on account of their outstanding service to virus taxonomy. They shall be elected by the full ICTV.

F. Members of the Bacterial Virus, Co-ordination, Fungal Virus, Invertebrate Virus, Plant Virus, Vertebrate Virus, and Virus Data Subcommittees

These shall be appointed by the Chairs of the Subcommittees and shall not require further approval by the ICTV.

G. Status of Study Group Members
Study Groups may be formed to examine the taxonomy of specialized groups of viruses. A Chair of a Study Group shall be appointed by the Chair of the appropriate Subcommittee and shall be a member of that Subcommittee ex officio and hence also a member of the ICTV.

Chairs of Study Groups shall appoint the members of their Study Groups. Members of Study Groups, other than Chairs, shall not be members of the ICTV, but their names shall be published in the minutes and reports of the ICTV to recognize their valuable contribution to the taxonomy of viruses. Study Group Chairpersons and Study Group members will be nominated for a period of three years. Their term of office will be limited to two consecutive periods of three years.

H. Finance Committee: A finance committee will be constituted comprising the Officers (President, Vice-President, the two Secretaries) and two nominated members. The nominated members will serve for a period of three years.

ARTICLE 5

MEETINGS

Plenary meetings of the full ICTV membership shall be held in conjunction with the International Congresses of Virology. Meetings of the ICTV Executive Committee shall be held in conjunction with the International Congresses of Virology. In addition, a mid-term meeting shall be held between Congresses.

ARTICLE 6

TAXONOMIC PROPOSALS

Taxonomic proposals may be initiated by an individual member of the ICTV, by a Study Group or by a Subcommittee member by sending it to the Chair of the appropriate subcommittee for consideration by that subcommittee. Taxonomic proposals approved by a subcommittee shall be submitted by its chair for consideration by the Executive Committee. Proposals approved by the Executive Committee shall be presented to the next plenary meeting of the full ICTV membership for ratification.

Separate proposals shall be required to establish a new taxonomic group, to name a taxonomic group, to designate the type species and the members of a taxonomic group or by circulation of proposals by mail followed by a postal vote.

Article 7

Voting

Decisions will be made on the following basis:

(i) At meetings, or postal votes, of the Executive Committee
A simple majority of the votes of those present, or those replying within two months of a questionnaire being sent out.

(ii) At plenary meetings, or postal votes, of the full ICTV membership
A simple majority of the votes of those present, or those replying within two months of a questionnaire being sent out. A quorum consisting of the President or Vice-President together with 15 voting members will be required.

(iii) Voting members of the full ICTV comprise the Members of the Executive Committee, the National Members, the Life Members and the Chairpersons of the Sub-committees.

In the event of a tie in (i) or (ii), the President shall have an additional casting vote.

Article 8

The Rules of Nomenclature of Viruses

The rules of nomenclature of viruses, and any subsequent changes, shall be approved by the Executive Committee and at a plenary meeting of the full ICTV membership or by circulation of proposals by mail followed by a postal vote.

Article 9

Duties of Officers

A. Duties of the President shall be:

1. To preside at meetings of the Executive Committee and plenary meetings of the full ICTV membership.
2. To prepare with the Secretaries the agendas for meetings of the Executive Committee and the plenary meetings of the full ICTV membership.
3. To act as editor for ICTV reports to be published after each plenary meeting of the ICTV.

B. Duties of the Vice-President shall be:

1. To carry out the duties of the President in the absence of the President.
2. To attend meetings of the Executive Committee and plenary meetings of the ICTV.

C. Duties of the Secretaries shall be:

1. To attend meetings of the Executive Committee and plenary meetings of the ICTV.
2. To prepare with the President the agendas for meetings of the Executive Committee and the plenary meetings of the ICTV.
3. To prepare the Minutes of meetings of the Executive Committee and plenary meetings of the ICTV and circulate them to all ICTV members.
4. To act as Treasurer of the ICTV. To handle any funds that may be allocated to the ICTV by the Virology Division of the IUMS or other sources.

5. To keep an up-to-date record of ICTV membership.

ARTICLE 10

PUBLICATIONS

No publication of the ICTV shall bear any indication of sponsorship by a commercial agency, or institution connected in any way with a commercial company, except as an acceptable acknowledgment of financial assistance. Furthermore, any publication containing material not authorized, prepared, or edited by the ICTV, or a committee or subcommittee of the ICTV, may not bear the name of the ICTV or the IUMS.

ARTICLE 11

ICTV STATUTES

The Statutes of the ICTV, and any subsequent changes, shall be approved by the ICTV Executive Committee, by a plenary meeting of the full ICTV membership and by the Virology Division of the IUMS.

ARTICLE 12

DISPOSITION OF FUNDS

In the event of dissolution of the ICTV, any remaining funds shall be turned back to the Secretary-Treasurer of the Virology Division of the IUMS.

THE RULES OF VIRUS CLASSIFICATION AND NOMENCLATURE, 1993

Substantial revisions have been made to the Rules of Virus Classification and Nomenclature recently with the intention of improving guidance to ICTV Subcommittees and Study Groups. Even so, it is unlikely that these Rules are the best that can be devised, so it is hoped that feedback to the ICTV from the virological community will result in further refinements. Comments are always welcome.

GENERAL RULES

1. Virus classification and nomenclature shall be international and shall be universally applied to all viruses.
2. The universal virus classification system shall employ the hierarchical levels of order, family, subfamily, genus, and species. To the extent that species designations are not yet complete, international vernacular names are used for many viruses.
3. The ICTV is not concerned with classification and nomenclature below the species level. Delineation of serotypes, genotypes, strains, variants, isolates, etc., is the responsibility of acknowledged international specialist groups.
4. Artificially created viruses and laboratory hybrid viruses will not be given taxonomic consideration. Again, delineation of these entities is the responsibility of acknowledged international specialist groups.
5. Taxa will be established only when representative member viruses are sufficiently well characterized and described in the published literature so as to allow unambiguous identification and discrimination from similar taxa. Likewise, nomenclature will only be recognized when pertaining to viruses that are sufficiently well characterized and described in the published literature so as to allow unambiguous identification and discrimination from similar viruses.

RULES PERTAINING TO NAMING TAXA AND VIRUSES

6. Existing names of taxa and viruses shall be retained whenever feasible.
7. The rule of priority in naming taxa and viruses shall not be observed.
8. No person's name shall be used.
9. Names for taxa and viruses should be easy to use and easy to remember. Euphonious names are preferred.
10. Subscripts, superscripts, hyphens, oblique bars and Greek letters may not be used.
11. New names shall not duplicate approved names. New names shall be chosen so as not to be closely similar to names in use currently or in the recent past.
12. Sigla may be accepted as names of taxa, provided that they are meaningful to virologists in the field and are recommended by acknowledged international specialist groups.
13. Any meaning imparted by a name of a taxon or a virus must avoid excluding viruses which are legitimate members of the taxon by alluding in the name to characteristics not possessed by all members or potential members, and not apply equally to a different taxon.
14. New names shall be chosen with due regard to national or local sensitivities. When names are universally used by virologists in published work, these or derivatives shall be the preferred basis for creating names, irrespective of national origin. In the event of the advance of more than one candidate name, the relevant Study Group or Subcommittee will make a recommendation to the Executive Committee of the ICTV, which will then decide among the candidates.
15. Proposals for new names and name changes shall be submitted to the ICTV in the form of taxonomic proposals.

RULES PERTAINING TO SPECIES

16. A virus species is defined as a polythetic class of viruses that constitutes a replicating lineage and occupies a particular ecological niche.
17. A species name shall consist of as few words as practicable.
18. A species name, usually together with a strain designation, must provide an appropriately unambiguous identification without mention of its genus or family name.
19. Numbers, letters, or combinations thereof may be used as species epithets where such numbers and letters already have wide usage. However, newly designated serial numbers, letters or combinations thereof are not acceptable alone as species epithets.
20. Approval by ICTV of newly proposed species, species names, and type species will proceed in two stages. In the first stage, provisional approval will be given. Provisionally approved proposals will be published in an ICTV Report; then, after a 3-year waiting period, if not withdrawn or modified, proposals will receive final approval.

RULES PERTAINING TO GENUS

21. A genus is a group of species sharing certain common characters.
22. A genus name shall be a single word ending in "...virus."
23. Approval of a new genus must be linked to approval of a type species.

RULES PERTAINING TO SUBFAMILY

24. A subfamily is a group of genera sharing certain common characters. It should only be used when needed to solve a complex hierarchical problem.
25. A subfamily name shall be a single word ending in "...virinae."

RULES PERTAINING TO FAMILY

26. A family is a group of genera (whether or not these are organized into subfamilies) sharing certain common characters.
27. A family name shall be a single word ending in "...viridae."
28. Approval of a new family must be linked to approval of a type genus.

RULES PERTAINING TO ORDER

29. An order is a group of families sharing certain common characters.
30. An order name shall be a single word ending in "...virales."

The Format for Submission of New Taxonomic Proposals

Contents

Over the last years the Executive Committee of ICTV has evolved procedures and rules to facilitate the processing and assessment of new taxonomic proposals for viruses. This section, which summarizes the present position, is provided to assist virologists wishing to make a contribution to the work of ICTV.

I. Initiation of New Proposals

The key units in the organization of the ICTV are the host-oriented subcommittees. Most of these subcommittees are organized into study groups of working virologists. New taxonomic proposals are usually initiated by these study groups, and less commonly by the subcommittees themselves.

It should be emphasized that, apart from the formal organization, it is perfectly in order for any individual virologist to initiate a new taxonomic proposal. Any such proposal should be in the format outlined below, and should be sent to the Chairperson of the appropriate subcommittee for consideration.

II. Processing of New Proposals

A taxonomic proposal originating in a study group or favorably considered by a study group after receipt from an individual virologist is forwarded to the appropriate subcommittee. If it is approved by the subcommittee, the proposal is then considered by the Executive Committee of ICTV. The Executive Committee of ICTV may approve a proposal, decline to approve, or send it back to the subcommittee for suggested changes.

Proposals approved by the Executive Committee go forward every 3 years to the plenary meeting of the full ICTV membership for final ratification.

III. Publication of New Proposals

Some new proposals pass through the ICTV and are approved without any prior publication. Such proposals then appear first in an official ICTV triennial report. Other proposals are published at an earlier stage in the Archives of Virology, which is the official journal of the Virology Division of the International Union of Microbiological Societies.

These publications may be enlarged presentations of taxonomic proposals being formally submitted by ICTV study groups. Two examples of this sort, published in Intervirology concern the family Caliciviridae (Schaffer et al., 1980) and the family Bunyaviridae (Bishop et al., 1980). A proposal for establishing the family Potyviridae family, comprising three genera has been published in Archives of Virology (Barnett, 1991).

Such publications allow individual virologists to scrutinize proposals and to make their views known to the appropriate ICTV subcommittee. It should be emphasized, however, that publication in itself does not give the proposals any status as far as ICTV is concerned.

IV. TIMING OF EVENTS IN THE PERIOD 1993-1996

There is a plenary session of the ICTV held every three years at the International Congress of Virology. The next plenary session will be held at the Xth International Congress in Jerusalem, Israel in August 1996.

There is no deadline for submitting proposals to the Executive Committee of the ICTV. Subcommittee chairs can send proposals to the ICTV Secretary for circulation to members before any Executive Committee meeting. New taxonomic proposals should be in the hands of the secretary before May 1996, so that the proposals can be circulated to the members before the Executive Committee of the ICTV during the Virology Congress of 1996.

V. STANDARD FORMAT FOR PRESENTING NEW TAXONOMIC PROPOSALS

Chairs of study groups and subcommittees should use the following guidelines and format in preparing new taxonomic proposals.

GUIDELINES:

1. Each individual taxonomic proposal should be submitted as a separate item (not mixed with explanatory or historical details). For example, a proposal to form a new family must be separate from a proposal for a new genus and separate from a proposal designating the type species for the genus.

2. Attention is drawn to rule N°20, which requires that approval of a new family must be linked with approval of a type genus and that approval of a new genus must be linked with approval of a type species.

3. Each proposal should contain information in the following format: Date.........

From the... ... ... Subcommittee or Study group
Taxonomic Proposal N°.:

1. Proposal: The taxonomic proposal in its essence, in a form suitable for presentation to ICTV for voting.

2. Purpose: A summary of the reasons for the proposal, with any explanatory and historical notes.

3. A summary of the new taxonomic situation within the family, group or genus (e.g. for a new genus- 'The family would now consist of the following genera:... ... ... ')

4. Derivation of any names proposed.

5. New literature references, if appropriate.

IV. Taxonomic Events in the Period 1993–1996

There is a plenary session of the ICTV held every three years at the International Congress of Virology. The next plenary session will be held at the Xth International Congress in Jerusalem, Israel in August 1996.

There is no deadline for submitting proposals to the Executive Committee of the ICTV. Subcommittee chairs can send proposals to the ICTV secretary for circulation to members before any Executive Committee meeting. New taxonomic proposals should be in the hands of the secretary before May 1996, so that the proposals can be circulated to the members before the Executive Committee of the ICTV during the Virology Congress of 1996.

V. Standard Format for Presenting New Taxonomic Proposals

Chairs of study groups and subcommittees should use the following guidelines and format in preparing new taxonomic proposals.

Guidelines

1. Each individual taxonomic proposal should be submitted as a separate item (not mixed with explanatory or historical details). For example, a proposal to form a new family must be separate from a proposal for a new genus and separate from a proposal designating the type species for the genus.

Attention is drawn to rule 17.2, which requires that approval of a new family must be linked with approval of a type genus and that approval of a new genus must be linked with approval of a type species.

Each proposal should contain information in the following format: Title

From the ... Subcommittee or Study group
Taxonomic Proposal No. ...

Proposal. The taxonomic proposal in its essence, in a form suitable for presentation to ICTV for voting.

[illegible] notes.

A summary of the new taxonomic situation within the family, group or genus (e.g., if accepted, the family would now consist of the following genera: ...

Derivation of any names proposed.

New literature references, if appropriate.

PART IV: INDEXES

AUTHOR INDEX

A

B

C

D

G

H

I

J

K

L

M

N

O

P

Q

R

S

T

U

V

W

X

Y

Z

VIRUS INDEX

Numbers

A

B

C

D

E

F

G

H

I

J

K

M

N

O

P

Q

R

S

T

Thosea asigna virus, ***Tetraviridae***, 374
Thosea baibarana NPV, ***Baculoviridae***, 111
Thosea sinensis GV, ***Baculoviridae***, 112
Thottapalayam virus, ***Bunyaviridae***, 309
Thylidolpteryx ephemeraeformis NPV, ***Baculoviridae***, 111
Thymelicus lineola NPV, ***Baculoviridae***, 111
Tibrogargan virus, ***Rhabdoviridae***, 286
Ticera castanea NPV, ***Baculoviridae***, 111
tick-borne encephalitis virus, ***Flaviviridae***, 419
tick-borne virus, ***Flaviviridae***, 419
Tillamook virus, ***Bunyaviridae***, 310
Tilligerry virus, ***Reoviridae***, 217
Timbo virus, ***Rhabdoviridae***, 285
Timboteua virus, ***Bunyaviridae***, 306
Tinaroo virus, ***Bunyaviridae***, 307
Tindholmur virus, ***Reoviridae***, 219
Tinea pellionella NPV, ***Baculoviridae***, 111
Tineola hisselliella NPV, ***Baculoviridae***, 111
Tipula paludosa NPV, ***Baculoviridae***, 111
Tiracola plagiata NPV, ***Baculoviridae***, 111
Tlacotalpan virus, ***Bunyaviridae***, 305
tobacco bushy top virus, ***Umbravirus***, 390
tobacco etch virus, ***Potyviridae***, 352
tobacco leaf curl virus, ***Geminiviridae***, 163
tobacco mild green mosaic virus, ***Tobamovirus***, 436
tobacco mosaic virus, ***Tobamovirus***, 436
tobacco mosaic virus satellite, Satellites, 489
tobacco mottle virus, ***Umbravirus***, 390
tobacco necrosis virus, ***Necrovirus***, 400
tobacco necrosis virus satellite, Satellites, 489
tobacco necrosis virus small satellite, Satellites, 491
tobacco necrotic dwarf virus, ***Luteovirus***, 382
tobacco rattle virus, ***Tobravirus***, 440
tobacco ringspot virus, ***Comoviridae***, 346
tobacco ringspot virus satellite, Satellites, 492
tobacco streak virus, ***Bromoviridae***, 454
tobacco stunt virus, Unassigned viruses, 505
tobacco vein banding mosaic virus, ***Potyviridae***, 354
tobacco vein distorting virus, ***Luteovirus***, 382
tobacco vein mottling virus, ***Potyviridae***, 352
tobacco wilt virus, ***Potyviridae***, 354
tobacco yellow dwarf virus, ***Geminiviridae***, 160
tobacco yellow net virus, ***Luteovirus***, 382
tobacco yellow vein assistor virus, ***Luteovirus***, 382
tobacco yellow vein virus, ***Umbravirus***, 390
tomato apical stunt viroid, Viroids, 496
tomato aspermy virus, ***Bromoviridae***, 456
tomato black ring virus, ***Comoviridae***, 346
tomato black ring virus satellite, Satellites, 490
tomato bunchy top viroid, Viroids, 496
tomato bushy stunt virus, ***Tombusviridae***, 395
tomato bushy stunt virus satellite, Satellites, 491
tomato golden mosaic virus, ***Geminiviridae***, 163
tomato leaf crumple virus, ***Geminiviridae***, 164
tomato leaf curl virus - Au, ***Geminiviridae***, 163
tomato leaf curl virus - In, ***Geminiviridae***, 163
tomato leafroll virus, ***Geminiviridae***, 161
tomato mosaic virus, ***Tobamovirus***, 436
tomato mottle virus, ***Geminiviridae***, 164
tomato pale chlorosis virus, ***Carlavirus***, 478
tomato planta macho viroid, Viroids, 496
tomato pseudo-curly top virus, ***Geminiviridae***, 161
tomato ringspot virus, ***Comoviridae***, 346
tomato spotted wilt virus, ***Bunyaviridae***, 314
tomato top necrosis virus, ***Comoviridae***, 346
tomato vein yellowing virus, ***Rhabdoviridae***, 284
tomato yellow dwarf virus, ***Geminiviridae***, 164
tomato yellow leaf curl virus - Is, ***Geminiviridae***, 164
tomato yellow leaf curl virus - Sr, ***Geminiviridae***, 164
tomato yellow leaf curl virus - Th, ***Geminiviridae***, 164
tomato yellow leaf curl virus - Ye, ***Geminiviridae***, 164
tomato yellow mosaic virus, ***Geminiviridae***, 164
tomato yellow top virus, ***Luteovirus***, 381
Tongan vanilla virus, ***Potyviridae***, 354
Tortrix loeflingiana NPV, ***Baculoviridae***, 111
Tortrix viridana NPV, ***Baculoviridae***, 111
Toscana virus, ***Bunyaviridae***, 312
Toxorhynchites brevipalpis NPV, ***Baculoviridae***, 111
Trabala vishnou NPV, ***Baculoviridae***, 111
Tradescantia/Zebrina virus, ***Potyviridae***, 354
Trager duck spleen necrosis virus, ***Retroviridae***, 198
Tranosema sp. virus, ***Polydnaviridae***, 145
tree shrew adenovirus 1, ***Adenoviridae***, 132
tree shrew herpesvirus, ***Herpesviridae***, 126
Triatoma virus, ***Picornaviridae***, 335
Tribec virus, ***Reoviridae***, 218
Trichiocampus irregularis NPV, ***Baculoviridae***, 111
Trichiocampus viminalis NPV, ***Baculoviridae***, 111
Trichomonas vaginalis virus, ***Totiviridae***, 249
Trichoplusia ni cypovirus 5, ***Reoviridae***, 231
Trichoplusia ni granulovirus, ***Baculoviridae***, 111
Trichoplusia ni MNPV, ***Baculoviridae***, 107
Trichoplusia ni Single SNPV, ***Baculoviridae***, 107
Trichoplusia ni virus, ***Tetraviridae***, 374
Trichosanthes mottle virus, ***Potyviridae***, 354
Triticum aestivum chlorotic spot virus, ***Rhabdoviridae***, 287
trivittatus virus, ***Bunyaviridae***, 306
Trombetas virus, ***Bunyaviridae***, 305
Tropaeolum virus 1, ***Potyviridae***, 354
Tropaeolum virus 2, ***Potyviridae***, 354
Trubanaman virus, ***Bunyaviridae***, 314
Tsuruse virus, ***Bunyaviridae***, 307
Tucunduba virus, ***Bunyaviridae***, 305
Tulare apple mosaic virus, ***Bromoviridae***, 454
tulip band breaking virus, ***Potyviridae***, 352
tulip breaking virus, ***Potyviridae***, 352
tulip chlorotic blotch virus, ***Potyviridae***, 352
tulip top breaking virus, ***Potyviridae***, 352
tulip virus X, ***Potexvirus***, 481

U

V

W

X

Y

Z

TAXONOMIC INDEX

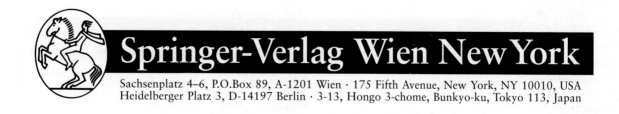

Springer-Verlag
and the Environment

WE AT SPRINGER-VERLAG FIRMLY BELIEVE THAT AN international science publisher has a special obligation to the environment, and our corporate policies consistently reflect this conviction.

WE ALSO EXPECT OUR BUSINESS PARTNERS – PRINTERS, paper mills, packaging manufacturers, etc. – to commit themselves to using environmentally friendly materials and production processes.

THE PAPER IN THIS BOOK IS MADE FROM NO-CHLORINE pulp and is acid free, in conformance with international standards for paper permanency.